Birkhäuser

Mathematics and the Built Environment
Volume 4

More information about this series at http://www.springer.com/series/15181

José Calvo-López

Stereotomy

Stone Construction and Geometry in Western Europe 1200–1900

Birkhäuser

José Calvo-López
Escuela Técnica Superior de Arquitectura y
Edificación
Universidad Politécnica de Cartagena
Cartagena, Spain

ISSN 2512-157X ISSN 2512-1561 (electronic)
Mathematics and the Built Environment
ISBN 978-3-030-43217-1 ISBN 978-3-030-43218-8 (eBook)
https://doi.org/10.1007/978-3-030-43218-8

This book is published under the imprint Birkhäuser, www.birkhauser-science.com by the registered company Springer Nature Switzerland AG
The registered company address is: Gewerbestrasse 11, 6330 Cham, Switzerland

Preface

One of the areas where mathematics interact with architecture and engineering is the geometrical definition and subdivision of construction elements. Although these issues arise with any material, they are particularly relevant in ashlar masonry. In contrast to concrete or brick, the general shape of vaults and arches built in hewn stone is materialised using special pieces, the voussoirs. On many occasions, these elements require complex forms. For example, an ordinary voussoir of a hemispherical dome has six faces, including two spherical portions, two conical ones and two planar shapes. All these surfaces must be controlled by templates or other instruments, specifically adapted to the form of a particular member. As a result, the three-dimensional definition of each different voussoir involves relatively advanced geometrical operations. From Antiquity to recent times, a rich set of graphical techniques, implemented in large-scale tracings drawn in floors and walls, has been gradually developed in order to solve these problems.

Full-scale drawings from Antiquity represent planar shapes, but not three-dimensional objects. From the Middle Ages on, masons used orthogonal projection to depict solids. Such knowledge was considered a valuable secret; stonemasons' lodges tried to keep it away from outside eyes, but dissenting masters brought this lore to the printing press. Later on, Renaissance architects and engineers with a background in construction rather than the figurative arts tackled a difficult problem: orthogonal projection accurately represents planar figures when they are parallel to the projection plane, but distorts oblique shapes. Although the social ascension of architects and engineers brought about the scorn of poets and generals and the ill will of traditional masons, they developed a rich set of methods for the reconstruction of the true size and shape of such figures. This research program was carried out on a mainly empirical basis, without strong ties with the learned geometry of this period, which showed no interest in projections. The next century saw the unexpected intrusion of clergymen into this artisanal realm, as well as a singular duel between geometricians and stonecutters. As a result, these practices coalesced into a discipline known as *stéréotomie*, the science of the division of solids, which metamorphosed during the years of the French Revolution into the descriptive geometry of Gaspard Monge, a general theory of orthogonal projection that provided the foundation of Projective Geometry.

Monge was the founder of the original École Polytechnique, an institution providing a general scientific basis for engineers in several branches of technology; later on, students would finish their training in several "application schools". This didactic program embodied the Enlightenment idea of "applied science": rational geometry furnished the basis for stonecutting, which was considered an application of this abstract science.

However, such a view does not fit the convoluted history of stonecutting. During the twentieth century, several scholars have argued for an opposing view, which can be defined as "formalised practice": it was not geometry which gave birth to stereotomy; rather, stereotomy fostered descriptive geometry. However, intensive research in the last decades has cast some doubts on this alternative view. Thus, throughout this book I will try to avoid the opposing, reductionist conceptions of "applied science" and "formalised practice" and gather facts from many angles, waiting until the final chapter to analyse in depth the issue of these interactions between artisanal procedures and learned science.

With this goal in mind, I will deal shortly in Chap. 1, Introduction, with a number of basic issues, such as the different strategies used in the formal control of concrete, brick and ashlar constructions; the connections of geometrical form and mechanical behaviour in pre-industrial architecture and engineering; the process of ashlar construction and the connections between actual stonecutting and preliminary full-scale tracings; the repertoire of architectural forms built in ashlar, the basic concepts of descriptive geometry and the primary methods of construction history. Since this book may interest readers coming from widely different backgrounds, I am afraid some of them will find these sections rather basic; of course, they may skip or skim through such passages.

In order to frame the methods and concepts of stonecutting in their historical context, before analysing them in depth, Chap. 2 will survey the biographies of the most relevant writers on the subject. Some of them, such as Philibert de l'Orme, Girard Desargues, Guarino Guarini and Gaspard Monge, are well-known figures in the history of architecture or mathematics; for other writers, such as Alonso de Guardia or Jacques Gentillâtre, we know little more than their names. While many treatises reached the presses and exerted wide influence, other texts, in particular Spanish ones, remained in manuscript form. On some occasions, these notebooks are rough or sketchy, but these traits make them remarkably interesting: they reflect workshop practice directly. The material in this chapter will be arranged in four sections, dealing with medieval masons, Renaissance architects, seventeenth-century clerics and Enlightenment military engineers, although, of course, no clear-cut division can be established between these social groups and historical periods.

Chapter 3 will deal in depth with the techniques used in the stonecutting process, from formal definition to actual carving and voussoir placement. It will analyse full-scale tracings, the planar and three-dimensional constructions used in these tracings, and the models used to verify the precision of geometrical procedures and to instruct apprentices. Of course, tracings are a preliminary stage for actual voussoir dressing. I will deal next with carving instruments, stressing the difference between mechanical ones such as the pick, the chisel or the axe and a set of geometrical instruments allowing the transfer of shapes from the tracing to the stone block, such as the ruler, the square, the arch square, the bevel and the templates. Next, I will analyse the basic dressing techniques, squaring and templates, as well as several hybrid methods. The final section of this chapter will deal with the transport, hoisting and placement of the voussoirs. Of course, these issues relate to mechanics, rather than mathematics, but they should not be left aside since they play an essential role in ashlar technology. In particular, the gradual rediscovery of powerful hoisting equipment in the Middle Ages led to the transition from Romanesque rubble to rough-hewn stone in Gothic severies and Renaissance ashlar, bringing about a remarkable evolution of stonecutting methods.

Chapter 3 deals with general procedures used with all kinds of constructive elements. However, a distinctive trait of stereotomy, compared with other applied geometrical fields such as perspective, is the striking variety of elements that can be built in hewn stone and the different control techniques that can be applied to each one. Thus, in Part II, the true nucleus of the book, I will survey these architectural and engineering types, including chapters on arches, trumpet squinches, rere-arches, stairs and vaults; the latter, considering their diversity and complexity, will be split across three different chapters, namely, cylindrical, spherical and ribbed and coffered ones. For each chapter, I will explain the tracing and dressing procedures covered in the most relevant stonecutting texts, stressing the nature of the geometrical problems posed by each element: oblique cylinders in skew arches; cones in trumpet squinches; cylinder intersections in lunettes; approximate spherical developments in domes and sail vaults; ovals and ellipses in oval vaults; orthogonal projections of sloping circles in torus vaults; distances and triangulations in coffered vaults and so on, stressing the evolution of solutions from the Renaissance to the Enlightenment. Such abstract, geometrical explanations risk removing these construction elements from their architectural contexts; to avoid this, I will mention a small number of representative examples of each element, showing that this seemingly infinite repertoire has its roots in the architectural intentions and challenges in each historical period.

Part II presents stereotomy as an isolated science, but this vision does not match the real nature of the discipline. To avoid this, a few short essays dealing mainly with the connections of stonecutting with other branches of knowledge will be included in Part III. First, I will address the connections of stereotomy with functional requirements such as circulation and lightning, the mechanical behaviour of constructions and the aesthetical ideals of each period briefly. However, we should not forget that caprice has led masons to seek unnecessary complications in order to show their skill and, indirectly, the wealth and power of their patrons.

This leads to the issue of the social standing of stonecutters or stereotomy theorists and the status of stereotomy as a branch of knowledge. Quite different groups have struggled for the control of the subject, from medieval masons to Renaissance architects, seventeenth-century clergymen, Enlightenment engineers and nineteenth-century scientists. These social shifts have brought about parallel mutations in the forms of dissemination of knowledge and the status of stonecutting, which evolved from an artisanal practice to a branch of mathematics. Thus, a neutral, academic view of the different stonecutting methods, such as the one presented in Part II, would risk missing the point completely.

The next section will deal with two interconnected issues, namely, the geographical distribution and the historical evolution of stereotomy. Jean-Marie Pérouse de Montclos stressed the concentration of classical ashlar masonry in France and Spain, identifying stereotomy as the critical factor or *pierre de touche* of the singularity of French architecture; at the same time, he placed its roots in Languedocian Romanesque. Later on, Joël Sakarovitch sought its origin in Christian Syria. Two or three decades later, we have in our hands a wealth of case studies that suggest a different, more complex and nuanced narrative. Thus, this chapter will present a polycentric vision of the origins of stereotomy, stressing its dependence on the availability of materials and a layered vision of its historical evolution.

Finally, I will come back to the central question posed before, the connections of stereotomy with learned science, dealing first with the relation between stonecutting and learned science in the Middle Ages, including both Euclidean and practical geometry. As Lon Shelby pointed out, the knowledge of Euclid between medieval masons was, for all practical purposes, nonexistent. Gradually, during the Renaissance and the seventeenth century, some concepts and methods of classical geometry, together with notions borrowed from other sciences, such as perspective, cosmography and cartography, were added to an otherwise empirical matrix, and the subject evolved slowly to the status of a branch of mathematics.

After addressing these issues, I will deal with the tricky question of the role of stonecutting in the formation of descriptive geometry. Such broad generalisations as Pérouse's statement about descriptive geometry as an offspring of stereotomy seem to focus on problems and methods; however, a science is also based on concepts. A systematic comparative analysis of both branches of knowledge shows that descriptive geometry, despite its purpose as a general tool, inherited a wide range of problems from stonecutting, due to the solid, spatial nature of this technology, as Sakarovitch explained convincingly. This does not exclude problems brought about by other areas, such as artillery or the theory of shadows. Methods pose subtle issues: for example, some procedures in sixteenth-century stonecutting, using orthogonals to a horizontal line passing through the projection of a point, may call to mind nineteenth-century rotations and rabatments; however, it is not easy to tell whether Renaissance writers were thinking specifically about rotations or applying an empirical procedure. In any case, it must be stressed that the modern concept of projection, including projectors and a projection plane, was explained by Alberti in the fifteenth century, but did not reach mainstream stereotomy treatises until the early eighteenth century.

It is important to remark that all these issues are open problems, fuelled by the large number of studies published in the last three decades; perhaps further research, or the emergence of still unknown manuscripts, will bring about significant developments in these issues. Thus, rather than putting forward a "definitive", closed conclusion, the book will end with a provisional summary of all the previous findings on the historical evolution of the discipline and the origin of descriptive geometry, stressing the areas where problems are still open.

Of course, this book is built on the foundation of the historical sources and many recent studies. Those quoted in the text are included on the final reference list; however, to avoid excessive length, other studies will be included in an independent bibliography in the making.

Finally, the reader may ask why this book focuses in Western Europe, or even consider such stance as parochial or Eurocentric. The answer is simple: as stated in Sect. 12.3.3, western stonecutting is only part of a large jig-saw puzzle, but the whole picture is too big for a single book and a single author. Hopefully, in the next decades, researchers from the Islamic countries, Armenia, India, or other locations will carry out studies on the stonecutting techniques of these cultures, expanding our insight of a complex phenomenon.

Cartagena, Spain
2019

José Calvo-López

Acknowledgements

This book has benefited from countless talks with Enrique Rabasa, José Carlos Palacios, Lázaro Gila, Santiago Huerta, the late Joël Sakarovitch, José María Gentil, Miguel Ángel Alonso, Ana López Mozo, Francisco Pinto, Eliana di Nichilo, Giuseppe Fallacara, Sergio Sanabria, Luc Tamboréro, Miguel Taín, Arturo Zaragozá, José Antonio Ruiz de la Rosa, Miguel Sobrino, Alberto Sanjurjo, Fernando Marías, Richard Etlin, María Mercedes Bares, Rafael Marín, Soraya Genin, David Wendland, Marco Rosario Nobile, Carmen Pérez de los Ríos, Rocío Maira, Rosa Senent, Pau Natividad, Benjamín Ibarra, Ricardo García-Baño, Rafael Martín Talaverano, Macarena Salcedo, Vincenzo Minenna, Fabio Tellia, Idoia Camiruaga, Antonio Luis Ampliato, Juan Clemente Rodríguez, Bill Addis, Emmanuela Garofalo, Javier Ibáñez, Agostino de Rosa, María Aranda, Carlo Inglese, Alessio Bortot, Paolo Borin, Giulia Piccinin, Francesca Gasperuzzo, Antonio Calandriello, Marta Perelló and many others.

Enrique Rabasa, Miguel Ángel Alonso, Miguel Taín and Clara Calvo have read portions of the book and given their valuable opinion. Alessio Bortot has provided useful help with images and other issues. Pau Natividad, Macarena Salcedo and Ángeles Fuensanta Martínez have provided valuable help in the editing stage. A very big thank you must go Sarah Goob and Sabrina Hoecklin from Birkhäuser and particularly to Kim Williams whose comments, suggestions and corrections have made this book a much better one.

Miguel Taín, Pablo Navarro, Pau Natividad, Macarena Salcedo, Enrique Rabasa, Miguel Alonso, Ana López-Mozo and Idoia Camiruaga have graciously contributed images for the book. Reproductions of treatises and manuscripts have been provided by the Library of the School of Architecture of Universidad Politécnica de Madrid, the Library of ETH Zurich, Universitätsbibliothek Bern, Stiftung Werner Öeschlin, Historisches Archiv Köln, Biblioteca Central Militar-Ministerio de Defensa de España, Library of the Università Iuav in Venice and Biblioteca de Cultura Artesana de Mallorca. Unless stated otherwise, other photographs, drawings, surveys, transcriptions and translations are by the author.

Funds for travel expenses and image rights have been provided by research grants BIA2006-13649, BIA2009-14350-C02-02, BIA2013-46896-P and HAR2016-76371-P, from the R+D Plan of the Spanish Government as well as projects 11988/PI/09 and 19361/PI/14 from Fundación Séneca. An invitation to stay for a term at Università Iuav from Agostino de Rosa and the IR.IDE research group has furnished a most inspiring physical and intellectual environment in the final phases of the book.

Contents

Part II Constructive Elements

About the Author

José Calvo-López is an architect. His Ph.D. dissertation on Renaissance stone-cutting techniques was awarded the Extraordinary Doctoral Prize of Universidad Politécnica de Madrid in 2002. He teaches in Universidad Politécnica de Cartagena, where he has been Dean of the School of Architecture and Director of the Masters' degree in Architectural Heritage and the Doctoral program in Architecture and Building Engineering. He is working presently on two research projects about the transition from Gothic to Renaissance in Spain and stone construction and geometry in the Roman and Early Medieval world. He has published research papers on stereotomy and spatial representation in *Archivo Español de Arte, Nexus Network Journal, Construction History, Informes de la Construcción, International Journal of Architectural Heritage* and *Architectura*. He has been a Visiting Professor in Universitá IUAV di Venezia and has participated in seminars or given lectures at the Max-Planck-Institut für Wissenschaftgeschichte, Centre Alexandre Koyré (CNRS), Centre d'Etudes Superieurs de la Renaissance (CNRS), Universitá di Palermo, Escuela de Patrimonio Histórico de Nájera, University of Texas at Austin, Politecnico di Torino and Universitá di Padova. He is a member of the Editorial Boards of *Revista EGA*, *Construction History*, *Nexus Network Journal* and *Revista de Historia de la Construcción* and the Scientific Committees of several congresses and conferences in Construction History, Artistic Theory and Architectural Drawing, as well as a member of the board of the Spanish Construction History Society.

List of Figures

Chapter 1
Introduction

Abstract The problem of the geometrical definition and subdivision of construction members is particularly relevant in ashlar masonry. In contrast with concrete or brick, the general shape of vaults and arches built in hewn stone is materialised using special pieces, the voussoirs. From Antiquity to recent times, a rich set of graphical techniques, implemented in large-scale tracings drawn in floors and walls, has been gradually developed in order to control the shape of these elements. In the Early Modern period, these methods crystallised in a branch of knowledge known as stereotomy, that is, the science of cutting solids, which furnished most problems and methods of descriptive geometry. The introduction of this book deals with a number of basic issues, such as the different strategies used in formal control of concrete, brick and ashlar constructions; the connections of geometrical form and mechanical behavior in pre-industrial architecture and engineering; the process of ashlar construction and the connections between actual stonecutting and preliminary full-size tracings; the repertoire of architectural forms built in ashlar, the basic concepts of descriptive geometry, and the primary methods of construction history.

1.1 Concrete, Brick, Ashlar: Three Different Approaches to Formal Control in Masonry Construction

When humankind began to settle and build in the Neolithic, it started showing a will to control the shape of the resulting product, even if tools for this task were rudimentary. Huts or stone circles in the period show approximately round, square or triangular plans, conical roofs or triangular cross-sections. As complex societies developed, with an ever-growing division of labour, specific tools for formal control gradually appeared; even the constructions of the first Egyptian dynasties show precise geometrical shapes. Later on, formal control of all kinds of masonry has played a central role in pre-industrial building, as attested by the extensive literature on this topic shown in Chap. 2.

It is easy to forget that these formal control processes can be carried out in three different ways, depending on material. In poured concrete, rubble masonry or pisé, the shape of the masonry element is guaranteed by the use of formwork (the provisional

J. Calvo-López, *Stereotomy*, Mathematics and the Built Environment 4,
https://doi.org/10.1007/978-3-030-43218-8_1

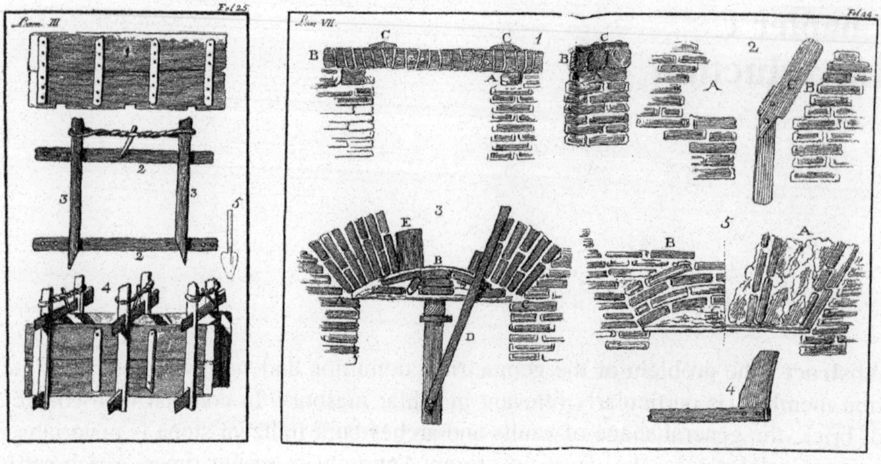

Fig. 1.1 Building in rubble and brick (Villanueva 1827: pl. 3, 7)

structure supporting an element under construction, which reproduces it shape) which should provide an exact negative of the enclosing surface of the element to be built (Fig. 1.1). Although it is necessary to conceive the shape of the element previously, either through a drawing or in the imagination of the builder, actual geometrical control is assured by the formwork.

Formwork is also used when building arches or vaults in brick, rough-hewn stone, or adobe. In this case, however, the primary purpose of centring and formwork is to support the structure until the mortar has hardened, rather than shaping the ensemble. In fact, it is evident that brick walls can be built without the use of formwork, by simply controlling the verticality of the wall with struts and the horizontality of joints with strings. In arches and vaults built with these materials, the form of the individual piece, either brick or block, does not conform to the general shape of the element under construction; thus, the thin, almost invisible, mortar wedges between bricks are responsible for the final form of the element (Fig. 1.1).

It is easy to overlook this idea, in particular when full formwork is used. In contrast, in structures built without continuous formwork, such as walls or timbrel vaults, some device must be used to control the general shape of the element under construction. Along with the struts and strings used in walls, a bent bar is used as a formal control device to guide simple timbrel vaults in the popular tradition of Extremadura (Fig. 1.2). The same idea appears in a passage by Fray Laurencio de San Nicolás, stating that

> When building a timbrel vault, formwork is not necessary. Thus, you should fix a strut with a rope tied around at the centre of the ring of the springings. The strut will be used as a guiding bar for the dome; it should have at its end a small board the size of a brick, so that each brick

Fig. 1.2 Controlling a timbrel vault. Team led by Manuel Fortea, Seville 2006 (Photograph by the author)

> rests on this board while you are placing the next one; doing the same for all courses, you will finish the dome with perfection[1]

Ashlar masonry is based on a different strategy. First, it is ideally possible to build a whole structure using a single stone, as in megalithic construction or the Mausoleum of Theodoric in Ravenna, which is covered by a single stone spanning a chamber 11 m wide. However, geological properties of the material, as well as available transport and hoisting equipment, impose practical limits on the size of stone pieces used in most buildings. Thus, to construct an ashlar structure, builders must start dividing ideally the portion of space occupied by each element in smaller pieces, called *blocks* when prism-shaped, or *voussoirs* when featuring curved faces. These pieces should fit snugly in order to materialise the final piece; that is, they must fill the space occupied by the element (Fig. 1.3). Thus, rather than using formwork as the primary formal control device, as in concrete, or entrusting this role to mortar, as in timbrel vaults, it is the shape of the elementary piece, either block or voussoir, which plays the central role in assuring the correct form of the resulting element.

[1]San Nicolás (1639: 93v): *Siendo tabicada no necesita de cimbra ninguna; y así, en el centro del anillo, a nivel del asiento de la media naranja, fija un reglón con un muelle que ande alrededor, y el reglón así fijo ha de servir de punto, o cintrel para labrar la media naranja, teniendo al fin de punto una empalma del grueso del ladrillo, para que en ella misma descanse cada ladrillo asentado, en el ínterin que otro asientas, y haciendo así en todas las hiladas, acabarás la media naranja con toda perfección.* Transcription and translation by the author.

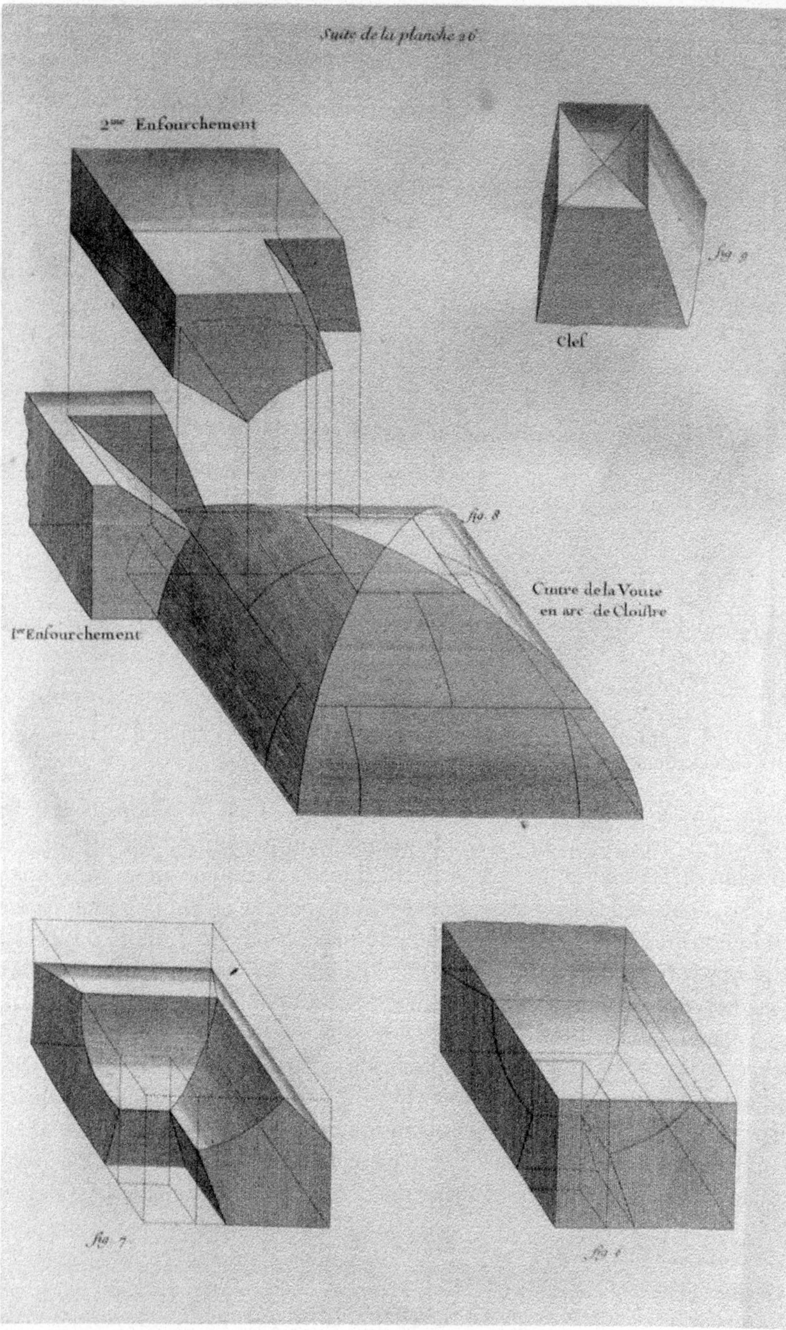

Fig. 1.3 Building a pavilion vault in ashlar (de la Rue 1728: pl. 26 bis)

These conceptions did not arise all at once. In Antiquity, Egyptian, Greek, and Roman architecture made good use of ashlar construction but, generally speaking, avoided complex geometrical problems (Pérouse [1982a] 2001: 181). In the first phases of Romanesque architecture, rubble masonry prevailed; gradually, medium-sized blocks were used, at least in some areas. However, Gothic architecture took another route: it used a two-tier system based in a network of linear elements, the *ribs*, while the spaces between the ribs were covered with portions of surfaces called *severies* resting on the ribs. In the first stages, particularly when building sexpartite vaults, voussoirs with parallel bed joints were used in vault ribs (Maira 2015: 166, 342). This means that mortar between voussoirs was responsible for the curved shape of the rib, as in brick. Later on, Rodrigo Gil (c. 1560: 24v-25v; see also Palacios 2006: 2420; 2009: 89–91) explained the role of a series of struts as a support for the ribs; afterwards, webs are built resting on the ribs. In this case, although the voussoirs are wedge-shaped, the struts guarantee the formal control of the rib, which in its turn controls the web. With the advent of the Renaissance, architects proposed and patrons demanded the transfer of Roman architectural elements, originally conceived in brick or concrete, to hewn stone, an aristocratic material; this required the quick implementation of new geometrical methods to tackle this challenge.

In the case of walls made of blocks, these procedures are relatively simple, since the result is in most occasions a cuboid whose dimensions may be set by the mason, within some limits. In contrast, in the case of some special arches, and in most vaults, as well as trumpet squinches, rere-arches or staircases, the problem is remarkably complex: voussoirs may assume a wide variety of shapes, usually with sloping bed joints and curved intrados and extrados faces (Fig. 1.4). Thus, the construction of an element in ashlar masonry involves at least six phases. First, the conception of the general shape of the element. Second, the choice of a scheme, called *bond*, for the

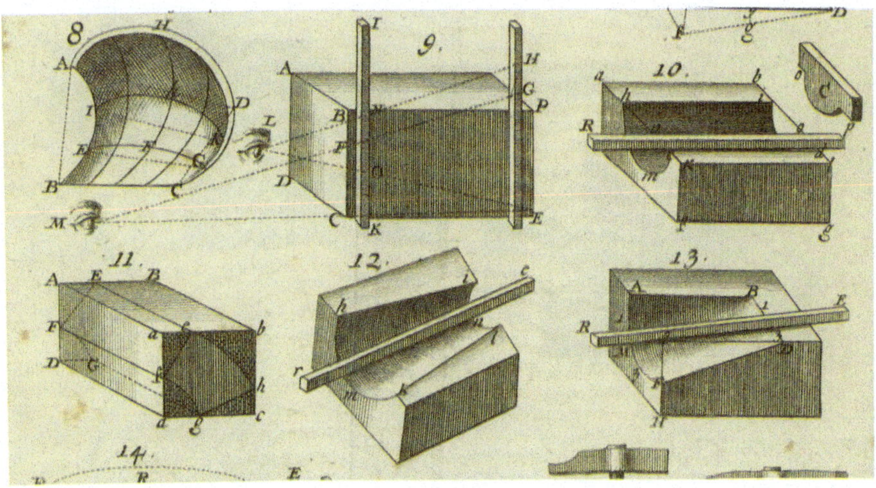

Fig. 1.4 Dressing blocks and ruled surfaces (Frézier [1737-1739] 1754-1769: pl. 28)

division of this element into pieces, either blocks or voussoirs. Next, the geometrical definition of the form of each voussoir, usually involving the shape of its faces and the angles between faces or edges. Fourth, the transfer of these geometrical data to the actual stone being shaped or *dressed*. Next, or usually at the same time, the mechanical task of dressing the stone with tools such as picks, axes and chisels; and finally, the placement of the voussoirs in their correct position in the building.

It is easy to understand that most of these processes—in particular, the definition of the shapes of the voussoirs—involve complex geometrical problems; the stability of the element usually demands the use of sloping bed joints, at least in some of the courses, as we will see in Sect. 1.2. As remarked by Palacios ([1990] 2003: 189) an apparently simple piece, a voussoir in a hemispherical dome, involves five different surfaces: portions of two different spheres in the inner and outer external surfaces; two parts of different truncated cones in in the upper and lower joints; and two identical planar shapes at the side joints (Fig. 1.5). In order to dress the voussoir, a mason usually starts by taking out stone with a chisel and a mallet until he arrives at a spherical surface, controlling its shape with a specific instrument in the shape of a curved ruler, the *templet* (not to be confused with a *template*). This operation does not by itself provide the edges of the inner faces of the voussoir, which are placed at parallels and meridians of the sphere. In order to set these lines, the mason should

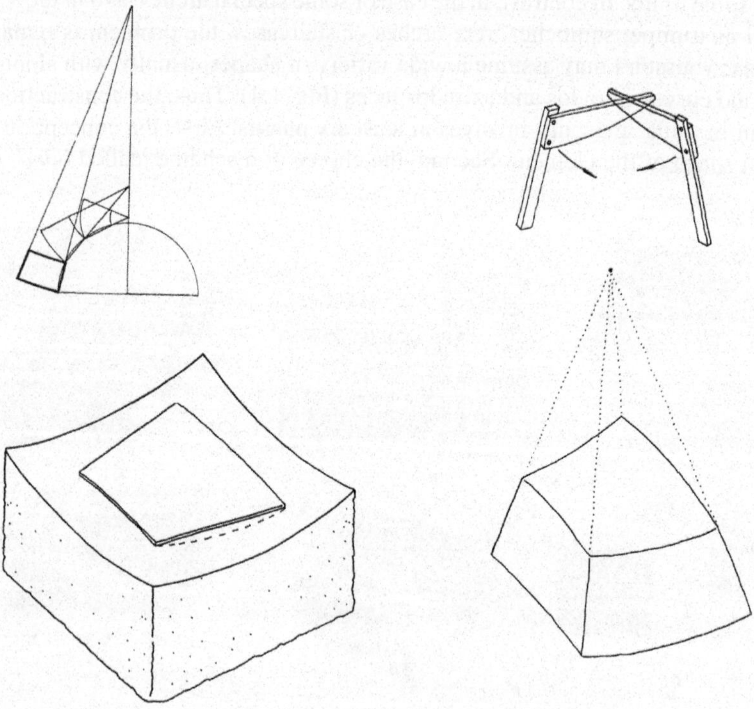

Fig. 1.5 Voussoir for a hemispherical dome (Drawing by Enrique Rabasa)

use a *template*, that is, a patch of paper or other flexible material representing the development of a cone which is inscribed in the interior of the sphere. The shape of the template needs to be defined in a preliminary full-scale tracing. Applying the template to the inner surface of the voussoir and bending it slightly to materialise a conical portion, the mason can mark the four edges of the voussoir face. Next, he can dress the joints of the voussoir using the arch square, another specialised instrument in the shape of a set square with a curved arm and a straight one. Since the straight arm of the arch square is orthogonal to the tangent to the curved arm at the junction of both branches, the mason can rest the curved side on the spherical surface, materialising normals to the spherical surface, specifically radii of the sphere. When the mason moves the arch square along the meridional edges of the inner face, the instrument materialises the surface generated by a radius moving along a meridian, which is a plane, allowing the mason to control the correct orientation of the side joints. In contrast, the surface generated by the arch square when moving along a parallel is a cone, furnishing the upper and lower joints of the voussoir.

This example shows clearly that Renaissance masons tackled quite complex geometrical problems using simple tools, an example of the smart use of scarce resources by pre-industrial societies. Rather than using equations, builders have addressed these problems by graphical means such as full-scale tracings and a set of simple, well-designed tools for the transfer of the geometrical constructions from the tracing to the stone being dressed. These techniques, developed empirically, have played an important, although by no means exclusive, role in the birth of orthogonal projection and descriptive geometry and such methods as changes of projection plane, rotations and rabatements. The historical development of these processes will be the subject of this book.

1.2 Strength, Stability, and Form

At this point, the reader may ask: what about the structural behaviour of masonry? In the industrial era, we tend to take two assumptions for granted: first, the main problems in building construction are related to the strength of materials; and second, almost any form can be built, using adequate materials and scientific computations. Also, industrial arrogance leads us to dismiss traditional, pre-industrial structural dimensioning based on arithmetic or geometrical rules or scale models as naïve or unscientific. However, what is naïve and unscientific is to extrapolate computing methods designed to analyse steel or reinforced concrete structures to masonry constructions.

Simply put, steel or reinforced concrete structural designers endeavour to optimise material using the smallest sections able to resist the loads resting on these structures, placing the material as close to their highest admissible stresses as safety factors allow. Also, both steel and reinforced concrete can withstand both compression and tension.

Masonry structures behave differently. First, brick and stone can resist high compressive stresses—up to 100 N/mm^2 for some kinds of stone, in contrast to around 25 N/mm2 for ordinary concrete—but their admissible tensile strengths are much lower. Second, as we have seen, masonry structures are usually divided into smaller pieces; joints between these pieces are usually filled with mortar; the adherence between mortar and stone is just a fraction of the already low tensile strength of stone; mortar itself may decay with time. Thus, it is advisable to discard these factors and assume that the admissible tensile strength of masonry, considered as a single, homogeneous "material", equals zero (Heyman 1995: 12, 14).

All this has led builders from many different cultures to use arches and vaults, where the material is subject almost exclusively to compression, and little or no tension appears. Doors and windows opened in walls may be solved using a *lintel*, a single horizontal element. However, lintels are subject to tension on their lower half; since stones used in construction do not resist tension well, there are practical limits to the span of an opening covered by a lintel. This has led builders to use arches, that is, linear elements with a curved directrix, usually divided into voussoirs. As a result of the shape of the wedge, the weight of the voussoir equals the resultant of two forces applied to the sides of the voussoir, known as *bed joints*, and thus the voussoir rests in equilibrium. Of course, if the bed joints were vertical, the forces on both sides of the voussoir would be horizontal, and they would be unable to compensate the weight of the voussoir; therefore, shaping the voussoir as a wedge is essential. Thus, each voussoir is subject to opposite, inward-pushing or *compressive* forces. This behaviour suits well the properties of stone, which is much better fitted to resist compression than tension. At the same time, compression holds voussoirs together; in theory, mortar between voussoirs is not necessary, although it is usually employed to smoothe the irregularities of bed joints. This system has a significant drawback: the compressive force in the lower bed joint of the entire arch cannot be exactly vertical. Thus, arches exert a horizontal force, known as *thrust*, at their lower end, known as the *springing*. Walls and piers are perfectly fitted to resist vertical forces since compressive strength in construction stones is quite high. In contrast, they cannot easily resist horizontal thrusts, which may topple them; this leads to the use of buttresses and flying buttresses. It is important to keep in mind that this phenomenon has nothing to do with the strength of the material, but rather with the stability of the element, and thus primarily with its shape.

Builders from earliest times, at least from Mesopotamia and Rome, learned empirically that for an arch or vault to be stable, it must comply with a minimal thickness. Starting from the work of Robert Hooke (1676: 31; see also Block 2006), it was proved that to avoid tensile stresses, the arch or vault should accommodate in its interior a curve, known as *pressure polygon* (Fig. 1.6), which takes the shape of a catenary in the case of a self-supporting arch. Since the admissible compressive strength of most stones used in masonry construction is rather high, failure by stone crushing rarely happens. In other words, the working stresses are far below the admissible ones for the selected material. This means that the dimension of the sections of the arch is constrained only by the need to accommodate the pressure polygon, and not by the need to resist compressive stresses. As a result, when designing masonry elements,

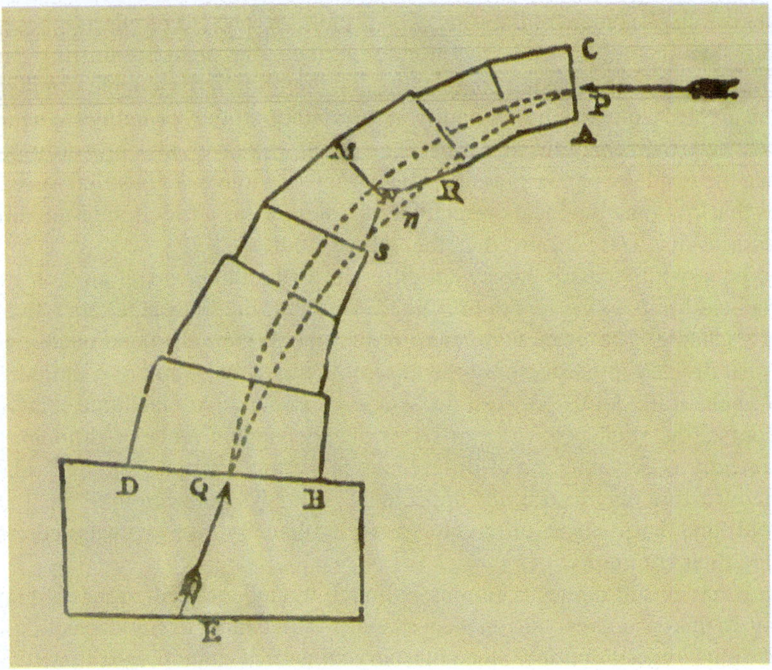

Fig. 1.6 Pressure polygon (Moseley 1843: 405)

it is usually safe to forget about admissible compressive strength and assume it is infinite (Heyman 1995: 14), as surprising as this may seem.

The consequences of such a principle are remarkable. The main structural problems of masonry constructions are not connected with stress or strength of materials, but rather with equilibrium and form. This justifies the use of arithmetic or proportional computations, graphic rules, or reduced-scale models, as suggested in many historical sources. These procedures were usually included in works dealing with architecture in general or with the geometrical problems posed by the definition of masonry elements, rather than specialised structural treatises, which appear only in the second half of the eighteenth century. Thus, I will mention these rules briefly when dealing with stonecutting theorists, but I will not analyse in depth these issues; the interested reader may consult the works of Jacques Heyman (1969; 1982; 1995) and Santiago Huerta (2002; 2004).

Thus, the problem with compressive stresses within the interior of the arch lies in their direction, rather than their magnitude. It is easy to understand that, in a stable arch, the pressure polygon loosely follows the shape of the arch. Thus, the resultant of the stresses is nearly horizontal at the top of the arch and increases its slope gradually until it reaches the springings. Ideally, bed joints should be orthogonal to these resultants to prevent the voussoir from shifting. In practice, two factors prevent sliding. Since bed faces are never perfectly dressed, their roughness causes some friction that opposes the displacement of the voussoirs. If mortar is used to regularise these rough surfaces, then the adherence of the mortar and the stone prevents shifting. None of these factors should be trusted completely, and therefore a reasonable orientation of joints in arches, from near horizontality in the vicinity of the springings to near verticality in the proximity of the keystone, is essential. However, absolute orthogonality between the resultant of compressive stresses and bed joints is not strictly necessary; studies that compute the ideal orientation of bed joints as a function of the direction of stresses (Aita 2003) are just that, theoretical studies.

The issue with vaults is somewhat different. If the thickness of an arch exceeds the practical limits to the length of a single voussoir, the mason adopts a two-stage division scheme. The piece is divided into horizontal *courses*, and each course is in its turn divided into voussoirs; the ensemble takes the shape of a cylindrical or *barrel* vault. If the joints between the voussoirs in a course match the joints in the next course, the vault behaves as a series of independent arches. Although this is mechanically acceptable, small differences in terrain strength, thermal dilatation and other factors may lead to the appearance of cracks between individual arches. To avoid this, the joints in different courses are usually laid in staggered fashion, breaking their continuity.

The behaviour of domes is similar, although the independent arches are laid out radially. In this case, even with an ideal shape, a vault cannot exclude tensile stresses, at least in its lowest sections. Since masonry structures cannot resist these stresses, as we have seen, cracks will appear. Usually, such cracks bring about much alarm and a large number of reports, as happened in Saint Peter's in Rome in the eighteenth century. However, the consequences are not so dramatic as it may seem at first view. Cracks divide the bottom sections of the vault in a series of arches, relieving tensile stresses; if the shape and the thickness of the vault are adequate, damage does not go beyond this point, and the structure is stable, as it seems to be the case with Saint Peter's.

Thus, the interactions between the formal control and the mechanical safety of masonry can be summarised in a few statements. While masonry cannot resist tensile stresses without cracking, it is quite well suited to withstand compressive stresses. Thus, when using masonry elements, builders have relied mainly on constructive types that are subject to little or no tensile stress, such as arches, vaults and domes. Generally speaking, what is relevant for the behaviour of such elements is the general shape and the placement and orientation of joints, rather than the properties of the material. In particular, the element should accommodate the pressure polygon, which can assume different shapes. The orientation of joints should approximate the orthogonals to the resultant of compression stresses on the joint, but strict perpendicularity is not necessary. What is essential is that adjoining voussoirs fit one another, that is, that the ensemble of the voussoirs fills the space occupied by the constructive element. Small voids between voussoirs can be filled with mortar, but intensive use of this practice would turn ashlar into rubble masonry. As a result, precise stonecutting is essential to guarantee efficient space filling.

Thus, structural safety places several constraints on the geometrical design of a masonry element, both at the general and the detail levels, but they are rather loose. In other words, since the stability of an arch or vault relies mainly on its capacity to accommodate the pressure polygon, many shapes can fulfil this condition: in fact, round, pointed, segmental, basket handle, elliptical, horseshoe, ogee and Tudor arches, as well as many other variants, have been successfully used in Western architecture. Also, in pre-industrial construction, both structural and formal control problems depend on geometry and are usually addressed using drawings or tracings. This explains why, during the Early Modern period, structural rules and stone-cutting problems appear in many occasions in the same treatises and manuscripts (compare for example Huerta 2004: 133–383, and Chap. 2 of this present book; see also Sakarovitch 2002: 592–598); a modern reader may be startled to find that, generally speaking, the space allocated to stereotomy in these texts is much larger than the one devoted to mechanical issues.

1.3 Dressing Techniques, Geometrical Methods, and Transportation Technology

Voussoirs of Gothic ribs are usually dressed starting from a template of the bed joint, also using a templet or rounded ruler to control the curvature of the rib. In contrast, the severies between ribs were executed in rubble, at least in the earlier phases of Gothic architecture, and thus did not require any specialised formal control tools other than the planks used to support them while mortar hardens. In fact, these severies do not need to adapt to a predefined shape. In the earliest phases, their form approximates a ruled surface passing through two or more perimetral ribs (Maira 2015: 166); later, they adopt the shape of a loosely defined double-curvature surface (Fitchen [1961] 1981: 98–99). These procedures are appropriate for Gothic vaults built as a network of ribs supporting severies. However, the Renaissance approach to ashlar construction is radically different: in contrast to the surfaces of Gothic webs, which are a result of the constructive process and do not require specific formal control methods, Renaissance surfaces must conform to predefined shapes, and thus need complex geometrical procedures in order to assure that the final element materialises this shape.

The new carving strategies devised by masons to cope with such requirements can be divided into two broad groups (Palacios [1990] 2003: 18–20; Sakarovitch 1993: 121–124; Sakarovitch 1997: 10–12). When dressing *by squaring*, the mason starts with a block in the shape of a cuboid which encloses the volume of the voussoir (Fig. 1.7). It is important to take into account that the mason assumes that the voussoir adopts in this phase its natural position (Sakarovitch 2005b: 48–49; Sakarovitch 2009b: 9–11). That is, a side face of the original cuboid, placed orthogonally to the floor or a bench, corresponds to features of the voussoir that should be placed vertically in the final element, such as the face of an arch. At the same time, other features that will be placed on a sloping position in the executed masonry, such as bed joints, are also placed on slanted planes in this phase. The advantage of such

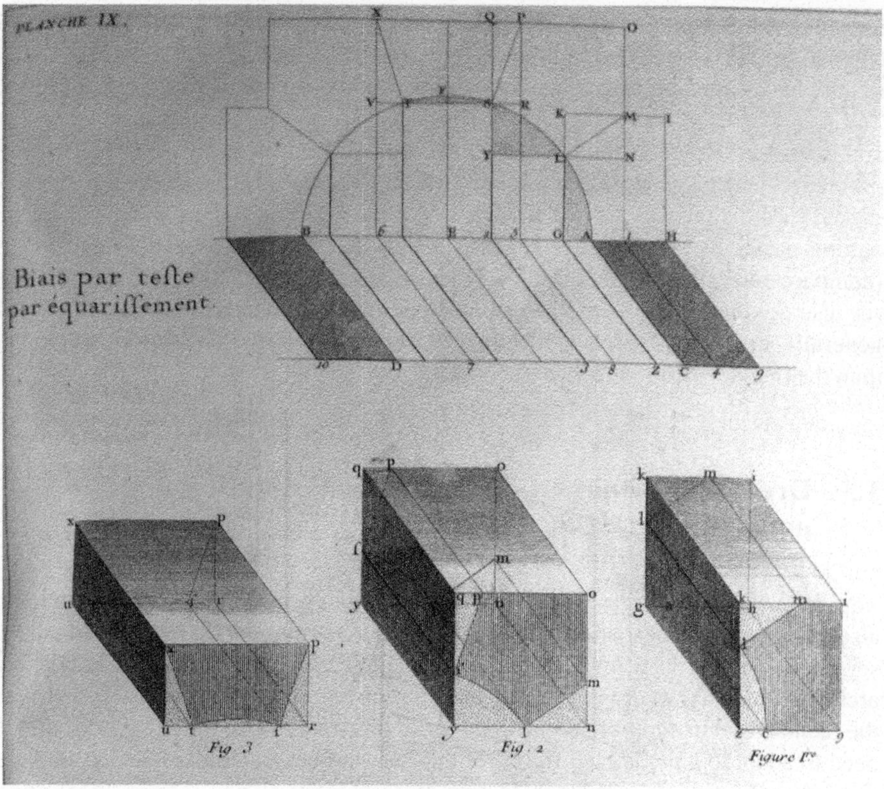

Fig. 1.7 Skew arch dressed by squaring, detail (de la Rue 1728: pl. 9)

approach is that the full-scale tracing (Fig. 1.8) needs to include no more than the horizontal and vertical projections of the voussoir faces. These projections may be drawn on block faces with a goose quill and diluted clay or with a metallic scriber; alternatively, these shapes may be transferred to templates, which are to be placed on the faces of the cuboid. The mason carves the block controlling the shape of the faces with a square; this is why this method is called "squaring". Thus, the resulting faces are either planes orthogonal to the faces of the original cuboid or cylinders whose generatrices are perpendicular to the original faces. The edges of the resulting voussoir are formed at the intersections of two faces, and, generally speaking, are not controlled directly. This notion was not universally known in the eighteenth century, and Amedée-François Frézier (1737-1739: II, 13) said they "appear at random". Thus, voussoir surfaces that do not correspond with planes or cylinders generated by the set square from orthogonals to the faces of the cuboid, such as intrados surfaces or the bed joints, should be dressed on a second stage, starting from the intersections between these planes and cylinders. Thus, the squaring method presents some disadvantages. First, it involves a two-stage dressing process. Second, if the mason is not careful, he can take off too much material, making the block useless, since the stonecutter

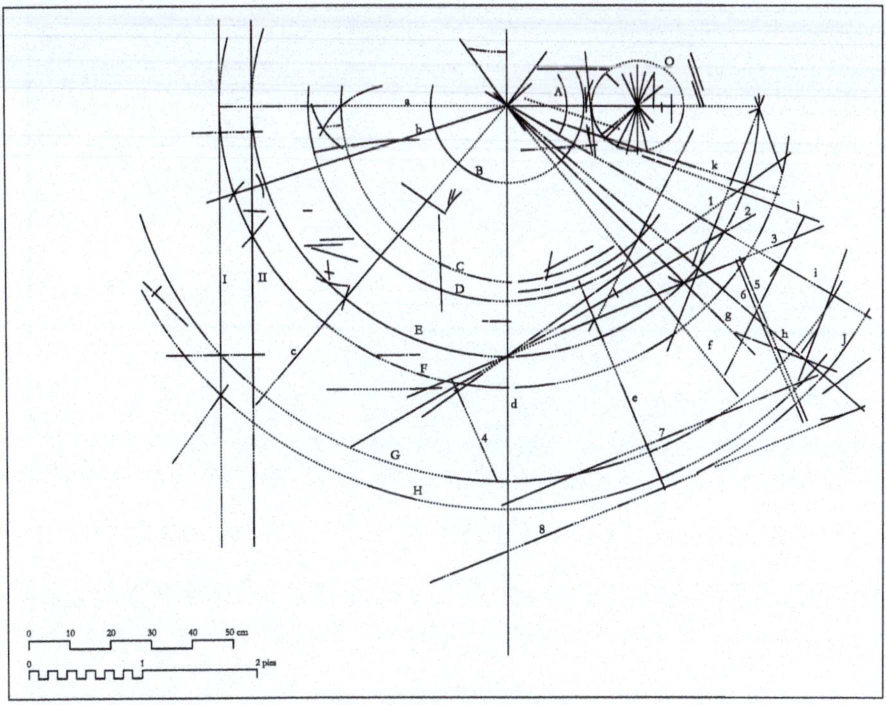

Fig. 1.8 Tracing for the vault in the chapel of Junterón in Murcia cathedral, probably prepared for dressing stones by squaring (Calvo et al. 2005a: 147)

cannot add material to the stone. Third, in most occasions, the initial block does not fit the finished voussoir snugly, leading to additional waste of labour and material.

To overcome these shortcomings, masons use the *template* or *direct* method when possible (Fig. 1.9). In this strategy, the stonecutter rotates the enclosing block ideally to fit the final voussoir as closely as possible. This is done generally aligning the largest face of the voussoir, usually that of the intrados, with a face of the initial block. Since intrados faces are usually sloping, the block does not adopt its "natural" position; the largest face, standing for a slanted plane, is placed on the floor or the bench in a horizontal position. The mason starts by marking on the faces of the initial block the shapes of the faces of the finished voussoir, such as that of the intrados, carving these surfaces directly; this explains the phrase "direct method". Since all reference to actual horizontals and verticals is lost, the mason should determine beforehand the true size and shape of the intrados and bed faces, which do not usually coincide with horizontal and vertical projections (Sakarovitch 2005b: 49–50; Sakarovitch 2009b: 11–12). Thus, in the preliminary tracing phase, the mason should perform several operations, constructing the shape of special-purpose templates starting from horizontal and vertical projections. This is why the method is also called "by templates". In the squaring method, templates may actually be used, but the mason obtains them directly from the projections; in contrast, in the templates method, the mason must

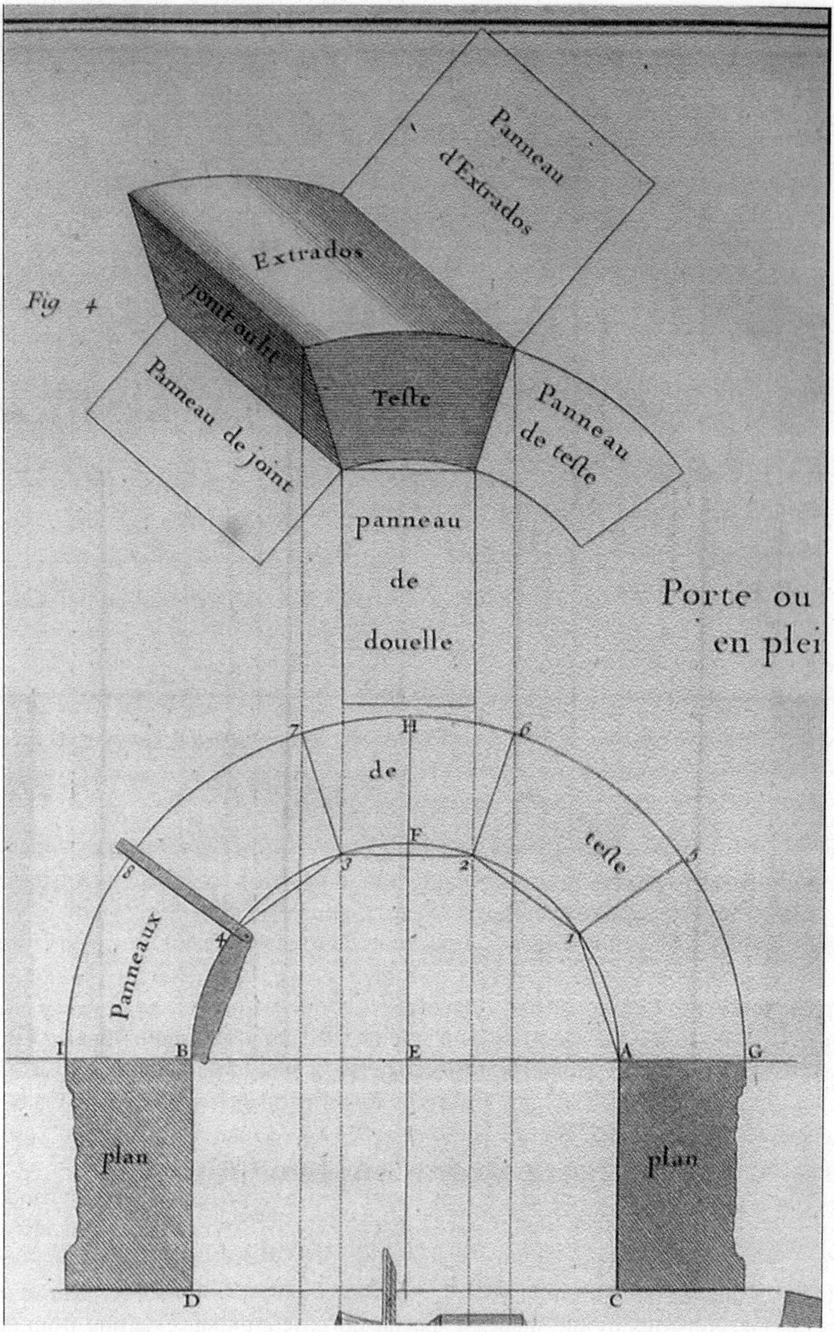

Fig. 1.9 Templates for a round arch, detail (de la Rue 1728: pl. 4)

perform a number of transformations in order to determine the true shape and size of the intrados and bed faces.

Several procedures are used to transform shapes laid on sloping planes, and thus not depicted in true shape in the plan or the elevations, into true-shape-and-size representations. This may be achieved by triangulation, that is, reconstructing the form of a planar figure from the length of its edges and diagonals. However, this method is slow, tiresome and prone to errors. Gradually, Renaissance masons developed an alternative method based on orthogonals or parallels to reference lines, starting from the projections of relevant points. Such an approach has points in common with the rotations introduced in nineteenth-century descriptive geometry by Théodore Olivier, as well as with the affine transforms of projective geometry. However, it is not easy to ascertain whether masons were thinking in terms of rotations; I will come back to this issue in Sect. 12.5.2. In any case, since both operations, triangulation and orthogonals, offer representations of planar faces, the resulting templates should be materialised in rigid materials, such as wooden boards; flexible materials, such as paper, cardboard, leather, cloth, tin sheets or other materials, can be used, provided they are applied on a planar surface. On other occasions, masons used templates obtained by conical or cylindrical developments (Fig. 1.10), either precise or approximate. Such templates must be flexible since they are applied on a previously carved surface, either spherical, conical, or cylindrical; the purpose of the template is to provide a representation of the edges of the surface in order to control the position of these lines.

The influence of transportation and hoisting technology on block and voussoir size and, indirectly, on dressing and setting out procedures, should not be left aside. Although this may seem counterintuitive, the specific surface of the ensemble of blocks used in a masonry element grows as block size diminishes, since new joints must be added in order to divide each block into smaller ones.[2] Thus, the building process with small stones is entirely different from the one used with large blocks or voussoirs. First, it is uneconomical to dress small stones precisely, since the face surface and the dressing effort would rise exponentially. Further, a small stone cannot be dressed with powerful tools such as the pick, the axe or the patent axe: it would be thrown away by the impact of the dressing instruments. Thus, small chisels must be used in this case, multiplying dressing effort. As a result, small stones are usually employed as rubble, that is, small formless pieces. The relative advantage of rubble is that it can be moved or lifted by a single worker, even with a single hand; a significant drawback is the need to use large quantities of mortar between the surfaces of irregular pieces. Large pieces of rubble are seldom used, since they would need wide mortar

---

[2]A 1 m $\times$ 1 m $\times$ 1 m cube can be built using 4 blocks measuring 1 m x 0.5 m x 0.5 m each. The combined surface of the six faces of each block equals $4 \times (1 \times 0.5) + 2 \times (0.5 \times 0.5) = 2.5$ m^2, and the total face surface of all blocks equals 10 m. If using blocks measuring 0.5 m $\times$ 0.25 m $\times$ 0.25 m, we will need 32 blocks, that is, eight times the number in the preceding example. Now, the combined surface of all the joints in a single block equals $4 \times (0.5 \times 0.25) + 2 \times (0.25 \times 0.25) = 0.625$ m^2, that is, a quarter of the total face surface for each block in the preceding example. However, since there are 32 blocks, the total face surface of the whole cube amounts to 20 m^2, that is, twice the surface of the cube built with larger blocks.

Fig. 1.10 Tracing for a half-dome in the parish church in Carnota, with flexible templates (Photograph by Miguel Taín)

joints, posing shrinking problems; also, it is difficult to carve channels for ropes, holes for pincers, or mortises for lewises into an irregular piece. The alternative to this approach is ashlar, that is, large, precisely dressed blocks or voussoirs. Dressing effort rests within reasonable limits, mortar is sparsely used or dispensed with, but powerful transportation and hoisting means are essential.

The importance of such issues in architectural and construction evolution, in particular in the Middle Ages, is usually overlooked. In the Roman world, powerful transportation and hoisting technology, together with good roads, made it possible to bring monolithic granite columns measuring 40 ft, that is, around 12 m high, from Aswan and the Mons Claudianus in Egypt to the porch of the Pantheon (Fig. 1.11) in Rome (Wilson Jones 2003: 208–210). Since technology is a social effort, not just a product of abstract knowledge, all this disappeared in the disruption of the Early Middle Ages; Western Europe had to settle for rubble or small stones. Gradually, hoisting equipment re-emerged during the Gothic period; in the early stages, it was used for rib voussoirs, while severies were still executed in rubble. In some Late Gothic schools, the web was also carefully carved; finally, this led to the blurring of the distinction between ribs and severies, paving the way for Renaissance construction. In any case, speed of execution demanded even larger pieces, and as a result more powerful equipment, as seen in the giant cranes depicted in the Hatfield drawing of the Escorial workplace (Navascués 1986).

Fig. 1.11 Rome, Pantheon, columns in the pronaos (Photograph by the author)

1.4 The Repertoire of Stereotomic Literature

In Sects. 1.4 and 1.5, I will present an elementary introduction to the geometric repertoire of ashlar construction and the concepts and methods of descriptive geometry, for readers without a substantial background in these matters. Of course, those acquainted with both subjects may skip or skim through both sections.

The simplest type of arch is the *round arch* (Fig. 1.12a). It follows the shape of a semicircle; all voussoirs are identical; its layout can be controlled with just a rope; joints are set out as radii of the semicircle. The lower surface of an arch is called *intrados*, while the upper one is known as *extrados*. The front and back surfaces of the arch are called *faces*. The surfaces where each *voussoir* leans on the preceding one are known as *bed joints*; on some occasions, for the sake of clarity, I will refer to the lower, visible edge of the bed joint as *intrados joint*, while the side edges will be called *face joints*. The lower surfaces of the entire arch are the *springings*; this term is also used for the edge of this surface belonging to the intrados. The first voussoir

Fig. 1.12 Arches, trumpet squinches and rere-arches; **a** Round arch; **b** pointed arch; **c** segmental arch; **d** basket handle arch; **e** skew arch; **f** splayed arch; **g** ox horn; **h** corner arch; **i** arch on a battered wall; **j** arch on a curved wall; **k** trumpet squinch; **l** trumpet squinch on a curved wall; **m** simple *capialzado*; **n** simple rere-arch; **o** rere-arch with arches on both faces (Drawing by the author)

of the arch is known as the *springer*, while the central one is the *keystone* since it usually the last one to be placed and is fundamental for the stability of the ensemble. The springer lies on the upper surface of a wall, pier, or column, known as the *impost*, while the edge of the wall lying below it is known as the *jamb* of the arch.

Despite its simplicity, the round arch has some limitations; in particular, it exerts significant thrust against its supports. The *pointed arch* (Fig. 1.12b) is made up from two branches in the shape of circular arcs. As Rabasa (2000: 43–47) pointed out, it is formally efficient: it can be adapted to different spans using identical voussoirs, it exerts less thrust and needs lighter centring than a round arch with the same span, particularly if the bed joints of the first voussoirs are horizontal; in this case, the ensemble of these voussoirs is called *springer* or *tas-de-charge*. However, the pointed arch is higher than a round arch with the same span; where a low-rise arch is needed, builders use the *segmental arch* (Fig. 1.12c), whose directrix is a single circular arc, shorter than a semicircle. The segmental arch exerts a strong thrust upon its supports; in order to reduce it, builders use the *basket handle arch* (Fig. 1.12d), whose directrix is an oval made up from three circular arcs, laid out so that tangents at the junctions are coincident, to avoid kinks.

In all these cases, the area covered by the arch is rectangular, since the springings are orthogonal to the face planes, that is, the outer surfaces of the wall in which the arch is opened. The intrados of the arch is a circular cylinder or a combination of several cylinders. These cases are considered elementary and thus, are absent or treated summarily in most stonecutting manuscripts and treatises. In some cases, for example, to enter a room from its corner, a *skew arch* (Fig. 1.12e) may be need. The area spanned by such arch is usually rhomboidal; that is, springings are oblique to the faces, although mutually parallel. The intrados of a skew arch is either a circular or an elliptical cylinder or a particular ruled surface, as we will see in Sect. 6.2. In other cases, springings are laid out as convergent segments, in order to achieve a broader opening in one of the faces of the arch, so the area is trapezial. Such arches are known as *splayed arches* (Fig. 1.12f), although a particular case, the right trapezium, is called in French as *corne de boeuf*, that is, "ox horn" (Fig. 1.12g). The intrados of the symmetrical splayed arch is part of a circular cone, although the apex is not materialised. The ox horn can be solved either as an oblique cone or, more often, as a particular ruled surface. On some occasions, builders place arches at the junction of two walls; the result, known as *corner arch* (Fig. 1.12h), is akin to a combination of two skew arches, while the intrados is materialised either by circular or elliptic cylinders, cones or ruled surfaces.

On some occasions, particularly in military architecture, openings must be pierced in battered walls; in these cases, the intrados of the arch or vault is usually a circular cylinder, but its intersection with the wall surface results in an ellipse (Fig. 1.12i). On other occasions, arches are opened in curved walls, either convex or concave (Fig. 1.11j). In these cases, the intrados is again a circular cylinder; however, the piece poses the problem of the intersection of two cylinders, the one in the intrados and the one on the surface of the wall, resulting in a warped curve. Arches may be also opened in vaults, in order to provide lighting from above. The problem is geometrically akin to the one posed by arches in curved walls; however, constructive solutions are entirely different; I will analyse them in Chap. 8 when dealing with cylindrical vaults.

Other pieces are conceptually similar to arches. *Trumpet squinches* (Fig. 1.12k) are used to bridge the span between two converging walls; they are like splayed arches, but in this case the apex of the cone is materialised. Also, in many occasions the open end of the trumpet squinch is curved, to support a cylindrical wall or even fancier shapes (Fig. 1.12l). *Rear arches* or *rere-arches* are openings in walls, akin to arches, where one of the faces reaches greater height than the opposite one. This effect may be brought about using two lintels placed at different heights (Fig. 1.12m), a lintel on one face and an arch in the other one (Fig. 1.12n), or arches with different rises, such as a round one on the front face and a segmental arch on the back (Fig. 1.12o).

Stairways may have round or straight plans. In the simplest case of first kind, *spiral stairs* (Fig. 1.13a), steps include a part of vertical straight strut, known as *newel*, or either rest on it. To provide a narrow opening, so the staircase may be lit from above, a helical support may be used (Fig. 1.13b). Of course, such opening cannot be very wide; when a larger well is needed, builders use cantilevered steps, starting from the external wall. In some rare cases, this layout is reversed, and steps are cantilevered from a central newel (Fig. 1.13c). When the diameter of the staircase is large, and builders do not intend to use cantilevered steps, *vaulted stairs* may be used (Fig. 1.13d). In this solution, several voussoirs laid out as an arch span the distance between the wall and the newel; then, another arch is placed at a higher level, in order to follow the rise of the stair, and so on.[3] This solution involves two concurrent problems, the radial layout of the voussoirs and the rise of the stairs: while cross-sections are semicircular, all intrados joints are helical.

In many other cases, staircases feature one or several *flights*, each one with a rectangular, rather than round, plan. However, the long edges of each flight, known as *strings*, are usually curved, to increase the stiffness of the flight. On some occasions, straight strings are employed as an alternative. Generally speaking, steps are set orthogonally to the wall, while intrados joints can be parallel or perpendicular to the wall. However, on other occasions, particularly in staircases with three or more flights, the steps, the intrados joints or both are set out radially; thus, the frontier between spiral and straight staircases is not as airtight as it may seem.

In any case, the most important position in the stereotomic repertoire is occupied by vaults and domes, known in some French treatises (Derand 1643: 329; La Rue 1728: 43) as *maîtresses-voûtes*, that is, "principal vaults". The simplest way to cover a rectangular area is the *barrel vault* (Fig. 1.14a), an extension of the concept of the round arch. As mentioned earlier, when the thickness of the arch exceeds practical limits, the piece must be divided into *courses* separated by intrados joints, which are generally parallel to the springings of the vault; on its turn, each course is divided into individual voussoirs. The cross-section of the barrel vault may be semicircular, pointed, or surbased, in the shape of a segmental or basket handle arch.

---

[3]This is a simplified explanation, based on the one given by most stonecutting treatises and manuscripts. In practice, voussoirs are laid out in staggered fashion in order to interlock each arch with the next one and avoid cracks between arches.

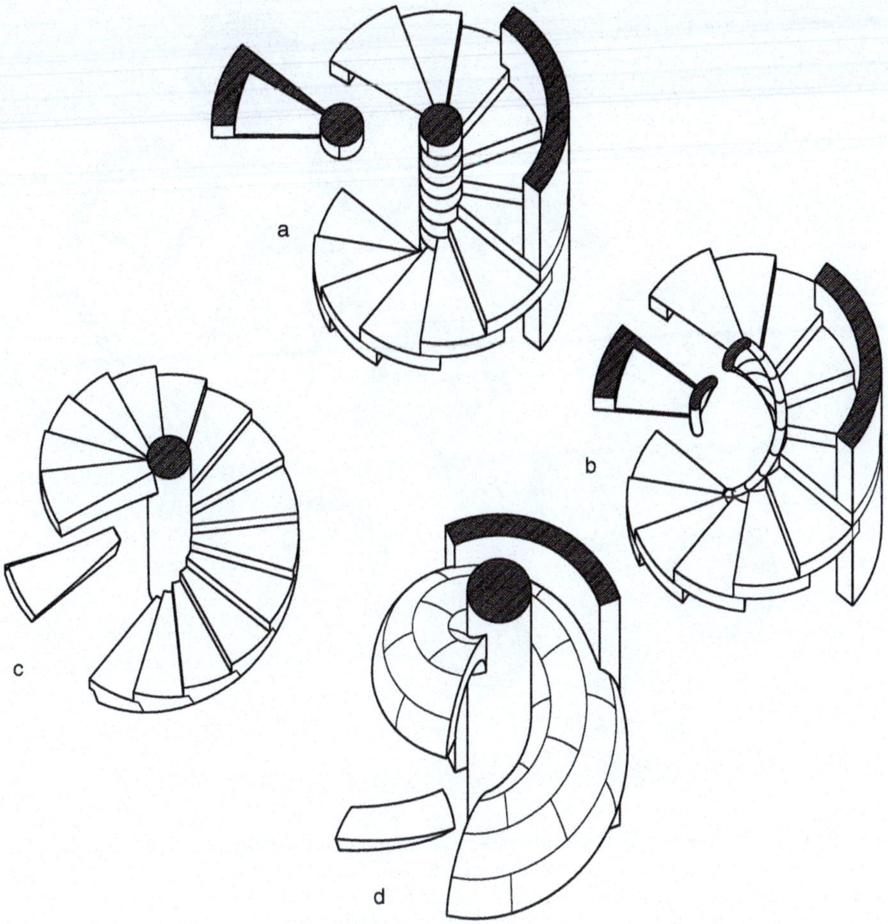

Fig. 1.13 Staircases; **a** straight-newel staircase; **b** open-well, helical-newel staircase; **c** cantilevered staircase; **d** vaulted staircase (Drawing by the author)

Barrel vaults can be grouped in several ways. Two vaults with equal radii and orthogonal axes may be laid out over a square area. In this case, the intersection of both intrados surfaces results in two half-ellipses, placed over the diagonals of the area; of course, the ellipses divide the area into four portions. Each of these portions is covered by two different cylindrical portions; the lower one starts at the springings of the vault, while the upper one starts from a semicircular cross-section of one of the vaults and extends through the horizontal generatrices of this upper cylinder until it meets the ellipses placed over the diagonals. Of course, these portions are redundant; builders use one of them and do not materialise the other one. If they choose the upper ones, with their generatrices perpendicular to the nearest side of the area, the result is a *groin vault* (Fig. 1.14b); semicircular openings can be placed over the sides of the area, allowing light to enter from above, as in Roman baths. When builders

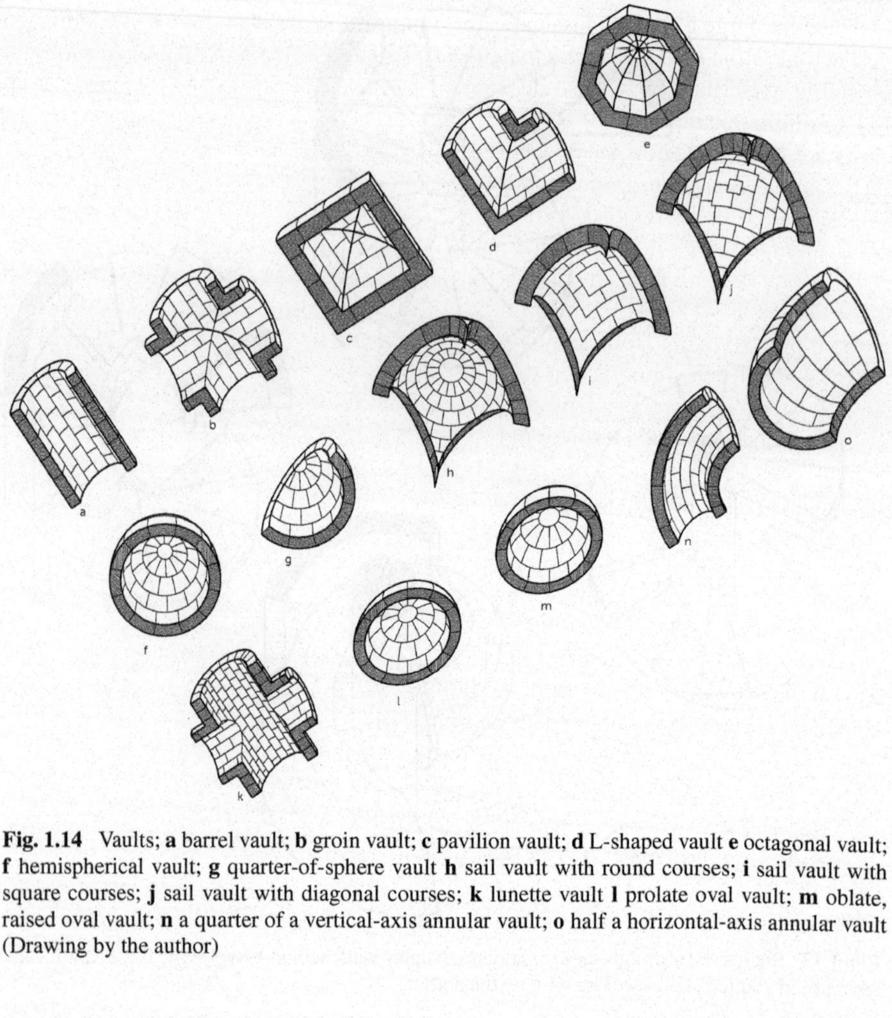

Fig. 1.14 Vaults; **a** barrel vault; **b** groin vault; **c** pavilion vault; **d** L-shaped vault **e** octagonal vault; **f** hemispherical vault; **g** quarter-of-sphere vault **h** sail vault with round courses; **i** sail vault with square courses; **j** sail vault with diagonal courses; **k** lunette vault **l** prolate oval vault; **m** oblate, raised oval vault; **n** a quarter of a vertical-axis annular vault; **o** half a horizontal-axis annular vault (Drawing by the author)

use the lower portions, with their generatrices set in parallel to the nearest side of the area, the result is a *pavilion* or *cloister vault* (Fig. 1.14c). Following the same method, three or four cylinders with the same radius and concurrent axes can be used, resulting of course in six or eight portions. Again, the builder can choose between upper and lower portions. In this case, the equivalent of the groin vault, using the upper portions, was rarely used in the pre-industrial period. In contrast, octagonal pavilion vaults (Fig. 1.14e) have been used widely, from Roman constructions such as the Domus Aurea or the so-called Temple of Minerva Medica to Romanesque buildings such as the Baptistery of Florence and several Renaissance examples, such as the pointed vault in the adjacent cathedral.

The other large group of classical vaults features a spherical intrados. The simplest case is that of the *hemispherical dome* (Fig. 1.14f), a half-sphere built over a circular springing, described by Alonso de Vandelvira (c. 1585: 60v) as "the starting point and foremost example of all Roman [i. e., classical] vaulting".[4] Usually, intrados joints are laid out as horizontal sections of the sphere, or parallels, while side joints are set out as meridians; other bonding schemes, probably derived from sail vaults, are mentioned in the treatises, although seldom used. Variants of the hemispherical dome include those with segmental and basket handle cross-sections.

On many occasions, a square area may be covered with a spherical vault, without using squinches. The problem can be solved cutting a hemispherical dome by four vertical planes, rising from a square inscribed into the circumference of the springings. The result is known as *sail vault* (Fig. 1.14h). It was built in brick in the Early Renaissance by Filippo Brunelleschi in the aisles of San Lorenzo and Santo Spirito, as well as in the Foundling Hospital; in contrast, Spanish and French builders frequently built this element in ashlar. Dividing a sail vault by the parallel-and-meridian scheme used in domes leads to odd-shaped voussoirs and much stone waste in the corners. To avoid this, another scheme, using two families of vertical planes, is frequently used (Fig. 1.14i, j). Putting the same idea in practice, triangular or polygonal vaults with a spherical intrados can be built, dividing them by vertical or—more rarely—horizontal planes.

Other types of vaults used in the Early Modern period include oval and torus-shaped pieces. In addition to surbased domes, where an oval can be used as a cross-section, a variant of the hemispherical dome may cover oval or elliptical plans. In these cases, the builder may choose to make the rise of the vault equal to the shorter axis of the area, resulting in a *prolate ellipsoid*, as in the church of Sant'Andrea in Via Flaminia by Vignola (Fig. 1.14l). He can also match the rise with the longer side of the springers' oval or ellipse, leading to an *oblate ellipsoid,* although this solution is not frequent (Fig. 1.14m); he can also use a choose a *scalene ellipsoid*, with three different dimensions for the diameters of the springing and the rise of the vault. Again, joints usually follow parallels and meridians; however, some solutions, particularly, a proposal by Gaspard Monge (1796), the founder of descriptive geometry, involve warped joints.

Torus or *annular vaults* are generated by the rotation of a circular generatrix around an axis. Such an axis can be vertical (Fig. 1.14n), allowing the vault to cover the ambulatory of a church or a gallery around a courtyard; in this case, only the upper half of the torus is used. Horizontal-axis torus vaults can be used to span a rectangular area with two semicircles at the extremes, using the top half of the outer part of the torus surface, that is, one quarter of this figure (Fig. 1.14o). In other applications, these vaults cover a rectangle with two quadrants at the ends, to materialise a niche, using an eighth part of the whole torus. In all cases, the surface is divided using the meridians (that is, the successive positions of the generating circumference), and the parallels (that is, sections by planes that are orthogonal to the surface axis).

[4] Alonso de Vandelvira (c. 1585: 60v): *Esta capilla es principio y dechado de todas las capillas romanas.* Transcription taken from Vandelvira and Barbé 1977. Translation by the author.

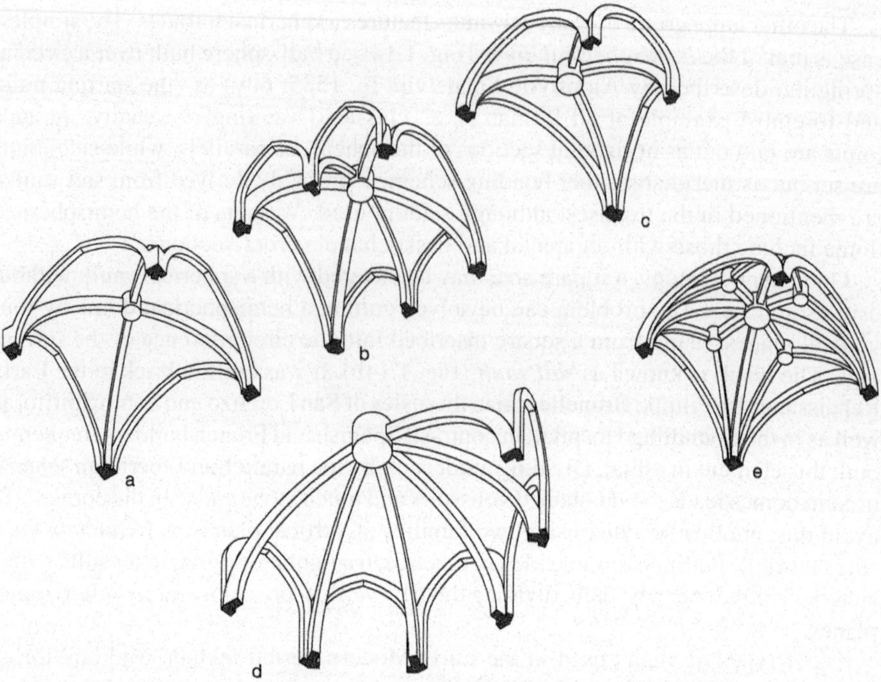

Fig. 1.15 Rib vaults; **a** quatripartite vault; **b** sexpartite vault; **c** trapezial vault; **d** semipolygonal vault; **e** star or tierceron vault (Drawing by the author)

As an alternative approach, vaults can be built using a network of linear elements, known as *ribs*, and then laying a continuous, comparatively thin surface, known as *web*, over the ribs. Each part of the web between a closed group of ribs is known as *severy*. Such a two-tier system is associated with the Gothic period. However, we should not forget that the system derives from Islamic sources, together with the pointed arch (see, for example, Bony 1984: 12–17), that Gothic was not exterminated by the Renaissance, but rather survived until the emergence of Neo-Gothic (see, for instance, Summerson [1953] 1969: 236–243) and that the rib-and-severies scheme was used for classical-looking *coffered vaults* during the sixteenth century in France and Spain.

In Western Europe, square or rectangular rib vaults can be *quadripartite,* featuring four perimetral arches and two diagonal ribs meeting at a central keystone (Fig. 1.15a), or *sexpartite* (Fig. 1.15b) including an additional transversal rib. While outer arches are usually pointed, diagonal ones are almost always semicircular, at least in twelfth- and thirteenth-century Gothic. Other rib vaults used in this period are *semipolygonal* (Fig. 1.15d) ones,[5] used in chancels and divided into triangular

[5]The name may be misleading. True semipolygonal vaults in the shape of half-pentagons or half-heptagons are used here and there. However, these vaults frequently include an additional rectangular section, placed over the junction of the chancel and the nave. Usually, this additional section is

severies by a series of radial ribs meeting at a single keystone; *trapezial vaults* (Fig. 1.15c), the standard solution for ambulatories; and *triangular vaults*, used in small spaces or as an alternative solution for deambulatories, involving rectangles and triangles.

Later, other ribs were added to create a tightly knit network, reducing the size of the severies. In a first step, each triangular section of a quadripartite vault was divided into three portions by two *tiercerons* and a *lierne*. Tiercerons start from the springings at the corners of the plan and rise to *secondary keystones*, placed in the middle of each quarter of the vault, where they meet the liernes, which join these keystones with the main one. This scheme (Fig. 1.15e) is the basis of most Late Gothic vaults; other ribs—in particular, curved ones—may be added to form quite complex vaults. While in Western Europe rib vaults usually maintain their individuality, in the German Empire transverse arches are frequently suppressed, and the whole nave of a church may be covered by a single vault with a multitude of ribs, known as *Netzgewölbe* or *net vault*.

As I have mentioned, ribs and severies were used in the sixteenth century in France and Spain to build Roman-inspired vaults, known as *coffered vaults*. The repertoire duplicates the catalogue of one-tier vaults, omitting some types that are ill-suited for the coffered system. For example, coffered barrel vaults, either semicircular or surbased, are frequent, while groin or pavilion vaults are all but nonexistent. Octagonal coffered vaults, in contrast, are more frequent, at least in the literature. Hemispherical coffered vaults, laid out using parallels and meridians, are also frequent in practice. Sail vaults, in contrast, are seldom solved with horizontal ribs; instead, two sets of vertical ribs are used. Oval and torus coffered vaults, although mentioned in manuscripts, are rarely used in practice.

1.5 Descriptive Geometry Concepts and Stonecutting

As we will see in Sect. 12.5, the concepts and methods of descriptive geometry took shape along a slow historical process, stemming from stonecutting and other sources. However, it will be useful here to present a few basic descriptive geometry concepts, in their "canonical" form, as established by the late-eighteenth-century systematisation by Gaspard Monge and his followers. Generally speaking, I will focus on the concepts that are relevant for our subject, using the vocabulary of English-language descriptive geometry manuals of recent decades, particularly Paré et al. (1996), which is different from the terminology of classic French treatises.

---

divided using radial ribs, as the rest of the semipolygonal vault, in contrast with the adjacent nave vaults, usually quadripartite or sexpartite.

As stated clearly by Monge (Laplace et al. [1795] [6] 1992: 305–306, 308–309), descriptive geometry has two aims. First, to represent in a sheet of paper—or the flat surface of a floor or wall—a three-dimensional solid. Second, to bring forward some geometrical properties which are implicit in these representations, particularly to determine the precise lengths of segments, angles between lines and shapes of planar figures when they are not rendered exactly by standard representation.

In order to achieve the first aim, descriptive geometry uses *orthogonal projection*, which is a particular case of *cylindrical projection*. To represent a solid (for example, a voussoir) its corners may be projected onto a plane using straight lines as *projectors*. This plane, materialised by the paper, the floor or the wall, is known as the *projection plane*. The intersection of the projector and the projection plane furnishes the representation of the point, known as the *projection* (Fig. 1.16a). It is easy to understand that the projector of a point must be determined unambiguously. Cylindrical projection uses parallel projectors, following a particular direction, known as the *direction of projection*. As said, orthogonal projection is a particular case of cylindrical projection in which the projectors are orthogonal to the projection plane.

This does not mean that other projections are useless. Oblique cylindrical projection (Fig. 1.16b) offers cavalier, military and transoblique or "Hedjuk's" perspective, while central projection (Fig. 1.16c) is the basis of linear perspective. All of these are present in historical stonecutting treatises and recent stereotomic studies as a vehicle for didactic illustrations.[7] However, in stonecutting the role of operative tool is reserved to orthogonal projection, and more precisely to the *orthographic* variant, where projection planes are either horizontal or vertical;[8] in the rest of this section, I will focus on this kind of representation.

In orthographic projection, the projectors of all points of a straight line or segment are parallel and intersect the same line and are thus in the same plane. The intersection of this plane with the projection plane is another line, the *projection of the line* (Fig. 1.16d). It can be easily proved that the projections of two convergent lines are convergent and the projections of two parallel lines are parallel (Fig. 1.16e, f). However, when two lines are skew (that is, neither parallel nor concurrent), *their projections* may be either parallel or convergent (Fig. 1.16g). As a result, even when the projections of two lines are parallel, we cannot take it for granted that the lines themselves are parallel; in the same way, the convergence of the projections does not guarantee the convergence of the original lines.

---

[6]This edition, taken directly by stenographers from Monge's lectures at the École Normale in 1795, will be quoted throughout this book, rather than the well-known 1799 edition; in fact, Sakarovitch (2005a: 225) lists the 1795 edition as first and the 1799 as second.

[7]To name just a few examples, Martínez de Aranda (c. 1600:114); Derand (1643: 35, 121, 165); Bosse and Desargues (1643a: pl. 5, 8, 9); la Rue (1728; pl. 9, 18, 20s, 26s, 31, 36s); Frézier (1737-1739: II, pl. 28, 29, 67; III, pl. 70, 71, 72, 73); Palacios ([1990] 2003: 38, 42, 118, 140, 188, 200, 290, 296, 366); Gelabert et al. ([1653] 2011: 327, 365, 385, 397).

[8]Orthogonal projection using an oblique projection plane is known as orthogonal axonometry, which is divided into isometrics, dimetrics and trimetrics. All these variants arose in the nineteenth century in England and Germany and were scarcely used in stereotomic treatises.

Fig. 1.16 Orthogonal projection; **a** orthogonal projection; **b** oblique projection; **c** central or conical projection; **d** projection of a straight line; **e** projection of convergent lines; **f** projection of parallel lines; **g** projection of skew lines; **h** and **i** projection of lines orthogonal to the projection plane; **j** projection of a line parallel to the projection plane; **k** projection of a line oblique to the projection plane; **l** projection of a shape parallel to the projection plane; **m** projection of a shape oblique to the projection plane; **n** and **o** projections of shapes orthogonal to the projection plane (Drawing by the author)

There is a notable exception to these concepts. If a line is orthogonal to the projection plane, the projector of any point in the line will be the orthogonal to the projection plane passing through the point. Since only one orthogonal to a plane may be drawn through a point, the projector of the point and the given line will overlap. The same idea can be applied to all points in the line; thus, the projectors of all points will overlap with the line, resulting in the same projection. Therefore, the projection of a line orthogonal to the projection plane, such as a vertical line in a horizontal projection, is a single point (Fig. 1.16h). The same idea applies to a line that is orthogonal to a vertical projection plane, known as a *line in point view* (Fig. 1.16i).

When a segment is parallel to the projection plane, the segment, its projectors, and its projection will form a rectangle. The length of the projection will equal the length of the segment, which may be measured in projection (Fig. 1.16j). In particular, the length of horizontal segments may be measured in a plan; segments belonging to planes parallel to a vertical projection plane, known as *frontal segments*, may be measured in the elevation. In contrast, if the segment is not parallel to the projection plane, it will form a trapezium with the projectors and the projection (Fig. 1.16k). Since the projection is orthogonal to both projectors, its length equals the shortest distance between projectors, while the segment itself will be longer than the projection. In other words, segments that are oblique to the projection plane are shortened in orthogonal projection and may not be measured directly. Further, it should be stressed that not all vertical planes are parallel, and thus, a segment belonging to a vertical plane which is not parallel to the projection plane cannot be measured in an elevation.

As a corollary, if a planar shape belongs to a plane parallel to the projection plane, it will preserve its shape in projection, since all edges and diagonals will maintain their lengths (Fig. 1.16l). If a shape lies in a plane oblique to the projection plane, some edges of the shape—or at least, some lines connecting points in different edges—may be parallel to the intersection of both planes; the length of these lines will be maintained in projection. In contrast, other lines in the figure will not be parallel to the projection plane, and will thus be shortened in projection. As a result, the shape of a figure oblique to the projection plane, such as the intrados of a voussoir in plan, or a slanted wall in elevation, will be distorted (Fig. 1.16m). An extreme case of such distortion arises when a figure is set on a plane orthogonal to the projection plane, such as a vertical plane in horizontal projection (Fig. 1.16n). All the projectors of the points of the figure will belong to this plane, and their projections will be placed at the intersection of the plane of the figure with the projection plane, which is a single straight line. The same idea may be applied to a figure set on a plane which is orthogonal to the projection plane of an elevation, known as a *plane in edge view*; the roof of the typical frontal view of a house drawn by children is a good example (Fig. 1.16o). For the same reasons, an angle set on a plane parallel to the projection plane may be measured directly, while an angle set on an oblique plane cannot be determined without using special operations.

A single orthogonal projection of a point is ambiguous since a given projector passes through infinite points, and all points in the projector share the same projection. Thus, to determine exactly the position of a point in space, at least two orthogonal projections, usually a plan and an elevation, must be used. Monge (Laplace et al. [1795] 1992: 310–312) stressed that two projections are enough for this task. If the projection of a point is known, its projector can be reconstructed easily as the orthogonal to the projection plane passing through the projection, which is, of

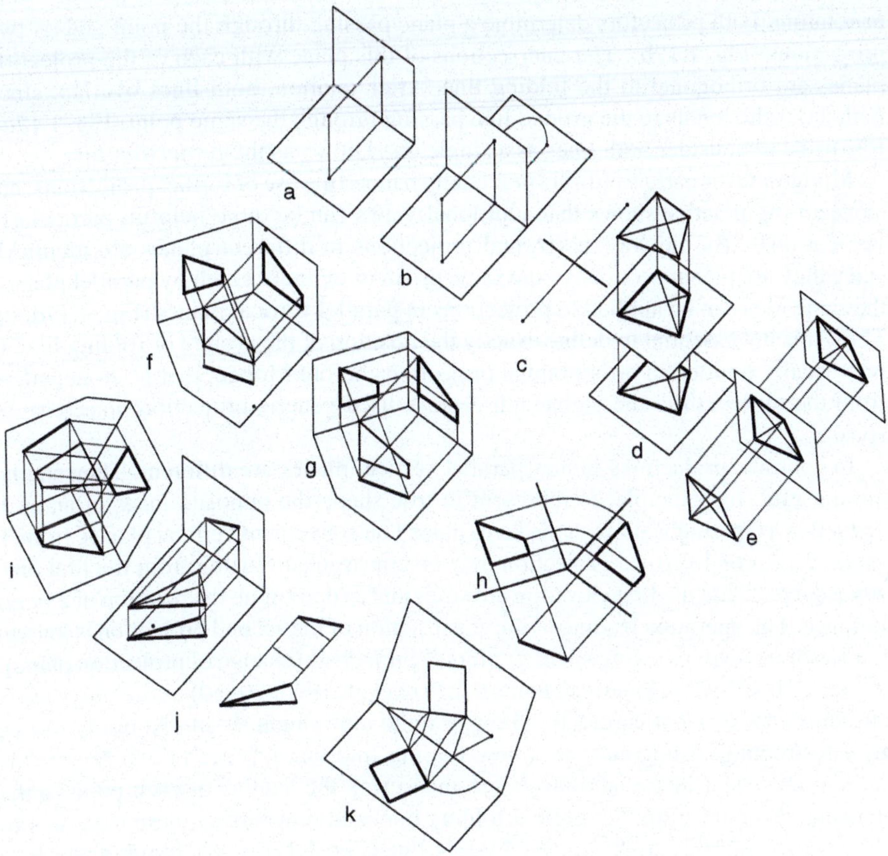

Fig. 1.17 Double orthogonal projection, auxiliary views and revolutions; **a** double orthogonal projection; **b** and **c** double orthogonal projection on a single sheet; **d** and **e** projections on several parallel planes; **f, g** and **h** change of vertical projection plane; **i** change of horizontal projection plane; **j** measuring the length of a segment through revolution; **k** rabatment of a square (Drawing by the author)

course, unique; this is one of the advantages of orthogonal projection.[9] Repeating the operation for the other projection, we get two projectors, which determine the position of the point in space, since their intersection is unique (Fig. 1.17a).

It is useful to draw both projections on the same sheet of paper. Monge assumed that one of the projection planes was rotated around the intersection of both planes, to bring them to a single leaf. The intersection of both projection planes is known in French descriptive geometry treatises as the *ligne de terre*, that is, *ground line*; recent American manuals (Paré et al. 1996: 9–11) use the term *folding line*. Before folding

[9]In oblique projection, the projector is also unique, and may be reconstructed unambiguously; however, the operation is cumbersome, since two angles are needed in order to define the direction of projection.

or rotating, both projectors determine a plane passing through the point and its two projections (Fig. 1.17b). The intersections of this plane with each of the projection planes are orthogonal to the folding line. After rotation, both lines overlap, since both are orthogonals to the ground line passing through the same point (Fig. 1.17c). It is usual to consider both lines as a single one known as the *projection line*.

Monge and his early followers practically banned the use of further projections, but stonecutting practice shows that additional views can be quite helpful, particularly vertical ones. By the way, horizontal projections in different planes are identical, since they are planar sections of the same prism of cylinder taken by parallel planes; the same idea can be applied to projections in parallel vertical planes (Fig. 1.17d, e). Thus, it is not essential to define exactly the position of the ground or folding line. It was usually omitted in orthographic projections before Monge; that is, stonecutters from the thirteenth to the eighteenth century used generic projection planes, so to speak.

In contrast, projections in non-parallel vertical planes are different. They can be quite useful, for example, to represent in true shape the opposite faces of an arch opened in a tapering wall, with the faces placed on convergent vertical planes. In such cases, the second vertical projection may be constructed starting from the first one, since the plan and the first elevation are sufficient to determine the position of a point in space. The operation leading to the construction of the second projection is known in French treatises as *changement de plan de projection* (change of projection plane); the result is called in American manuals (Paré et al. 1996: 20–30) as *auxiliary view*, stressing that it is not essential, although quite convenient. It can be easily drawn by constructing orthogonals to a new folding line through each point that needs to be represented in the auxiliary view, measuring the heights of each point in the existing elevation from the original folding line, and transferring these measures to the parallels starting from the new folding line (Fig. 1.17f, g, h). In other words, a new elevation is drawn using the parallels to determine the horizontal projection of each point and transferring their heights from the old elevation to the new one.

The same idea can be applied to construct a new projection on a sloping plane; that is, substituting the sloping plane for the horizontal one. This may be done by drawing parallels from an elevation and measuring horizontal distances from the old folding line, rather than measuring heights; the operation is called *changement de plan horizontal* (Fig. 1.17i). However, this idea was too abstract for stonecutters. It was neither used in practice nor explained in the technical literature, although a singular stonecutting method put forward by Girard Desargues (1640) and his follower Bosse (1643a) has some points in common with this operation; as we will see in Sect. 2.3.2, Desargues's proposal was violently rejected by practising stonemasons.

Rather than using this method to determine the true shape of figures set in sloping planes, stonecutters used several methods implicitly involving the rotation of figures rather than projection planes. The simplest instance of these methods is the computation of the length of a segment using a right triangle, with the horizontal projection of the segment and the difference in heights between their ends as catheti. Of course, the hypotenuse of the triangle gives the length of the segment; the operation can be understood as the rotation of the segment around its horizontal projection (Fig. 1.17j).

This procedure may be repeated as many times as necessary to determine the true forms of all kinds of polygonal shapes by triangulation; in particular, stonecutters used the edges and the diagonals of voussoir faces to construct true-shape templates (see for example Vandelvira c. 1585: 27v; Palacios [1990] 2003: 96–105).

Planar figures may also be rotated around a horizontal axis to bring them to a horizontal plane. In this operation, known as *rabatment*, from the French "rabattement", that is "act of pulling down", points move along circumferences whose centre is on the axis (Fig. 1.17k). More precisely, the circle belongs to a plane orthogonal to the axis, while its centre is given by the intersection of the axis with a line orthogonal to the axis passing through the point. Since the axis is horizontal, the circle is placed on a vertical plane, and thus it is represented in horizontal projection by the orthogonal to the axis drawn through the projection of the point. However, to locate the point exactly, we need to use an additional constraint. If we know the distance of the point to the axis, we can transfer it to the orthogonal to place it. Alternatively, we can measure the distance of the point to an arbitrary point of the axis and transfer it to the rotated figure. Nineteenth-century manuals use both approaches. Stonecutting manuals use variants of these methods; however, we should ask ourselves whether stonemasons were thinking in terms of rotation when they applied these methods; I will deal with this issue in Sect. 12.5.2.

Another descriptive geometry concept anticipated by masons is the opposition between developable and warped surfaces. Straight lines can be drawn on some surfaces, such as cylinders and cones, passing through any point; these surfaces are known as *ruled surfaces*, and the straight lines are known as *ruling lines*. In contrast, no straight line is entirely contained in a sphere or a torus; this kind of surfaces are known as *non-ruled surfaces*. If each ruling line in a ruled surface is parallel to an infinitely close line, the surface may be unfolded or *developed* without extending or cutting it; the result of the operation is a planar figure that may be bent again in order to apply it to the figure, as the templates used by stonecutters for cylindrical surfaces. The same operation can be carried out when two infinitely close ruling lines intersect, as in a cone. In contrast, when two infinitely close ruling lines are skew lines, as in a hyperbolic paraboloid or an ox horn, the surface cannot be developed without extending or cutting it. In three-dimensional space, non-ruled surfaces, such as the sphere, are non-developable (see Lawrence 2011). Of course, this fact brings about the central problem of cartography: the surface of the earth must be developed approximately in order to depict it on a plane.

1.6 Research: Sources and Methods

As any survey book, this present volume does not present original primary research, with a few exceptions. Instead, it is based in a large number of studies carried out during the last thirty years by a large number of researchers. Those quoted explicitly in the text are listed in the final reference list. However, the standard format of this book series limits the bibliography to only those references which are effectively cited in the text; a more inclusive online bibliography on stonecutting and stereotomy is in the making.

These studies explore three main groups of information. First, as in any work in the fields of architectural or construction history, extant constructions play the role of primary sources. However, construction elements do not by themselves explain the methods used for their execution. For example, direct inspection of an arch or vault does not usually allow us to tell whether its voussoirs were dressed by squaring or with true-shape templates. To gain an insight into the methods used in stonecutting, the construction historian must turn his or her eyes to technical literature. In our field, these secondary sources assume many forms, from medieval manuscripts to nineteenth-century scientific journals. As we will see in Chap. 3, the information provided by these sources, at least from the sixteenth century on, is plentiful and generally precise. However, when reading treatises, or manuscripts prepared for the press, the seasoned scholar suspects now and then that some concepts and methods have been polished with didactic purposes. Do these treatises and manuscripts reflect actual workshop practice? As an alternative to more formal sources, the scholar can turn his or her attention to personal notebooks: they are rougher and less systematic, but closer to actual practice. Also, full-scale tracings offer rich insights into the real procedures used by stonecutters. While many of them have disappeared due to loss of scaffoldings or renovation of wall facing or pavements, the few extant examples bring about valuable information about execution procedures, at least for the setting-out phase. Modern surveying technologies allow checking whether a tracing matches a particular element; if the answer is affirmative, we may assume that the procedures outlined in the full-scale drawing were used in the execution of the element. In any case, the information provided by these tracings about dressing procedures, not to mention transportation, hoisting, and placement, is indirect. Thus, in order to increase our knowledge of the technology of ashlar construction we need to use built examples, archival documentation, technical literature, and traces of execution, all at the same time.

Extant constructions. Of course, the study of a particular architectural work or constructive element must begin with the collection of data about the building itself and its historical context. Paradoxically, architectural historiography from the Renaissance on, and particularly in the twentieth century, has placed more stress on the context than on actual constructions. Studies about design and execution circumstances, including date, authorship, and patronage, based on archival documents, have flourished during the twentieth century. The techniques used in these studies—in particular, palaeography, the historiographical science that allows reliable transcription of documents according to set rules—are well known, so I will not deal with them in depth in this section.

In contrast, convenient and reliable techniques for building surveying, allowing precise measurement of non-accessible construction elements, have arisen only in the last decades, so I will explain them with some detail in the next paragraphs. In the 1990s, a researcher wishing to precisely document an arch or vault, up to the level of the individual voussoir, needed to resort either to tape measures taken from a scaffolding or to analogical or analytic photogrammetry. The first method is quite cumbersome and expensive; archaeologists employ it for façades, but it is not used widely for vaults. Analogical photogrammetry requires physically huge and prohibitively expensive equipment; as a result, its use was restricted to a few national heritage documentation centres. The introduction of analytical, computerised photogrammetry at the end of the twentieth century removed these barriers; however, the procedures were still cumbersome and the precision, both for analogical and analytical photogrammetry, was not outstanding, at least by present-day standards. Further, all these methods favoured drawing lines on horizontal or frontal planes, for example, contour lines, which are quite useful in topography, but not as much in stereotomic studies.

Around 2000, the introduction of total stations furnished with laser distance meters brought about a significant leap forward in architectural heritage surveying. These instruments provide a convenient and efficient way to gather the coordinates of physically inaccessible points, provided they are visible from the station point. Although such a method may be used to determine contour lines, it focuses on known points carefully chosen by the operator. Such an approach is particularly appropriate for stereotomic studies. Most ashlar masonry can be modelled efficiently starting from key points, such as voussoir corners. Usually, the lines joining these points are either straight ones or circular arcs, with a few exceptions such as ellipses or the warped surfaces of rib vault severies. Thus, joints in the masonry can be easily reconstructed starting from the points gathered with a total station (Fig. 1.18); in their turn, these lines provide a set of generatrices and directrices for the outer surfaces of the constructive element under study.

Although quite precise, data gathering with laser total stations is slow and tiresome, at least in comparison with alternative methods.[10] An interesting alternative, quite popular in the 1990s and the first decade of the twenty-first century, is crossing-image photogrammetry, which uses several digital photographs of the relevant element.[11] A single photograph is an instance of conical or central projection.[12] As with orthogonal projection, any point in a photograph shares its projector and its projection with other points; thus, the precise location in space of a point cannot be determined from a single image.[13] However, using two or more photographs taken

[10] A detailed, voussoir-by-voussoir survey of a complex vault may need about 5,000 points taken with a laser total station, typically needing a working day for its completion. In contrast, taking about one hundred photographs of the same vault, using a tripod, which should be sufficient for a crossing-image or automated photogrammetry survey, may take about an hour of field work.

[11] In this technique, each point in the object must be covered at least in two photographs, but it is advisable to include each key point in at least six photographs. As a result, an arch, rere-arch or squinch may be surveyed using 10 or 20 photographs, while a vault may need 50 or 100, depending on size, complexity and other circumstances.

[12] This assumes that lens distortion is suppressed or, at least, reduced to a point where it is irrelevant for practical purposes. This is assured by lens calibration, which is carried out either automatically or manually, with tools included in photogrammetry programs.

[13] Some techniques for the reconstruction of objects with straight edges and right angles have been put forward; see for example (Mula 2016). However, when dealing with pre-industrial artefacts with significant tolerances, these assumptions lead to coarse approximations.

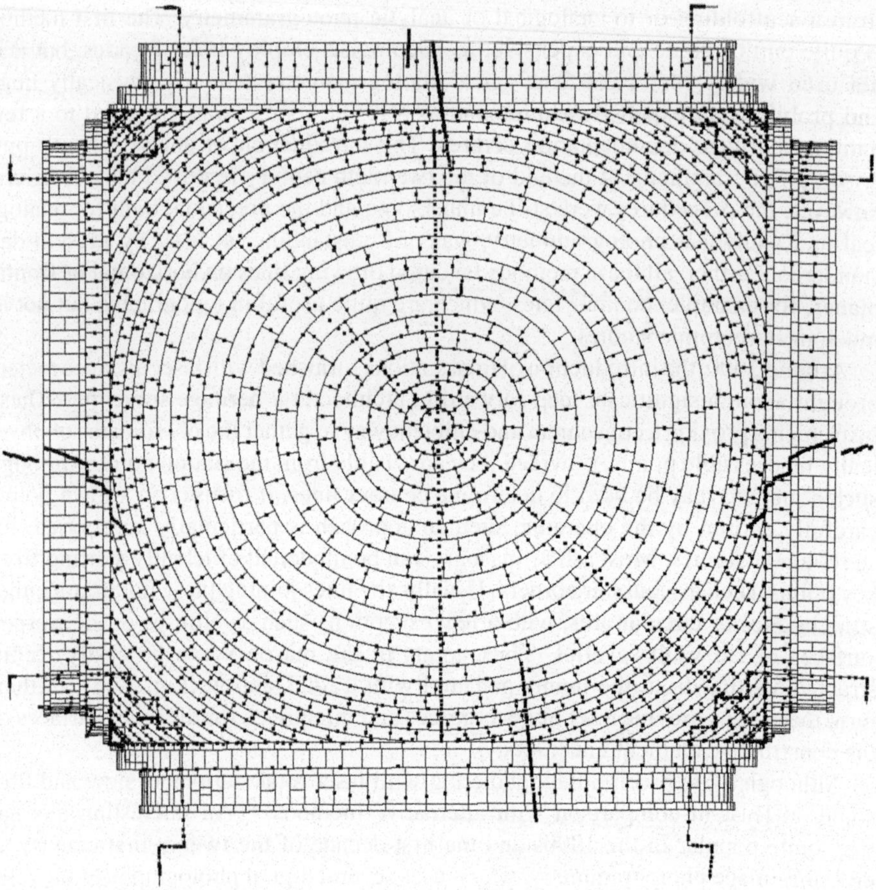

Fig. 1.18 Plan of a vault in the Hospital de las Cinco Llagas in Seville, based on data gathered with a total station (Natividad 2017: II, 390)

from known points, the projectors of a point may be reconstructed, and the point may be placed at the intersection of both projectors. Thus, the operator should mark the position of all relevant points in the architectural element in all photographs where each one is visible.[14] Some of these points may be used also by a specialised program to perform *relative orientation*, that is, to determine the successive positions of the camera unless they are determined by GPS or a similar technique. In any case, it is easy to understand that crossing-image photogrammetry trades short data gathering times for intensive office work. As with laser total stations, these points can be connected by lines and surfaces, either in the photogrammetry program or in an

[14]This means that if there are, say, 2,000 relevant points in a vault and 50 photographs are used, the surveyor must mark 100,000 points in the photographs. In practice, not all points will be visible in all photographs, and the actual number of points will be reduced, but it is still usually quite large.

external Computer Aided Drawing (CAD) program. A final step is lacking, however: photographs, like linear perspectives, do not furnish scale information. To overcome these shortcomings, the surveyor must perform *absolute orientation*, taking the coordinates of three points with a total station or an equivalent instrument and using them to rotate and scale the model in a CAD program.

Another data-gathering method that allows reduced operating times, both in the field and the office, is the 3D laser scanner. However, it must be stressed that it embodies a different paradigm. Rather than searching for significant points, laser scanning performs a massive bombardment of rays, giving as a result the coordinates of a large set of randomly placed points (a "point cloud") on the external surface of the measured object. Such approach is quite convenient for terrains, sculpture, or the analysis of degraded portions of masonry in conservation work. However, it is not the ideal one for stereotomic studies, since voussoir corners and other key points need to be cherry-picked from the huge point cloud furnished by the scanner; the sheer size and structure of the point cloud makes the extraction of these relevant points difficult.

An interesting alternative to laser scanning is automated photogrammetry. As in crossing-image photogrammetry, the surveyor starts by taking a fair number of photographs of the object. However, when arriving in the office, the location of corresponding points for relative orientation is performed by the program; in theory, the surveyor does not need to mark particular points in the photographs. Once this phase is over, the program keeps searching for corresponding points in the photographs, building a dense point cloud. In the next stages, the program overlays a polygonal mesh on the point cloud and projects portions of the photographs onto this mesh, obtaining a three-dimensional, textured model of the arch or vault under study. The result is similar to the product of laser scanning (Fig. 1.19), but the resulting point cloud is better suited to manipulation in CAD programs; this allows easy placement of significant points, or even "drawing in space" from the point cloud. Another advantage of this technology is that the surveyor can mark relevant (as opposed to random) points in one photograph, and the program will place them on the mesh. The method is also quite flexible since the surveyor can increase the density of photographs in essential areas, such as the ribs of a Gothic vault.

Full-scale tracings. As we have seen, preserved full-scale tracings offer information about stonecutting methods that cannot be deduced from built elements. Most tracings match more or less tightly actual built elements; this guarantees that the tracing shows actual constructive practice, rather than being embellished in any way. In any case, before dealing with methods for checking the degree of similarity between tracings and built constructions, we should analyse the different kinds of traces left by the construction process, to avoid some confusions.

Usually, the earliest of these marks—and the most significant for our purposes—are full-scale tracings. These tracings typically include a prefiguration of the shape of the elements the mason intends to build, drawn in orthographic projection, either horizontal, vertical, double or multiple (Fig. 1.20). From the Renaissance on, tracings

Fig. 1.19 Diagonal section of the vault of the chapel of Galliot de Genouillac in Assier, prepared by means of automated photogrammetry (Survey by Pablo Navarro-Camallonga and the author)

may include true-size-and-shape templates for voussoir faces. Those executed in scaffoldings or under walls, column bases and the like, have disappeared or are not visible; others, placed under stairways or other out-of-the-way places, have been documented in recent decades (see Sect. 3.1.1).

There are, however, other traces of the construction process that should not be mistaken for this kind of tracings. Once the full tracing was executed, the shape of templates was copied onto a template and applied to a stone block to control its dressing, marking the outline of the template on a face of the block. Again, these marks have usually disappeared, since they coincide with the edges of a block, rib or base. Only when builders changed their mind during execution, as is probably the case in Bylands Abbey, are these marks visible (Fergusson 1979). Construction processes left other marks on stone surfaces; for example, position marks were used to guarantee the correct placement of blocks or voussoirs. Frequently, short lines were inscribed on the upper faces of two contiguous blocks or voussoirs, crossing the joint between them; the blocks in the next course usually hide these marks. Position marks may be quite sophisticated. For example, the voussoirs in the extrados of the vault in the Chapter Hall of Seville cathedral are numbered; this hints strongly that the vault is elliptical rather than oval, since an oval needs only two different kinds of voussoirs, while all pieces in a quarter-ellipse are different (Gentil 1996: 142, Fig. 32, n. 5).

Much has been made of the so-called masons' marks, left on the visible faces of blocks or voussoirs, in order to quantify the work carried out by a mason or team and pay them accordingly. They are usually rather simple, including two or three chisel incisions. On some rare occasions they are quite elaborate, fostering fanciful theories about their geometrical composition and cosmic symbolism. They furnish

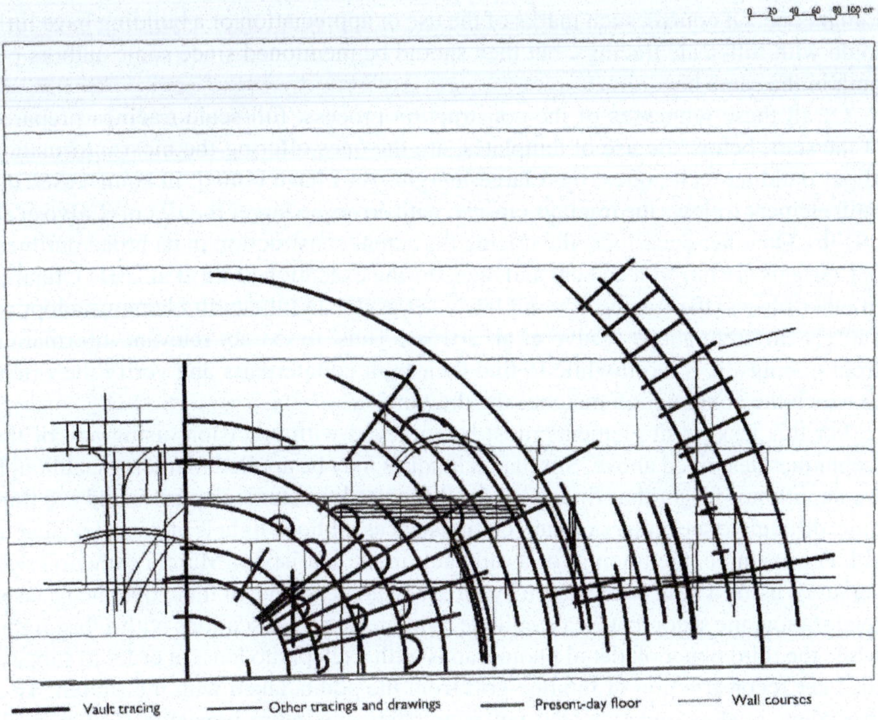

Fig. 1.20 Tracing for a sail vault in Murcia cathedral (Topographic survey by Miguel Ángel Alonso, Pau Natividad and the author)

an indirect route to authorship, especially for periods or buildings where archival documentation is not available. However, these marks do not provide information about the building process itself, in contrast to full-scale tracings. It is a bit surprising to see that a whole science, called *glyptography*, has been built around this issue, and even some serious studies about full-scale tracings (Bessac 1984) have been published in the proceedings of conferences in this field.

Other traces on building surfaces have didactic purposes. It is well known that the education of masons was carried out at the workplace. Elementary geometrical constructions, such as a rose with six petals, a good exercise to gain skill in the use of the compass and understand, at least empirically, that the side of the hexagon equals the radius of the circumscribed circle, are quite common (Calvo et al. 2013b: 288). Other diagrams, such as the spirals on top of a capital in Chartres Cathedral, presented by Branner (1960b) as operative tracings, may also be exercises on the use of the compass.

Finally, once the buildings are finished, their use leaves all kind of marks, and even when they are abandoned, visitors leave graffiti, ranging from the famous one by Lord Byron in the temple of Poseidon in Sunion to the not-so-well-known ones left by stonecutting apprentices in the spiral staircase at the Abbey of Saint-Gilles in

Languedoc. Of course, such marks of the use or appreciation of a building have little to do with full-scale tracings, but they should be mentioned since some authors put both in the same bag.

Of all these witnesses of the construction process, full-scale tracings prepared at the start, before the use of templates, are the ones offering the most information about building techniques, by a large margin. As I have hinted, in some cases the built element follows the tracings closely, with errors no larger than 2 cm (Calvo et al. 2013b). On other occasions, the tracing fits actual construction in its broad outlines, but variants are introduced here and there during execution (Taín et al. 2011); finally, in other places, the tracing is completely set aside, and the built elements adopt an entirely different shape (Calvo et al. 2013a). Thus, to extract relevant information from tracings, it is worthwhile to find their built counterparts and verify the extent to which the construction matches the drawing.

For this task, built elements must be surveyed with precision, using any of the techniques described above. Similar techniques may be applied to tracings, although we should not forget that lines are usually incised on stone or plaster and are thus quite difficult to see under ordinary circumstances; oblique light is often an invaluable aid. For example, the tracing for a sail vault in the sacristy of Murcia cathedral was surveyed using a laser total station by three researchers. One of them operated a laser total station, the second one cast oblique light on particular sections with a flashlight, while the third one took detail photographs with a telephoto lens, in order to support the later reconstruction of the line grid from the points taken with the station. This procedure is slow and tiresome but guarantees maximum precision (Calvo et al. 2013b; see Fig. 1.20).

Alternatives to this approach are based on the direct use of photographs. Careful shooting techniques should be used to guarantee that subtle incised lines remain distinct: the use of the tripod to avoid vibration is mandatory, focusing should be as precise as possible and optimal aperture is advisable. Since tracings are usually placed on a flat surface, photographs can be rectified, taking the coordinates of several fixed points with a laser total station, and using specialised programs to apply a homographic transformation, either to the photograph itself or to a CAD drawing traced from the picture (Irles and Maestre 2002). Of course, it is essential to reduce the distortion of the photograph to irrelevant levels before applying the transform. Another approach is the use of photogrammetry, either crossing-image or automated. However, the application of this technique to the tracings in the church of Saint Clare in Santiago de Compostela has shown that this method is convenient and reliable when the tracings are incised clearly, but is useless in the case of subtle, almost worn-out lines.

Treatises and manuscripts. Research methods for technical literature are of course completely different. First, for manuscripts and older books, say up to the sixteenth century, any serious study must be based on a transcription of the source. In fact, a first stage of the renewed interest in stereotomic literature arising in the 1980s was based on facsimile editions and transcriptions (Vandelvira and Barbé 1977; de L'Orme and Pérouse 1988; García et al. 1991; Martínez de Aranda and Bonet 1986).

Transcriptions may be literal, conforming to established rules set out by palaeographic literature; however, on some occasions, modernised transcriptions, according to present-day spelling, are useful. Quotations from sixteenth-century treatises and manuscripts throughout this book use modern spelling, while those from later periods are literal.

In any case, to gain an in-depth knowledge of the complex concepts and methods embodied in these sources, a detailed geometrical analysis is essential. Since stonecutters solve these problems through tracings, graphical methods have proved powerful tools for this task. In the 1980s, the resources of traditional drawing were used to illustrate the complexities of stereotomic literature and intuitively present the results (for example, Sanabria 1984: III, ill. 11, 12, 89; Palacios [1990] 2003: 366). Military perspective, that is, cavalier perspective with a horizontal projection plane, or its variants, such as transoblique or "Hedjuk's" perspective, were used frequently, since they preserve the true shape of the plan while giving an intuitive rendering of the volume of the depicted element (Sanabria 1984: III, ill. 12, 90; Palacios [1990] 2003: 80, 92).

Later, three-dimensional models, including both the constructive element, the full-scale tracing or even the stages of the dressing process, were built by means of CAD programs (Trevisan 1998; Calvo 2000a; Fallacara 2003a; Fallacara 2003b: II; Salvatore 2011a; Salvatore 2011b; Gelabert et al. [1653] 2011[15] and many later works). This has brought about powerful tools for checking the precision of traditional stonecutting procedures. Moreover, 3D models may be used to present research results in many ways, from shaded and textured renderings to actual models embedded in PDF files, used in public presentations, or spread through Internet.

Manuscripts include hidden information furnished by paper texture, watermarks, and other signs, as well as the structure of the sheets of manuscripts, all of which are essential to date anonymous works or analyse the accumulation of sheets of different origins. These issues can be analysed through well-known methods, including cold-light sheets or ultraviolet photography. Other items of hidden information are specific to technical sources. In addition to text and visible drawings, most stereotomic manuscripts present compass marks and lines drawn with a drypoint or stylus in the intermediate stages of drawing preparation. These marks can be found by traditional means, such as loupes or linen testers, or using high-resolution digital images, enhancing their contrast to very high levels. This furnishes crucial information about the geometrical methods used in the preparation of manuscript drawings (Fig. 1.21), as well as the textual history of a particular work, and indirectly on those employed in full-scale tracings, although prudence is advisable.[16]

[15]The drawings for the 2011 Gelabert edition were prepared by Yuka Irie around 2000, using a Japanese CAD program that furnished military perspectives directly.

[16]For example, in some drawings in the manuscript of Alonso de Vandelvira (c. 1585: 70r) in the library of the School of Architecture of Universidad Politécnica de Madrid, drypoint lines are quite detailed, making it possible to follow the steps of the construction of the drawing. In contrast, in the neater drawings of the manuscript in the National Library of Spain (Vandelvira 1646: 126), dry-point lines are limited to a few axes or essential lines, as if the drawing was traced from a

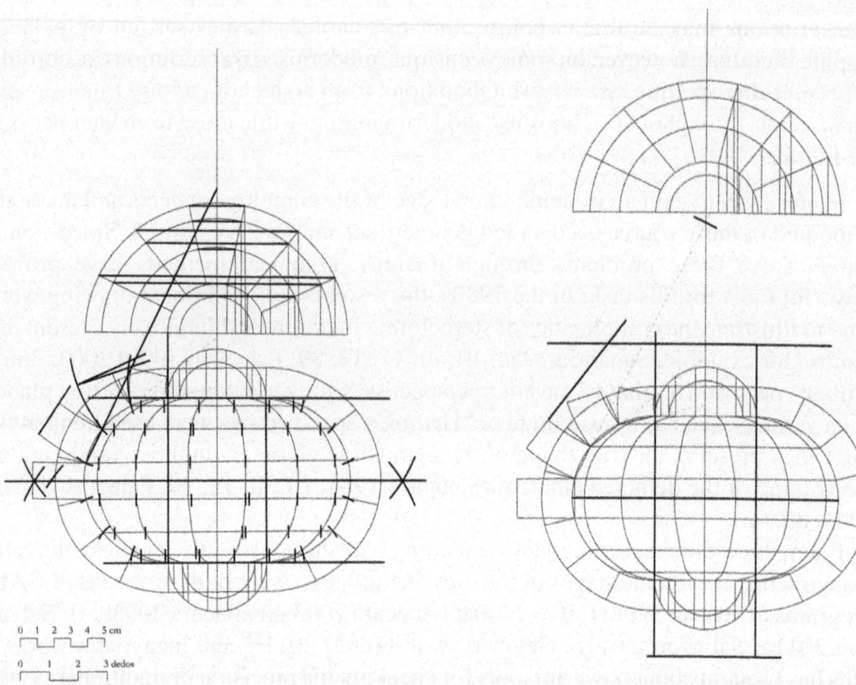

Fig. 1.21 Hidden lines in two manuscripts of *Libro de trazas de cortes de piedras* (Calvo et al. 2005a: 243, 245). Left, Vandelvira (c. 1585: 70r); right, Vandelvira and Goiti ([c. 1585] 1646, 126)

This bring us to the context of treatises and manuscripts. Of course, information about the circumstances of the preparation of a particular text, the life and career path of its author, as well as details about the buildings he was involved in, are essential for framing this literature. Such information can be found usually in historical works based on archival documentation; methodologies in this field are well known, so there is no need to deal with them in detail here. Further, no analysis of a stereotomic source is complete unless it is compared with other similar sources and built examples. CAD modelling provides a powerful tool for this end, allowing comparisons of theoretical solutions in the treatises and practical examples.

Historical research and digital innovation. In any case, CAD models do not allow direct testing of the dressing procedures suggested or implied in technical literature. Researchers have addressed this task using experimental archaeology techniques, that is, reproducing the dressing process. This can be done literally, using the tools

similar one. Further, both drawings are identical in size. This strongly suggests that the drawing in the National Library was traced from the one in the School of Architecture or another close one. See Calvo et al. 2005: 240–242.

of the craft in hewn stone,[17] working with plaster as in Renaissance models (Palacios et al. 2015, in particular 30–32) or using state-of-the-art 5-axis machines or robotic diamond wire cutting machines (Fallacara 2003b: I, 194–227; Parisi and Fallacara 2009: 290–308; Colella 2014; Barberio 2014). Following the old tradition of stereotomic models, such experiments are usually carried on at a reasonably reduced scale (see in particular Tamboréro and Sakarovitch 2003: 1900).

This approach exceeds the strict boundaries of historical research. It leverages synergies between construction history investigation, architectural education and practical application, as a recent special issue of the *Nexus Network Journal* shows (Fallacara and Barberio 2018a). In addition to its didactic potential (Sakarovitch 2003b: 75–78; Sakarovitch 2008; Rabasa 2005; Rabasa 2007b), a group formed around Claudio D'Amato in the Politecnico di Bari[18] has stressed its practical advantages both as a generator of new architectural forms through topological transforms and as a sustainable construction process; only time will tell if the availability of renewable energy will confirm this last promise.

The approach of another group at ETH Zurich is somewhat different. Rather than starting from historical examples and subjecting them to topological transformations, they begin with free-form surfaces, alter them so they are subject only to compression, excluding tension; next, they tesselate these surfaces, perform "stereotomy" (that is, they generate bed joints and extrados surfaces), check their structural viability and finally, produce a prototype by means of a CNC machine and the required formwork, all through digital processes (Lachauer et al. 2010; Rippmann and Block 2010a; Rippmann and Block 2010b; Rippmann and Block 2011; Rippmann and Block 2018).

All this leads to the application of stereotomical methods to other materials and problems, following the road of Frézier and Monge (see Chap. 6). In addition to hewn or milled stone, recent literature presents experiments in post-tensioned masonry (Todisco et al. 2018), concrete and stone powder (Azambuja and Sousa 2018), glass-fiber reinforced plastic (Diles 2018), wood and glass (Tellia 2018), weathering steel (Tibuzzi 2018). All this has been applied or proposed to solve acoustical problems (Azambuja and Sousa 2018), or ephemeral architecture for the London 2012 Olympics (Tibuzzi 2017), rather than the usual problems of preindustrial architecture. However, this book cannot deal in any depth with "Stereotomy 2.0" as Fallacara and Barberio (2018a; 2018b) call it, since it focuses on old-fashioned Stereotomy 1.0; the interested readers may consult the special issue of the *Nexus Network Journal* (Fallacara and Barberio 2018a).

---

[17] Sakarovitch (2003b: 75–78); Rabasa (2005); Sakarovitch (2006a: 2777–2782); Rabasa (2007a); Rabasa (2007b: 89–96); Sakarovitch (2008); Rabasa (2008: 127–128); Palacios and Maira (2014); see in particular Clifford et al. (2018) for the Prehispanic use of stone tools.

[18] Fallacara (2003b: I, 103–125); Fallacara (2006); Fallacara and Tamboréro (2007: 82–145); Etlin et al. (2008); Fallacara (2014a); Fallacara (2014b); Fallacara and Stigliano (2014); Errede (2014); Calabria (2014); Cascione (2014); Boccadoro and Barberio (2014); Gadaleta (2018); see also Potié (2005: 78).

Part I
Writers and Techniques

Chapter 2
Writers

Abstract In order to frame the methods and concepts of stonecutting in their historical context, this chapter surveys the biographies of the most relevant writers on the subject. The material in this chapter is arranged in four sections, dealing with a number of social groups that fought for the control of this branch of knowledge, including medieval masons, Renaissance architects, seventeenth-century clerics, and Enlightenment military engineers. Many of these authors, such as Philibert de l'Orme, Girard Desargues, Guarino Guarini, or Gaspard Monge, are well-known figures in the history of architecture or mathematics; for other writers as Alonso de Guardia or Jacques Gentillâtre, we know little more than their names. While many treatises reached the presses and exerted wide influence, other texts, in particular, Spanish ones, remained in manuscript form. Some of these notebooks are rough or sketchy, but these traits lend them a particular interest, since they reflect workshop practice directly, without the refinements of published treatises.

2.1 The Late Middle Ages and the Masons

2.1.1 The Sketchbook of Villard de Honnecourt

The National Library of France holds a unique document about medieval architecture and decoration (Villard c. 1225). It may be dated between 1220 and 1240 and includes 33 parchment sheets filled with drawings on both sides, together with short explanatory captions in Picard, some of which are translated into Latin. A text on the verso of the first sheet states "Villard de Honnecourt greets you … and (asks you) to pray for his soul … in this book you can find sound advice in the great techniques of masonry and the devices of carpentry (and) the technique of representation as the discipline of geometry requires and instructs".[1] Since this artifact includes a number

[1] Villard (c. 1225: 1v): *Wilars dehonecortot vos salve et si proie … quil proient por sarme … Car en cest livre puet on trover grant consel de le grant force de maconerie et des engiens de carpenterie. Et si troveres le force de le por … traiture, les trais ensi come li ars de iometrie ….* Transcription and translation taken from Villard/Barnes 2009.

J. Calvo-López, *Stereotomy*, Mathematics and the Built Environment 4,
https://doi.org/10.1007/978-3-030-43218-8_2

of leaves under the same binding, it has been classified as an *album*, that is, a notebook that once held blank pages. Some authors, however, have pointed out that the sheets were assembled after use, and called the document a *portfolio*.[2] Given that a fair number of the drawings, but by no means all, represent architectural subjects, nineteenth-century authors (Quicherat 1849: 65, 67, 71; Villard/Lassus 1858: 45, 47, 51) surmised that Villard was an architect or mason, and thus the document was a lodge book or *Bauhuttenbuch* (Villard/Hahnloser [1935] 1972: 220–221, 238). Along the twentieth century, other specialists have remarked that Villard was quite proficient at figurative drawing, at least for the period, while he made some blunders in architectural drawing (see, for example, Branner 1963: 135–137 and Ackerman 1997: 44–45, for a different, nuanced opinion, see Clark 2004). The most recent and authoritative study (Villard/Barnes 2009: 221–222, 223–225, 229–230) posits that the main author was a lay agent for Cambrai cathedral and travelled to many places to gather models for this building; in other words, the document is an artist's sketchbook.

Schneegans (1901: 47–48; see also Villard/Hahnloser [c. 1225] 1972: 194–200) pointed out that some of the drawings and texts were not prepared by Villard, but rather by two anonymous scribes, identified by Schneegans as ms. 2 and ms. 3.[3] Following this path, Barnes (Villard/Barnes [c. 1225] 2009: 11–14) has identified no fewer than eight hands in addition to Villard, named as Hand I through Hand VIII. In particular, a cleric seems to have added Latin annotations, including the famous one stating that Pierre de Corbie and Villard designed a church *inter se disputando*, while Villard himself had described the conception of the design in more sober terms as "a double-ambulatory church devised by Villard de Honnecourt and Pierre de Corbie".[4]

Two groups of drawings in the portfolio are particularly interesting for us. The plans of the church designed with Pierre de Corbie, the cathedral of Meaux and the abbey of Vaucelles, show objects placed at different levels, from column sections to rib layouts, in correct orthogonal projection. At the same time, two elevations for Reims cathedral, interior and exterior (Fig. 2.1), depict the nave also in orthogonal projection, although in this case there are some mistakes or licences. After eight centuries of orthogonal projection, this does not strike us as extraordinary; however, to thirteenth-century eyes, this kind of architectural representation was quite innovative. Architectural drawings from Classical Antiquity, such as the well-known tracing of the pediment of the Pantheon in the Mausoleum of Augustus (Haselberger 1994a; Haselberger 1994b; Inglese 2013), always on rigid supports and at large or full scale, usually depict objects in the same plane, either vertical or horizontal. High Middle Ages miniatures or the plan of Saint Gall follow the same practices. Only in some twelfth-century miniatures can we find walls and battlements passing in front of other

[2]Trying to be neutral, I will refer to this document as a "sketchbook", since it includes many drawings and most texts are intended as captions.

[3]Both Schneegans and Hahnloser mention two hands in addition to Villard, but Hahnloser reverses the chronological sequence and thus the names of the authors.

[4]Villard/Barnes (c. 1225: 15r, 95): *une glize a double charole, k[e] vilars dehonecort trova et pieres de corbie*. Transcription and (free) translation taken from Villard/Barnes 2009.

Fig. 2.1 Elevation and section of Rheims cathedral nave (Villard c. 1225: 31v)

walls. The drawings of Reims cathedral in Villard's manuscript may be connected in part with this tradition since modern scholarship sees the main author as an agent for Cambrai cathedral. However, Ackerman (1997: 42) remarked that the cathedral had only risen to the triforium level when Villard visited Reims; thus, he took it for granted that Honnecourt had gathered information from drawings prepared by or for the cathedral architect. I will come back to this issue in Sect. 12.5.2, but as far as I know, Villard's plans and elevations are the oldest extant examples of such a method applied to actual architectural drawing.

This new representational system was used by another draughtsman, Barnes's Hand IV, to address some construction problems in a series of small schemata. No fewer than 18 diagrams are packed into fol. 20r. All of them are first-rate sources for the history of medieval construction and surveying technology; however, I must focus here on three of them. The first one, (Villard c. 1225: 20r, dr. 9-i) represents a skew arch with parallel springings, a set square aligned with one of the imposts, and several simple marks. It carries the legend *[P]ar chu tail om vosure besloge*, that is, "In this way you may dress the voussoir for a skew arch". Although several interpretations of this simple scheme have been put forward,[5] the problem is still open, as we will see in Sect. 6.2.2. However, it seems likely that the marks are connected with the dressing process for the voussoirs of the arch, which cannot be addressed without an elevation. In this case, this scheme could represent a very old witness of double orthogonal projection, which is an essential tool of the stonecutting technique, as we have seen in Sect. 1.3. Other schemes (Villard c. 1225: 20r, dr. 18-r, 8-h; see also Bechmann [1991] 1993: 175–180), also by Hand IV, represent a splayed arch and an opening in a curved wall. The latter includes a curved straightedge or *cherche ralongée*, which seems to be used as a reference plane in order to measure the positions of voussoir corners and thus dress the voussoirs by squaring; again, an elevation is needed in order to control the shape of the voussoir, but it is nowhere to be seen in the sheet.[6]

2.1.2 *Mathes Roriczer, Hannes Schmuttermayer and Lorenz Lechler*

The sketchbook of Villard de Honnecourt is the only extant document from Early and Middle Gothic that resembles, albeit vaguely, a modern treatise on architecture or construction; to find another similar document, we must wait until the late fifteenth century. In part, such historiographical desert derives from the tradition of secrecy of masons' lodges. In any case, an interesting episode shows that these secrets were

---

[5]Branner (1957); Villard/WG/Bucher (1979: 120); Lalbat et al. (1987); Sanabria (1984: I, 38); Bechmann ([1991] 1993: 169–175); Villard/Barnes (2009: 133).

[6]Barnes (Villard/Barnes 2009: 133) posits here that the legend below this drawing *[P]ar chu tail om vosure des tor de machonerie roonde* is mismatched, and corresponds that the preceding drawing; thus, he translates the text not-so-literally as "By this means one cuts a voussoir of a cylinder of round masonry". Also, he ascribes to this drawing the legend below the preceding scheme, *[P]ar chu fait om choir deus pires a un point si lons no seront*, that is, "By this means one makes two stones fall at one point, if they will not too far apart", concluding that the drawing represents an arch about to be closed, lacking just the keystone. However, he does not seem aware of the fact that in later periods, *tour ronde* designates an arch opened in a curved wall, and thus, Bechmann's interpretation, although a bit far-fetched in the details, is generally sounder: the legends are not misplaced and the scheme represents the plan, not the elevation, of an arch opened in curved wall. Transcriptions and translations taken from Villard/Barnes 2009.

not airtight. In 1459, delegates from the lodges of the Germanic Empire met at Regensburg to approve a set of common rules governing masons' practice and create a network to enforce them. The lodges of Bern, Vienna, Cologne and, most of all, Strasbourg, were placed at the summit of the organisation. One of the rules approved in the meeting stated that no mason should teach anyone who was not a mason or a formal apprentice how to extract elevations from a plan (Frankl 1945: 46–47). It may be surmised that Regensburg masons, who had hosted the meeting and were left out of the top tier of the network, were not happy, and probably did not feel bound by these statutes (Roriczer, Schmuttermayer and Shelby 1977: 46–61).

Later on, Mathes Roriczer, a prominent Regensburg stonemason, brought to the presses three short booklets, one on pinnacles (1486), one on gables (c. 1490b), and one on geometry in general (c. 1490a), explaining masons' secrets. Although only six pages long, the booklet on geometry is of the utmost interest for us since it shows the extent of masons' geometrical knowledge at the end of the fifteenth century. For example, it includes an incorrect construction for the regular pentagon, a problem that was correctly solved in Euclid's *Elements* (c. -300; see also Meckspecker 1983); we will come back to these issues in Sect. 12.4.1. There is also a most interesting passage in the booklet on pinnacles (1486: 5r-5v), where Roriczer explains how to construct an elevation starting from the plan (Fig. 2.2). In essence, he draws an axis of the elevation; he measures distances to the axis in the plan and transfers them to orthogonals to the axis placed at relevant levels in the elevation. However, a

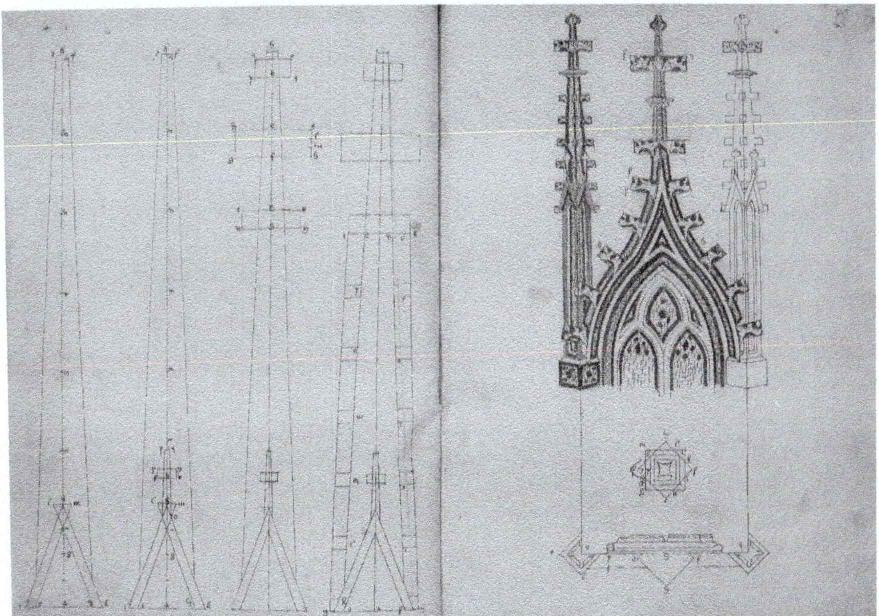

Fig. 2.2 Pinnacles and gablet (Facht 1593: 39v-40r, after Roriczer 1486 and Roriczer c. 1490b)

twentieth-century architect or engineer would be surprised to see that Roriczer does not use projection lines connecting plan and elevation (see Sect. 3.1.3).

This procedure must have been common knowledge among late-fifteenth-century masons, at least in Germany, and probably in all Europe. About two years later, Hannes Schmuttermayer, a Regensburg goldsmith, working independently from Roriczer, published another booklet on pinnacles using the same method, although the explanations are less clear (Schmuttermayer c. 1500).

Another interesting text from the period is Lorenz Lechler's *Unterweisungen und Lerungen für seynen Son Moritzen* (1516), which is a series of instructions from a stonemason to his son, who was starting to work in the trade. The original manuscript is lost, but the text and a few drawings of rib and respond sections have been preserved in another manuscript by Jacob Facht von Andernach (1593: 41r-61r); there is also another copy prepared in Bern around 1600 (Shelby and Mark 1979: 113–114). These notes deal mainly with structural dimensioning of the main elements in a church and, generally speaking, are out of the scope of this present book. What is more relevant for our purposes is the design of rib and respond sections (Fig. 2.3), a problem that was addressed, or at least alluded to, in Villard de Honnecourt's templates (c. 1225: 32r). Lechler is more specific: as a rule, he starts from a square whose side is related to the width of the church walls and then inscribes rotated squares inside the first one (Facht, Roriczer and Lechler [1516] 1593: 41r, 42r-42v), a technique used also by Villard (c. 1225: 19v, 20r) as well as Hand IV and Roriczer (1486: 3r-3v; 1490b: 1r-1v). However, he divides segments into equal parts, embellishes profiles with curved

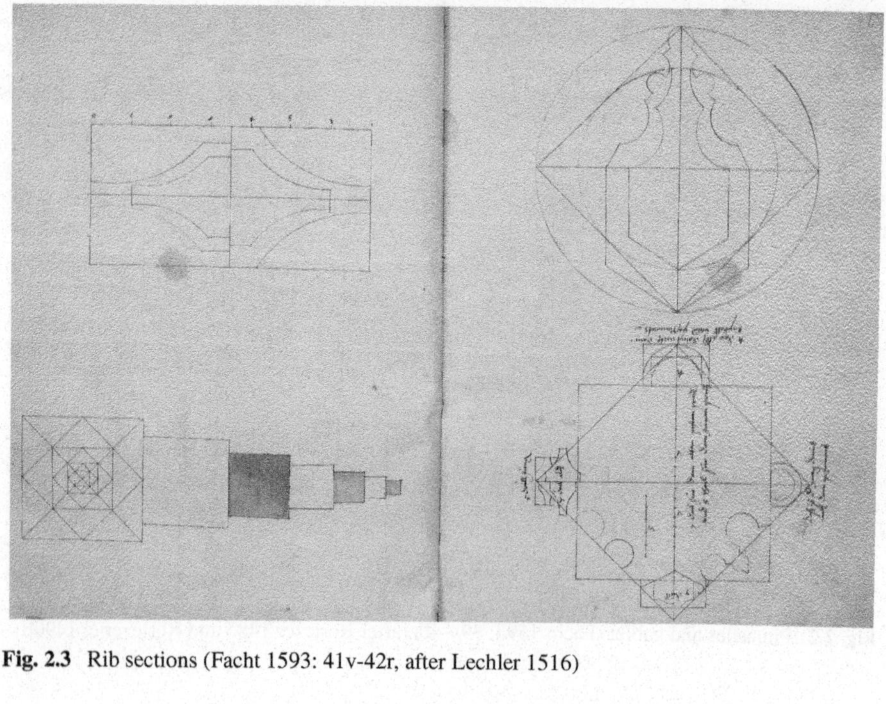

Fig. 2.3 Rib sections (Facht 1593: 41v-42r, after Lechler 1516)

sections, or leaves aside the initial square, starting with a rectangle (Facht, Roriczer and Lechler [1516] 1593: 41r-42v).

Up to this moment, we have not seen actual stonecutting problems,[7] except in the most schematic drawings of Hand IV. German manuscripts at the turn of the sixteenth century, such as those by Wolfgang Rixner (1467–1500) and Hans Hammer (c. 1500) finally address the spatial problems posed by rib vaults (see Sect. 2.1.6). However, these texts deal with complex *Netzgewölbe* or net vaults, while the texts of Pedro de Alviz, Rodrigo Gil, Hernán Ruiz and Josep Gelabert focus on simpler cases for didactic reasons; thus, I will survey these manuscripts in the next pages and come back to the German masters in Sects. 2.1.6 and 2.1.7.

2.1.3 *Pedro de Alviz and Rodrigo Gil de Hontañón*

The first known document resembling a modern stereotomy manual was probably the original of MS 12.686 in the National Library in Madrid, written around 1540. The late sixteenth-century copy includes a fair number of drawings of trumpet squinches, splayed and skew arches, arches opened in corners or round walls, sloping vaults, rere-arches, barrel, groin, and pavilion vaults, as well as rib vaults, some of them with accompanying text; the absence of staircases and spherical vaults is remarkable. Some pieces are solved by squaring, other by templates. The frequent use of triangulation connects the manuscript with a later one by Alonso de Vandelvira (c. 1585).

Javier Gómez Martínez (1998: 31–32) remarked that three rib vaults shown at the end of the manuscript are not simple didactic exercises; since one of them includes a scale and the dimensions of the other two are consistent with the scaled one (Fig. 2.4), they seem to be prepared for a particular construction. He went further and posited that the building that best fitted these drawings was the parish church of Garcinarro, near Cuenca. For Gómez Martínez, these details date the manuscript to the 1540s and connect it with Pedro de Alviz, a mason in the entourage of Francisco de Luna, who was a prominent figure in the area in the mid-sixteenth century and the grandfather of another writer in the field, Alonso de Vandelvira. Later, Ricardo García-Baño (2017) analysed the handwriting and the codicological structure of the notebook, prepared a survey of the vaults in the Garcinarro church and compared them with the drawings in the manuscript. This led him to confirm that the rib vaults are working drawings for the Garcinarro church. However, he has shown that the manuscript in the National Library is not in the handwriting of Pedro de Alviz, but rather seems to date from the late sixteenth century; it appears to be a copy of an earlier one dating from the mid-sixteenth century, perhaps by Alviz or somebody in his entourage. This would explain why some solutions are somewhat crude, compared with late

[7]Shelby and Mark (1979: 124–126) take it for granted that Lechler knew Facht von Andernach's techniques for the control of rib curvatures and keystone heights. Such position is defendable, since Hammer (c. 1500) and Rixner (1467–1500) present early, rough versions of these methods; however, for the sake of clarity, I will deal with Hammer, Rixner and Facht von Andernach separately.

Fig. 2.4 Plan of a rib vault for the parish church at Garcinarro (Alviz, attr. c. 1544: 28v)

sixteenth-century texts; at the same time, this hypothesis fits well with the frequent use of triangulation, anticipating Vandelvira's approach.[8]

A particular detail connects this manuscript with another one by Rodrigo Gil de Hontañón: the rib vaults in both manuscripts are inscribed in a circle, although only

[8]Taking all these issues into account, I will refer to this manuscript throughout this book as "Alviz c. 1544", although the reader should remember that it is the original of the manuscript of the National Library that is attributed to a mason in the entourage of Alviz.

a quarter or half the circumference is depicted, in order to fit into the available space of the page (Alviz c. 1544: 28v, 29v; García and Gil de Hontañón 1681: 25r). In both drawings, the semicircle or quadrant seems to be used to control the layout of tiercerons. However, Gil de Hontañón reuses it as a cross-section of the vault and furnishes valuable information lacking in the operative drawings of the Alviz manuscript.

Gil was master mason of the cathedrals of Salamanca, Segovia and Plasencia (Hoag 1958; Casaseca 1988), and a prominent figure in mid-sixteenth architecture in Central Spain. The original of his manuscript is now lost, but the whole text or part of it was included in another one, *Compendio de arquitectura y simetría de los templos*, written by Simón García (1681), an obscure late-seventeenth-century architect from Salamanca. Most scholars dealing with this subject (Menéndez Pelayo 1883–1889: II, 570, note 1; Camón 1941: 305; Hoag 1958: 404–410; Sanabria 1982: 282) identify Gil's contribution with the first six chapters of the *Compendio*, although there are some disagreements about the details (see in particular Gómez-Moreno 1949: 11–12 and Bonet [1979] 1991). Hoag (1958, 410) surmised that Rodrigo started planning his manuscript in the early 1560s, while Sanabria (1982: 283) argued for an earlier date, between 1544 and 1554.

The first five chapters in the *Compendio* are an assortment of architectural problems, dealing with plans and dimensions of churches, structural rules (both graphical and arithmetical), windows, towers, and staircases; he takes some material from Vitruvius and focuses on anthropomorphic measurements (Chanfón [1979] 1991; Huerta 2002; Calvo and Salcedo 2017). Although there is a drawing of a helical-newel stairway or *Caracol de Mallorca* (Fig. 2.5), according to Sanjurjo (2015: 66–67) it is included here more as an illustration of anthropomorphic concepts than as a real explanation of a constructive technique. Thus, for our purposes the most interesting section of the *Compendio* is the sixth chapter, which deals with rib vaults; it includes the drawing mentioned before and furnishes valuable data about the geometric layout of these constructions, its tracing methods and even the execution of the centring. For example, full-scale tracings prepared on scaffoldings, exactly under the vault, allowed masons to know the curvature of the ribs, the height of the secondary keystones and the height of the struts supporting the ribs. Also, they were used to control execution after placing the voussoirs, checking that plummets hanging from the keystones overlapped their theoretical position in the tracing (Gil de Hontañón c. 1560: 24v-25r).

2.1.4 Hernán Ruiz II

Hoag (1958: 406) suggested that Simón García had probably polished the style of Rodrigo Gil's contribution to the *Compendio* and structured it into chapters. If this is true, Rodrigo's original manuscript would have resembled a personal notebook, with

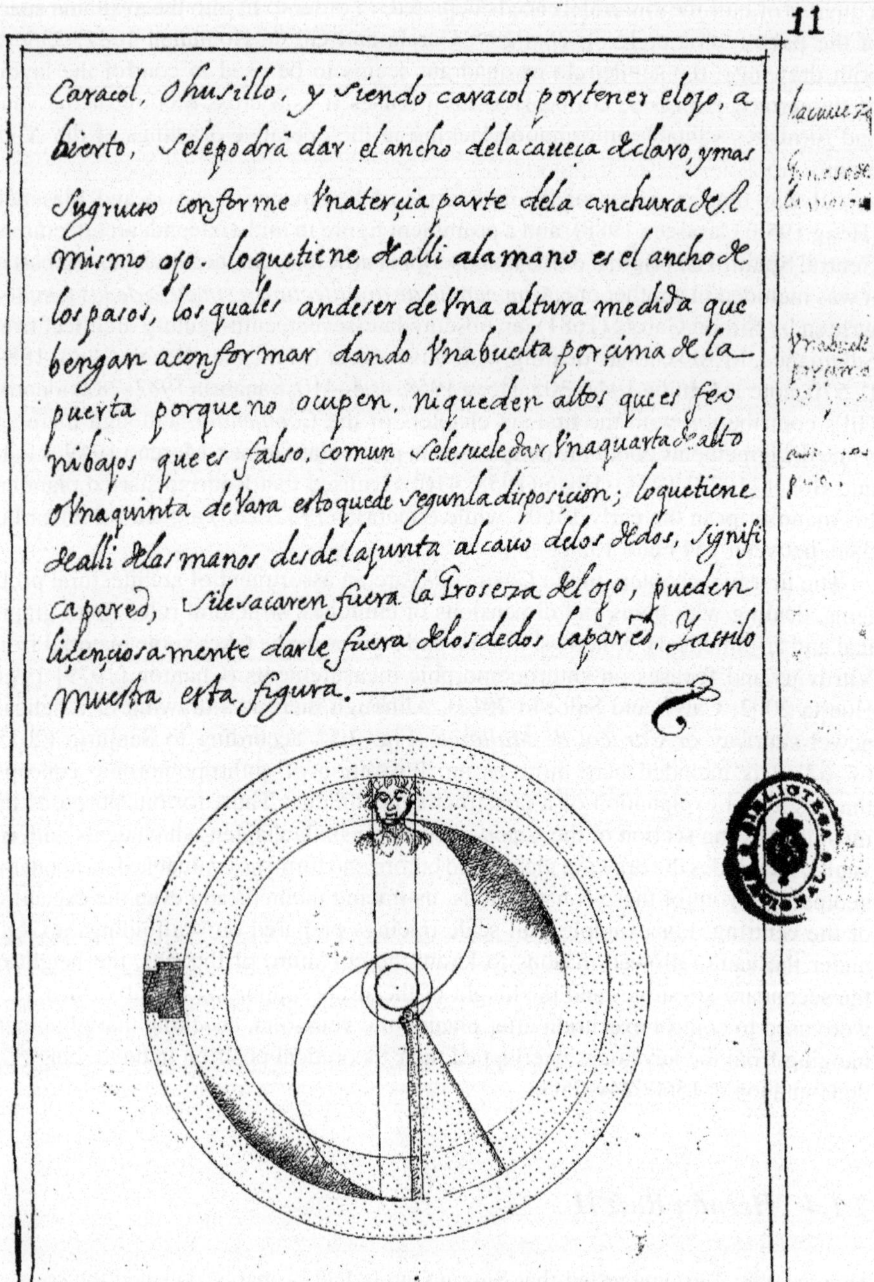

11

Caracol, Ohusillo, y siendo caracol portener elojo, a
bierto, Selepodrá dar el ancho delacaueça declaro, ymas
sugrueso conforme Vnaterçia parte dela anchura del
mismo ojo. loquetiene dalli alamano es el ancho de
los pasos, los quales andeser de Vna altura medida que
bengan aconformar dando Vnabuelta porçima dela
puerta porque no ocupen, niqueden altos que es feo
nibajos que es falso. comun selesuele dar Vnaquarta de alto
oVnaquinta de Vara estoquede segunla disposicion; loquetiene
dalli delas manos desdelajunta alcauo delos dedos, signifi
capared, silesacaren fuera la groseza del ojo, pueden
liçençiosamente darle fuera delos dedos lapared. Yassilo
muestra esta figura.

Fig. 2.5 Spiral staircase (García 1681: 11r after Gil de Hontañón c. 1560) a looser structure. Hernán Ruiz II, also known as Hernán Ruiz the Younger,[9] prepared such a manuscript, preserved in the library of the School of Architecture of Madrid (Ruiz c. 1560; see also Ruiz/Navascués 1974; Banda 1974; Jiménez 1998a; 1998b). After being apprenticed with his father, he inherited the post of master mason of the cathedral of Córdoba, which was being built inside the Great Mosque; although utterly preposterous, such undertaking demands great technical competence. In any case, a simple bishopric was too small for his ambitions, so he moved to Seville, the hub of commerce with the Americas, where he secured first the post of master mason of the Hospital de las Cinco Llagas and, afterwards, the Metropolitan cathedral, the widest church in Christendom, rightly known as Magna Hispalensis. Again, he tackled apparently unsolvable problems: he finished the ill-conceived Royal Chapel, built a Christian bell tower over the Almohad minaret, and started the construction of the Chapter Hall, the first elliptical space built from the ground up in the Renaissance (Banda 1974; Morales 1996; Gentil 1996).[10]

At his death in 1569, the Chapter Hall was unfinished. The chapter entrusted the post of master mason to Pedro Díaz de Palacios, leaving aside Hernán Ruiz III, the son of the master, who fled to Málaga carrying the drawings and templates prepared by his father for the elliptical vault. Díaz was unable to complete the vault without the templates and was expelled from the post of master mason. The cathedral chapter consulted with highly-skilled stonemasons and architects, such as Francisco del Castillo, Andrés de Vandelvira, Juan de Herrera and Juan de Minjares, and finally entrusted the direction of the works to Juan de Maeda, master mason of Granada cathedral, who measured up to the task and carried on with the execution of the vault, which was finally completed by his son, Asensio de Maeda (Gentil 1996).

Such a convoluted history shows that tracings and templates were an essential tool of Renaissance stone construction, and also that medieval secrecy was still alive in the second half of the sixteenth century. Although the sketchbook in Madrid does not include drawings about the Chapter Hall, it offers valuable information about the skills and interests of an architect and stonemason of the period. Several authors have put forward different opinions about the date of the manuscript; however, the most recent and authoritative study, by Alfonso Jiménez (1998a; see also 1998b), places it between 1555 and 1567. Again, it is a mixed bag of drawings and texts, including a translation of the first book of Vitruvius, a large number of drawings on architecture

---

[9]Most of the studies about the master mason of Seville refer to him as *Hernán Ruiz el mozo*, or *el joven*, that is, "the younger". However, his son Hernán was also an architect; he is usually called "Hernán Ruiz III". Thus, it seems clearer to mention the master of Seville as "Hernán Ruiz II", as done by Banda (1974).

[10]The Chapter Hall was begun in 1558, after the construction of Sant'Andrea in Via Flaminia in 1550-c. 1553 by Vignola, a rectangular-plan church covered by an oval dome. It was finished in 1592, before the completion of Sant'Anna dei Palafrenieri, also by Vignola, with an elliptical plan, which was begun in 1565; however, Sant' Anna was not covered until the eighteenth century. See Lotz (1955: 52); Lotz ([1974] 1995: 119–120); Gentil (1996: 108–110, 120–127).

and ornamentation, some isolated structural rules, idiosyncratic perspective methods, many geometrical schemes, and several stonecutting drawings.[11]

Most of these stereotomic problems, dealing with vaults, staircases, rere-arches and arches, are solved by squaring. Except for a remarkable rere-arch and a tierceron vault, they are rather simple but furnish an interesting, albeit slightly outdated, picture of stonecutting methods in the first half of the sixteenth century. The stonecutting tracings in Murcia cathedral, dating from the 1520s and perhaps the 1530s (Calvo et al. 2005a; Calvo et al. 2013b), lack templates, implying that voussoirs were carved by squaring. In contrast, a tracing from around 1543, using templates based on cone developments, has been found in the rooftops of Seville cathedral (Ruiz de la Rosa and Rodríguez Estévez 2002), and can be safely ascribed to Hernán Ruiz II's forerunner in the post of master mason, Martín de Gaínza. Remarkably, Ruiz does not include any tracing for hemispherical domes.

In contrast, the tracing for a rib vault in Hernán Ruiz II's sketchbook, without any accompanying text, offers a rather complete and sophisticated system for the control of these vaults, considering its early date. The solution provided by Ruiz, based in an assortment of true-shape representations of ribs in different orientations, reappeared in seventeenth- and eighteenth-century treatises such as those of François Derand (1643) and Amedée-François Frézier (1737–1739), although they leveraged the resources of learned geometry in order to control the tangencies of the tiercerons and liernes, as we will see in Sect. 10.1.4.

2.1.5 *Josep Gelabert, Josep Ribes and the Tornés Family*

Gothic architecture survived for the whole Early Modern period, until the emergence of Neogothic in England in the mid-eighteenth century (Summerson 1953 [1969]: 236–243; Gómez Martínez 1998). In fact, some of the most detailed explanations about rib vault construction methods can be found in *Vertaderes traces de l'art de picapedrer*, a manuscript written in 1653 by Josep Gelabert, a stonemason living in Majorca. Perhaps the isolation of this territory favoured the survival of Gothic techniques; remarkably, all constructions mentioned by Gelabert are built in the city of Majorca, in contrast to Vandelvira, who cites directly or indirectly several examples or archetypes located in Andalusia, but also others in Murcia, Mallorca, Cuenca and France (Gelabert and Alcover 1977; Gelabert and Rabasa [1653] 2011: 12–13; Vandelvira c. 1585: 14r, 21r, 52v, 53r, 55v, 59r, 69v, 83v, 97v, 119v).

Gelabert was born in 1622, seems to have worked in the renovation of the nave of the cathedral of the city and died in 1688 after falling from a scaffolding.[12] The manuscript is preserved in the Biblioteca de Cultura Artesana in the city of Majorca.

[11] See Álvarez (1998); Jiménez (1998c); Huerta (2004: 148–151); Gentil (1998); Calvo (2019); Ruiz de la Rosa (1998), Pinto (1998).

[12] An archival document transcribed by Gambús (1988: 773–774) states that *treballant de son art Joseph Gelabert picapedrer ... estant a un bastiment, es trenca y es rompe y caygue al payment de la sala que havia dos canes de alt, de que li sque gran copia de sanch per lo nas y boca sens parlar ni*

Although it is not divided into parts or chapters, each problem carries a heading, a written explanation in Catalan and a neat line drawing, with few exceptions; the general structure is rather clear. It includes sixteen different rib vaults (Gelabert 1653: 131v-147r),[13] while Hernán Ruiz II (c. 1560: 46v), Alonso de Vandelvira (c. 1585: 94r-97r) or François Derand (1643: 392–395) limited themselves to one standard type, the square-plan tierceron vault. Further, Gelabert does not focus exclusively on the ribs, but explains in detail all the key elements of the vault: springers, keystones and, in a world first, the web (Gelabert and Rabasa 2011: 412–419). Except for severies, the rest is supported by didactic drawings (Fig. 2.6) showing the profile of the keystones, the width of the ribs, and the layout of bed joints in springers. Ribs are quite wide; it is not easy to tell if Gelabert was exaggerating for didactic reasons. He also deals with other typically Gothic elements, such as pointed arches, twisted piers, and spiral staircases with helical newels; the latter two were innovations introduced by Guillem Sagrera in the Merchants' Exchange of the city.

Such close ties with the medieval tradition do not mean that Gelabert leaves aside classical forms: he deals with round and basket handle arches; groin vaults, particularly a rectangular-plan variant that is found frequently in Majorca; "Roman" doorways, with a presumably rusticated lintel; round and oval windows. His explanation of the hemispherical dome (Gelabert 1653: 50v-51r) is remarkable; he basically follows the standard procedure as described by Vandelvira (c. 1585: 60v-61v) although the graphical presentation is different, and he explains the dressing process in detail, stressing that templates should be flexible.

Even richer in Gothic vaults is the manuscript *Llibre de trasas de bias y muntea* by Josep Ribes i Ferrer, a Catalan mason (1708). The manuscript includes no fewer than forty different rib vaults; in addition to the basic types, such as the quadripartite and the tierceron vault, it presents many elaborate combinations of tiercerons and liernes (Tellia and Palacios 2015; see also Tellia and Palacios 2012 for trumpet squinches). There are some points of contact with Gelabert: there are no chapter divisions in either manuscript, although problems are arranged systematically, except for a few placed at the end. Ribs, keystones, and springers are extremely wide, perhaps because both masters wanted to show their intersections clearly. However, while Ribes draws ribs directrices in true size and shape, stemming from a common springer, following

donar señals de confession sino lo aliento que dura per spay de mitge hora, de que morí. Bastiment means both "building" and "scaffolding" in Catalan; however, the meaning "scaffolding" fits better in the context. Thus the fragment may be translated as "while Joseph Gelabert, stonemason, was working in his craft at a scaffolding, it was cut and broken, and he fell on the pavement of the hall from three meters high; he bled copiously from the nose and mouth; he did not talk or give signals of confession, except his breath, which lasted for half an hour, and after that he died". (See also Gambús 1988: 778; Gelabert/Rabasa 2011: 7–8).

[13]The manuscript includes a number for each drawing, written in ink in the upper right corner of the sheets, as well as smaller numbers, written in pencil in the lower left corner, for each folio. In order to avoid confusion, all references to the text and the drawings are based on the smaller folio numbers; drawing numbers are recorded in Gelabert/Rabasa (2011).

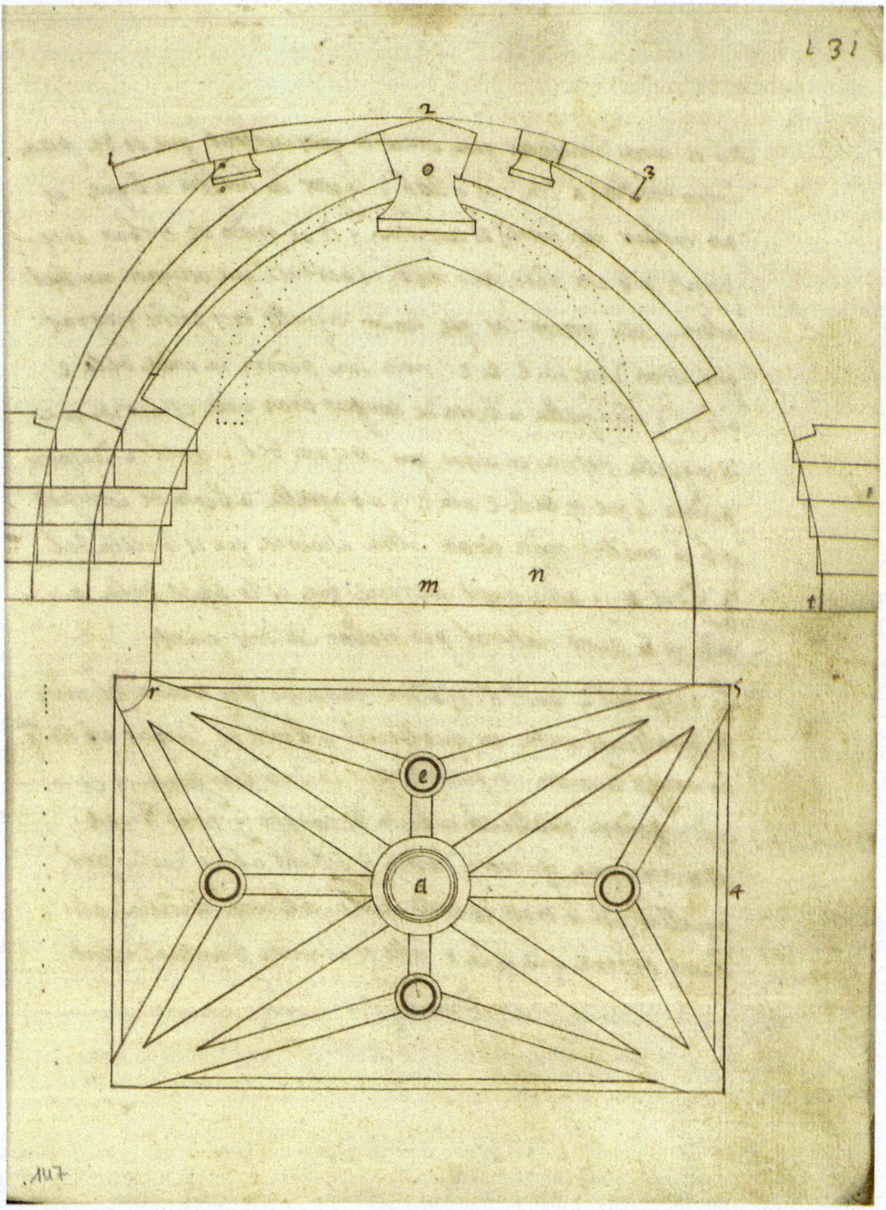

Fig. 2.6 Rectangular star vault (Gelabert 1653: 147r)

the usual practice, Gelabert inverts the scheme and joins ribs by the central keystone; since their horizontal projections are not equal in length, they depart at the springers.[14]

The similarities between both manuscripts can be explained, of course, by the common adscription of Majorca and Catalonia to the Crown of Aragon. However, two manuscripts written and drawn by several members of the Tornés family give a good picture of the state of events after the War of the Spanish Succession in another domain of the same realm, Aragon itself. The Tornés family was a respected dynasty of masons and architects based in Jaca, a historically significant spot in the Camino de Santiago, which had fallen into provincial isolation in the slopes of the Pyrenees at that time. Both manuscripts are a striking mixture of family records and love poetry with architectural, stonecutting and fortification drawings, some of them with explanatory text, some without. It has been shown (Tornés and Juan [c. 1700] 2013: 35–38, 40–41; see also Tornés and Juan [c. 1700] 2015) that although no fewer than six hands were involved in the preparation of the notebook, all stonecutting drawings seem to be by the hand of Antón Tornés Grasa. Most of them (Tornés and Juan [c. 1700] 2013: 58r, 59r; see also Juan 2014 and Juan 2015) are almost literal copies of the printed treatise of Fray Laurencio de San Nicolás (1639: 100r, 103r). Other drawings (Tornés and Juan [c. 1700] 2013: 49v, 50v) represent the angle between intrados and face joints in skew and corner arches using a technique employed by Jean Chéreau (1567–1574: 113r) and explained by Mathurin Jousse (1642: 10–11, 18–23). However, the most interesting drawings in the notebooks are a pair of rib vaults (Tornés and Juan [c. 1700] 2013: 56r, 56v). As in Gelabert's elevations, diagonal ribs and tiercerons meet at the central keystones; in contrast, they are drawn as single lines, as in Hernán Ruiz II (c. 1560: 46v). All this shows that at this stage, Aragon was looking to both Castile and France, but its shared heritage with Catalonia and Majorca was still present at deep levels.

2.1.6 Hans Hammer, Wolfgang Rixner and Master WG

As we have seen, most Spanish manuscripts and French treatises[15] explain the layout of ribbed vaults using as a basic example a tierceron vault, on some occasions with the addition of a few curved ribs; they seem to take it for granted that the reader will extrapolate this method to complex vaults. Such an approach was insufficient to address German net vaults, and thus writers in the Empire followed a different route.

The sketchbook of Hans Hammer (c. 1500), like the one by Villard de Honnecourt and many other Early Modern personal notebooks, is a mixed bag. It includes a

[14]Compare for instance Ribes (1708: 99, 103, 107, 108); Ruiz (c. 1550: 46v); Vandelvira (c. 1585: 96v) and Derand (1643: 393), with Gelabert (1653: 135r, 136r, 138r, 142r). Sheets in the Ribes manuscript are unnumbered; page numbers are taken from a PDF reproduction furnished by Biblioteca Nacional de Catalunya.

[15]Alviz (c. 1544); García y Gil de Hontañón (1681); Ruiz (c. 1560); Vandelvira (c. 1585); Guardia (c. 1600); Derand (1643); Gelabert (1653); Ribes (1708); Tornés and Juan ([c. 1700] 2013: 56r, 56v); Frézier (1737–1739); see also Sect. 10.1.4.

most interesting section on elevation equipment; crosses akin to those in Roriczer's *Wimperbuchlein* (Hammer c. 1500: 2v); spires whose construction method recalls Lechler, many vault plans (Fig. 2.7) from the simplest ones to whole churches; staircases; oblique pointed arches; and some puzzling geometrical schemes (Hammer c. 1500: 23v, 24r). Comparing the latter with the manuscript of Facht von Andernach, it seems clear that they are true-shape representations of the curvature of vault ribs. For example Facht (1593: 24r), shows a semicircle representing the standard curvature of most ribs in a complex vault; however, not all ribs can adopt the same radius, so there is an auxiliary arch furnishing the curvature of a non-standard rib. Vertical lines make it possible to measure the height of the keystones (see Sect. 10.1.5).

Fig. 2.7 Rib vaults (Hammer c. 1500: 26r)

The manuscript by Wolfgang Rixner (1467–1500), preserved at the Albertina gallery in Vienna, is again a conglomerate of architectural problems, including elaborate rib vault plans in the German tradition, decorative motifs such as animal drawings or traceries, and geometrical schemes for the tracing of regular polygons. Most interesting for our purposes are a series of sheets cut with a knife so that the remaining sections represent traceries or ribs (Rixner 1467–1500: 78, 81, 85, 87, 89, 100, 102, 104, 106, for example). Even more significant for our aims are some schemata of vault elevations (Rixner 1467–1500: 41; see also 34, 94, 96, 97). They include both the intrados and the extrados of some ribs, as well as bed joints, the enclosing solids of some voussoirs and several vertical lines connecting specific points in the ribs with the horizontal plane at the springing level; in essence, the concept is the same one used in Hammer (c. 1500: 23v, 24r).

Another manuscript including vault schemes is preserved in the Städelsches Kunstinstitut in Frankfurt; the initials "WG" and the dates 1560 and 1572 are inscribed in the binding. The same initials, a mason mark and the year 1572 appear on several occasions in the body of the manuscript. These dates are consistent with the watermarks of the paper, found in Bavaria, Salzburg and Hungary from 1527 to 1581; Bucher surmises that the book was physically made in 1560, and was then slowly added to until its completion in 1572. Also, according to Bucher (Villard/WG/Bucher 1979: 195), the same masons' mark appears in the church of Saint Leonard in Stuttgart together with that of Abelin Jörg. This connects the author of the manuscript with a well-known family of masons working around Munich; however, since no other data about the author are known, he is identified as Master WG.

In the initial sections and a few final fragments (WG 1572: 1–48, 279–280, 313–314), basic constructions were inscribed with a stylus and a divider; next, the draughtsman went over them with a pencil in order to arrive at the finished drawing, as usual in other manuscripts. In contrast, in the rest of the book, Master WG (1572: 49–276) includes cut sheets, as did Rixner, in order to show building elevations, traceries and, more frequently, rib vaults. Most of these sheets, representing vault plans, do not feature pencil lines, just stylus lines on the axes of ribs. All this suggests that Rixner and master WG used these cut-outs as showpieces for patrons, folding or bending them to adopt the three-dimensional shape of a real vault.[16]

Although these sheets are quite striking, some drawings in the initial section are also quite interesting for our purposes (WG 1572: 37–47). As in Hernán Ruiz II and Derand, there are several rib elevations, drawn as circular arcs. With two exceptions

[16]Bucher (Villard/WG/Bucher 1979: 196) states that "Like Lechler and, later, Simon Garcia, the author must have been fully conscious that Gothic architecture had come to an end, but that its principles and richness were worth preserving for future generations". Later on, he adds that "The most fascinating aspect of the manuscript is its late date. It is nearly inconceivable that a volume so steeped in Gothic nostalgia should have been painstakingly assembled at a time when Filarete and Alberti were passé, Philibert de l'Orme had already published his *Premier Tome de l'Architecture* ...". Such "nostalgic" view of WG's intentions must be put into context. Bucher was writing in 1979; of course, he could not know the studies by Gómez Martínez (1998) or Rousteau-Chambon (2003) about Spain and France; however, he overlooked Summerson's ([1953] 1983, 397–407) pages about Early Modern Gothic in England.

(WG 1572: 21, 39), they are drawn separately from plans, which are present a few sheets before (WG 1572: 4, 10, 14, 16, 23, 29); however, it is not easy to tell whether any of these separate plans match the elevations. In any case, we may surmise safely that the elevations represent ribs in true size and shape. Thus, Master WG follows basically the same approach as Hernán Ruiz II (c. 1560: 46v), although he usually duplicates some ribs, perhaps the perimetral ones, to show their layout as pointed arches.

2.1.7 The Codex Miniatus, Jacob Facht von Andernach, and Bartel Ranisch

The Codex Miniatus 3 in the Austrian National Library, named by Bucher (1972) "Dresden sketchbook of vault projection" on account of its watermarks, sheds more light on these issues. Each of the 21 sheets representing vaults includes a plan and an elevation. However, Bucher's denomination is slightly misleading, because while the plan is actually projected, elevations depict oblique ribs in true size and shape. Plans are neatly drawn, including most arrises in the ribs; these are represented in the elevation with two lines for each rib, standing for the intrados and the extrados. Each elevation includes several numbered vertical lines; the same numbers represent keystones in the plan. This suggests that vertical lines are used to measure the height of the keystones, as in Rodrigo Gil (c. 1560: 24v-25v). In some cases (Codex Miniatus c. 1570: 1v, 2v, 3r, 3v, 5r, 6r, 6v, for example), the elevation of a single rib is enough. We may safely assume that in these examples, rib curvature is standardised, simplifying the preparation of formwork, centring and arch squares; the single rib in the elevation, known as *Prinzipalbogen* in modern studies (Müller 1990: 168–183; Tomlow 2009, 197–199; Pliego 2007), furnishes both the curvature of all ribs and the height of the keystones. However, such a neat system has a drawback. In general terms, each keystone can be reached by several paths; if all paths feature the same length, the height to the keystone is univocally determined. In contrast, when some keystones may be reached through paths with different lengths, such system cannot be applied strictly, as we will see in Sect. 10.1.5; in this case, the draughtsperson may include additional ribs in the elevation, in order to adjust curvatures and reach secondary keystones precisely (Codex Miniatus c. 1570: 2r, 4r, 5v); alternatively, masons may manipulate rib intersections.

Similar techniques can be seen in the drawings of the manuscript by Jacob Facht von Andernach (1593). Again, vaults are neatly drawn in plan and elevation, usually with the edges and axis of each rib, depicting the intrados and extrados. Instead of joining plan and elevation, as in the Codex Miniatus, both drawings of each vault are placed in facing pages (Fig. 2.8). Generally speaking, designs are a bit more complex, and elevations with several ribs (Facht 1593: 3v, 4v, 7v, 8v, 10v, etc.) are more frequent, although the strict *Prinzipalbogen* method is used where suitable (Facht 1593: 5v, 11v). Bed joints, present here and there in the Codex Miniatus (c. 1570:

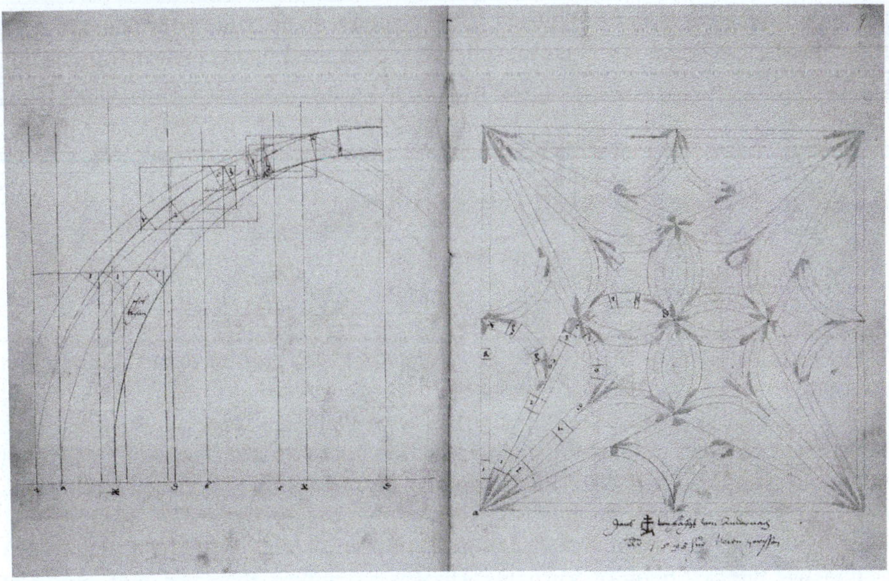

Fig. 2.8 "Elevation" and plan of a ribbed vault (Facht. Roriczers and Lechler 1593: 8v-9r)

5v, 8v), are now represented systematically; voussoirs are enclosed in rectangles, as in Renaissance drawings illustrating the squaring method (Facht 1593: 3v, 4v, 5v, etc.). This connection is also evident in two pages representing an oblique splayed arch, a sloping vault and a curved arch, together with Tuscan, Doric and Ionic orders and a classical cul-de-lampe (Facht 1593: 21v, 22r, 28r, 34v, 35v, 37r). On the other hand, fols. 38v-42v are taken from Roriczer and Lechler. In other words, Fach von Andernach was trying to assemble in a single book a wide variety of subjects in the architectural culture of his time, both Gothic and Renaissance.

A different method is presented in Bartel Ranisch's book (1695) about the churches of Danzig, now Gdansk. In contrast to the rest of the texts I have analysed in this chapter, this is not a manual or a personal recompilation of stonecutting methods, but rather a description of several extant churches. However, they are explained in the utmost detail, with a separate rib curvature diagram for each vault. Ranisch's idiosyncratic method has been analysed by Pliego (2017). As we will see in Sect. 10.1.5, he imposes a rigid standardisation; that is, he sets as a rule that the curvature of all ribs should be equal. In the first example, the vault of the church of Saint Mary, he draws a quadrant standing for the diagonal rib, reusing it to compute the height of the keystone of the perimetral arch over the larger side of the plan. Then, in a separate drawing, he applies a singular procedure: he measures the distance from the meeting point of the tiercerons placed near the short side of the area to the central keystone in the plan and applies it *twice* in the elevation. After a complex procedure, he places the centre of the ribs below the springing line. That is, he does not impose the condition of a vertical tangent in the springing, in contrast to the Codex Miniatus,

Facht von Andernach and Ruiz. Generally speaking, Ranisch uses the same method for other vaults (Pliego 2017); as we will see in Sect. 10.1.5, this leads to remarkable problems.

2.2 The Renaissance and the Architects

2.2.1 Philibert de l'Orme

We know little about the life of Villard de Honnecourt and nothing about Hand IV, but the background of the rest of the preceding writers is strictly artisanal. They belong to long dynasties of masons: Lechler writes for his son; Roriczer is a relative of the Junkers or Eagles of Prague, the members of the famous Parler family; Rodrigo Gil was an illegitimate son of Juan Gil, master of the cathedrals of Salamanca and Seville. Although Hernán Ruiz II fits in this model, his interests go far beyond this medium. As we have seen, he was interested in the theoretical problems of perspective and tried to translate the first book of Vitruvius. He was a complex character, half medieval mason and half Renaissance architect. He was fully aware of his role as a designer, and in fact, he presented drawings and models frequently to the Chapter of Seville, but, in contrast to some architects coming from the figurative arts, he did not give up his role as construction supervisor.

Catherine Wilkinson (1977b) labelled these architect-builders as "new professionals". She presented Philibert de l'Orme as an archetype of this social group, which plays a central role in French and Spanish sixteenth-century architecture. De l'Orme was the son of a well-known builder; if we believe his own words, he commanded 300 men when he was 15 years old, probably when building the ramparts of Lyon (de l'Orme 1561: 35r; see Bonnet 1993: 25–26; Pérouse 2000: 22–23). Later on, he completed his education measuring ruins in Rome; his thoroughness in this task brought him to the attention of his first protector, Cardinal Jean du Bellay. He was perhaps in contact with Antonio de Sangallo the Younger during the construction of Palazzo Farnese (Bonnet 1993: 32–68). After this, he worked for Henri II of France in Fontainebleau, putting under his command Gilles le Breton, a mason who had refused to obey Serlio's orders. As a reward for his services, de l'Orme was chosen as Superintendent of the King's Works and Abbot of Saint Serge of Angers (Blunt 1958: 30–79; Potié 1984: 28–30; Potié 1996: 33–52; Pérouse 2000: 47–71; Pérouse 2001: 74–75). He also worked for Diane de Poitiers, lover of Henri II, at the Chateau d'Anet. He built there two striking pieces, the chapel vault with interlacing spiral ribs, and the trumpet squinch under the king's studio (Sanabria 1984: 248–249; Evans 1988; Sanabria 1989; Evans 1995: 180–189; Potié 1996: 92–106; Trevisan 1998; Trevisan 2000; Fallacara 2003a; Lenz 2009).

His fame was so enormous that Rabelais referred to him as Great Architect of King Mégiste (Rabelais [1534] 1978: 32; Potié 1996: 81). However, his luck changed suddenly. A courtesan hit Henri II in the eye with a spear in a tournament; he died within a few days. De l'Orme was expelled from the supervision of the King's works, his post was entrusted to Francesco Primaticcio, and he was attacked by jealous masons on the outskirts of Paris. All this shows the difficult position of the "new professionals", opposed not only to traditional artisans but also to figurative artists coming from Italy. Significantly, when charged with having embezzled the King's funds, he boasted that he had expelled from France "the barbarian fashions and the thick joints"; that is, he stressed not only his elegance as a designer but also his skill as a builder (Pérouse 1988: 21).

In any case, he survived all these attacks and, what is more striking, landed a big commission from the Queen Mother, Cathérine de Medici, the apparent rival of Diane de Poitiers: the vast Palais des Tuileries in central Paris. However, the scorn against the mason turned courtier had not ended. It has been told many times that Ronsard, fuming after having been kept waiting in the Queen's anteroom for a long time, hung on the closed door a notice with the words *Fort reverend habe*, which can stand for a misspelt transcription of *Fort reverend abbé*, that is, "Very reverend abbot". However, the phrase can also be read as an abbreviation of *Fortuna reverend habet*, a fragment from an epigram from Ausonius, referring to Agathocles, tyrant of Syracuse, son of a potter and potter himself in his youth, reminding him of the need to carry his good fortune with modesty (Potié 1996: 45–46; Ceccarelli 1996: 31–33; Pérouse 2000: 75–76).

De l'Orme did not follow Ronsard's advice. When explaining the *vis de Saint Gilles* in his landmark *Premier Tome de l'Architecture* (1567), he states that:

> I saw in my youth that anyone who understood the layout of the vis de Saint Gilles was held in high esteem among masons … At that time masons tried to understand it and dress the voussoirs using templates …. Some workers dressed them by squaring; however, there is no ingenuity in this, and much material is lost[17]

Of course, de l'Orme sided with neither such low-ranking apprentices, nor with the good workers that dressed the voussoirs by templates. Instead, he proposed a different method:

> If I were to direct its construction, I would not care to dress it by templates, and of course not by squaring; there is not so much difficulty as workers think. It is quite easy to dress the voussoirs using arch squares and bevels; having all the templets that are needed, it is quite easy to trace all stones.[18]

---

[17]De l'Orme (1567: 123v-124r): *Jay vu en ma jeunesse que celui qui savait la façon du trait de la dite vis Saint-Gilles, et l'entendait bien, il était fort estimé entre les ouvriers... en ce temps là les ouvriers travaillaient fort à l'entendre et principalement pour la faire par panneaux.... On en rencontrait quelquesuns qui la faisaient par équarrissage, mais en cela n'y a guère d'esprit ne d'industrie, et y faut perdre beaucoup de pierres.* Transcription taken from http://architectura.cesr.univ-tours.fr; translation from the author.

[18]De l'Orme (1567: 124r): *Si je l'avais à conduire je ne me soucierais guère de la faire par panneaux, ni moins par équarrissage, vous avisant qu'il n'y a point tant de peine, ni tant de difficulté que les ouvriers le pensaient pour lors, et que plusieurs encore le pensent, pour ne le savoir. Il est aussi fort aisé et facile de la faire avec des buveaux et sauterelles, car en ayant les cherches rallongées qu'il y faut, et leurs équerres, il est facile d'en tracer justement toutes les pierres.* Transcription taken from http://architectura.cesr.univ-tours.fr. Translation from the author.

However, comparing de l'Orme's solution with those of Vandelvira (c. 1585: 52v-53r) and Martínez de Aranda (c. 1600: 230–233), who make no bones about dressing the voussoirs by squaring, the differences are not remarkable. In fact, the method used by the Spanish masters is not so simple as it may seem, since it involves the use of an auxiliary elevation. Comparing this solution with that of de l'Orme, it is evident that the basic technique, stemming from orthogonal projections, is the same in all three authors; the enclosing solids of the voussoirs, depicted only in its lower parts, as usual in De l'Orme, make it clear that he is going to dress the voussoirs by squaring. The only difference between de l'Orme and the Spaniards is his use of templates; however, since the intrados surface of the piece is clearly warped, such templates cannot be used directly; as we will see in Sect. 3.2.5, these templates are probably used to control the shape of an intermediate stage of the voussoir in the squaring method.

Such boastful attitude on de l'Orme's part (see Pérouse 2000: 83) can be explained in part by his career. Both the *Nouvelles Inventions pour bien Bâtir a Petit Frais* (1561), a carpentry manual, and *Le Premier Tome de l'Architecture* (1567), the first architectural treatise written by a Frenchman,[19] were published after Henri II's death; they can be read as a complement for his plea in the embezzlement case and, at least for the *Nouvelles inventions*, as a promotional portfolio for the Tuilleries commission. All this suggests that de l'Orme intended to showcase his great knowledge and ingenuity (see Fig. 2.9), rather than explain each problem "in such a detailed and easy way as I can",[20] as he states in another passage.

For example, when dealing with hemispherical vaults, he starts directly with a sail vault with diagonal courses (de l'Orme 1567: 111v-112v, 113r; Fig. 1.14j). Such a piece is a derivative of the hemispherical vault cut by four vertical planes passing through the sides of a square area; thus, four spherical caps, which are different due to the rectangular shape of the plan, must be left aside. If these complications are not enough, in this variant the problem is not solved by dividing the vault by horizontal planes, which would give round courses separated by joints following the parallels of the sphere. Instead, the courses are divided by joints placed at the intersections of the sphere with two families of vertical planes; to make matters worse, such vertical planes are parallel to the diagonals of the plan. This cavalier attitude contrasts with the didactic approach of Alonso de Vandelvira and other Spanish manuscripts, as we will see in Sect. 2.2.3.

In other words, de l'Orme prefers narrative brilliance to didactic clarity (see Calandriello 2020). This also explains the awkward succession of stonecutting problems. Throughout the book, he uses as a leitmotiv the construction of a house. Books III and IV are devoted to the geometrical problems of stone construction, presenting a relatively large catalogue consisting of 32 different problems. This general scheme

[19]Other architectural treatises had been published in France before, but they were translations of Vitruvius, Alberti, Diego de Sagredo and Sebastiano Serlio.

[20]De l'Orme (1567: 74v): *la façon du trait, lequel je décrirai le plus particulièrement et simplement que je me pourrai aviser*. Transcription taken from http:// architectura.cesr.univ-tours.fr; translation from the author. (For a different interpretation, see Galletti 2017b, 157–160).

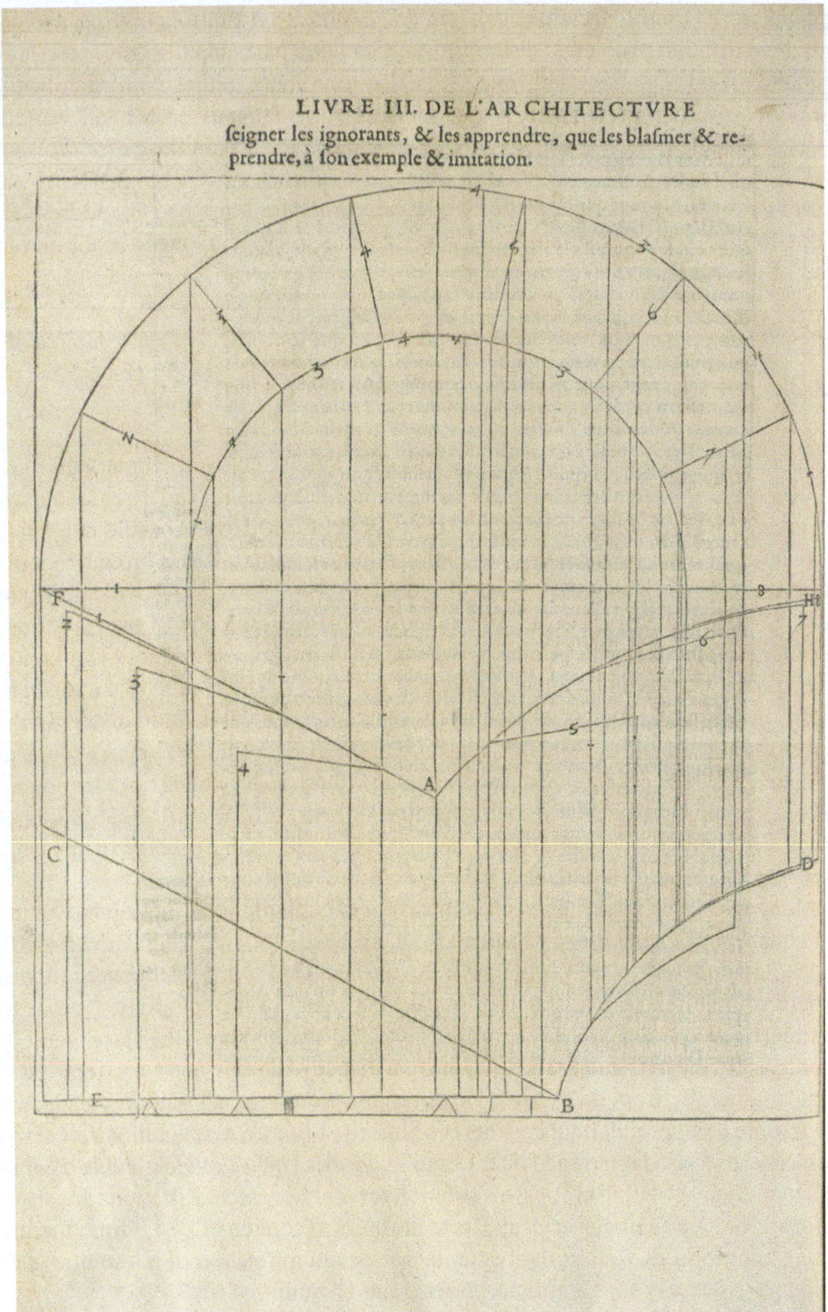

Fig. 2.9 Arch in a corner between a straight and a convex wall (de l'Orme 1567: 81v)

explains de l'Orme's decision to begin his explanation of stonecutting problems with three difficult problems: a sloping vault and two particular kinds of rere-arches frequently used in cellars. However, it is also true that together with such brilliant and not-so-useful pieces as the Anet squinch or the diagonal-course sail vaults, de l'Orme explains other sensible, down-to-earth pieces, such as the *porte biaise par teste*, a skew arch with parallel springings (de l'Orme 1567: 71r-72r). This piece tackles a problem in spatial geometry that the learned tradition had not addressed up to this moment: the construction of true-size-and-shape templates for the intrados sides of the voussoirs.

2.2.2 *Jean Chéreau and Jacques Gentillâtre*

Despite its not-so-didactic character, de l'Orme's treatise enjoyed a remarkable circulation and exerted great influence in France in the late sixteenth and early seventeenth centuries; it was reprinted in 1568, 1576, 1603, 1626 and 1648. This did not help the publication of other treatises until an outburst of new materials was brought about by the appearance of Desargues's disrupting leaflet in the 1640s, as we will see in Sects 2.2.6, 2.3.2, and 2.3.3. However, two manuscripts offer a picture of the evolution of stereotomic knowledge in France between de l'Orme and Desargues. The first one bears the name of Jean Chéreau, builder of the nave of the church of Saint-Jean in Joigny, in Burgundy. It was written a few years after the publication of de l'Orme's treatise: not before 1567, since it takes much material from the *Premier tome*, and not after 1574, since it includes a dedication to Charles IX, who is not mentioned in the text but can be identified from a portrait (Pérouse [1982a] 2001: 95–96; Lemerle 2016). By a curious coincidence, it is now preserved in the Library of the Polish Academy of Sciences in Gdansk, the city of Bartel Ranisch.

Along with other materials, such as drawings of capitals, lions, labyrinths, sundials and square root calculations (Pérouse [1982a] 2001: 96, 138, 185; Trevisan 2011, 9–10), it includes about 35 problems in stereotomy. De l'Orme's influence, although evident, should not be overstated. It is true that many of the solutions, such as two versions of the *vis de Saint Gilles* with round and square plans, the spiral vault, the sloping vault, the triangular sail vault, the fan-shaped sail vault, and some rere-arches, are taken literally from de l'Orme. However, there are original sections, such as the remarkable lunette vault built by Chéreau himself at Joigny, new solutions for several squinches, and an adaptation of de L'Orme's *corne de vache double* (double ox horn), which prefigures Mathurin Jousse's solution for the skew arch. Rib vaults are absent. All in all, Chereau's manuscript attests to the quick reception of De L'Orme's treatise amongst masons, as well as the beginning of a reinterpretation of his solutions that would pave the way for Gentillâtre, Jousse and Derand.

The second author, Jacques Gentillâtre, is well known through a set of almost 300 drawings preserved at the library of the Royal Institute of British Architects in London (Coope 1972). He was born in 1578, apprenticed in the workshop of Jacques II du Cerceau and worked in cities held by Protestants in the period, such as Sedan and Montbéliard. However, what is more important for our purposes is an anonymous manuscript in the National Library of France, attributed to Gentillâtre on the basis of similarities with the London drawings (Gentillâtre c. 1620; see Chatelet-Lange 2006). It includes no fewer than 594 sheets with text and drawings on both sides. In Vitruvian fashion, it is divided into ten books and provides material on arithmetic, geometry, fortification, war machines, civil architecture, the orders, stonecutting, carpentry, optics and mechanical arts. The two books devoted to stereotomy (Gentillâtre c. 1620: 406r-451v) include skew arches, rere-arches, arches in curved walls, hemispherical and surbased domes, groin vaults, trumpet squinches, straight and spiral staircases, an early solution for the famous "Rere-arch of Marseille" (Fig. 2.10), and a fair number of rib vaults. Along with diagrams in plan and elevation and explanatory texts, Gentillâtre inserts here and there perspectival renderings to suggest to the reader the complex spatial layout of some pieces, such as squinches, rere-arches and staircases.

Although some content is taken from Palladio, such as the well-known drawings of the Chambord staircase, and de l'Orme, as in the sail vault resembling a fan vault in plan,[21] other solutions are more innovative, in particular the complex squinches, which prefigure solutions by Jousse and Derand, and the rib vaults (Gentillâtre c. 1620: 450r-451v). There are twelve different plans of complex rib vaults, including examples with curved ribs, double liernes, or groups of tierceron vaults. However, of greater interest is another drawing which includes a rectangular-plan rib vault divided into triangles; the presence of circles and semicircles suggest that the grid is projected onto a spherical surface. Thus, the result has some traits in common with German *Netzgewölbe*; at the same time, it recalls some vaults in the upper storey of the Merchants' Exchange in Seville and suggests connections with cartography (see Sects. 10.1.5 and 12.4.3).

2.2.3 *Alonso de Vandelvira*

Alonso de Vandelvira's *Libro de trazas de cortes de piedras* is the first complete, systematic text dealing exclusively with stereotomy.[22] Villard, Ruiz, Chéreau and

[21] Cf. Palladio (1570: I, 65) with Gentillâtre (c. 1620: 444v), and de l'Orme (1567: 114r), with Gentillâtre (c. 1620, 419r).

[22] Two manuscripts of this work have survived. The older one belongs to the library of the School of Architecture of the Polytechnic University of Madrid (MS R-10); its date and scribe are unclear. It bears the title *Exposición y declaración sobre el tratado de cortes de fábrica que escribió Alonso de Vandelvira por ... Bartolomé de Sombigo y Salcedo, Maestro Mayor de la Santa Iglesia de Toledo* (Explanation of the Treatise on Stonecutting Written by Alonso de Vandelvira by ... Bartolomé de Sombigo y Salcedo, Master Mason of the Holy Church of Toledo). However, Bermejo (1954: 292–293) and Barbé (1977: 21–23) showed that the handwriting of the manuscript proper does not belong to Sombigo and the manuscript predates Sombigo's tenure as master; that is, Sombigo is neither the author nor the copyist of the School of Architecture manuscript.

The other manuscript is preserved in the National Library of Spain, also in Madrid (MS 12.719), with the title *Libro de Cortes de Cantería de Alonso de Vandelvira, Arquitecto. Sacado a luz, y*

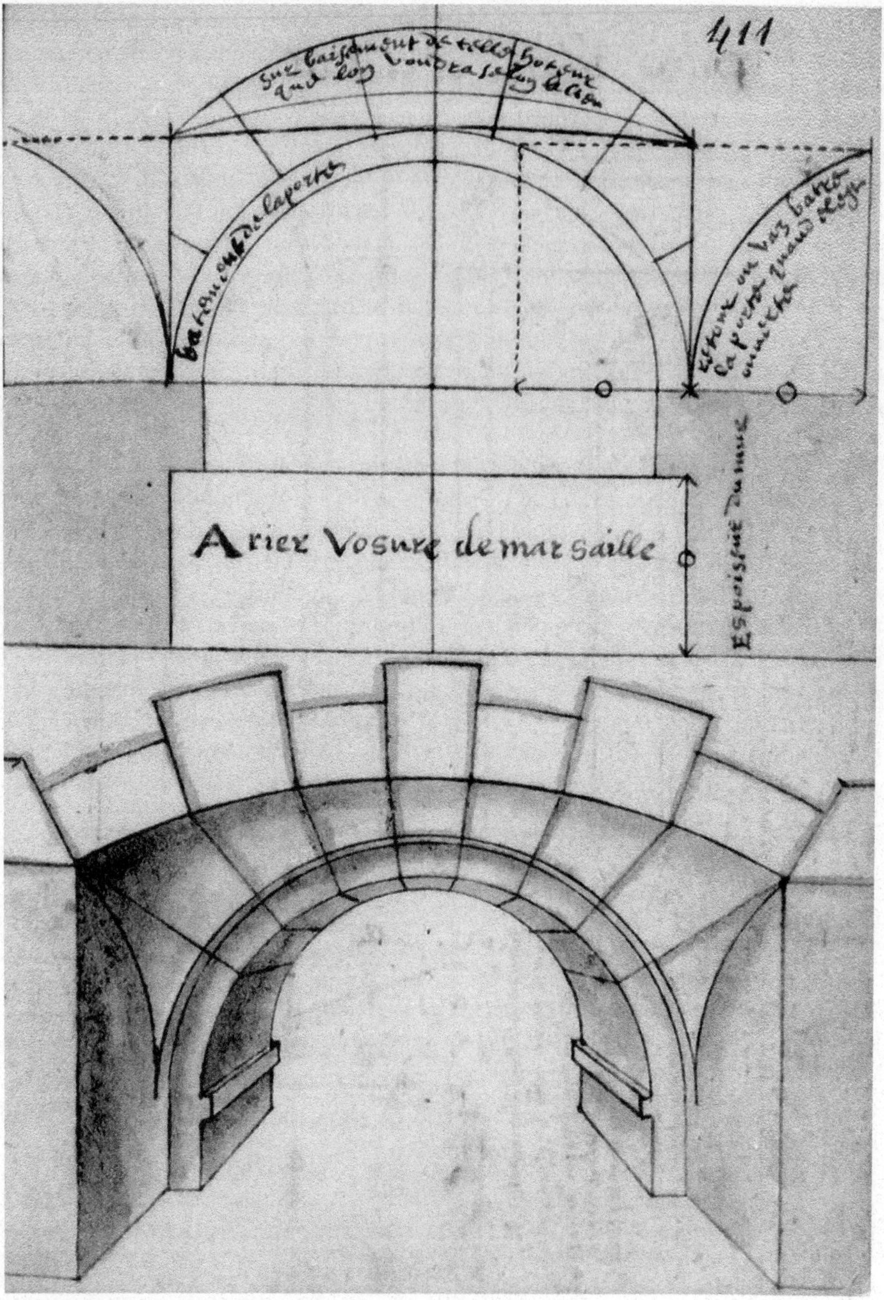

Fig. 2.10 Rere-arch of Marseille (Gentillâtre c. 1620: 441r)

Master WG had assembled personal notebooks dealing with many different issues. De l'Orme's *Premier Tome* is clearly structured, but also involves a wide range of subjects; although the third and fourth books are devoted almost exclusively to stonecutting, the selection of problems is far from complete (de la Rue 1728: preface). In contrast, Vandelvira's manuscript, written between 1578 and 1591,[23] presents a systematic catalogue of almost every stonecutting problem known in the period, starting with squinches and proceeding through arches, sloping vaults, rere-arches, stairways and vaults, reaching a brilliant finale with ribbed and coffered vaults (Fig. 2.11). This sets it apart from personal notebooks and suggests it was prepared for the presses; in fact, two passages in Fray Laurencio de San Nicolás (1665: 155, 217–218) imply that there were attempts to publish it in the seventeenth century, many years after its actual preparation around 1585.[24]

The life of Vandelvira offers an interesting example of the changing fortunes of stonemasons and architects in the sixteenth and seventeenth centuries. He was the grandson of Francisco de Luna, who worked for the Order of Saint James and exerted an overwhelming influence across a broad territory stretching from Cuenca to Murcia. Even more relevant in Spanish architecture is his father, Andrés de Vandelvira, master mason of the Cathedral of Jaén. The prominence of Andrés has led some scholars to

---

aumentado por Philipe Lazaro de Goiti, Arquitecto, Maestro Mayor de Obras de la Santa Iglesia de Toledo ... Año de 1646 (Book on Stonecutting Tracings by Alonso de Vandelvira, architect. Brought to light and completed by Philipe Lázaro de Goiti, Architect, Master Mason of the Holy Church of Toledo ... Year of 1646). There is no reason, as far as I know, to doubt about Goiti's role as a copyist and the date of the manuscript. We have shown (Calvo et al. 2005: 240-242) that the dimensions of some drawings in this manuscript exactly match the corresponding ones in the manuscript held by School of Architecture. However, the drypoint marks in the National Library manuscript suggest the drawing was traced from another drawing, while the same marks in that of the School of Architecture hint that the drawing was constructed in its own sheet. Thus, it seems that the National Library manuscript was copied either from that of the School of Architecture or from one very close to it. This casts doubts about Goiti's "completion".

Further, the title *Tratado de arquitectura* given by Barbé does not seem to fit this work well, since it was not published until the twentieth century and its scope is not the entire field of architecture. Most scholars use the title *Libro de trazas de cortes de piedras* mentioned by Fray Laurencio de San Nicolás (1665: 217–218) or *Libro de cortes de cantería* written in the cover and the frontispice.

[23]The precise date of the original manuscript is not known. It seems to have been prepared between 1578, when the staircase of the Chancillería or Courthouse in Granada (mentioned in the text as built), was completed, and 1591, the date of the death of Juan de Valencia, a master in the Escorial circle who had a copy of the manuscript in his possession (see López Martínez 1932: 166–167; Barbé 1977: 18-19). However, the possibility of multiple versions of the manuscript cannot be discarded.

[24]San Nicolás (1665: 155): *... la Arquitectura, como penden de estampa y ni en España ay quien las abra, ... por la costa de las planchas ... esta ataja a los que viven con ansia de escribir; y así dejan mano escritos muchos papeles; yo he visto algunos, particularmente de cortes de cantería, que los ay en España muy curiosos, y ingeniosos* (in Architecture, since [writers] need plates, and nobody can engrave them in Spain ... due to their high cost, those who want to write are deterred by expenses; I have seen some interesting and ingenious books, in particular on stonecutting ...). The only stonecutting book mentioned by name by San Nicolás (1665: 217–218) is that of Vandelvira; the neatness of the Goiti manuscript fits well with an original prepared for print.

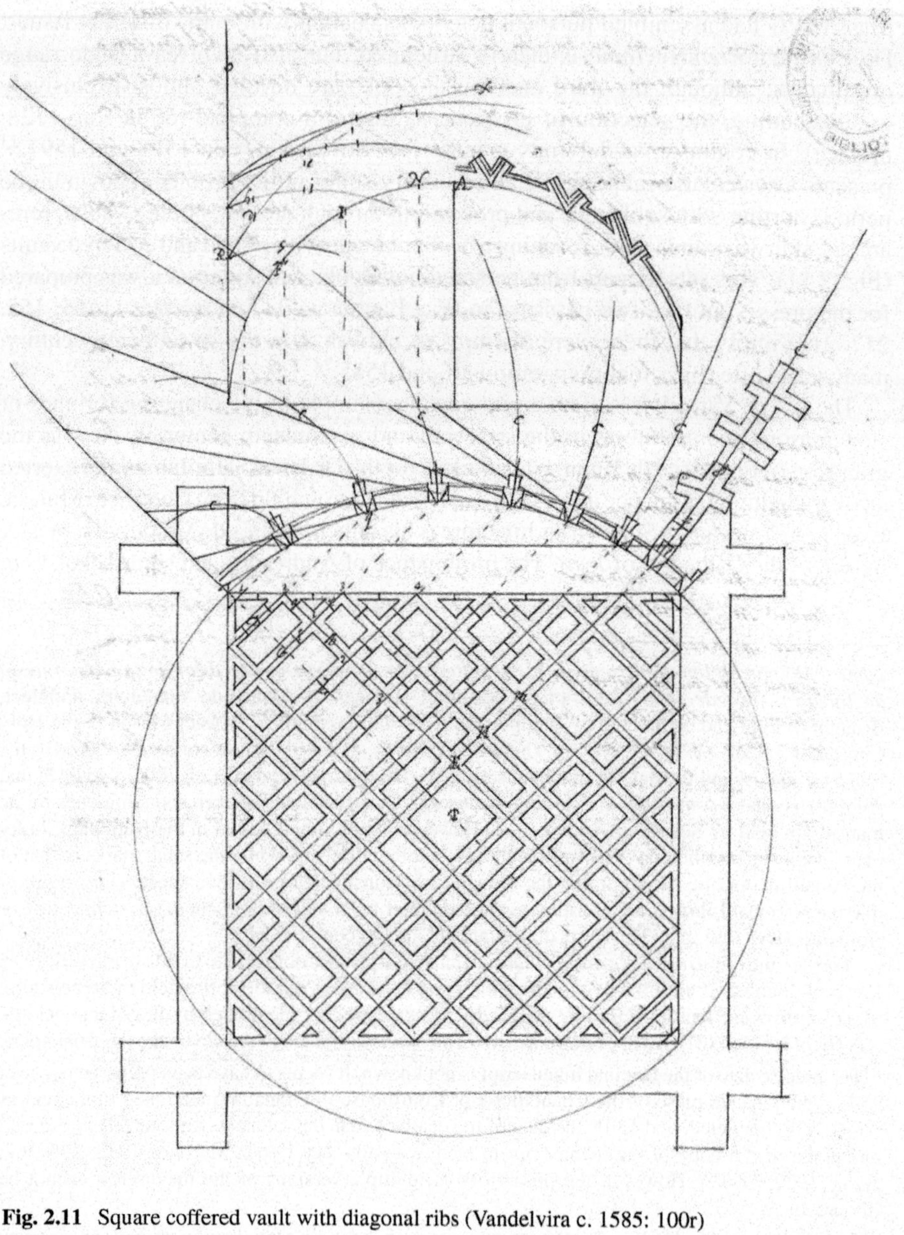

Fig. 2.11 Square coffered vault with diagonal ribs (Vandelvira c. 1585: 100r)

attribute the manuscript to him; however, the factual evidence contradicts this possibility, since the text mentions him as deceased, while the staircase in the Chancillería in Granada and other examples cited in the manuscript are dated after his death.

Although Alonso was trained as a stonemason, probably under Hernán Ruiz II, he did not practice this craft for some years; he was married to the daughter of a wealthy landowner and sat on the town council of Sabiote, near Úbeda. At the same time, his father recommended Alonso Barba, his long-time assistant, as master mason of Jaén Cathedral. However, Alonso de Vandelvira lent wheat from the municipal granary to some friends who could not return it and was imprisoned as a result. Once freed, he did not pursue his political career and returned to stonecutting, first in the church of Saint Peter in Sabiote. However, the prospects in the kingdom of Jaén were not brilliant, and he moved to Seville, which was booming as the centre of the commerce with the Americas. He carried out several important commissions and was offered the post of master mason of the Merchants' Exchange, begun with plans by Juan de Herrera in a severe late-sixteenth-century style. Some drawings in Vandelvira's manuscript (c. 1585: 115v, 118v) bear some resemblance to the vaults in the upper storey of the Merchants' Exchange, although on close inspection the similarity is not literal (Natividad 2017: I, 312–314). In any case, funds for the completion of the building were lacking, construction stalled, and Vandelvira was underpaid. Thus, he left for Sanlúcar de Barrameda, where he built several churches, while the upper storey of the Exchange was built finally by Miguel de Zumárraga (Plegezuelo 1990; Cruz Isidoro 2001: 96–100). Later on, Vandelvira ended his remarkably long career in Cádiz, where he presented his candidacy for the post of master of the city works. Although another stereotomic author, Cristóbal de Rojas, remarked that even if Vandelvira were in Rome, the city ought to call him,[25] the post was initially granted to another stonecutting writer, Ginés Martínez de Aranda, who left for Santiago de Compostela and left the job free for Vandelvira (see Sect. 2.2.4).

In addition to its complete and systematic nature, Vandelvira's manuscript shows a didactic quality which is lacking in de l' Orme. As we have seen, when dealing with spherical vaults, De l'Orme begins with one of the most complex spherical vaults, namely a diagonal-course sail vault (1567: 111v-112v). Next, he adds another whimsical sail vault with the layout of an English fan vault (de l'Orme 1567: 113r) and comes back to diagonal-course vaults with the rectangular variant of this problem, introducing further complexities (de l'Orme 1567: 115v). In contrast, Vandelvira starts his section on vaults with the hemispherical dome (c. 1585: 61r-62v). After dealing with other vaults, he addresses sail vaults beginning with the basic type, the square vault with round courses; next, he introduces the rectangular variant. After this, he deals with the square vault with vertical courses set in parallel to the sides of the area and its rectangular variant; and only when he has explained these problems thoroughly does he deal with the irregular, triangular and diagonal-course vaults, finally proceeding to surbased sail vaults (Vandelvira c. 1585: 82r-93v). Thus, Vandelvira graduates the difficulties didactically, clearly explaining the crucial steps

[25] Cámara (1981: 259–260): *este hombre … fue considerado por Rojas tan competente que aunque hubiera estado en Roma habría que haberlo llevado a Cádiz …* See also Cámara (2015: 18).

in the tracing process, although many passages are somewhat synthetic in order to avoid repetition. Instead, dressing details are seldom explained in detail; he seems to think that an experienced stonemason would understand them with just a few hints. Other texts in the personal notebook tradition, such as the one of Alonso de Guardia (c. 1600) fill this gap.

2.2.4 *Cristóbal de Rojas, Ginés Martínez de Aranda and Alonso de Guardia*

Cristóbal de Rojas was an engineer serving Philip II of Spain in Cádiz, a key point for the Spanish Crown. The city controls the entrance to a bay close to the estuary of the Guadalquivir river, the point where the ships returning from the Americas finished their oceanic journey, and was thus subject to frequent English and Dutch attacks led by such controversial figures as Sir Francis Drake. However, this was merely a phase in Rojas's long career, which took him to Brittany as part of the Spanish support of the Catholic League and the Duc de Mercoeur in the Wars of Religion. Apart from Pedro Luis Escrivá, he was the first high-profile Spanish military engineer; before him, Philip II employed Italians such as Tiburzio Spanocchi or Giovanni Battista Antonelli (Mariátegui 1880: 15–38; Checa 1989: 324–332; Cámara 1998). Like de l'Orme with poets, Rojas was despised by noblemen, at this moment in full command of the military profession; he retorted with jokes about the ignorance of generals about scales, stressing his experience in construction since he was trained as a mason. In any case, he asked in several occasions to be awarded the title of captain; finally, he was granted it only on an honorary basis, so that he could inscribe it in the frontispiece of his *Teórica y práctica de fortificación* (1598), the first treatise in military engineering written in Spain and the second printed work dealing with stereotomy in Europe (Rojas 1598: 30v; Mariátegui 1880: 36–37, 40–41; Cámara 2014: 137–138, 145–146).

This book starts with an interesting introductory section, stressing that the engineer should master geometry, arithmetic, and the ability to choose fortification locations, a skill that cannot be learned in books, but rather by practical, on-site experience. After this, Rojas includes a selection of Euclidean propositions, followed by a section on arithmetical rules. Next, the main part of the book is devoted to fortification, with a broad range of plans and explanations of fortresses, while the shorter third part deals with construction issues, including only nine pages of stereotomic schemes (Rojas 1598: 97r-101v; see also Calvo 1998), without individual explanatory texts. Significantly, in the general introduction to this section, Rojas justifies the lack of these texts arguing that "to know how to make a vault you need much experience, so I will not include an explanation, since this issue is deeply hidden; however, anybody with some knowledge in this field, will understand architectural issues from images".[26] Notwithstanding that, the thirteen stonecutting schemes included in this section bring about some interesting innovations, particularly a skew arch which seems solved using orthogonals to oblique intrados joints (Fig. 2.12, below) instead of the traditional solution, which leads to elliptical face joints (Fig. 2.12, above); later, Ginés Martínez de Aranda would offer a clear explanation of this problem.

The career of Martínez de Aranda has some points of contact with Alonso de Vandelvira. He was born in Baeza, near Úbeda; he belonged to a long dynasty of masons, but the late- sixteenth-century crisis led him to seek work in a small village, Castillo de Locubín. Some years later, he moved to a larger town, Alcalá la Real, where he executed plans from other masters, such as Ambrosio de Vico, in the abbey church of Santa María de la Mota. However, in a stroke of good luck, Martínez de Aranda met Abbot Maximiliano de Austria, the illegitimate son of a member of the Habsburg family. Within a few years, Don Maximiliano was made bishop of Cádiz, a city that was devastated after an Anglo-Dutch attack in 1596; he entrusted to Martínez de Aranda the reconstruction of the damaged cathedral. In this period, he won a competition for the post of master mason of Cádiz, winning over Alonso de Vandelvira. However, Don Maximiliano was promoted to the position of Archbishop of Santiago de Compostela, and Martínez de Aranda followed him again, while Alonso de Vandelvira took the post of master of Cádiz, as we have seen before. In Santiago, Martínez de Aranda was involved in important works, such as the staircase that descends from the West front of the cathedral to the Plaza del Obradoiro, as well as the monastic church of San Martín Pinario. He seems to have distanced himself from the archbishop and Santiago around 1608, although there is no evidence about the particular circumstances. It appears that he had invested the profits from his activity in Cádiz and Santiago in a lucrative, quiet business in Castillo de Locubín, while maintaining some constructive activity, probably to help his relatives and disciples, such as Juan de Aranda Salazar (Galera 1978; Gila 1988; Gila 1991: 51–87, 265–287; Antón Solé 1975; Bonet 1966: 115–130; Vigo 1996).

Martínez de Aranda wrote another systematic, thoroughly complete stonecutting manual, under the title of *Cerramientos y trazas de montea* (c. 1600); as far as I know, only a portion of it has survived. According to its introduction, it was divided into five parts, dealing with arches, rere-arches, staircases, squinches and cylindrical vaults, and spherical and octagonal vaults. The first two parts and a portion of the third one are preserved in a copy by Martínez de Aranda conserved in the Instituto de Cultura e Historia Militar in Madrid.[27] At least one other version of the same work seems to have been prepared by the author; one of the copies was addressed to

[26] Rojas (1598: 101r): ... *consiste el saber hacer las bóvedas en el mucho uso y experiencia que se tendrá de ellas, y así no diré su declaración, por ser materia que la tiene dentro en sí muy escondida, aunque fácil de comprenderla al que tuviere algunos principios, con los cuales conocerá por la pinta todas las cosas de arquitectura.* Translation by the author.

[27] The manuscript in the Instituto de Historia y Cultura Militar (MS 457) includes some notes that connect it with members of the Churriguera family; one of them states that Alberto de Churriguera wrote it. However, the handwriting does not match that of Alberto and, in contrast, is quite similar to the signature of Ginés Martínez de Aranda in the period 1600–1620. About this issue, see Calvo (2009) and, for a different opinion, Bonet (1986).

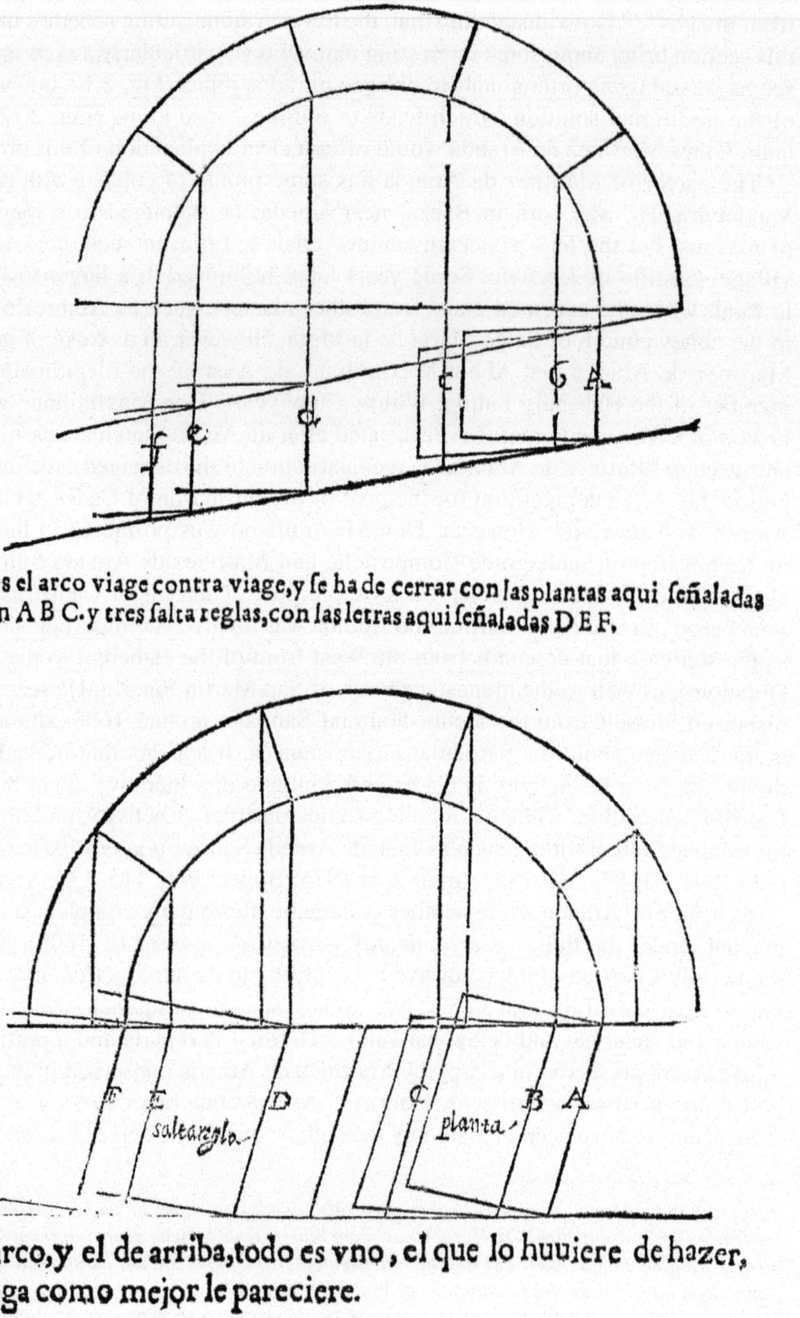

Fig. 2.12 Schemes for skew arches (Rojas 1598: 99v)

Don Maximiliano, but the dedication is lacking in the manuscript in Madrid. In any case, the approach of the preserved parts is encyclopaedic, including no fewer than 70 problems about arches and 51 about rere-arches (Fig. 2.13), in addition to several introductory propositions and a most interesting prologue where Martínez de Aranda puts forward his opinions about the nature of stonecutting, its role in architecture and the social standing of masons and their lore. Although the structure is quite systematic, the body of the text shows clear connections with the oral tradition; for

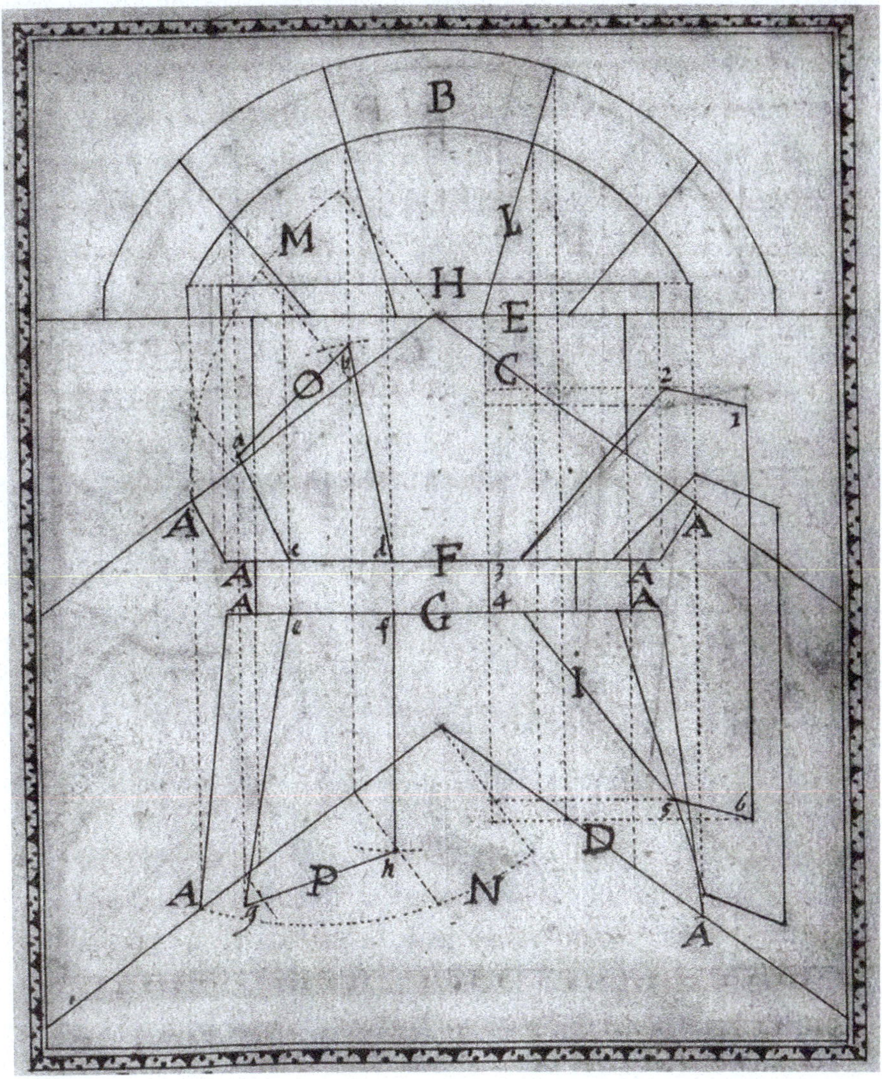

Fig. 2.13 Combination of rere-arches in a corner door (Martínez de Aranda c. 1600: 208)

each problem, the tracing procedure is explained in remarkable detail, in a repetitive fashion. In contrast, dressing techniques are only mentioned in the most complex cases. As with Vandelvira, the dedication, the neat structure, and the prologue, under the heading "To the reader", hint that Martínez de Aranda intended to publish his manuscript (Calvo 2009).

Several text passages and setting out procedures in the manuscript, such as the construction of templates for skew arches using orthogonals to oblique intrados joints, recall the treatise of Cristóbal de Rojas. The documents regarding the selection process for the post of master mason in Cádiz discovered by Alicia Cámara (2015: 18) cast light on this issue. Such passages may be part of an effort to flatter Cristóbal de Rojas, who favoured Alonso de Vandelvira, and win him to Martínez de Aranda's side; the dedication to Don Maximiliano may also have been written with this intention. In other words, the original manuscript may have been prepared as a contribution to Martínez de Aranda's portfolio for the competition; the copy in the Instituto de Cultura e Historia Militar, without the dedication, seems to be a later version of the original manuscript.

Another manuscript (Guardia c. 1600; see also Marías 1992: 352–353, 357–359, and Calvo 2015a) shares some traits with those of Rojas and Martínez de Aranda, at least regarding vocabulary and setting out techniques. However, it is quite different from these works in other aspects. Rather than a printed treatise or a systematic compilation, it is a hotchpotch of hastily-drawn sketches in the blank pages of a factitious copy of two emblem books by engraver Battista Pittoni (1562; 1566) and poet Lodovico Dolce; even the free space of the printed sheets is used here and there (Fig. 2.14). The name of Alonso de Guardia is inscribed in some pages; as far as I know, no mason with this name is mentioned in studies about Early Modern Spanish architecture. In any case, its haphazard quality lends it a remarkable interest: it is nearer to actual workshop practice than the academic compilations of Vandelvira and Martínez de Aranda. In particular, it explains dressing processes for basic pieces, such as splayed arches, deals in detail with the tracing and carving process of rib vaults, and includes pieces that seem too simple to mention for Vandelvira, such as the hemispherical dome dressed by squaring.

2.2.5 *Juan de Portor y Castro and Francisco Fernández Sarela*

Another manuscript in the wake of Martínez de Aranda is the late one by Juan de Portor y Castro, a mason from Santiago de Compostela, although he also had some connections with Granada. In addition to similarities of vocabulary and stonecutting techniques, its connections with Martínez de Aranda are evident in a passage discussing the tracing procedure for a cantilevered spiral staircase, stating that:

> Although this procedure is not used often, you may present it in the competition for a position. It was one of the tracing methods that were put forward by Juan de Aranda Salazar

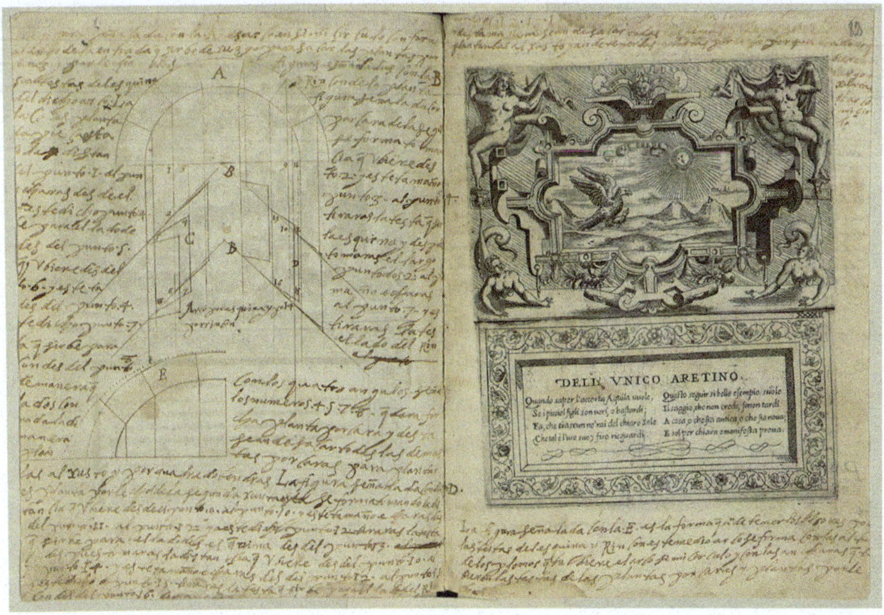

Fig. 2.14 Corner arch (Guardia c. 1600: 81v-82r)

in his contest against Bartolomé Fernández Lechuga for the job of master mason of Granada Cathedral.[28]

Aranda Salazar was the nephew of Ginés Martínez de Aranda; a few years after this contest, held in 1631, he moved to Jaén, where he was charged with the construction of the vaults of the cathedral, begun by Andrés de Vandelvira in the preceding century (Gómez Martínez 1998: 38–39; Galera 1977: 109–111).

Portor's manuscript includes the date 1708, and a large part of it may have been written around this date. This section resembles Martínez de Aranda not only in vocabulary and general methods but also in details such as a penchant for applying templates to non-developable surfaces and even in the shape of the step used in the cantilevered staircase mentioned in connection with Aranda Salazar (Fig. 2.15). This does not mean that this section of the Portor manuscript is a slavish copy of Martínez de Aranda. Quite the contrary, it comprises solutions for pieces that were included in neither Vandelvira nor in the preserved portions of Martínez de Aranda, such as cantilevered staircases, pointed lunettes, or some idiosyncratic solutions for rib vaults. In contrast, other parts of the manuscript, with smaller drawings and handwriting, seem to have been put together as late as 1717–1718; they mimic solutions by

[28]Portor (1708: 22r): *Que aunque no es traza que se ofrezca muchas veces ocasión para ejecutarla, es traza para valerse de ella en una oposición. Como ya ha sucedido porque fue una de las trazas que valieron a un Maestro mayor en la Santa Iglesia de Granada, Juan de Aranda Salazar en la oposición que tuvo con el maestro Bartolomé de Lechuga en dicha Iglesia de Granada.* Translation by the author.

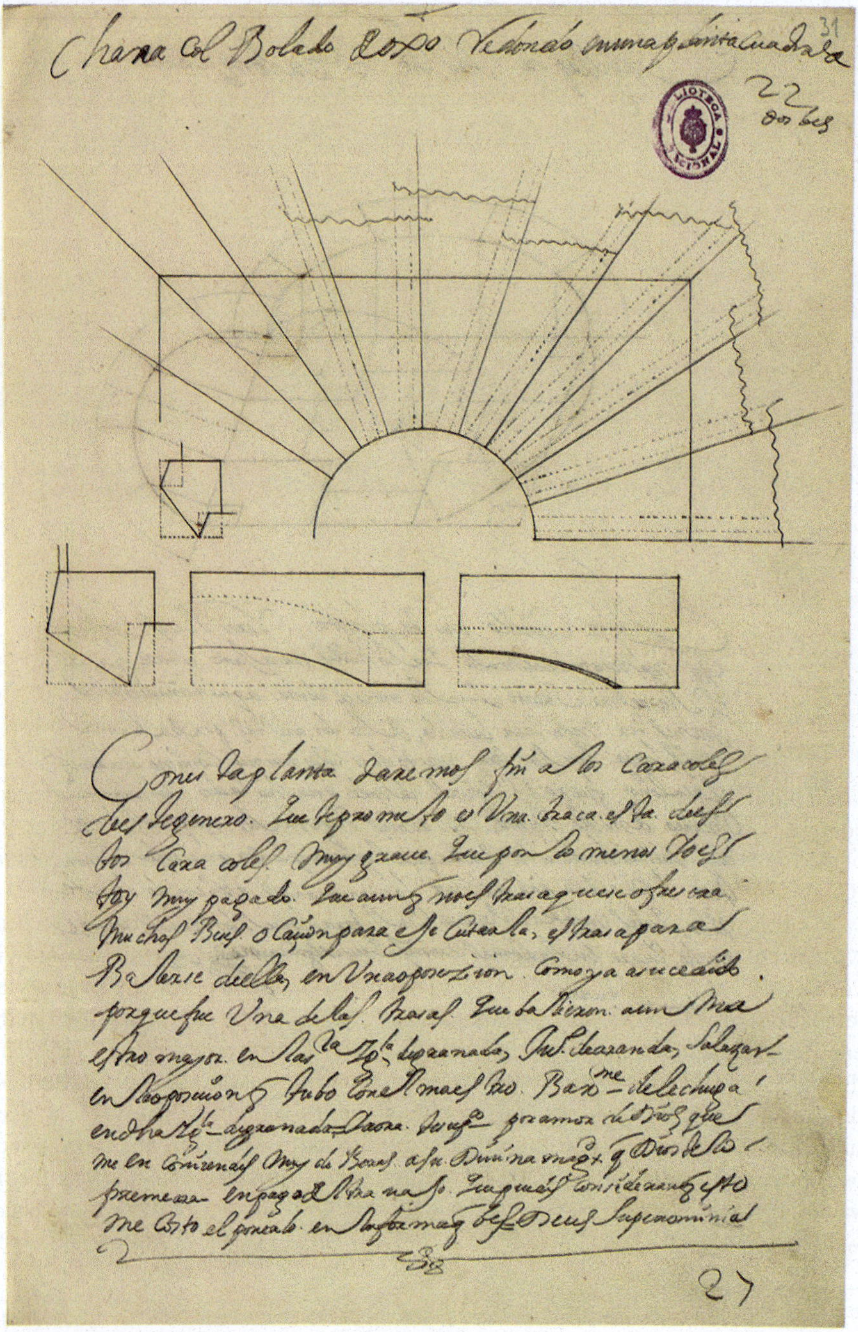

Fig. 2.15 Cantilevered staircase (Portor 1708: 22r)

Tomás Vicente Tosca, the author of a treatise (1707–1715) on mathematics popular in eighteenth-century Spain (Carvajal 2011a; Carvajal and Cortés 2013).

Still another late offspring of the Martínez de Aranda line is the manuscript of Francisco Antonio Fernández Sarela, dated 1740 and preserved in the library of the convent of Saint Francis in Santiago de Compostela (Carvajal and Cortés 2013; Fernández Sarela and Cortés 2015: 67–79). It shows many similarities with Portor, such as the pointed lunettes; this is not surprising since both masters had worked together in the cathedral workshop. There is also material taken from Fray Laurencio de San Nicolás and Juan de Torija, as well as a drawing resembling some singular trefoil arches in the convent of Saint Clare in Santiago.

2.2.6 *Mathurin Jousse, Juan de Aguirre and Fray Francisco de Santa Bárbara*

In France, after a bibliographical desert lasting for more than 70 years, the publication of a short leaflet on stereotomy by Girard Desargues in 1640 brought about much turmoil and a host of publications, ranging from pamphlets to full-blown treatises, as we will see in Sect. 2.3. Mathurin Jousse brought to the presses a work prepared some years before, the first printed book dedicated entirely to stonecutting; in fact, the royal privilege was granted in 1635 (Jousse 1642: viii; Pérouse 1982d; Babelon 2006). Recent studies (Le Boeuf 2001a; Le Boeuf 2001b) have shown that the author was neither an architect nor a mason, but rather an ironsmith who had previously published one book on locks and grilles and another in carpentry. Quite probably, he was trying to profit from the sudden interest in stonemasonry brought about by the controversy triggered by Desargues to complete a collection of books on the building trades, stressing that he was revealing the most crucial secret in architecture.

Jousse was connected with the college at La Flêche, a key centre of the educational network of the Society of Jesus in France. The publication of his book did not please Father François Derand, former professor of mathematics at the college, who was also preparing a treatise on stereotomy in the heat of the debate between Desargues and the Paris stonemasons. It has been posited (Pérouse 2009) that Jousse had borrowed from Derand's teachings at the college, or even plundered the manuscript of his work while the Jesuit professor was writing it. However, this is not what Derand says:

> It is true that a longer writing about the same subject, published around six months ago, under the title *Secret de l'architecture*, has surprised me in the middle of my printing. However, I have found it false in many points and lacking the finest problems, and the wealthiest practices of this art. I see that its author has not reached his goal, and he should present his work in better shape, if he wants it to be taken as legitimate and useful, in accordance with our great hope of finding in it the finest secrets of Architecture. Anyone taking the pains to study what this book includes will see that I am telling the truth. Even joining the three works [De l'Orme, Desargues and Jousse] on condition that the *Secret d'Architecture* will

be freed from its errors, this work on vaults cannot be considered as finished, only sketched at most...[29]

Thus, Derand accused Jousse of sloppiness, rather than plagiarism. Comparing both works, as I will do in Chap. 4, we get the impression that Jousse transmits more or less literally the workshop practices of the early seventeenth century, while Derand is trying to polish them with a veneer of mathematics. Probably the ironsmith Jousse sought the assistance of a professional mason, but this advisor was not Derand. In any case, the *Secret d'Architecture* is an invaluable document, the witness of Early Modern tradition in France. A mason should understand Jousse's language without difficulties, although some erudite terms such as *hemycicle* arise here and there. Plates are placed systematically on even-numbered pages, and the corresponding text is set at the next odd-numbered sheet, so the reader can easily follow the written explanation with the help of the figure. It shares many traits with Vandelvira and Martínez de Aranda while being more advanced in some points, such as the systematic use of cylindrical developments for intrados templates (Fig. 2.16). This attests to the interchange of stereotomical knowledge on both sides of the Pyrenees and the dynamic, evolutionary nature of this field.

Its comprehensive and systematic nature enhances its value. He starts with arches, including skew ones and those opened in corners and battered or round walls; next, he deals with sloping barrel vaults, trumpet squinches, spherical and cylindrical vaults, spiral staircases and annular vaults, lintels, and rere-arches. Thus, Jousse established the standard table of contents of French stereotomic treatises, which remained unchanged up to the nineteenth century.

Two manuscripts attest to the influence of Jousse's manual on Spanish masons. One of them (Manuscrito de arquitectura y cantería, c. 1600), preserved in the National Library of Spain in Madrid, bears the name of Juan de Aguirre, but this unknown figure does not seem to be the author. It is divided into two distinct parts, although none is marked by a particular heading. The first one deals with stonecutting problems, while the second one includes several architectural drawings, most of them orders taken from Vignola, with two staircases copied from Palladio, including his enhanced version of that of Chambord (Salmerón 2015). Stereotomic drawings share many traits with Vandelvira, such as the use of bevel guidelines to compute angles between face and intrados joints; however, other traits, such as the use of cylindrical

[29]Derand (1643: ii r): *Il est bien vray qu'vne plus grande piece concernant le mesme suiet, & mise au jour depuis six mois en çà ou enuiron, sous le titre Secret de l'Architecture ... m'a preuenu & surpris au milieu de mon Impression. Mais comme iel'ay reconnu fautiue en beaucoup de chefs, & destituée d'ailleurs des plus beaux Traits, & des plus riches pratiques de l'Art; i'ay iugé que son Autheur n'auoit aucunement atteint au but, & qu'il sera obligé de donner vne meilleure forme à son ouurage, s'il veut qu'il passe pour legitime, & qu'il nous soit autant vtile, comme est grande l'esperance qu'il pretend que nous conceuions, d'y trouuer les plus beaux secrets de l'Architecture. Ceux qui pendront la peine de parcourir ce qu'elle contient, verront comme ce que i'en dy, ne deroge en rien à la verité. Estant donc ainsi, que iognant mesme le trauail de ces deux (ce qui se entend à condition que celuy du Secret de l'Architecture &c. soit purgé de ses fautes,) à ce que Philebert de Lorme, l'unique que ie sçache qui les a precedé en cette entreprise, en a escrit dans ses oeuures, cet ouurage des voûtes ne peut au plus estre dire qu'ebauché, mais non acheué ...* Translation by the author.

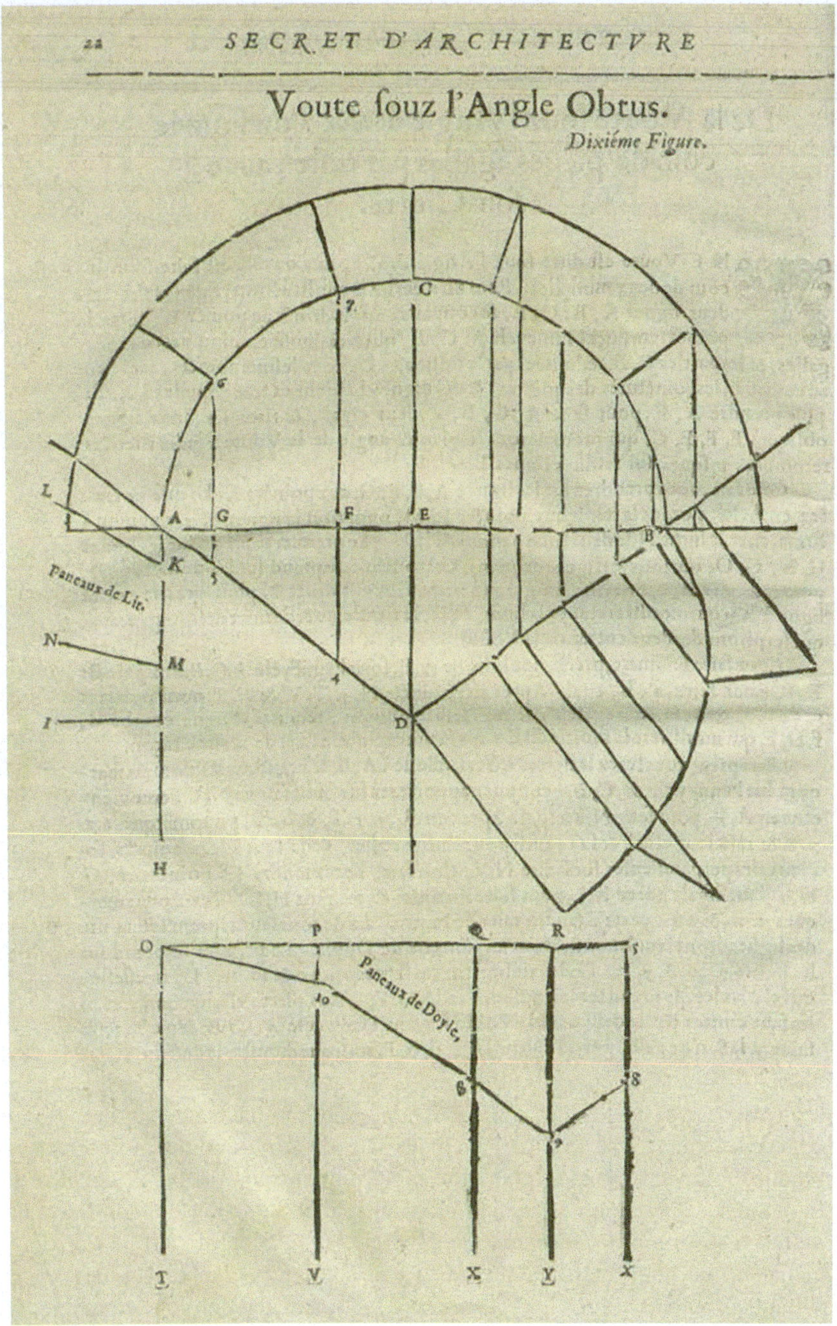

Fig. 2.16 Corner arch (Jousse 1642: 22)

developments for intrados templates, recall the drawings of Jousse and Derand. Also, pointed lunettes, included by Jousse, appear for the first time in Spanish literature.[30]

In this case, the influence of Jousse is indirect, subject to interpretation and compatible with other Spanish influences. In contrast, a manuscript by Fray Francisco de Santa Bárbara, dating from 1766, preserved in the Municipal Archive in Xàtiva, combines direct influence from Jousse, which is visible in the title, rendered in Spanish as *Secretos de Arquitectura,* and many drawings, with other problems that are not taken from Jousse, but rather derived from the Spanish tradition (Bérchez 2005–2006: 196–197). The author and the fact that it was written in the Valencian Monastery of Saint Michael of the Kings connect it with the many stonecutting treatises written by clergymen in the seventeenth century, as we will see in Sect. 2.3.

2.3 The Seventeenth Century and the Clergymen

2.3.1 Fray Laurencio de San Nicolás and Juan de Torija

Clergymen showed an increasing interest in stereotomy along the seventeenth century. While in the preceding centuries religious orders had generally depended on secular architects, from the sixteenth century onwards they began to train architects from their own ranks. One of them was Fray Laurencio de San Nicolás, the son of a mason who entered the order of Augustinian Recollects after the death of his wife while working in Jarandilla. His son left for Madrid, where he worked as a builder; later on, he entered the Augustine order, following a successful career as an architect. He erected churches and monasteries for different orders, particularly the convents of Augustinian Recollects in Salamanca and Talavera and the parish of Saint Martin, a church for the Benedictinian nuns of Saint Placidus and another one for the female branch of the military order of Calatrava, all in this latter city; he also sat on an advising panel of the Madrid town hall (Martín González 1989; Díaz Moreno 2008: xxiii-xlvi).

San Nicolás wrote a general treatise on architecture, *Arte y uso de arquitectura,* in two parts, appearing in 1639 and 1655. George Kubler (1957: 80) assessed it as "the best book on architectural instruction ever written"; such a judgement may be a bit excessive, but in any case, it is a comprehensive, sensible, and practical educational manual. The first part includes introductory sections in geometry and arithmetic, followed by an explanation of the main traits of the classical orders. Next, he presents a section on arches and vaults, alternating masonry and carpentry, as well as a section in façades and the general composition of churches. From this moment on, the book includes staircases and some topographical issues, such as levelling instruments, which leads to another section in geometry; he concludes with

[30]There is a crude drawing in Alonso de Guardia (c. 1600: 76v) that may represent pointed lunettes, but it seems to be unfinished. See Calvo 2015: 450–451.

a chapter recommending princes not to grant the posts of master mason to figurative artists, but rather to building specialists.

This first part was the fiercely attacked by Pedro de la Peña, who tried to prevent its distribution on several grounds that seem quite irrelevant for a modern reader, such as small discrepancies on measurements of churches or whether Pythagoras invented the square root and the right angle (San Nicolás 1665: 4–21). San Nicolás counterattacked in the second part of his treatise with accusations of plagiarism against de la Peña and Juan de Torija, saying that they had copied a manuscript by Alonso de Vandelvira (San Nicolás 1665: 155–156). Thanks to this episode, we know that the stonecutting text by Vandelvira was known as *Libro de trazas de cortes de piedras* and some attempts at its publication were made in the mid-seventeenth century. As for the second part of *Arte y uso de Arquitectura* itself, it begins with a lengthy section on the classical orders, explaining the differences in the conceptions of Vitruvius, Serlio, Palladio, Viola Zanini, Arfe y Villafañe, Vignola and Scamozzi, together with some remarks on Sagredo, Cattaneo, Labacco and Rusconi; he stresses that this section is illustrated with copper engravings, in contrast to the woodcuts of the first part. After some additional information about wooden roofs, he embarks on an attack on *Breve tratado de todo género de bóvedas* (Short treatise on all kinds of vaults) by Juan de Torija, clearly stating that it is "a book on vault measurements", giving alternative solutions. This paves the way for another lengthy section on geometry and proportions, spanning no fewer than 140 pages, followed by a few chapters about the building regulations of Toledo and a final autobiographical chapter explaining his trajectory and the inspiration for entering the Augustinian order.

As for the strictly stereotomic matter, what is more remarkable is that San Nicolás mixes stonemasonry, bricklaying and carpentry in an assortment of chapters about the building trades (1639: 64–108); on some occasions—for example in the discussion of lunettes (San Nicolás 1639: 103–105)—, it is not easy to understand whether he is speaking about brick, ashlar or both. There are sections on round, segmental, basket handle and pointed arches, skew and corner arches, rere-arches, barrel, pavilion (Fig. 2.17), groin, hemispherical, sail and annular vaults: a reasonable, although synthetical, repertoire of stonecutting problems.

As we have seen, Torija's short treatise (1661) deals with vault measurements rather than strictly stereotomic issues; for example, it does not include a single bed joint template or bevel guideline.[31] In particular, it seems that its main goal is to measure the surface area of vaults in order to determine the cost of renderings or

[31] Marías (1988); Rabasa (2000: 235); for a different point of view, see Barbé (1981) and Pérouse (1982b). Moreover, it is interesting to note that surface computations and actual stereotomy are included in different chapters in the general, elementary manual of Juan García Berruguilla (1747: 84–98, 99–110). In spite of this, some authors assume that Torija's *Breve tratado* was copied from Vandelvira's *Libro de trazas de cortes de piedras*. This notion seems to be based on a passage by Fray Laurencio de San Nicolás which states that Torija copied Vandelvira, but he does not go as far as saying that it was the *Breve tratado de todo tipo de bóvedas* the book he copied from Vandelvira. It may be another lost book, perhaps one of these books mentioned by San Nicolás (1655: 155–156) that did not reach the presses as a result of high costs. Another manuscript that includes a chapter on stonecutting including drawings without templates or the voussoir enclosures for squaring is the one by Andrés Julián de Mazarrasa (c. 1750); see Mazarrasa and Fernández Herrero (1988).

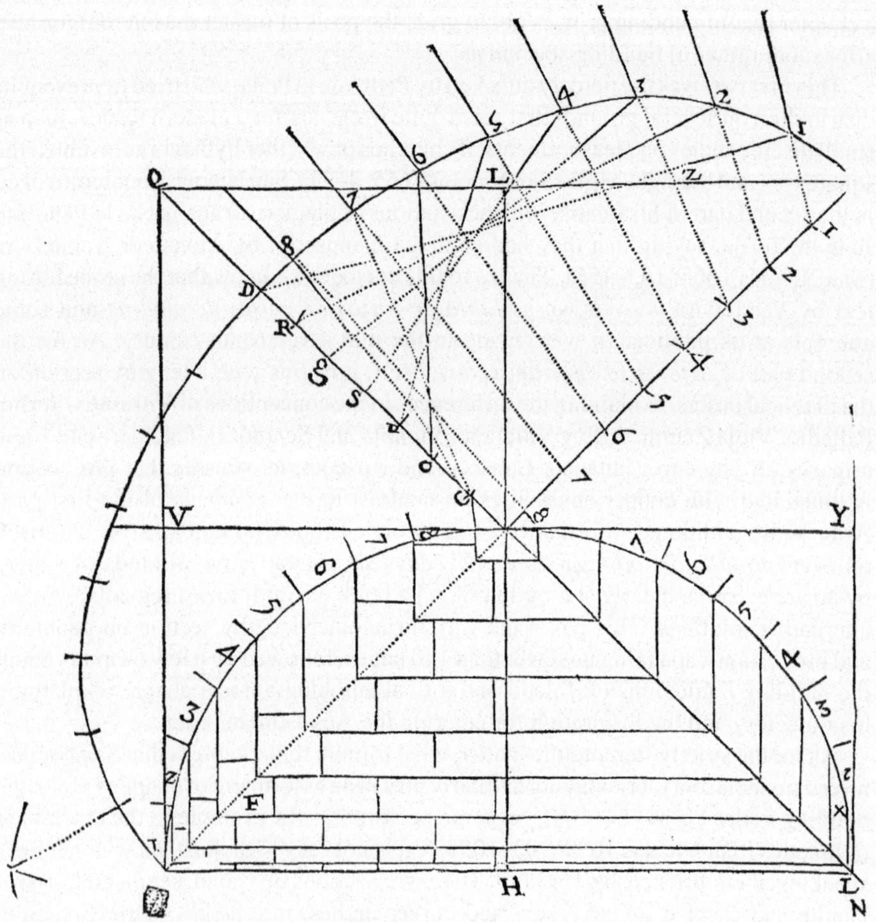

Fig. 2.17 Pavilion vault (San Nicolás 1639: 100r)

surface treatments of ashlar masonry; perhaps this explains the mention in the title of "all kind of vaults", including both brick and stone. Quite significantly, it includes only cylindrical vaults and a splayed arch, with the exception of a flawed calculation of the area of the pendentives under a hemispherical vault. The intrados surfaces of the cylindrical vaults and the splayed arch are correctly computed, or at least approximated, by developments (Fig. 2.18). Therefore, at first glance these schemes resemble the flexible intrados templates of Jousse (1642: 20–21, 64–65) and Derand (1643: 171); on closer inspection, Torija actually uses flat templates, as did Martínez de Aranda (c. 1600: 24–26), but groups them in chains, as Jousse and Derand did. This poses an interesting question: did Torija reuse stereotomic techniques for surface calculations, or did knowledge travel the other way around? For the moment, it seems that the idea passed from stonecutting to quantity surveying, since all these texts

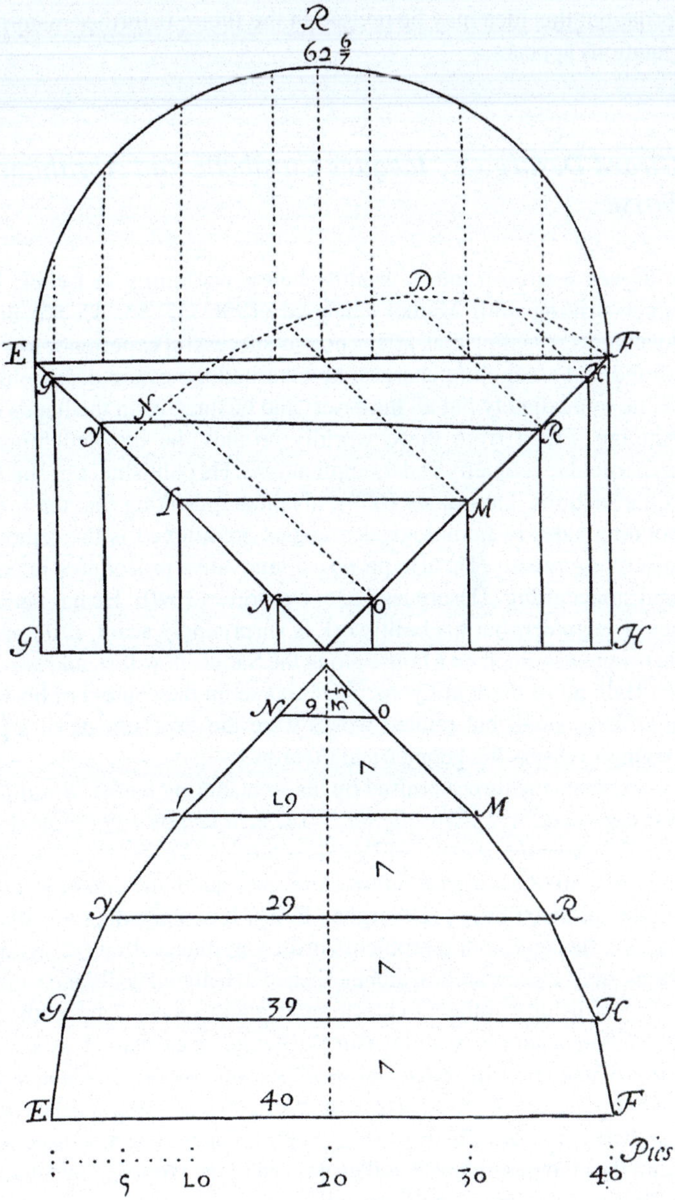

Fig. 2.18 Development of a pavilion vault for quantity surveying purposes (Torija 1661: 11)

predate Torija; but this idea may be revised in the future if further material on vault area computations appears.

2.3.2 *Girard Desargues, Jacques Curabelle and Abraham Bosse*

Girard Desargues was born into a wealthy bourgeois family in Lyons. His career seemed rather striking until Marcel Chaboud (1996: 35–36, 47–50) unearthed a series of documents revealing that, after a not-so-successful experience as a merchant, including fraud and smuggling charges, he inherited the estate of his older brother Christophe; he immediately put all his assets and businesses in the hands of another brother, Antoine. Freed from these worldly errands, he devoted himself to his passions, in particular geometry and its applications. He published a ground-breaking work on conic sections (Desargues 1639); his theorem stating that the intersections of corresponding sides of homological triangles are aligned is the remote foundation of projective geometry. He also proposed innovative procedures in perspective, sundials, and stonecutting (Desargues 1636; Desargues 1640). He has been taken for an architect or engineer, but his built work is surprisingly scant, little more than a remarkable trumpet squinch on a house along the Saone, now lost, and two staircases in the Town Hall, all of them in Lyons.[32] Of course, in the context of his biography, it seems that these small but refined works were the experiments of a gentleman geometer trying to assess his geometrical theories.

This also explains the turmoil raised by his stonecutting leaflet, *Brouillon project d'exemple d'une manière universelle du S.G.D.L. touchant la practique du trait a preuues pour la coupe des pierres en l'architecture ...*, (Draft project as an example of a universal system of Monsieur Girard Desargues about the practice of tracing for stonecutting in architecture ...) (Desargues 1640). It explained a radically different way of tackling the stereotomic problems involving intersections of cylinders with oblique planes, that is, skew arches, arches opened in battered walls and sloping barrel vaults. Instead of using parallels to a reference plane, as usual up to this moment, he devised an idiosyncratic system using different axes named as *essieu*, *sous-essieu*, *contre-essieu* and *traversieu* and involving the use of two slanted projection planes (La Gournerie 1855: 10–12, 44; Schneider 1983: 59–104; Sakarovitch 1994b; Sakarovitch 2010; Boscaro 2016: 51–87). Such an operation has some features in common with the changes of horizontal projection plane proposed by Olivier (1843–44:1: 18–22); however, Desargues's procedures, based on triangulation, are different from nineteenth-century solutions.

[32]See Saint Aubin (1994); Boscaro (2016: 105–133, in particular 121–123). The house along the Saone is not documented, but rather attributed to Desargues by nineteenth-century sources. In contrast, the involvement of Desargues in the Hôtel de Ville is attested by some documents; however, it is not clear whether his role went farther than that of an external advisor.

Professional stonecutters were infuriated by this unexpected intrusion in their field by an idle bourgeois. Jacques Curabelle, known as the best stonecutter of the period in Paris, published several pamphlets with such Baroque titles as *Foiblesse pitoyable du sr G. Desargues employée contre l'examen fait de ses oeuures* (1644a) and a more substantial *Examen des oeuvres du Sieur Desargues* (1644b) in 81 pages. Desargues answered with *Recit au vray de ce qui a este la cause de faire cet escrit* (1644). Both sides went so far as to paste posters about the issue in the streets of Paris, agreeing finally on a singular challenge. Two teams of masons, instructed by Desargues and Curabelle, were to erect arches according to the theories of their respective leaders; the winner was to be awarded a prize of the substantial sum of 200 pistoles. Ultimately, the duel did not take place since the contenders could not agree on the choice of a jury. It seems that Curabelle took it for granted that the panel of judges would be composed of masons, but Desargues retorted bluntly that "the geometers ... should not go to the school nor the lesson of the masons; quite the contrary, masons ... should go to the school and the lesson of geometers; that is to say, the geometers are the masters, and the masons are the pupils".[33] This incident marks a crucial paradigm shift in the field: for Desargues actual arches or vaults are no longer the criterion of validity of stereotomic procedures; stonecutting methods should be tested against the abstract concepts of rational geometry.

Desargues's original leaflet includes four drawings and four pages of fine print; its circulation seems to have been quite limited, and his position did not gain immediate acceptance.[34] Abraham Bosse, an engraver and faithful follower of Desargues, expanded the minimalist booklets of Desargues into full-fledged books, dissecting the techniques of his master in sequential steps. Though his profession fostered a pedagogical, visual approach (Fig. 2.19), he did not gain wide acceptance: although accepted as a professor in the Royal Academy of Painting and Sculpture, his teaching of Desargues's perspectival methods raised much opposition and led to his dismissal. His stonecutting treatise (Bosse and Desargues 1643a) is quite interesting as a detailed explanation of Desargues's obscure stereotomic procedures; however, it exerted little influence on later literature. As for Curabelle (1644b: 19, 21, 22, 27, 31, 55, 63, 78, 81), his booklet is the first one to introduce the word *stéréotomie*, that is, "section of solids" into the debate: the stonecutter, not wishing to be set back by the scientists, resorted to learned language. The controversy caused a sudden renewal of interest in the matter, fostering the publication of a remarkable number of books along the seventeenth century, starting with Jousse, as we have seen, and Derand, the subject of Sect. 2.3.3. These events started a slow drift of the subject into the area of speculative geometry. However, Desargues's methods were not followed by these stonecutting

---

[33]Desargues (1648: [iii]): ... *les Geometres ... ne vont ny à l'Ecole ny à la leçon des Massons, mais au contraire, les Massons ... vont à l'Ecole & à la leçon des Géomètres, en quoy de mesme, les Geometres sont maistres, & les Massons disciples ...* This passage is found at the third page of a section with unnumbered pages at the start of Bosse 1648, with the title "Reconnoissance de Monsieur Desargues", probably written by Desargues himself. See also Sakarovitch (1994b) and Le Moël (1994).

[34]Only a few copies have been found; the only complete one is preserved in the library at Quimper (Pérouse [1982a] 2001); see Sakarovitch (2010) or Boscaro (2016) for a reproduction.

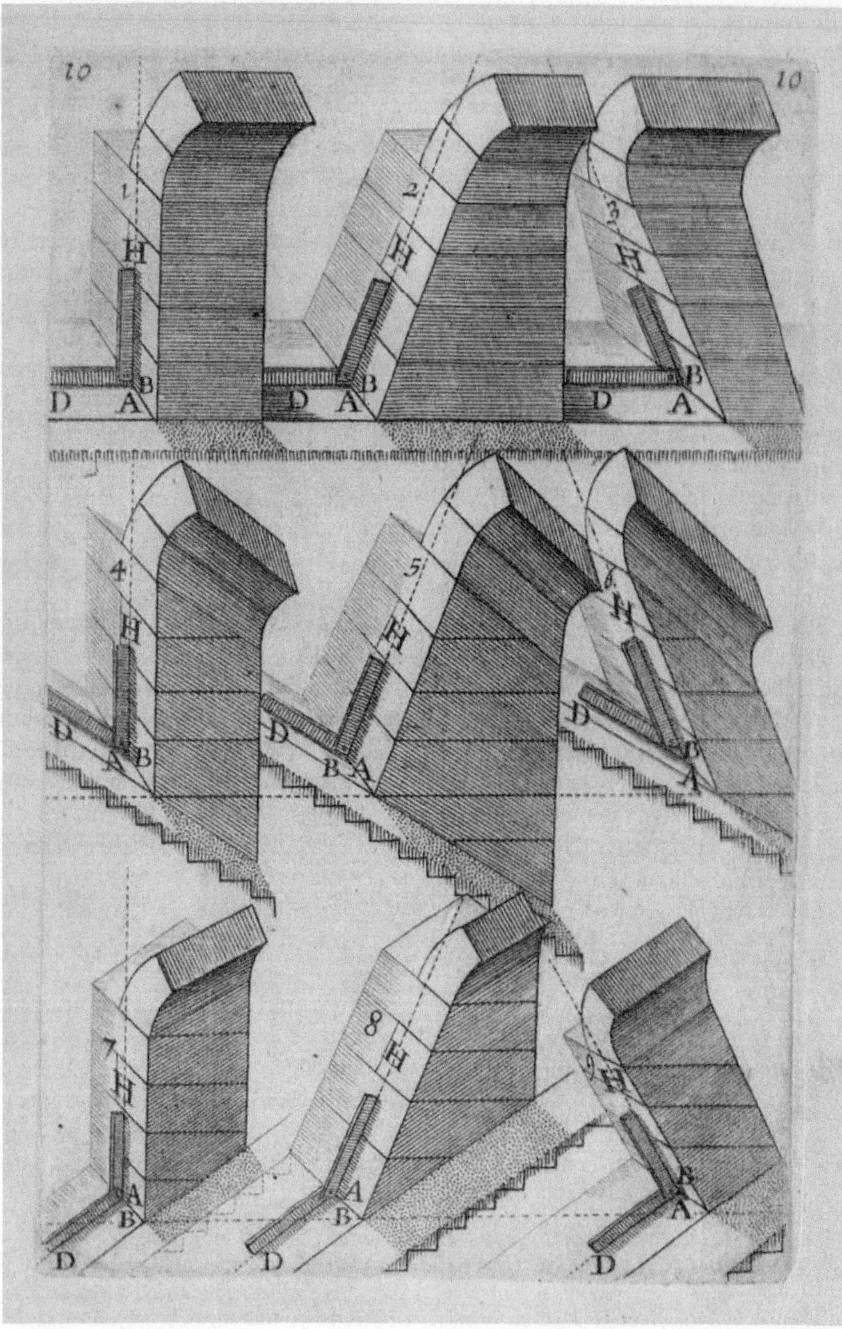

Fig. 2.19 Battered and sloping arches and vaults (Bosse and Desargues 1643a: pl. 10)

theorists; the depiction of Philippe de la Hire and Amedée-François Frézier as disciples or spiritual sons of Desargues (Pérouse [1982a] 2001: 100) should be understood at a rather abstract level.

2.3.3 *François Derand*

While Desargues had tried to put stonecutting under the command of geometry, Derand's approach was subtler. As a member of the Society of Jesus, he taught mathematics at the prestigious college of La Flêche, the alma mater of Descartes. He also acted as an architect for the Society; his most relevant work is the main church of the Jesuits in Paris, presently known as Saint-Paul-Saint-Louis, which exerted a remarkable influence in the buildings of the Society in France (Moisy 1950; Moisy 1952; Pérouse 2009). It has been argued that he was no real architect and that he needed the help of a professional builder; were we to apply this criterion to the entire history of architecture, Alberti, Bernini and Wren would be disqualified from the profession. Moreover, the vaults of Saint-Paul-Saint-Louis use an ingenious solution for the rectangular-plan groin vault, employing raised ellipses for the arches in the short edges of the plan, the typical approach of a geometer to this problem, which is of course included in his treatise (Derand 1643: 329–335).

Rather than starting from new foundations, as Desargues did, Derand built his treatise on the time-honoured traditional methods of the masons, refining them with the concepts, and especially with the systematic nature, of erudite geometry (Pérouse 1982e). In his own words,

> I have followed this route in this treatise, in particular in the problems that I have taken from old practices accepted from long ago by the masters, taking into account the sound effects that result, rather than these small [geometrical] subtilities, which encumber the masons rather than making their works more polished and stable.[35]

Derand's rib vault (1643: 392–395) offers a clear example of this approach. His method is generally similar to the solution adopted by Hernán Ruiz II (c. 1560: 46v): he draws a group of true-size-and-shape representations of all ribs, including perimetral arches, diagonal ribs, tiercerons and liernes over the edge of the plan. However, while Ruiz had placed the centres of these arches by trial and error, as has been shown by Rabasa (1996a: 428), Derand constructs the bisector of the line joining both ends of the tierceron and places the centre of the tierceron at the intersection of this bisector with the springings plane; this assures that the tangent at the start of the rib is vertical. He also recommends the reader to place the centres of the liernes at the vertical line passing through the central keystone, to "meet in a more agreeable

---

[35]Derand (1643: [v]): *Ainsi me suis-ie comporté en ce Traité, és Traits particulierement que i'ay tiré des pratiques anciennes receües de long-temps parmy les Maistres, ayant eu plus d'égard aux bons effets qui en resultent, qu'à ces petites pointilles & subtilitez, qui sont plus propres à embarasser les ouuriers qu'a rendre leurs ouurages plus polis & solides.* Translation by the author.

way".[36] That is, he means to draw both liernes with horizontal tangents at their ends, avoiding a salient point at the central keystone. In both cases, Derand is using the erudite concept of tangent to polish a traditional stonecutting construction, rather than starting from scratch.

The contents of the treatise are as comprehensive as these of Vandelvira and Jousse. However, while de l'Orme (1567), Vandelvira (c. 1585) and Jousse (1642) arrange their subject matter as a continuous string of stonecutting problems, with a heading for each section, but no intermediate divisions, Derand organises his neatly into five parts. Like de l'Orme, he starts with sloping barrel vaults, an awkward decision, criticised by later theorists (de la Rue 1728, v–vi). From this point on, he follows with skew arches, rere-arches and lunettes, all included in the first part of the book. The second one is devoted to *portes*, a category encompassing arches in curved walls and a few other skew arches. The third part deals with trumpet squinches, while the fourth includes *maitresses voûtes* (that is, vaults proper), and the fifth closes the treatise with stairways, both spiral and straight.

Derand makes every effort to facilitate the reader's understanding of the subject, up to the point of making the treatise somewhat boring. The book is printed in folio format, measuring 39 cm high, in contrast to 33 cm for Jousse's quarto, best fitted for the workshop. Fine copper engravings substitute Jousse's woodcuts. Derand eschews geometrical demonstrations, arguing that

> intentionally, I have not added the proofs of these stonecutting problems to the practical solutions I am proposing, in order not to confuse the matter, which is complex enough without the demonstrations.[37]

With this exception, everything is explained in detail, both tracing methods and dressing procedures (Fig. 2.20); variants are given when appropriate. As a result, the text runs up to several pages of fine print for many sections, while Jousse had limited his explanations to one page per problem; in most occasions, the large sheets are used to accommodate two or more stonecutting drawings. This does not allow the careful symmetry between text and figures in the *Secret d'Architecture*, but Derand, not to be surpassed by Jousse, applies a radical solution: he duplicates the sheets where necessary to allow the reader to follow the explanations easily.

---

[36]Derand (1643: 394): ... *se trouuent d'vne plus agreable rencontre.*

[37]Derand (1643: preface, [viii]): ... *par dessein formé nous n'auons pas voulu ioindre les demonstrations de nos Traits, aux pratiques que nous proposons en ce Traité, afin de n'embarrasser par trop ces matieres, qui sans cela ne le sont desia que trop en soy.* Translation by the author. See also la Gournerie (1855: 9–10).

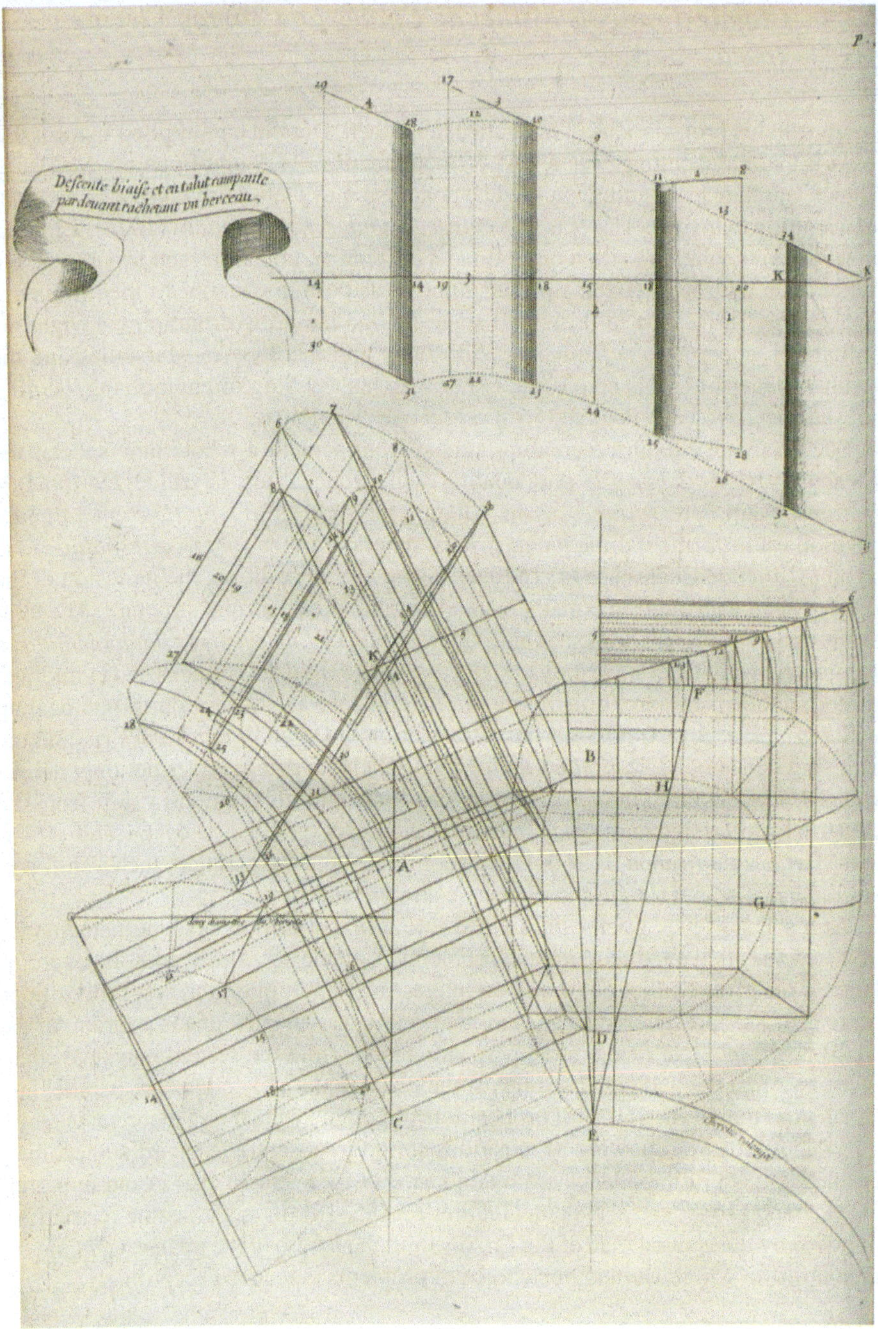

Fig. 2.20 Skew sloping vault on a battered wall intersecting a barrel vault (Derand [1643] 1743: pl. 39)

2.3.4 *Claude-François Milliet de Chales and Tomás Vicente Tosca*

Despite all his protestations of a practical approach, Derand had implied in a passing remark that stonecutting was a part of mathematics.[38] Claude-François Milliet de Chales, another Jesuit, drew the logical conclusion of Derand's stance and included stonecutting as treatise XIV of his *Cursus seu mundus mathematicus* (1674), a general work on mathematics written on Latin and, of course, not addressed to masons, but rather to an educated readership. The author was born in Chambéry, then in Savoy, and died in Turin, also in this small state; he had taught hydrography, navigation, military engineering, mathematics, philosophy and theology in Marseilles and the Jesuit colleges of Lyons and Turin. Other than his teaching of engineering, he does not seem to have been involved with actual construction.

Such a treatise required creating almost from scratch a whole new specialised vocabulary (Fig. 2.21): the part dealing with stonecutting is called *De lapidum sectione*, vaults are *fornices* as in Alberti, rere-arches are *posticæ arcuationes*, pavilion vaults *fornices arcuati claustrales*, spherical vaults *sphericæ testudines*, and so on (Milliet de Chales [1674] 1690: II, 619–620). In addition, the inclusion of the subject in an encyclopaedic mathematical work imposed a rigid structure. The treatise is divided into five books: an introductory one or *Liber primus fundamentalis*; a second one with the title *De arcubus, & fornicibus cylindricis* (arches and cylindrical vaults); another one on *Fornicibus conicis* (conical vaults); the fourth one dealing with *Fornicibus principalibus*, a hasty translation of Derand's *maitresses voûtes*; and a closing one about *Testudinibus helicoiedibus*, a label that includes not only helical vaults, but also other kinds of staircases. At first sight, this implies a shift from the constructive classification of architectural elements established by Derand (1643) and Martínez de Aranda (c. 1600), to a geometrical one; in practice, such a shift was not radical, since "helical vaults" include straight staircases.

At a second level, books are divided into propositions, as befits a mathematics treatise, classified into theorems and problems, with the occasional addition of corollaries. This gives Milliet de Chales an interesting opportunity: the first propositions of treatise XIV are theorems in descriptive geometry. The first one is a variant of the well-known theorem of the three perpendiculars, while the second one states that, given two parallel and equal segments, their horizontal projections are parallel and equal. And that is all. The third proposition, although labelled as a theorem, is really a problem; all the following propositions in the treatise are standard stonecutting problems, most of them taken from de l'Orme (1567), Jousse (1642) and in particular Derand (1643; see also de la Gournerie 1855, 13). In any case, the restrictions imposed by the general plan of the *Cursus* limit the length of the treatise to 74 pages, in contrast to 453 in Derand; the choice of problems is reduced accordingly.

---

[38]Derand (1643: [ii]): ... *comme en toutes les sciences, & particulierment és Mathematiques, donce celle-cy fait parti* ... (as in all sciences, including mathematics, of which [stonecutting] is a part). The phrase may seem ambiguous at first sight, until the reader understands that *celle-cy* refers to *la science des traits des voûtes*, included in the preceding sentence.

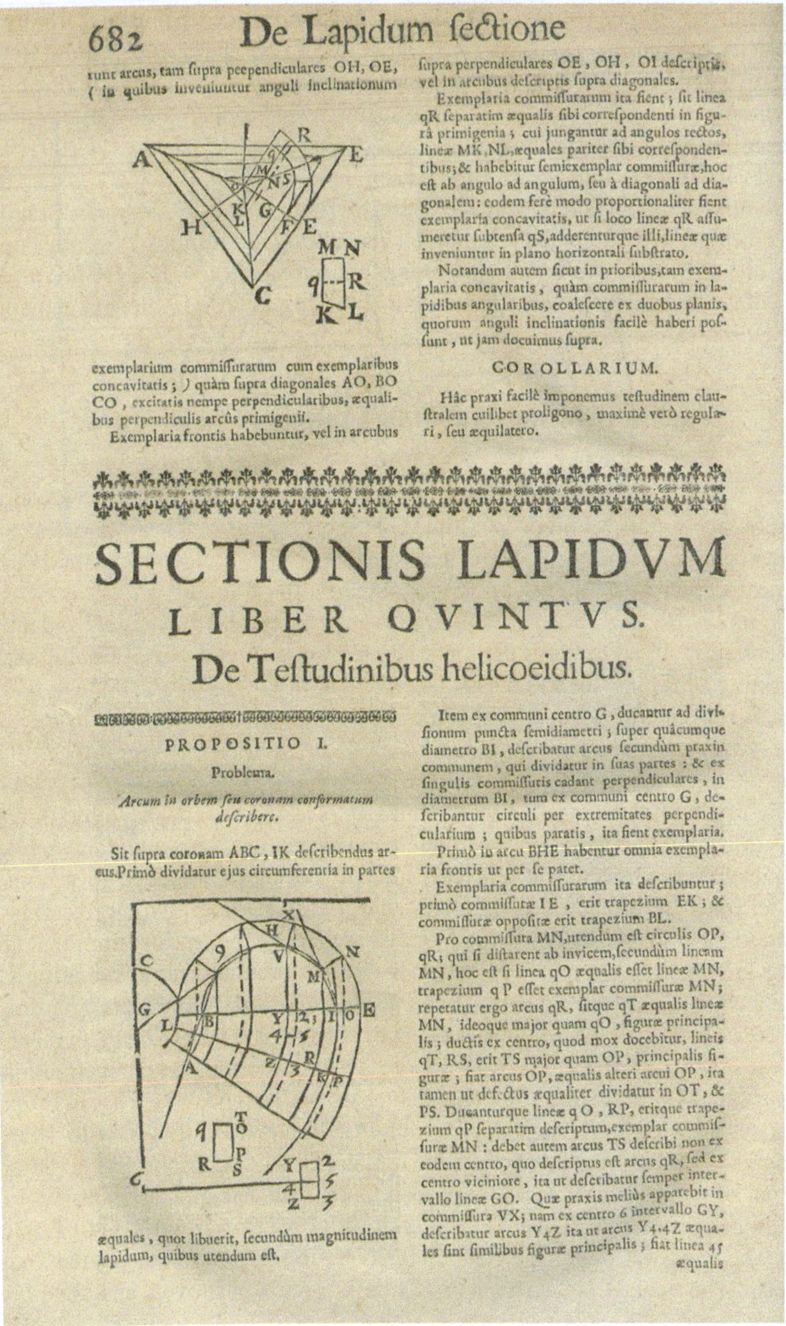

682 De Lapidum ſectione

runt arcus, tam ſupra peependiculares OH, OE, (in quibus inveniuntur anguli inclinationum

exemplarium commiſſurarum cum exemplaribus concavitatis ;) quàm ſupra diagonales AO, BO CO , excitatis nempe perpendicularibus, æqualibus perpendiculis arcûs primigenii.

Exemplaria frontis habebuntur, vel in arcubus ſupra perpendiculares OE , OH , OI deſcriptis, vel in arcubus deſcriptis ſupra diagonales.

Exemplaria commiſſurarum ita fient ; ſit linea qR ſeparatim æqualis ſibi correſpondenti in figurâ primigenia ; cui jungantur ad angulos rectos, lineæ MK ,NL, æquales pariter ſibi correſpondentibus; & habebitur ſemiexemplar commiſſuræ, hoc eſt ab angulo ad angulum, ſeu à diagonali ad diagonalem : eodem ferè modo proportionaliter fient exemplaria concavitatis, ut ſi loco lineæ qR aſſumeretur ſubtenſa qS, adderenturque illi, lineæ quæ inveniuntur in plano horizontali ſubſtrato.

Notandum autem ſicut in prioribus, tam exemplaria concavitatis , quàm commiſſurarum in lapidibus angularibus, coaleſcere ex duobus planis, quorum anguli inclinationis facilè haberi poſſunt , ut jam docuimus ſupra.

COROLLARIUM.

Hâc praxi facilè imponemus teſtudinem clauſtralem cuilibet proligono , maximè verò regulari , ſeu æquilatero.

SECTIONIS LAPIDVM

LIBER QVINTVS.

De Teſtudinibus helicoeidibus.

PROPOSITIO I.

Problema.

Arcum in orbem ſeu coronam conformatum deſcribere.

Sit ſupra coronam ABC , IK deſcribendus arcus. Primò dividatur ejus circumferentia in partes æquales , quot libuerit, ſecundùm magnitudinem lapidum, quibus utendum eſt.

Item ex communi centro G , ducantur ad diviſionum puncta ſemidiametri ; ſuper quâcumque diametro BI , deſcribatur arcus ſecundùm praxin communem , qui dividatur in ſuas partes : & ex ſingulis commiſſuris cadant perpendiculares , in diametrum BI , tum ex communi centro G , deſcribantur circuli per extremitates perpendicularium ; quibus paratis , ita fient exemplaria.

Primò in arcu BHE habentur omnia exemplaria frontis ut per ſe patet.

Exemplaria commiſſurarum ita deſcribuntur ; primò commiſſuræ I E , erit trapezium EK ; & commiſſuræ oppoſitæ erit trapezium BL.

Pro commiſſura MN, utendum eſt circulis OP, qR; qui ſi diſtarent ab invicem, ſecundùm lineam MN , hoc eſt ſi linea qO æqualis eſſet lineæ MN, trapezium q P eſſet exemplar commiſſuræ MN; repetatur ergo arcus qR, ſitque qT æqualis lineæ MN , ideoque major quam qO , figuræ principalis ; ductis ex centro, quod mox docebitur, lineis qT, RS, erit TS major quam OP , principalis figuræ ; fiat arcus OP, æqualis alteri arcui OP , ita tamen ut defectus æqualiter dividatur in OT , & PS. Ducanturque lineæ q O , RP, eritque trapezium qP ſeparatim deſcriptum, exemplar commiſſuræ MN : debet autem arcus TS deſcribi non ex eodem centro, quo deſcriptus eſt arcus qR, ſed ex centro viciniore , ita ut deſcribatur ſemper intervallo lineæ GO. Quæ praxis meliùs apparebit in commiſſura VX; nam ex centro 6 intervallo GY, deſcribatur arcus Y4Z ita ut arcus Y4.4Z æquales ſint ſimilibus figuræ principalis ; fiat linea 45 æqualis

Fig. 2.21 Triangular-plan pavilion vault and vaulted spiral staircase (Milliet [1674] 1690: II, 682)

Probably, Milliet's treatise did not exert much influence on masons, architects and engineers. In contrast, a Spanish derivative, the *Compendio matemático* by Tomás Vicente Tosca (1707–1715), an Oratorian father from Valencia, member of the pre-Enlightenment circle of the *novatores*, seems to have played an important role in the training of Spanish military engineers (Capel 1988: 221; Marzal 1991: II, 989; Galindo 1996: 177; Calvo 2007: 174). Like that of Milliet, this work is also divided into treatises; one of them, dealing with *Montea y cortes de cantería* (tracing and stonecutting), was published separately (Tosca 1757). It mimics Dechalles up to the point of using the same notation on some occasions, although there are also some interpolations, such as a passage about the crossing tower of Valencia cathedral.

2.3.5 *Guarino Guarini and Juan Caramuel y Lobkowitz*

Milliet was not the only one, or even the first one, to include stereotomy in a mathematics treatise written in Latin. Guarino Guarini, a Theatine priest born in Modena and, of course, one of the leading figures in seventeenth-century European architecture, wrote a lengthy book on geometry, the *Euclides adauctus et methodicus mathematicaque universalis* (Euclides augmented and methodical and universal mathematics). In addition to the material stemming directly from Euclid, there are interesting additions, such as those on conic sections, logarithms, geodesy, and, most relevant for our purposes, a treatise on projections and another one in *De superficiebus corporum in planum redigendis* (About the reduction of the surfaces of solids to the plane).

The first half of treatise XXVI (Guarini 1671: 445–452) deals with orthographic projection, explaining that lines parallel to the projection plane maintain their length, oblique lines are shortened, and perpendicular lines are reduced to a point; that parallel lines are projected as parallels; that figures parallel to the projection plane maintain their shape, while oblique figures are subject to deformation; that oblique circles are projected as ellipses, and so on. Even more interesting is treatise XXXII, dealing with cylindrical, conical, spherical, and toroidal developments (Guarini 1671: 572–596). Although the whole subject is presented as a purely mathematical issue, the text includes a passing remark about its application to stone construction.[39] In contrast, most figures show diagrams in the style of Derand, with a series of intrados templates and bed joint templates hanging from them. Thus, it is easy to deduct that Guarini's abstract procedures may be applied to skew, corner and curved arches, trumpet squinches with straight or curved faces, sail, and annular vaults.

All these problems reappear in *Architettura Civile* (Guarini [c. 1680] 1737: 191–265). The book was published posthumously by Bernardo Vittone and the Theatines of Turin. The final 34 plates, depicting Guarini's built, destroyed, and unbuilt works

[39]Guarini (1671: 573): *Si annulus solidus esset diversis segmentis compositus, vt solent portarunt arcus lapidei* ... (if solid rings are divided into several segments, as usual in doors [solved] with stone arches ...).

and projects are frequently reproduced in survey books about Baroque architecture. In contrast, the first 41 plates and the 316 pages of text, dealing with geometry, stereotomy, the orders and oblique architecture, are seldom mentioned. In fact, the book is divided into five treatises, dealing respectively with architecture in general, horizontal projections, orthogonal vertical projections, *ortografia gettata* or theory of developments, and geodesy. This does not mean other subjects are not present; for example, the orders are included in the third treatise, dealing with orthogonal elevations, while stereotomic problems are presented in the fourth treatise, which focuses in developments.

At the beginning of the fourth treatise, Guarini ([c. 1680] 1737: 191) contrasts *ortografia retta* (the theory of projections) to *ortografia gettata* (that of developments, in his own parlance)

> This orthography is in contrast to the preceding one [right orthography] in its title and its operating method. In right orthography, planar surfaces are elevated with orthogonal lines, in order to create solid bodies and give form to the building, while in this [extended orthography] spatial bodies are reduced to a plane using perpendicular lines to extend their surfaces.[40]

Next, he stresses the utility of the theory of developments for stonecutting in these terms:

> However, this [extended orthography] is also quite useful, and absolutely necessary for the architect, although not well known in Italian architecture, and only put in practice wonderfully in many occasions by the French. In order to dress stone, and find its exact form, it is necessary to know the shape of its surfaces; stones dressed in these shapes, when placed in the building, will stand in their place and meet each other. This is why this orthography has been found. It extends these surfaces in the planes ...[41]

From this moment on, there are no explicit references to arches, vaults, templates or dressing procedures; everything is explained in terms of cylindrical and spherical sections and developments. However, looking at the plates (Fig. 2.22, 2.23), the application of these abstract problems to construction elements is not hard to guess: there are corner arches, arches in round walls or lunettes (Guarini [c. 1680] 1737: tr. IV, pl. III), sloping vaults (pl. IV), skew arches (pl. VI), trumpet squinches (pl. VIII), spherical vaults (pl. XII) and finally, an annular vault (pl. XIV). That is, Guarini has advanced a remarkable distance towards Desargues's goal of putting stonecutting under the rule of geometry. However, rather than starting from the ground, as Desargues had tried, he builds on a long tradition, for many of his solutions can be

---

[40]Guarini ([c. 1680] 1737: 191): *Questa Ortografia, siccome é opposta nel suo titolo all'antecedente [ortografia retta], così anche nel suo modo di operare; perchè là dove inquella le superfizie piane s'innalzano con linee perpendicolari, per dare a loro corpo, e formare la Fabbrica, questa per lo contrario i corpi in alto sospessi con linee perpendicolari riduce in piano per istendere la loro superficie.*

[41]Guarini ([c. 1680] 1737: 191): *Non è però questa di quella meno utile, anzi chè assolutamente necessaria all'Architetto, abbenché poco conosciuta dalla Italiana Architettura, solamente dalla Francese in molte occasioni egregiamente adoperata. Perchè adunque per tagliare le pietre, e ritrovare le giuste forme è necessario sapere, quali siene le loro superfizie, acciocchè fatte, e tagliate secondo quelle, quando si pongono in opera, si assettino al suo luogo, e convegnano colle altre, perciò e stata ritrovata questa Ortografia, che appunto mette le loro superfizie in piano ...*

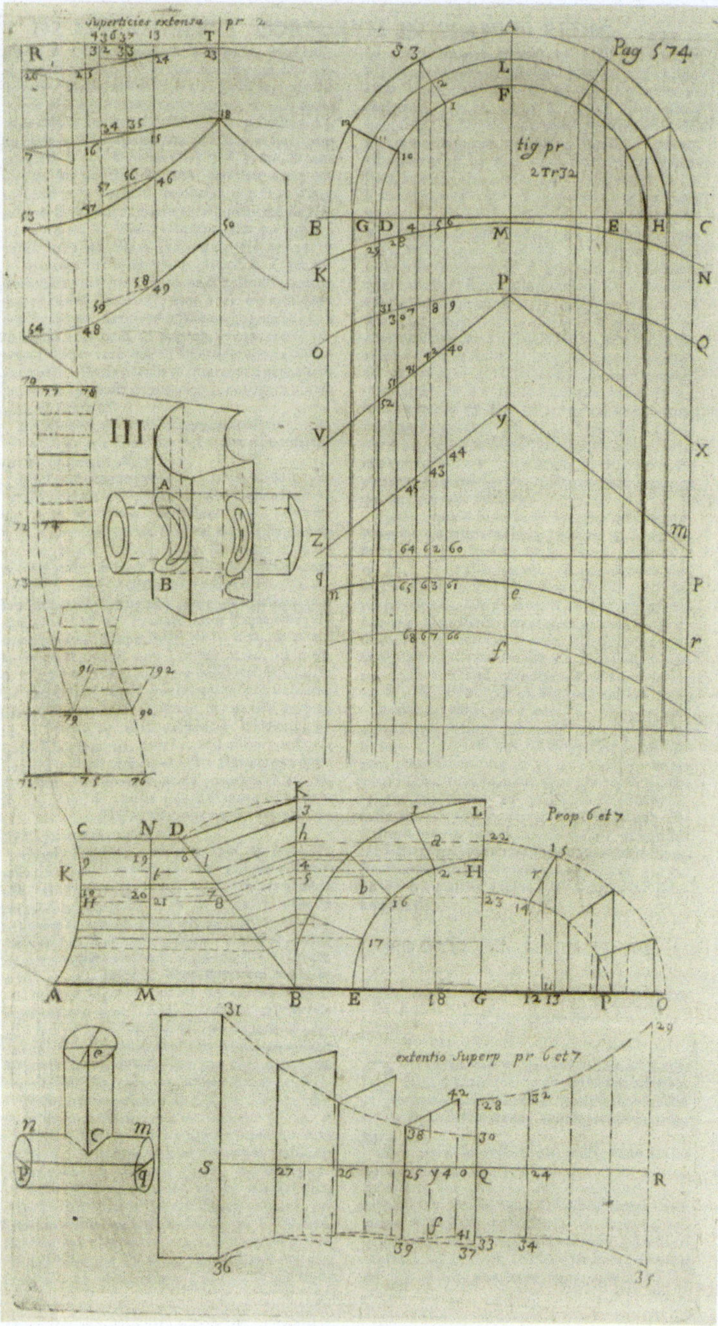

Fig. 2.22 Schemes for corner and curved-face arches (Guarini 1671: 574)

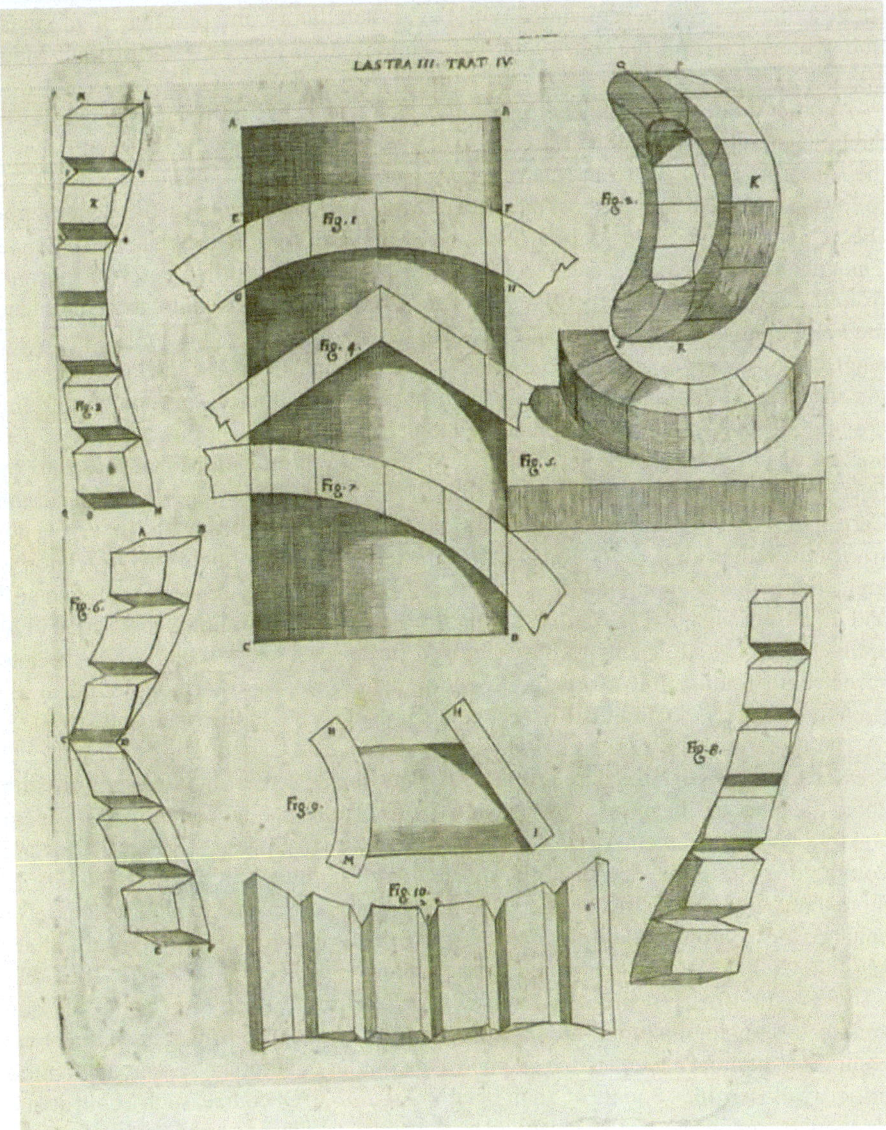

Fig. 2.23 Schemes for corner and curved-face arches (Guarini [c. 1680] 1737: treatise 4, pl. 3)

traced back to Derand or, even earlier, to the sixteenth century. This poses, of course, the question of his sources. He may have known Castilian stonecutting during his hypothetical trip to Lisbon to build Saint Mary of the Divine Providence in Lisbon, or its influence in Southern Italy while staying in Messina to direct the construction of Santa Maria Annunziata. However, the remarkable similarities with Derand suggest that he became acquainted with the French theory and practice of stonecutting while in Paris for the construction of Saint-Anne-la-Royale (Coffin 1956; Meek 1988: 12–40).

Another interesting issue in the *Architettura Civile* is its treatment of the theory of "oblique architecture" put forward by Juan Caramuel y Lobkowitz. This author was the son of a gentleman from Madrid and a lady from Frisia; a Cistercian monk, he followed an ecclesiastical career as abbot of Melrose and Disemberg, the Benedictine abbey in Vienna and Our Lady of Emaus of Montserrat in Prague, bishop of Misia, Campagna and Vigevano, and elected archbishop of Otranto. He defended Prague from the Swedish in the Thirty Years' War, fortified Leuven against the Dutch and the French, upheld the rights of the Spanish monarchy to the Kingdom of Portugal, taught theology in Alcalá de Henares and Leuven, held a debate with Gassendi about the satellites of Jupiter, proposed for the first time the binary numeration system, prefigured fuzzy logic, worked on geographical latitudes confronting Mersenne, was the first one to point out the heretical nature of several propositions in *Augustinus* by Jansenius and was accordingly attacked by Pascal in *Les Provinciales*, negated the Cartesian theory of the *turbillons*, reformed the square by Bramante in Vigevano, wrote more than fifty works on theology, cryptography, political law, astronomy, logic, mathematics, combinatorics, and several branches of natural science, and still had time to publish *Architectura civil recta y oblicua* ..., (Straight and oblique civil architecture), which seems to have enjoyed an appreciable circulation in the late seventeenth-century Italy, to judge by the fierce attacks by Guarino Guarini in his *Architectura Civile* (Caramuel 1678; see also Meek 1988 and Gasperuzzo 2019).

The three volumes of the book, lavishly produced in Caramuel's own Episcopal Presses in Vigevano, are an assortment of many subjects, starting with a discussion about the number of ages of the world and the description of the Temple of Jerusalem and following with extensive treatises on arithmetic and geometry. Only in the second volume does he start with an actual discussion of architecture, which occupies no more than a third of the treatise since the final volume is devoted to plates. He tries to enhance the catalogue of the classical orders with the addition of the Hierosolymitan, Mosaic, Gothic, Attic, Atlantic and Paranymphic genres (Caramuel 1678: II-V: 42–79).[42] But the focus of the whole work is the attempt to build a theory of oblique architecture, using distorted orders to accommodate them to stairs, as well as oval columns to fit more closely to the geometry of circular or elliptical spaces, an implicit criticism of Bernini's Square of Saint Peter's (Fig. 2.24). Of course, such a bold attack on papal patronage needs to be justified by the highest authority, and Caramuel resorts to no less than the skew windows of the Temple of Jerusalem (Caramuel 1678: II-VI, 3–4).

Rather than the short and outdated section on stonecutting included in the book (Caramuel 1678: II-VI, 20–22), what is interesting for us is the connection of these issues with a theory of geometrical transformation in architecture. Article XIII in treatise VI deals with skew and splayed arches and sloping vaults, including four

---

[42]Caramuel's treatise page numbering starts anew with each treatise, not only with each volume; thus, "II-V:42-79" means "volume 2, treatise 5, pages 42–79".

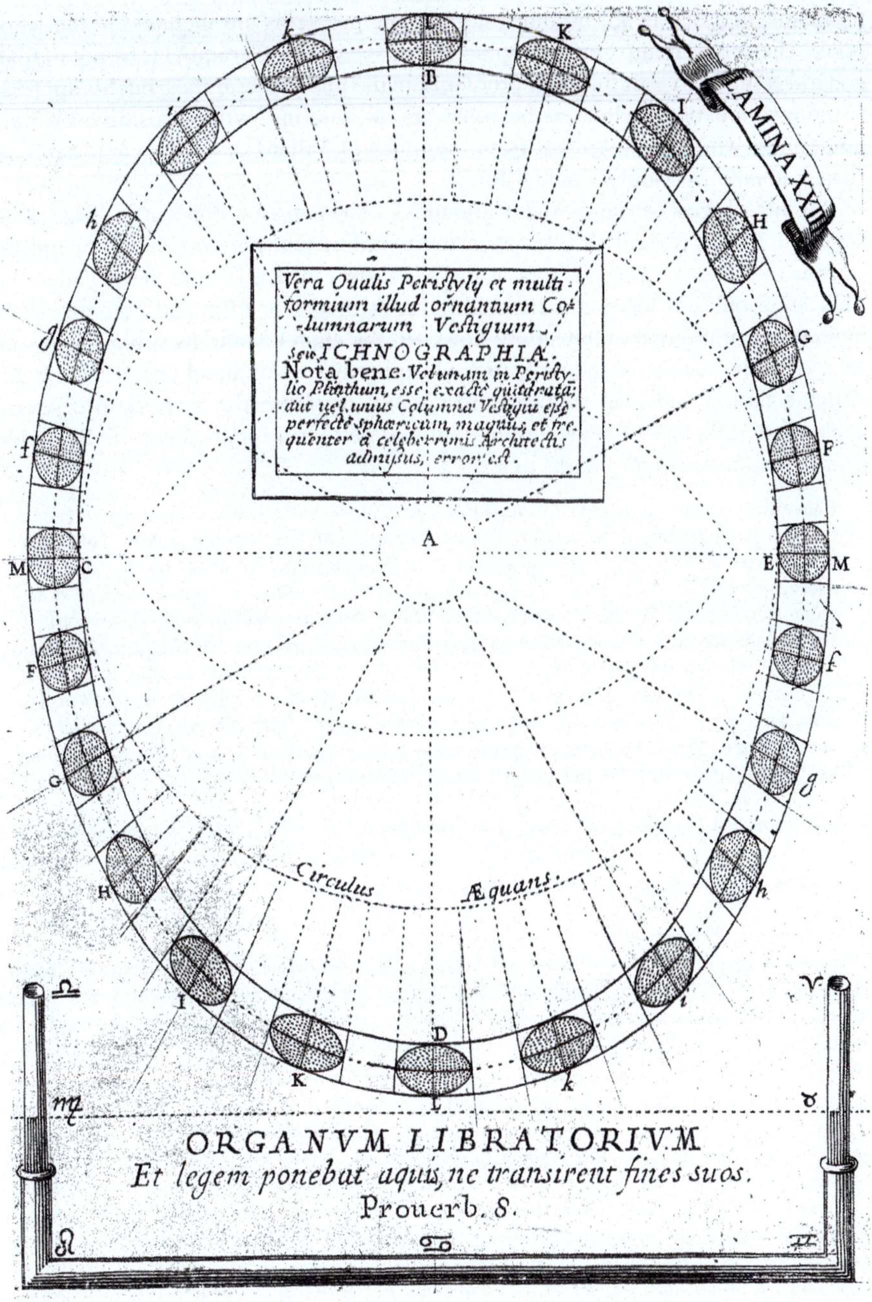

Fig. 2.24 Plan of a square with elliptical-section columns (Caramuel 1678: III, pl. 23)

problems, and that is all; no other stereotomic problems are included in the whole book. The solutions are clearly impractical. It seems that Caramuel is trying to find a general solution to four different problems, in the line of Desargues, but his approach is quite simplistic: he dresses the voussoirs by squaring, without using even bevel guidelines, which were known from the times of Villard de Honnecourt and used systematically by Vandelvira and Jousse.

All this suggests strongly that Caramuel's knowledge of stonecutting was rather superficial and amateurish. However, he mentions that his knowledge of oblique architecture stems from his stay at the Monastery of La Espina, near Valladolid. This huge building does not include rampant orders or elliptical columns,[43] the main icons of oblique architecture; however, the church contains several skew and splayed arches, some of them mentioned explicitly by Caramuel (1678: II-6:2, 21). Oblique motifs and oval supports are directly or indirectly connected to several sophisticated themes of stonecutting. Philibert de l'Orme appreciates the Belvedere stair by Bramante, but remarks that

> I have seen a vault … in a place named Belvedere, close to the palace of the Pope in Rome … Near this building, there is a large round staircase with a well in the middle. This vault is supported at both sides; the stairway it is not built on steps, but rather on the vault that ascends around the columns … Over them there is a brick vault … it is a nice and well-done work. However, if the architect who has directed its construction had understood geometrical tracings, he would have used a slanted shape for all members, even the bases and capitals, which are designed orthogonally, as if the vault were to be placed at the same level; over the capitals and under the bases, he has placed wedges in order to adapt to the slope of the stair. This shows that the worker who has made the staircase does not understand what the architect must know. Because, instead of making the vault in brick, he may have built the vault in ashlar, placing sloping arches between the columns …[44]

He does not furnish a drawing, but Vandelvira (c. 1585: 54r-55r) included these rampant orders in the *Caracol de emperadores* (Emperors' staircase), an elaborate construction including a vaulted stairway on the exterior and a smaller, open-well one

[43] In this paragraph, the terms "ellipse" and "oval" are used deliberately as if they were interchangeable. This is of course geometrically incorrect, but most seventeenth-century architectural writers did not differentiate between them. In fact, Caramuel's stance derives from the fact that he did not understand, or did not wish to understand, that Saint Peter's Square is traced as an oval, not as an ellipse.

[44] De l'Orme (1567: 124r-124v): *J'ai vu une vis quasi semblable … au lieu nommé Belvédère près le palais du Pape à Rome, … Tout auprès y a quelque bâtiment ayant une vis ronde assez grande, et à jour par le milieu … La dite voûte est portée sur des colonnes du côté du jour, et de l'autre côté sur des murailles, n'ayant point de marches, sinon la voûte qui rampe tout autour desdites colonnes, … Par le dessous y a une voûte de brique … et se montre l'œuvre fort belle et bien faite. Mais si l'architecte qui l'a conduite eût entendu les traits de géométrie, desquels je parle, il eût fait tout ramper, je dis jusques aux bases et chapiteaux, qu'il a fait tous carrés, comme s'il les eût voulu faire servir à un portique qui est droit et à niveau ; par le dessus des chapiteaux, et au-dessous des bases du côté de la descente, il a mis des coins de pierres pour gagner la hauteur du rampant. Laquelle chose montre que l'ouvrier qui l'a faite n'entendait ce qu'il faut que l'architecte entende. Car au lieu qu'il a fait la voûte de brique, il l'eût faite de pierre de taille, et d'une colonne à autre des arcs rampants.* Transcription is taken from http://architectura.cesr.univ-tours.fr; translation by the author. See also Calvo (2002) and Camerota (2005b).

in the interior. Such examples can be traced further back in time, to the balusters of the Blois or Chambord staircases or those in the Hospital of the Holy Cross in Toledo or the Chapter Staircase in León cathedral; furthermore, such affine deformations are present also in skew ribs in Late-Gothic vaults, known in Spanish as *revirados*. Oval columns are used in Spanish corner arches of the sixteenth century, such as those in the Pimentel Palace in Valladolid, the Palace of the Guzmanes in León or the entrance to the sacristy of the Chapel of El Salvador in Úbeda. Such an apparently arbitrary choice is justified by the adaptation to the contradictory geometry of corner arches, where the faces of the arch lie on the surfaces of intersecting walls, while the springings of the arch are parallel to the axis of the ensemble, usually placed at the bisector of the wall planes; thus, for maximum consistency, the section of columns placed before the arch should be tangent both to parallels to the springings and the faces; since these lines are not orthogonal, the construction leads to elliptical sections for the columns (Calvo 2002).

Caramuel's criticisms of Bernini raised much debate. He was probably granted the position of bishop of Vigevano in order to remove him from Rome; if so, the move was ill-advised, since it placed the Episcopal press in his hands. Guarini did not appreciate this foreign intrusion in an Italian debate and had bitter words to say against "a certain someone who has written in Spanish about architecture".[45] His hostility led him to underrate Caramuel's contribution to oblique architecture and ascribe the invention to Serlio:

> In his Book I, Chapter 6, Serlio gives some details of [oblique] architecture; and Caramuel has published an entire treatise with many figures, and an architecture, where he strives not only to diminish but also to enlarge the cornices of any given design proportionally, so it is useful for the architecture of staircases and their vaults.[46]

In this case, Guarini's remarks are a bit unfair. It is true that Serlio had explained a method that allows changing the scale of a complex figure drawing parallels, intersecting them with an oblique line and tracing a new set of parallels from the intersections, while keeping the proportions of the original figure.[47] However, Hernán Ruiz II (c. 1560: 39v-40r) expanded the use of this method in order to perform affine transformations, so to speak, particularly changes of scale with different factors in two original axes. Later on, Ginés Martínez de Aranda used similar transforms in the balusters of the staircase of the Obradoiro Square in Santiago de Compostela, arriving at a solution that anticipates some plates in Caramuel (1678: II, part IV, pl. I) and Guarini ([c. 1680] 1737: treatise III, pl. XV, figure 8). Going back, such

---

[45]Guarini ([c. 1680] 1737: 71): *Questa osservazione milita contro un certo, che ha scritto nella Favella Spagnuola di Architettura.*

[46]Guarini ([c. 1680] 1737: 169): *Il Serlio nel Lib. I al cap. 6 da qualche insegnamento di questa Architettura; ed il Caramuel ne fa un Trattato intiero con molte figure, ed è un'Architettura, che si adopera non solamente a diminuire, ovvero accrescere le cornici proporzionatamente, e qualsisia dato disegno, ma serve anche all'Architettura delle Scale, ed a' suoi Volti, e però dovendo noi trattare delle Scale è conveniente proporre questa cognizione.*

[47]Serlio (1545: 9r): ... *cioe d'vna piccola farne una magiore proportionatamente* ... The French translation in the same volume is even clearer: *agrandir vne cornice, & garder la proportion en toutes ses parties* (to enlarge a cornice, and maintain the proportion in all its parts).

solutions seem to derive from Gothic geometrical methods, particularly affine transformations in the profiles of the ribs, known in Spanish as *revirado* (Rabasa 2000: 106–111). Thus, what Caramuel and Guarini are trying to do is to reuse a long tradition of geometrical transformations in order to set a firm foundation for Baroque architecture.

2.4 The Enlightenment and the Engineers

2.4.1 Philippe de la Hire and Jean-Baptiste de la Rue

Neither Guarini's and Milliet's inclusion of stonecutting in mathematical treatises nor the use of Tosca's treatise as a manual in military academies were isolated phenomena. During the eighteenth and nineteenth centuries, innovation in stonecutting techniques stemmed mainly from some French educational institutions, particularly the Royal Academy of Architecture, the School of Bridges and Roads, the Military Engineering School at Mézieres, the École Polytechnique and the Central School of Arts and Manufacturing, in chronological order. When the architectural school was created in 1671, the post of director was entrusted to François Blondel, a naval, civil and military engineer and professor of mathematics and fortification at the Royal College in Paris. He was involved with the supervision of fortifications in Britanny and Normandy, the construction of the rope factory in Rochefort and a bridge at Saintes; he also refurbished some city gates in Paris and built anew the well-known Porte Saint-Denis (Perdue 1982; Lemonnier 1911–1929: I, ix-xi, xxi-xxxi).

His lessons at the Academy were published in the widely influential *Cours d'Architecture*. At the start, he reformulated the Vitruvian list of the sciences the architect should know, replacing letters and history with mechanics, hydraulics and stonecutting, which were taken for granted as parts of Mathematics:

> … taking into account that knowledge of architectural precepts is not sufficient in itself to make an architect, since this quality involves many other skills, His Majesty desired that other subjects that are absolutely necessary to architects, such as geometry, arithmetic, mechanics or moving forces, hydraulics or the movement of waters, gnomonics or the art of making sundials, military architecture or fortifications, perspective, stonecutting and other parts of mathematics were taught in the second hour of lessons in the the Academy…[48]

---

[48]Blondel (1675–1783: preface [iii]): … *comme il est vray que la connoissance des preceptes de l'Architecture ne suffit pas toute seule pour faire un Architecte, cette qualité supposant beaucoup d'autres lumieres; sa Majesté a voulu que pendant la seconde heure des leçons de l'Academie, l'on enseignât publiquement les autres Sciences qui sont absolument necessaires aux Architectes comme sont celles-cy, la Geometrie, l'Arithmetique, la Mechanique c'est a dire les forces mouvantes, les Hydrauliques qui traittent du mouvement des eaux, la Gnomonique ou l'art de faire les Quadrans au Soleil, l'Architecture militaire des fortifications, la Perspective, la Coupe des Pierres & diverses autres parties de Mathematique* … Translation by the author.

Stonecutting was not included in the *Cours*, which is based on the lessons taught during the "first hour", that is, architecture proper; the book deals mainly with the classical orders, although there are sections on arcades, doors, triumphal arches, bridges and even formwork. However, Blondel also published a large volume with the title *Résolution des quatre principaux problèmes d'architecture* (Solution to the four main problems in architecture) (Blondel 1673; see also Gerbino 2005). These crucial problems are the geometrical definition of the entasis or contour of classical columns; the construction of a conic section which is tangent to three straight lines, applied to rampant arches; the determination of the bed joints of such arches; and the line according to which beams should be cut in order to make them as strong as possible; thus, the word "problem" should be understood in the mathematical sense. The first three sections are related to stereotomy, although not central to it.

In any case, Blondel did not write a comprehensive text on stonecutting. His appreciation of Desargues is reflected in his notes to the second edition of Savot's *Architecture Française*, remarking that

> I am surprised to see that Mr Desargues's rule, explained in the book by Mr Bosse, is seldom used, since it is infallible and that it can be used in all cases.[49]

Does that mean that Desargues's theory of stonecutting was taught at the Royal Academy? The available evidence points in another direction. If we want to grasp a notion of the teaching of stonecutting in the Royal Academy, we should look at four manuscripts by Philippe de la Hire in the Institut de France, the Library of the School of Bridges and Roads in Paris, and the Municipal Libraries in Rennes and Langres (de la Hire 1688a; 1688b; c. 1688c; 1688d; see Tamboréro 2013: 195–196). The author was the son of Laurent de la Hire, a painter in the entourage of Desargues and Bosse. Philippe started training in painting, following the steps of his father, but he soon switched to science; the renderings in the first part of the Ponts et Chausées manuscript show his competence as a draughtsman. He was involved with the Royal Academy of Sciences, and in fact he was made Director of this institution a few months before his death. He wrote on conic sections, gnomonics, astronomy, mechanics, surveying, acoustics and optics; he also worked on a general map of France and carried out levelling for the construction of canals (Becchi and Foce 2002: 29–31; Sakarovitch 2013: 9–11; Pinault-Sørensen 2013; Rousteau-Chambon 2013).

He was also a professor in the Royal Academy of Architecture; the unpublished manuscripts on stonecutting probably derive from his lessons in the "second hour". The manuscript in the Institut is, according to Tamboréro (2013: 195) a draft for the one in the École des Ponts et Chausées. The first part of the latter deals mostly with arches and cylindrical vaults, although it includes a section on the *vis de Saint-Gilles*, including careful renderings that recall Bosse's engravings. The second part deals with groin, pavilion, and hemispherical vaults, sloping vaults and trumpet squinches.

---

[49] Savot/Blondel ([1624] 1685: 352): *... je suis étonne que la Regle universelle de Monsieur Dezargues expliquée dans le Livre du Sr Bosse, soit si peu en usage, veu, qu'elle es infaillible dans la pratique & qu'elle peut servir à tous les cas.*

The drawings are much sketchier, although the handwriting is still careful; again, it seems a preliminary draft. As for the content of this work, de la Hire introduced several significant innovations; for example, he used for the first time dihedral angles, in contrast to the stonecutters' traditional use of the bevel to measure angles between edges. His approach to sloping vaults is also quite interesting. Like Desargues, he searches for an "ordinary method" capable of solving all instances of cylindrical vaults; however, instead of using Desargues's abstract procedure, he performs a series of transformations or *rabattements*, in Tamboréro's words (2013: 196–197), reducing complex cases to the simplest one: the horizontal axis vaults.

Jean-Baptiste de la Rue was also a member of the Royal Academy of Architecture in the early eighteenth century. He entered the corporation as second-class architect and dealt with the problems of development and quantity surveying of surbased vaults. However, he seems to have also worked in the field of civil engineering, since he delivered a presentation to the Academy about the formwork for the bridge of Gien and a machine for the removal of foundation piles (Lemonnier 1911-29: V, 28, 34, 42–43, 46, 270, 276, 338; see also Pérouse 1982f). The draft of his *Traité de la coupe des pierres* (1728) was also presented to the Academy, receiving a positive evaluation (Lemonnier 1911-29: V, 2–4). It has been praised many times for its high-quality graphics. While Derand (1643: 35, 121, 165) had included a few oblique projections in order to show some details in the dressing technique, de la Rue made lavish use of carefully hatched and shadowed axonometrics, linear perspectives and oblique projections, sequential dressing schemes and transparencies (Fig. 2.25)[50]; he even included folded paper models when he felt that drawings were insufficient (Bortot and Calvo 2019).

Such graphical virtuosity has overshadowed other strengths of the treatise. Pérouse ([1982a] 2001: 100; see also 1982f), while remarking his visual quality and didactic clarity, included him among the last masons, saying that the subject matter had shrunk in comparison with Derand. This judgment is unfair; it may apply to a typical masons' manual derived from his treatise, such as the one by Simonin (1792) but not to De la Rue's treatise in itself. First of all, de la Rue clearly and didactically classifies the subject into arches and rere-arches, *maitresses voûtes* or full-blown vaults, trumpet squinches, sloping vaults and staircases. With 184 pages, it is more manageable than Derand, but it leaves out few essential issues. Second, despite Perouse's judgement, he brought about remarkable innovations. He put forward new methods for tracing the face joints of oblique trumpet squinches by cutting a symmetrical squinch. He also remarked that templates used up to his day for the intrados of spherical vaults, based on cone developments, were inexact since the generatrix of a cone is not coincident with the meridian of a sphere. Thus, he devised an alternative method, placing the corners of the voussoir on a dressed spherical surface with a compass using triangulation; this amounts to an idiosyncratic way of drawing in space (de la Rue 1728: 50–52; for a different opinion, see Sakarovitch 1998: 143). When necessary, he explains the dressing procedure in detail, using sequential drawings

[50]De la Rue (1728; pl. 5, 9, 11, 12, 18, 24, 26, 27, 29, 30, 31, 34v, 36v, 47v; see also Rabasa 2000: 245; Alonso et al. 2011: 659–668.

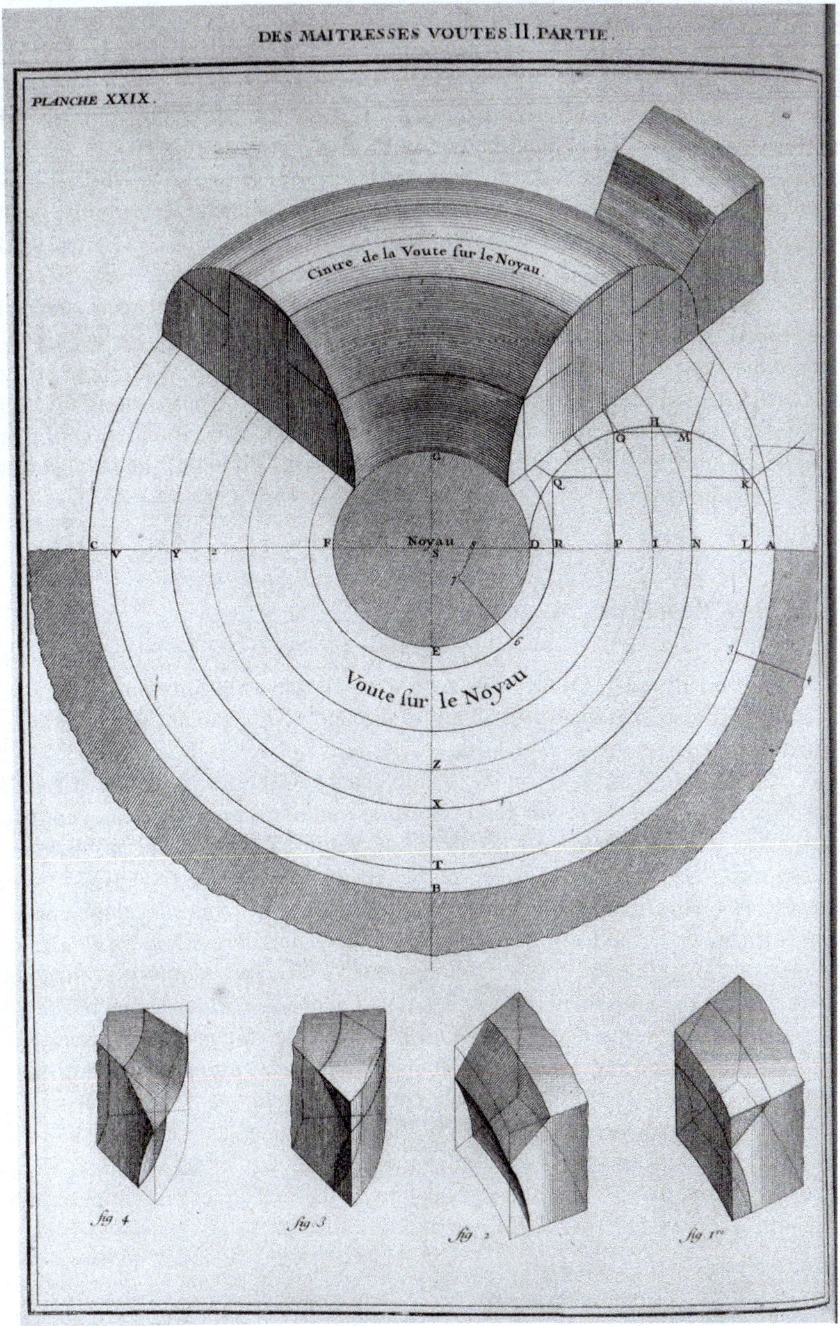

Fig. 2.25 Annular vault (de la Rue 1728: pl. 29)

that describe each of its phases. It is no wonder that Gaspard Monge used it as a textbook at the original École Polythechnique, although bowdlerising it, as we will see.

However, de la Rue's most significant innovation is the *Petit traité de stereotomie* (Short treatise on stereotomy) at the end of the volume (de la Rue 1728: 163–180; see Sakarovitch 1998: 143; Rabasa 2000: 239). The author unearthed the neologism coined by Curabelle to title an eighteen-page appendix on abstract geometry, dealing with the sections and development of the cylinder, the oblique cylinder, the circular cone and the sphere. It is not clear whether the inclusion of these pages was an afterthought or if was conceived from the start as an independent work since it is not included in the general table of contents of the book; it also includes a short *Avant-propos* or prelude on its own. However, as we will see in the next section, Amedée-François Frézier greatly expanded the *Petit traité,* transforming it into the full-fledged first volume of his treatise, providing the scientific foundation of stone-cutting technique, and paving the way for the final conversion of the subject into a branch of mathematics.

2.4.2 Amedée-François Frézier

Like Rojas and Blondel, Frézier was a military engineer; by a remarkable coincidence, he was also stationed in Brittany at the end of his career. Much earlier, he had been sent on a scientific expedition (or perhaps a covert military mission) to the coasts of Chile and Peru; later on, he was in charge of the fortifications of Landau, in the Palatinate, then in French hands. Despite building a hospital and twenty-four forts, he had enough leisure to write the three-volume *La théorie et la pratique de la coupe des pierres et bois … ou traité de stéréotomie* …(Frézier 1737–1739), the longest work on the field.[51] Reversing de la Rue's approach, this encyclopaedic treatise devotes the entire first volume to abstract issues, including *Tomomorphie*, that is, the shape of the sections of solids by planes (Fig. 2.26); *Tomographie*, or the operations that allow drawing them in true shape on a plane; the orthographic projections of these solids, including plans or *Ichnographie* and elevations or *Ortographie*, and finally the true-shape representation of angles or *Goniographie*. As Marta Salvatore (2011a; 2011b) has stressed, Frézier's approach prefigures Monge's descriptive geometry: an abstract geometry based on drawing that applies not only to stonecutting but to many other crafts. Only after these issues are sorted out, does Freziér deal with *Tomotechnie*, that is, the application of these abstract principles to actual stonecutting (Frézier 1737–1739: I, viii-ix).

[51] Frézier also published the *Eléments de stereotomie* …, (1760) an abridgement of his main treatise. Instead of reducing the theoretical introduction and keeping or increasing the practical cases, as usual with such introductory manuals, he did exactly the opposite. He kept the theoretical foundation and reduced drastically the repertoire of practical problems. This suggests he was following the prevailing trends in engineering schools of the period. See Sakarovitch (1995) and Sect. 12.2.

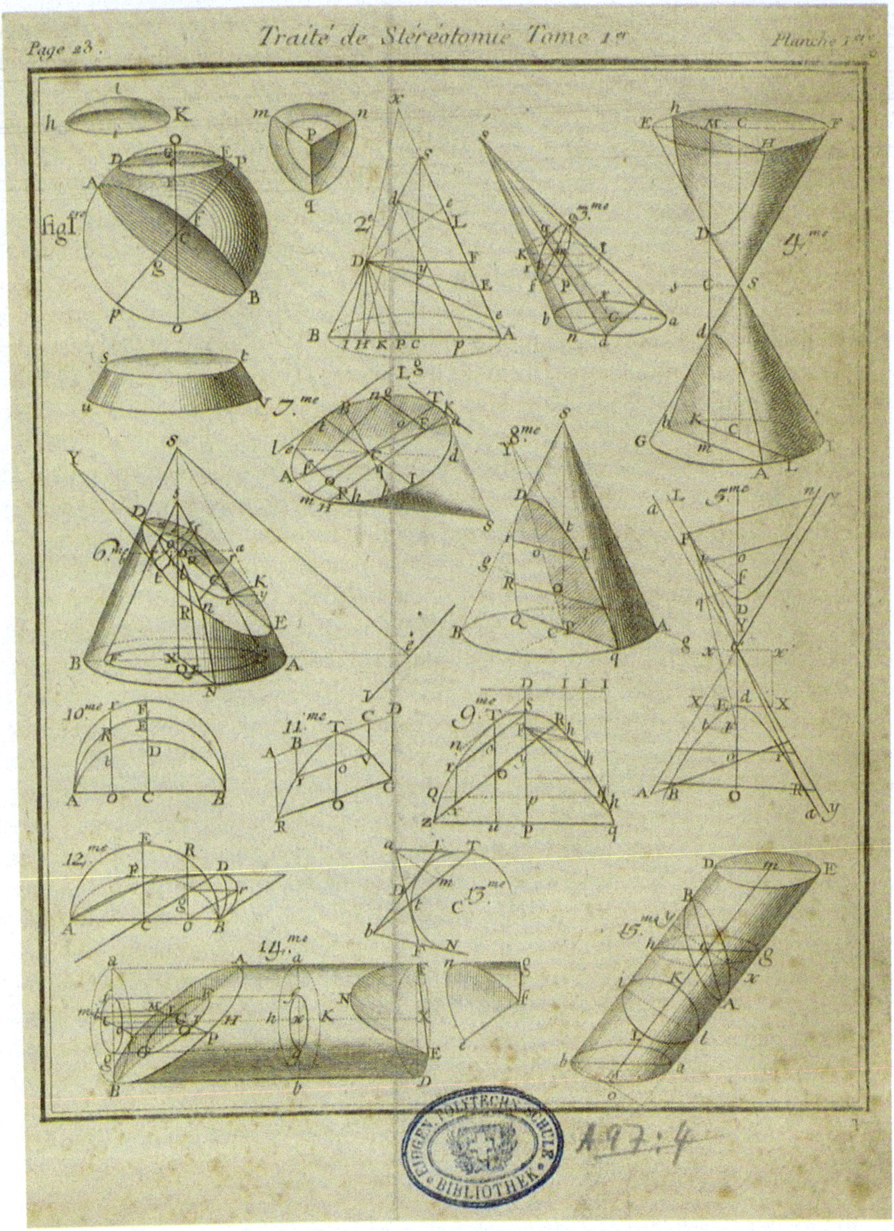

Fig. 2.26 Spherical, conical and cylindrical sections (Frézier [1737-39] 1754–1769: pl. 1)

In accordance with the encyclopaedic conception of the book (see Potié 2008: 156–158), the second and third volumes chart almost all known solutions for each problem. For example, when dealing with hemispherical domes, Frézier mentions four ways of laying out the courses: horizontal, vertical, sloping and a combination of all the above. Then he gives no fewer than four different solutions for the horizontal-course variant, while most of the authors before him had settled on a single solution (Frézier 1737–1739: II, 310–331). He first explains a solution using spherical segments, which he admits to be based on the "drawing in space method" pioneered by de la Rue (1728: 50–52; see also Sect. 2.4.1) and then the usual solution by cone developments. Next, he solves the problem by reducing the sphere to a polyhedron, that is, using rigid templates instead of the flexible ones of the preceding method; and finally, he addresses the issue by squaring. Then, he goes ahead with the vertical-, sloping- and combined-course variants, and at last deals with incomplete spherical vaults, for a grand total of no fewer than fifty pages.

This approach enables Frézier to explain, dissect and attack the methods proposed by earlier writers. On other occasions, however, he puts forward new, idiosyncratic methods. A good example is his preferred solution for skew arches (Frézier 1737–1739: II, 139–140; see also Sect. 6.2.2). A host of authors had endeavoured to use either developable surfaces or bed joints orthogonal to the face plane of these arches. Frézier tries to fulfil *both* requirements simultaneously; this leads to an elliptical cylinder whose axis is oblique to the face plane. In order to generate bed joints, the cylinder is cut by a fan of planes, whose common line is perpendicular to the faces of the arch. However, since this line is not parallel to the axis of the cylinder, the intersections of the sheaf of planes with the cylinders are elliptical arcs, and Frézier makes no bones about using them as bed joints.

Thus, his desire to explain all known solutions to a given problem does not imply that Frézier's attitude is eclectic or neutral: he stresses over and over the *fausseté de l'ancien trait*, that is, the "falsehood of old tracing methods" or similar expressions. In particular, he takes exception to the use of "irregular", that is, non-developable, surfaces (Frézier 1737–1739: II, 140); he remarks that the sections of cones through planes that are perpendicular to the generatrices are not circular (Frézier 1737–1739, II: 231); he points out that if a sail vault with vertical courses is dressed using templates constructed by cone development, the voussoirs at the diagonals are formed by two different cones, and so on (Frézier 1737–1739: II, 344). Other corrections are subtler: he does not accept Derand's rectification of a circular arc using two chords, but in practice, he admits some simplification:

> it is common knowledge that this difficulty [the rectification of the circle] is geometrically unsolvable; however, in practice, it is sufficient to take several small chords, whose length does not differ much from their arcs and place them on a straight line.[52]

We can see he is trying to place stonecutting under the rule of geometry, as Desargues did; he dares, for the first time in almost a century, to explain the method of

[52]Frézier (1737-39: III, 43–44): *Cette dificulté qui est, comme l'on sçait, Geometriquement insurmontable, ne tire à aucune conséquence pour la pratique, òu il suffit de prendre de suite plusiers petits cordes qui different peu des arcs, & les ranger sur une ligne droite.*

the *Brouillon projet* ..., although he presents it as a general method and offers alternative procedures (Frézier 1737-39: II, 191–192). This strategy is also shown by section headings, which give systematically the geometrical definition of a stonecutting operation or element and its name in stonecutters' jargon, taking the form, for example, "To pass a planar surface through three known points of a solid. In terms of the art, to dress a face"[53]; that is, Frézier (1737-39: II, 15, for example).

Another interesting point is given by Frézier's assessment of masons' graphic procedures:

> The confusion that arises in the drawings of the books that deal with stonecutting frequently originates in the multiple kinds of representation that are put together in the same tracing; on many occasions, the plan is joined to the profile, sometimes even to the elevation; all them are mixed without divisions ... Frequently, vertical objects are turned around, as if they went down instead of rising; on some occasions they lean on their sides, when they should be vertical ... The need to join several objects in a small plate makes this encumbrance almost unavoidable; it is also useful to indicate their relations more sensibly [...] Although it would be more natural to place each kind of drawing separately, it is however true that such simplicity would show less clearly the relations of lines, and this layout would be less convenient than joining, and even mixing, the *plans, profiles* and *elevations*.[54]

2.4.3 *Gaspard Monge*

Freziér's tolerance would last only for a few decades. At the end of the century, a scientist connected with military engineers, Gaspard Monge, launched an attack against such an unruly system of projections. Frézier was born into an old aristocratic family; his surname was granted when a distant ancestor presented a plate of strawberries, *fraises* in French, to Charles III. In contrast, Monge was a commoner; this excluded him from the higher ranks in the army, reserved for noblemen and the sons of glass traders. However, at 18 he prepared a plan of his hometown, Beaune, with instruments he had made himself. In 1764, Captain Antoine-Nicolas-Bernard du Vignau, an engineering officer passing by the town, was struck by his ability and brought him to the Military Engineering Academy in Mézières, where he worked in the drawing, models, and stonecutting workshop. After two years, he was promoted

---

[53] Frézier (1737-39: II, 15): *Par trois points donnez, dans un solide faire passer une surface plane. En termes de l'Art. Dégauchir un Parement.*

[54] (Frézier 1737-1739: I, 271–272): *La confusion que l'on trouve dans les desseins des Livres qui traitent de la coupe des Pierres, vient souvent de la multiplicité des especes de representations que l'on rassemble dans la même Epure; car souvent on y joint le plan au Profil, quelquefois encore à l'elevation, & l'on mêle les uns avec les autres sans divisions ... Souvent les objets verticaux sont renversez, comme si au lieu de monter ils tomboient du haut en bas; quelquefois ils sont placez de côté, quoiqu'ils doivent être verticaux ... La necessité de rassembler plusiers objets dans une petite planche rend cet embarras presque inévitable; d'autant plus qu'il a son utilité pour indiquer plus sensiblement leur rapports ... Quoiqu'il soit plus naturel de mettre chaque espece de dessein à part; il est cependant vrai que cette simplicité d'objet indique moins sensiblement les rapports des lignes, & que l'on trouve en cela moins de commodité qu'à rassembler, & même quelquefois à meler les* Plan, Profil & Elevation. See also Sakarovitch (1992a: 532).

to the post of *répetiteur*, that is, assistant teacher of mathematics and later on he replaced the full professor, Abbé Bossut, as the actual instructor. Subsequently, he was involved with the teaching of physics, drawing, perspective, shadows, stonecutting and topography; finally, in 1775 he was granted the post of Royal Professor of Mathematics and Physics (Taton 1950: 4–5; Belhoste and Taton 1992: 270–271; Sakarovitch 1998: 223–227; Sakarovitch 2005a: 227; Sakarovitch 2007: 48–51). It was probably in those years that he conceived the notion of a generalisation of stonecutting and topographical methods, later known as descriptive geometry. In fact, one of the reasons for his success at Mézières was that he found a simple way to solve the problem of *défilement*, that is, the height of a wall built to protect a position from enemy gunfire; later on, he would claim that he was not allowed to reveal these secrets (Dupin 1819: 11; see also Sakarovitch 2003b: 71; Lawrence 2011: 705–706; and Carlevaris 2014: 637–650). At the same time, he was connected with the Royal Academy of Sciences, where he presented eight papers on analysis and differential geometry. Later on, he taught hydraulics at the Louvre; he finally left Mézières in 1784.

With the onset of the French Revolution, Monge, who probably had withstood much humiliation at Mézières on account of his humble origins, was an ardent Jacobin; he was made minister of the Marine for a few months and took part on the scientific expedition that followed Napoleon to Egypt. But his main occupation in these years was the organisation of two famous schools, imitated by many others around the world: the École Normale and the École Polytechnique.[55] The former was created for the education of high-school teachers, while the latter was intended to provide a primary scientific training to engineers in many disciplines, who would go on to complete their training afterwards in specialised *écoles d'application*. Freed from the constraints of the Áncien Régime, Monge taught his descriptive geometry in both the Normale and the Polytechnique, but the approaches were different. In the first one, although stressing their utility for artisans, Monge focused on the capacity of graphical procedures to solve abstract geometry problems. His lessons were recorded in shorthand and published by his pupils (Laplace et al. [1795] 1992: 305–453; see also Belhoste and Taton 1992: 279–289; Sakarovitch 2005a: 228–235); this has led to an overly abstract conception of this science. In contrast, his teaching at the École Polythechnique, where descriptive geometry was allotted half the time of the first year, is not so well-known. Only the general outline of the course and some exercises by pupils have survived; it seems that it focused broadly on the same subject matter as the lessons at the Normale, but much emphasis was placed on practical drawing exercises in stonecutting, topography, carpentry and shadows (Belhoste and Taton 1992: 277–279, 289–299; Sakarovitch 1994a: 77–82; Sakarovitch 1998: 151, 242–268; Sakarovitch 2005a: 226).

[55] The École Polytechnique was born as the École Centrale des Travaux Publics in 1794 and renamed the next year. Although there were some subtle differences between this first Centrale and the Polytechnique, they are not relevant for our purposes so that I will refer to both schools as a single one. About this, see Belhoste and Taton (1992: 275–277).

In any case, Monge in his first lesson at the École Normale stressed that two projections, horizontal and frontal, are sufficient to represent exactly a point in space (Fig. 2.27). He considered projection planes as fixed; this allowed him to use their intersection, known as the ground line[56] for basic tasks such as finding the intersections of lines with the projection planes and determining whether a line belongs to a plane. Also, he represented ordinary planes using their intersection with projection planes. In a word, the new system was quite well fitted to the solution of abstract geometrical problems, while its advantages in practical issues such as stonecutting and topography were not so clear (Laplace et al. [1795] 1992: 308–317, 328–331; see also de la Gournerie 1855: 25–27; de la Gournerie 1874: 114).

Monge used de la Rue's plates as didactical material in the École Polytechnique, but he eliminated cavalier perspectives (Rabasa 2011: 732). This implies a crucial change in the use of orthographic projection. In stonecutting treatises, from the times of de L'Orme (1567: 69), this method of spatial representation had been reserved for operational (as opposed to representational) purposes, mainly the determination of templates and bevel guidelines. Plans and elevations were presented generally without hatches, shading or shadows. When writers had a particular interest in showing the spatial configuration of a difficult-to-understand element, they resorted to shadowed axonometrics or perspectives.[57] However, on many occasions, eighteenth-century military engineers dispensed with linear perspective and tried to enhance the representational strength of plans and elevations using shades and shadows (Sakarovitch 1998: 85–89, 90, 92; Sakarovitch 2007: 49–51; see also Muñoz 2015: 73–75).

Despite his involvement with stonecutting in the school of Mézières, Monge left only one paper dealing indirectly with stereotomy. In his article on the lines of curvature of the ellipsoid (Monge 1796; see also de la Gournerie 1855: 27–28; de la Gournerie 1874: 126–135; Sakarovitch 1992a: 536–539; Sakarovitch 2009c), he studied the lines along which the surface formed by the normals to another surface is developable. In the case of the sphere or the ellipsoid of revolution, parallels and meridians fulfil this condition, since the resulting surfaces are planes and cones, respectively. In contrast, in the scalene ellipsoid, (that is, the one with three different axes), lines of curvature do not match parallels; instead, in an ellipsoid with two horizontal axes, they go up and down. Passing abruptly from abstract geometry to practical construction, Monge takes it for granted that in an ellipsoidal vault, bed joints should be materialised as the developable surfaces generated by the normals to the curvature lines of the ellipsoid. Compared with the usual solutions to the problem, such an approach has advantages and disadvantages. On the one hand, surfaces generated by normals allow the use of the square to control the dressing of the bed joints; the use of templates based on developable surfaces facilitates this

---

56 Nowadays, the intersection of both projection planes is known as *ligne de terre* (ground line), a term borrowed from linear perspective. However, Monge did not use it; it was introduced by Louis-Léger Valleé ([1819] 1825: 5); see Rabasa (2011).

57 This is particularly true in Bosse and Desargues (1643a: pl. 2, 3, 4, 5, 6, 8, 10, 11, 67, 78, 79), De La Rue (1728: pl. 12, 17, 18, 21, 24, 26, 27, 29, 30bis, 31, 32, 34, 36bis, 40bis, 61bis), or Frézier (1737–1739: pl. 54, 62, 73, 74).

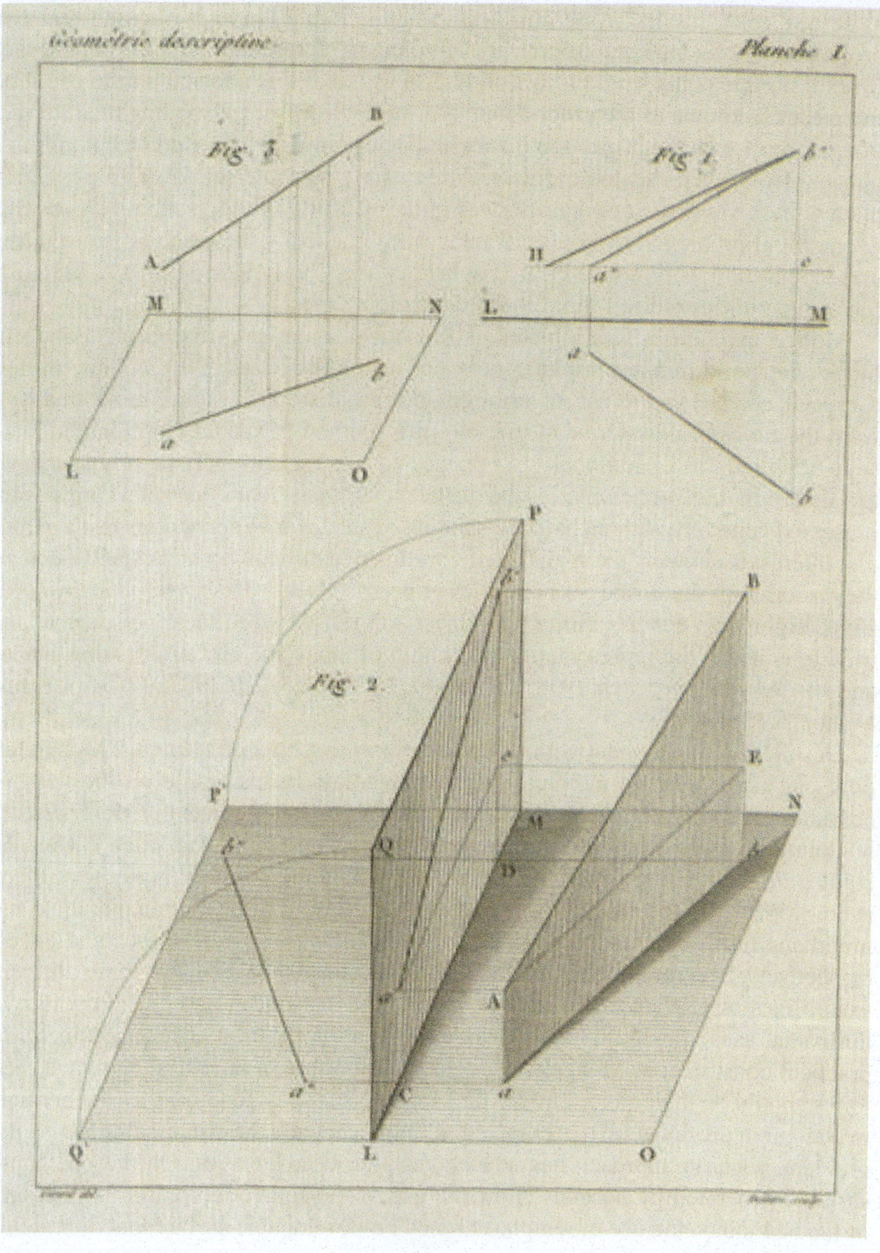

Fig. 2.27 Projections of a straight line (Monge 1799: pl. 1)

task.[58] On the other hand, the construction of the awkward rising and descending intrados joints results in a complex tracing, where any error would compromise the precision of the result; in other words, the ease of dressing comes at the expense of difficulties in setting out. Furthermore, all these problems do not arise in an ellipsoid of revolution; several authors (Vandelvira c. 1585: 73v-76v; Derand 1643: 398–400) had used this shape,[59] laying out the intrados joints as parallels and meridians. In this case, the normals along parallels generate elliptical cones, which are developable. In any case, Monge was not afraid of the complications posed by the scalene ellipsoid; he went so far as to propose this shape for the covering of the National Assembly that was being built at that precise moment.

Furthermore, he recommended placing the speaker under one of the umbilic points of the ellipsoid, including a tribune for the public along the edge of the ellipse, and materialising the vault with ribs following curvature lines, adding that:

> All these ribs, which are vertical at their springing, should curve around one of the umbilic points and go down vertically to the columns in the opposite side of the hall; they will cross orthogonally with other ribs following the other family of curvature lines. The space between ribs could be left open, either to cast light on the hall or to allow ventilation; they would form a transparent cover less fantastic than the rose windows in our Gothic churches.[60]

In short, what Monge (1796) proposed was a temple to geometrical Reason, placed at the political centre of the Nation. Stonecutting here goes much further than the solution of a practical problem: its role is that of an illustration, even an emblem, of the power of Science.

2.4.4 *Jean-Nicholas Hachette, Charles Leroy and Jean-Baptiste Rondelet*

The approach of Monge characterised the host of books on descriptive geometry and stereotomy published in the nineteenth century.[61] I will mention only three of them. Jean-Nicholas Hachette, Monge's closest pupil, published a treatise on descriptive

---

[58] In any case, La Gournerie (1855: 28, 32) asserted that only four out of twelve of the stonecutting sheets in the collection of in the École Polytechnique, prepared or at least supervised by Monge, follow this rule.

[59] This assertion should be understood in a broad sense. Neither Vandelvira nor Derand use the term "ellipsoid"; moreover, they do not clearly distinguish ellipses from ovals. See Sect. 9.4.2.

[60] Monge (1796: 162–163): *Tous ces nervures, verticales à leur naissance, se courberaient autour de l'un ou de l'autre ombilic, redescendraient ensuite à plomb sur les colonnes opposées, et elles seraient croissés perpendiculairement par d'autres nervures pliées suivant les lignes de l'autre courbure. Les intervalles de ces nervures pourraient étre a jour, soit pour éclairer la salle, soit pour donner des issues a l'air, et formeraient un vitrage moins fantastique de les roses de nos églises gothiques.*

[61] By that time, these terms had undergone a series of semantic shifts. For De La Rue (1728: 163–180) "stereotomy" was an abstract science dealing with the sections of solids. This branch of knowledge is known as "tomomorphie" in Frézier (1737-39: I, viii-ix); in this period, "stéréotomie"

geometry with a series of appendices on perspective, shadow theory and stereotomy. The section on stonecutting, although based on the general principles on descriptive geometry set out in the main part of the book, stands well as a short treatise or manual in the craft. In fact, the plate headings are labelled *Cours de stéréotomie par M. Hachette*, in contrast to those on perspective or shadow theory, which are marked as belonging to the general course in descriptive geometry. This set of plates, with their explanatory texts, includes skew arches, arches on round walls, a special rere-arch known as the *arrière-voussure de Marseille*, hemispherical and ellipsoidal vaults, pavilion vaults, annular vaults, sloping vaults, trumpet squinches, and staircases, including the *vis de Saint-Gilles*, with a total of 20 plates; it represents a short but reasonably well-balanced selection. Despite the inclusion of some traditional examples such as the *vis de Saint-Gilles*, the reader gets the impression that the purpose of this section is not to instruct the reader on stonecutting techniques, but rather to show the power of descriptive geometry. For example, Monge's treatment of curvature lines and its application to the division of ellipsoidal vaults into voussoirs is explained in detail (Hachette 1828: 320–322, pl. 6); however, it is illustrated only with a plan of the vault, without the addition of an elevation; thus the reader is at pains to grasp the most striking trait of Monge's proposal. The treatment of the *arrière-voussoure de Marseille* (Hachette [1822] 1828: 317, pl. 3) is another case in point. The long section on the ruled surfaces of this rere-arch dutifully explains the problem of the continuity of the surfaces of the intrados of the piece, using tangent planes. Ultimately, it refers the reader to a passage in the general descriptive geometry course, where Hachette ([1822] 1828: 96) explains that if two ruled surfaces have a straight line in common and, further, the plane tangent to both surfaces is identical at three different points of the straight line, then both surfaces share the same plane tangent at all points in the straight line. As Sakarovitch (1992a: 534–536; 1998: 307) and Rabasa (1996b: 32; 2000: 278–286) have remarked, stonecutters had built similar rere-arches for centuries without resorting to such concepts of higher geometry.

The relatively small number of stonecutting examples in Hachette's appendix is a result of the conception of the work, a section in a general treatise on descriptive geometry. However, even in the specialised treatise on stereotomy by Charles Leroy (1844), no more than half the plates of the book deal with actual stonecutting; in fact, the title page makes it clear that "stereotomy" means "applications of descriptive geometry, including shadows, perspective, gnomonics, stonecutting and carpentry". There are a few more stonecutting examples than in Hachette, but the repertoire is by no means complete; Leroy seems to be thinking that once the reader masters a few procedures—in particular changes of the projection plane—he will be able to solve any problem in this field. In this case, Monge's ellipsoid is explained in detail (Leroy [1844] 1877: pl. 43–44) although the author also offers an alternative solution

---

is a broad term encompassing not only solid sections but also projections and actual stonecutting, which is designated by the neologism "tomotechnie". Further on, the apparition of the term "Descriptive geometry", which means "Geometry of (or by) Drawing" occupied the semantic field of "stereotomy", which on its turn moved on to occupy the meaning of "tomotechnie", that is, actual stonecutting. Such displacement has gone so far that some Spanish authors allude to "blocks with careful stereotomy", meaning, of course, proper dressing technique.

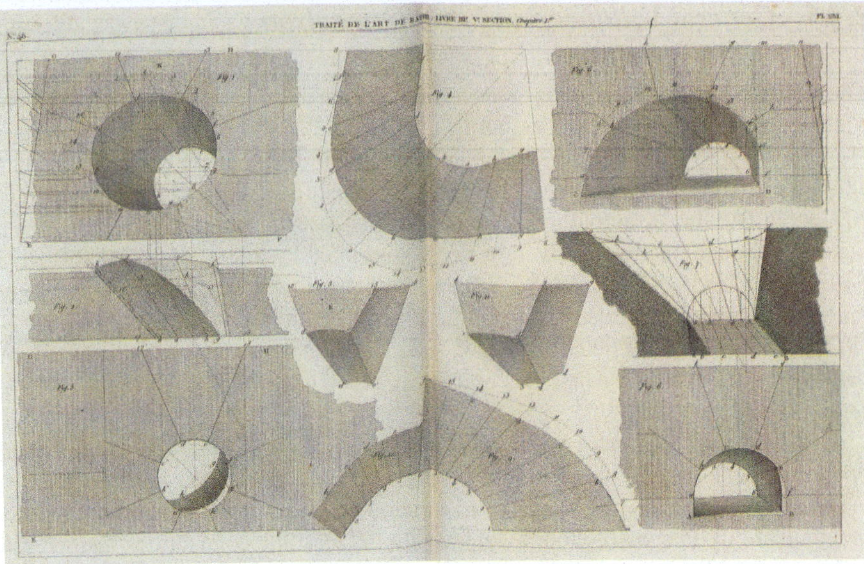

Fig. 2.28 Oculi and gunports (Rondelet 1824: pl. 46)

with horizontal courses. In short, rather than being treated as a practical discipline in the field of building, stonecutting seems to be included as a provider of academic exercises on descriptive geometry, a complement of Leroy's (1834) general book on the subject.

Another interesting example is the *Traité de l'art de bâtir* by Jean-Baptiste Rondelet (1802–1817), a general construction treatise, in contrast to those of Hachette and Leroy, which are sections or derivatives of descriptive geometry literature. A lavish work in eight volumes, it enjoyed a wide circulation due to its clarity of exposition and the quality of the plates (Fig. 2.28), following the line that started with Bosse and de la Rue. However, many illustrations adhere to a systematic, almost religious, use of orthographic projection. A good example of this is his treatment of ribbed vaults. Derand (1643: 393) and Frézier (1737-39: 2, pl. 71) had eschewed the construction of a full orthographic elevation of the ribbed vault, perhaps because diagonal ribs would be shown obliquely, and such representation would be useless when executing rib vaults; both authors focus on true-size-and-shape representations of all ribs, including diagonals and tiercerons. In contrast, following Monge's tenets, Rondelet ([1802–1817] 1834: pl. 41, 42) draws careful shadowed elevations of rib vaults, with diagonals and tiercerons shown in oblique view.

2.4.5 General Vallancey and Peter Nicholson

The author of the first treatise on stonecutting written in English was General Charles Vallancey (or Vallancy). He was born in Windsor, the son of a French Protestant. He joined the Royal Engineers and was stationed in Ireland. He prepared plans to counteract the Irish Rebellion of 1798 and built Queen's Bridge, now Liam Mellow's Bridge, in Dublin. He wrote many books on Irish issues, ranging from grammar to history, although his theories were generally rebuked by later authors (Webb 1878: 540; Moore 1885–1900: LVIII, 82–83). In his professional field, he translated essays on fortification and engineering from the French and wrote *A practical treatise on stonecutting* (1766). As Peter Nicholson (1827: v) remarked, the treatise was to have had five parts, but only the first one, borrowing much material from de la Rue (1728), reached the presses.

Nicholson was the son of a Scottish stonemason; in eighteenth-century Scotland, operative stonemasons' trade organisations were still associated with speculative Freemasons' lodges. When he moved to London in the last decade of the century, his practical ability in construction, together with his masonic connections, helped him to launch establish himself as a collaborator of several architects, in particular with Robert Smirke, the designer of the British Museum. He started developing his own system of spatial representation when he heard about Monge's theory. The result was a hybrid system combining orthographic projections and axonometrics, in an attempt to compensate for the lack of intuitive spatial representation of descriptive geometry proper. This "British system of projection" enjoyed a warm reception in London architectural schools for some time; probably it was understood as a bulwark of English empiricism against Napoleonic abstraction (Lawrence 2003: 1274–1276).

Nicholson also published many books about the building trades, including *A Popular and Practical Treatise on Masonry and Stone-Cutting* (1827) and *The Guide to Railway Masonry* (1839). The first starts with a chapter on general issues on geometry, such as the positions of lines, points and planes, although it includes a section on projections on a cylinder; this is a first hint of the interest of the treatise in oblique bridges. Next, the second chapter deals with projections and developments. The third chapter addresses the actual construction of templates for built pieces, beginning with skew arches, as we would expect, and following with arches opened in round walls. The next chapters deal with niches, domes, conical roofs, groin and rib vaults, mouldings, lintels, ending with a drawing of Waterloo Bridge and its centring. Thus, many traditional pieces of the repertoire of stonecutting, such as corner or sloping arches, rere-arches, squinches or oval vaults are left out; this manual focuses on the most common jobs for a London stonemason of the period.

The second treatise explains in great detail an alternative solution to the orthogonal quartering of skew arches, known as helicoidal bond. As we will see in Sect. 8.1.2, the old problem of skew arches had resurfaced in the nineteenth century in connection with railroad bridges. While Adhémar (1861), Dupuit (1870) and many other French engineers tried to find an ideal solution using bed courses perpendicular to compressive stresses inside the masonry, leveraging descriptive geometry, Nicholson (1839:

1–20, plates 17-24, 27, 31-31-34; see also La Gournerie 1855: 32–36; La Gournerie 1874: 122–125) took a more practical route, using helical bed joints. The idea may seem far-fetched, but helices are represented as straight lines in a development of the intrados cylinder, thus easing the construction of templates and the dressing of the voussoirs, in contrast to sophisticated French solutions to this problem.

2.4.6 *Théodore Olivier and Jules Maillard de la Gournerie*

The last stand in defence of stonecutting as a practical, empirical discipline of the building trades was fought by Jules Maillard de la Gournerie, the successor of Leroy in the descriptive geometry chair at the École Polytechnique, in a bitter debate with Théodore Olivier. Ironically, Olivier saw himself as a champion of Monge's practical stance. He had arrived at the École Polytechnique one year after Monge left actual teaching at the school for more urgent duties; however, Olivier was introduced to Monge by Hachette and revered him as a secular saint. After some years as assistant teacher in the School of Artillery at Metz, he felt uncomfortable with the direction taken by the École Polytechnique under the leadership of Laplace, Cauchy and Poisson, "who are unable to understand anything but Algebra"[62]; as a result, he accepted an offer by Bernadotte, the Napoleonic general who became King of Sweden and Norway, to act as instructor to the Crown Prince and teach descriptive geometry at the Military Academy at Marienberg. Back in France, in 1829, together with the physics professor Péclet, the chemist Dumas and the capitalist Lavallée, he founded the École Centrale des Arts et Manufactures, a new kind of engineering school. While the Polytechnique and the École des Ponts et Chausées were state schools entrusted with the education of civil engineers and military officers, the Centrale was created to instruct *ingeniéurs civils* at the service of the emerging French industry. The title may be misleading; the pupils were trained to be the *médecins de les usines et fabriques* (the doctors of factories and industries), and the professional title may be translated as mechanical or industrial engineer (Comberousse 1879: 18–30; Sakarovitch 1992b: 589; Sakarovitch 1994c).

Olivier published a host of treatises, textbooks and papers on descriptive geometry. However, he is best remembered for the systematisation of two basic procedures in the discipline, changes of projection plane and rotations, proposed in the final sections of a memoir in an apparently independent subject, the construction of the tangents to a curve in a double or multiple point (Olivier 1832: 348–350; Olivier 1843–1844: I, 18–22).[63] Monge had used both, and in fact, such ideas may derive indirectly from the old practices of stonecutters.

---

[62]Olivier (1851: xi-xii;) ... *ces hommes [comme Laplace, Cauchy et Poisson] incapables de comprendre autre chose que l'algèbre* ... See also Sakarovitch (1994c: 327).

[63]In this period, rotations applied both to lines and planes. Later on, rotations were differentiated into rotations proper, when they are applied to lines, and rabatments, which are performed on figures.

This proposal raised a bit of turmoil. In particular, Jules Maillard de la Gournerie delivered several ironical attacks against Olivier; he was perhaps striking out indirectly at Monge and even Desargues. In contrast to Monge and Olivier, who had little or no experience in practical construction, de la Gournerie was an able engineer: he built the lighthouse in Bréhat, once again in Brittany, on a small rock that barely emerged in low tide. Such practical experience led him immediately to question Olivier's insistence on changes of the horizontal projection plane, which he found unnatural:

> M. Olivier has proposed a method to solve the problems of descriptive geometry that has been received with approval by several distinguished professors, and which is taught at some schools ... for each problem, [this method] eschews the projection planes that have been chosen for the representation of the geometric system under consideration, following the nature of the givens, and replaces [these projection planes] with others where the magnitude of unknowns is immediately shown. This change of projection planes is performed, either by successive rabatments of auxiliary projection planes keeping the system fixed, or through rotations of the system around the first planes ... In the ordinary procedures in Stereotomy, the horizontal projection of objects is invariably maintained; however, a number of different elevations are used, including oblique projections. Thus, the method proposed by M. Olivier does not lead to new tracings, except when it requires the change of both projection planes, and in this case, it is not convenient for applications, particularly in Stereotomy.[64]

De la Gournerie did not stop here. Later, he stated bluntly that

> In Monge's teaching ... stereotomy ... is reduced to its geometrical aspects.
>
> Monge was dominated by this incomplete approach to the question, to the extent that he tried to solve, using Geometry alone, the problem of the division of a vault into voussoirs.
>
> Thus, the famous geometer set aside the central givens of the problem; he was not concerned by the position of a vault with regard to gravity, the placement of its springings, or the existence of unevenly distributed loads ... only the geometrical nature of the surface was taken into consideration[65]

---

[64]La Gournerie (1860: vi): *M. Olivier a proposé pour résoudre les problèmes de la Géometrie descriptive une méthode qui a reçu l'assentimet de plusieurs professeurs distingués, et qui est suivie dans quelques établissements ... consiste à abandonner, dans chaque problème, les plans de projection qui, d'après la nature des données, ont été choisis pour la représentation du système géométrique que l'on considère, et à en prendre d'autres sur lesquels les grandeurs des inconnues sa manifestent immédiatament. Ce changement de plans coordonnés est obtenu, soit par des rabatements succesifs de plans de projection auxiliaires devant le système fixe, soit par des rotations du système devant les premiers plans ... Dans les procédés ordinaires de la Stéréotomie, on conserve invariablement la projection horizontale des objets, mais on en fait quelquefois plusiers élévations, et même des projections obliques. La méthode proposée par M. Olivier ne conduit donc a des tracés nouveaux que lorsqu'elle exige le changement des deux plans de projection, et dans ce cas elle convient peu aux applications, notamment à la Stéréotomie.*

[65]De la Gournerie (1874: 114): *Dans l'enseignement établi par Monge, ... la Stéréotomie ... est reduit a sa partie géométrique.*

Monge s'était laissé tellement dominer par cette manière incomplète de considérer la question, qu'il a entrepris de résoudre, par les seules ressources de la Géométrie, le problème de la décomposition d'une voûte en voussoirs ...

Le célèbre Géomètre faisait ainsi complétement abstraction des données essentielles du probème; il ne s'inquiétait ni de la situation de la voûte par rapport à la direction de la pesanteur,

This also provided an occasion to attack Leroy, who had praised Frézier for having used orthogonal projections before Monge. However, de la Gournerie made it clear that all earlier stonecutting authors had used this method, remarking that de la Rue had extended his application to the solution of abstract geometry problems. Of course, he did not miss the opportunity to redouble his invectives against Olivier, remarking that if he had read Frézier well, he would have known that changes of horizontal projection plane are useless in practical operations and particularly in stereotomy; in his own words:

> If Olivier had followed the wise advice of Leroy, that is, if he had read Frezier's observations about Desargues's methods, he would not have proposed the change of projection planes as a general procedure for the solution of descriptive geometry problems without first refuting the criticisms raised against this method.[66]

Twenty years before, La Gournerie had delivered a speech at the beginning of his courses at the Conservatoire des Arts et Métiers, another school connected with Olivier. He presented a detailed account of the succession of stonecutting treatises, from de l'Orme to Frézier, in order to support his views, stressing rightly that Monge had built his descriptive geometry on a long-standing tradition, but had tried to generalise it in order to solve abstract problems, introducing folding lines and the representation of planes. The publication of the speech (de la Gournerie 1855) marks the beginning of the historiography of stereotomy, just at the moment when practical stonecutting was fading away under the pressure of the Industrial Revolution.

---

ni de la position des points d'appui, ni de l'existence possible de surcharges diversement réparties … la nature géométrique de cette surface était seule prise en considération.

[66]De la Gournerie (1874: 153–154; see also Rabasa 2011: 718): *Si Olivier avait suivi le sage conseil de Leroy, s'il avait lu les observations de Frézier sur la manière de Desargues, il n'aurait pas proposé le changement des plans de projection, comme procédé général de solution pour les problèmes de la Géométrie descriptive, sans chercher à réfuter les critiques dont cette méthode avait été l'objet.*

Chapter 3
Techniques

Abstract This chapter will deal in depth with the techniques used in the stonecutting process, from formal definition to actual carving and voussoir placement. It analyses full-size drawings, the planar and three-dimensional constructions used in these tracings, and the models used to verify the precision of geometrical procedures and to instruct apprentices. Next, it deals with carving instruments, stressing the difference between mechanical ones such as the pick, the chisel or the axe and a set of geometrical instruments allowing the transfer of shapes from the tracing to the stone block, such as the ruler, the square, the arch square, the bevel and the templates. The next sections analyse the basic dressing techniques, squaring and templates, as well as several hybrid methods and the procedures used to control the shape of cylindrical, spherical and warped surfaces. The final section of this chapter will deal with the transport, elevation, and placement of the voussoirs. Although these issues do not seem to be connected with geometrical problems, the use of powerful elevation tools allowed the transition from Romanesque rubble to rough-hewn stone in Gothic severies and Renaissance ashlar, bringing about a remarkable evolution of stonecutting methods.

3.1 Setting Out

3.1.1 Large-Scale Tracings

Antiquity. We are prone to project our nineteenth- and twentieth-century technical habits on past periods, taking it for granted than small-scale drawings on flexible supports have been the norm during the entire history of architecture. However, the main flexible support in Antiquity, papyrus, reaches at most a width of about 40 cm, and thus is not suitable for detailed drawings. In contrast, a fair number of full- and large-scale architectural tracings[1] from Antiquity, inscribed on walls and floors,

[1] "Trace" and "tracing" are used throughout this book with the meaning "to draw, sketch, outline, etc.", in particular in large- or full-scale drawings. On some rare occasions, using them with the meaning "to copy (a drawing, etc.) by following its lines on a superimposed transparent sheet", I will state that I am using this sense of the word.

J. Calvo-López, *Stereotomy*, Mathematics and the Built Environment 4,
https://doi.org/10.1007/978-3-030-43218-8_3

have been preserved throughout Europe and the Middle East. Most scholars dealing with the passage where Vitruvius (c. -25: I.2.2) mentions the three "species of disposition"—ichnography (plans), orthography (elevations) and scenography (perspective)—do not stress that the author is talking about full-scale tracings, although he mentions *in solis arearum descriptiones*, which I would translate as "drawings in the area below a future building".[2] Usually, such *descriptiones*, used as modern layouts, have remained hidden. Only in some particular examples, such as the Roman temple in Évora, the degradation of stone in column bases has unveiled the tracings (Pizzo 2016).

In other cases, tracings represent vertical figures, such as arches, pediments or sections of bases and columns, rather than horizontal layouts (see Inglese 2000: 91–108). Several tracings in the temple of Apollo in Didyma represent the complex bases of some columns of this structure. However, the most interesting one seems related to the formation of the entasis of a column through a remarkable method; we will come back to this issue in Sect. 4.2.2 (Haselberger 1983; Ruiz de la Rosa 1987: 124–128; Wilson Jones 2009: 99). Another remarkable tracing from Antiquity is located near the mausoleum of Augustus in Rome; it represents the pediment of the Pantheon, about 1 km away (Haselberger 1994a; Haselberger 1994b; Inglese and Pizzo 2000: 31–94; Inglese 2013; Inglese 2014: 31–94). Other large drawings on rigid supports, such as the well-known *Forma Urbis Romae*, a huge plan of Severan Rome, initially set in a wall of Vespasian's Templum Pacis, are drawn to scale, not at full size. A remarkable feature of these drawings from Antiquity is that they do not represent objects lying in clearly different planes; at most, the tracing shows the cornice and the back surface of the pediment of the Pantheon or the different levels of Roman hills and valleys. In other words, the concept of projection, the representation of clearly different planes, is still lacking.

Middle Ages. Later, parchment substituted papyrus as a writing support. It was quite expensive: the well-known plan of Saint Gall is drawn on the hides of no fewer than five animals. Thus, the Middle Ages inherited large-scale tracings, overlaying them with allegorical connotations of the act of laying out the foundations of a building. Paintings and miniatures depict Pope Liborius marking with a hoe the outline of the basilica of Saint Mary Major in miraculously fallen snow in the Roman summer, or Saint Peter, Saint Paul and Saint Stephen furnishing ropes to Abbot Gunzo (Fig. 3.1) in order to trace the foundations of the church at Cluny (Miscelanea … cluniacensis c. 1200: 43r; Colombier [1953] 1973: 86). Almost no examples of full-scale tracings from the High Middle Ages have survived in the West, although a small number of large-scale drawings have been preserved in the Byzantine world, such as those at the sixth-century church of the Holy Cross at Resafa (Ousterhout [1999] 2008: 64–65). This has led some scholars to posit that Romanesque constructions

[2]Quirico Viviani's (1830–32: I, 58) translation, *le descrizioni delle forme nei suoli delle aje*, gives a rich insight on this issue, provided that we keep in mind that one meaning of *descrizione*, nowadays rare but still common in the early nineteenth century, is "drawing" and that *aja* means "a zone or area of levelled ground".

Fig. 3.1 Saint Peter, Saint Paul and Saint Stephen bring ropes to Abbot Gunzo for the layout of the church at Cluny (*Miscelanea … cluniacensis* c. 1200: 43r)

were carried out without any kind of drawing, through *imago in mente conceptum*; of course, we should be careful about such arguments from silence.

In contrast, a large number of full and large-size tracings from the Late Romanesque and Gothic periods have been found, analysed and even forged (Villard/Barnes [c. 1225] 2009: 224–226). First, as stated in Sect. 1.6, we must differentiate simple marks resulting from the application of templates, such as those found at Bylands Abbey (Fergusson 1979), or axes and auxiliary notches made during the dressing process (Willis [1842] 1910; Bessac 1984; Rabasa 1996a; Ruiz de la Rosa 1996) from full-fledged tracings, which involve usually more than one voussoir or block and are executed *before* preparing the templates.

Complex tracings have been preserved in the tribunes of Soissons Cathedral (Brunet 1928; Barnes 1972), the flat roofs of Clermont-Ferrand (Claval 1988), representing gables, windows and flying buttresses, the triforia in the transepts in Reims cathedral (Erlande-Brandenburg 1993: 78), the walls of the crypt of Roslyn Chapel in Midlothian, Scotland (Shelby 1969: 545–546; Erlande-Brandenburg 1993: 79) and the church of Santa Maria in Ponte near Cerreto di Spoleto (Inglese 2000: 115–116). In England, perhaps due to inclement weather, tracings were prepared frequently in specialised rooms, such as those preserved in York Minster and Wells Cathedral (Harvey 1968; Colchester and Harvey 1974; Holton 2006). Archival documents state that Thomas of Canterbury, master mason in charge of the Saint Stephen chapel in the palace of Westminster was *in trasura super moldas operanti*, or *operanti in trasura et moldas* de novo *reparanti* (working on templates in a tracing room and repairing templates).[3] Unfortunately, few tracings for vaults and other three-dimensional pieces have been preserved.

The Early Modern period. Much less attention has been devoted to Renaissance and Baroque tracings. It seems that an old prejudice opposes the "anonymous" medieval master, working without drawings and employing tracings, to the Renaissance artist, prone to use drawings on paper. Neither of these visions fit the available

[3]Hawkings and Smith (1807: 172–173); Hastings (1955: 58, 159, 174); see alsoand Shelby (1964: 394), about the *trasyng hous* of Exeter cathedral and the *trasour* at Windsor Castle.

evidence. Medieval masons were not anonymous, as Harvey's dictionary (1954) has shown; they used *both* drawings and full-scale tracings, just as their Early Modern counterparts did.

Most studies on Early Modern tracings come from a few locations in Spain, such as the basement corridors in the Escorial (López-Mozo 2008), the sacristy and the chapel of Junterón in Murcia Cathedral (Calvo et al. 2005a: 145–50; Calvo et al. 2013b; see Fig. 1.20), the rooftops of Seville Cathedral (Ruiz de la Rosa and Rodríguez 2002; Pinto and Jiménez 2016), the Cathedral of Jaén (Gutiérrez 2017) and a large number of locations in Galicia (Taín 2003; Taín 2006; Taín and Natividad 2011; Taín et al. 2012; Calvo et al. 2013a; Calvo et al. 2016; Cajigal et al. 2016; see Fig. 3.2). There are documentary references to the tracing houses of Granada Cathedral and The Escorial (Gómez-Moreno 1963: 90; Bustamante 1994: 209, 228) and even the actual tracing room of Seville Cathedral, dating from the sixteenth century, has been preserved (Pinto and Jiménez 1993). However, there is no reason to suppose such tracings are a Spanish speciality. Of course, tracings do not appear unless somebody is looking for them; perhaps increased awareness of these issues will bring forward discoveries in the decades to come.

Locations and purposes. On some occasions, tracings were prepared exactly below the element under construction, in order to control the placement of voussoirs. Generally speaking, in the Gothic period they were executed on scaffoldings placed under vaults, at springer level, while in the Early Modern period, they were laid

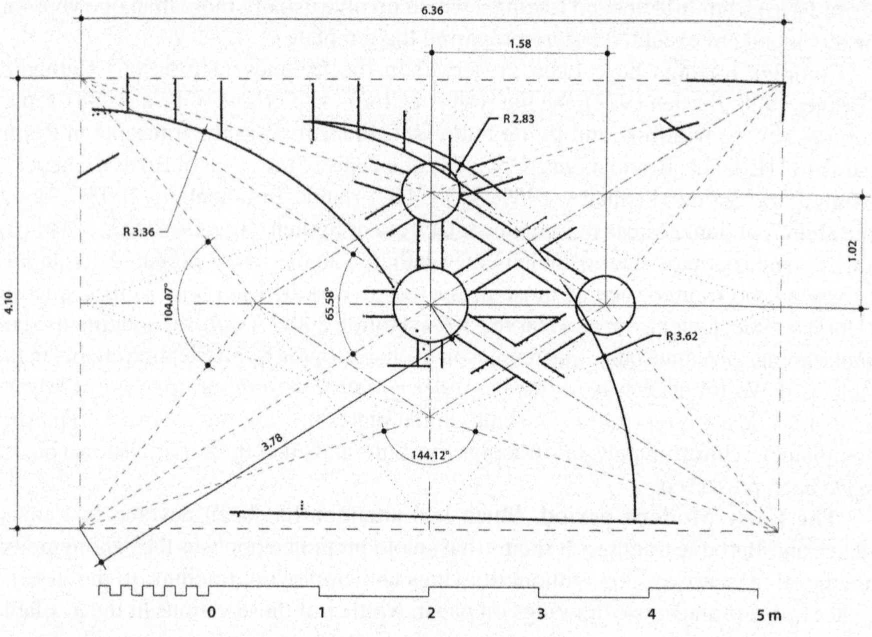

Fig. 3.2 Tracing for a vault in the sacristy of Tui cathedral, inscribed on the floor of the chapel of Saint Catherine (Drawing by Miguel Ángel Alonso and the author)

directly on the floor (Garcia and Gil de Hontañón 1681: 24v-25v; Vandelvira c. 1585: 23r, 23v). We will come back to these issues when dealing with placement control methods in Sect. 3.3.4. Here, it is important to summarise that tracings were prepared in three different kinds of locations: directly under the element under construction; in specialised tracing houses; or in secluded places such as tribunes, rooftops, socles in less frequented aisles, cloister walls, basements or spaces below stairways.

The contrast between the huge, complete tracing in the parish church of Szydłowiec, (Fig. 3.3) and the small, fragmentary one at Tui cathedral is quite striking. The former represents the vault in the church choir, filling an entire wall, with a dimension of 8.5 × 12.7 m (Brykowska 1992), while the plan of the vaults of the sacristy in Tui (Fig. 3.2) measures 2.4 × 2.1 m, including a single instance for each different kind of rib (Taín et al. 2012). This suggests that the Szydłowiec drawing was executed to secure the client's approval, while ordinary tracings, such as the one at Tui, were prepared for the masons' use; as a result, they are extremely economical (see Sakarovitch 1992a: 532). It is easy to understand that tracing on all fours or from a scaffolding is quite taxing, so the Tui masons left out any unnecessary line. This also explains why such tracings are usually difficult to understand. They are not representations in the ordinary sense of the word, designed to convey orders from the designer to the executors, as modern blueprints are. Rather, they are

Fig. 3.3 Tracing for the choir of the Szydłowiec parish church on a wall of the nave (Photograph by the author)

private drawings, prepared by the master mason for his own use; they were employed to assure formal control of the piece under construction or to determine shapes not deriving directly from plans and elevations, such as true-size-and-shape templates for the intrados and the bed joints of voussoirs.

Tracing instruments. The primary tool for these tracings is the rope supplied by the saints to Abbot Gunzo. A tight one, held between two fixed points, can be used to draw a line with the help of a scriber or stylus. Fixing the rope at one end with a nail and tying the scriber to the other end, the mason can draw circles and arcs. Also, ropes can be used to draw gardener's ellipses, as we will see in Sect. 3.1.2. However, Philibert de l'Orme remarked that ropes made from hemp are not dependable, since they do not keep their length when wet; it is advisable to use either ropes made from the inner bark or phloem of lime trees or rulers three, four or six fathoms long (about 6, 8 or 12 m). Piercing two holes at the ends, these straightedges can also be used to draw circles or arcs.[4]

When trying to translate Vitruvius, Hernán Ruiz II was at pains with the term *area*; he defined it as "a flat surface, levelled and prepared to trace with ruler, compass, and square".[5] Such tools were enlarged versions of the ones used by architects and engineers in the twentieth century. A drawing illustrating Matthieu Paris's *Lives of the Offas* shows the king's architect holding a large compass reaching up to his waist and an unusual square with two convergent arms (Paris c. 1250–1259; 23v; see also Colombier [1953] 1973: 99; Gimpel [1958] 1980: 36). The huge compass can be used to draw medium-sized arcs; large ones require ropes or rulers, as remarked by de l'Orme (1567: 33v). For example, Ginés Martínez de Aranda (c. 1600: passim; see also Calvo 2000a: I, 136) mentioned the compass 86 times when dealing with 51 instances of rere-arches, which are usually small pieces, while he included only 20 references to this instrument when explaining 70 problems dealing with arches.

As for the two-armed square, its main advantage is its versatility: it can be used in both the setting out and dressing phases. The convergent sides of each branch have led to different interpretations, even involving the Golden Mean (Morgan 1961: passim; Sené 1970); more likely, they were used when drawing the angles between the bed joints of a voussoir (Branner 1957: 65–66; Shelby 1965: 247–248; Shelby 1969). In accordance with Renaissance ideals of exactitude, de l'Orme brings back the three-sided square (Fig. 3.4), a drawing instrument known in Antiquity (Vitruvius c. -25: IX.0.6; Faventino [c. 250] 1540: 21r; Isidore c. 630: 19.18.1), advising the

[4]De l'Orme (1567: 33v): *Pour donc bien équarrir un fondement, vous prendrez une ligne ou cordelle qui soit faite d'écorce d'arbre, comme de tille (pour autant que la ligne de chanvre ne retient sa mesure quand elle est mouillée) et la ferez de telle longueur que vous voudrez, lorsqu'on ne peut avoir un si grand compas qu'il serait de besoin. Au lieu de la dite ligne on pourrait user de longues règles et étroites en forme de compas, le tout selon la commodité du lieu où vous serez. Soit en une sorte ou en l'autre, vous prendrez ladite ligne ou règles de trois, quatre ou six toises (la plus longue a le plus de jugement) et en userez ainsi que si c'était un compas, ayant une broche ou pointe à chacun bout*. Transcription is taken from http://architectura.cesr.univ-tours.fr; translation by the author.

[5]Hernán Ruiz II (c. 1560: 13r): *Area es un planiz aparejado y nibelado para delinear con regla y compas y escuadra*.

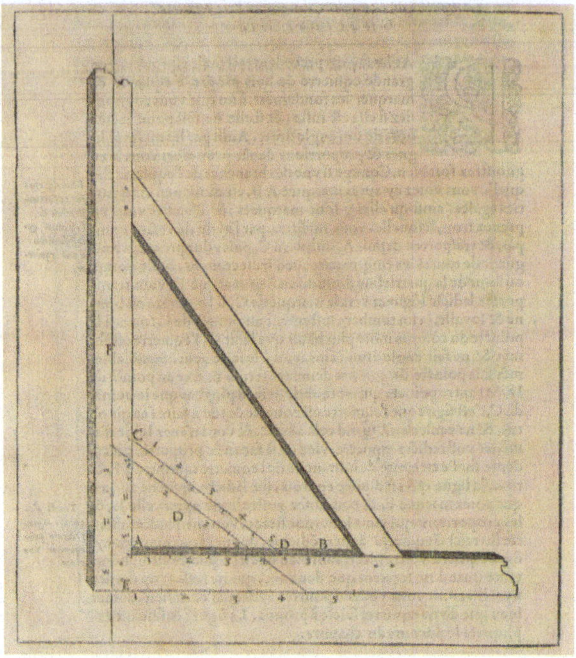

Fig. 3.4 Three-sided square (de l'Orme 1567: 36v)

reader to use the Egyptian triangle with sides measuring 3, 4 and 5 units in order to check it and correct it if necessary (de l'Orme 1567: 36r-38v).

All in all, there are two main reasons to use full- or large-scale tracings. The former obviate the need for scale changes, which were problematical in the Middle Ages and most of the Early Modern period. We should recall that decimal notation was popularised in Europe by Simon Stevin in 1585, and even then, its use was not widespread between masons. However, this does not explain the frequent use of reduced-scale drawings in walls and floors. Parchment was extremely costly in the Middle Ages (Erlande-Brandenburg 1993: 74); in the Early Modern period, paper was still expensive. Thus, reduced-scale drawings on parchment or paper were used generally as presentation drawings for the client, while tracings on floors and walls, either full- or at a reduced scale, were used by masons for their own needs.

3.1.2 Planar Geometry Methods

There is a remarkable body of literature about the amazing geometrical knowledge of medieval masons. However, such writers often fail to study actual medieval masons' texts, such as those by Villard de Honnecourt (c. 1225) and other scribes in his portfolio, the anonymous Regius (c. 1390) and Cooke (c. 1400) manuscripts, the printed booklets of Mathes Roriczer (1486; c. 1490a; c. 1490b) and Hannes Schmuttermayer

(c. 1500), and the sketchbooks by Wolfgang Rixner (1467–1500) and Hans Hammer (c. 1500), to mention only those prior to 1500 (see Sects. 2.1.1, 2.1.2 and 2.1.6). In this section I will present a factual study of these scant remaining primary sources of medieval masons' geometrical knowledge, together with several Early Modern sources written by masons, architects or engineers or widely circulated among them[6]; however, I will leave interpretations for Sect. 12.4.1.

Orthogonals and parallels. First, we should try not to project onto medieval and Early Modern full-scale tracings the usual practices of nineteenth- and twentieth-century technical drawing on paper. Drawing parallels directly on floors or walls, by sliding a triangle along another one, is utterly impractical; also, long orthogonals drawn with the square are unreliable. Thus, several sources hint that auxiliary orthogonals were used to draw parallels, and vice versa. Remarkably, no particular method for the tracing of perpendiculars, other than the use of the square, is explained by either Villard or other scribes in his portfolio, or by Roriczer. Tracing the foundations of a cathedral using only a square, however large, would lead to considerable errors. Abbot Suger stated that he had laid out the foundations of Saint-Denis Abbey using geometrical and arithmetical instruments. The problem was rather complex, since he had to connect the new east and west ends across the existing nave; modern plans show clearly that he did not succeed. Later on, the layout of many Gothic constructions is rather acceptable. This hints strongly that medieval builders must have known a method for the layout of orthogonals with ropes; neither squares nor de l'Orme's long rulers can solve the problem for a whole cathedral.

De l'Orme explains two suitable methods for this purpose (1567: 33r-35v). One of them is again based on the Egyptian triangle. Although the other one is not so well known, it is quite practical for the laying out of perpendiculars in large tracings, for example when opening foundations. De l'Orme explains it with rulers, probably in order to show the advantages of this instrument; however, it can be easily extrapolated to ropes to construct orthogonals in large tracings. The builder should begin tracing an equilateral triangle (Fig. 3.5); next, he should construct an isosceles triangle using one side and the prolongation of another side of the equilateral triangle. The base of the isosceles triangle is orthogonal to the third side of the equilateral triangle. Both of these methods can be used with long ropes, tracing orthogonals as long as a cathedral.

A few years later, Alonso de Vandelvira (c. 1585: 3v) included the well-known method for bisecting a segment using four arcs and an orthogonal in the introduction of his stonecutting manuscript, calling the bisector *esquadría*. This hints that he considered this procedure to be a means for tracing perpendiculars rather than, or in addition to, bisecting the original line. The placement of these explanations amongst the rather theoretical introduction, taken in part from Juan Pérez de Moya (1568; see Barbé 1977: 26) a popular Spanish mathematician of the period, casts some doubt

[6]See Alviz c.1544; Ruiz c. 1560; Gil de Hontañón c. 1560; de l'Orme 1567; Chéreau c. 1567–1574; Vandelvira c. 1585; Martínez de Aranda c. 1600; Derand 1643; Frézier 1737–1739; as for texts written by non-masons but widely circulated among them, see Dürer 1525; Serlio 1545; Jousse 1642.

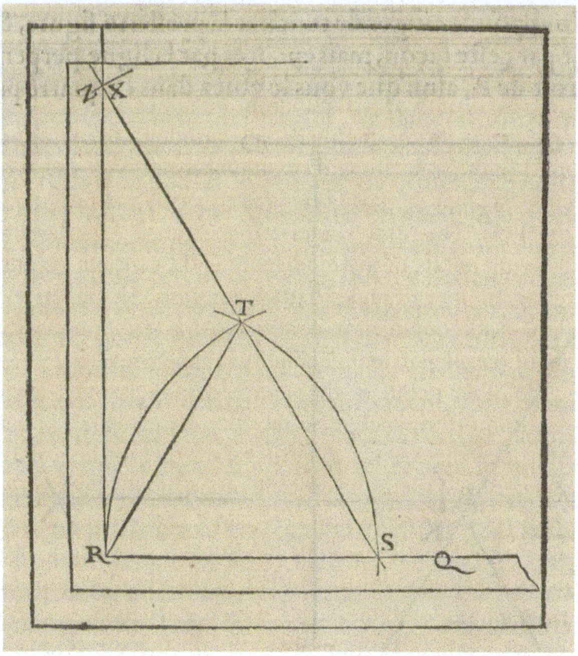

Fig. 3.5 Method for constructing orthogonals based on isosceles triangles (de l'Orme 1567: 34v)

on the practical use of this procedure. In any case, this detail shows how the methods of learned geometry penetrated into builders' circles along the sixteenth century, although this influx should not be overstressed (Rabasa 2015a: 464–466). As we may expect, Vandelvira's explanation is devoid of any proof; it curiously resembles nineteenth-century drawing manuals rather than sixteenth-century translations of Euclid (1576: 17r; see also 12v), which use equilateral triangles, and implicitly full circles, but not the short arcs drawn by Vandelvira, which befit the economical nature of full-scale tracings.

The layout of parallels is a different issue. The classical Euclidean construction (Euclid -300 : I, 31), based on a segment forming equal angles with the given line and the parallel, is utterly impractical for masons' needs. Equal angles may be constructed using a bevel (the masons' protractor), but the errors induced by this method when applied to a large-scale tracing may be large. Cristóbal de Rojas (1598: 7v), quoting the same Euclidean proposition, explains a different method, based on two circles with their centres placed on the given line; the parallel should be tangent to the circles (Fig. 3.6). Again, a slight error in the appreciation of tangency may lead to remarkable errors in the direction of the parallel line.

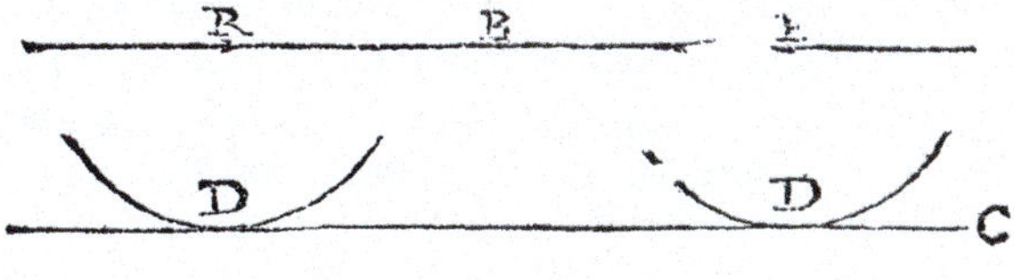

Fig. 3.6 Tracing parallels using circular arcs (Rojas 1598: 7v)

A subtle detail in the stonecutting manuscript by Ginés Martínez de Aranda (c. 1600: 16–17) hints that masons constructed parallels using auxiliary orthogonals, and vice versa. When he is about to draw a number of parallels representing the intrados joints of a skew arch, he states that the base line of the elevation serves as a *juzgo* for the parallel lines; although the word is not common in Spanish, it is used in other sections of the manuscript with the meaning of "auxiliary line". This suggests that the base line is used as an ancillary construction for the tracing of the parallels. This may sound rather strange to an architect or engineer trained in the drawing habits of the nineteenth and twentieth centuries, when a triangle slid easily along another triangle to draw parallels. However, such technique is not practical in large-scale tracings; in contrast, using a common orthogonal in order to construct a set of parallels is both economical and precise.

Another detail in the same passage is even more remarkable: when trying to draw an orthogonal to an intrados joint, which is perpendicular to the *juzgo* itself, Martínez de Aranda does not draw the orthogonal to the intrados joint with the square; he explains his tracing procedure as *tirarás la linea en blanco 1 galgada con* [*el juzgo*]. Since *galgar* refers to a procedure carried out with a gauge, we should understand that the author measures a distance along one of the orthogonals to the *juzgo* and carries it to another perpendicular; joining the resulting points, he may draw a parallel to the *juzgo*, which is of course orthogonal to the intrados joints.

Such practices may also explain the elevation-drawing methods in Roriczer's *Buchlein den fialen gerechtigkeit* (1486: 5r-5v; see Sect. 3.1.3). Rather than using projection lines, Roriczer first draws a vertical axis of the elevation; next, he constructs several orthogonals to the axis, representing horizontals at different heights; he then transfers horizontal measures taken from the plan to the orthogonals in the elevation. Although with a different intention, a similar technique may have been used in the remarkable set of stonecutting tracings in the convent church of Saint Clare in Santiago de Compostela (Calvo et al. 2016: 60–63; Calvo and Taín 2018a: 67–68). A line at least 16.20 m long runs through the whole ensemble of tracings without an apparent function (Fig. 3.7); however, most tracings in the entire church floor, covering an area of roughly 15 × 15 m, have their bases, axes, or springing lines drawn orthogonally to the long line, which plays the role of an axis of the whole set; a shorter line is drawn orthogonally to the axis at one of its end, as a base for the ensemble. Thus, the general axis serves as an organising device of the whole group, acting as an aid to the tracing of perpendiculars and parallels.

Division of segments and circles; regular polygons; circle centres. The approach of Renaissance architects and masons to learned geometry is exemplified by Philibert de l'Orme's (1567: 38v-39v; see also Manceau 2015) method for the division of a segment in an arbitrary number of equal parts (Fig. 3.8). He uses Thales' theorem, but he adapts it to the constraints of full-scale tracings. To divide the segment *AB*, he uses a number of parallel lines. Although such lines are oblique to *AB*, he does not use an orthogonal to the lines, as in Martínez de Aranda's construction, but rather two orthogonals to *AB*; this may seem a waste of labour, but we should take into account that the first orthogonal is necessary in order to apply Thales' theorem.

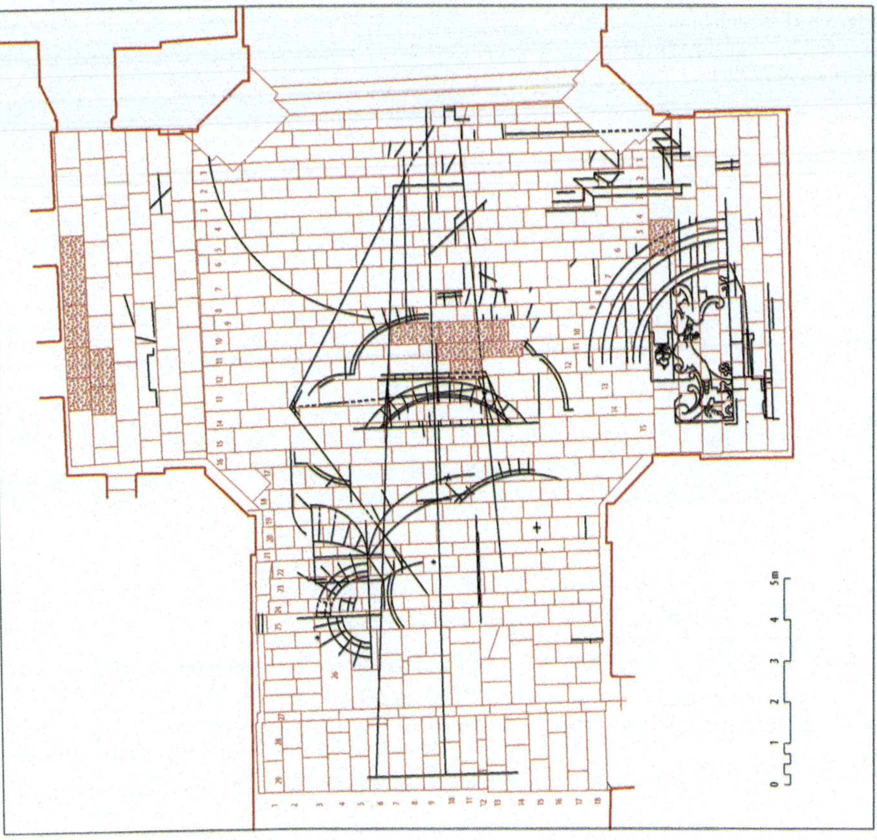

Fig. 3.7 Tracings on the floor of the church of Saint Clare in Santiago de Compostela (Drawing by Idoia Camiruaga, Miguel Taín and the author)

In contrast, masons' approach to the division of circles shows clearly the practical, rough-and-ready nature of these methods. Such an operation, of course, is essential in stonecutting, for example to set out an arch or vault with voussoirs of the same size; further, it is associated with the construction of regular polygons. Dividing a circle in two, four, eight or sixteen parts using the bisector of a segment, a technique explained by Vandelvira (c. 1585: 3v), is relatively easy. However, arches are usually divided into an odd number of voussoirs, in order to use a single member as a keystone. Roriczer uses an incorrect construction for the regular pentagon in his *Geometria Deutsch* (c. 1490a: 2r; see also Dürer 1525: Eiii bis r[7]; Meckspecker 1983) (Fig. 3.9). First, he draws an edge *a-b* of the pentagon. Next, he constructs two full circles with their centres at both ends of the edge, with their radii equal to the length of the side.

[7] Some folios in this book are numbered, others are not. Further, numbers take the form "Aiii", for example, meaning "third numbered folio in group A" and unnumbered sheets are designated as *bis*, *tria*, etc. Thus, "Eiii bis r" means "*fourth* folio in group E, recto face".

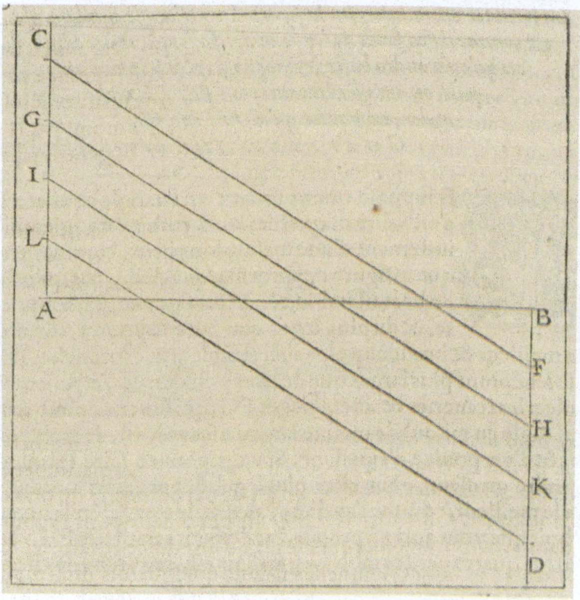

Fig. 3.8 Division of a segment into equal parts (de l'Orme 1567: 39r)

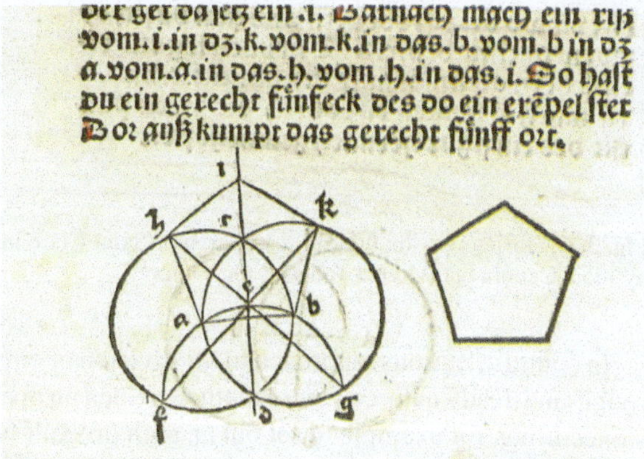

Fig. 3.9 (Incorrect) construction of a regular pentagon (Roriczer c. 1490a: 2r)

This construction provides, of course, the bisector of the edge, but Roriczer uses it with another purpose. He draws a third circle with its centre at the intersection *d* of the first two circles, passing through *a* and *b*; of course, its radius will equal again the edge of the pentagon. Next, he traces a line passing through the intersection *f* of the first and third circles, and the intersection of the third circle with the mediatrix, *e*; he extends this line until it reaches the second circle at *k*, placing a corner of the polygon at this point. Of course, the length of *b-k*, the new edge of the pentagon, equals the original side, but nothing assures that the angle *a-b-k* measures 108°, as in a regular pentagon; in fact it measures 108.37°. Rorizcer repeats the same construction at the other side, getting a new corner *h*, and then closes the pentagon at *i* so that *h-i* and *k-i*

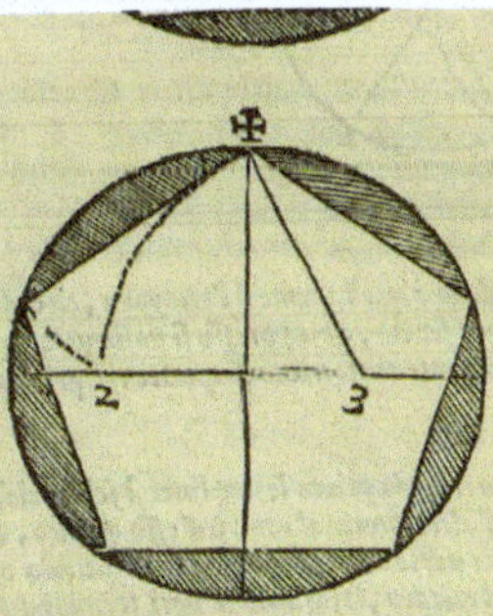

A formar questa figura pentagona, cioè di cinque lati, non è cosi facile come l'altre, per essere di lati dispari, & di più numeri che tre, nondimeno à farla teoricamente, si farà cosi. Fatto vn cerchio perfetto, dentro di quello si farà una croce, cioè una linea piana che sia il diametro, & sopra il diametro caderà una linea a piombo. Poi dal lato sinistro sia diuiso il mezo diametro in due parti uguali, che sarà 3. & da quello alla sommità, che sarà una croce, sia allargato il compasso, & dalla croce in giù fin sopra il diametro sia circuito, non mouendo però la punta delle seste dal 3. & doue caderà la linea curua partita dalla croce sopra il diametro, quãto sarà dal 2. alla croce, quello sarà giustamente un de' cinque lati del pentagono. In questa figura si trouano anchora le dieci faccie, percioche dal centro al numero 2. sarà un lato delle dieci faccie, & più questa figura ne reca anchora il lato di sedici faccie, & però dalla circunferentia al 1. verso il centro al punto 2. lì si trouarà uno de' lati delle sedici faccie.

Fig. 3.10 Construction of a regular pentagon (Serlio 1584: 18v) reprinted from Serlio 1545

should equal in length *a-b*. Thus, all edges of the pentagon are equal in length, but not all internal angles are equal; in fact, they measure 108.37°, 107.04° and 109.19°. This is striking, since Roriczer could have used the exact and straightforward Euclidean solution to this problem (Euclid c. -300: 4.11). Thus, we are led to conclude that he did not know Book IV of the *Elements*, which was available in this period (Euclid 1482).

It is true that the Euclidean method is rather cumbersome when used in full-scale tracings; however, a few decades later, Serlio (Fig. 3.10) included in his First Book (1545: 20r; compare with Euclid c. -300: 4.11) an exact construction for the regular pentagon: he takes two orthogonal diameters of a circumference, divides a radius in halves, constructs a right triangle with half a radius and a full one as catheti, draws an arc with its centre in one of the corners of the triangle, finds the intersection of this arc with a diameter and draws another arc that furnishes a corner of the pentagon. He presented also a simple and exact method to inscribe an octagon inside a square: the builder draws an arc with its centre in a corner of the square, equalling the radius with the length of half a diagonal; the intersections of the arc with the sides of the square should furnish two corners of the octagon. Repeating the construction four times, the mason can find all four corners and draw the octagon (Serlio 1545: 19r).

In contrast, Serlio's construction for the nonagon or a polygon with a larger number of sides is difficult to understand:

> The figure shown will be very useful for those who need to divide a circle in equal parts, no matter how many; however, in order not to confuse the reader with a great number of parts, I will suppose by way of example that we want to divide a circle in exactly nine parts; we will take the fourth part of the whole circle and divide it in nine parts and four of these parts will surely make a nineth part of the whole circle ...[8]

[8]Serlio (1545: 20v): *La figura quì sotto dimostrata sarà di gran giouamento a tutti quelli, a cui bisognara diuidere alcune circonferentie in quante parti gli accadera quantunque fossero gran numero, imo dispari; ma essempi gratia per non confondere il lettore in gran numero di parti, vorremo fare un circolo perfetto diuiso in noue parti giustamente: prenderemo adoncha la quarta*

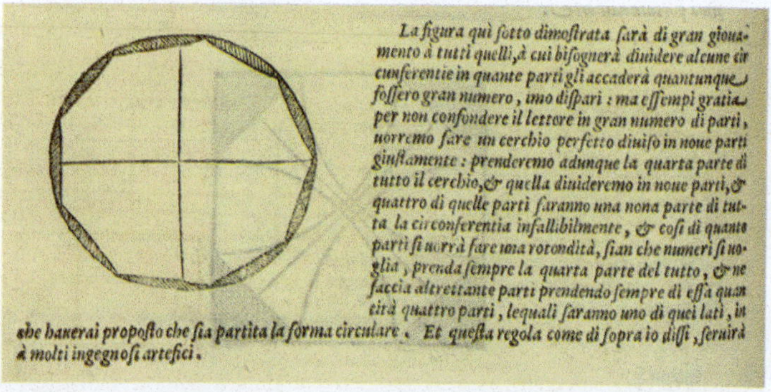

La figura quì sotto dimostrata sarà di gran giouamento à tutti quelli, à cui bisognerà diuidere alcune circonferentie in quante parti gli accaderà quantunque fossero gran numero, ımo dispari: ma essempi gratia per non confondere il lettore in gran numero di parti, uorremo fare un cerchio perfetto diuiso in noue parti giustamente: prenderemo adunque la quarta parte di tutto il cerchio, & quella diuideremo in noue parti, & quattro di quelle parti faranno una nona parte di tutta la circonferentia infallibilmente, & cosi di quante parti si uorrà fare una rotondità, sian che numeri si uoglia, prenda sempre la quarta parte del tutto, & ne faccia altrettante parti prendendo sempre di essa quantità quattro parti, lequali faranno uno di quei lati, in che hauerai proposto che sia partita la forma circulare. Et questa regola come di sopra io dissi, seruirà à molti ingegnosi artefici.

Fig. 3.11 Division of the circle into nine "equal" parts (Serlio 1584: 18v) reprinted from Serlio 1545

In other words, to divide a circle into nine equal parts, first divide a quadrant into nine equal parts; then, multiply by four the distance between two consecutive points; this will be the distance between two consecutive points dividing the full circle into nine parts. This is incorrect; in fact, taking the radius of a circle as the unit, the distance between the points that divide a quadrant into nine points is 0.174 units; applying Serlio's rule, the points that divide the full circle into nine parts would be set 0.696 units apart, while in fact the distance between the corners of a regular nonagon is 0.684 units. Moreover, Serlio does not explain how to divide the quadrant in nine parts; thus, the whole page (Fig. 3.11) reeks of circular reasoning.[9]

Stonecutting manuscripts provide some clues about the methods used to divide arches into voussoirs. Hidden lines with geometrical constructions can be seen in the manuscripts of Hernán Ruiz II, both copies of Vandelvira, and the one attributed to Pedro de Alviz (Rabasa 1996a; Calvo et al. 2005a: 240–242; García Baño and Calvo 2015); however, as yet no occult lines have been found with an explicit construction for the division of arcs into equal parts. Again, a tip is given by a subtle detail in Martínez de Aranda (c. 1600: 96–97). He explains the construction of a complex piece, a double-splayed arch with a groin in the middle plane in the shape of a pointed arch. It is easy to check on the drawing that the distances between corners of the voussoirs lying at the groin are equal, including the distance between two points that lie at opposite branches of the pointed arch. A similar situation arises in a skew arch

---

parte di tutto lo circolo, & quella diuidiremo in noue parti & quattro di quelle delle parti faranno una nona parte di tutta la circonferentia infallibilmente ....

[9]There is an alternative interpretation of the passage, but it is not likely. Serlio may be thinking about dividing a circunference by means of a rope wounded around it, for example around a column drum. In this case, the mason can wind a rope around a quarter of the drum, mark on it the points that correspond to the ends of the quadrant, divide this portion of the rope in nine parts by trial and error, and apply four times the ninth part of the quadrant in order to get a ninth part of the full circumference. However, as we will see in the next paragraph, both Martínez de Aranda and Jousse divide pointed arches in equal parts taking into account the length of the chords, not the circular segments.

in Jousse (1642: 14–15). Both faces of the arch are semicircular, but their projection onto a single plane results in a pointed arch, and Jousse instructs explicitly the reader to divide the projected arch into five parts. Neither this problem nor the one posed by Martínez de Aranda's actual pointed arch can be solved through compass-and-ruler constructions. Thus, we must conclude that both Martínez de Aranda and Jousse addressed these issues by trial and error; the same method may have been applied to simpler problems such as the division of a round arch into five, seven or nine parts. This would explain in part Serlio's idiosyncratic construction for the nonagon: rather than dividing the whole circle in nine parts, he divides one quadrant by trial and error; next, he multiplies the distance between points by four to divide the entire circle.

Another problem regarding circles arising frequently in stonecutting practice is the determination of the centre of a circle given three points. In his carpentry manual, de l'Orme (1561: 13r-13v; see also Serlio 1545: 11v-12r) explains a classical graphical construction under the colourful name *les trois points perdus* (the three lost points). As usual in learned geometry treatises, de l'Orme determines the bisectors of two segments joining these points; the centre of the circle is placed where both bisectors meet. This procedure is also explained in the stonecutting chapters of his architectural treatise although the issue is less clear, since he mixes his explanation with the description of the *cerche ralongée*, (extended circle). His exposition is rather puzzling:

> In order to find the stretched templet [derived from] a circumference and explain it adequately, it cannot be drawn with a single stroke of the compass, nor from a single centre, but rather with several centres and lines, curved and round.[10]

Ovals, ellipses, and other curves. A comparison with Spanish texts can shed some light on de l'Orme's baffling passage; however, in order to understand it better, we should deal first with the methods for constructing ovals and ellipses. Medieval masons do not seem to have used real ellipses; when using a non-pointed surbased arch, they gave it the shape of a three-centre oval.[11] Four well-known constructions (Fig. 3.12) for such ovals are included in Serlio's First Book (1545: 17v-18v; see also Duvernoy 2015: 430–434).[12] All of them are based in a geometrical kernel, in the shape of a rhombus or a square; this nucleus allows the construction of four symmetrical radii passing through the meeting points of the four arcs that make up the oval, that is, the points where the curvature of the oval changes suddenly. Two centres are placed at each of these radii; this guarantees tangency between each consecutive pair of arcs. The second, third and fourth ovals have fixed proportions. In contrast, the first one can be constructed with different ratios between axes; however, it cannot be

[10]De l'Orme (1567: 55r-55v, 56r): *Pour trouver donc promptement la cherche rallongée d'une circonférence, et la donner bien à entendre, elle ne se peut trouver ou prendre tout d'un coup avec le compas, ni d'un seul centre, mais bien avec plusieurs centres et plusieurs lignes, courbes ou rondes* ... Transcription taken from http://architectura.cesr.univ-tours.fr; translation by the author. See also Sanabria (1984: 203–204).

[11]Of course, full ovals require four centres, while oval arches can be traced using three.

[12]In the first edition, fol. 18r is erroneously numbered as fol. 20r.

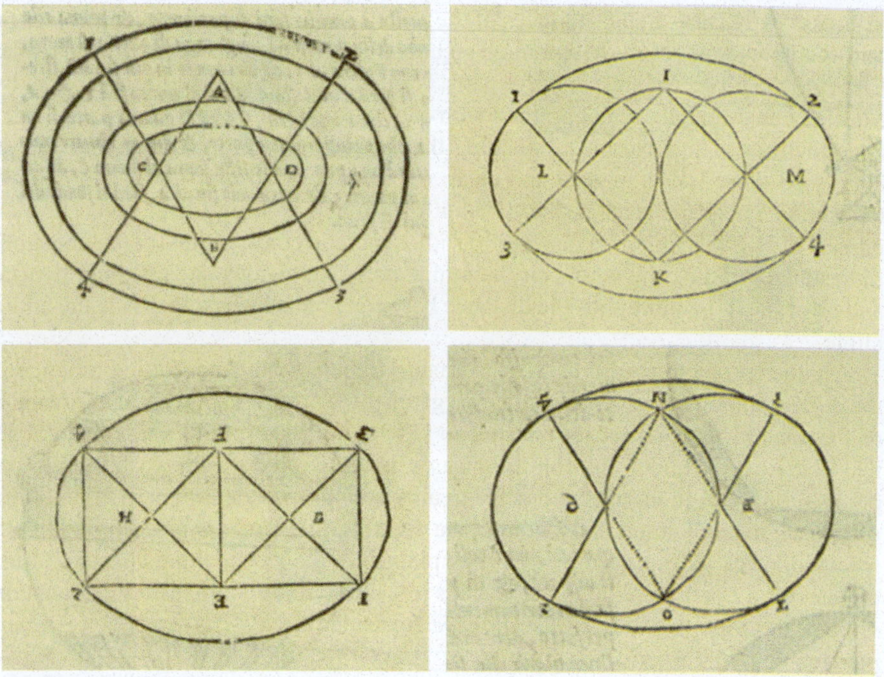

Fig. 3.12 Four methods for oval construction (Serlio 1584: 13v-14r) reprinted from Serlio 1545

adapted to a particular dimension except by trial and error (Gentil 1996). Vandelvira (c. 1585: 18r) used a variation of this approach furnishing three-centre oval arches; again, his variant cannot be used to construct an oval arch of a given span and rise. Another variant is the one followed by Vignola in Sant'Anna dei Palafreneri, starting with a kernel in the shape of four Pythogorean triangles and arriving to a 4:3 ratio between the axes (Duvernoy 2015: 443–444).

Ana López-Mozo (2011) has advanced the hypothesis that the Escorial masters may have known a layout for ovals of any given dimension at the end of the sixteenth century. Other masters use an alternative approach, leading to an approximation of the ellipse. Quoting Serlio, Alonso de Vandelvira states

> if you want to raise or lower the basket handle arch you can use this method, which is included in Sebastiano Serlio's First Book on Geometry … and you should use this rule to build an oval vault according to the available area; joining two arches in this shape you will get an oval figure[13]

[13]Vandelvira (c. 1585:18v): *Si quisieres subir o abajar el arco carpanel lo podrás hacer por esta traza, la cual pone Sebastiano Serlio en su Primero Libro de Geometría … y esta regla se tenga para trazar una capilla oval conforme te pidiese el sitio, porque otro arco de esa otra parte de éste será figura oval*. Transcription by Vandelvira/Barbé 1977. See also Serlio 1545: 13v-14r.

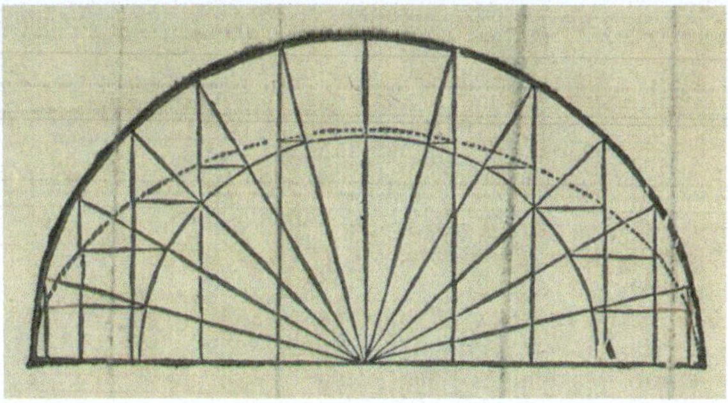

Fig. 3.13 Construction of a semi-ellipse by points (Serlio 1600: 11v) reprinted from Serlio 1545

The construction (Fig. 3.13) is well known; it furnishes the position of several points of an ellipse. The mason or draughtsperson constructs two concentrical semicircles whose diameters equal the larger and lesser axes of the ellipse. Next, he draws a common radius of both semicircles, as well as a horizontal line starting from the intersection of this radius with the lesser semicircle, and a vertical line going down from the meeting point of the radius and the larger circle. A point of an ellipse can be placed where both lines meet. Repeating the construction for a suitable number of radii, Serlio and Vandelvira arrive at a set of points of the resulting curve. However, Serlio states that "a curved line should be drawn; it cannot be made with the compass, it should be drawn with a knowing and practical hand".[14] He seems to be thinking about drawings on paper; however, Vandelvira gives another option, suitable for full-scale tracings: "these are the points that the arc should cross; you should join them with the compass in groups of three, or by hand if the points are close".[15] Later on, Martínez de Aranda generalises the procedure for raised arches: tracing a vertical line from the intersection of each radius with the lesser semicircle and a horizontal one from the larger circumference, he gets points of a different ellipse, raised rather than surbased. Moreover, he does not consider the option of tracing the curve by hand: "where (both lines intersect) you should mark points; next, you should take the points in groups of three and the softened circumference will be formed".[16] This is an instance of the use of the *cherche ralongée* method: points in a complex curve are grouped in sets of three; the centre of the arch joining each group is determined by the *trois points perdus* procedure; finally, a curve made up from a series of circular arcs, passing through all points, is constructed.

All this means, of course, that masons are not thinking about an ellipse, but rather about a transformation of the circle; even in the seventeenth century, the learned

---

[14]Serlio (1545: 13v): *sia tirata una linea curua, la quale non si puo fare col compasso, ma con la discretta, & pratica mano sara tirata.*

[15]Vandelvira (c. 1585: 18v): *aquellos sean los puntos por do ha de ir el arco, los cuales irás adulciendo con el compás de tres en tres puntos o con la mano si fueren los puntos espesos.* Transcription taken from Vandelvira/Barbé 1977.

[16]Martínez de Aranda (c. 1600: 1–2): *adonde tocaren harás unos puntos después cogerás los dichos puntos de tres en tres y quedará formada la circunferencia indulcida G.* Modernised transcription from the autor; a literal version can be found in Calvo (1999: III).

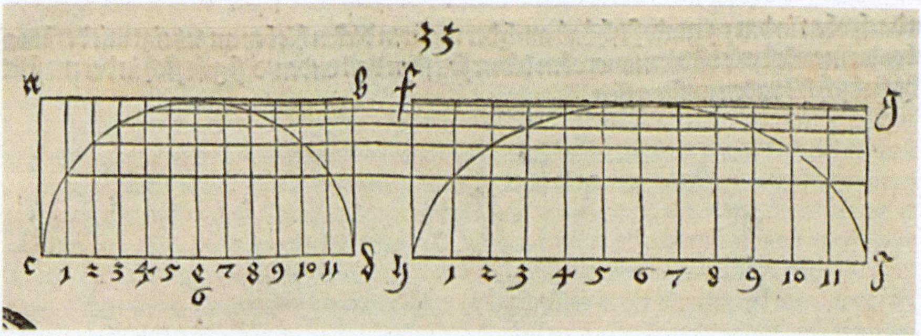

Fig. 3.14 Construction of a semi-ellipse by points (Dürer 1525: C iii v)

Jesuit François Derand (1643: 294–296) includes a short chapter entitled *Du compas à ovale: ou pour former des ellipses* (About the oval compass; or to draw ellipses). The result of Serlio, Vandelvira and Martínez de Aranda's construction is a hybrid figure: it is composed of circular arcs, but the ends of these arcs are placed along an ellipse (for a different outlook on the issue, see Huerta 2007).

Other writers explain alternative constructions for surbased arches: for example, Albrecht Dürer (1525: Ciii v) draws a single semicircle (Fig. 3.14); then, he divides a horizontal diameter into a number of equal parts and constructs verticals through the division points in order to measure the height of their intersections with the circle. Next, he reconstructs the circle using these heights, but the distances between the verticals are multiplied by a given factor. Thus, the circle is subject to a transformation and the result is, of course an ellipse. In any case, he does not mention the method used when actually tracing the curve. It is worthwhile to remark that Dürer ascribed this construction to stonemasons, and he included it just before the well-known constructions of conic sections, where he actually uses the term *Die linie ellipsis* (the line ellipse); I will come back to this issue in Sect. 3.1.3.

Hernán Ruiz II (c. 1560: 41v) includes in his *Libro* a transformation that uses basically Dürer's procedure, although the graphical means are different, since he uses radial lines rather than verticals. At least, this shows a good understanding of Dürer's method; he also masters Dürer's procedure for the determination of conic sections, getting better results than Dürer or his printers (Ruiz c. 1560: 28r-29r). All this makes it clear that Renaissance stonemasons' practical methods were not so distant from Dürer's abstract concerns.

Alternatively, masons could use the well-known "gardener's method" in full-scale tracings. Some discussions about this issue have arisen in the last decades.

The allusions to this technique in Serlio are ambiguous,[17] but Pietro Cattaneo or Ambroise Bachot describe the method clearly.[18]

The *cherche ralongée* method can be used to draw all kind of curves, not only ovals and ellipses. De l'Orme (1567: 55r-55v, 56v, 120v-121r) substitutes a series of circular arcs for helixes, while Martínez de Aranda (c. 1600: 4) does the same for the three-dimensional curve resulting from the intersection of two cylinders. All this explains why de l'Orme connects the "three lost points" with the issue of the *cherche ralongée:* any conceivable curve can be drawn by points, joining them in groups of three. All this implies a deviation from learned geometry: although sophisticated methods are used to find points belonging to an ellipse, helix or lunette, these points are joined simply by means of circular arcs.

We have therefore seen that learned geometry, either theoretical or practical, is all but non-existent in medieval masons' texts; later on, a number of Euclidean constructions find slowly their way into builders' manuals, although they are generalised, particularised or otherwise adapted to fit masons' needs, and almost always deprived of any kind of proof; we will come back to these issues in Sect. 12.4.1.

3.1.3 Spatial Geometry Methods

This section poses a particular problem. I will endeavour to explain spatial geometry methods used by Late Medieval and Early Modern stonemasons for twenty-first-century readers as clearly as possible. The easiest way to do this is through the concepts of descriptive geometry shown in Sect. 1.5, such as projection, development, and rotation. However, such an approach begs a question, both tricky and central in this book: did stonemasons really think in the abstract terms of descriptive geometry? In order to address this issue as objectively as possible, I will explain here the essential traits of stonecutters' methods grouping them under descriptive geometry terms for

[17]Serlio (1545: 13v) … *auenga che molti muratori hanno una certa pratica, che col filo fanno simili volte le quali veramente corrispondono al ochio, & si acorda anchorauo con alcune forme ouali fatte col compasso,* (… many masons in practice make vaults with a string, which at first sight look like oval shapes drawn with the compass). This sentence has been interpreted as a reference to a gardener's ellipse, although Kitao (1974: 71–72, 107, note 124, Fig. 3.46) and Huerta (2007: 230) posit that it describes a particular type of oval construction. There is, however, no substantial evidence for the ellipse or the oval theory; thus, we should consider Cattaneo as the first instance of a clear mention to a gardener's ellipse in architectural or constructive literature.

[18]Cattaneo (1567: 158): *Piglisi con la corda o filo, la distantia, che è dalla intersegatione de le due linee a ciascun capo de la linea A B che tal corda uerrà a esse per la metà di tal linea A B & in tal parte de la corda si fermi un ponto, o si facci un nodo, piantisi dipoi l'altro capo de la corda in ponto C, ouero in ponto D & con tal Corda arcuando si uegga in qual parte di de la linea A B batte il ponto, o nodo di tal corda, che in nostra batte da ogni banda in ponto E & in ciascuno di questi due ponti. E si ficchi un chiodo o polo, dipoi si douerà sempre per regola general addopiare la corda, o filo cuanto gliè la linea A E e questa corda addopiata serà guida di tale ouato, però che arcuando dentro a quella, con un chiodo o altro stiletto, si uerrà facilmente a causar la figura ouale non diminuita.* See also Bachot (1598, page without number at the end).

the sake of clarity. Next, throughout Part II of this book, I will deal with a fair number of examples of such methods applied to particular constructive problems. Finally, in Sect. 12.5 I will come back to the central question, trying to assess the role of stonecutting in the formation of descriptive and projective geometry.

Orthogonal projection. While in planar geometry stonecutters used here and there some classical geometry methods, the picture in spatial geometry is entirely different. Simply put, the fundamental spatial control method of stonecutters, orthogonal projection, is nowhere to be seen in the learned geometry of Antiquity and the Middle Ages. As a result, masons developed several specific methods based on orthographic projection, which evolved gradually during the Late Middle Ages and the Early Modern period, up to the point that Albrecht Dürer (1525: Ciii, Oiii ter r) ascribed double orthographic projection and an antecedent of affine transforms to stonemasons.

As we have seen when dealing with full-scale tracings, orthographic projection is all but absent in Antiquity; of course, the representation of objects lying in the same plane cannot be taken as an evidence of projection. Such artefacts as the *Forma Urbis Romae* or the tracing for the Pantheon pediment do not depict clearly objects in different planes. The well-known plan of Saint Gall is basically a layout scheme representing the axes of the walls of the monastic buildings; when trying to represent objects in different planes, for example in the cloister arcades, the author drew additional elevations (Horn and Born 1979: 241–249; Galtier 2001: 348–349); the result is akin to the *centripetal perspectives* used in Early Medieval miniatures (Galtier 2001: 42–44, 115–116, 348–350).

Some progress in these issues arose in miniatures dating from the eleventh or twelfth centuries, such as those depicting the tower of the Tábara monastery, including a cross-section of some elements, or some miniatures of *In Ezechielem* by Richard of Saint Victor, showing clearly arcades passing in front of battlements (Galtier 2001: 374–389; Cahn 1994: 60–62). It is worthwhile to remark that Richard was a student of Hugh of Saint Victor, that some manuscripts of his commentary on Ezekiel are preserved in the library of Cambrai, and that recent research such as the Villard de Honnecourt edition by Barnes ([c. 1225] 2009: 229–230) posit that Villard was an agent for Cambrai cathedral. However, there is a clear difference between the hesitancy of the Tábara and Ezekiel miniatures and some drawings in the sketchbook of Villard de Honnecourt (c. 1225: 14v, 15r, 31v; see also Ackerman 1997). Both the internal and the external elevations of the nave of Reims cathedral (Fig. 2.1) show clearly objects placed on different vertical planes, namely the aisle windows on the one hand and the high window, triforium and arcades on the other; both planes are set apart by the width of the aisle. Plans, particularly the well-known ones for the cathedral of Meaux and the church designed with Pierre de Corbie (Fig. 3.15), represent different levels as well, from the columns to the vault ribs; in contrast, the depiction of ribs in the plan of Cambrai cathedral is still rather unsure.

These plans, elevations and sections are general architectural drawings, prepared by Villard de Honnecourt, who was not an architect or mason (Villard/Barnes 2009: 229–230). However, Hand IV scribbled a number of diagrams about practical geometry problems on the same sketchbook (Villard c. 1225: 20r, in particular dr. 8 and

Fig. 3.15 Plans of a church designed with Pierre de Corbie and the cathedral of Meaux (Villard c. 1225: 15r)

9), including what seems to be a skew arch and an arch opened in a curved wall. The identification of the problems is not clear, and the analysis of the solutions is problematical. In any case, it seems clear that both schemata deal with spatial problems and the hastily drawn marks in the skew arch probably represent voussoir corners.[19]

In the decades that followed, starting with the Reims palimpsest (Branner 1958) and continuing with a number of elevations of Strasbourg cathedral (Recht 1995: 47, 51; Recht 2014), the competence of architectural draughtspersons increased quickly,

[19]See Branner (1957); Lalbat et al. (1987, 1989); Bechmann ([1991] 1993: 169–180). Barnes (Villard/Barnes 2009: 133) posits that drawing 8 represents a keystone rather than a complete arch. This interpretation is based on a supposed mismatch between the captions of drawings 7 and 8 and involves a dog-tooth joint between the keystone and the underlying voussoirs. This kind of joint is used for example in Theodoric's Mausoleum in Ravenna but is quite uncommon in the West during the Late Middle Ages. In my opinion, Bechmann's ([1991] 1993: 169–180) interpretation, although not free of problems, is more likely. In any case, this issue again shows the problematic interpretation of these diagrams.

reaching virtuosity in the plans and elevations in the collection of the Academy of Fine Arts in Vienna (Böker 2005) or the drawings for Siena cathedral and baptistery (Recht 1995: 57–63; Ascani 1989: 266–272).

In contrast, stonecutting diagrams all the way across Europe, from the German *Bogenaustragungen* (see Sects. 2.1.6 and 2.1.7 for manuscripts and Böker 2005: 227, 237–238, 252, 254, for independent drawings) to the Spanish manuscript of Hernán Ruiz II (c. 1560: 46v) depict vaults in correct horizontal projection, but do not include actual vertical projections of the ensemble of ribs. Rather, masons use *independent* true-shape representations of individual ribs; this enables them to control the curvature of these members, the correct union of tiercerons and liernes at secondary keystones and even the slope of bed joints (Rabasa 1996a; see also Tomlow 2009). In contrast, when dealing with other pieces such as skew arches or rere-arches, Ruiz (c. 1560: 47r, 47v) uses correctly drawn vertical projections as well as plans.

All this brings us to the issue of *double* orthographic projection. Some authors have pointed out a window in Villard de Honnecourt's sketchbook (c. 1225: 10v) featuring the horizontal section of a thin column superimposed on the elevation, as an early example of double projection (Sakarovitch 1990: 70). However, the detail is quite small and, what is worse, inconsistent with the elevation. Remarkably, similar details, such as the profiles of pilasters in some Strasbourg drawings, superimposed in an otherwise "correct" elevation, have been pointed out as a proof of the non-systematic nature of Gothic orthographic drawing (Sakarovitch 1990: 70). Thus, the issue demands a nuanced approach: the small details in plan or profile superimposed onto the Villard or Strasbourg elevations, dating from the thirteenth century, may be understood as a larval stage of double projection, which appears in full-fledged form in the centuries that followed. A remarkable example is a Sienese elevation of a bell tower, perhaps a copy of Giotto's design for the Florence bell tower (Ascani 1989: 266–268). Although the plan is lacking, the oblique sides of the octagonal top storey are rendered in a remarkably correct vertical projection; it is quite difficult to draw the crowning with such precision without taking measurements from a plan. It may be argued that such representation comes from an artist, not from a mason, although the authorship of the drawing is still unclear. However, in the same period an explicit representation in double orthogonal projection, albeit not entirely consistent, can be found in a drawing for another bell tower, this time in Freiburg im Breisgau (Sakarovitch 2005b: 45–46); a few decades later, a drawing for the cathedral of Milan represents the plan and the elevation of the cathedral, superimposed in order to save paper, as in full-scale tracings (Colombier [1953] 1973: 82–83; Sakarovitch 1990: 70–72).

Sources about the particular methods used to construct these early double-projection drawings are scarce. As I have anticipated in Sect. 2.1.2, Roriczer does not use projection lines connecting plan and elevation, as in modern multiview drawings (Fig. 2.2). Rather, he explains how to construct an elevation starting from the plan in these terms:

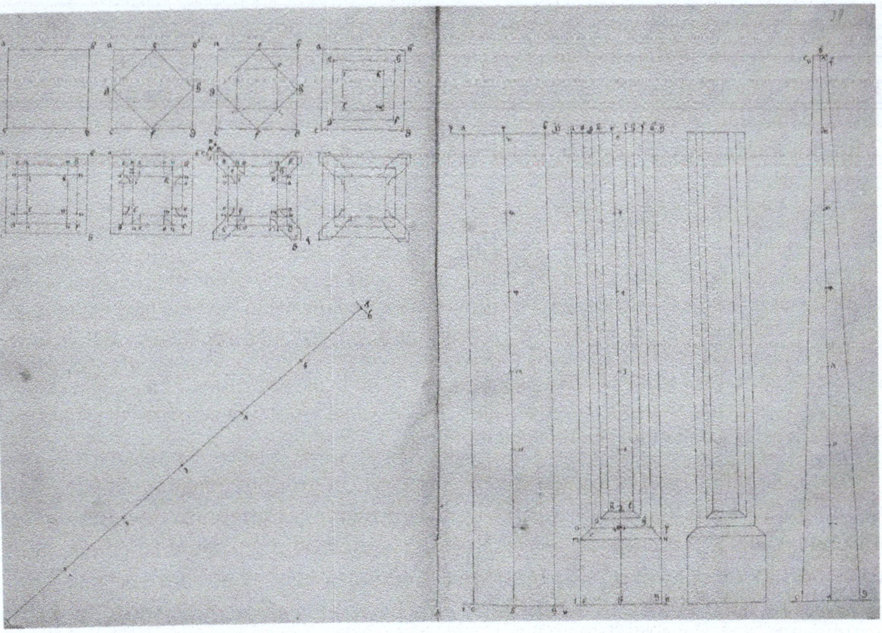

Fig. 3.16 Plan and elevation of a pinnacle (Facht 1593: 38v-39r, after Roriczer c. 1486)

> If you want to extrapolate [from] the base plan of the pinnacle, then draw a long, vertical line … Then draw the nearest figure … Make a [horizontal] line above through *r*, and mark there two letters, *x* [and] *y*. Do the same below through *s* … Then place one [leg of the] the dividers on the centre line [that bisects] line *a-b* of the base figure … and open the dividers out to *a* or *b*. Whatever that distance is, take the same unchanged dividers and set [one leg] at *r* on the centre line of the extrapolation device [With the other leg] make a small point on the line *x-r-y* on the side of *x*, [and] mark an *a* there. Then swing the dividers around toward *y* and score *a-b*.[20]

Thus, Roriczer (1496: 5r-5v) brings the width of the pinnacle to the elevation, placing it along a horizontal line (Fig. 3.16). This guarantees the consistency of horizontal and vertical projections; if we were to draw parallels through *a* and *b*, they would join the corresponding points in plan and elevation, just as projection lines do in descriptive geometry schemes or as twentieth-century architects did when using tracing paper[21] in order to construct an elevation starting from a plan. However, these projection lines are remarkably absent in Roriczer; this leads him to use a procedure that, although tiresome to our eyes, is quite practical when executing full-scale tracings.

The next step in the emergence of projection lines is the well-known letter to Pope Leo X by Raphael Sanzio and Baldassare Castiglione. When explaining how to

[20]Translation by Lon Shelby, taken from Roriczer/Schmuttermayer/Shelby 1977.

[21]Exceptionally, I am using here "trace" meaning "to copy (a design) by drawing over the lines visible through a superimposed sheet of transparent paper".

construct an elevation starting from the plan, Raphael advises the reader to measure frontal distances in the plan and transfer them to the elevation, as in Roriczer. Also, he constructs the axis of the elevation, although it is not used in order to guarantee the consistency of plan and elevation. Rather, two different vertical lines are used with this purpose, probably in order to improve the precision of the horizontals in the elevation:

> After having drawn the plan … you should draw … a line with the length of the base of the entire building; from the middle of this line, you should draw another straight line, making right angles at both sides which will represent the entrance to the building. From both ends of the base line, you should draw two parallel lines, orthogonal to the base line; both lines should have the height of the building … Afterwards, between these two lines … you should bring the measure of the columns, pilasters, windows and other ornaments drawn in the middle of the plan of the building, and from any point in the extremities of the columns or pilasters and windows, or from the ornaments of the windows, you should draw all the lines, always using parallels to the two end lines.[22]

The general procedure recalls that of Roriczer: in order to construct an elevation, the authors advise the reader to start by first drawing a horizontal line as a general base and then a vertical axis. As in Roriczer's method, the draughtsperson should draw horizontals at different heights; marking these heights at both verticals at the sides, he or she can control the tracing of the horizontals more precisely. The axis is still useful when transferring measures from the plan to the elevation, as in Roriczer; however, since windows in a classical composition will usually be aligned, Raphael advises using additional verticals in order to control the placement of windows and ornaments: such verticals suggest projection lines, although their presence is not evident.

This step was taken in the very next years. A detailed tracing in the sacristy of Murcia cathedral (Calvo et al. 2013b) depicts the plan and the cross-section of a sail vault, including bed joints in both projections (Figs. 1.20 and 3.17). These joints are evenly spaced in the cross-section, but their horizontal distance increases as they approach the keystone. It is quite difficult to compute these distances directly from the section, but the analysis of the tracing, as well as its comparison with the built element, shows that the projections were drawn with remarkable precision. All this suggests that projection lines must have been used in the preparation of this drawing. However, only one line that can be taken as a projection line can be seen, in an otherwise well-preserved tracing. Thus, the most likely hypothesis is that projection lines were materialised with a rope or another instrument that did not leave any traces.

---

[22] Sanzio and Castiglione/Di Teodoro (2003: 79–80): *Designato che si ha la pianta … devesi tirare … una linea della larghezza delle basi de tutto lo edificio e, dal punto di meggio di questa linea, tirarai un' altra linea dritta, la quale faccia da l'un canto e da l'altro dui anguli retti: e questa sia la linea della intrata dello edificio. Dalle due estremitati di la linea della larghezza tiraranssi due linee parallele perpendiculari sopra la linea della base, e queste due linee siano alte quanto ha da esser lo edificio … Dippoi, tra queste due estreme linee … se pigli la misura de le colonne, pilastri, finestre et altri ornamenti dessignate nella metà della pianta de tutto lo edificio dinanti e, da ciaschun punto de le estremitati delle collonne o pilastri e vani, overo ornamenti de finestre, farai il tutto, sempre tirando linee paralelle di quale due estreme.* Transcription taken from Sanzio/Castiglione/Di Teodoro (2003); translation by the author. See also (2003: 142–143).

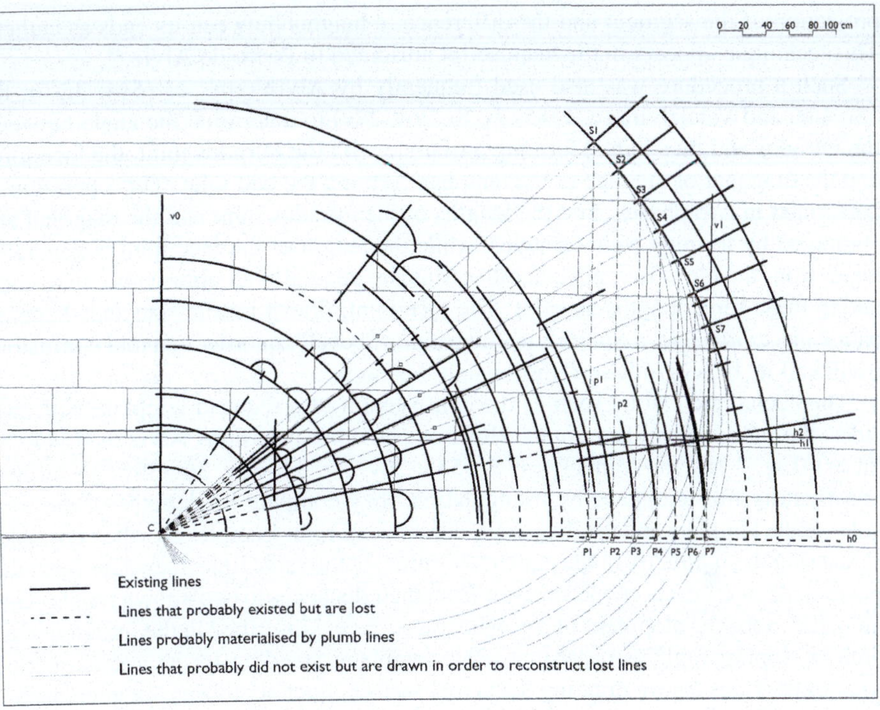

Fig. 3.17 Tracing for the vault over the sacristy of Murcia cathedral (Survey by Miguel Ángel Alonso, Pau Natividad and the author)

In the same year he Murcia vault was completed, 1525, Albrecht Dürer published his *Underweyssung der Messung*, a remarkable compilation of all kinds of geometrical problems solved by graphical means. As we saw in Sect. 3.1.2, double orthographic projection appears twice: in connection with conic sections and with the shadowed perspective of a cube (Dürer 1525: Ciiii r, Ciiii v, Ciiii bis v, Ciiii ter r, Oiii ter v, Piii). In the second passage, Dürer (1525: Oiii ter r) ascribes it to stonemasons; he also mentions masons when explaining how to locate the points of an ellipse using what is nowadays known as an affine transform, in a passage (Dürer 1525: Ciii) placed just before the pages on conic sections, as we have seen in Sect. 3.1.2. Projection lines, explicitly connecting both views, appear only in the preparatory drawings for the perspective of the cube. However, from this moment on, projection lines are ubiquitous in stonecutting treatises and manuscripts.[23]

Segment lengths and triangulations. Several specific methods allow masons to determine the true size and shape of voussoir edges, faces and angles. The length of a segment may be determined by forming a right triangle with the horizontal

23 Alviz (c. 1544: 2r, 3r, 6r); Ruiz (c. 1560: 47v, 147r-152r); de l'Orme (1567: 72r, 74 r, 79v); Chéreau (c. 1567–1574: 109r, 112r, 113r); Vandelvira (c. 1585: 8r, 10v, 11v); Rojas (1598: 98v-100v); Martínez de Aranda (c. 1600: 3, 7, 9).

projection of the segment and the difference in heights between its ends as catheti. This technique appears in the manuscript attributed to Pedro de Alviz.[24]

Such a procedure was also used frequently by Alviz (attr., c. 1544: 3r, 7r, 8r, 13r, 14r) and Vandelvira (c. 1585: 7v, 8r, 26v, 27v) to determine the angle between the intrados and face joints, forming a triangle with the intrados joint, the face joint and the diagonal of the face of the template forming the bed joint. This operation is performed in two phases: first the lengths of the intrados joint and the diagonal are computed by forming right triangles, while the length of the face joint is generally taken from the elevation; next, another triangle, generally an obtuse one, is formed taking these three lengths as sides. The face joint, drawn in true size and shape, is called the *saltarregla*, a Spanish word meaning "bevel"; in order to avoid confusion, I will call it "bevel guideline" throughout this book.

This technique can be seen as the construction of a reduced template with two sides, the intrados and face joints. However, on other occasions Alviz (c. 1544: 3r, 7r, 8r) (Fig. 3.18) and Vandelvira (c. 1585: 26v-28r, 31r-35v, 36v-39v) go further, constructing full templates by triangulation: they compute the lengths of the four sides of the template and the diagonals, either using right triangles or, when possible, measuring them directly in the plan or elevation; at the same time, they determine the position of each corner of the voussoir from their distances to two known points. This amounts to the development of a polyhedral surface[25] inscribed in the intrados of an arch or sloping vault, in order to prepare templates standing for an approximation to intrados faces *before* dressing the actual surface. Such templates are quite useful, since they represent intrados joints and the four corners of the intrados face; once the planar surface determined by the four corners has been materialised, it is quite easy to carve the final surface, hollowing the intrados face. Other writers apply this technique to trumpet squinches, such as the famous one at Anet (de l'Orme 1567: 92r-99v) and rere-arches (Martínez de Aranda c. 1600: 116–118, 121–123, 128–136).

Auxiliary views. This procedure is used in the manuscript attributed to Pedro de Alviz (c. 1544: 9r, 11r, 21r, 22r), Hernán Ruiz II (c. 1560: 47v), de l'Orme (1567: 127v), Chéreau (c. 1567–1574: 104v) and Vandelvira (c. 1585: 10v, 11v, etc.). However, sixteenth-century authors limit themselves to auxiliary views based on vertical projection planes (Fig. 3.19). As we have seen, Desargues (1640) published a leaflet putting forward a radical revision of most stonecutting methods, using slanted planes as a reference; however, these techniques were not adopted in actual practice (Sakarovitch 1994b; Sakarovitch 2009a). Later, auxiliary views based on sloping planes were proposed by Théodore Olivier (1843–44: I, 18–22) as an abstract descriptive geometry method; it is quite interesting to know that de la Gournerie (1860:

[24] Alviz (c. 1544: 3, 6, 7, 8, etc.), de L'Orme (1567: 92–96); Vandelvira (c. 1585: 7v, 8r, 8v, 9v, 26v, 27v, etc.). See also Palacios ([1990] 2003: 30–37, 96–105) and García-Baño (2017:109–112, 157–161).

[25] The term "polyhedral surface" is used here and throughout the book in its broadest sense, meaning "a surface composed of planar polygons, regular or irregular". Such a definition does not require the surface to be closed; in fact, all instances of these surfaces used in stereotomy are open, and thus do not enclose a polyhedron. This may seem counterintuitive, but still this use of the term "polyhedral surface" is etymologically correct and consistent with other terms such as "dihedral" or "trihedral".

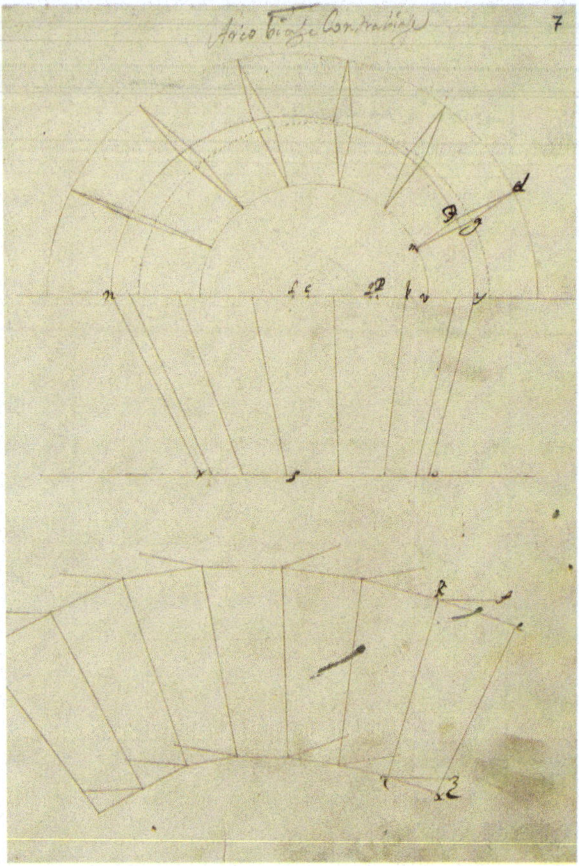

Fig. 3.18 Splayed arch with intrados templates and bevel guidelines constructed by triangulation (Alviz, attr. c. 1544: 7r)

vi) attacked the idea, stressing that the sensible, practical art of stonecutting had eschewed such changes of the horizontal projection plane (Rabasa 2011: 717–718).

It remains open to discussion whether auxiliary views in de l'Orme, Vandelvira and Martínez de Aranda imply real changes of projection plane. Usually, sixteenth-century masons chose as the main vertical projection the most complex view of a given piece; thus, auxiliary views usually depict figures in a single plane and, strictly speaking, are not changes of projection. In contrast, Derand (1643: 49, 63, 67, 95, 223, 343) made lavish use of this technique, depicting even complex figures laid out in non-developable surfaces by means of real changes of projection; I will come back to this issue in Sect. 12.5.2.

Measuring angles. As we have seen, Alviz and Vandelvira measure the angle between intrados and face joints by triangulation. Martínez de Aranda uses the same method on rere-arches; however, on other occasions, (c. 1600: 11–12, 15–16, 40–41) he computes the angle between the intrados and face joints in some arches using a simple construction (Fig. 3.20, right). He draws in plan an orthogonal to both faces of the arch through the front end of the intrados joint. Next, he measures the apparent

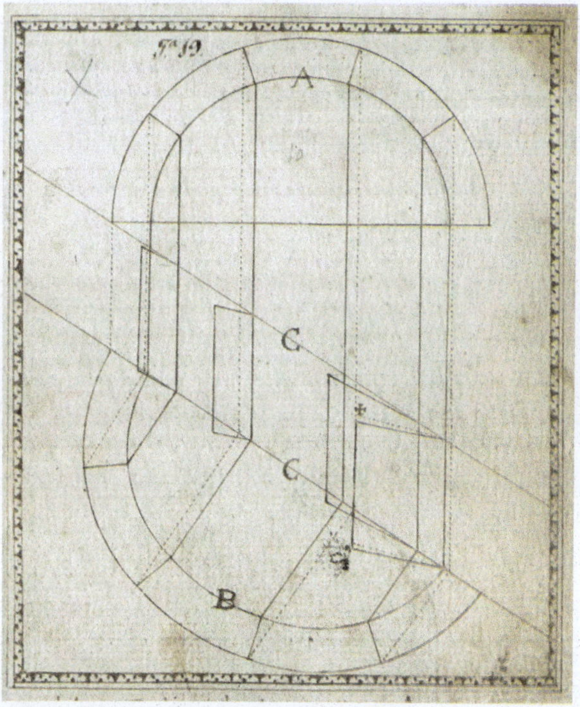

Fig. 3.19 Skew arch with a rhomboidal plan, showing a cross-section and an auxiliary view (Martínez de Aranda c. 1600: 9)

distance between both ends of the intrados joint in the elevation and transfers it to the plan, starting from the intersection of the orthogonal with the back face, in order to locate the rear end of the intrados joint. Then, he draws a true-shape representation of the intrados joint connecting both ends; the angle between the intrados joint and each of the face joints may be measured from the true-shape representation and the corresponding arch face. It is easy to prove that the construction is exact if we assimilate it to a revolution around the orthogonal, which represents a line in point view. The front end of the intrados joint, being on the rotation axis, does not move; the back end moves on a plane orthogonal to the axis, that is, the back-face plane, maintaining its distance to the axis, which can be measured in the elevation. Since the face joint and the vertical projection of the intrados joint are drawn so that their extensions pass through the centre of the arch, the face joint is coplanar with the intrados joint and the line in point view; this assures that it will reach the horizontal plane at the same time as the intrados joint.

Did stonemasons really think in such abstract terms? Martínez de Aranda's solution is an exception. De l'Orme uses a similar procedure (1567: 70v) in an assembly of two skew arches (Fig. 3.21); however, he groups the lines that make it possible to measure angles between the intrados and face joints at the springers, rather than

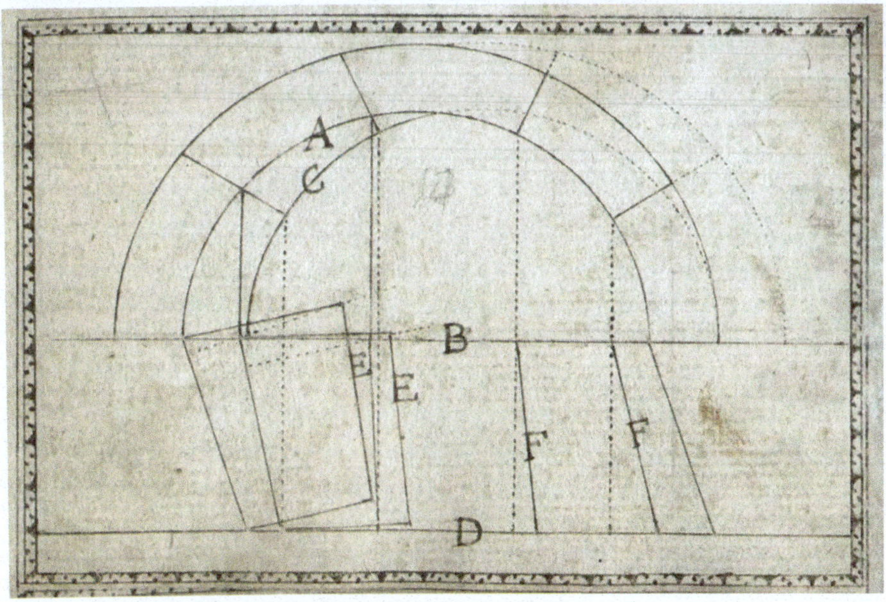

Fig. 3.20 Skew arch solved by templates, at the left; by squaring, using bevel guidelines, at the right (Martínez de Aranda c. 1600: 15)

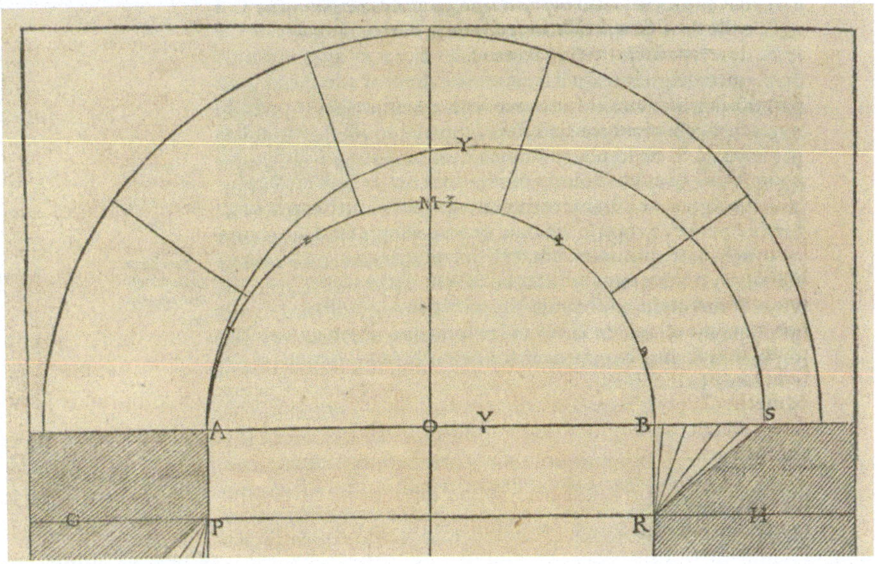

Fig. 3.21 Double ox horn, detail (de l'Orme 1567: 70v)

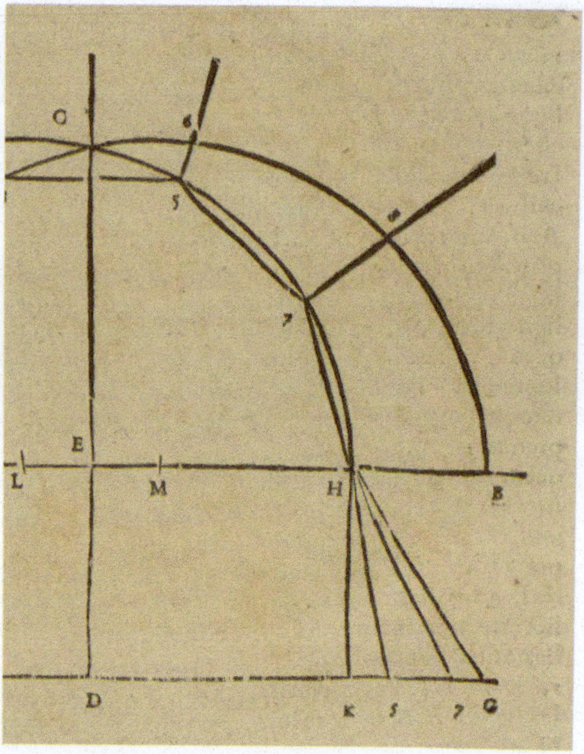

Fig. 3.22 Skew arch with rhomboidal plan and orthogonal bed joints, detail; intrados joints are drawn at the springers (Jousse 1642: 14)

placing them at the horizontal projection of the actual intrados joint. This solution appears in Guardia (c. 1600: 70v), Jousse (1642: 14–17) (Fig. 3.22) and Derand (1643: 124–126); it may even have been used by Hand IV (Honnecourt c. 1225: 20r, dr. 9; see Lalbat et al. 1987, Lalbat et al. 1989) although the interpretation of such schematic diagrams remains problematic. Thus, it seems that the procedure originated simply as a true-shape construction, unconnected to the idea of rotation; later Martínez de Aranda placed it under the intrados joint for didactic reasons.

Drawing true shapes. Similar techniques are used by Martínez de Aranda (c. 1600: 41–42) in order to construct full templates when both faces of the piece are mutually parallel and orthogonal to the springings. In contrast, when intrados joints are parallel to the springings and oblique to the faces (Fig. 3.23), masons revert to a slightly different technique, using projection planes orthogonal to the intrados joints.[26] In order to construct true-shape depictions of the quadrilateral joining the four corners of a voussoir, they maintain the lower intrados joint of a voussoir in their original position. Next, Vandelvira (c. 1585: 19v, 20v) draws an orthogonal to this joint through the horizontal projection of one or both ends of the upper intrados joint. Since both joints are lines in point view and thus orthogonal to the

[26] Alviz (c. 1544, 9r, 11r); de L'Orme (1567: 72r, 74r); Vandelvira (c. 1585: 19r, 19v, 20v); Martínez de Aranda (c. 1600: 9, 10, Fig. 3.19); in contrast, the technique is absent in Ruiz (c. 1560).

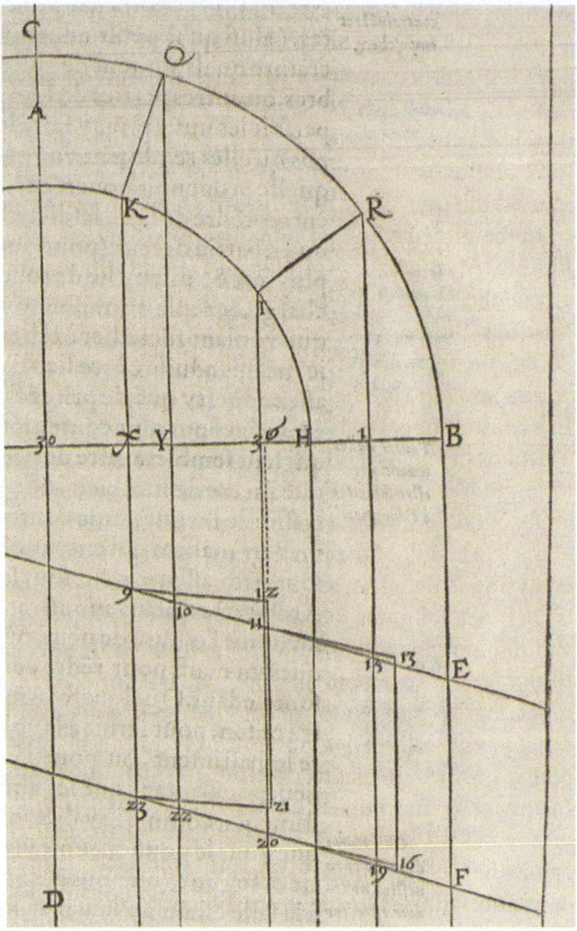

Fig. 3.23 Skew arch with rhomboidal plan and circular cross-section, detail (de l'Orme 1567: 72r)

projection plane, their distance may be measured directly from the elevation and transferred to the orthogonal in order to place the true-shape representation of the upper joint. In addition to allowing the computation of the distance between intrados joints, the orthogonals mark the position of the ends of the upper joint. It may be proved that the operation is exact by arguing that the whole template for the intrados face of the voussoir rotates around the lower intrados joint, which does not move in the revolution; the ends of the upper intrados joint move along circumferences placed on planes orthogonal to the lower intrados joint; since the intrados joint is horizontal, these planes are vertical and thus are depicted as straight lines—in particular orthogonals to the lower intrados joints—and that the upper intrados joint preserves its parallelism and its distance to the lower one in the revolution.

However, a crucial question remains: did Vandelvira think in these abstract terms? Although orthogonals are explicitly represented in the drawings, Vandelvira does not mention them in the text. De l'Orme (1567: 71r-72r) (Fig. 3.23), other authors dealing with the same problems (Martínez de Aranda c. 1600: 6–11, 71–73; Jousse 1642: 10–11, 20–23; Derand 1643: 153–161, 167–172) or even Vandelvira himself (c. 1585: 19r) use a procedure that Martínez de Aranda calls *galgar*, from *galga,* meaning "gauge", as we saw in Sect. 3.1.2. Generally speaking, these authors transfer the distance between the intrados joints to a reference line and draw a parallel to the lower intrados joint from this point, furnishing the position of the upper intrados joint in the true-shape diagram; next, they transfer the distance to the reference line from the projected upper intrados joint to its true-shape depiction. For economy, de l'Orme, Martínez de Aranda and Derand use the springing line in the elevation as a reference line, but any orthogonal to the intrados joint would perform the same function. Such a procedure recalls a diagram by Hand IV (Villard/Barnes [c. 1225] 2009: 20r, drawing 8) representing an arch in a round wall. A singular templet is used, with a straightedge laid tangentially to the wall and *two* arcs in the opposite edge. Everything suggests that the templet was used as a reference to compute the relative positions of voussoir corners. All these methods lead to the same result as orthogonals to intrados joints, but the antecedent of the templet in Villard's sketchbook and the connections with the gauge, used mainly during the dressing process, hint at an empirical origin of the procedure. I will come back to this issue in Sect. 12.5.2, after analysing in detail several instances of the use of these methods throughout Chap. 4.

In other solutions to the problem of the skew arch, faces act as projection planes and the intrados joints are oblique to them. In this case, Vandelvira (c. 1585: 27v) resorts to the development of polyhedral surfaces by triangulation, a tiresome technique that is prone to the accumulation of errors. This suggests that Vandelvira did not know an alternative method used by Martínez de Aranda (c. 1600: 16–17) and perhaps also by Cristóbal de Rojas (1598: 99v) for skew arches with round faces. In this case, the projection plane is parallel to the faces of the arch, rather than orthogonal to the intrados joints. As a result, the distance between *intrados joints* cannot be measured in the elevation, but the distance between *voussoir corners* can be determined directly. Since voussoir corners rotate in a plane orthogonal to the axis of revolution—specifically the lower intrados joint—the mason can draw an orthogonal to this joint starting from the original position of the upper corner of the voussoir; this will represent the trajectory of the top corner during a revolution. It is worthwhile to remark again that Martínez de Aranda actually draws the orthogonal to the intrados joint, in contrast to other diagrams in his manuscript, but he explains the operation as *galgar,* that is, to draw a parallel to the springings line in a cross-section, which is used as a *juzgo* or auxiliary construction, as we have seen in Sect. 3.1.2. In the next step, the actual distance between two corners of the voussoir, placed at the lower and upper joints, may be taken easily from the elevation; the mason can draw an arc with this radius and its centre at the lower corner; the intersection of this arc with the orthogonal starting from the projection of the upper corner will give the position of this corner in the true-shape template. In other words, Martínez de Aranda, and perhaps Rojas, are generalising the use of measurements from a reference plane to the case where the intrados joints are oblique to the vertical projection plane. Although the procedure is quite economical, nothing suggests that either de l'Orme (1567:

67v-69r) or Vandelvira knew it; the former eschews the issue altogether, while the latter uses a tiresome and unreliable method based on triangulation.

Orthogonals to intrados joints or measurements from a reference plane are used in another technique for the computation of the angle between face and intrados joints. When dealing with skew arches or rere-arches, Chéreau (c. 1567–1574: 113r), Guardia (c. 1600: 71v), Jousse (1642: 10–11, 18–33) and Tornés ([c. 1700] 2015: 49v, 50v) transfer the horizontal projection of the lower end of face joints to the springings, and then draw the joint as a line passing through this point and the intersection of the face plane with the arch axis. The operation can be understood as a rotation around the arch axis. The low end of the face joint moves on a plane orthogonal to the axis, but this is not enough to determine its final position; another constraint is needed. Thus, the joint is virtually extended until it reaches the axis; this point does not move, of course, and thus provides another point which makes it possible to trace the joint. Once again, it is not clear whether these authors were thinking in such abstract terms; quite significantly, most of them do not try to find the upper end of the face joint, since their primary goal is to determine the angle between the face and intrados joints.

Hernán Ruiz II (c. 1560: 46v) depicts the diagonal ribs and tiercerons in a star vault in true shape, to avoid deformation by projection. This technique may be understood at first glance as a rotation around a vertical axis, bringing diagonals and tiercerons to the plane of the outer arches. However, nothing suggests that the author was thinking in these terms; in fact, de l'Orme (1567: 108v) presents diagonal ribs and tiercerons without any connection with the plan, as we will see in Sect. 10.1.4, while actual stonecutting tracings, such as the one in Tui cathedral (Taín et al. 2012), suggest this was the usual practice up to the eighteenth century. The idea of ribs revolving around a vertical axis does not appear explicitly until Frézier (1737–39: III, pl. 71); he includes in his diagram arcs suggesting the rotation of diagonals and tiercerons around a vertical line passing through the springings.[27]

Cylindrical and spherical developments. As we have seen, many authors use developments of polyhedral surfaces in order to control the execution of arches or sloping vaults; once a portion of this surface is materialised in an individual voussoir as a planar shape, masons continued to carve without using templates until they reached a cylindrical surface. In contrast, Mathurin Jousse (1642: 46) and Derand (1643: 171, 173, 177), dress cylindrical surfaces directly, using an approximation to a cylindrical development by dividing into two parts the portion of a circle that belongs to a voussoir and using both chords; this procedure seems to be implied in de l'Orme (1567: 73v-77r) and Vandelvira (c. 1585: 21v-22v).

Developments of spherical surfaces appear in the literature with de l'Orme (1567: 111v-115r); however, they are anticipated in actual practice by a tracing in the rooftops of Seville Cathedral dating from 1543 or 1544 (Ruiz de la Rosa and Rodríguez 2002). De L'Orme, the Seville masters and many others inscribe a set of

[27]Vasari (1568: I, 269) remarks that Paolo Uccello *Trovò similmente il modo di girare le crociere e gli archi delle volte* (he likewise found the way to rotate the ribs and the arches of the vaults); however, the meaning of this phrase remains unclear.

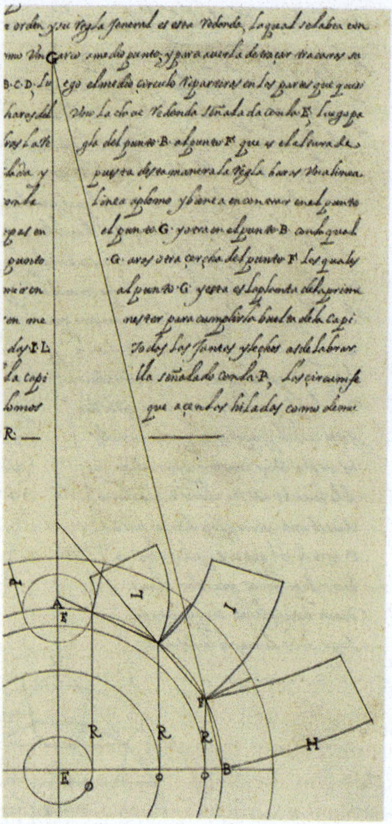

Fig. 3.24 Hemispherical dome, detail ([Vandelvira c. 1585: 61r] Vandelvira/Goiti 1646: 118)

cones in the intrados of a spherical vault and develop it using the well-known procedure that was taught at elementary schools up to a few decades ago (Fig. 3.24).[28] As we will see in Sect. 12.4.3, this method may have been taken from cartography or cosmography. Of course, a spherical surface cannot be developed exactly. Such conical templates do not coincide exactly with the spherical surface, but actual replication of the procedure by Enrique Rabasa (2003: 1681; see also Rabasa 2000:

[28]As stated in Sect. 2.2.3, there are two manuscripts of Alonso de Vandelvira's *Libro de Trazas de Cortes de Piedras*, one in the School of Architecture of Technical University of Madrid and another one in the Spanish National Library in Madrid. Text quotations refer to the manuscript in the School of Architecture, as in the standard edition of Geneviève Barbé (Vandelvira/Barbé 1977). Figures are generally taken from the manuscript in the National Library, which is usually neater. However, some drawings are lacking in the National Library manuscript, while in some cases the available reproductions are clearer in the School of Architecture manuscript. In order to enable the reader to connect the figures with the text, figures taken from the National Library manuscript are referred to those in the School of Architecture manuscript. Thus, the phrase in the caption ([Vandelvira c. 1585: 61r] Vandelvira/Goiti 1646: 118) should be interpreted as follows: "Image from the copy by Goiti, in the National Library, page 118; corresponds with drawing in the School of Architecture, folio 61 recto".

174–175; Rabasa 2007a: 22–23) has shown that the difference is negligible, even in small vaults. The method poses another problem: in order to compute the length of the template for a voussoir of given length, it is necessary to rectify the circumference.[29] However, sixteenth-century masons did not use π; in fact, numerical dimensions are seldom used in stonecutting. A copy of the Vandelvira manuscript interpolates a striking remark about this issue: "you will close both arcs [which represent developed directrices of the cone] at will".[30] In many cases, such a light-hearted approach to the issue was practical, since stone usually arrived from the quarry in different-sized blocks. Only when precise control of the length of the voussoirs was necessary, did the mason have to resort to other methods, such as the direct determination of the distance between two corners along a horizontal joint in plan.

Intersections of cylinders and projections of circles. In Sect. 3.1.2, we have seen a number of two-dimensional procedures providing points of ellipses. In addition, many treatises and manuscripts present several methods addressing three-dimensional problems that give ellipses as a result, such as intersections of cylinders with oblique planes and projections of circumferences lying on sloping planes. Up to the eighteenth century, these problems are never treated as abstract issues, and the word "ellipse" seldom appears explicitly. Quite to the contrary, the problem is solved using joints in the masonry as auxiliary lines. As with many other issues, the technique appears in the manuscript attributed to Alviz (c. 1544: 21r, 22r, 23r; see also Ruiz c. 1550: 47v; de l'Orme 1567: 127r-128v; Vandelvira c. 1585: 23v; Martínez de Aranda c. 1600: 3, 7, 9); the problem is addressed using the intersecting plane as a projection plane and constructing an auxiliary view.

The same technique is used when computing the projection of a circle on an oblique plane (Martínez de Aranda c. 1600: 72, 76) or, inversely, when a circumference lying in a sloping or oblique plane is projected onto a horizontal or frontal plane (Vandelvira c. 1585: 70r; Alviz c. 1544: 11r; see García Baño 2017: 210–220). In these cases, projectors are drawn from all corners of the voussoirs until they meet the horizontal or frontal plane, furnishing the coordinates of the projection (Fig. 3.25). In fact, authors do not differentiate clearly between the two operations; they may even combine both methods: a circle is projected into an oblique plane, generating at the same time a cylinder that intersects another oblique plane (Martínez de Aranda c. 1600: 9, 10); in extreme cases, it is not clear whether a diagram represents a projection or an intersection (Martínez de Aranda c. 1600: 3).

---

[29] This problem involves computing the length of the circumference of a circle. Thus, it is different from the problem of the squaring of the circle, although both involve π.

[30] Vandelvira/Goiti ([c. 1585] 1646: 118): *las quales dos çerchas cerraras por do quisieres*. This phrase is not included in the manuscript Raros 31 conserved in the Library of the School of Architecture of the Polytechnic University of Madrid.

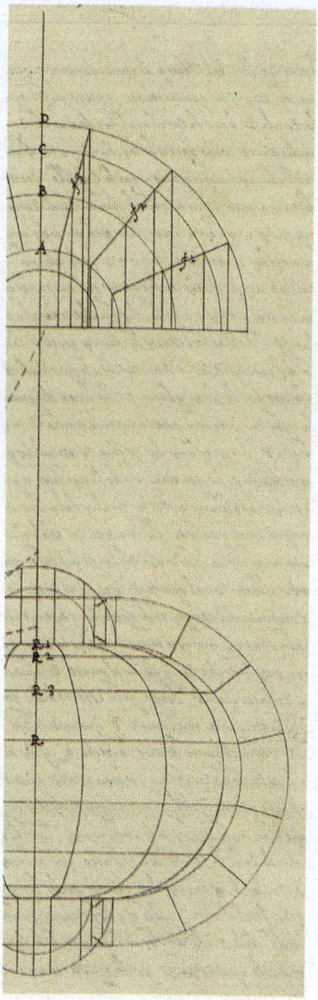

Fig. 3.25 Horizontal-axis annular vault, detail ([Vandelvira c. 1585: 70r] Vandelvira/Goiti 1646: 126)

3.1.4 Models

Architectural models. From de l'Orme's *Le premier tome de l'Architecture* (1567: 21v-24v, 76v), the first printed treatise on the subject, to Derand's *L'architecture des voûtes* (1643; 449) stonecutting treatises and manuscripts mention or depict reduced-scale models (see also Vandelvira c. 1585: 22r; Rojas 1598: 80v, 88v-89r; Martínez de Aranda c. 1600: i-iii; San Nicolás 1639: 69r, 70v, 91r, 95v) (Fig. 3.26). Further, Spanish archival documents attest to the use of models in actual construction or stonecutters' instruction. In particular, some of the Spanish sources make it clear that, in addition to full-scale tracings, stonecutters used frequently reduced models to verify the validity of their tracings. This is quite striking to our ears, since it amounts

Fig. 3.26 Stereotomic models (Derand [1643] 1743: pl. 205)

to no less than an empirical approach to geometry; we will come back to this matter in Sect. 12.2.

In any case, before dealing specifically with stereotomic models, it is worthwhile to comment on a few key facts about Early Modern architectural models. First, documentary evidence showing the use of three-dimensional models during the sixteenth century in Spanish cathedrals, royal palaces, hospitals, fortifications, granaries, gardens and cranes is almost infinite.[31]

Pierre du Colombier and Franz Bischoff have stressed that there is no evidence for the use of models as project or construction tools in the Gothic period outside Italy.[32] In contrast, John Fitchen ([1961] 1981: 5–6) or Lon Shelby (1964), while accepting the lack of documentary evidence for medieval architectural models outside Italy, still take it for granted the existence of actual models. In any case, a small detail hints that architectural models were an innovation in sixteenth-century Spain. The Count of Tendilla, a central figure in the introduction of the Renaissance in Spanish architecture, told Cardinal Cisneros in 1509 that his monastery of San Antonio de Mondéjar was so small that it "seems to be made as a *modelo* (as they say in Italy) of a bigger one";[33] this suggests that the Spanish word is of Italian origin and was

[31] See Marías (1989: 505); Gómez-Moreno (1941: 60, 75, 185–186); Rosenthal (1961: 17–18, 178–181); Rosenthal (1985: 25, 42–43); Wilkinson (1977a: 107, 108, 118, 123, 128, 138); Marías (1983–86: II, 6; III, 253–254, 256, 258, 266–267, 280; IV, 62, 66); Kubler (1982: 74, 77–78); Bustamante (1994: 26–27, 43–44, 112, 118, 136, 147, 149, 188, 443, 648); Cámara (1998: 102–103, 132, 134); García Tapia (1990: 182–183); Zaragozá (1997: 29), to cite just a few.

[32] Colombier ([1953] 1973: 95–96); Bischoff (1989: 287–295) See also Kostof ([1977] 1986: 74–75); Ettlinger ([1977] 1986: 109); Millon (1997: 19); Erlande-Brandenburg (1993: 72–73).

[33] Gómez-Moreno (1925: 24): *no paresce syno que se hizo para modelo (como dizen en Italia) de otro mayor.*

not in frequent use in Spain at the beginning of the sixteenth century. The use of models in Valencian Late Gothic construction—for example in the funerary chapel of Alphonse V in Valencia—may be explained by the ties of the city and the patron with Italy.

The choice of materials also argues for the Italian origin of this technique. Spanish models in this period were usually made in plaster, wood and clay. Plaster derives probably from Italian sources, since earlier Italian models, such as the one built by Francesco Talenti and Andrea Orcagna in 1357 for the columns of Santa Maria del Fiore in Florence, were made in this material (Colombier [1953] 1973: 96). However, gypsum was used in Spain for models much longer than in Italy or France. The evidence for the use of this material in Italian architectural models after 1400 is scant; de l'Orme recommends the use of wood, paper, or cardboard for models (1567: 22v). Thus, earlier Spanish plaster models may have derived from Italian sources, but the longevity of the technique seems be justified by the availability of gypsum in Spain and the use of models in stereotomic testing, as we will see later.

Models were usually large. For example, the one for the completion of the bell tower of Seville Cathedral was as high as a person, roughly on a scale of 1:50. In the Escorial the model of the monastery was stored in the attic, to free some rooms so that a mason or carpenter of some rank could live in them (Morales 1993: 342–343; Morales 1996: 24–25; Bustamante 1994: 188). Wood models, such as the one for the lantern of the dome of Santa Maria del Fiore in Florence made by Brunelleschi, frequently included movable pieces. In a model of Valladolid cathedral, façades, towers, and stairs are detachable, and the whole model can be split in half to allow viewers to experience the interior as well as the exterior of the building. In this way, large models could be carried to distant places. Juan Bautista de Toledo prepared in Madrid a general wooden model of the project or *traza universal* for the Escorial; when it was needed, the model was dismantled, packed, and sent to the building site (Vasari 1568: I, 317; Carazo 1994: 96–97; Bustamante 1994: 26–27, 44). Such practice may have fostered the use of stereotomic models with independent voussoirs.

Renaissance models played a number of different roles (Cabezas 1992: 5, 8). First, models were used by the architect in the conception of the project, as a well-known passage from Alberti attests ([1485] 1991: 33–34). Second, they were frequently submitted to the patron to obtain approval. In 1544, Alonso de Berruguete made two models of the church of the Hospital of Tavera, probably from designs by Alonso de Covarrubias, one of them on a round plan and the second one on a square one. The patrons had still not chosen a plan for the church and the models were built to help them make a decision (Wilkinson 1977a: 107–108, 138; Marías 1983–1986: III, 253–254, 280). Third, models were used to convey design intent from the architect to the builders. In France, a carefully made model is known still in our days as a *chef d'oeuvre*, which may be literally translated as a "foreman of the works". This makes it clear that on some occasions Renaissance or Baroque models played the role of modern-day working drawings (Potié 1996: 57–59; Truant 1995: 49; Lawrence 2003:

1272–73).[34] In the early phases of the construction of the Escorial complex, Philip II stressed that the execution of the general model of the Escorial should not begin until he had approved the drawings of the complex. That is, in this case the model was not used to secure approval for the work, which was granted with the drawings, but rather to transmit orders to the executors: Juan Bautista de Toledo was required to bring the model to the construction site "since it will be necessary to consult the model every hour for doors and windows and other things".[35] In other words, models mirror drawings in their design, presentation and execution roles.

Stereotomic models. There is also substantial evidence for other technical uses of models in sixteenth-century Spain, in particular for the preparation of templates and the determination of angles between edges of voussoirs. Many treatises, manuscripts and archival documents confirm the use of models to test these geometrical procedures. At first glance, these methods may startle a modern engineer, architect, or mathematician: they amount to an empirical approach to geometry. However, a closer look at these Spanish models reveals that there is more sense in these methods that may seem at first sight.

Francisco de Goycoa, master mason, signed a contract in Cuenca in 1568 for the training of Bartolomé de Anchia; he agreed to teach his pupil to

> make models for different kinds of building [elements] in small volumes cutting plaster and other materials, teaching him the arch squares and stonecutting methods needed to construct any classical or Gothic vault, and also trumpet squinches and rere-arches[36]

This short passage offers useful information about several aspects of models. First, Goycoa was using plaster for stereotomic models in the late sixteenth century; in this period wood prevails in architectural models in the Escorial. Second, taking also into account a drawing by Derand (1643: 449), it is safe to surmise that the model-making procedure reproduces the stonecutting technique in plaster (Fig. 3.26), starting from a block to arrive at an individual model of a voussoir, cutting each of its faces at the right angle with the help of bevels. We may also assume that in a second phase these voussoirs were joined together to verify that the pieces of the arch or vault fitted together. Such models recall those for the Florence lantern or the large ones in the Escorial or Valladolid cathedral, made of a number of movable pieces. However, the main point of interest of this document is the stress on the didactical use of models:

---

[34]The first meaning of *chef d'oeuvre* in Académie Française (1986) is *Ouvrage probatoire qu'exécutaient les ouvriers dans la corporation où ils voulaient passer maîtres* (test piece executed by workers in the guild where they want to be accepted as masters). Of course, its most frequent meaning in these days is *Ouvrage parfait, très beau* (perfect, very beautiful work; masterwork; masterpiece), the second meaning of the word in the same dictionary.

[35]Bustamante (1994: 26): *para lo que tocare a puertas y ventanas y otras cosas que será menester mirar en él cada hora.* See also Cabezas (1992: 5).

[36]Marías (1989: 455–456): *contrahacer diferentes trazas de edificios en volumenes pequeños cortando en yeso y en otras cosas enseñandole los baiveles y cortes de piedras para cerrar cualquier buelta, así de capillas romanas como de obra francesa y pechinas y capialçados*". *Edificio* means literally "building"; however, the context suggests that Goycoa is talking about vaults and other construction elements.

models are used to show an apprentice, in a most intuitive manner, the complex problems involved in the construction of arches, vaults and squinches.

This idea appears frequently in later Spanish stonecutting treatises and manuscripts. De Rojas says that "it is impossible to understand the division of an arch in voussoirs without making a clay or plaster model … in my youth I made many models of different kind of vaults".[37] This sentence recalls a passage from de l'Orme where the author warns the reader that stonecutting "is difficult in itself, and cannot be easily understood, unless by those that … understand the tracings taking the pains of making models, cutting small pieces in wood or stone".[38]

An interesting passage hints that models are not only used to instruct apprentices in stereotomical techniques, but also to verify geometrical constructions. When explaining the procedure to dress the voussoirs of an arch opened in a round wall, Vandelvira (c. 1585: 22r; see Palacios [1990] 2003: 80–87; Wendland et al. 2015: 1788–1789 and Sect. 6.4.1) puts forward two alternative methods. In the first method, he instructs the reader to dress the voussoirs with the aid of templates. This involves a fairly complex geometrical construction. Templates are drawn using orthogonals to a line in point view to show the true shape of the voussoir edges. In addition, Vandelvira introduces a new construction to determine the curvature of the edges of the templates, but this procedure leads to a strange result; in Vandelvira's words: "Now you will ask how it is possible that, the arch being convex and concave, the templates are reversed; they are convex in the concave side, and they are concave on the convex side".[39] To attest to the exactitude of his method, Vandelvira suggests that the reader prepare a model of the arch by the second method, squaring; once this is done, the reader should apply the templates to the model to verify that they represent the true shape of the voussoir's faces: "if you want to prove it, make a model by squaring, as I will teach you further on, and apply the templates to the model, and you will find that the templates are correct".[40]

As he had announced, three sheets later he explains how to dress the voussoirs by squaring. Vandelvira almost excuses himself for explaining such a simple arch by this method: "I will show now the arch in a round wall by squaring, because I had promised to do so, and also to cast light on other pieces that can be done only by

[37]Rojas (1598: 88v-89r): *no se puede saber perfectamente el cerramiento de un arco, si no es contrahaciéndolos por sus piezas de barro, o de yeso … en tiempo de mi mocedad me ocupe en contrahazer, y levantar modelos de muchas diferencias de cerramientos de capillas*. See also Martínez de Aranda (c. 1600: [i]).

[38]De l'Orme (1567: 78v): *… la chose est difficile de soi-même, il est aussi malaisé qu'elle se puisse entendre, sinon par ceux qui ont … intelligence des traits avec la peine qu'ils prendront de les contrefaire, coupant de petites pièces de bois ou de pierre …* Transcription taken from http://architectura.cesr.univ-tours.fr. Translation by the autor.

[39]Vandelvira (c. 1585: 22r): *Dirás ahora cómo, siendo el arco torre cavado y torre redonda, las plantas van al contrario que las primeras, van redondas a la parte del torre cabo y a la parte del torre redondo van cavadas …* Transcription taken from Vandelvira/Barbé 1977. Translation by the author.

[40]Vandelvira (c. 1585, 22r): *… y si lo quisieres probar contrahaz un arco de éstos por robos, como te enseñaré adelante, y luego planta estas plantas y harás la prueba ser éstas ciertas*. Transcription taken from Vandelvira/Barbé 1977. Translation by the author.

squaring".[41] Such concern likely arises from a notion that was widely spread between Renaissance masons: as we will see in Sect. 3.2, squaring involves a great loss in labour and material, so masons usually resorted to an intensive use of templates in the dressing process. It seems that Vandelvira applies the squaring method to this arch because he acknowledges its strong empirical evidence and didactic potential, leveraged by means of models.

Martínez de Aranda assured the reader that "I have always been careful to make models of these stonecutting tracings before writing about them; I did as much as possible to cast light on these difficult terms consulting with wise men and eminent stonecutters".[42] This hints that models for Martínez de Aranda are not only didactic resources, but also research tools; they are used to test the correctness of the tracings. According to the copy of the Vandelvira manuscript by Felipe Lázaro de Goiti, the copyist has tested most of the stonecutting problems in the book by means of plaster models, since stereotomical knowledge cannot be acquired only by reading; it is necessary to make many models. Doing so, Goiti claims to have tested his manuscript in the touchstone, and contrasted it with experience, which is the mother of all knowledge; in this way, the text will achieve the esteem it deserves for its correctness and importance.[43] All this recalls the empirical approach to geometry upheld by Curabelle in his confrontation with Desargues; models and actual constructions furnish methods both for research and verification in stonecutting. The implications of these issues are crucial for the epistemological standing of stonecutting; we will come back to the problem in Sect. 12.2.

---

[41] Vandelvira (c. 1585: 24v): *Porque dije de enseñar el arco torre cavado y redondo por robos y por-que también sea lumbre para entender otras trazas que no se pueden hacer si no es por robos, pongo ahora éste.* Transcription taken from Vandelvira/Barbé 1977; folio number follows the original numbering in the manuscript, transcribed by Barbé in round brackets. Translation by the author.

[42] Martínez de Aranda (1600: [i]): *siempre tuve cuidado y principal intento de contrahacer las dichas trazas y ponerlas por modelos antes de ponerlas por escritura cuanto pude hice por sacar a luz la grande obscuridad que los términos de ellas tienen consultándolas con hombres doctos y personas eminentes y tracistas.* See also Cabezas (1992: 10).

[43] Vandelvira and Goiti (1646: 3r): … *e gastado muçho tiempo en … prouar los mas de los Cortes haçiendo sus pieças de yeso: porque este Arte no se alcança con sola la letura, sino que es necesario modelar y contrahaçer vna y muçhas veçes. Lo qual a sido como examinar los dichos papeles en la piedra del toque, y registrarlos en el contraste de la experiençia, madre de todas las facultades* … All this does not seem to be factually true, since he had reproduced literally the designs of another manuscript; however, it reflects clearly the conceptions of masons about the empirical nature of their craft. In particular, *la piedra del toque* (the touchstone) was a special piece of dark siliceous stone, such as basalt or jasper, used to gauge the purity of gold and silver by the colour of a stroke left by a piece of gold or silver on this stone.

3.2 Dressing

3.2.1 Geometrical Instruments

Between the basic processes of *setting out*, that is, the geometrical definition of a masonry element, and *dressing*, the mechanical execution of the units that are to be integrated into the masonry, it is necessary to *transfer* the geometrical shapes resulting from the tracing process to the blocks being dressed. This process demands several specific instruments. Some of them are lineal, such as the straightedge, the *gauge*, a ruler used to transfer measures by making notches on its edges, and the *templet,* a straightedge with a curved edge. Others have two operating sides, such as the squares used in the dressing process, the *bevel* or stonecutters' protractor, and the *arch square*, that is, a square with a curved arm. Finally, the most visible of these instruments is the *template*, usually with four operating sides.

We should not forget that transfer and dressing processes overlap in practice. Typically, the mason prefigures the shape of the voussoir or block using geometrical instruments; then, he cuts some stone with mechanical tools; in the next steps, he controls the resulting form with geometrical instruments again, cuts some more stone and so on, in a slow iterative sequence. Notwithstanding that, stonecutters' jargon makes it clear that geometrical tools embody basic geometric concepts, as we will see: straightedges for planarity, squares for orthogonality, templets for curvature, bevels for angles, and templates for complex shapes.

The straightedge, the gauge and the templet. In stonecutting parlance, the ruler stands for the straight line, the plane or even the lintel, as shown by such designations as *arrière-voussure bombée et reglée* (hollowed and ruled rere-arch) (Derand 1643: 148–150) or *capialzado a regla* and *escalera adulcida a regla,* (respectively, "ruled rere-arch" and "softened ruled staircase") (Vandelvira c. 1585: 44r; 59v-60r). While the connection between the ruler and the straight line is obvious, its identification with the plane stems from its use in the dressing process. When explaining a skew trumpet squinch with an oval profile, Vandelvira states that "you should dress the voussoir face with the ruler, then you should place the template on this surface";[44] this means the mason should dress a flat surface, verifying its planarity with the ruler, and then lay a rigid template on it. However, the ruler can also be used to dress non-planar surfaces. Dealing with the same squinch, Vandelvira remarks a few lines later that the intrados should be hollowed in order to materialise its final shape: "you should place the templet on the face of the squinch, then you should dress the intrados from the templet to the vertex of the template using the ruler".[45] In other words, when

[44] Vandelvira (c. 1585: 8v): *primero se ha de labrar el paramento de la dovela a regla, luego se ha de plantar la planta en el dicho paramento.* Transcription is taken from Vandelvira/Barbé 1977. Translation by the author.

[45] Vandelvira (c. 1585: 8v): *… háse de plantar la cercha en la cabeza, luego desde la cercha a la punta de la planta se ha de labrar a regla.* Transcription is taken from Vandelvira/Barbé 1977. Translation by the author.

dressing the intrados of the piece, which is an assembly of three different conical surfaces, the mason should use the ruler to assure that generatrices are straight.

When explaining a similar operation for splayed arches, Martínez de Aranda (c. 1600: 105, 107) advises the reader to dress the voussoirs *a regla y borneo* before hollowing them. This colourful expression means the mason should verify the planarity of the dressed face using a straightedge while closing an eye; that is, he should check that two edges of the surface appear to overlap from a single point of view and thus the surface is effectively planar.

In contrast, on other occasions, masons use the ruler intentionally to dress a warped, non-developable surface. For example, when dressing a rere-arch, Martínez de Aranda recommends that the reader "dress the intrados from one end to the other with the shape of arches *A C* placing the straightedge squarely from one side to the other so that the intrados will be warped".[46] That is, after dressing auxiliary lines with the shape of arches, the mason should dress the intrados leaning the ruler on these arches and keeping it on a plane orthogonal to the face of the piece; however, since arches *A* and *C* are not parallel, the result will be a non-developable or warped surface.

At the beginning of the Enlightenment, Frézier (1737–39: II, 15–16, 18–20, 21–22, 35–37) explained in detail the dressing of planar, cylindrical, conical, and warped surfaces with the help of the ruler. In all cases, the method is basically the same as in the preceding paragraph: after carving two auxiliary lines playing the role of directrices, the mason should dress a ruled surface, controlling its generatrices with the help of a straightedge. Our modern notion of ruled surface and even the distinction between developable and warped surfaces seems to stem from this masons' practice, as I will discuss in Sect. 12.5.3.

A wooden straightedge may be used as a gauge, *galga* or *échasse*, marking on its edge the distance between two points with notches in order to transfer it. Such operation may be performed both in the setting out phase, as we have seen in Sect. 3.1.2, and in the dressing process, as explained by Derand:

> Gauges are flat wood rulers on which notches or mortises are made, in order to score on one side the length and in the other one the width of stones … To return a stone is to give it a second bed or face, opposed to the first one … so that the second one is parallel to the first one. It is the same as workers mean when they talk about gauging a stone; however, to return a stone, it is necessary that the second bed or faces … has the same marks than the first one, so that one set of marks is in front of the other one; this is not necessary when gauging a stone.[47]

---

[46]Martínez de Aranda (c. 1600: 46): *labra los bolsores por una parte y otra con la forma que tuvieren los bolsores de los arcos* A C *y de unas testas a otras las labraras a regla plantando la regla de quadrado que vengan a quedar por las caras engauchidos.*

[47]Derand (1643: 5): *Les échasses, son des bois plats en forme de regles, sur lesquels on fait des creus ou entilles, pour marquer en l'vn des costez des iceux la longueur, & en l'autre la hauteur des pierres …/Retourner une pierre c'est luy donner un second lit ou parement, opposé en sorte à un premier … que le second soit parallele au premier. C'est cela mesme que veulent signifier les ouvriers, quand ils disent iauger une pierre; excepté que pour retourner une pierre, il faut que le second lit ou parement … porte les mesmes marques qui se trouvent sur le premier, & ce en telle*

Another tool is the *templet*, a ruler with a curved edge. We saw in Sect. 3.1.2 the methods used by de l'Orme (1567: 55r-55v) and other authors in order to draw *cherches ralongées*, that is, representations of a wide variety of curves using a series of circular arcs; the term refers both to the geometrical construction and the instrument. However, not all templets had these complex shapes. In its simplest form, this edge is just a circular arc; it can be used to draw arcs on stone, for example when dressing the blocks for a curved wall. This simple instrument can be employed in more sophisticated ways: taking into account that the sphere has the same curvature in all directions, the dressing of the intrados face of the voussoirs for a spherical dome can be controlled revolving a templet around its axis (Frézier 1737–39: II, 25–37, pl. 29). As we will see in Sects. 3.2.6 and 9.1.1, this is a crucial phase, since the templates for such voussoirs must be placed on a spherical surface.

The square. As we have seen, the square is shown in a drawing of King Offa with his architect (Paris 1250–1259: 23v) as a symbol of the medieval mason, together with the compass. With the arrival of the Renaissance, it was subject to substantial transformations. The medieval square is used both as a tracing instrument and as a dressing tool, checking the orthogonality between two faces of a block or voussoir: in the stained-glass window of the History of Saint Chéron in Chartres cathedral, two squares are placed between two half-dressed blocks (Colombier [1953] 1973: 17). By the sixteenth century, the square with divergent arms was no longer useful in the tracing process; it was replaced by another variant, with two arms, parallel edges and bevelled ends, represented in the Golden Staircase in Burgos cathedral, from the decade of 1520 (Marías 1991: 249, 259). A few decades later, the three-sided square of Antiquity reappears as a specialised tracing tool, as we have seen (de l'Orme 1567: 36r-38v); from this moment on, the square with two arms is confined to the role of a dressing tool.

While the ruler stands for the straight line or the plane, the square embodies the concept of orthogonality; for example, Fray Laurencio de San Nicolás (1639: 66v) states that the intrados and extrados of three-centre arches are dressed *a escuadra*, that is, orthogonally to the faces. This shows that the square is used not only to verify the perpendicularity of two planar surfaces but also to control the generation of ruled surfaces: placing an arm on a planar surface, the other arm will be used to check the correct direction of the generatrices of the resulting surface. In particular, the square, in French *équerre*, gives its name to one of the primary dressing methods, *équarrisement* or squaring, which relies mainly on orthogonal projections, while the alternative method, *par panneaux,* or by templates, uses true-shape constructions and developments (Frézier 1737–39: II, 11–12).

The bevel. This basic instrument, known in French as *sauterelle* from its similarity with the grasshopper's leg, is an articulated square, similar in appearance to a compass, used "to take a measurement from the tracing, or from the executed work, in order to cut a stone by its end, or any other part, while being in the workshop,

---

sorte que les unes soient directement opposées aux autres: ce qui n'est pas neccesaire quand on ne fait que iauger la Pierre.

before placing it".[48] It does not appear in medieval sources; in Villard's times, angles were measured by gradients and transferred taking the measures of the catheti of a right triangle, perhaps using the square.[49]

Several Spanish manuscripts include lines known as *saltarreglas*, the Spanish for *sauterelle*, in order to control the angle between the face and the intrados joints, although methods differ between authors. Martínez de Aranda (c. 1600: 11–12, 15–16, 40–41) uses the *saltarregla* in order to improve precision when dressing voussoirs by squaring. In this method, the voussoir is left in its natural position in most phases of the dressing process, and the outline of the voussoir face is scored on the stone from the start. Therefore, the author knows the position of the face joint beforehand and the line known as *saltarregla* represents the intrados joint; as we have seen in Sect. 3.1.3, the angle between both joints may be determined through a construction anticipating nineteenth-century revolutions. In contrast, Vandelvira (c. 1585: 47r-49r) uses the *saltarregla* here and there in elements solved by squaring but also in a large number of pieces solved using intrados templates (Vandelvira c. 1585: 7v, 8r, 8v, 9v, 10r, 26v-28r). In these cases, the natural position of the block is manipulated during the dressing process, and the mason does not have a reference for horizontal and vertical planes; in contrast, he knows the position of the intrados joint, which is marked on the template. Thus, in Vandelvira's method, the line in the tracing known as *saltarregla* represents the position of the face joint, not the intrados joint.

The arch square. This instrument, known in French as *biveau*, is similar to the bevel, although one or both arms are curved, either convex or concave. Perhaps this has fostered the appearance of the English term "bevel" for the *sauterelle*, not the *biveau*.[50] Like most stonemasons' instruments, it seems to have medieval antecedents. It appears in the stained-glass window of the History of Saint Sylvester in Chartres cathedral (Fig. 3.27), with a bent arm and a straight one; it can be used to dress concave surfaces, such as the intrados of an arch or rib. However, Hand IV in Villard de Honnecourt's sketchbook does not seem to know this instrument, since he uses a combination of a *cerce* and a square instead of it; this raises doubts about its actual use in the Middle Ages (Colombier [1953] 1973: 30; Shelby 1965: 247; Villard c. 1225: 20r, in particular dr. 6). In any case, the instrument was widely used in the sixteenth century, both in France and Spain (Ruiz c. 1560: 29v; Bustamante 1994: 156, 158, 215, 227, 231; de l'Orme 1567: 54v-55r, 56v; Wilkinson 1991: 268; Vandelvira c. 1585: 7v).

Generally speaking, French treatises present the arch square with an articulation (Fig. 3.28), just like the bevel (de l'Orme 1567: 54v, 56v.; de la Rue 1728: 2, pl. 2); in the words of Father Derand,

---

[48]De l'Orme (1567: 55): *La sauterelle est quasi semblable au Buveau, fors qu'elle est toute droicte, & s'ouure & ferme comme l'on veult, pour prendre vne mesure sur le traict, ou sur l'oeuvre, à faire couper vne pierre par le bout, ou autrement, estant sur le chantier, premier que de la mettre en oeuvre*. See also Derand (1643: 4).

[49]Villard (c. 1225, 20v, dr. 7). See also Shelby (1965: 246–247); Shelby (1969: 539); Shelby (1971: 145); Barnes/Villard (2009: 143–144).

[50]According to the Merriam-Webster dictionary, "bevel" derives from the medieval French word *bevel*, from Old French **baivel*; remarkably, *baivel* is the Spanish name for the *biveau*.

Fig. 3.27 Stained-glass window of the History of Saint Sylvester. Chartres Cathedral (Photograph by the author)

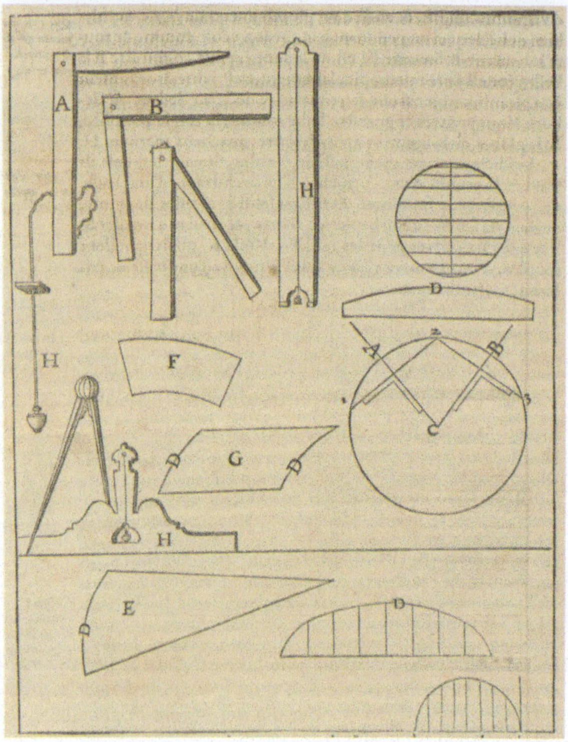

Fig. 3.28 Stonecutting instruments (de l'Orme 1567: 56v)

> The arch square shares with the bevel the articulation of its branches; however, its arms are not straight; sometimes both are convex, on other occasions they are curved and concave, in other one is bent and the other straight; or either both are concave, and half a branch is straight, as needed.[51]

In contrast, there are no references to articulated arch squares in Spanish stonecutting texts of the sixteenth and seventeenth centuries: Vandelvira (c. 1585: 7v) represents a fixed version at the beginning of his manuscript. We should bear in mind that the articulated arch square is useful when a large number of voussoirs with the same curvature and different angles between the bed joints and the intrados surfaces are to be dressed; when this is not the case, a fixed instrument, or a set of these tools, which are of course easier to manufacture than the articulated arch square, can solve the problem more efficiently.

Sanabria (1984: 190–191) pointed out that de l'Orme prefers bevels and arch squares to templates, in particular when dealing with the *vis de Saint-Gilles*, the mythical acme of medieval and Renaissance stonecutting; thus, he surmised that the use of the arch square was a recent evolution of the template method. Barbé (1993: 131) went further and took it for granted that the use of arch squares and bevels was an improvement over templates. At first sight, de l'Orme's text seems to support this view:

> … masons found the invention of not limiting themselves to making a window, but rather placing a full quarter of the stairway on the air, using a design called 'cantilevered quarter of stairway', which is done in different ways; some masons dress the voussoirs by squaring, others do it using templates. As for myself, I do not need anything but an arch square or a bevel, together with a square; after having constructed the templet, I will build the cantilevered quarter of stairway in any fashion that is needed …[52]

However, we should not forget the theatrical tone of many passages in de l'Orme (see Sect. 2.2.1). A short passage in Fray Laurencio de San Nicolás shows that *reglas-cerchas*, that is, arch squares, can be conceived as partial templates and there is not much difference in use with actual full templates: "prepare arch squares, according to *A-Y-N* for the inside, and another arch square according to *B-O-M*, or full templates, which are the same thing; with these you should dress the bed joints

---

[51]Derand (1643: 4): *Le buveau conuient avec la sauterelle en la mobilité de ses branches, mais il differe d'elle en ce que ses branches ne sont point à droite ligne; mais quelquefois toutes les deux sont rondes & bombées, quelquefois au contraire elles sont courbes & creuses au dedans, d'autrefois l'vne est ronde & l'autre droite; ou bien toutes les deux estans creuses, la moitié de l'vne se trouue droite ainsi que l'on peut auoir affaire.*

[52]De l'Orme (1567: 120v; see also 123v-124r): … *les ouvriers ont trouvé l'invention de ne se contenter seulement d'y faire une fenêtre, mais bien de mettre tout un quartier de vis à jour, et en faire un trait qu'ils appellent le quartier de vis surpendu, lequel se fait en différentes sortes, car les uns le font par équarrissage, et les autres par panneaux. Quant à moi, je ne voudrais sinon qu'un buveau ou sauterelle avec une équerre, de sorte qu'après avoir tiré la cherche rallongée, je ferais le quartier de vis rampant en toutes sortes* … Transcription is taken from http://architectura.cesr.univ-tours.fr. Translation by the author.

in both voussoirs".[53] Thus, the difference between full templates and arch squares is not so relevant; a mason who knows how to construct a full template can prepare an arch square by simply using two sides of the full template. As a result, de l'Orme's assertions can be understood under a quite different light: boasting of virtuosity, he states he can dress the voussoirs of the "cantilevered quarter of a stairway" with just an arch square, in contrast to beginners, who had difficulties in executing this piece with full templates; in other words, the arch square demands more dexterity from the mason.

Templates or molds. The stained-glass windows of the History of Saint Sylvester and the History of Saint Chéron in Chartres cathedral include templates, also called molds, representing the section of vault ribs (Colombier [1953] 1973: 17, 30). The central role of this simple instrument in Gothic construction is clear from the start: when William of Sens begun the construction of Canterbury Cathedral at the end of the twelfth century, he sent templates to Caen, in order to get stones correctly dressed and shipped across the English Channel. Later, canvas templates were also sent to Caen in order to dress stone for Westminster Palace. Thus, templates allow the separation of the tracing, dressing and placement processes in stone construction (Colombier [1953] 1973: 24; Shelby 1964: 394, note 33; Ruiz de la Rosa 1987: 291–293; Palacios [1990] 2003: 18–20; Zaragozá 1997: 29) offering, together with full-scale tracings, the most important method of formal control in Gothic construction.

This is shown by the medieval tradition of master masons handing templates to the workers (Colombier [1953] 1973, 98; Shelby 1964, 393; Kostof [1977] 1986: 88–89; Burns 1991: 202) and the importance given by Villard de Honnecourt (2009, 32r; see also 30v-31v, 32v and Alexander 2004) to the templates of Reims cathedral, shown in connection with its sections and elevations (Fig. 3.29). This central role is perpetuated in the Renaissance. As I have noted in Sect. 2.1.4, after the death of Hernán Ruiz II his son fled to Málaga with the templates for the Chapter Hall in Seville Cathedral, and Pedro Díaz de Palacios was unable to complete the vault (Morales 1996: 47, 152; Gentil 1996: 123–127; see also Erlande-Brandenburg 1993: 69). In the Escorial, many contracts mention a copy of the templets, arch squares and templates delivered to contractors, while the original remained in the hands of the King's Surveyors; if the contractor needed more than one copy, he had to prepare it at his own expenses (Bustamante 1994: 233, 238, 242, 360, 381). All this stresses the central role of templates, in full or simplified forms, in the geometrical control of the works and the execution of contracts.

More specifically, Escorial contracts differentiate between *plantas*, *moldes* and *contramoldes* (Bustamante 1994: 147, 155, 156, 239, 360, 369, 381). *Contramoldes* or *contramotlles* (Bustamante 1994: 147, 155, 156, 239; Zaragozá 1992: 99; 1997: 29, 31) seem to be negative templates, placed outside the member being dressed. The difference between *plantas* and *moldes* reflects an important conceptual distinction.

---

[53] San Nicolás (1639: 73v): *haz reglas cerchas, según A Y N para la parte de adentro, y otra regla cercha según B O M o plantillas enteras, que lo mismo es lo uno que lo otro, y con ellas se han de ajustar los paramentos por la parte de sus lechos y sobrelechos.*

Fig. 3.29 Templates for Rheims cathedral (Villard c. 1225: 32r)

In sixteenth-century Spanish, templates associated with complex geometrical tracings are called *plantas*, while simpler mouldings are called *moldes* (Vandelvira c. 1585: 19v; Wilkinson 1993: 267; Bustamante 1994: 240, 242). The same distinction exists in Catalan, between *patró* and *motlle* (Martín 1981: 209, 212; Zaragozá 1992: 99) or in French, between *panneau* and *moule* (de l'Orme 1567: 57r, 77r; Sanabria 1984: 207; Potié 1996: 68–69). It may seem surprising, but *molde, moldura, moulure,* or the English words "moulding" and "mold", derive from the *moule* of William of Sens and Villard de Honnecourt, while the neologisms *panneau* or *planta* were created in the Renaissance for complex stereotomic shapes. As an exception, "mold" is frequently used in English as a synonym of "template".

Other terms also reflect conceptual differences. Most stonecutting problems in Martínez de Aranda (c. 1600: 6–11, 16–17, 18–34, for example) are solved by dressing the stones *por plantas al justo.* This term is used for templates representing the shape of a voussoir face in true shape, placed on a planar surface to mark the edges of the voussoir, as opposed to templates in orthographic projection. The author differentiates systematically between intrados and bed joint templates, called *plantas por cara* and *plantas por lecho*, respectively. Usually, Vandelvira (c. 1585: 9v, 26v-27r, for example) follows a different approach, although the distinction is more apparent than real. He constructs full templates only for the intrados faces and calls them simply *plantas*; the shape of bed joints is controlled through *saltarreglas*, that is, segments standing for face joints, making it possible to measure the angle between the intrados joint and the face joint; I will translate *saltarreglas*, when used in this sense, as "bevel guidelines". However, the full template for the bed joint can be constructed following the same method used for the bevel guideline; in fact, Vandelvira (c. 1585: 19v) does so when he needs extra precision, for example in skew arches with face mouldings.

The template repertoire is apparently larger in de l'Orme (Fig. 3.30): in addition to *panneax de doyle par desoubz*, (intrados templates), and *paneaux de joint*, (bed joint templates), we find *paneaux de teste* and *paneaux de doyle par desus* (respectively, face templates and extrados templates) (de l'Orme 1567: 77r; see also Derand 1643: 3 and de la Rue 1728: 1–2). Templates for the front faces of arch voussoirs are also present in Vandelvira (c. 1585: 12r) and Martínez de Aranda (c. 1600: 4, 178, 180) although they are usually not shown as individual shapes, but rather in a development, presented as a set of templates and called *cimbria.* As for the extrados templates, it is safe to surmise that this surface was usually left rough, either to improve the adherence with the overlaying masonry in the case of arches or just because extradoses are only visible from the secluded spaces under the roofs, in the case of vaults. Thus, down-to-earth Spanish manuscripts generally eschew extrados templates.

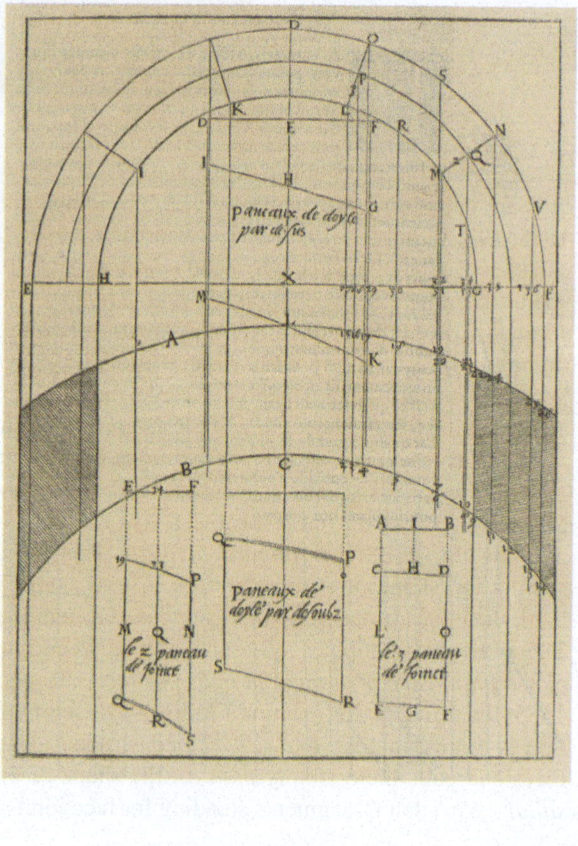

Fig. 3.30 Arch in a round wall with different kinds of templates (de l'Orme 1567: 77r)

A remarkable variety of materials are used in the preparation of templates. In medieval England, masons use wood, canvas, linen cloth and parchment (Shelby 1964: 394; Shelby 1971: 142–143). During the construction of the Basilica at Vicenza, Palladio used tin, iron, paper, and wood templates (Burns 1991: 204, 207). According to de l'Orme, *moules* can be made in copper, wood, tin or cardboard; [54] there is no reason not to extrapolate this to *panneaux*. In any case, he later remarks that paper or cardboard templates are boiled in order to reduce their rigidity and glued to a wooden base: "However, I do well to inform my readers that I never find my figures carved exactly as I have drawn them, since masons usually moisten, and even boil the paper of the drawing a little before glueing it to a board, in order to carry out dressing".[55] Similar techniques were put into practice when building the Gates of Quart in Valencia in the fifteenth century (Zaragozá 1992: 99); starched paper sheets, a pot and glue are bought in order to paste the *motles* or templates, probably on a wooden base. A striking detail shows the importance of templates. The main vault in the Gates is a skew one; thus, it is not strictly symmetrical, and a voussoir in the outer left quarter has its counterpart in the inner right section. Natividad (2010: 125) has shown that a small error is mirrored exactly in the diametrally opposed voussoir; such coincidence may be explained only by an error in a template used in both voussoirs. In the Royal Chapel of Alphonse V of Aragon, in the same city, it seems that paper templates were used directly, without using a wooden base (Zaragozá 1997: 33); however, the use of starched paper suggests they were conceived as rigid templates. Wood prevailed in the Escorial complex: stonecutting supervisors requested a carpenter in order to prepare templates, while other managers argued that these craftsmen should be under the command of carpentry supervisors. According to de l'Orme, mouldings were drawn on stone with a *broche d'acier*, that is, a steel scriber or stylus. This suggests the use of *moules* in wood again; the idea can be extrapolated to stereotomic templates.[56]

In any case, the rigid or flexible nature of the templates is more important for our purposes than material. Paper templates glued on a wooden base or spread on a planar surface count as rigid templates, while paper, cloth or tin templates adapted to a conical, cylindrical, or spherical surface act of course as flexible templates, as explained by Frézier:

> These templates are made of rigid material such as wood boards when dressing a planar surface; or sometimes in flexible materials, such as cardboard, tin, or plumb sheets when

[54]De l'Orme (1567: 56r): *Et se font les dits moules de cuivre, de bois, de fer blanc, ou papier de carte, et servent à mouler et marquer les pierres pour les tailler*. Transcription is taken from http://architectura.cesr.univ-tours.fr. Translation from the author.

[55]De l'Orme (1567: 106v): *Toutefois je veux bien avertir les lecteurs que je ne trouve mes figures si justement taillées que je les avais portraites, pour autant que les tailleurs ont coutume de mouiller, et quelquefois faire un peu bouillir le papier de la portraiture, premier que de le coller sur la planche, pour la conduite de leur taille*. Transcription is taken from http://architectura.cesr.univ-tours.fr. Translation from the author. See also Sanabria (1984: 205–206).

[56]De l'Orme (1567: 57r): *Quand les pierres sont équarries et jaugées, on les moule et trace avec une petite broche d'acier sur les moulures des oeuvres qu'on veut tailler à la pierre*. Transcription is taken from http://architectura.cesr.univ-tours.fr. Translation from the author.

dressing a curved surface, whose contour is determined by way of development, which is the less frequent solution in execution.[57]

This issue has fundamental geometrical consequences since rigid templates were obtained through true-shape constructions or triangulation of polyhedral surfaces, while flexible templates were the result of developments of cylindrical or conical surfaces, sometimes approximating spherical ones. However, before analysing with this issue in depth, I will deal with mechanical instruments and the basic dressing methods used to carve planar faces, returning to cylindrical, conical, and spherical surfaces in Sect. 3.2.6.

3.2.2 *Mechanical Instruments*

Bessac ([1986] 1993) has published a full book about all kinds of stonecutting instruments, analysing their construction, function, characteristic marks, evolution from Antiquity to the twentieth century and even their names in the different regions of France. It is impossible and unnecessary to deal here with this subject in such detail; the interested reader may consult Bessac, Warland ([1929] 2015) or Rockwell (1993). Rather than following this path, I will focus on three primary instruments, mentioning their variants briefly. Father Sigüenza, stressed the expenses brought about in the Escorial worksite by "… the repairs of the tools, the picks, the axes and their handles, chisels and mallets, that are worn out continuously …";[58] he does not mention other tools.

These instruments have been in use in Europe almost continuously from Antiquity to our times. In particular, the pick and the axe are to be found in stonecutters' tombs in the Isola Sacra in Ostia or the Museum of the Berry in Bourges (Adam 1984: 35–36), while chisel marks are frequent in Greek and Roman construction. Still in the eighteenth century, Frézier (1737–1739: II, 17; see also pl. 28) mentions combinations of the pick, the axe, and the chisel, with the accompanying mallet, but not other tools such as the bush hammer, the mitre or the saw, which appeared on the eve of the Industrial Revolution or later.

The pick and its variants. The pick (Fig. 3.31) is used to remove larger irregularities in the faces of the blocks, both in the quarry and on arrival in the worksite. This operation can be performed as a first stage before the use of the axe or as a final dressing for parts left in the rough, such as the extrados of many vaults, or surfaces that will be rendered or plastered (Bessac [1986] 1993: 5). In the Escorial complex, contracts differentiate clearly between parts that are to be dressed *a picón* (with the

[57]Frézier (1737–39: II, 12): *Ces modeles se sont sur des matieres infléxibles comme des planches, losqu'il s'agit de la formation d'une surface plane, & quelquefois sur des matieres flexibles, comme du Carton, du fer-blanc, ou des lames de plomb, lorsqu'il s'agit d'une surface courbe, dont on cherche le contour par la voye du dévelopement qui est la moins ordinaire dans l'execution.*

[58]Sigüenza ([1605] 1907: II–III, 440): … *el adobo de las herramientas, picos y escodas, y sus astiles, cinceles y macetas que se gastan a cada paso* ….

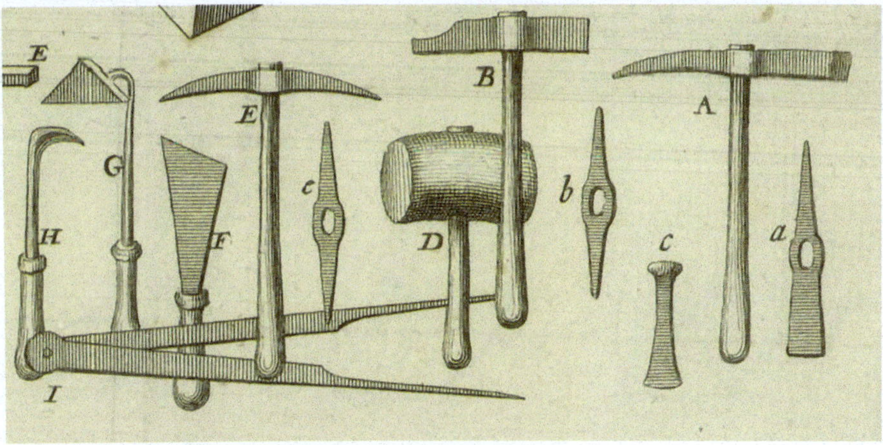

Fig. 3.31 Picks, bevel, mallet, and other stonecutting instruments (Frézier [1737-1739] 1754-1769: pl. 28)

pick) and those that are to be left *bien escodadas* (thoroughly dressed with the axe) or *trinchantadas* with a small axe.[59] The use of the pick is attested to in Egypt, Greece and Rome (Bessac [1986] 1993: 21–24). A clear example, with pointed ends in both sides, was included in the Herrad von Landsberg manuscript, dating from 1181–1185. However, this tool appears frequently as a part of a combined instrument, with a pick in one end and an axe in the other. It can be seen in this form in the stained-glass windows of Chartres, or in a miniature by Jean Fouquet, which shows a stonemason using the pick side to dress a pier (Colombier [1953] 1973: 17, 23, 30, 31, 37, 43, 48; Gimpel [1958] 1980: 36).

An instrument performing similar functions is the stonecutter's hammer, known as *marteau têtu* in French; in our times, it is used to take out irregularities along the edges of the initial block, before removing material from the faces with the pick. It seems that it was used in the Early Middle Ages to break, rather than dress, stones into rubble (Erlande-Brandenburg 1993: 97). For convenience, it is frequently combined with the pick. There is evidence of its use in Greece, Rome and perhaps Egypt. From this point on, proof of its use is somewhat scant, although it is represented in medieval miniatures, in combination with the axe, and has in fact survived up to the twentieth century (Frézier 1737–39: II, 17, pl. 28; Warland [1929] 2015: 5–6; Bessac [1986] 1993: 25–38).

The axe and its variants. However, the main stonecutters' tool, up to the eighteenth century, is the axe. Although it may be used tangentially, as in carpentry, it is frequently used orthogonally to the dressed surface, when dressing blocks or voussoirs (Fig. 3.32). Thus, it usually leaves a footprint with small parallel traits (Bessac

[59]Bustamante (1994: 309): *La piedra … labrada a picon excepto el fajón que será escodado* (stone … should be hewn with the pick except for the strip that will be dressed with the axe). Transcription by Bustamante 1994, modernised by the author. See also p. 155, 156, 213, 221, 233, 258.

Fig. 3.32 Using the axe. Stonecutting workshop of the School of Architecture of Madrid (Photograph by the author)

[1986] 1993: 39–49). The axe is particularly useful for dressing planar surfaces, but it can be used for the intrados of voussoirs; its edge can be straight or clawed. Proof of the use of this instrument can be found in Egypt and Greece, and quite frequently in Rome, particularly in gravestones. It is less frequent in the Middle Ages; an instrument with an axe at both ends appears in the *Biblia Sancti Petri Rodensis* (c. 1020: III, 89v). Other representations, such as the mosaics at Monreale cathedral, also from the twelfth century, also include an axe with a single end (Colombier [1953] 1973: 19, 57, 102, 105; Gimpel [1958] 1980: 34, 43, 47, 53). Later on, the axe is found usually in combination with the pick, for example in the stained-glass windows at Chartres, as we have seen. The same instrument is to be found in the manuscript of *Los veintiún libros de los ingenios y las máquinas* (Lastanosa c. 1570: 252r), in a passage about extraction in the quarry. A particularly interesting representation of an axe can be found in the Grandes Chroniques de France, from the fourteenth century, which shows, despite its diagrammatic depiction, the instrument used in a planar surface (Gimpel [1958] 1980: 38).

Another instrument with a linear edge is known in French as the *polka*: one or both of the cutting edges are perpendicular to the handle, rather than parallel, as in the traditional axe. The variant with two perpendicular edges is scarce; the instrument with a perpendicular and a parallel edge, that is, the combination of the standard axe and the polka, is more frequent. Although the name is relatively recent, the instrument can be traced to the Middle Ages, where it replaces to a certain extent the standard axe (Bessac [1986] 1993: 49, 58–59). In some cases, it is difficult to tell whether a miniature or drawing represents a *polka* or a combination of a pick and an axe (Colombier [1953] 1973: 43; Erlande-Brandenburg 1993: 103). The popularity of the *polka* in this period may be explained by two factors: its convenience, since it serves both as a pick and an axe, and its fitness for dressing small stones.

The chisel, the mallet and their variants. Both the pick and the axe are, in Bessac's terminology, thrown percussion tools; their movement furnishes the kinetic energy needed to break the stone and remove unnecessary material. In contrast, the *chisel* is a posed percussion tool; kinetic energy is brought about by the *mallet* or the *dummy* striking the handle of the chisel. It is used for careful work, for example when dressing preliminary lines to define the edges of a planar surface, the ribs of Gothic vaults or sculptoric work (Bessac [1986] 1993: 121–133). As a result, it occupies a central place in Late Middle Ages ashlar construction. Gervase of Canterbury remarked that the old piers in the eponymous cathedral were not dressed with the chisel; this hints that the standard carving procedure from the thirteenth century on involved this tool. In the earliest depictions, such as the manuscript of Herrad von Landsberg and the stained glass at Chartres, the cutting end of the chisel appears near the edges of the member being dressed. This suggests that the chisel was used mainly to dress preliminary lines (Fig. 3.33), known as *marginal drafts*; as a second step, a flat surface inside the marginal drafts was to be dressed with the axe or, from the eighteenth century on, with the bush hammer. Later on, during mature and late Gothic, the chisel was used for most dressing tasks in ribbed construction, as Gervase's remark and graphic evidence shows; with the advent of the Renaissance, it was again restricted to preliminary lines and, of course, delicate tasks such as mouldings and sculptural work (Bessac [1986] 1993: 133–137; Colombier [1953] 1973: 31).

There is a wide variety of derivatives of the chisel. The working end of the chisel can range from a simple *point*, known as *punch* in heavier instruments, to a wider edge, as in the *driver* or the *boaster*; between them stands the *drafter*, used for

Fig. 3.33 Using the chisel. Course "El arte de la piedra". Universidad CEU-San Pablo (Photograph by the author)

marginal drafts. As in the axe, the edge of the chisel can be straight or toothed, as in the *waster*. Furthermore, there are curved chisels, known as *gouges*, in a large range of shapes and radii, used for mouldings. The handle can be wooden or metallic, forming a single piece with the cutting end (Bessac [1986] 1993: 108–120, 138–153; Warland [1929] 2015: 2–5). Another variant, probably brought about by the appearance of the bush hammer in the eighteenth century, is known in France as *ciseau boucharde*; it features several rows of teeth, instead of a single one as in the waster (Bessac [1986] 1993: 155–157).

The tool used to strike the chisel or its variants against the stone is called a *mallet* when made of wood, or a *dummy* when metallic. In both instruments, the working mass or *beat* can adopt a number of different shapes: cuboid, cylinder, truncated cone or, frequently, a bell shape; needless to say, beats in mallets are larger than the ones in dummies (Bessac [1986] 1993: 159–171; Warland [1929] 2015: 6–7).

The Industrial Revolution and beyond. Other instruments appeared in the eighteenth century, with the onset of the Industrial Revolution and the greater availability of metal tools. The most relevant was the *boucharde* or *bush hammer*, used for dressing planar surfaces: it takes the form of a hammer with a number of diamond shaped points (Fig. 3.34). Usually, the head of the tool is interchangeable, so the mason can start with a head with a few, large diamond points, say a 2×2 array, to dress a surface roughly; next, he can proceed gradually with other heads with a larger number of smaller points, up to 100 points in a 10×10 pattern, in order to get a smoother finish. This instrument was used by marble workers, in particular in Carrara, from the seventeenth century on. However, in the nineteenth century, the tool was adopted by stonecutters because of its adaptability and efficiency, replacing the axe, which was relegated to a secondary role. The bush hammer leaves a point grid pattern on the dressed surface, quite different from the short parallel lines of the axe; it is easy to differentiate these historical finishes, featuring a more or less irregular series of short marks, from the regular patterns resulting from the renovations of the last centuries, using the bush hammer. A variant of this instrument is the *patent axe* or *patente*, which features a series of linear edges, so to speak a group

Fig. 3.34 Using the bush hammer. Stonecutting workshop of the School of Architecture of Madrid (Photograph by the author)

of chisels, rather than the array of points of the bush hammer. The footprint of this tool is similar to that of the axe, although on close inspection the parallelism of the edges differentiates it from historical finishes (Bessac [1986] 1993: 77–91; Warland [1929] 2015: 1).

Another instrument that appeared during the Industrial Revolution was the *chemin-de-fer*, literally "railroad", known in English as "French scraper", a series of sturdy metal blades used to achieve a smooth surface in soft limestones. *Drags* are toothed steel plates used to the same end. Other tools used in soft or medium-hard stones are *files* and *mitres* and *rifflers*, that is, special files with a curved tip used to clean difficult sections, in particular in mouldings, and several varieties of *saws*, used to cut soft stones (Bessac [1986] 1993: 188–230; Warland [1929] 2015: 6–9).

One of these varieties is called the *hand saw* since it can be operated with one hand, while other ones such as the *cross-cut saw* and the *whipsaw* require the use of both hands. Mechanical saws arose in the twentieth century, often using industrial diamonds or silicon carbide. Other machines employed widely from the twentieth century on are *planers* and *canting* machines and *polishers*. The pre-industrial antecedent of these machines is the *lathe*, used to execute balusters and other pieces involving surfaces of revolution. (Bessac [1986] 1993: 253–261; Warland [1929] 2015: 10–15). Although these machines are highly efficient, they cannot solve all dressing problems, particularly the complex ones addressed by traditional stonecutting. Thus, during the twentieth century, *pneumatic hammers* have been used by stonecutters and sculptors. Workshops using this technology are fitted with air compressors and piping to distribute pressurised air to the work posts. Pipes end in nozzles designed to receive small heads in the shape of tiny bush hammers or short chisels, greatly increasing the efficiency of the carving process. However, the typical stonecutting instrument of the early twenty-first century is the 3- or 5-axis *machine tools* operated by *computer numerical control* (CNC); the main role of the stonecutter is to prepare CAD models that are translated to machine control language (Fallacara 2003b: 194–198, 213–218; Colella 2014). Obviously, such technology is quite far from the historical processes covered by this book; so, I will not deal with it in any depth, focusing on the next sections in traditional techniques.

3.2.3 Dressing by Squaring

As we have seen, stonecutting treatises and manuscripts from the Renaissance to the Enlightenment classify dressing methods in two broad areas. Both take their names from geometrical instruments: the square - *équarrissement, escuadría* - and the template - *panneaux, plantas* - and thus, are tied closely to geometrical concepts. The square is applied when tracings rely on orthographic projection, while the template is used where geometrical constructions give true-shape representations of the faces of dressed stones. Thus, both methods anticipate key concepts in descriptive geometry: double orthographic projection for the squaring method, and developments and true-shape constructions for templates.

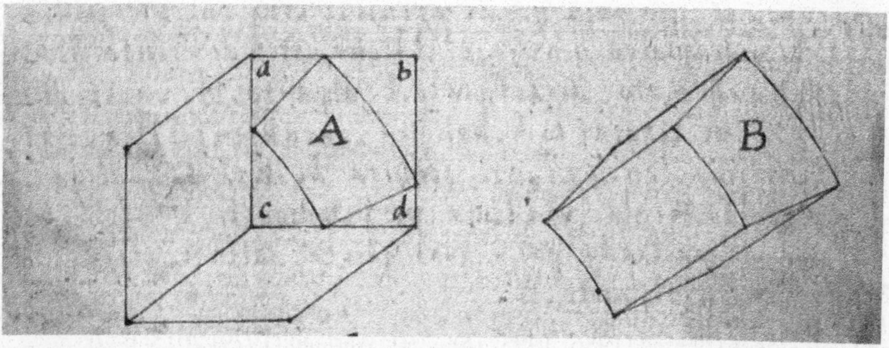

Fig. 3.35 Dressing by squaring (Martínez de Aranda c. 1600: 114)

However, the picture is not so simple. It has been pointed out (Rabasa 2000: 158–160) that modern stonecutting mixes both methods, while Frézier, suggests dressing some pieces by a method known as *demi-équarrissement* or *media escuadría* (Frézier 1737–39: II, 13, 115–116; see also Bails [1779–1790] 1796: 433–435). For the sake of clarity, I will address these hybrid methods in Sect. 3.2.5, while dealing in what follows with simple squaring.

The concept of squaring. When dealing with skew arches with elliptical faces, Derand remarks that

> although this technique, which is executed with the ruler, the square and the compass, may be considered as squaring, workers do not give it this name, since they want everything that is done by squaring to be traced only by transfer of heights and cantilevered spans, taken from the tracing, and brought to the stone the masons means to dress, and convert into a voussoir.[60]

This implies that the canonical squaring method eschews templates; de la Rue (1728: 2) is more explicit: "To trace by squaring is a way of tracing stones starting from measurements taken from the full-scale drawing, without using templates".[61]

Thus, squaring involves scoring the orthogonal projections of the edges and faces of the voussoir on a rectangular cuboid and cutting away several wedges of redundant stone from the block with the aid of the projections (Fig. 3.35), until the voussoir reaches its definitive form (Sakarovitch 1993: 121–124, 135). This is explained by Martínez de Aranda in simple, empirical terms:

> I suppose that figure A is the voussoir you want to dress by squaring, and this square has four angles *a*, *b*, *c*, and *d*; you should mark the edges of this voussoir and cutting away *a* in

[60]Derand (1643: 160) *Bien que cette pratique, qui s'execute auec la regle, l'équaire, & le compas, puisse entrer au rang de celles qui se font par équarrissement; si est-ce neantmoins que les ouuriers ne luy donnent point ce nom, parce qui'ils veulent que tout ce qui se fait par dérobement, ou par équarrissement, se trace seulement par transport des hauteurs, & des auances des retombeés, prises dans le trait, sur la pierre que l'on veut tailler & reduire en voussoir ....*

[61]De la Rue (1728: 2): *Tracer par équarrissement ou dérobement, est une manière de tracer les Pierres par des mesures prises sur l'épure, sans se servir de Panneaux.*

the upper joint, and *b* in the extrados, and *c* in the intrados, and *d* in the lower bed joint, and passing these cuts from one face to the opposite one, this voussoir will take shape as shown in figure B.[62]

Later on, Frézier's explains the operation using geometrical concepts:

... before forming a figure in the shape of an oblique solid, you should start by forming with the square a cube, or parallelepiped, capable of containing it. Next, you will trace on each of the planar surfaces, supposed to be in a horizontal or vertical position, the projection of a surface of the body you intend to materialise, and remove from the parallelepiped everything that exceeds the outline of each projection, removing unnecessary stone; since the surfaces of this body are squares or rectangles before being dressed, the method using them is called *by squaring.*[63]

Although Aranda and Frézier connect the terms *cuadrado* and *équarrissement* with the square and the rectangle, the name of the method may also derive from *équerre,* the French name for the square. Once the shape of the faces has been drawn on stone, this instrument plays an essential role in this procedure. This tool allows the stonemason to transfer the shape and position of the surplus wedges from one face of the block to the other; when one arm of the square rests on a flat face of the voussoir, the other arm materialises lines perpendicular to the planar surface. In this way, the square can generate planes at right angles to the faces of the voussoir, such as the joint planes, or cylinders with generatrices orthogonal to face planes, such as the intrados surface of a round arch or a barrel vault.

This procedure has a clear geometrical meaning. Architectural drawing handbooks and even descriptive geometry texts use the word "projection" for a process that involves passing a line through a given point and finding the intersection of the line with a plane. In this sense of the word, the projection of a point is a point, the projection of a straight line is a straight line, and the projection of a curve is generally a curve. This is what stonecutting writers do when reproducing the front of an imaginary voussoir on the stone surface. Nevertheless, in the rigorous vocabulary of projective geometry, the whole process involves two steps, projection and section; strictly speaking, the projection is only the first phase. Hence, the projection of a point is a line, the projection of a line is a plane, and the projection of a curve is,

---

[62]Martínez de Aranda (c. 1600: 113–114): *Supongo que la figura* A *es el bolsor que quieres entrar en cuadrado y el dicho cuadrado son los cuatro ángulos* a b c d *con el cual dicho cuadrado cogerás los extremos del dicho bolsor y robándolo por el lecho alto con el robo* a *y por el tardos con el robo* b *y por la cara con el robo* c *y por el lecho bajo con el robo* d *y pasando los dichos robos de una testa a otra quedará formado el dicho bolsor como parece en la figura* B.

[63]Frézier (1737–39: II, 11–12): *... avant de former une figure de solide oblique, on commence par en former une de cube, ou de parallelipipède à l'equerre, capable de la contenir. Ensuite on trace sur chacune des surfaces planes supposées en situation Verticale ou Horisontale, la projection des surfaces du corps qu'on se propose de former, & l'on retranche du parallelipipede tout ce qui excede les contours de chaque projection, en abattant la pierre superflue; & parce que les surfaces de ce corps sont ou en quarré, ou en quarré long, avant d'etre taillées; on appelle la méthode qui en suppose de telles, par* Équarrissement. The expression *quarré long* means "rectangle", since the starting block is usually a rectangular cuboid; however, I have translated it as "long square" in order to keep Frézier's (debatable) argument, which derives the name of the technique from the shape of the faces of the starting block, rather than the use of the square.

generally, a surface; thus bed joints are created as projections of straight lines, while the intrados surface is generated as an orthogonal projection of the face arc by means of the square.

Complex pieces. As explained by Martínez de Aranda and Frézier, this method seems quite straightforward; however, Vandelvira remarks that "all pieces solved by squaring are more difficult than those dressed by templates and bevels".[64] The reason for this apparent contradiction is that Martínez de Aranda's description is merely didactic, so he takes as an example the voussoir of a semicircular arch. In this case, all voussoirs are identical to the keystone; any skilled mason would dress all stones using the projections of the keystone to minimise the volume of the enclosing block. Doing so, there is little difference between squaring and templates since the horizontal projection of the keystone intrados is the same as the intrados template.

Moreover, in actual practice, ordinary pieces are dressed by templates or hybrid methods, to prevent waste of labour and material. As de l'Orme remarked, in squaring there is a *gran perte de pierres* (a significant waste of stone) (de l'Orme 1567: 73v; see also Sanabria 1989: 276–277; Palacios [1990] 2003: 18–20; Sakarovitch 2005b: 53). Thus, for simpler pieces, the use of templates was preferred; in contrast, the squaring method was reserved for a few complex pieces that could not be solved in practice using true-size templates, such as rere-arches (Vandelvira c. 1585: 46r).

A good example of such pieces is explained by Martínez de Aranda when dealing with the *rincón de claustro*, a reinforcement arch placed under the junction of two orthogonal barrel vaults (Fig. 3.36); its outer half takes the shape of a diagonal strip of a pavilion vault while its inner half takes that of a strip of a groin vault:

> I suppose you want to dress the second voussoir in the outer section *E*, you should take from the plan of this arch its intrados template which is between the four angles *a*, *b*, *c*, and *d* and with this template you should dress the voussoir by squaring so its height equals the height of the envolving prism from point *e* to point *f*; after dressing it by squaring as I have said you should take material from both faces according to the enclosing shape *E*, so the voussoir will be dressed with the shape between the angles *g*, *h*, *i* and *l*, and it will form a crease in the intrados; the voussoirs on the other side of the arch will feature a groin. In this way, you should dress all the voussoirs in the arch except the keystone, which should be dressed so that half the voussoir should include a crease and the other half a groin.[65]

Thus, the mason should begin dressing a prism with the total height of the voussoir, using the plan of the voussoir as the base. Next, he should cut four wedges to shape

[64] Vandelvira (c. 1585: 26r): *Toda traza que es por robos es más dificultosa que la que se labra por plantas y saltareglas.* Transcription is taken from Vandelvira and Barbé 1977.

[65] Martínez de Aranda (c. 1600: 85–87): *supongo que quieres labrar la pieça segunda de haçia el Rincon E tomaras en la planta del dicho arco su planta por cara que esta entre los quatro angulos a b c d y con esta dicha planta labraras de cuadrado la dicha pieça que tengo de alto lo que tubiere de alto el quadrado de su bolsor desdel punto e al punto f y despues de labrada de quadrado con forma aRiba dicha la Robaras por entramas testas con los Robos que tubiere el quadrado E que benga a quedar la dicha pieça por entramas testas como pareçe el bolsor entre los angulos g h i l y por la cara quedara labrada en Rincon y si fuera del lado contrario quedara labrada la dicha cara en arista y desta manera se an de yr labrando todas las demas pieças deste arco eçeto la clabe que a de yr labrada de forma que la mitad quede por la cara en arista y la otra mitad en Rincon que por enmedio benga a estar en cuadrado.*

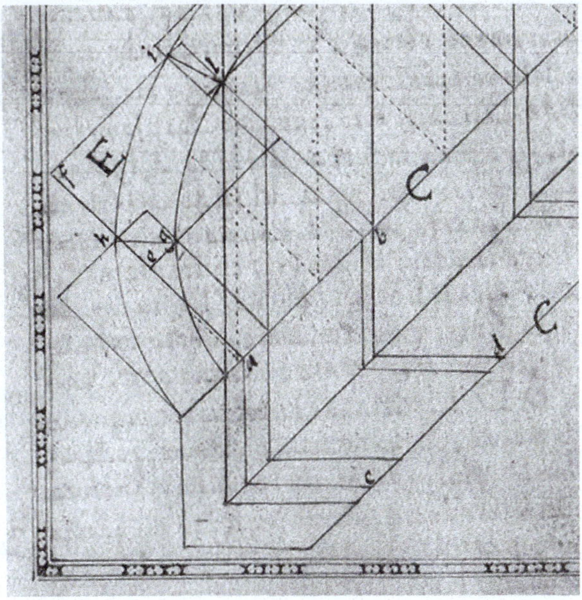

Fig. 3.36 Reinforcement arch for an L-shaped vault, detail (Martínez de Aranda c. 1600: 86)

the bed joints, the intrados, and the extrados. This phase is explained in further detail by Derand (1643: 322) when dealing with the groin vault; he stresses that the mason should take horizontal and vertical distances to the edges of the cuboid, transfer them with a gauge and join the resulting points with a templet or arch square in order to define the wedges that should be taken off.

However, Martínez de Aranda's problem is more complex, since he is dealing with a combination of a groin and a pavilion vault. In order to tackle the problem, the mason should mark on each of the front and back faces the shape of the voussoir taken from the elevation *g-h-i-l*. To take a wedge off and materialise the lower bed joint, the mason should start from one face of the voussoir, leaning the square in *g-h*; upon reaching the symmetry plane, the mason should begin in the opposite front and repeat the operation. Martínez de Aranda stresses that a crease appears in the outer half of the arch while the other half features a groin; that is, the arch includes a section of a groin arch, but also a part of a pavilion vault. This simple remark has important consequences: in the inner half of the arch, the mason may keep carving happily until he surpasses the groin. In contrast, in the outer section, he should be careful to avoid crossing the symmetry plane, that is, the virtual position of the crease. If not, he will go beyond the intrados surface, and the whole stone will be spoilt: in traditional stonecutting, a mason can remove stone but he is unable to add material.

Moreover, precision is essential, since the mason has no specific instruments to control the shape of the groin-turned-crease; in fact, it appears automatically, almost miraculously, as Frézier will remark much later in a passage that summarizes the advantages and disadvantages of the squaring method:

The advantage of the squaring method consists in, first, when the sections of a vault are not circular, it avoids the preparation of a great number of templates, since each voussoir needs a different template.

2nd, in that it is not necessary to know the curved lines formed at the intersection of curved surfaces; they arise at random, taking off the stone on the intrados using a ruler carried along the cross-section of the element.

Its *disadvantages* are 1st it consumes a great quantity of stone in pure loss since it is necessary to find sloping surfaces between horizontals and verticals … The *second* disadvantage is that it is necessary not only to dress needlessly the surfaces of a parallelepiped that is to be cut afterwards, but also on some occasions another second set of surfaces, which are also useless, and are needed only to dress the third set, which is the only one that subsists when the member is finished, …The *third* disadvantage is that if angles are somewhat altered by execution, and the squaring process does not achieve the exact transfer of the angles between two different surfaces, either for errors in the squares and arch squares or from the hand of the worker that executes the piece, important errors can arise, and edges will have an irregular and ill-shaped contour.[66]

3.2.4 Dressing with True-Shape Templates

As a consequence, stonecutting treatises and manuscripts, starting with de l'Orme, Vandelvira and Martínez de Aranda, address most problems using templates. Leaving flexible templates aside for the moment, these methods are based on true-shape constructions, either triangulation or several specific stonecutting methods, as we have seen in Sect. 3.1.3.

Stand-alone templates. De L'Orme explains the use of templates for a trumpet squinch, constructed by triangulation:

I want to warn the reader that the stones of all kind of squinches are more challenging to dress than the stones in other pieces, since after dressing the intrados of the stone, you can mark on it its template exactly, but for the others, such as the bed joint templates, face templates and also extrados templates, you should avoid using them to cut the stone in a single pass, because in this case you will waste it, and it will be useless. Thus, you should cut some

[66]Frézier (1737–39: II, 13–14): *L'avantage de la méthode par équarrissement consiste 1° en ce que l'on s'épargne la peine de faire un grand nombre de panneaux pour la construction d'une voute, lorsque ses ceintres se sont pas circulaires; parce qu'il faut changer à chaque voussoir/2° En ce qu'il n'es pas nécessaire de connoitre les·lignes courbes, qui se forment par l'intersection des surfaces courbes; on les forme par une espece de hazard, en abatant successivement la pierre d'une Döele à la régle trainée sur un Arc-Droit/Ses* Désavantages *sont 1° qu'elle consomme beaucoup de pierre en pure perte; car puisqu'il faut chercher des surfaces inclinées entre des verticales & des horisontales … Le* second *Désavantage est, qu'il faut non seulement faire inutilement les surfaces d'un Parallelepipede qu'il faut recouper, mais souvent des secondes surfaces, qui sont encore inutiles, & qu'il ne faut supposer que pour trouver les troisèmes, qui doivent subsister quand l'ouvrage est achevé, … Le* troisième *Désavantage est, que si les angles sont un peu alterez par l'execution, & que l'équarrissement ne soit pas exact dans les renvois, que ces angles font d'une surface à une autre, soit par la faute des Équerres ou des Biveaux, ou de la main de l'Ouvrier qui s'en sert, il peut en résulter des erreurs sensibles, & des arétes d'un contour irrégulier & mal formé.* See also (Sakarovitch 2003b: 72) and Trevisan (2011: 14).

> stone in the side of a joint, a little bit more stone from the side of the face of the arch, do the same thing from the extrados side … and not everything at once, but rather cutting skillfully, surrounding the stone with templates all around until they meet exactly, and they touch one another by all their edges, for the bed joints and the intrados as well as in the front face, because if you are not careful, your stone will be immediately wasted, and it will be useless.[67]

We should remember that de l'Orme is talking about a piece that involves a specific difficulty in the acute vertex of the triangular shape of the intrados. Applying this method to the voussoir of an arch or vault, the mason can carve a provisional flat surface for the intrados of the voussoir and score the shape of the intrados template on it with a scriber. Next, he can start gradually taking stone off from *two* adjacent sides of the voussoir, say the face and the upper bed joint, dressing two planes passing through the intrados joint and the chord of the face arc. As stressed by de l'Orme, it is essential to check the result at intervals with bed joint and front face templates until both templates meet each other and the intrados plane. When this position is reached, the edge between the front face of the voussoir and the bed joint, as well as the planes of the face and the joint are fixed in space, so the mason can dress both planes easily, checking their planarity with the ruler.

For a modern architect or engineer, this procedure sounds unnecessarily complicated. In the early eighteenth century, Frézier (1737–39: I, 372–374), addressed the problem in a more efficient way, using the dihedral angle between the intrados and joint planes; that is, the angle between the intersections of both planes with a third plane orthogonal to their common intersection, in this case the intrados joint. Apparently, this idea was too abstract for de l'Orme or Spanish manuscripts of the sixteenth century; none of them mentions it. However, the later steps in the dressing process are much more straightforward. Once the planes of the face and the joint are dressed, the stonemason can mark on them the outline of their respective templates. After this, the mason can dress the plane of the lower joint with the aid of a ruler leaning in the intrados joint and the face joint, inscribe the joint template on it, and dress the rear face by making the ruler rest on both face joints and the chord of the face arc.

As we have said, no Spanish manuscript or treatise describes this method explicitly. However, Martínez de Aranda explains the construction of intrados and joint templates in almost every section of his manuscript (c. 1600: 6–10, 16–7, 20–33,

---

[67]De l'Orme (1567: 99r): *Mais je veux bien avertir le lecteur que les pierres de toutes sortes de trompes sont plus difficiles à tailler que de beaucoup d'autres sortes de traits, pour autant qu'après avoir fait un parement à la pierre pour la doile de dessous, vous pouvez bien tracer son panneau justement, mais pour les autres, comme pour les panneaux de joints, panneaux de tête, et aussi panneaux de doile par le dessus, gardez vous bien de les tracer pour couper la pierre du premier coup, car vous la gâteriez, et ne pourrait plus servir. Il faut donc ôter un peu d'un des joints, et puis un peu du côté de la tête, semblablement du côté de la doile de dessus, et ainsi conséquemment un petit de l'un et petit de l'autre, et non point tout à un coup, mais coupant si dextrement le tout que vous puissiez armer votre pierre de panneaux tout autour qui se rapportent justement et se touchent l'un l'autre par toutes leurs extrémités, tant par les joints que par les doiles et par le devant, où est le panneau de tête, car si vous n'y preniez garde, votre pierre serait incontinent gâtée, et ne pourrait servir.* Transcription is taken from http://architectura.cesr.univ-tours.fr. Translation is from the autor. See also Calvo 2003: 464–465, and Trevisan 2011: 15.

49–65, etc.). Face templates are usually present, either included in the elevation or grouped in a *cimbria* or set of templates (Martínez de Aranda c. 1600: 6–11, 85–90; see also de l'Orme 1567: 100r-103r; Vandelvira c. 1585: 10v-12v, 15r, 17r-17v, 23r-23v). This strongly suggests that Aranda considered this procedure to be the canonical method in stonecutting. Perhaps for this reason he never explained the dressing technique explicitly, while he remarked the details diverging from this paradigm, as we have seen for the groin vault.

Templates and bevel guidelines. In any case, some evidence suggests that everyday dressing methods in sixteenth-century Spain were simpler ones. Alonso de Guardia (1600: 82v) explains the dressing of a voussoir of the splayed arch using an intrados template, a joint template and the arch square of the face (Fig. 3.37). As we have seen, Alonso de Vandelvira usually explains the construction of intrados templates, calling them merely *plantas*, but he generally skips joint templates, the *plantas por lecho* of Martínez de Aranda. Instead, Vandelvira usually constructs *saltarreglas*, that is, lines standing for the face joint, which allow him to measure the angle between the intrados and face joints. Some details suggest that the ensemble

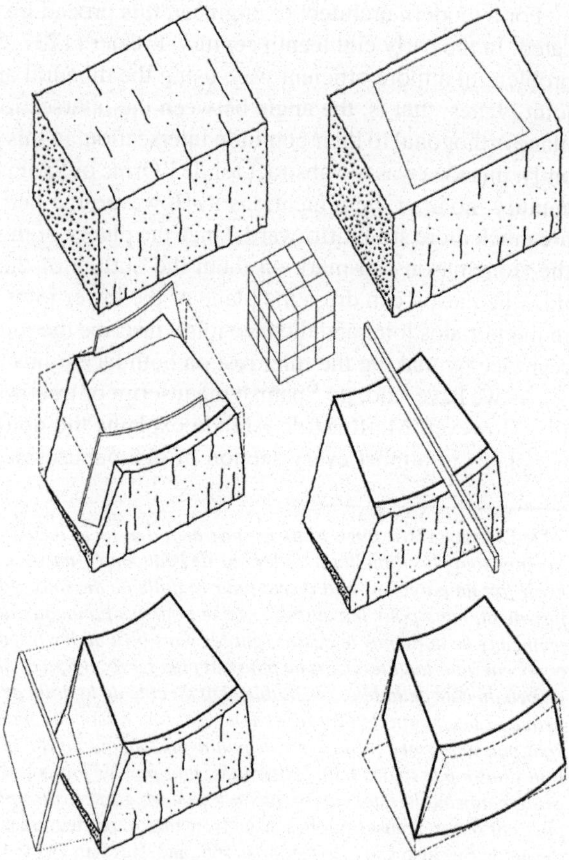

Fig. 3.37 Dressing a voussoir for a splayed arch according to Alonso de Guardia (Calvo 2000a: 236)

of these *saltarreglas*, or bevel guidelines, and the intrados joint plays the role of a simplified bed joint template. In a skew arch with semicircular faces, Vandelvira explains the construction of *plantas* and *saltarreglas*, adding that if the mason is to carve mouldings, he should distort the templates in order to adapt them to the bevel guidelines.[68] The result is the most complete joint template one can conceive, representing not only its four edges but also a highly detailed section of the mouldings.[69] Vandelvira gives few hints about the way templates and bevel guidelines are used in this particular case; however, the consideration of the *saltarregla* as a simplified template suggests a variation of de l'Orme's method. After dressing the plane of the intrados side of the voussoir and scoring the intrados template on it, the stonemason can gradually take material from the front face of the voussoir, until the face template and the bevel, opened in the angle marked by the guideline, assemble in the face joint. This method is less cumbersome than the use of joint and face templates since the bevel can be used even if the joint is not yet dressed. Once the first face joint is fixed in space, the dressing process can go on as we have seen before.

Again, Frézier explains the advantages and disadvantages of the full use of templates systematically in these terms:

> First, it is evident that the operation is more direct. Thus it is shorter. Second, since there are fewer preliminary flat surfaces, it is easier to use stones of a smaller volume. Third, since there is less material to take off, it brings about an economy in the supply of stone. 4° That since the operation is based on the shape of surfaces and their outlines can be drawn exactly following the rules of stonecutting tracings, it is carried out much more surely, and as a result, it is necessarily more precise. To summarise, it is the most learned method and the main object of study in stonecutting, and thus the authors in this field focus on it ... The only disadvantage is a greater need of instruments, if this name can be applied to templates.[70]

In other words, dressing with templates requires less time, labour and material. Since this method is based on true-shape depictions of the faces of the voussoirs, it is more precise. As a result, this method is preferred by most stonecutting authors, despite the need for a large number of templates.

---

[68]Vandelvira (c. 1585: 19 v): *si quisieres echar molduras has de extender los moldes en las saltarreglas*. Transcription is taken from Vandelvira and Barbé 1977.

[69]In many sections of his manuscript, Martínez de Aranda (c. 1600: 16, 19, 25) represents the three edges of the joint template corresponding to the intrados joint, and both face joints in solid line, while he renders the extrados joint in dashed lines. This graphic treatment seems to suggest the reader he can choose between applying the intrados and face joints as *saltarreglas* and using four joints as a full template.

[70]Frézier (1737–39, II: 14–15): *Premierement, il est visible que l'operation étant plus immédiate elle doit étre plus courte/Secondement, qu'y ayant moins de supposition de surfaces planes à faire préceder, il y a plus de facilité à faire servir des pierres de moindre volume/Troisièmement, qu'y ayant moins à retrancher, il se trouve une plus grand oeconomie dans la consummation de pierre/4° Que l'operation étant fondée sur l'etenduë des surfaces, dont on a pù exactement tracer les contours par les régles de l'epure, on y est conduit beaucoup plus sûrement, & par conséquent elle en doit étre plus exacte/Enfin c'es la plus sçavante méthode & le principal objet de l'etude de la Coupe des Pierres, don't les Auteurs qui ont traité ont fait le plus de cas ... Le seul Désaventage qu'on y trouve, est un plus grand attirail d'instrumens, si l'on peut appeller les panneaux de ce nom.*

3.2.5 Dressing by Hybrid Methods

While stonecutting treatises stress the opposition between squaring and true-shape templates, they explain on many occasions several methods standing between both paradigms, using orthogonal projections and more or less complete templates. These hybrid methods are not afterthoughts of learned theorists; quite to the contrary, they are present in the earliest phases of the development of stonecutting techniques.

Bevel guidelines scored on stone faces or drawn in tracings. One of the classical solutions for the skew arch, known in French as *biais passé* (Fig. 3.38, top), involves bed joints orthogonal to face planes. In order to control the shape of the voussoirs, Derand instructs the reader to use this technique:

> One begins first by tracing the voussoirs all alike, as if the arch were straight, not skewed; the line *A-B* should be scored on the lower bed joint of the first voussoir, numbered in the tracing with number 1; next, the line *C–D* should be marked on the upper bed joint, and the templet *B–D* should be laid on the points *B* and *D*; and the straight line *D–L* should be traced from the front of the stone to its back, starting in *D* and finishing in *L*; another similar line should be traced in the same fashion in the lower bed joint from *B* to *O*; by this means will be had the proper surface of the voussoir which must be cut.[71]

That is, the reader should dress a voussoir for a round arch by squaring; once this is done, he should mark on both bed joints the actual position of each voussoir corner, scoring a guideline representing the intrados joint; this allows measuring the angle between intrados and face joints. Also, this guideline furnishes a reference for the correct position of a templet representing the edge of the back face.

No actual template is used in this method, which may date from the times of Villard de Honnecourt (Villard c. 1225: 20r, dr. 9); in fact, the procedure amounts to drawing the intrados joints directly on the stone. However, other writers (Martínez de Aranda c. 1600: 15–16; Guardia c. 1600: 70v; Jousse 1642: 14–15) use these guidelines in the preliminary tracings on floors or walls. Derand himself (Fig. 3.38, bottom) included the guidelines in the tracing for the ox horn, another piece with a trapezial, rather than rhomboidal plan, drawn in the lower section of the same sheet (Derand 1643: 124–126; see also de l'Orme 1567: 69v-70v). Such bevel guidelines may be understood as utterly simplified templates, as we saw in the Sect. 3.2.1. However, the road leading from these embryonic templates to actual hybrid dressing methods is fairly complex;

---

[71] Derand (1643: 124): *Il faut en premier lieu tracer les voulsoirs tout de mesme, comme si la porte estoit droite & sans biais; & choisissant celuy qui doit seruir le premier, on portera sur iceluy le dérobement ou retombée A B, qui se placera sur le deuant du lit inferieur du voulsoir, representé sous le trait & marqué du chifre 1, selon qu'il s'y voit repairé par les mesmes lettres A B: puis retournant sur le trait sera pris C D, & porté au lit superieur du mesme voulsoir, au lieu où sont placées les mesmes lettres C D; & la cherche D B estant couchée sur les repaires B & D; & la ligne droit D L estant tirée du deuant au derriere de la pierre, commencant en D, & finissant à rien au point L; & vne autre estant tirée de mesme sorte, au lit inferieur de B, au point O: on aura par ce moyen le dégauchissement du vousoir, & ce qu'il faudra couper.* The term *dégauchissement* may be misleading; it means literally, "to avoid being warped", although it is normally used with the meaning "to dress a planar surface". This does not fit with the context since the final surface is actually a warped one; all this suggests that Derand is using the word, by extension, with the meaning "giving a surface its proper shape".

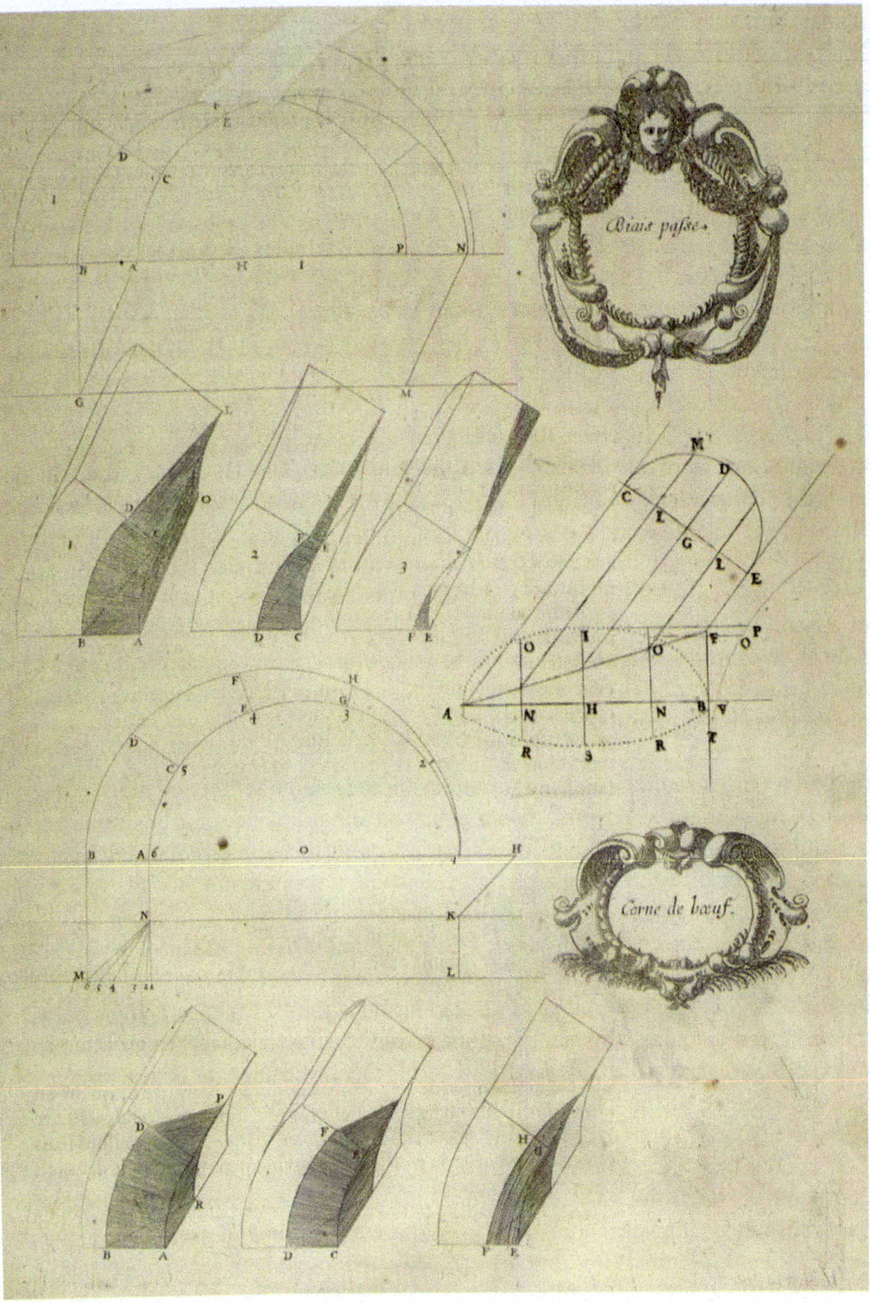

Fig. 3.38 Above, dressing the voussoirs of a skew arch, marking bevel guidelines directly on the bed joints. Below, dressing the voussoirs of the double ox horn, marking bevel guidelines in the tracing (Derand [1643] 1743: pl. 60). See also details in Figs. 6.10 and 6.25

it may be followed through several examples in the encyclopaedic manuscript of Martínez de Aranda and other treatises (see also Calvo 2003: 465–470).

Auxiliary templates in squaring. Although for Derand the canonical squaring method excludes the use of templates, he accepts the use of simplified templates or *échasses* to avoid much direct tracing on stone surfaces. For example, in the groin vault (Derand 1643: 334–335), instead of scoring horizontal and vertical distances on the edges of a cuboid and joining them with a templet, the mason may use a simplified triangular template, with two straight sides equalling horizontal and vertical distances and a third, curved side standing for the curve represented by the templet in the standard procedure. Also, for Frézier (1737–39: II, 12–13, 108–109; see also Pérouse de Montclos 1982a: 90; Palacios 1986: 102), the squaring method does not exclude the use of templates.

Moreover, Frézier explains a hybrid method known as *demi-équarrissement*, that is, "half-squaring". In order to avoid dressing a preliminary cuboid and marking horizontal and vertical distances on it, Frézier (1737–39: II, 115–116) advises the mason to dress a face plane and use on it a bevel to construct the angle between the chord of the voussoir and a horizontal line. In the next step, the mason should dress a plane orthogonal to the face plane and apply on it an intrados template. Finally, he can use the arch square to score the side joints on the face plane, using the square to control the carving of the bed joints. Frézier explains the system taking a round arch as an example, but he admits that its advantages in this case are not remarkable; in contrast, it simplifies the dressing of other complex pieces, since the mason may dispense with the preliminary cuboid.

Martínez de Aranda (c. 1600: 40–41) explains the carving of the voussoirs of a rere-arch with a semicircular arch in the front face and a segmental arch in the rear. Before dressing the voussoirs, the stonemason should make a fairly simple tracing, depicting the plan of the arch, the elevation, with the semicircular and segmental arches, and the division of the arch into voussoirs. It is interesting to note that Aranda does *not* represent the horizontal projection of the intrados joints because he does not need to use it in the carving process. Thus, the lines crossing the intrados in the plan are not *projections* of the intrados joints, but rather true-shape constructions used to measure the angle between face joints and intrados joints. Like Derand, he measures the projected distance between two corners of the voussoir; however, while Derand marked this distance directly on the voussoir face, Martínez de Aranda transfers it to the back face of the arch in the full-scale tracing. This operation allows him to construct bevel guidelines furnishing the angle between intrados and face joints.

The dressing process is fairly simple and more economical in labour and material than ordinary squaring. The stonemason begins by carving a mixtilinear block whose base reproduces the vertical projection of the voussoir and whose height equals the depth of the rere-arch. After that, he can score the angle between intrados and front joints in the bed joint side of the block, using the bevel. Repeating the operation on the other side, he can place the four corners of the intrados face of the voussoir. Next, he may take off material from the block until the intrados face is correctly dressed (Fig. 3.39). In the *Arco capialzado viaje por cara,* which is a biased variation of the preceding rere-arch, Martínez de Aranda advises the reader to use a ruler to check the

Fig. 3.39 Dressing a voussoir for a rere-arch with the bevel, according to Ginés Martínez de Aranda (c. 1600: 40-41) (3D modelling and rendering by the author)

dressing of the intrados. He stresses that the ruler should move on planes orthogonal to the front face, generating a warped surface: "you should dress the intrados face passing the ruler squarely from the front to the back so the intrados will we warped".[72]

Using the same technique in more complex pieces, he arrives gradually at the combined use of templates and squaring. For example, he addresses the problem of an arch formed by two rere-arches joined back to back, known as *Arco por arista,* literally "groin arch". The result is an arch with a V-shaped section, with two semi-circular faces and a groin in the form of a segmental arc, placed on a vertical plane which is parallel to the planes of the two faces. Martínez de Aranda repeats the method used in the preceding example, but instead of a single bevel guideline, the result is a template representing the V-shaped cross-section (Fig. 3.40). The description of the stonecutting operations makes it clear that the dressing method stands half-way between squaring and templates. Martínez de Aranda advises the reader to mark in both bed joints of the block the shape of the V-shaped cross-section of the arch. Since this section is not a single line, as in the preceding example, Martínez de Aranda designates it as *planta por lecho*, or joint template, while using the traditional word *robos* for squaring:

[72]Martínez de Aranda (1600: 46): *de unas testas a otras las labrarás a regla plantando la regla de cuadrado que vengan a quedar por las caras engauchidos.*

Fig. 3.40 Dressing a voussoir for a double rere-arch or "groin arch" with bed templates, according to Ginés Martínez de Aranda (c. 1600: 46-47) (3D modelling and rendering by the author)

> You should dress it by squaring by both faces, taking off the wedge between the numbers 1 2 5 6, so that after squaring in the lower bed joint it will assume the shape of the joint template F, and in the upper bed joint that of template *G*.[73]

A similar technique is illustrated quite clearly by de la Rue (1728: 50, pl. 27), with the help of excellent engravings (Fig. 3.41). The author takes exception to the use of templates based on conical developments for spherical vaults, and thus offers two alternative methods. The second one is remarkable since it is based on drawing directly on a spherical surface. The first one, although not as innovative, is also quite interesting. De la Rue dresses the voussoirs of the vault by squaring, starting from the plan of the voussoir, which takes the form of a wedge with two planar faces at the sides and two arcs at the front and the back. Next, he places a true-size-and-shape template, which can be obtained easily from the elevation, on both sides. In addition to the edges of the side joints, the template provides voussoir corners; the edges of the bed joints can be controlled easily starting from the corners with templets of the appropriate radius.

Auxiliary squaring in the true-shape method. In the preceding examples, the stonecutting procedure is based mainly on orthographic projections, although Martínez de Aranda, de la Rue, and Frézier use true-size templates as auxiliary

[73]Martínez de Aranda (1600: 46–47): *la robarás por entrambas testas con el robo que parece entre los números 1 2 5 6 que venga a quedar después de robada por el lecho bajo con la forma que tuviere la planta por lecho F y por el lecho alto quedará con la forma que tuviere la planta por lecho G.*

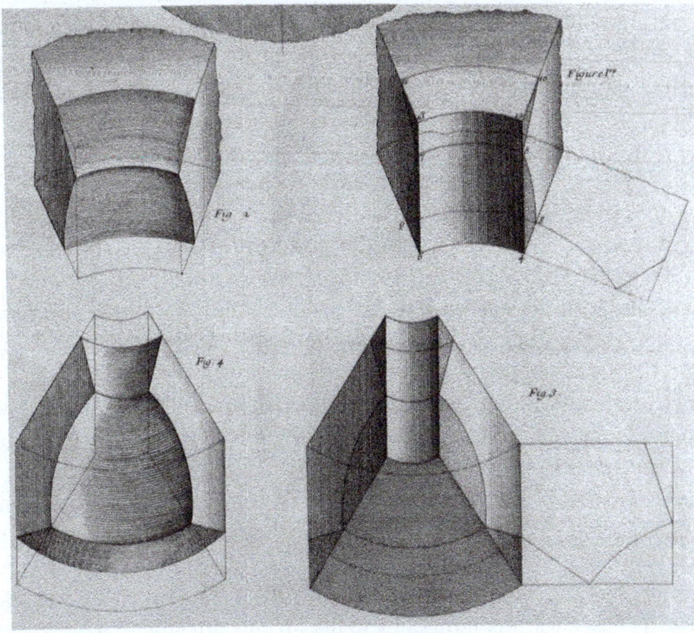

Fig. 3.41 Dressing a voussoir for a hemispherical dome by squaring with true-shape templates of the side joints, detail (de la Rue 1728: pl. 27)

devices. Inversely, in a skew arch with semicircular face arcs, the dressing of the voussoirs is performed with the aid of true-shape templates; however, Martínez de Aranda (c. 1600: 16–17) suggests that the reader use the templet obtained through orthogonal projection, which may be considered a one-sided template, to dress the intrados of the voussoir. As always, the stonemason is to prepare a full-scale tracing, in plan and elevation. After this, he should construct true-shape templates for the intrados and bed joints (Fig. 6.24). Additionally, the mason may prepare face templates, taking them directly from the elevation of the arch. Thus, when the mason starts to dress the voussoirs, he has an intrados template, two joint templates and, if necessary, two face templates.

This seems more than sufficient to employ the procedure suggested by De L'Orme (1567: 99r), enclosing the voussoir in the five templates and cutting it gradually until all templates match. However, Martínez de Aranda instructs the stonemason to use also the cross-section of the arch by a vertical plane perpendicular to the arch axis, obtained as an auxiliary view. As a result of the obliquity of the arch, the face arch is semicircular, and the cross-section is a raised ellipse, which Martínez de Aranda calls *arco encogido* or shortened arch. Once this is done, he instructs the reader to "dress the voussoir intrados squarely with the shape of the shortened arch".[74]

[74]Martínez de Aranda (c. 1600: 17): *por las caras de los bolsores se han de labrar de cuadrado con la forma que tuviere el arco encogido.*

Such an operation can be performed with the aid of a templet carried along the intrados, keeping it parallel to the plane of the cross-section and orthogonal to the intrados joints. Doing so, the intrados surface, an elliptical cylinder, is generated as the projection, in the strict sense of the word, of the "shortened arch" or raised ellipse.[75]

It is interesting to note that for Martínez de Aranda the molds used in the template method are *plantas al justo*, which can be translated as "exact templates"; however, in other examples where templates are used in connection with the squaring method (c. 1600: 193, 197, 201, 229, 242, 245, 249), the operation is called *plantar de cuadrado*, a term that means "orthogonal" in sixteenth-century masons' jargon, and comes from the same root as *escuadra, escuadría, équerre, équarrissement* and *square* (Martínez de Aranda c. 1600: 7, 12, 85).

Dihedral angles. Another example of the combination of squaring with true-shape templates can be found in Jousse's and de la Rue's solutions for groin vaults. As we have seen before, Martínez de Aranda (c. 1600: 85–87) dresses by squaring the *rincón de claustro*, which is a combination of a groin and a pavilion vault, while Derand (1643: 329–335, 346–348)[76] applies the squaring method both to groin and pavilion vaults.

In contrast, Pérez de los Ríos and García (2009) have pointed out that Jousse (1642: 156–157) uses folding templates in the groin vault (Fig. 3.42). As we will see in Sect. 8.2.2, no other template is shown, while the elevation includes the typical portions of the enclosing blocks under each voussoir. This suggests that he intends to dress the voussoirs at the groin by squaring, although he uses a folding template to control the execution of the intrados face. At the end of the section, he says "If you want to dress the voussoirs by squaring you should take points A, 1, 2, 3, 4, 5, 22, C, & 23".[77] That is, he gives two solutions, the traditional dressing process by squaring, akin to the one used by Martínez de Aranda, along with an alternative based on squaring complemented by folding intrados templates.

De la Rue (1728: 44–46; see also 68, 77–79, plates 36, 40) goes a step further. First, the mason should dress two planar surfaces, so that the dihedral angle between both planes fits the one the mason has computed in the tracing; this is a recent innovation, introduced in stereotomy by de la Hire (c. 1688a: 71r-74r; see also Tamboréro 2008: 73–74; Tamboréro 2009: 94–97). Next, the mason can apply a folding template on both faces, scoring its outline. Then, he can easily dress the side joints between the voussoir crossing the groin and ordinary ones, since they are orthogonal to the intrados joints. After this, the mason can score the template on the

[75] In actual practice, the cross-section was quite probably drawn with a mixed technique. First, the mason places the points of the ellipse through the intersection of the oblique cylinder of the intrados with the cross-section plane; second, he joins these points using circular arcs (Martínez de Aranda c. 1600, 1–4; see also Sect. 3.1.2).

[76] The actual page 346 is misnumbered as 344; thus, the section on pavilion vaults starts on the second instance of page 344.

[77] Jousse (1642: 157): *Si l'on veut faire les pieces par dérobement, il faut prendre aux poincts A, 1, 2, 3, 4, 5, 22, C, & 23.*

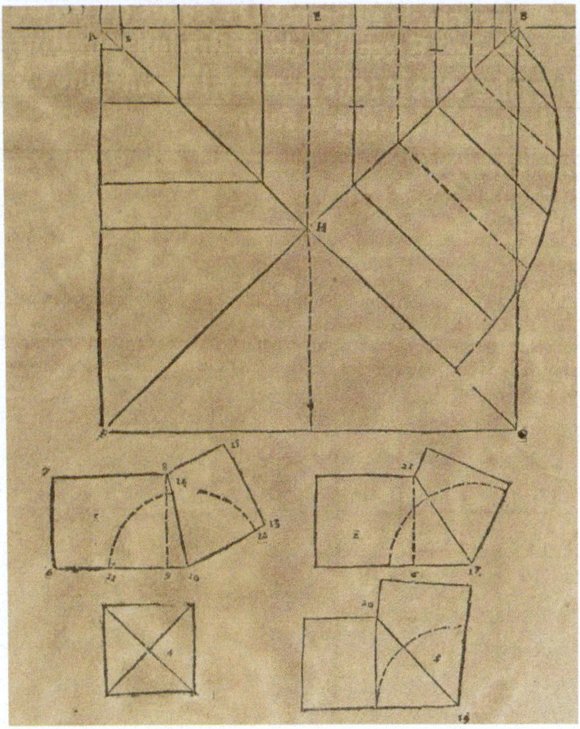

Fig. 3.42 Groin vault, showing folding templates, detail (Jousse 1642: 156)

side joint, dressing the intrados and bed joints easily, since both are orthogonal to the side joints (Fig. 3.43).

Oblique projections. Martínez de Aranda (c. 1600: 47–48) also includes in his manuscript a skew variation of the "groin arch", that is, two skew rere-arches placed back-to-back. He starts by tracing a doorway as a semicircular arch. Since the plane of the doorway is oblique to the axis of the arch, the extrados surface will be an

Fig. 3.43 Controlling a dihedral angle with the bevel. Course "El arte de la piedra". Universidad CEU-San Pablo (Photograph by the author)

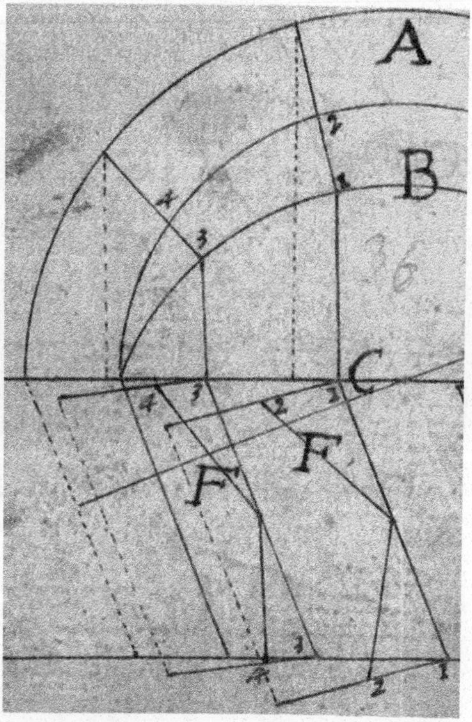

Fig. 3.44 Skew "groin arch" depicted using oblique projection, detail (Martínez de Aranda c. 1600: 47)

elliptical cylinder. As in the "groin arch", the intrados is formed by two ruled surfaces passing through their respective face arcs and a middle groin. The elevation included in Martínez de Aranda's manuscript is not an orthogonal projection, but rather an unusual oblique projection (Fig. 3.44; see also Martínez de Aranda c. 1600: 37, 45). The picture plane is parallel to the arch faces, but the projectors are parallel to the axis of the arch; hence, they are horizontal, but not perpendicular to the projection plane. In this way, both face arcs, and even the start of the groin or *arista*, are superimposed in this idiosyncratic elevation. Martínez de Aranda starts the dressing process by carving a block defined by two lines in point view passing through the voussoir corners at the groin and the extrados joints. However, this operation furnishes only the face side of the voussoir. He cannot carve the voussoir squarely as he has done in the previous examples since the joints are not orthogonal to the face planes. Therefore, to carry the face template from front to back, as he did in his theoretical explanation of the squaring method, he needs the intrados template of the block. According to the manuscript, the stonemason should construct the intrados template using a construction in true shape based in orthogonals to a horizontal oblique line, as in the preceding example. In addition to this, he will also mark the *arista* or groin in the template and transfer it to the stone. The result of this operation is a full template representing the true size and shape of the intrados of the intermediate block from which the voussoir will be carved, but not of the intrados of the definitive voussoir.

When tracing the bed joint templates, the stonemason should first construct the joint template of the enclosing block, using orthogonals to the intrados joints as before. Since the intrados template does not represent the final voussoir, it is also necessary to take away two wedges below the ruled surfaces passing through the face arcs and the groin. To do that, the mason may measure in the oblique elevation the distance between the projection of the groin and that of the face arc. Transferring this distance to the representation of the face joint in the joint template, he obtains a corner of the joint template of the shaped stone. Aranda's text is short but unambiguous: "you should construct the bed joint templates again and place them squarely as shown in the bed joint templates *F* and you should remove material as done in the preceding example".[78] That is, the mason is to construct a second V-shaped joint template to cut the voussoir to its definitive form, removing material it as in the basic orthogonal variant of this piece.

3.2.6 Dressing Cylindrical, Spherical and Warped Surfaces

As we have seen when dealing with materials in templates, most templates are rigid. This is obvious when the templates are made from wood, but even the paper templates mentioned by de l'Orme (1567: 106v), glued on a wood base, fall into this category. This fact has important geometrical consequences: on many occasions, the template for a non-planar face, such as the intrados of a voussoir, does not represent the *development* of a face, but rather its *projection* onto a plane passing through three or four of its corners. Frézier explains their use in these terms:

> … before dressing a curved[79] surface, one must first place the corners at their precise distance; these corners are the solid angles in the voussoir. Of these corners, at least three can be placed on a surface that is planar, conical or cylindrical; one can also place on the same planar surface two opposing straight sides. Thus, having formed a planar surface, or as the masons say, *dressed a face*, one can mark a large portion of the contour of the voussoir …[80]

[78]Martínez de Aranda (c. 1600, 48): *en las plantas por lechos formarás segunda vez las plantas por lechos para plantarlas al justo como parece en las plantas por lecho F y les robarás las piezas conforme se hizo en el arco por arista en la cara.* The idiosyncratic phrase *formarás segunda vez las plantas por lechos* seems to allude to the construction of the V-shaped template inside a enclosing rhomboidal one.

[79]In some descriptive geometry manuals, "curved surface" means "double curvature surface" or "non-ruled surface". However, here Frezier is using *surface courbe* with the meaning "non-planar surface".

[80]Frézier (1737–39: I, 310): … *avant que de creuser une surface courbe, on en doit premierement situer les bornes dans leur juste distance; ces bornes sont les angles solides des voussoirs, desquels il y en a au moins trois qui peuvent étre appliquez à une surface plane, & ordinairement conique ou cylindrique, on peut placer sur la méme surface plane les côtez opposez qui sont droits; de sorte qu'ayant formé une surface plane, ce qu'on appelle en termes de l'art* dresser un Parement, *on y peut placer une grande partie du contour d'un voussoir ….*

However, there is another alternative: to prepare templates that do not represent the projection of the intrados on a plane, but rather its development so that the template can be applied directly to the finished intrados surface.

Flexible templates for cylindrical surfaces. In the seventeenth century, Derand advocates the use of flexible templates:

> These templates are simply the form and the figure of the faces of the voussoirs, transferred into a thin material, which must also be flexible since the templates are applied to the concavity or the convexity of said voussoirs. To this end can be used thin wood boards or sheets of lead, copper sheets, cardboard, tin, etc.[81]

Derand (1643: 171, 173, 177) and Jousse (1642: 46–47) apply these templates to cylindrical surfaces, drawing a long series of connected templates (Fig. 3.45) which include, for example, all the intrados templates for an arch opened in a corner or round wall.[82] In another passage, Derand (1643: 159) uses for these templates the expression *coucher dans son creux*, literally "place in its hollow". Some details hint that flexible templates for cylindrical surfaces were not unknown in the sixteenth century. As we have seen, de l'Orme (1567: 74v-77r) draws both intrados and extrados templates for an arch opened in a curved wall. On close inspection, the distance between joints in the extrados template is slightly larger than the one in the built arch, while the width of the intrados template equals the one in the real member. This makes much sense since the intrados template can be applied to a planar surface, as explained by Frézier; later, the mason can keep carving to materialise a cylindrical surface. In contrast, a planar surface passing through the extrados joints would penetrate the mass of the voussoir.

All this suggests that de l'Orme is thinking about rigid templates for the intrados and flexible ones for the extrados. He uses three points *O*, *S*, *N*, to determine the curvature of the template edges (Fig. 3.46); this suggests he is adding the lengths of the chords *O–S* and *S–N* to get the total width of the template. It seems that Vandelvira uses the same technique since lines dividing each voussoir in half are also present in his manuscript (c. 1585: 21v, 22r, 22v; compare with 21r, for instance). However, the issue is far from clear, since De l'Orme's text is ambiguous: "you will take the length of the three points *O S N* and draw separately three parallel lines of the same length".[83] Moreover, he states in the text that the same method is to be used for intrados templates, but in this case, the distance between joints is the same in the built arch and the template.

---

[81]Derand (1643: 3): *Ces panneaux, à bien dire, ne sount autre chose que la forme & la figure des costez des voussoirs, transferée sur quelque matiere mince et deliée, laquelle doit etre aussi flexible, quand les panneaux sont pour estre appliquez dans la concauité, ou sur la conuexité desdits voussoirs. A cela donc pourront seruir les ais de petite épaisseur, les lames de plomb, ou de cuiure, le carton ou le fer blanc, &c.* See also Frézier (1737–1739: II, 12).

[82]MS 12.744 in the National Library of Spain, connected with the name of Juan de Aguirre, uses the same method; it has been dated to the mid-seventeenth century, although the date is based mainly on the presence of this technique.

[83]De l'Orme (1567: 75v): *Vous prendrez donc la largeur des trois points OSN, et en tirerez à part trois lignes de même largeur, qui seront parallèles …* Transcription is taken from http://architectura.cesr.univ-tours.fr.

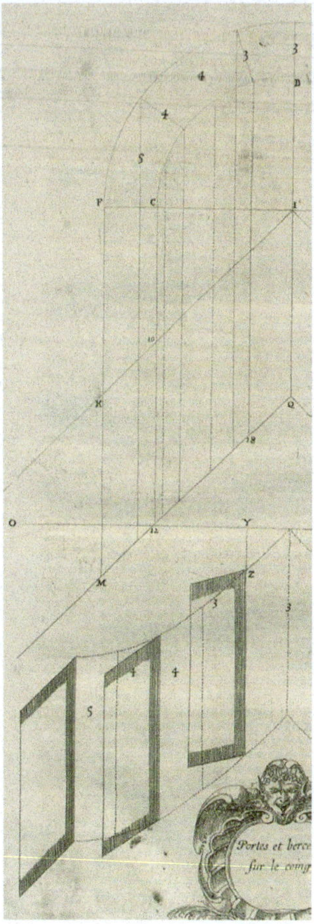

Fig. 3.45 Corner arch, showing flexible templates for cylindrical surfaces, detail (Derand [1643] 1743: pl. 81)

Templates for spherical surfaces. Remarkably, flexible templates were also used in the sixteenth century for spherical elements, although the sphere is not a developable surface. As we will see in Sect. 9.1.1, the procedure starts by carving a portion of a spherical surface with the aid of the templet, taking into account that the sphere has the same curvature in all points and directions; next, the mason should inscribe cones, which are easily developable, in the spherical intrados. Derand explains the use of these templates in these terms: "on the hollowed surface, you should place the template for the course of the voussoir you are dressing; and after placing its sides on the concave surface, you should mark the outline of their ends and bed joints ...".[84] Josep Gelabert states clearly the need to use flexible templates in this dressing procedure: "... these are the intrados templates; I should make it clear that they cannot

[84]Derand (1643: 356): *sur le parement creusé, comme dit est, se couchera le paneau de l'assise, à laquelle appartient le voulsoir que vous auez en main; & ayant repairé ses costez dans la doüele creuse de vostre voulsoir, vous en tracerez les ioincts des bouts & des lits.*

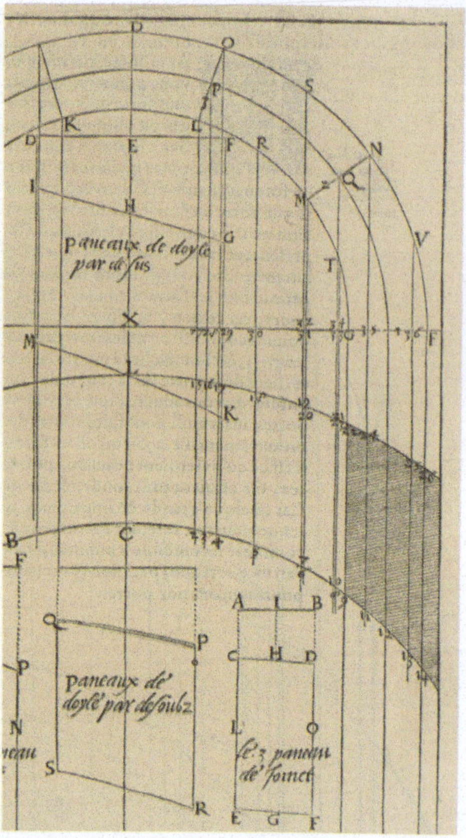

Fig. 3.46 Intrados and extrados templates for an arch in a round wall, detail (de l'Orme 1567: 77r)

be made from wood, but rather from cardboard, double paper or something that can be bent".[85] Of course, when applying these conical templates to spherical surfaces, the fit will not be perfect, but the practical implications of this fact are negligible, as Rabasa (1996a; 2000: 174–175) has verified in practice.

Templates and warped surfaces. There are other, more problematic, instances of the use of templates for non-developable surfaces, in such elements as skew and splayed arches, rere-arches or helical staircases. A paper template cannot represent such surfaces without deformation; in fact, a property of ruled warped surfaces is that two adjacent generatrices are neither parallel nor convergent (Bosse and Desargues 1643: 30–31). Thus, two consecutive intrados joints will not be coplanar, and no template can lean on both intrados joints; in Frézier's words,

> … surfaces with more than three edges can have their angles in different planes since they can be divided into triangles; thus, a template with four sides can be divided into two triangles

[85]Gelabert (1653: 50v) *axo son las plantas de duella advertint que nos poden fer de post sino que an de ser de carto o de paper dobla o de qualsevol altra cosa sols que es pua doblegar*. Transcription is taken from Gelabert/Rabasa 2011.

> … a hollow roof tile, in spite of having a conical curvature, adapts so well to a board that its four corners touch it. A part of a cylinder, a portion of a sphere, such as the faces of the voussoirs of regular vaults, has the same property. It is not the same with a portion of the Marseille rere-arch or the rere-arch of Saint Anthony …[86]

Nevertheless, sixteenth-century Spanish manuscripts include on many occasions templates for warped voussoir faces, known as *engauchidas*, from the French *gauche*. Alonso de Vandelvira uses them only as an auxiliary step to construct the *saltarreglas*, that is, the segments representing the face joints in order to determine their angles with the intrados joints: "Rere-arches are always dressed by squaring; although templates are constructed, they are only used to obtain the *saltarreglas*".[87] However, other authors, from Alviz (c. 1544: 15r, 16r, 20r) to Juan de Portor y Castro (1708: 4v, 7v, 9r) use templates for warped faces without such warning. In particular, Martínez de Aranda (c. 1600: 119–120; see also 45–46, 113, 222–223) acknowledges clearly that the intrados of a skew rere-arch is *engauchida*, that is, warped; however, he is not deterred by this fact and constructs templates boldly. Portor (1708: 28v) is even clearer: he includes in his manuscript a *Capialzado engauchido por plantas*, that is, "Warped rere-arch solved by templates".

Martínez de Aranda's procedure (c. 1600: 100–105) may be interpreted, albeit anachronically, in the following terms: he rotates a warped intrados face around one of its sides, in order to bring it to a horizontal plane, obtaining a result a template that seems to represent at first sight the intrados in true size and shape. Since two corners of the voussoir are on the axis of revolution, they will not move; the position of the third corner can be determined taking into account that it will rotate in a plane that is perpendicular to the axis and its distance to the first corner may be taken from the elevation. However, this procedure cannot be used for the fourth corner, since it is not coplanar with the other three. In fact, when the third corner reaches the horizontal plane, the fourth one will not have reached it, or else it will have passed along it. Thus, Martínez de Aranda (c. 1600: 82–85; see also Calvo 2000a: I, 247–252) seems to compute the position of the fourth corner simply by measuring their distances to the second and third corner and transferring them to the intrados template. Such a template represents in true size and shape the four sides of the voussoir intrados and one diagonal while distorting the other diagonal; although it does not represent the true size and shape of the intrados of the voussoir, it can be useful in the dressing process. It makes it possible to mark the first three corners, dressing the plane that passes through these three points; once this is done, the mason

---

[86]Frézier (1737–39: I: 311): … *les surfaces de plus de trois côtez peuvent avoir leurs angles en differens plans; puisqu'elles peuvent être divisées en triangles; ainsi une Doele plate de quatre côtez peut être divisée en deux triangles … une tuile creuse, quoique d'une courbure Conique, s'adapte si bien sur une planche que ses quatre angles la touchent. Une portion de Cylindre, une portion de sphère, telles que sont celles des voussoirs des voutes régulieres, a la même proprieté. Il n'en est pas de même d'une portion d'Arriere-Voussure de Marseille ou de St. Antoine* …. For the Marseille and Saint Anthony rere-arches, see Sects. 7.3 and 7.2.3, respectively.

[87]Vandelvira (c. 1585: 46r): *Los capialzados todos son por robos que aunque están aquí en los demás las plantas sacadas sólo sirven para que por ellas se saquen las saltareglas.* Transcription taken from Vandelvira/Barbé 1977. Translation by the author.

can rotate or fold the template around this diagonal to dress the other triangle in the intrados and materialise the fourth corner.

All this recalls the use of folded templates in Jousse's solution for a groin vault (1642: 156–157; see Pérez de los Ríos and García Alías: 2009). This piece does not feature a warped intrados since the web surfaces are cylindrical, but the voussoirs that cross the groin overlap two different intrados surfaces. Jousse uses V-shaped templates for these voussoirs; by folding the template along a line coincident with the groin, the mason can control the dressing of both faces.

Later on, templates for warped surfaces are used by Frézier, although there is no reason to suppose he knew the manuscripts of Martínez de Aranda (c. 1600) and Portor (1708); the available evidence shows both were circulated locally (Calvo 2000a: I, 111–112, 251). Moreover, his approach is quite different. The first step in Frezier's method (1737–39: I: 310–311, II: 35–37, 445–447) involves a plane passing through three corners of the intrados of the voussoir, as in Martínez de Aranda. However, he does not rotate or fold the template by its diagonal; rather, he projects the fourth corner onto the plane determined by the other three vertexes, since he considers that

> warped surfaces cannot be defined by other means that their distances to planar surfaces, measuring the lengths of lines perpendicular to this plane, or lines whose slope is known, ending on different points of the warped surface to which it is compared".[88]

This leads Frézier to a precise dressing method: after marking the intrados template on a planar surface, he dresses the face of the arch using the square, draws the face joint on it using the bevel, and measures along this line the distance of the fourth vertex to the planar surface on which the template has been scored. Next, he can dress the bed joint and materialise the intrados using the ruler, since the surface, although warped, is a ruled one; he goes as far as dividing into equal parts the edges where the ruler leans, in order to assure a correct execution.

3.3 Placement

3.3.1 Transportation

It is too easy to forget that transportation of materials from the extraction point to the worksite posed formidable challenges for pre-industrial societies. In the twelfth century, eight statues of oxen were placed in the highest sections of the towers of Laon cathedral in remembrance of a miraculous event. The roofing of the cathedral required long beams, but the city is located on a high plateau. It seemed impossible

---

[88]Frézier (1737–39, I: 310–311): *aussi on ne peut connoitre les surfaces courbes, qui ne sont pas régulieres, que par leurs distances à des surfaces planes, en mesurant les longueurs des lignes perpendiculaires à ce plan, ou don't l'inclinaison est connue, terminées à differens points de la surface courbe, à laquelle on la compare.*

to carry the logs to the acropolis until many large oxen appeared suddenly as if sent by the angels and raised the load to the top of the hill. The legend highlights the difficulties faced by medieval builders when carrying materials to the worksite in a world where broad roads, suitable for carts, were all but inexistent.

Thus, on many occasions builders resorted to transportation by boat. Stone was brought across the English Channel from Caen to Canterbury cathedral, Battle Abbey, the Tower of London or the Westminster palace; different materials were brought from different locations in the Crown of Aragon to the Castel Nuovo in Naples, including stone from Santanyí, in Majorca. In these cases, geography made the use of boats unavoidable. In contrast, when the Chapter of Seville cathedral decided to build their temple entirely in stone, in a city lying in the middle of the Guadalquivir valley, they had to employ stone extracted from several underground quarries in Puerto de Santa María, in the bay of Cádiz. Rather than using the impracticable roads of the period, they brought the stone along the bay and the Atlantic coastline to Sanlúcar de Barrameda, and then up the course of the river to Seville (Erlande-Brandenburg 1993: 106; Filangieri 1937: 303, 306; Rodríguez Estévez 2010: 118–128).

Even in a well-financed Renaissance worksite, such as the Escorial, transportation was a major economic and organisational issue, placing strong constraints on stonecutting procedures. A remarkable debate about whether stone should be dressed at the quarry or the building site was held in 1576, as several authors have shown (Sigüenza ([1605] 1907: III, 440–441; Kubler 1982: 27, 37–38, 55–56, 80–81; Wilkinson 1985: 236, 239; Wilkinson 1993: 270–272; Bustamante 1994: 411–413). At the suggestion of Juan de Herrera, King Philip II himself decided that the stone was to be dressed at the quarry to speed construction. However, the masonry foremen, Pedro de Tolosa and Lucas de Escalante, procrastinated arguing that they did not understand the instructions given by Juan Bautista Cabrera, supervisor of the oxcarts, who did not belong to the building trades. Cabrera's intervention, although not usually stressed, hints that the bottleneck lay not only on the stonecutting procedure, but also on the capacity of the carts. On such a vast working site, with ten different contractor teams working in the main church, available oxcarts were unable to feed the huge construction organisation. Thus, bringing dressed stone rather than rough blocks reduced the load, optimising transportation; also, stone was picked directly from the carts by the cranes, minimising the need for hoisting equipment. The main problem raised by the new system was the damage to dressed blocks brought about by the bad state of the roads, stressed by Tolosa and Escalante. Thus, it was decided that stone would be carried to the construction site with the addition of *un grueso de cordel*, that is, the width of a cord, over visible faces. Even with the new system, transportation still demanded huge resources. Large expanses of grassland were set aside for the 600 oxen hauling the carts. Some pieces posed almost insurmountable challenges, such as the jambs of the main entrance, carried by 30 pairs of oxen (Sigüenza [1605] 1907: III, 451, 457; Bustamante 1994: 295, 427; for other instances of oversized blocks, see Fitchen [1986] 1989: 155–157).

Thus, roads, carts, oxen, and pastures were essential elements of pre-industrial stone construction technology. The available auxiliary means influenced the choice of stone size: for example, the less-than-ideal state of this factor in the High Middle

Ages led to the widespread use of rubble in Romanesque architecture. Dressing techniques are directly related to the constraints set by oxen and carts, as shown clearly by the Escorial example. However, other elements in the system, such as hoisting means, falsework and placement control, are also essential, influencing the choice of stone sizes and dressing methods; I will deal with them in what follows.

3.3.2 Falsework

Father Sigüenza described the removal of the formwork for the central dome in the Basilica of the Escorial in these terms:

> … the crossing dome, so full of timber, scaffolds, cranes, formwork, boards and beams, so thick and so dense that the bond and strength of so much wood was admirable; everything was necessary for the safety of such a large and heavy machine … Everybody was afraid to remove the centring, scaffolding, cranes and all the timber in the church; at first sight, it seemed something big, intricate, difficult, dangerous, nobody dared to address the issue … Friar Antonio, the foreman, to whom God had given the light to go out of this darkness, managed to clear away everything easily and safely and removing such host of beams, struts and boards appeared … a shining temple, which rejoiced the souls with its greatness, proportion and beauty.[89]

Such elaborate Renaissance falsework was the result of a gradual evolution during the Middle Ages. Although definitions fluctuate, I will differentiate between *formwork* or *shuttering*, that is, the artifact that replicates, either literally or schematically, the intrados shape of the element under construction; *centring*, the horizontal structure supporting the formwork; and *shoring* or *underpinning*, the vertical or slanting struts that transmit the load of the element to the ground or previously built elements; I will use *falsework* for the ensemble of the three categories. A full study of the subject would be enough for several books; a systematic summary can be found in Fitchen ([1986] 1989: 85–129, 155–187). I will deal here with just a few examples in order to suggest how these apparently ancillary issues influenced architectural form and constructive methods, particularly in the Middle Ages.

Formwork. In contrast to hoisting devices or even scaffoldings, graphical sources about formwork are quite scant. However, a few instances of High Middle Age formwork have been preserved in place in Spain; all of them are placed in small, out-of-the-way spaces where nobody bothered to remove these auxiliary structures. The oldest of these is in a tower in the church of San Millán in Segovia, dating from

---

[89] Sigüenza ([1605] 1907: III, 464): … *la cúpula del cimborrio, y por de dentro tan llena de madera, de andamios, grúas, cimbras, tablados y vigas tan gruesas y tan espesas, que ponía admiracion y era de ver la trabazón y la fuerza de tanto enmaderamiento; todo era menester para la seguridad de tan grande máquina y peso … Había puesto mucho miedo el quitar las cimbras, andamios, grúas, y todo el enmaderamiento de la iglesia; mirado así a bulto espantaba, parecía una cosa grande, intrincada, dificil, peligrosa, no se atrevía nadie a entrar en ello … El obrero fray Antonio, a quien había Dios dado claridad para salir de estas oscuridades, lo hizo quitar con harta facilidad, sin peligro, y apareció luego en quitando tanta multitud de vigas, maderos y tablas, … un templo clarísimo, que alegró al alma con su grandeza, proporción, hermosura.*

the tenth or eleventh century, probably the oldest bell tower in Castile and perhaps the first in the Iberian Peninsula. The tower, covered by a pavilion vault executed with ribs in ashlar and severies in concrete, retains portions of the original formwork. Building on a study by Merino (2005: 776), Sobrino and Bustos (2007: 908, 909) have presented a reconstruction of these temporary structures, including full wooden ribs at the axes of the vault and a continuous web under the vault surfaces, as befits a concrete vault.

The tower in the church of Aldeaseca, in central Spain, dating from the twelfth or thirteenth century, includes several pointed barrel vaults. Although the formwork has disappeared completely, some boughs have been preserved within the rubble masonry; according to Sobrino and Bustos (2007: 908–909), they are remnants of a vegetal layer placed between the formwork and the actual vault in order to ease the removal of formwork. The castle in Molina de Aragón, a town which is actually in Castile, although close to the Aragonese frontier, retains remnants of formwork in two towers. A pointed barrel vault with transversal arches covers the tower of the Velador. There are footprints and actual remnants of the boards going from one arch to the next one; this suggests that the transversal arches acted as centring, supporting the boards of the formwork. In the case of the Homage Tower, the vault is quadripartite; again, the ribs act as centring, and the remnants of the formwork are laid in a rough and ready way; some sections near the corner go from the walls to the ribs, while the centre sections lean on the corner ones (Sobrino and Bustos 2007: 909–910).

Of course, such simple devices are not sufficient to support structures as large as the great Rhineland cathedrals, Cluny III, Saint-Sernin in Toulouse or the cathedral of Santiago de Compostela. Extrapolating the Castilian examples, some hypotheses may be put forward. Barrel vaults, either pointed as in the Cluny school or round, as in the Santiago group, may have been supported by a continuous series of planks (Bechmann [1981] 1996: 141), particularly where small rubble masonry or concrete was used. In theory, carefully hewn stones, such as those used in some Languedocian and Provençal examples, allow the use of a discontinuous set of planks, but the formidable weight of such vaults argues against this solution.

Gothic architecture is tied to a remarkable change to these systems. Bechmann ([1981] 1996: 25–29, 42–51, 132–142) has remarked that the depletion of forests in the twelfth century compelled builders to use lighter formwork; this led to, or at least fostered, the use of the rib vault. Although there is very little hard evidence about Gothic formwork and centring, Fitchen and Bechmann have advanced a series of interesting, although debatable, hypotheses. First, severies may have been supported by a discontinuous set of horizontal elements, either single planks or trusses with a curved upper component, a lower straight one and connecting pieces. This fits well the nature of the surfaces of mature Gothic webs, which are generally double curvature surfaces; however, Maira (2015: 40, 118, 122, 142, 166, etc.) has shown that in most sexpartite vaults, in both France and Spain, the intrados surface is a ruled one, with a few exceptions. Although the use of trusses with parallel sides cannot be discarded, solid planks fit these single-curvature surfaces better.

Two alternatives to this system have been proposed: the last course in a severy under construction may have been supported by stones hanging from ropes, stabilising them and allowing the masons to dispense with any formwork (Fitchen [1961] 1981: 180–189; Bechmann [1981] 1996: 186). There are some problems with this hypothesis. First, it requires regularly hewn stones; thus, it cannot be applied to Early Gothic vaults, whose severies were built in rubble. Second, this method may be efficient in middle courses; however, it is unnecessary in the lower courses, where bed joints are almost horizontal, and ineffective in the upper courses, where bed joints are almost vertical. Third, it is better fitted to the Angevin system of domical vaults, using diagonal and axial ribs, where the length of the upper courses is shortened. In contrast, in the Île-de-France system, which dispenses with axial ribs, the upper courses in the severies are the longest ones. It should also be stressed that the main evidence put forward by Fitchen ([1961] 1981: 180–183) to argue for the historical use of this system is a communication to a British journal in 1831 by the German architect Lassaulx (1830–1831: 226), who reported the use of this procedure in contemporary Vienna. He stated that the adherence of mortar helped the system; In any case, neither Lassaulx nor Fitchen discuss whether mortars used in nineteenth-century Austria were similar to slow-setting medieval ones. All in all, Fitchen himself ([1961] 1981: 187) admits that this system may have been complemented by light formwork in the higher sections of rib vaults.

Bechmann ([1981] 1996: 185–186) probably had these problems in mind when he suggested that masons could have used the surplus of gypsum in the Paris basin to prepare formwork in this material, making good use of its short setting times. This may explain the contraposition between the long spans between ribs in Île-de-France and Picardy and the shorter ones in Angevin and Aquitanian vaults. However, such a hypothesis raises a new problem. As we have seen in Sect. 1.1, when building in brick or ashlar, formwork provides support but also acts as a resource for general formal control. By its own nature, plaster of Paris cannot fulfil this task, so another formal control device, such as a set of light boards, must have been used.

The revival of single tier construction, brought about by the aesthetical ideals of the Renaissance, once again required the use of full or half-full formwork and corresponding centring and shoring again; however, both the scale and the complexity of buildings had generally increased, posing new challenges for builders. Documentary sources, up to the twentieth century, are still quite scarce and, moreover, the few available images are concerned with centring, while formwork is represented in edge view, so to speak. In any case, a drawing by Jacques Gentillâtre (c. 1620: 465v) shows full formwork. Later, two sheets in Rondelet ([1802–17] 1834: pl. 127–3, 4; 128–3), show spaces between planks, but they are no wider than the planks themselves. Twentieth-century restoration practices on Renaissance or Baroque structures (Huerta and Rabasa 2001: 69; Fitchen [1986] 1989: 104–105) point in the same direction: full or half-full formwork is the norm.

Centring. We have seen that in Romanesque construction, transverse arches frequently support formwork; however, they must be supported while being built, so they require their own centring. Extrapolating from later examples, this centring may have consisted in a frame made up from a series of planks following the outline of the arch, a straight beam joining its springings and a set of struts connecting the upper and lower parts in the form of a truss. If the breadth of the arch demands it, double frames are used, and their top edges are joined by a series of short boards known as *lagging*.

However, for larger arches, centring may have begun much above the springers, as suggested by projecting stones in the Pont-du-Gard, a remarkable Roman bridge near Avignon. The rationale of this system takes into account that friction between voussoirs is usually sufficient to keep them together until the slope of bed joints reaches about 30º. Thus, centring may have been supported by the projecting stones, in the Pont-du-Gard and in many other constructions as well. Generally, such protrusions were removed upon completion of the structure; however, they were left in place in the Pont-du-Gard for unknown reasons (Fitchen [1986] 1989: 161–163).

Gothic ribs also require strong centring, as do transverse arches in large Romanesque constructions. Early Gothic vaults probably inherited the same system, using double frames in the shape of a truss, joined by laggings at their upper edges; this device fitted well the rectangular sections of Early Gothic ribs. A late image of such a centring system can be found in a fourteenth-century French miniature (Colombier [1953] 1973: 46),[90] showing a mason placing the voussoirs of a round arch. The centring adopts the form of a round string of planks in the upper edge, a straight plank connecting both imposts, and a post and several braces joining both. Although the drawing is not completely clear, it seems that the post and two braces converge in the midpoint of the horizontal plank, while another pair of braces meets the horizontal plank at other points at springing level; thus, the truss is not completely triangulated. In mature and late Gothic, laggings were spaced more widely and transferred to the lower edge of the upper string of the truss; this evolution fostered the emergence of V-shaped rib sections, with lateral portions resting on top of the planks (Fitchen [1961] 1981: 151–153).

It has been suggested that the efforts to make the radii of diagonal and transverse ribs equal may have stemmed from a desire to reuse centring. This raises a problem, since round and pointed arches may be built with the same radius; however, their lengths are generally different. Thus, the centring for a semicircular diagonal rib may be used for a perimetral pointed arch, rotating it to place the surplus length under the springer. Moreover, from the thirteenth century on, Gothic builders used the *tas-de-charge,* that is, the practice of carving the initial sections of all ribs converging on a single springing—initially transverse and wall arches and diagonal ribs, later tiercerons—with horizontal bed joints. This allowed them to begin the centring at about one third of the length of each rib, as in the Pont-du-Gard, avoiding at the same

[90]British Library, Royal MS 14 E III, fol. 85v. The manuscript includes three different texts from the Lancelot-Grail Prose Cycle: "Estoire del Saint Graal", "La Queste del Saint Graal" and an abridged version of the "Morte Artu". The first miniature in fol. 85 v, the one that is relevant to us, depicts a king supervising the building of a church. See http://www.bl.uk/manuscripts/.

time the problems posed by the difficult intersections of both ribs and centres at the sprigers.[91]

With the advent of the Renaissance, builders returned to top lagging for arches, or formwork placed directly on independent centres for vaults without transverse arches. Sources are a bit more frequent; a general impression of the shape of centring may be grasped from Gentillâtre (c. 1620: 465v-466r), Jousse ([1627] 1650: 147–150) Zabaglia ([1743] 1824: plates 4, 5) or Rondelet ([1802–17] 1834: plates 126–128). The layout of the braces is quite diverse, verging on the fantastic, but there seems to be a trend toward triangulation, although there are exceptions. An interesting novelty is the application of structural calculations to centring by Rondelet ([1802–17] 1834: III, 169–173).

Another crucial aspect of centring is its removal. If taken away suddenly, the structure may suffer dynamic stresses leading to excessive deformation or collapse. For small arches, simply pushing a slightly oversized keystone between the supporting voussoirs will cause a small lateral expansion of the arch that will be sufficient to separate it from the centring (Fitchen [1986] 1989: 100). For larger pieces, wedges were usually placed at key points between the formwork and the centring, or between the underpinning and the centring. Once the voussoirs were placed over the formwork, the wedges were removed slowly in order to lower the formwork, allowing the element to adapt gradually to mechanical tensions and assume a slightly different shape (Fitchen [1986] 1989: 101–102; Palacios and Martín 2009: 57). Even with this system, accidents could occur (Huerta and Rabasa 2001: 69). An interesting alternative is the use of small bags filled with sand; pinching the sack, a jet of sand will project as a result from pressure so that the element may settle gradually. All this may explain the space between formwork and centring in some drawings by Gentillâtre (c. 1620: 466r), Zabaglia ([1743] 1824: pl. 4) and Rondelet ([1802–1817] 1834: pl. 128-1, 2).

Shoring, scaffolds and ladders. Unless centrings are supported by projecting stones, in Pont-du-Gard fashion, a vertical structure known as *shoring* or *underpinning* must bring the load of the element, the formwork and the centring to the ground. There is still less available evidence about these systems in the Romanesque period. Gothic builders also tried to reduce the need for timber in temporary structures below the formwork, in different ways. First, the general lightening of vaults brought about by the use of ribs and webs implied a remarkable reduction of the need of timber in underpinning; moreover, the use of the *tas-de-charge* limited the load that had to be supported by these structures. Second, Fitchen ([1961] 1981: 189–190) suggests that, instead of raising the provisional structures from the ground up, in many cases they started from triforium or gallery level with canted struts; for greater safety, they were then tied to the piers in the clerestory, before stained-glass windows were put into place. For smaller structures, they may also be tied to holes opened in the masonry, which are left open in many medieval constructions.

[91] Willis ([1842] 1910: 4); Choisy (1899:II, 272); Fitchen ([1961] 1981, 135–138); Bechmann ([1981] 1996: 210–214). See also Maira (2015: 60, 100, 106, 118, 126, 132, 146, 159, 180), for examples of sexpartite vaults with and without *tas-de-charge*.

Scaffolds are also built in many occasions with putlogs embedded in these holes or hung from belfries, spires or flying buttresses; another set of protrusions in the spandrels of the Pont-du-Gard may have served for this purpose (Fitchen [1986] 1989: 87, 163–164). However, Fitchen ([1986] 1989: 85–86) stresses the difference between *masons' scaffolds*, with a double row of posts, one placed against a wall and the other at the open side of the structure, and *bricklayers' scaffolds*, which dispense with the wall posts, relying on the union of the putlogs with the wall. However, he remarks (Fitchen [1986] 1989: 87) that medieval builders eschewed scaffolding when possible, using the building itself as an alternative. This is confirmed in many medieval images (Colombier [1953] 1973: 28, 32, 33, 46, 48) showing ladders in the forefront, almost as an icon of building in progress, while the scaffolding that was to be reached through these ladders is set back or absent. Even when using scaffolds, where possible builders avoided placing heavy hoisting equipment on them (Fitchen [1986] 1989: 85, 90); we will see an interesting example in the next section.

3.3.3 Hoisting

Antiquity. Once the stones are dressed, either at the worksite or the quarry and brought to the construction site, they must be hoisted to carts or their final position. Again, these issues may seem a non-problem after the Industrial Revolution, but they posed formidable challenges before mechanisation. Vitruvius mentions several machines designed to solve these problems, such as a *gin* or small crane formed by one or two canted struts held in place by three or more tight ropes. Such a device was complemented with some machines to increase its lifting power: a windlass or capstan, a wheel and a tryspast or polyspast. The wheel includes a large cylinder and an axis connected with a pulley or tryspast; in this way, the device reduces the movement of the ropes and increases the lifting power of the ensemble (Fleury 1993: 98–112; for a medieval version of the instrument, see Fitchen [1986] 1989: 92–93). A large version of such cranes is depicted in the well-known relief of the Hateri (Fleury 1993: 124–127; Fitchen [1986] 1989: 94), showing several workers inside the wheel, treading on it to exert rotating power. Such depiction seems to be an artist's impression rather than a realistic description; later representations of these wheels, such as those in Breughel paintings, make it clear that no more than two workers can cooperate in this effort.

Heron of Alexandria offers additional data on these issues, in particular about the tryspast and the polyspast; he also mentions the instrument known as three-legged lewis or Saint Peter's keys. Large pincers, fastened to holes in the sides of stones, were known in Antiquity; however, they are dangerous when dealing with soft stone. In this case, masons can carve a dovetail-section mortise, broader at the bottom and narrower at the opening, in the upper side of a block. Then, an ingenious device, the three-legged lewis, formed by three separate pieces, can be inserted inside the box, placing the side pieces first and the centre one last; once this is done, a rope can be passed through a hole in all three sections of the device in order to lift the stone.[92]

Middle Ages. Such sophisticated mechanisms seem to have been forgotten in the disruption of the High Middle Ages. Workers appear in the *Biblia Sancti Petri Rodensis* or the manuscript of Herrarde de Landsberg, both from the twelfth century, carrying stones in their shoulders; in the stained-glass windows of Châsse de Mozac, Beauvais and Chartres, they carry them in a handbarrow held by two workers (Colombier [1953] 1973: 19, 31, 33, 55; Gimpel [1958] 1980: 30, 35, 38, 43, 44; Erlande-Brandenburg 1993: 102, 109). Of course, such constraints justify the small size of stones used in Early Romanesque architecture. Later on, capstans and windlasses, usually placed in scaffoldings, reappear; in a further refinement, hoists with large handspikes, often in the shape of a cross, set at ground level, are connected to pulleys placed over scaffoldings or in simple L- or T-shaped cranes (Colombier [1953] 1973: 19, 25, 28, 29, 31, 48, 50, 99, 110; Gimpel [1958] 1980: 34, 37, 79; Erlande-Brandenburg 1993: 24, 100, 118, 122; Fitchen [1986] 1989: 92–94); of course, such technical advances, or rather re-discoveries, underpin the gradual use of larger stones in the Gothic age.

In the later phases of this period, construction technology seems to borrow many concepts from maritime and mining trades. The crane in the chapel of Alphonse V in Valencia was set up by mariners; also, ropes were bought from maritime suppliers. A few decades later, the Chapter of Murcia cathedral asked the Marquis of Vélez for a blacksmith from his alum mines in Mazarrón to help in the cathedral crane, probably making fastenings, pulleys or tryspasts (Zaragozá 1997: 29; Gutiérrez-Cortines 1987: 117; see also Fitchen [1986] 1989: 94). The cranes in Breughel's Towers of Babel also seem to derive from maritime technology, in particular for the closed construction of the jib of the crane, which resembles the existing Hanseatic crane in Gdansk harbour. In any case, both the wheel in the Tower of Babel and the crane in Gdansk show that only one or two workers can walk inside the wheel; their treading actions cause the wheel to rotate, while the large radius of the wheel brings about a remarkable reduction of the force needed to hoist loads (see also Colombier [1953] 1973: 26–28; Erlande-Brandenburg 2003: 111, 122, 128).

The manuscript by Hans Hammer (c. 1500: 8r-9v, 13r, 14r-15v) includes several drawings of cranes (Fig. 3.47), reflecting further developments. Instead of placing the jib of the crane starting from ground level, most cranes include a frame at the base in the shape of a pyramidal frustum, furnished with a windlass with pairs or crosses of handles; none of them includes a treading wheel. The jib is a developed version of medieval ones; two strongly slanting struts pass between another pair of bars, holding a pulley. Even more interesting is a series of drawings depicting a series of tryspasts and polyspasts.

The Early Modern period. The Renaissance rediscovered the technical solutions of Vitruvius and Heron, albeit with some misunderstandings. Fra Giocondo takes

[92]Fleury (1993: 119–121); Warland ([1929] 2015: 72); Martines (2016); Peroni (2016); Carriero and Sabbadini (2016); see also Bruno (2016) about actual holes for this device in the stones of the Colosseum.

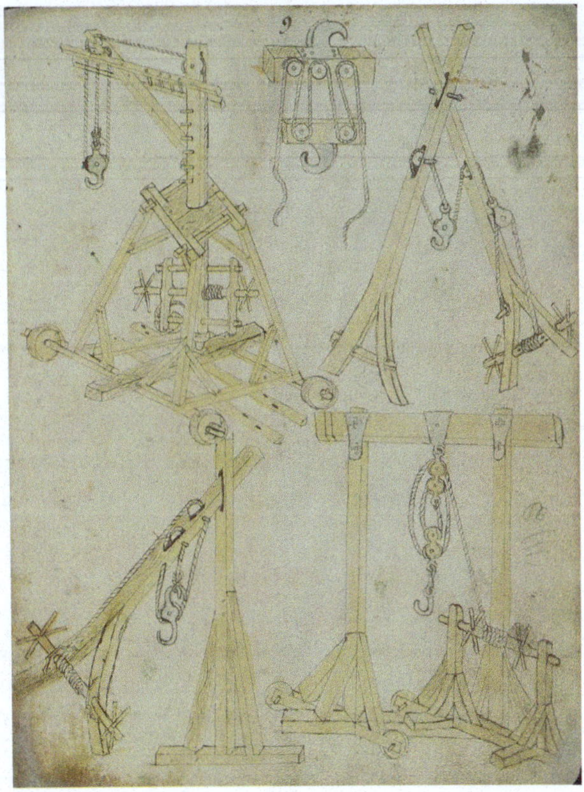

Fig. 3.47 Cranes, pulleys, polispasts and other hoisting devices (Hammer c. 1500: 9r)

it for granted that Vitruvian cranes include three slanted struts, forming a pyramid (Fig. 3.48); of course, this version of the instrument is rather cumbersome and almost useless for the task of lifting a stone and placing it on a cart, not to say at the upper part of a wall or a vault (Vitruvius/Fra Giocondo 1511: 95v). What is more shocking is that Lázaro de Velasco repeats the mistake; in a separate drawing, he includes drawings of pulleys with one, two and three sheaves; those with two and three wheels seem to belong to a single pentapast, but Velasco represents both separately. All this suggests that Velasco knew of the existence of tryspasts and pentapasts but could not describe the concept easily (Vitruvius/Velasco c. 1564: 146v). It seems that such misunderstandings came to an end in the late sixteenth century. Barbaro (Vitruvius/Barbaro 1556: X, 262) deals with a one-masted derrick, which he probably knew in the Arsenal of Venice, although he also represents a three-strut gin (Fig. 3.49). The Vitruvian gin or *cabrilla* used in the Escorial was improved by Juan de Herrera, who was also the author of a manuscript on *Architectura y machinas*, dealing with pulleys and tryspasts (Sigüenza [1605] 1907: III, 441; Llaguno 1829:II, 129; Kubler 1982: 27; Herrera c. 1575).

In any case, the *cabrilla* or gin must have been limited in the Escorial to ancillary tasks, such as lifting stones in order to place them on carts. The well-known Hatfield

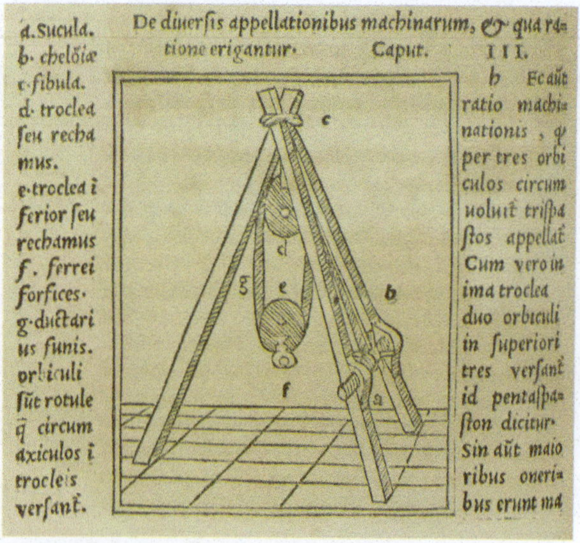

Fig. 3.48 Gin (Vitruvius/Giocondo [1511] 1523: 174v)

House drawing shows no fewer than twelve large cranes, analysed by Íñiguez (1963a, 1963b), García Tapia (1990: 175–181) and Lorda (1997, 2000). The larger ones, used by the ten contractor teams of the Basilica or Main Church, include a large two-storey wooden frame; a large wheel, akin to those of Gdansk or Brueghel's pictures, is housed in the second storey. On top of this storey, a cantilevered triangular frame goes out of the footprint of the cage, allowing the hoisting of stones; the drawing also shows a long row of two-wheeled carts leaving stones in the floor; from this point, they are picked by the large cranes.

The notes and comments in Lázaro de Velasco's translation of Vitruvius provide further information about these hoisting machines. For example, he instructs the crane operators to moisten the ropes with vinegar or seawater, since wet ropes "endure the fire that originates as the rope winds [around the wheel]".[93] He prefers the three-piece lewis to pincers; in fact, about 1575 at the Escorial the pincers were being replaced by Saint Peter's keys (Vitruvius/Velasco c. 1564: 147r-147v; see also Fleury 1993: 119–121; Lorda 1997: 91–92).

He also describes (Vitruvius/Velasco c. 1564: 156r) a windlass supported by two pairs of timbers, eight feet long. Each pair of these struts lies on a crossbar. A smaller crossbar also joins these timbers and, in turn, both pairs of struts are joined by another pair of timbers. The smaller triangle between both struts and the smaller crossbar forms a socket supporting the end of the windlass, with two pairs of handspikes allowing workers to operate the machine.

---

[93] Vitruvius/Velasco (c. 1564: 147): *resiste al fuego que se causa del rodear de la soga.* See also Calvo (2006).

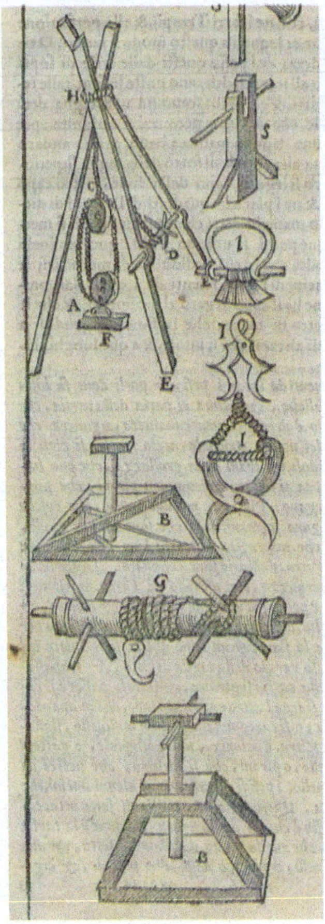

Fig. 3.49 Gin, windlass, pincers and other hoisting equipment (Vitruvius/Barbaro 1567: 446)

Along with an ordinary hoist, Velasco describes a capstan, the *torno encarcelado*, which may be literally translated as "prisoner windlass". The machine is built on two planks assembled in a T-shape and securely fastened to the ground. From each of the three arms of the T starts a strut; all three meet the crossing point of the T. The capstan is fastened by one end to the crossing point of the T and by the other end to the three struts; it carries two handspikes for workers to rotate the hoist. Velasco makes it clear that, after winding around the capstan, the rope must pass through two pulleys. The first of these pulleys must be fixed to the ground, far from the capstan, while the other pulley should be set on a high place. Thus, the hoist can be set in the centre of the courtyard of a building under construction, lifting loads to any location in the building. However, he also mentions another use of this device: the prisoner hoist can be used in a *suelo de bobeda*, literally "the floor of a vault"; taking into account the passage by Rodrigo Gil we have seen in Sect. 3.3.4, this "floor of the vault" must be a platform in a scaffolding (Vitruvius/Velasco c. 1564. 156r; see also Lastanosa c. 1570: 143v-147v; García 1681: 24v-25v).

He also includes quite a detailed description of a large crane (Vitruvius/Velasco: c. 1564: 154r-155v; see also Calvo 2006), even giving precise instructions for assembling and disassembling it. Although he furnishes no drawing, the description is so detailed that the machine can be reconstructed certainly; it has many features in common with drawings in Hammer (c. 1500: 3r-9v, 13r, 14r-15r) and Gentillâtre (c. 1620: 556r-556v). It is built around a pyramidal frustum with two platforms and four struts, in contrast to the prismatic frames of the Escorial cranes. A post or *mastel* is attached to the lower platform using an iron socket, allowing the post to rotate around its own axis, like a modern tower crane. The jib of the crane is joined to this post by another socket. The jib is supported by a strut, joined to it by its edge and several crossbars. The jib is 3 feet longer than the strut; it seems that the jib slopes downwards, in the manner of the beak of a crane. Inside the frame, between both platforms, there is a wheel 15 feet in diameter and 3 feet wide, allowing a man to go inside the wheel to exert traction to lift the weights. Velasco even goes into the detail of mentioning the joining pieces between the planks of the treading section, which allow the man to exert his traction more effectively and safely. The axis of this wheel is called the *mastel*, just like the post, since it rotates, acting as the windlass of the crane. From the windlass, the rope goes to two pulleys. Of course, we can assume one of the pulleys goes at the end of the jib, but the location of the other pulley is not clear. Velasco states that these two pulleys help each other, and the weight is divided between both pulleys and the jib is put sideways; thus, the rope is not subjected to great stress and works more efficiently than a horizontal or "perpendicularly straight" rope. All this suggests that Velasco is again struggling with the description of a tryspast. If this is true, the lifting power of Velasco's crane would be quite remarkable. The thrust of the treading man can be estimated at about 60 kg; however, it is multiplied by the ratio of the diameters of the wheel and the hoist, 15 feet divided by less than 1 foot. That makes the ratio more than 15 and the lifting power more than 900 kg. Again, the lifting power of the machine is multiplied three times by the tryspast; thus, the crane can lift stone blocks around 2,400 kg, taking into account a 20% power loss by friction (see Fleury 1993: 107, 132–135).

3.3.4 Placement Control

Needless to say, the correct geometrical execution of an element in ashlar masonry depends on two factors: the geometry of the individual voussoirs or blocks and the final location and orientation of these voussoirs and blocks in relation to each other and to the horizontal plane. The second factor is guaranteed by a correct placement procedure, which is as crucial as the setting out and dressing processes.

The plumb and the plumb rule. In the pre-industrial era, the primary tools used for placement control are the plumb and its derivatives, including the plumb rule and the *archipendulo* or string level.[94] The *plumb* itself is merely a mass, called *plummet* or *bob*, usually made from lead, hanging from a string. The instrument is known from Egyptian times; it appears, for example, in the *Biblia Sancti Petri Rodensis* and in the manuscript of Herrarde de Landsberg. While the plummet in the last example is tiny, the one in the *Biblia Rodensis* (c. 1020: III, 89v) is huge; it was perhaps made of wood, due to the scarcity of metal (Shelby 1961; Colombier [1953] 1973: 19, 31; Gimpel [1958] 1980: 35, 43).

This raises a significant problem: given a cylindrical plummet, as usual, the operator had to keep the upper end of the string at a distance to the wall equal to the radius of the plummet; of course, it is quite challenging to do this precisely. This problem was addressed in different ways. In Egypt, there are instruments including a plumb, a ruler to be applied against a wall and two projecting bars. It may be surmised that when the ruler was applied to a backwards leaning wall, the string rested firmly on the lower projecting bar; in contrast, if the wall leaned to the front, the string would not touch the lower bar; the method seems quite sophisticated, but difficult to operate. In the Holkham Bible, the plumb is used in a different fashion. The operator holds the string with the right hand and a small template in the left one; the template includes a substantial mortise in the centre. It may be surmised that the depth of the mortise equals the radius of the plummet; then, if the string touches the end of the mortise, it is parallel to the wall, and the operator can assume that the wall face is vertical (Shelby 1961: 128–129); again, this set of instruments does not seem easy to operate. The plumb rule solves the problem efficiently, offering greater precision and convenience. In this instrument, the string of the plummet is fixed to a straightedge with a nail; a notch is marked on the end of the ruler so that the line joining the nail and the notch is parallel to the edge of the straightedge. If the edge of the ruler is placed against a wall or arch face and the string of the plummet passes through the notch, the wall or face is vertical (de l'Orme 1567: 56–57; see also Derand 1643: 4) (Fig. 3.28).

The string level. This tool, known in Italian as *archipenzolo*, is also a derivative from the plumb. It is a three-arm square in the shape of an isosceles triangle, with a plummet attached to the corner on the axis of symmetry; a notch is scored in the middle of the opposite side. If this side is placed on a level surface, the plumb will materialise a perpendicular (in both senses of the word) to the opposite side, verifying the horizontality of the surface. In the Egyptian version of the tool, the lateral branches of the square protrude from the opposite side; thus, the instrument gauges the horizontality of the segment connecting the ends of the lateral branches, rather than the surface. In contrast, other examples, such as the one in the stained-glass window of Saint Sylvester in Chartres cathedral and the one drawn by de l'Orme, the horizontal side is continuous, thus making it possible to check a surface. This does not mean that the protruding sides are forgotten; they appear in the relief of the

[94]The drawing in the manuscript of Herrad von Landsberg represents a worker wielding a square near a wall; however, it is not clear if he is using it to check the horizontality of the bed joints of a wall. Since the manuscript was destroyed in 1870 in the French-Prussian war and is known only through copies, it is impossible to reach a firm conclusion.

Quattro Santi Coronati, the patron saints of artists and builders, in the Orsanmichele in Florence by Nani di Banco, at the beginning of the Quattrocento. A huge version of this instrument, used in topography, is shown in different treatises, in particular in Cristóbal de Rojas, who devised a procedure to compute the difference in heights between the ends of the sides and, repeating the operation as many times as needed, of a whole road or canal (Shelby 1961: 127–128; Colombier [1953] 1973: 30, 131; de l'Orme 1567: 56–57; Rojas 1598: 83r-84v; Derand 1643: 4; Esteban 1991).

Tracings as placement control tools. As I have mentioned in Sects. 2.1.3 and 3.1.1, full-scale tracings were sometimes executed precisely below the element under construction. In addition to employing them to prepare templates or bevel guidelines, these tracings were used as placement control tools. In the words of Rodrigo Gil de Hontañón (Fig. 10.12).

> m.m are struts, used to place the keystones before any rib; in order to cut them to the required height, you should proceed this way: the scaffolding should be prepared at the level of the springing of the vault, that is, the diagonal GC of the plan. However, in this position, it will be too low, since the springers are higher, and you will not be able to place the ribs over it. Thus, a second scaffolding such as 5 should be prepared. It should be covered with sturdy planks so that all the ribs can be drawn and scored on it, as you can see in the plan. Once this is done and the keystones are marked on the planks, plumb lines should be hung from the keystones to the tracing, that is, the keystones on the diagonal ribs.[95]

In other words, to control the layout of the ribs the mason should draw a full-scale tracing on a platform made of strong planks and placed on a scaffolding. This drawing includes a plan of the vault, set exactly beneath the piece under construction so that the mason can control the placement of voussoirs using a plumb line. The drawing also features an elevation furnishing the height of the primary and secondary keystones. For large vaults, Rodrigo Gil recommends that the reader place the board higher than the vault springings, at the point where the diagonal ribs and tiercerons separate. This suggests that the springers were controlled using an independent method, such as the system based on horizontal, stretched templates described by Rabasa (1996a; see also Pérez de los Ríos and Rabasa 2014); however, Rodrigo Gil does not mention this.

These practices were still used in the Early Modern period. When dealing with arches in battered walls or lunettes, Alonso de Vandelvira (c. 1585: 23r, 23v) recommends drawing the plan of the arch face directly beneath the element under construction; once the voussoirs are in place, the masons should hang plumb lines from the corners of the voussoirs, checking that they are placed over their theoretical positions.

---

[95]Gil de Hontañon ([c. 1560] 1681: 24v-25r): *… las m.m. son las mazas o pies derechos, para asentar las claves antes que crucero alguno. Para cortarlas al alto que requieren, se les toma en esta manera: el andamio se hace al nivel de donde comienzan a mover las vueltas, que significa la diagonal de la planta GC. Y porque allí estará bajo, por hallarse los jarjamentos con sus avanzamentos más altos, y no se alcanzará a asentar los cruceros sobre ellos, se hará otro segundo andamio como 5. Y este tan cuajado de fuertes tablones, que en ellos se pueda trazar, delinear, y montear, toda la crucería ni mas, ni menos de lo que se ve en la planta. Esto hecho y señaladas todas las claves en su lugar sobre los tablones dejar caer perpendículos, de la vuelta a ellas, esto es para las que están en los cruceros o diagonales.* Transcription is taken from García et al. 1991, modernised by the author.

Thus, the tracing serves a two-fold purpose: before dressing the voussoirs, it is used to prepare templates to control the carving process; after the voussoirs are ready, it allows the masons to check the precision of their placement. Vandelvira mentions this procedure in connection with two particular elements which feature a complex shape for the horizontal projection of the arch faces; of course, in most arches, the projection of the arch face is merely a line. However, such practice was also used for staircases, as shown by the tracing for the remarkable triple staircase in the convent of Saint Dominic in Bonaval, in Santiago de Compostela (Taín 2006).

Part II
Constructive Elements

Chapter 4
Simple Elements

Renaissance stonecutting literature pays little attention to the most frequent elements in ashlar construction: walls, piers, columns, lintels, and round, pointed or basket handle arches. For example, Martínez de Aranda explains that the first part of his manuscript deals with "difficult arches", loftily ignoring the simpler types.[1] Stereotomy is, up to a certain extent, a science of deformation; Martínez de Aranda himself alludes to this in the introduction to his manuscript "all figures causing any alteration of the removed material and extension of lines and circles".[2] However, the reader of the present book may expect an explanation of the simpler types before dealing with the more complex ones; in fact, from the seventeenth century on, treatises include at least schematic explanations about the more straightforward types, as I will do in this section.

4.1 Blocks and Walls

4.1.1 Straight Walls

In pre-industrial masonry construction, even the simplest element—the straight wall—involves not-so-elementary geometrical concepts. To dress the cuboid blocks that make up the wall, the mason should start from a flat surface, called the *surface of operation*; usually, the upper bed joint plays this role. Without the mechanical tools of the Industrial Revolution, the task is not as easy as it may sound. First, the mason should carve a straight line or *marginal draft* with the chisel, checking it with the straightedge or, for smaller pieces, with an arm of the square. Next, he can open

---

[1]Martínez de Aranda (c. 1600, [ii]): … *arcos dificultosos* ….

[2]Martínez de Aranda (c. 1600, [iii]): *Traza es toda figura que en su distribucion causare alguna alteración de robos y extendimiento de líneas y circunferencias.*

J. Calvo-López, *Stereotomy*, Mathematics and the Built Environment 4,
https://doi.org/10.1007/978-3-030-43218-8_4

another marginal draft at right angles with the preceding one, verifying its orthogonality with the square. No templates are needed to control these operations; the width and the depth of the block can be marked on these marginal drafts with the help of a gauge (when taken from a tracing), or a measuring rod (when stated in some unit of measurement).

So far so good, but the third draft poses a complex problem. Two orthogonal, convergent drafts lie on the same plane; however, nothing guarantees that a third one will belong to that plane. To assure this, the stonemason must use a specific procedure, called *boning*. The mason or an assistant should place a straightedge on the *first* draft, and another one in the *third* draft in the making. If both drafts are in the same plane, and we consider another point of the same plane as the centre of a conical projection, the images of both drafts will overlap. Stonemasons found, quite probably by empirical means, a way to leverage this geometrical property to verify the coplanarity of both drafts. They move their heads around until both rulers overlap to the eyesight; this assures the drafts are parallel or at least convergent. This operation is carried out more easily by closing an eye; in fact, the method is called *bornoyer* in French and *bornear* in Spanish, both from the French *borgne,* that is, one-eyed (Fig. 1.4). If there is no way to overlap the rulers to the mason's eyes, the drafts are skew lines, rather than parallel or convergent ones, and should be reworked. When working on hard stones, such as granite, the mason can lean the straightedge on *boning blocks* rather than the marginal draft; this amounts to using points in place of lines (Frézier 1737–1739: II, 15–17; Warland [1929] 2015: 81–84; see also De la Rue 1728: 2–3).

Once the third draft is correctly placed, it is easy to carve a fourth marginal draft, closing the outline of the operating surface, which should be dressed with the usual tools: pointer, axe, bush hammer or, for small blocks, tools in the chisel family. In order to guarantee the planarity of the operating surface, the mason should lean a straightedge (or, for smaller blocks, a square), against two marginal drafts. This practice is consistent with the proposition from Euclid (c. -300: XI.2) which states that if two lines lying on the same plan converge with a third line, the third line lies on that plane. Of course, this does not mean that masons knew and applied Euclid's Book XI; as we have seen in Sect. 2.1.2, Roriczer was not at ease with some propositions from the first books of the *Elements*.

Once the operating surface is hewn, the mason can dress four faces orthogonal to the surface of operation. For easier operation, each of these faces may be placed at a horizontal position during dressing, rotating the block with levers if necessary. Next, the mason can open drafts starting from the corners of the surface of operation. Two of these drafts, together with one of the edges of the surface of operation, will furnish the sides of the front face, the joints or the back of the block. Once again, the planarity of these faces should be verified using the straightedge, resting on two drafts. Using the boning method at this stage is not essential, although it is advisable as a safety check. Repeating the same procedure for the lower bed joint, the mason may finish the carving of the block.

4.1.2 *Curved Walls*

Although blocks for curved walls can be dressed using basically the same procedure, they introduce new geometrical concepts. First, the shape of the block must be scored on the operating surface, either with the help of a template or a combination of gauges and templets. That is, the mason should start preparing four preliminary drafts with the outline of a quadrilateral, dress the operating surface, mark onto it the outline of the template and start opening drafts orthogonal to the surface of operation, as in the simple block. Generally speaking, the side joints of the block should be planar, since there is no reason to dress curved inner joints, while the front and back faces should be cylindrical. These faces cannot be dressed with the axe or the bush hammer, except when the radius of curvature is quite large; instead, they should be executed with chisels or, in the case of very small radii, with gouges, that is, special chisels with a round section. It is essential to control the execution of curved surfaces by leaning one arm of the square on the operating surface and materialising the generatrices of a cylinder with the other arm of the square.

4.2 Piers and Columns

Needless to say, the blocks for a rectangular pier or the sections of a strictly cylindrical column, called *drums*, can be dressed following the outlines of the preceding section. However, more complex forms were frequently used in pre-industrial architecture, such as the compound bundled piers and *cantonnée* pillars of the Romanesque and Gothic periods or the subtly shaped columns of Classical Antiquity and the Early Modern period. Bundled piers were usually constructed by an assemblage of medium-sized stones for every course, and thus the dressing problem is conceptually similar to the one posed by blocks for curved walls.

4.2.1 *Twisted Columns*

The twisted supports used in the Merchants' Exchanges of the City of Majorca and Valencia (Fig. 4.1), as well as other locations, pose an interesting geometrical problem. Josep Gelabert (1653, 41v-42r) a mason from Majorca, gives fairly straightforward directions for the dressing of the drums of such piers (Fig. 4.2). After carving a rough block and two opposing flat bed joints, the mason should place on the bed joints an unusual template, with eight arrises and eight channels between them. It is essential to assure that the arrises of both templates are placed along lines perpendicular to the bed joints. Next, he should dress roughly a cylinder enclosing the arrises. Then, he should mark a line connecting each arris in the lower bed joint with the next arris in the upper bed joint, rather than the one placed directly above the starting

Fig. 4.1 Twisted column. Palma de Mallorca, Merchant's Hall (Photograph by the author)

one. This procedure results in a series of slanting lines drawn on the lateral surface of the cylinder. If these lines were drawn with a straightedge or a square, they would take the shape of a crooked line once the entire pier is assembled. A drawing in Gelabert suggests the use of flexible templates wound around each drum to control the shape of helixes (Gelabert / Rabasa 2011: 104); once the cylinder is developed, the helix takes the shape of a straight line. This is consistent with the documentation of the Valencia Merchants' Exchange, which mentions spear shafts used to control drafts.[3] Once these segments are drawn in each drum, the mason should dress channels between the arrises, using a chisel or gouge. When placed one on top of another, the arrises in Majorca took the form of a beautifully shaped helix; quite probably, masons finished those surfaces in place, to guarantee such an exquisite appearance. In contrast, those ones in Valencia include a round moulding over a flat base, creating an almost baroque effect.

[3] Aldana (1988: II, 65, 267) quotes an archival document which mentions *quatre astes de llances per atar les tirades des pilars* (four spear shafts to tie the drafts in the piers).

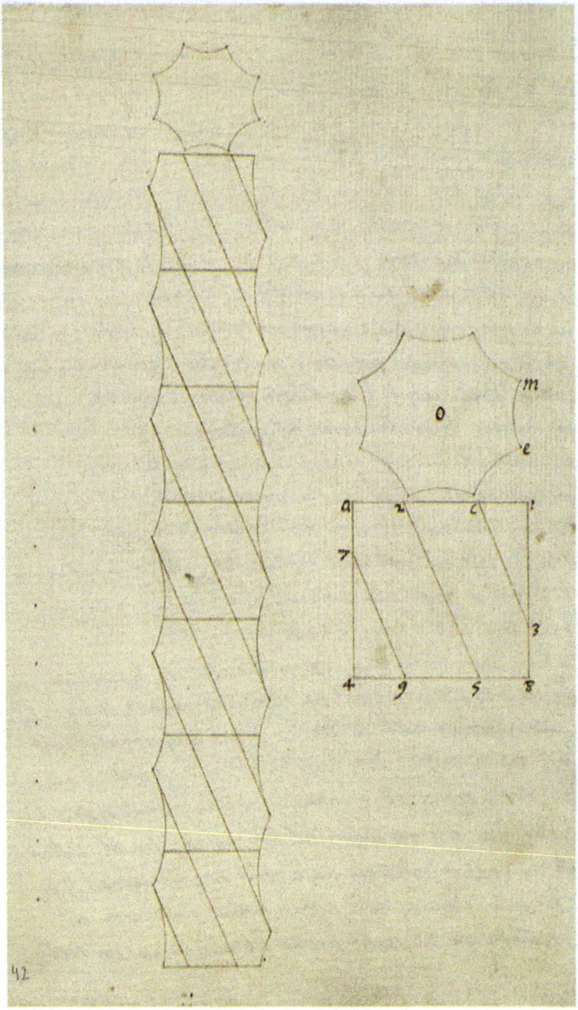

Fig. 4.2 Twisted column (Gelabert 1653: 42r)

4.2.2 *Classical Columns*

Columns in Greek, Roman, Renaissance, Baroque or Neoclassical architecture require different dressing and control techniques. Except for the Doric order, they include a *base* at their lower end, made up from a series of surfaces of revolution such as *toruses* when convex and *scotias* where concave; these details are carved with the help of an inverse template and, more efficiently, with a lathe. At the upper end, they include a *capital*, which is divided into an *echinus* and an *abacus*. In the Doric order, the former is a surface of revolution, while the latter is a square-plan block; in the Ionic and Corinthian orders, the capital is sculpted to include volutes or acanthus leaves, which is beyond the scope of this book.

The section between the base and the capital, taking up most of the length of the column, is called the *shaft*. In classical architecture, its enclosing volume is never exactly a cylinder, but rather a surface of revolution whose generatrix is a curve called *entasis*. Widely different methods for the construction of this curve have been proposed from Antiquity to the Early Modern period. A tracing in the theatre of Aphrodisias, now in Turkey, depicts a column shaft enclosed by a line parallel to the axis for the third part of its length and a slightly tapering line for the rest of the piece. Several Roman examples, for instance, one in the Temple of Hadrian in Rome, seem to follow the Aphrodisias method, although measurements are not entirely reliable as a consequence of degradation. In contrast, other examples seem to be formed by a combination of straight and curved sections (Wilson Jones 2009: 99–102).

This method may seem a bit crude; however, the crooked line may be smoothed using a flexible lamina. This is suggested by a possible interpretation of a passage in Alberti ([1485] 1991: 185–186; see also Becchi 2009: 279–281). He states that the profile of the column is composed of many lines, some straight and some curved (Fig. 4.3). In particular, it is controlled by five key points: the projections (the upper and lower sections of the column); two recessions near these points; and the belly, which is the widest section of the shaft, placed at one-third of its height. Next, he instructs the reader to prepare a full-scale drawing on a floor or wall, starting with a straight line representing the axis. The diameter on the base should be in a given

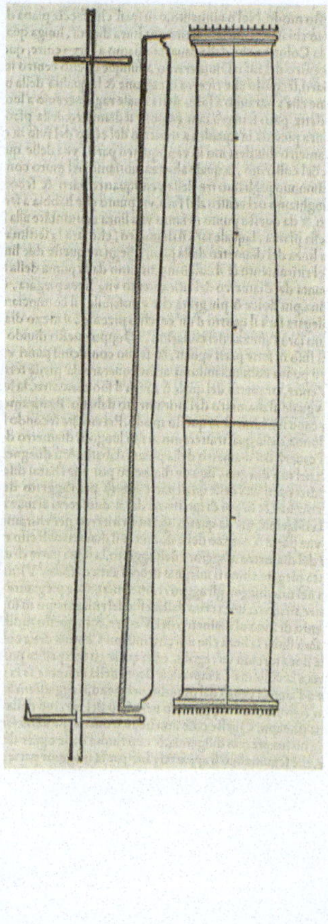

Fig. 4.3 Entasis (Alberti/Bartoli [1485] 1550: 198)

ratio to the length of the axis, depending on the order of the column. Then, 3/24 of the diameter will give the height of the bottom retraction; the diameter at this retraction should be 1/7 smaller than the diameter at the base. The whole length of the axis should be divided into seven parts, and the belly should be placed at the third division point.[4] The diameter of the belly equals that of the retraction at the base, as does the diameter of the top surface, while the diameter at the top retraction is 1/9 smaller. Surprisingly, Alberti instructs the reader to trace two straight lines, from the bottom retraction to the belly and from the belly to the top retraction; however, he adds that a *tabula gracilis* (thin board) should be used to draw the resulting section of the column. Cosimo Bartoli (Alberti/Bartoli [1485] 1550: 197) translates this "thin board" as *regolo,* which may be interpreted as an ordinary template with a curved, elegant shape. There is an alternative interpretation: the *tabula gracilis* may be a thin, flexible lamina used to draw a smooth curve passing through the recessions and the belly (Becchi 2009: 281).

Later on, a variant of this method was put forward by Pietro Cattaneo (1567: 131) and Andrea Palladio (1570: I, 15). A thin lamina or ruler was fixed by its ends to the upper end of the shaft and to a point at one third of its height, using nails; then, the lamina it was pushed using another nail until it adopted a graceful curve, and the profile of the lamina was marked with a pen or a pencil.[5] All this is consistent with the interpretation of Alberti's *tabula gracilis* as a flexible lamina; however, Alberti mentions clearly tracings on a floor or wall, while Cattaneo is thinking about drawings on paper and Palladio does not comment on this detail, but he uses the word *riga* (ruler).

Instead of these empirical methods, other authors have put forward geometrical tracings. In his translation of Vitruvius, Cesare Cesariano rendered "entasis" as *tumefactione* (swelling), and recommended the reader to trace it with a compass or an *asta longa*, that is, a ruler used to draw a circular arc, as suggested by de l'Orme for other purposes (1567: 33v; see also Sect. 3.1.1); Dürer (1525: Giiii v-Giiii bis r; see also Biiii ter r) refined Cesariano's procedure explaining, given three points, how to construct the circular arc.

A tracing in the temple of Apollo in Didyma, also mentioned in Sect. 3.1.1, offers interesting information about another method for entasis construction. It is drawn at full scale along the horizontal or radial axis; in contrast, along the vertical axis, the scale is reduced by a ratio of 1/16. The section of the column is depicted as a circular arc, intersected by many horizontal lines. This may seem striking, but we should take into account that the scale change should be reversed when dressing the stone for the column. Thus, the distances between horizontal lines are be multiplied by 16, while the distances from the intersections of the circular arc to the axis of the

---

[4]It is important to remember that Alberti counts as division points the ends of the line, so when he states that the belly should be placed at the fourth point from the base, he is actually talking about the third division point.

[5]This explanation is not included in the first edition of Cattaneo's treatise, published in 1554. Although the second edition of the treatise was published before Palladio's one, the latter stated in his book that he had shown this method to Cattaneo. As far as I know, this statement has not been disputed. See Becchi (2009, 287).

column are preserved; thus, the circular arc is transformed into an ellipse, furnishing the shape of the entasis (Haselberger 1983; Wilson Jones 2009: 99).

Serlio (1537: 7r) proposes a method leading to similar results. He divides the axes of the upper two-thirds of the shaft and an auxiliary semicircle into several equal parts, drawing horizontals through them; he then draws verticals from the intersections of these horizontals with the semicircle until they reach the horizontals drawn from the shaft axis, giving a series of points for the entasis. The result of the operation is an extremely elongated half-ellipse, although the upper part is not materialised and only two arcs are used; as in the Didyma tracing, the process is based on a change of scale of the half circle along an axis, while keeping the scale in the other direction unchanged; it can be considered as a particular case of an affine transform, although nothing suggests that the Didyma builders or Serlio were thinking in these terms.

Later on, Giacomo Barozzi da Vignola (1562: 32; see also Becchi 2009) proposed, as an alternative to Serlio's method, an elaborate procedure, which may have been inspired by a similar one applied by Dürer to the profile of fortification walls. The builder should draw the axis of the column and the necking (Fig. 4.4). Next, he should construct a right triangle with the radius of the shaft at the necking as a cathetus and the radius at the belly as hypotenuse, extending this line until it reaches a horizontal line drawn through the belly at point *E*. In the next step, a series of lines are drawn from *E*, and the radius of the belly is marked on each line starting from the axis, furnishing points for the entasis. Although these segments are equal, those between *E* and the axis are not, and thus the entasis is not a circular arc, but rather a Nicomedean conchoid. About a century later, François Blondel took up this method, including it among the four most important problems in architecture, and presenting a specially designed compass to implement it. To summarise, no fewer than five basic methods to trace the entasis were proposed: the simple scheme including two segments; the refinement of this scheme using a flexible lamina; the circular arc; the elliptical arc drawn by means of an affine transform; and the Nicomedean conchoid.

4.3 Round, Segmental, Pointed and Basket Handle Arches

4.3.1 Round Arches

Like the classical column, the round arch (Fig. 1.12a) is deceptively simple. In its canonical form, the directrix of a round arch is a semicircle divided into an odd number of equal portions. Of course, learned geometry offers methods to inscribe a decagon in a circle or a round arch into five voussoirs. However, only small arches are split into five pieces; for larger arches, this would lead to unmanageable voussoir sizes and very large curve segments. Books usually found in Early Modern stonemasons' libraries do not include procedures to divide a semicircle into seven, nine or a larger, odd number of portions. As we have seen in Sect. 3.1.2, some details in Martínez de Aranda (c. 1600: 96–97) and Jousse (1642: 14–15) hint that this problem was

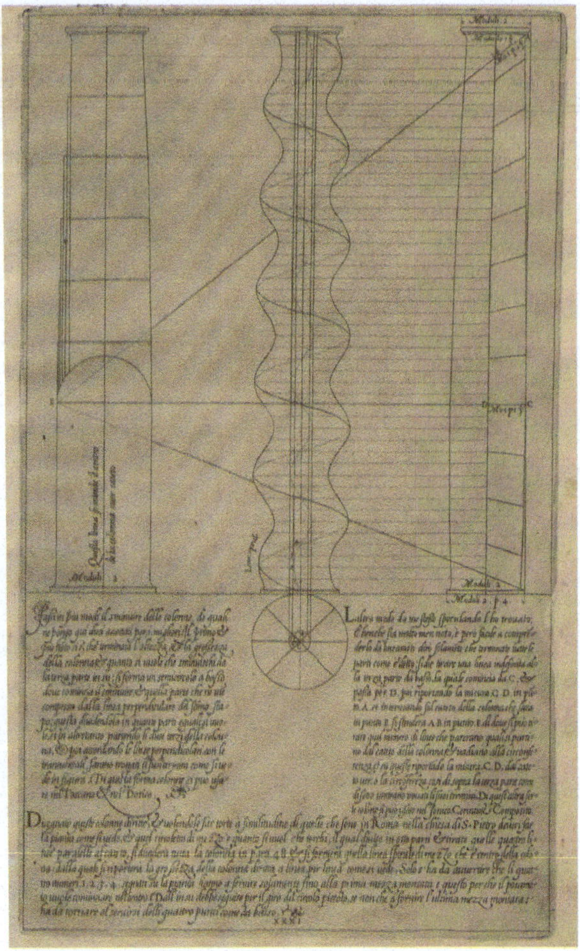

Fig. 4.4 Entasis (Vignola 1562: 32)

usually addressed by trial and error. This method is less tiresome if the keystone is larger than the ordinary keystones: in this case, a small error can be masked easily.

Once this difficult step has been overcome, the rest is more straightforward. The dressing procedure is explained by Derand (1643: 18–22). He appears almost ashamed to deal with such a simple issue, justifying himself by remarking that it is desirable to introduce the solution to complex problems through simpler ones. Thus, to dress a voussoir for a round arch, the mason can use the method we have seen for blocks used in curved wall, using as the operation surface *the face, not the bed joint*. That is, he should dress a flat face, scoring the outline of the face template on it and dress four surfaces, two for the upper and lower bed joints, another one for the intrados and, if necessary, another one for the extrados. However, the latter can be left in the rough, since it will be used to support the overlying masonry. Of course, all four surfaces should be orthogonal to the face plane and thus can be controlled with the square, following the generatrices of the intrados and extrados cylinders. Derand insists that the face template should be placed as carefully as possible, to prevent any waste of stone. However, he also explains an alternative method based on squaring. After dressing the front face, in order to avoid the use of the face template, the mason should draw the face outline directly on the stone; in particular, he should score on a dressed surface two orthogonal lines and transfer the width and height of the voussoir intrados to these lines with a gauge, marking two corners of the voussoir. Next, he should use an arch square to draw a circular arc passing through these points, as well as the upper bed joint. However, the enclosing rectangle needed in this method is larger than the one used in the first procedure.

For the sake of greater precision, de la Rue (1728: 8–9, pl. 3–4; see also Frézier 1737–1739:II, 107–116) suggests that the reader use the most complete set of templates (Fig. 4.5), including face templates, taken directly from the outline of the arch; bed joint templates, which are simple rectangles with the width and thickness of the arch as sides; and *flexible* intrados templates, where the length of the circular

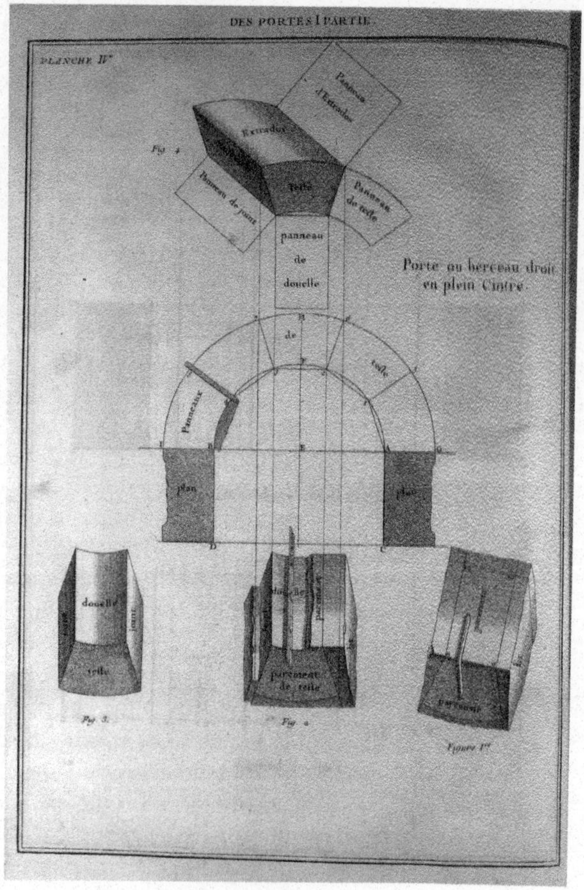

Fig. 4.5 Round arch dressed by templates (de la Rue 1728: pl. 4)

arc at the intersection of the intrados and the face is approximated by dividing the arc into two halves and adding them up. Of course, this amounts to a coarsely simplified rectification of the circle. Although the error involved is small, Frézier took exception against this method, advising the use of rigid templates, called *panneaux de doële plate*, which should be laid on a planar surface before hollowing the intrados. As an alternative, de la Rue remarks that voussoirs may be dressed using only the face template and the width of the arch.

4.3.2 Segmental Arches

When the rise of an arch is limited by other factors, such as the floor of an upper storey that would intersect with a round arch, segmental arches (Fig. 1.12c) may be used. They are set out joining both imposts with a circular arc with its centre placed below the springing line; the rise of such arches is shorter than half the span. According to Fray Laurencio de San Nicolás (1639: 64v-65r), these arches rise from a special piece, the *salmer* or *sommier*, which may be translated as "springer" (Fig. 4.6); it solves the transition between a horizontal joint at the springings, which does not pass through the centre of the arch, and the first ordinary bed joint. Except for this factor, the mason can apply the same setting out and dressing procedures as in the round arch. In any case, a striking comment by Gelabert (1653: 25v-26r), presented as a personal invention, shows that the voussoirs of a segmental arch can be rearranged

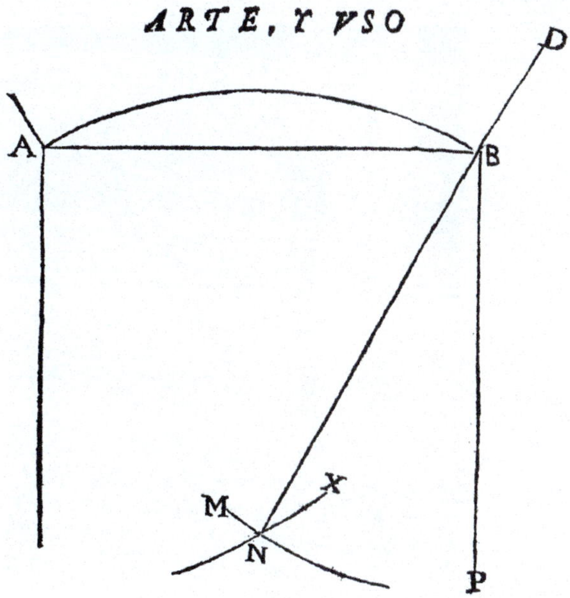

Fig. 4.6 Segmental arch starting from a springer (San Nicolás 1639: 64v)

as a pointed arch; in particular, both springers, cut obliquely, can be joined at the keystone of the pointed arch, avoiding the need to dress a V-shaped piece.

4.3.3 Pointed Arches

At first sight, the pointed arch (Figs. 1.12b and 4.7) seems more complex than the round arch. However, in practice, it avoids some of the problems raised by the semicircular one; this explains its fast spread once imported in Western Europe from Islamic sources (Bony 1984: 12–17). First, the need to find the length of the voussoirs by trial and error in the round arch can be avoided using standard round-arch voussoirs of reasonable length, placing them over the formwork until both arms of a pointed arch approach the apex and using two ordinary voussoirs in lieu of the keystone, cutting them at a skew angle so that they fit against one another at the plane of symmetry of the arch, as remarked by Gelabert (1653: 16v-17r; see also Rabasa 2000: 43–44). V-shaped keystones, although frequent in nineteenth-century Neo-Gothic, are all but non-existent in medieval Gothic.

All this makes the rise independent of the span, in contrast to the round arch. This enables Gothic builders to raise the keystones of the perimetral arches of ribbed vaults, avoiding the domical form which characterises Angevin construction; this leads to almost horizontal courses in the severies. Of course, when covering rectangular bays, builders can adjust the shape of pointed arches to reduce or nullify the difference in heights of the short- and long-side perimetral arches. Moreover, pointed arches

Fig. 4.7 Pointed arches. Lisboa, Convento do Carmo (Photograph by the author)

exert less thrust than round ones and require not-so-robust centring; both are crucial factors in tall Gothic structures (see, for example, Rabasa 2000: 44–46).

Voussoirs may be dressed using any of the procedures we have seen for the round arch. However, it seems that medieval masons used simpler methods. First, the thorough survey of sexpartite vaults carried out by Maira (2015: I, 167, 342) proves that most ribs in twelfth-century sexpartite vaults were built using voussoirs with parallel bed joints, relying on the mortar in bed joints to materialise the curvature of the arch, that is, employing stone as brick. For the moment, this conclusion may be extrapolated to other Early Gothic elements such as quadripartite vaults and independent pointed arches, although surveys of these elements as careful as Maira's one would be quite useful. From the thirteenth century on, masons seem to have computed the angle between upper and lower bed joints of each voussoir by using squares with divergent arms (Shelby 1969); this was one of the reasons for the standardisation of rib curvature and voussoir size in Gothic vaulting, since the angle depends on both factors.

4.3.4 Basket Handle and Tudor Arches

The obtuse angle between the springing of a segmental arch and the underlying vertical door or window jamb can be aesthetically unpleasing and is mechanically inefficient, since it exerts strong thrusts on the abutments. In order to prevent these effects, from the Late Gothic period on, masons and architects used basket handle arches, generally based on three-centre ovals. As we have seen in Sect. 3.1.2, most Renaissance masons and architects knew the four methods for the tracing of ovals explained by Serlio (1545: 17v-18v); other variants were used by Vignola (see Kitao 1974: 34, Fig. 51, note 128) and Vandelvira (c. 1585: 18r). However, none of these procedures may be easily applied to a basket handle arch of a given span and rise. The Tudor arch (Fig. 4.8), with four arcs and a pointed apex, furnishes a way to perform slight adjustments in the rise, since the centres of the pair of interior arcs can be displaced along the common radius at the meeting point between the inner and outer sections; however, it is of course eschewed in classical architecture. López-Mozo (2011) has argued that the architects and masons working in the Escorial complex possibly knew a procedure to construct a three-arc oval of a given span and rise. From the mid-sixteenth century on, architectural treatises explain the use of the gardener's ellipse to tackle this problem (Cattaneo 1567: 158; Bachot 1598; San Nicolás 1639: 67r-67v) or simple ellipsographs (Derand 1643: 294–296; San Nicolás 1665: 200–204; see also Maltese 1994).

Fig. 4.8 Tudor arch. Cambridge, King's College (Photograph by the author) Notice the—very slight—change of direction at the apex

4.4 Lintels and Flat Vaults

4.4.1 Lintels

An opening for a window or door can be spanned by a *lintel*, that is, a horizontal piece with a rectangular section. However, there are limits to the size of a lintel made of a single stone. First, the lintel works mechanically as a beam, and is thus subject to compression in the upper section and tension on the lower one. Stones used in construction can withstand a certain amount of tension, but they are really not well suited to this end; moreover, the amount of tensile stress on a beam is proportional to the square of the span. In addition, very large lintels raise difficult problems of transportation and hoisting. Thus, long lintels are usually divided into several wedge-shaped pieces, acting in fact as an infinite radius arch; such arrangement is called in some Spanish treatises as *arco degenerante* (degenerate arch). Its mechanical behaviour is startling. It cannot fail unless crushed by compressive stress and, in any case, this situation would arise only from a huge amount of load. However, it exerts more thrust in the buttresses than any other type of arch.

The wedge-shaped pieces of this element can be dressed easily using face templates, prepared from a simple tracing including only the elevation of the piece; no plan is strictly necessary. The lower edge of the lintel is divided into an odd number of parts, to avoid a vertical joint in the axis of symmetry. Next, face templates can be drawn passing through each of these points and a single centre, extending them until they reach the upper edge. Once this is done, the mason can dress a flat face for each voussoir, score on it a face template, and dress four planes orthogonal to this surface, starting from the sides of the template. Two planes will furnish the bed joints, one will stand for the lower, visible face of the lintel and the fourth one will be used as the upper face, which is hidden within the interior of the masonry.

An interesting variant of this kind of lintel is the *portal de apotecari*, literally "pharmacist's portal", shown by Gelabert (1653: 93v-94r). He does not approve it, warning that it wastes much stone, although he includes it in his manuscript in accordance with his encyclopaedic intentions. The bed joints in the outer face are strictly vertical, while those in the back face, which may be hidden in the masonry or in a dark hallway, are wedge-shaped (Figs. 4.9 and 4.10); thus, the back face supports the front. Gelabert stresses that both the vertical and the slanted joints should be included in the tracing, but he does not say a word about the dressing process, which raises difficult challenges: the portion belonging to the back face extends beyond the front one at one side, while the front portion exceeds the back one at the other side. Further, the transition between face and front should be placed on the same plane for all wedges. A possible dressing procedure may start by choosing a block large enough to enclose the profiles of both faces; dressing a flat surface for the back side of the voussoir; marking on it the back template; opening two marginal drafts orthogonal

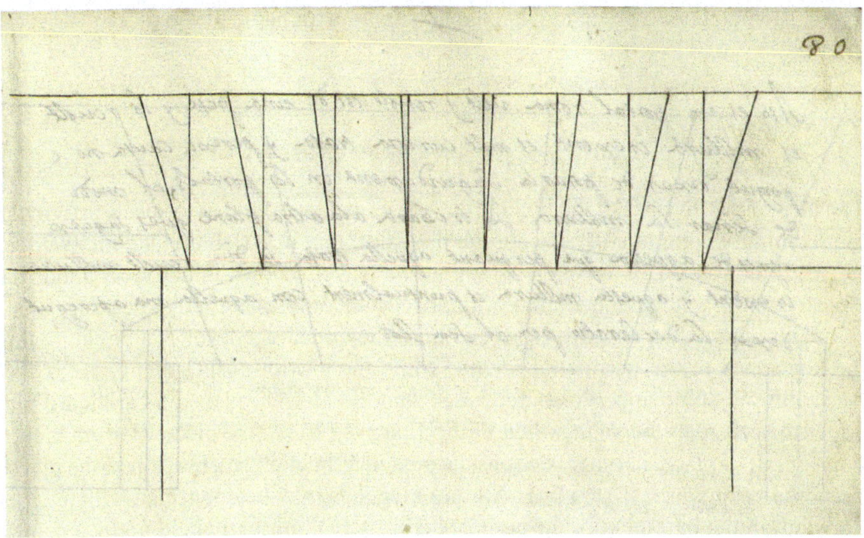

Fig. 4.9 *Portal d'apotecari* (Gelabert 1653: 94r)

Fig. 4.10 *Portal d'apotecari.* Exhibition *Arqueología Experimental* (Photograph by Enrique Rabasa)

to the back face plane at the lower corners of the back template, connecting it to the lower corners of the front face by lines in point view; dressing the front face orthogonally to the marginal drafts; scoring on it a rectangle with the dimensions of the front face; and carving from both faces very carefully until reaching the transition plane; such elaborate procedure explains the colourful name of the piece.

4.4.2 *Flat Vaults*

The idea of the division of the lintel into wedges can be extended to space, arriving at the seemingly contradictory idea of the flat vault; that is, a slab divided into wedges in two directions. This layout was applied in a well-known "flat" vault under the elevated choir of the main church of the Escorial complex. Strictly speaking, it is a sail vault where the pendentives ascend until they reach the level of a circle joining the keystones of the perimetral arches, while the section inside this circle is flat. According to some apocryphal accounts, Juan de Herrera placed a cardboard pier under the centre of the vault; when Philip II arrived to see the piece, Herrera kicked the pier out of place; in other versions, he passed a sheet of paper on top of the pier to show the vault was self-standing. All these legends may be connected with another flat vault in the basement of the complex (Fig. 4.11), presently held up by a central pier and four supporting arches, all in stone. The unusual shape of the piece suggests it was initially designed as a flat vault; perhaps it failed or developed cracks, leading

Fig. 4.11 Reinforced flat vault. The Escorial, monastery (Photograph by the author)

to the addition of the pier and the reinforcing arches. In contrast, the vault under the choir is placed between the lower stages of two bell towers, whose weight counteracts the remarkable thrust exerted by the vault. While the lower surfaces of both vaults are visible, the upper ones are covered by pavements, which have not been renovated in the last centuries; thus, there is no evidence at all for the internal divisions. However, in both cases, the lower surfaces are divided using circular and radial joints. This suggests that internal joints may take the shape of conical bed joints and vertical joints between the stones in the same course, translating a common scheme for sail vaults (López Mozo 2009: 356–367).

Other examples of these singular vaults were built in France. The space between the columns and the wall in Perrault's colonnade on the eastern façade in the Louvre is covered by a series of medallions with the emblem of Louis XIV. Seen from the outside, these pieces seem small, but they actually measure around 5 × 5 m, so they are solved as flat vaults (Fig. 4.12). In this case, the upper surface is visible; its central section is divided by circular and radial joints, and the enormous thrusts of these vaults are counteracted using crossing iron bars. By the eighteenth century, the expertise on flat vaults had reached high levels. In the Grand Theatre in Bordeaux, a columnar screen covered by flat vaults surrounds the front and sides of the building; the vaults at the corners are divided by joints expanding radially from the inner corner. In the narthex of the church of Saint Sulpice in Paris, there are two flat vaults; one of them is divided as a spiral vault, while the joints of the other one form the letters "SS".

Strikingly, these remarkable types are not included in most stonecutting treatises and manuscripts. However, two idiosyncratic methods for the division of flat

Fig. 4.12 Flat vaults. Paris, Louvre, Service gallery over the Grande Colonnade (Photograph by the author)

vaults were presented in 1699 to the Royal Academy of Science in Paris by the engineer Joseph Abeille and Father Sébastian Truchet (Gallon 1735: 159–164; see also Frézier 1737–1739: II, 71–72). Both Abeille and Truchet used identical wedge-shaped pieces, half of them set in one direction and the other half in the orthogonal direction (Fig. 4.13). Truchet's solution is more flamboyant, since it involves curved joint surfaces, while in Abeille's method all faces are flat. Later on, Frézier (1737–1739: II, 72–81) added new variants, including pieces with faces in the shape of irregular octagons and dodecagons. These methods had no antecedents in practice, as far as I know; however, some Spanish applications of Abeille's system in Lugo Cathedral and the Canal de Isabel II, near Madrid, have been analysed by Rabasa (1998) and De Nichilo (2002). In both cases, thrusts are not crucial, due to the small size of the pieces and, in the case of Lugo, the placement of the vault between two bell towers. Another recent example is the conversion of a Truchet vault into a skew arch by Giuseppe Fallacara (2009b; Parisi and Fallacara 2009: 290–308), a tour-de-force carried out with the help of advanced CAD and CNC techniques.

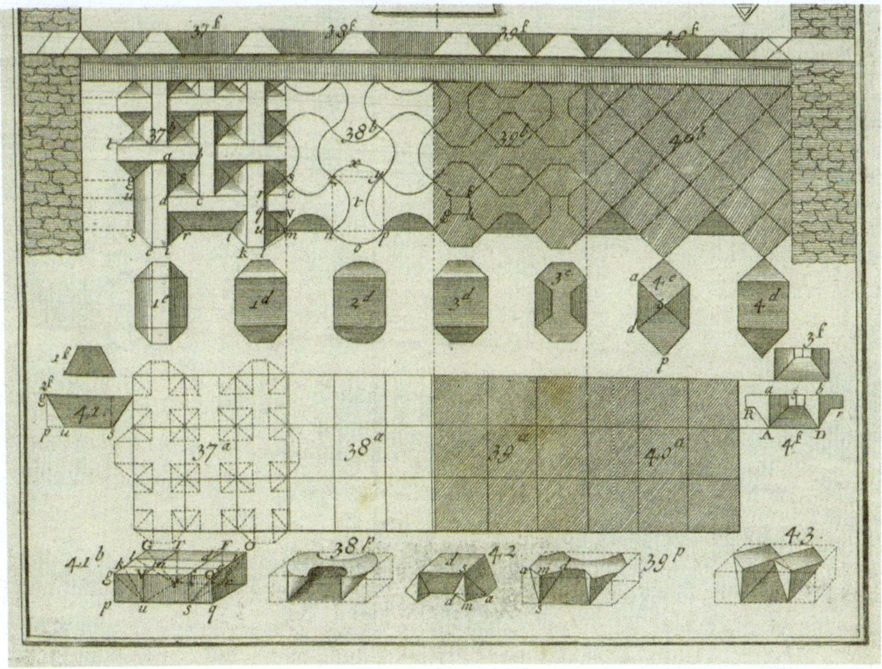

Fig. 4.13 Flat vaults invented by Abeille, Truchet and Frézier. (Frézier [1737–1739] 1754–1769: pl. 31)

Chapter 5
Trumpet Squinches

Abstract This chapter deals with trumpet squinches. These pieces bridge the gap between two converging walls; they are similar to arches, but their intrados continues up to the intersection of the walls, and thus it is strictly conical. The simplest type is the symmetrical, front-faced squinch, which ends at the vertical plane connecting the ends of both walls. Further complexities appear when the squinch is not symmetrical, or when the squinch is cantilevered beyond the line that joins the ends of the walls, either with a planar or curved face. The apex of such complications is the Trompe d'Anet designed by Philibert de l'Orme, whose directrix is a rampant arch and whose face follows a complex curve with several projections and recesses. As in other chapters, this one explains the different solutions offered by historical treatises, stressing that the complex types are solved using a combination of the methods used in the simpler ones, and thus, there is not a substantial reason to regard them as extraordinary.

5.1 Flat-Faced Trumpet Squinches

5.1.1 Non-Cantilevered Trumpet Squinches

Trumpet squinches (Figs. 1.12.k, 5.1) span the triangular area between two intersecting walls, either on a building's exterior, in order to increase usable room at a particular point, or in the interior, for example to turn a square area in the crossing of a church into an octagonal cornice, upon which the circular springer of a vault can start.

Symmetrical trumpet squinches with a round face arch. In its simplest form, the face of the squinch is a round arch set on a plane orthogonal to the bisector plane of both intersecting walls, and the intrados surface is half a cone of revolution. The apex is placed along an orthogonal to the face plane passing through the centre of the face arch. Since the intrados is a surface of revolution, intrados templates are identical, and the angle between intrados and face joints equals that between the intrados and face sides of the bed joint at the springer.

J. Calvo-López, *Stereotomy*, Mathematics and the Built Environment 4,
https://doi.org/10.1007/978-3-030-43218-8_5

Fig. 5.1 Trumpet squinch. Avignon, Saint-Bénézet bridge (Photograph by the author)

Vandelvira (c. 1585: 7; see also Palacios [1990] 2003: 24–29) gives a simple solution to this problem. After constructing the plan of the piece and the elevation of the face arch, the mason should draw an arc with its centre at the apex and a radius equal to the length of the intrados joint, which can be taken directly from the plan (Fig. 5.2). Next, he will draw another arc from the end of the springing whose radius equals the chord of the first voussoir edge; he can then draw a line from the apex to the intersection of both arcs, standing for first intrados joint. Thus, the apex of the cone, the end of the springing and the end of the chord provide the corners of the intrados template for the first voussoir.[1] The rotational symmetry of the piece allows Vandelvira to use this template for all voussoirs; also, he can employ the angle between the face and the springing as a bevel guideline representing the angle between any face joint and the corresponding intrados joint.

Jousse (1642: 78–79) uses exactly the same method (Fig. 5.3); in particular, he does not seem to attempt a cone development, since he uses no intermediate point and says "Afterwards you will take one of the portions of the half circle *A-F* in order

[1]At first sight, the circular arc at the short side of the template may suggest that Vandelvira is attempting a cone development. However, he does not use an intermediate point as he does in the approximate cylindrical developments used in arches in curved walls, as we will see further on. Moreover, he does not say the template is formed by the arc and both lines; instead, he refers to the template by the letters *A– B–D* marked at its corners. Thus, the arc is used only to transfer the length of the springer to the first intrados joint; the analysis of other similar pieces, where Vandelvira does not use arcs at all, supports this interpretation.

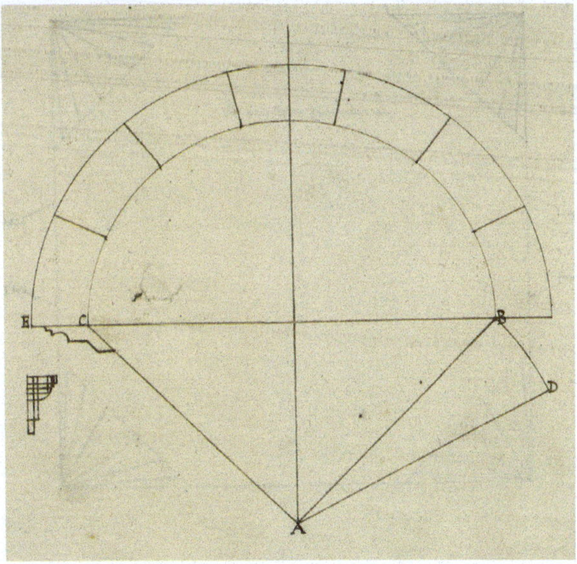

Fig. 5.2 Symmetrical trumpet squinch ([Vandelvira c. 1585: 7r] Vandelvira/Goiti 1646: 2)

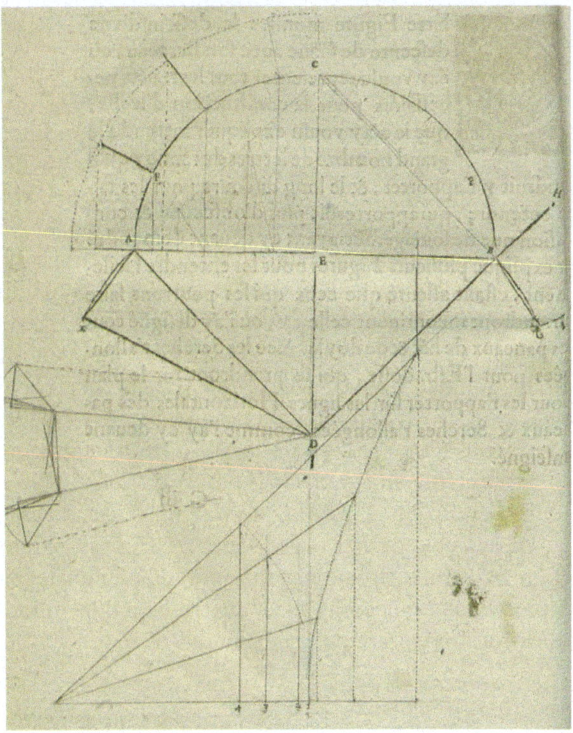

Fig. 5.3 Symmetrical trumpet squinch (Jousse 1642: 78)

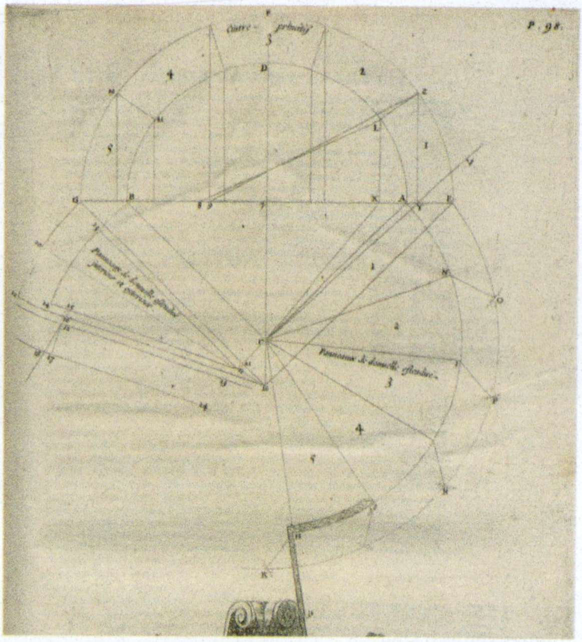

Fig. 5.4 Symmetrical trumpet squinch (Derand [1643] 1743: pl. 98)

to equal *A-K* and *B-G* to it".[2] Derand (1643: 207–212) includes an elaborate tracing, constructing all intrados templates using the same method (Fig. 5.4). In order to draw the face joint guidelines, he adds a circular arch with its centre on the vertex of the squinch, passing through the start of the extrados. Drawing arcs from the corners of the intrados templates with a radius equal to the thickness of the face arch, he locates the other end of the face joints at the intersections of these arcs with the larger one. The construction seems sophisticated at first sight until we notice that Derand is just rotating the angle between the springing and the face plane, a simple planar geometry construction. In any case, he adds afterwards that "It is worthwhile to remark that in this squinch it is not necessary to construct all bed joint templates, since the first one, *CAE*, is enough for all voussoirs, for they are all equal; this can also be applied to intrados templates, as it is evident from the tracing".[3] Thus, Derand's insistence in constructing all templates is purely didactic: he wants to show empirically that all templates are equal, paving the way for the explanation of complex squinches.

De la Rue repeats Derand's elaborate construction, adding two interesting points (Fig. 5.5). First, it is clear that intrados templates are flat and rigid (that is, they represent a polyhedral surface inscribed on the intrados of the squinch). The author does not attempt to develop the intrados cone, since he states: "the intrados templates

[2]Jousse (1642: 78–79): *En aprez l'on prendra l'vne des portions de l'Emicycle A, F, pour faire A, K, & B, G, son égal.*

[3]Derand (1643: 208): *Où il est bon de remarquer, qu'en cette trompe il n'est pas necessaire de tirer tous ces paneaux de ioint, le premier C A E estant suffisant pour tous, veu qu'ils sont tous égaux; comme le sont aussi tous les paneaux de doüele, ainsi qu'il se voit evidemment sur le trait.*

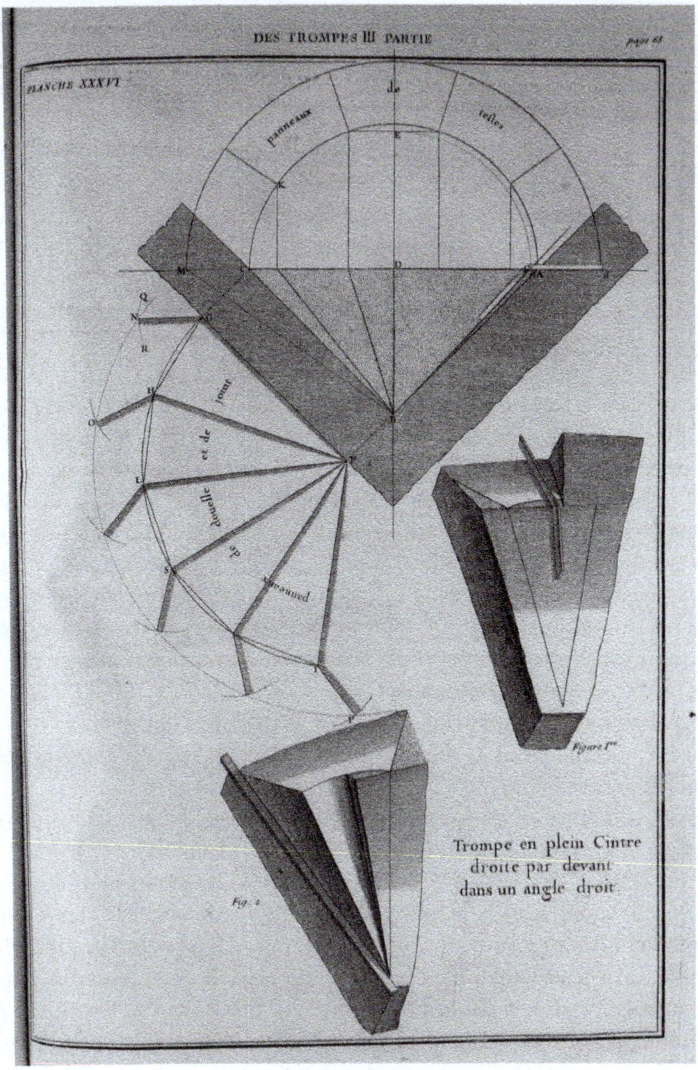

Fig. 5.5 Symmetrical trumpet squinch (de la Rue 1728: pl. 36)

AEC should be extended, taking the distances *GH* equal to *CK*, *HL* equal to *HG*, etc.";[4] *GH* stands for the short side of the template, while *CK* is the chord of the voussoir edge. Second, he adds an interesting explanation of the dressing procedure. As a preliminary step, the mason should draw in the elevation a line joining both face corners of the keystone, determine its intersection with the symmetry plane

[4]De la Rue (1728: 68): *on étendra les douelles de l'arc AEC, saisant les distances, savoir, GH égale à CK, HL égale à HG, &c.*

of the squinch and bring it to the springing. This operation departs from previous practice and seems rather strange at first; however, it is consistent with de la Rue's choice of rigid templates, since it furnishes in the plan the dihedral angle between the polyhedral face inscribed inside the intrados of the keystone and the face plane. After this, he explains the actual dressing method: the mason should prepare a flat surface, scoring the intrados template on it; next, he will transfer the dihedral angle between this surface and the face plane to dress the face. Then, he will score the face outline and dress the bed joints, controlling their planarity using a straightedge leaning on the intrados and face joints. Nothing is said about the dressing of the intrados, although the explanatory scheme suggests that it can be dressed easily using the straightedge to materialise cone generatrices.

Basket handle and asymmetrical trumpet squinches. All the solutions to the symmetrical squinch rely on the rotational symmetry of the piece, solving the problem with a single intrados template and a single bevel guideline, although Derand and de la Rue draw all templates and guidelines for didactic reasons. However, when the squinch does not show this symmetry, all templates should be constructed independently. This problem arises in squinches with basket handle face arches, and also in pieces where the face plane is not orthogonal to the bisector of the springings. In these cases, the apex of the intrados surface is not placed along an orthogonal to the face plane passing through the centre of the face arch—or in the basket handle case, the circular arcs that make up the face. Therefore, the intrados surface is not a cone of revolution. An interesting exception is de la Rue's (1728: 69–70) solution to the skew trumpet squinch, which uses a cone of revolution; however, in this case, the face plane is not orthogonal to the axis of the cone, and rotational symmetry alone does not solve the problem.

Vandelvira offers solutions for a squinch with a basket handle face arch (Fig. 5.6), an asymmetrical squinch, and a combination of both (Vandelvira c. 1585: 8r-10r; see also Palacios [1990] 2003: 30–37). The procedure is an extension of the one used in the symmetrical squinch. When constructing intrados templates, Vandelvira cannot take the length of intrados joints from the plan. Instead, he computes it by forming a right triangle, taking as catheti the horizontal projection of the intrados joint and the height of its upper end over the springing line; in this case, the lower end is always placed at the apex of the squinch, at impost level. In typical stonecutting fashion, this operation is carried out without drawing a single line. For the first voussoir, Vandelvira rotates the projection of the first intrados joint, bringing the apex to the springing line and marking this point, let us call it *X*, with two lines. At the other end of the rotated intrados joint there is a projection line joining the vertical projection of the upper end of the intrados joint, *C*, with the horizontal one, *B*. Thus, line *X*-*C* is the hypotenuse of a triangle whose catheti equal the horizontal projection of the intrados joints and the difference in heights between its corners. Therefore, *X*-*C* represents the real length of the intrados joints, as in nineteenth-century descriptive geometry manuals. Next, he constructs a triangle with the length of the intrados joint, the springing and the chord of the first voussoir edge, as in the symmetrical squinch; such a triangle provides the intrados template for the first voussoir. Since the lengths of the intrados joints are different, Vandelvira joins their ends with a

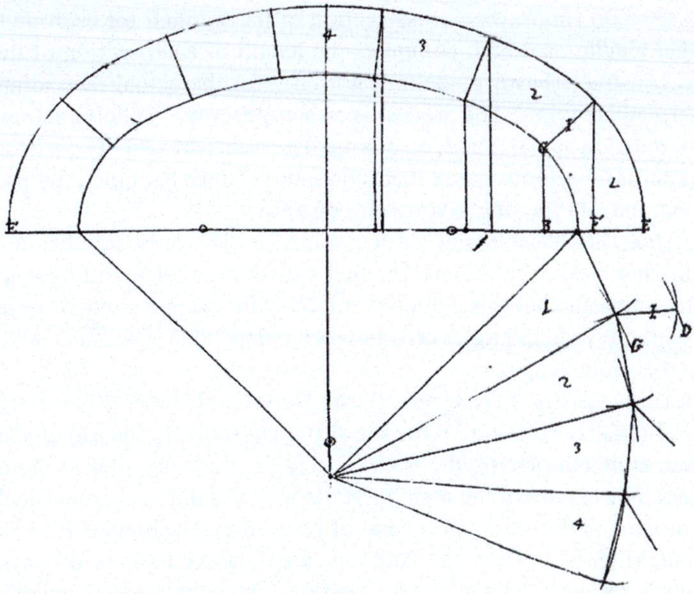

Fig. 5.6 Basket handle trumpet squinch (Vandelvira c. 1585: 7v)

line, rather than an arc; this makes it clear that he is trying to develop the polyhedral surface inscribed on the intrados surface, not the intrados cone. In the symmetrical basket handle squinch (Vandelvira c. 1585: 7v) bilateral symmetry allows reusing the templates for one half of the squinch in the opposite side; in contrast, in the asymmetrical squinch, either round or basket handle (Vandelvira c. 1585: 8v, 9v), he repeats this procedure for all the voussoirs. To reduce the work of tracing, he uses the upper intrados joint of each voussoir as the basis of the next one. This procedure seems recursive but does not lead to an accumulation of errors since each new intrados joint is computed from the start; in fact, he could have separated the templates if necessary.

As for the bevel guidelines, Vandelvira cannot take them from the springer, due to the lack of rotational symmetry in these variants. Thus, he constructs them by forming a triangle with the corresponding intrados joint, computed as above; the length of the face joint, taken directly from the elevation; and the diagonal of the bed joint, that is, the distance between the vertex of the squinch and the upper end of the face joint. Again, this distance cannot be measured directly. Vandelvira computes it forming a right triangle, with the horizontal projection of the diagonal and the height of the upper end of the face joint as catheti; the hypotenuse furnishes the length of the diagonal. This operation requires tracing a projection line from the upper end of the face joint but, other than that, it is carried out without the need to draw further lines.

Vandelvira (c. 1585: 8r, 10r) also includes some variants for large squinches where each course or section between two generatrices needs to be divided into two or more voussoirs. He draws a cross-section of the squinch for each dividing line, either basket handle or round; computes the length of each section of the intrados joints; and constructs bevel guidelines not only for the actual face joints but also for the intermediate joints. The process is rather tiresome. Vandelvira draws a single cross-section, but in actual practice, intermediate joints are usually set in at least two different planes, in order to break their continuity. Other than that, the procedure is the same that the one for single-voussoir courses.

Jousse (1642: 80–85) explains similar solutions. The most significant difference is that in the multi-course variants, the first course is smaller, and its templates are lacking; this suggests that this course is solved with a single stone or *trompillon*, to avoid the extremely sharp angles arising when independent voussoirs are carried all the way to the squinch apex.

In contrast, de la Rue (1728: 69–70; see Bortot and Calvo 2020) introduces an interesting alternative solution for the skew trumpet squinch. Instead of starting with a round face arch and placing the apex outside the perpendicular to the face plane drawn through the centre of the arch, he starts with a standard symmetrical squinch, with the intrados in the shape of a cone of revolution, but he cuts it by an oblique vertical plane (Fig. 5.7). Thus, the voussoirs are identical to those of a symmetrical round squinch, except for the cut-away portion. The solution is as ingenious as the approach: after drawing the elevation of the face arch and the plan of the squinch, including the oblique cutting plane, de la Rue transfers all voussoir corners to the right springing, using orthogonals to the axis of the squinch. Next, he constructs a single template of an uncut voussoir as in the symmetrical, round squinch. Up to this moment, the right-hand corners of the voussoirs are in their proper place, since the right-hand side of the template overlaps the springer. The left-hand corners of the voussoir must be transferred to the other side of the template, but this operation can be performed by simply drawing an arc with its centre in the squinch apex. Joining the left- and right-hand corners of each voussoir, the mason can draw the intrados templates, which appear grouped near the springer. As for the bevel guidelines, they can be constructed easily by taking into account that they all converge in a single point, namely the intersection of the oblique face plane with the axis of the cone of revolution. The lower ends of the face joints can be transferred to the springers through the same procedure used for the intrados templates.[5] Thus, the mason knows two points of each bevel guideline: the common intersection point of all guidelines and the lower end of each face joint, so he can join them easily to construct the guideline. In any case, de la Rue's neat drawing makes it clear that the mason should use as a guideline the portion of this line going up from the voussoir corner, rather than the other half going down to the common intersection point.

[5]In theory, the same points could be used, although de la Rue transfers them to the left springing for the sake of clarity.

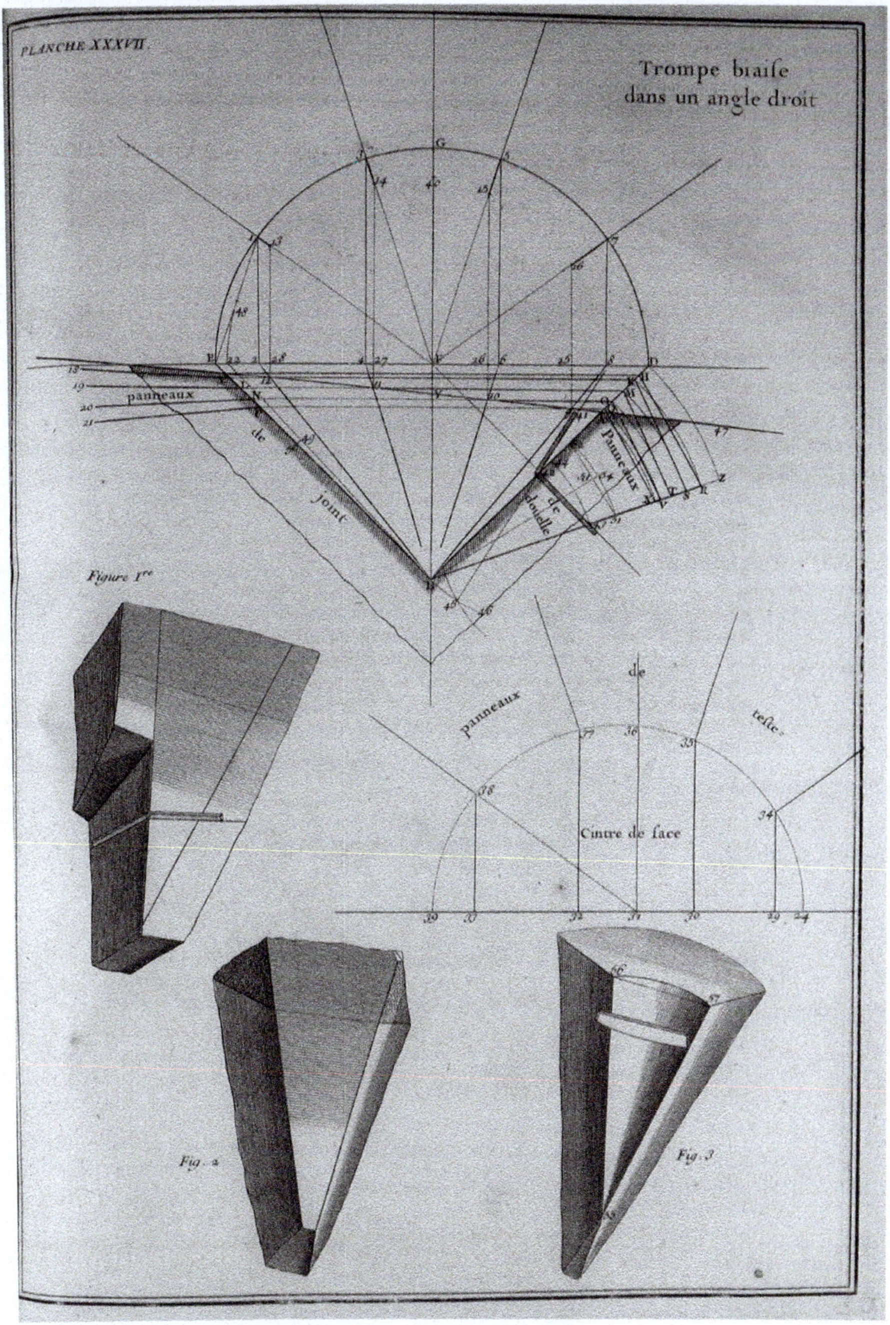

Fig. 5.7 Asymmetrical trumpet squinch (de la Rue 1728: pl. 37)

5.1.2 Corner Trumpet Squinches with Flat Faces

In the squinches we have seen so far, the piece does not go outside the triangle defined by both springings and the line joining their ends. From this point on, we will deal with several variants where the intrados surface is cantilevered beyond this line (Fig. 5.8). Since stone does not resist bending stress well, cantilevered elements are not frequent and may be seen as a constructive stunt; this is reflected in a number of remarks in French treatises that describe such pieces as *suspendu dans l'air* (hanging in the air) or similar expressions (see Etlin et al. 2012: 20–21). The foremost example of such literature is, of course, Philibert de l'Orme's lengthy description of the squinch he built at Anet castle. In typical fashion, de l'Orme (1567: 88r-99v) presents this complex squinch *first* and, in the next section, he explains a simple corner squinch (de l'Orme 1567: 100r-103r). For the sake of clarity, I will leave the Anet *trompe* for the next section, together with squinches with curved faces, while introducing the subject in this section with the simple corner squinch.

Fig. 5.8 Corner trumpet squinch. Santiago de Compostela, Cathedral, Concha de las Platerías (Photograph by the author)

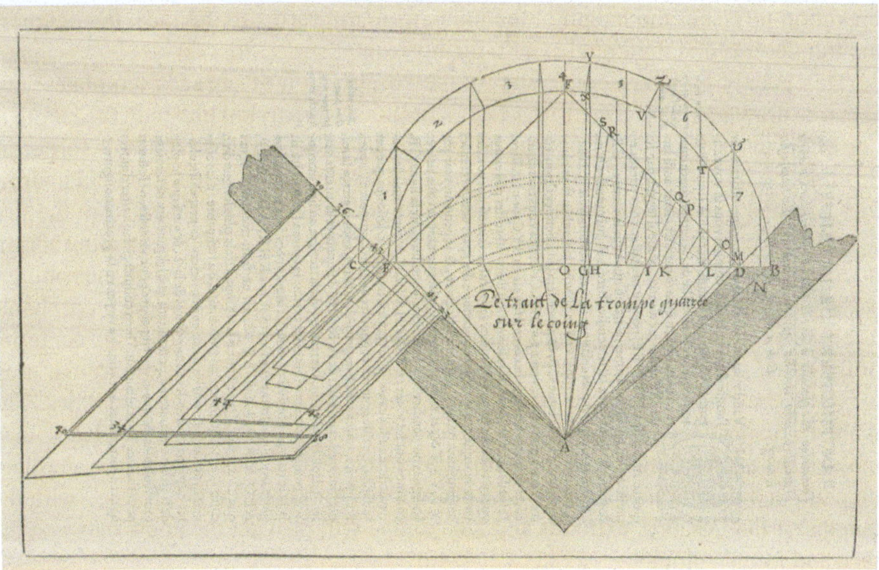

Fig. 5.9 Corner trumpet squinch (de l'Orme 1567: 100v)

Parallel projection. De l'Orme starts by drawing the plan of the squinch in the shape of a rotated square; two adjacent sides provide the springings, while the others stand for the faces, so the squinch is cantilevered from the diagonal joining the edges of the springers (Fig. 5.9). Next, he draws a round generating arch spanning the distance between springing ends. This arch will not be materialised in the built squinch; however, it is the key to the solution of the problem, as we will see. In the following step, he draws projection lines from the generating arch in order to locate voussoir corners, L, K, etc.; he then draws lines from the apex, passing through the projections of the corners of the generating arch, until they reach the horizontal projection of the face at points O, P, etc.

As a next step, de l'Orme constructs an auxiliary diagram in order to determine the length of the intrados and extrados joints. First, he brings to the left springing the intersections of the joints and the generating arch, *L*, *K*, etc., as well as their ends at the face side, *O*, *P*, etc. Next, he raises perpendiculars to the springing from each of these points, transferring the heights of voussoir corners to the relevant orthogonal line. This enables him to draw lines from the apex to the intersections of the intrados joints with the generating arch, extending them until they reach the face. Although he does not draw the section of each line between the apex and the intersection with the generating arch, this step is crucial, since at this stage he does not know the height of the face end of intrados joints. The result is not a projection, but rather a series of rotations of each intrados joint around a vertical axis passing through the apex until they reach a vertical plane passing through the springing. In any case, this operation furnishes the length of the joints from the vertex to the face, the height of their endpoints and their angles with the horizontal.

This is only an intermediate step in the construction of intrados templates. In a separate drawing (de l'Orme 1567: 102v), he plots a horizontal line with the length of one of the faces. Next, he brings the horizontal distances between voussoir corners to this line, both for the intrados and the extrados. Again, he draws perpendiculars from these points, marking on them the heights of voussoir corners, taken from the first diagram. We should expect two pairs of elliptical arches as a result, standing for the intersection of the intrados and extrados cones of revolution, with two planes oblique to the axis of the cone. This is not completely clear in de l'Orme's woodcut, probably as a result of errors in the execution of the print, dilatation of paper over the centuries or the usual practice of representing ellipses and ovals with a series of circular arcs. Next, he draws face joints, in order to get a continuous set of face templates. The diagram also offers essential data for the construction of intrados templates: they may be drawn forming a triangle with two intrados joints, taken from the auxiliary diagram and the intrados edge of a face template. Another triangle, formed with the lengths of the intrados, extrados and face joints, will provide the angle between the face and intrados joints.

For once, de l'Orme's solution is presented with didactic clarity. However, it does not seem to reflect stonecutters' practices; drawing so many lines at full scale in floors or walls would be tiresome and confusing. Vandelvira offers two different solutions to this problem. In the first one (c. 1585: 10v; see also Palacios [1990] 2003: 38–41) he draws the plan of the squinch and an elevation of a generating round arch, set on a plane parallel to a diagonal passing through the intersection of both faces (while de l'Orme had placed it exactly on the diagonal). This detail is not crucial; more remarkably, Vandelvira uses a parallel projection to transfer the arch to the faces. Since each half of the generating arch is projected onto one of the face planes, which are oblique to the projectors, the final shape of the projection will be given by two half-ellipses. Vandelvira constructs one of them in true shape, taking the heights of each voussoir corner from the generating arch and their horizontal projections from the plan. The result is called the *cimbra*; this Spanish word usually stands for a set of face templates, while on other occasions it also means "centring"; thus, the half-ellipse should be understood as a simplified set of face templates, giving a profile that may be used to control the centring. Next, Vandelvira will proceed as he did in the preceding squinches. By taking the length of an intrados joint from the plan, bringing it to the *cimbra* and measuring the hypotenuse of the triangle formed with the height of the corresponding voussoir corner, he may compute the length of each intrados joint. Repeating the operation for the next intrados joint and measuring the chord of the voussoir from the *cimbra*, he can draw the first intrados template. In a similar way, he can construct the first bevel guideline forming a triangle with the diagonal of the bed joint and the length of the face joint, all taken from the *cimbra*.[6]

[6]In theory, the height of the voussoir corner in this step may be taken from the elevation, rather than the *cimbria*, reducing the error brought about by the transfer of this height to the *cimbria*. However, we should recall that masons used full-scale tracings and that the length of the chord must be taken

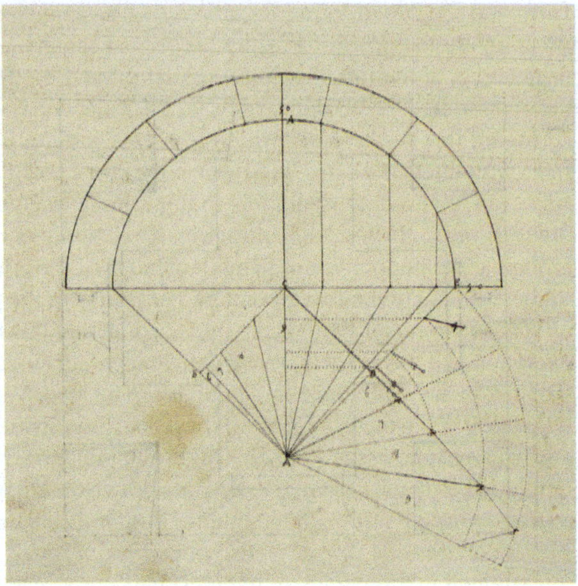

Fig. 5.10 Corner trumpet squinch, second solution ([Vandelvira c. 1585: 14v] Vandelvira/Goiti 1646: 18)

Central projection. Vandelvira's second solution (c. 1585: 14r-15r; see also Palacios [1990] 2003: 54–57) is quite ingenious. Again, he draws a square for the plan and a generating round arch set on a vertical plane passing through the furthest corner of the squinch, with its ends placed exactly at the prolongations of the springings of the squinch. In the next step, rather than using a parallel projection, the generating arch is projected *centrally* (Fig. 5.10). As in perspective projection, the inner and outer edges of the generating arch, as well as its joints, are projected onto the face planes using projectors converging in the apex of the squinch, as in de l'Orme. The intrados surface is generated by these projectors, extended from the projection to the apex. Since these lines, standing also for generatrices, pass through the round arch,[7] the intrados surface is a portion of a cone of revolution, as in de l'Orme.

Vandelvira's solution to the problem is quite elegant; it was used also by Alonso de Guardia (c. 1600: 29v; see also Calvo 2015a), while similar procedures may be found in French treatises. He starts by developing the cone of revolution, or rather the polyhedral surface inscribed on its interior, tracing an arc whose radius equals the distance between the apex of the cone and the start of the generating arch, and marking on this arc the chords of the voussoirs, taken from the generating arch.

from the *cimbria* since the elevation does not represent it in true size. Thus, taking the height from the elevation would have involved much going back and forth, bringing on larger errors. The same can be said about the heights of the upper ends of the face joints, which are taken from the *cimbria* in order to construct the bevel guidelines.

[7] I am using here the expression "generating arch" for an arch that is included in the preparatory tracing but not materialised in the built element, as on other occasions. However, in this case, from the standpoint of cone geometry, the generating arch plays the role of a directrix of the cone, since the generatrices are the projectors and intrados joints.

However, the actual intrados surface of the squinch is only a portion of this cone. To compute the positions of the corners of the actual voussoirs, he first draws the projections of the intrados joints, using projection lines drawn from the corners of the generating arch and joining the apex of the cone with their intersections with the springing line of the arch, which acts as a folding line. The intersections of these lines with the face planes will furnish the corners of the actual voussoirs. In the next step, Vandelvira draws parallels to the folding line passing through these points, in order to bring them to the springing. This amounts to a revolution of these points around a line in point view, namely the axis of the cone. Then, he will transfer each of these points to the corresponding generatrix in the polyhedral development using an arc; however, we should take into account that this last phase is a planar geometry operation and the arc is merely a convenient, didactical means to perform a distance transfer. In any case, once these points are located, Vandelvira can draw the intrados templates; given the symmetry of the piece, he constructs only those for the right side and the keystone of the squinch. Bevel guidelines are rotated around the squinch axis until they reach the springings; they are drawn taking into account that the extensions of face joints pass through the intersection of the squinch axis with both face planes.

Vandelvira (c. 1585: 15r) also constructs a set of face templates or *cimbria* for this squinch in a separate drawing. This is not as easy as it may seem: at this stage, he knows the length of the intrados joints, but not the height of the actual voussoir corners. In contrast to the elegant constructions in the preceding phases, he painstakingly draws an auxiliary elevation including all the directrices and generatrices of the cone passing through voussoir corners, just to measure the heights of three corners. Using these and the distances along the face plane taken from the plan, the mason can construct the true-shape representation of one of the face arches, which is an arc of an ellipse. Since the face plane cuts the cone of revolution obliquely, the centre of the conic section will be at the opposite side of the axis, the actually built portion will be less than a quarter of an ellipse, and the ensemble of both sides will resemble a pointed arch.

Jousse (1642: 86–87) explains a similar solution. Instead of placing the generating arch on a plane passing through the furthest corner of the square, he sets it on a plane rising from the diagonal where the cantilevered part begins, as did de l'Orme. However, such an arch is a directrix of the cone, and thus it is homothetic with the larger arch used by Vandelvira. As a result, the intrados joints are identical in both solutions, and this detail has little effect on the general solution. Another point is more relevant, although it does not affect the essence of the method: in Jousse's own words, "you should draw the half-circle *A*, *F*, *B*, and divide it in five *unequal* parts, making those in the lower sections larger than those in the upper part".[8] From this point on, Jousse follows Vandelvira's second method, with minor variations; for example, when constructing bevel guidelines, instead of tracing perpendiculars to the axis, he measures their positions along the face plane and transfers them to an extension of the springing, which is symmetrical with the face plane. In order to construct

[8] Jousse (1642: 87): …*faites l'Emicycle A, F, B, qu'il faut divider en cinc parties* inégales, *faisant celles du bas plus grandes que celles du haut* … (my emphasis).

Fig. 5.11 Corner squinch with uniform voussoirs (Jousse 1642: 88)

the *cintre* or true-shape representation of the face, he draws an utterly simplified orthographic elevation of the face arches, which contrasts sharply with Vandelvira's complex projection of all the cone generatrices; next, he takes the heights of the relevant voussoir corners from this diagram and their horizontal positions from the plan.

Another variant by Jousse (1642: 88–89), the *Trompe sous le coin de pieces égales* (corner squinch with uniform voussoirs) introduces an interesting issue, the regularisation of the chords of the voussoirs (Fig. 5.11). Unfortunately, the text does not explain the procedure,[9] but it can be safely deduced from the drawing. The mason is to construct the plan, the generating arch and the *cintre* as in the preceding variant; after this, he should divide the *cintre* into equal parts. Of course, such operation

[9]Jousse (1642: 89). The *full* text of this section is *Cette Trompe se fait comme la précedente; tous les Paneaux & Cintre se prennent comme i'ay enseigné, sans en faire vne plus ample décription, d'autant que ce ne seroit qu'vne répetition, veu que le dessein l'enseigne assez intelligiblement* (This squinch is traced as the preceding one; all templates and the *cintre* are constructed as I have shown, without need for a longer explanation, which would be a repetition, since the design shows it rather clearly).

(dividing an elliptical or oval[10] arc of arbitrary length), was most probably carried out by trial and error since there is no way to perform it using constructions with straightedge and compass. Next, the heights and positions of the division points are carried to the generating arch. It is worthwhile to stress that this arch is *not* divided into equal parts in this step, since Jousse's goal is to divide into equal parts the *cintre*, not the generating arch, and central projection on oblique planes does not preserve the shape of the arch. From this point on, the mason can construct intrados templates and bevel guidelines as in the preceding example.

Derand (1643: 212–217) follows the essential steps in Jousse's method, including the placement of the generating arch on the diagonal of the square, the first method with the unequal division of the squinch, and the inverse procedure, dividing the *cintre* into equal parts and applying them to the faces. However, he uses the midpoints of the voussoirs in order to construct an approximate development of the intrados cone. De la Rue (1728: 76–79) follows similar lines, but he does not use voussoir midpoints. He presents an enticing visual and textual explanation of the dressing method, which involves the use of the bevel to control dihedral angles between the undressed planar intrados surfaces and the squinch face, and the use of a templet (see Sect. 3.2.1) in order to control the shape of the final conical surface; this use of polyhedral surfaces in the dressing process explains the absence of voussoir midpoints and cone developments.

5.2 Trumpet Squinches with Curved Faces

5.2.1 Trumpet Squinches with Convex or Concave Faces

Parallel projection. An alternative approach for cantilevered trumpet squinches, the mason may dress the face with a round shape (Fig. 5.12). Vandelvira explains the simplest solutions; as in corner squinches, he offers two sets of solutions, based on parallel and central projection. The first group begins with a symmetrical squinch opened in a convex surface. Vandelvira (c. 1585: 11r; see Palacios [1990] 2003: 42–45) starts by tracing the plan with the springings and the generating arch. Next, he draws a circular arc standing for the face cylinder and projects the generating arch onto the cylinder using projectors parallel to the axis of symmetry of the squinch. From this point on, he follows the same procedures used in the first, parallel-projection based, solution for the corner squinch, using voussoir midpoints to control the curvature of face joints; he remarks that their curvature decreases as they approach the keystone (Fig. 5.13). On the pages that follow, he presents a series of squinches with skew convex, symmetrical concave and skew concave layouts, following the same principles (Vandelvira c. 1585: 11v-12v; see Palacios [1990] 2003: 46–47).

[10] In fact, a mixture of both, since the points of the *cintre* are those of an ellipse but were joined by circular arcs in practice.

Fig. 5.12 Trumpet squinch with a curved face. Lyon, Hotel Boullioud (Photograph by the author)

The *trompe de Montpellier*. Another set of alternatives is based on central rather than parallel projection. It appears for the first time in the manuscript of Jean Chéreau (c. 1570: 105v-106r) under the name *trompe de Montpellier*; he includes three separate drawings but no real explaining text (Fig. 5.14). However, the setting out procedure may be reconstructed by comparison with later solutions. After drawing the plan of the squinch and the elevation of the generating arch, Chéreau (c. 1570: 105v) projects the generating arch centrally onto the convex surface, including the midpoints of voussoir edges. Next, he transfers these projected points to the springings, using them to construct bevel guidelines, labelled *paneau de joingt*, and drawn as circular arcs. This is, of course, a simplification, since the face joints are sections of the wall cylinder through oblique planes, and thus elliptical arches. Although geometrically incorrect, such an approach has some practical sense, since the difference between the circle and the ellipse would be barely noticeable in practice; in any case, it is not easy to say if Chéreau was aware he was performing a simplification. Then, in two separate drawings (c. 1567–1574: 106r), he presents the intrados templates and the development of the face, called *cerche rallongée ou cintre*. It may

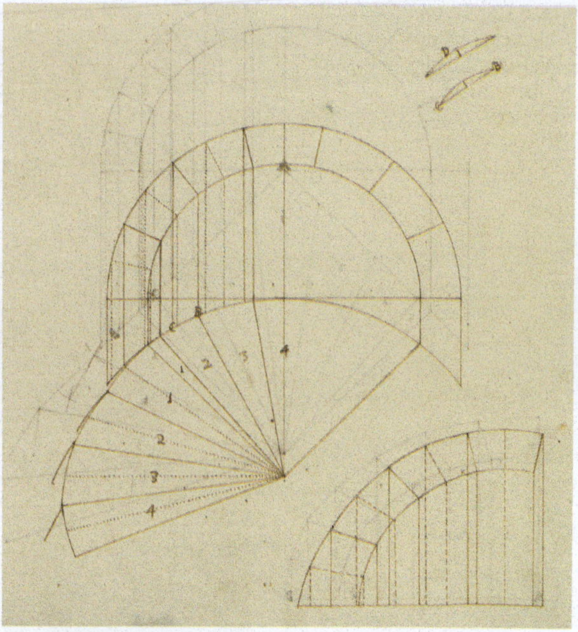

Fig. 5.13 Trumpet squinch with a curved face ([Vandelvira c. 1585: 11r] Vandelvira/Goiti 1646: 10)

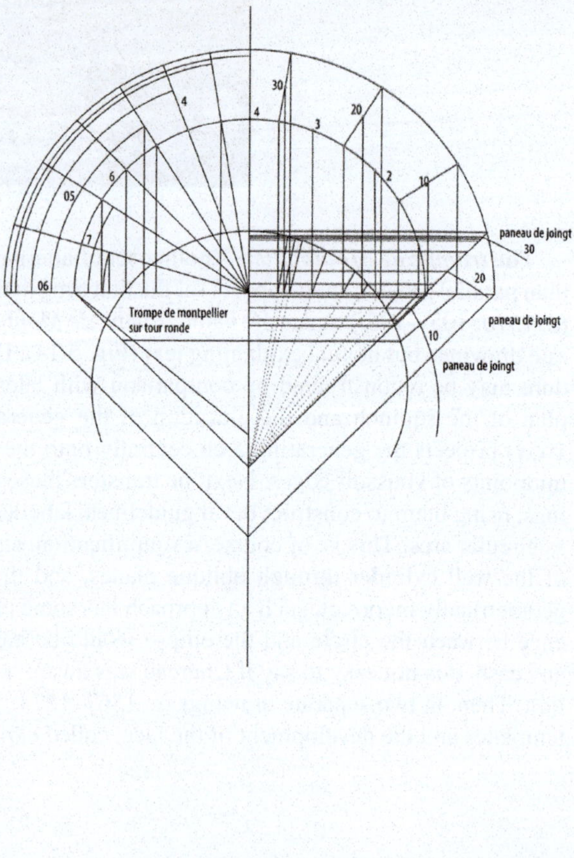

Fig. 5.14 Trompe de Montpellier (Chéreau 1567–1574; 105v, redrawn by author)

be surmised that the intrados templates are constructed by triangulation, using the midpoints of voussoir edges; in fact, some drypoint marks attest to the use of these points. Several straight lines seem to play the role of alternative bevel guidelines. The development of the face is drawn by starting with a horizontal line; horizontal distances between voussoir corners and midpoints are transferred to this line; next, perpendiculars are raised, and the heights of midpoints and corners are transferred to the orthogonals. In contrast to usual practice, the extrados is lacking.

Vandelvira (c. 1585: 15v-16r; see Palacios [1990] 2003: 58–59 and Aranda 2017) follows basically the same procedure in the *trompa de Mompeller*, although the generating arch is not secant to the wall cylinder, but rather tangent, and the intrados templates are grouped, starting from the right springer. The face development, however, is presented on a separate page and includes the extrados, in contrast to French treatises. Jousse (1642: 108–109) includes such a long explanation that the printer was forced to use a special small font in the lower half of the page to fit it onto a single sheet. Again, he follows Chéreau's procedure, but he advises the reader to divide the generating arch into *unequal* parts, placing the larger ones near the springers. He uses three quadrants in the generating arch, standing for the intrados, the extrados and an intermediate, auxiliary arc; as a result, he divides the face joints, as well as voussoir edges, into halves. The intrados quadrant is reused as the plan of the cylindrical surface. In addition to intrados templates and the face development, he constructs face templates, avoiding the use of the curved bevel guidelines of Chéreau and Vandelvira (Aranda 2017: 34). Derand (1643: 256–257) remarks that the distinctive feature of the *trompe de Montpellier* lies in the radius of the wall face, which should equal that of the intrados of the generating arch. In fact, Jousse uses the same solution, but Derand manages to arrive at a much simpler tracing, although he divides the voussoir edges and face joints into halves. He solves the problem using only intrados templates and a face development, eschewing bevel guidelines.

5.2.2 The Trompe of Anet

Geometrical studies of the stonecutting treatises of Jousse (1642) and Derand (1643) and large portions of de L'Orme (1567), de la Rue (1728), and Frézier (1727–39) and are almost non-existent. In contrast, the wealth of papers and book sections dealing with the squinch built by de l'Orme in the castle of Anet for Diane of Poitiers, in order to support a small study for the private use of Henri II, is really startling.[11] Perhaps the presence of the king or the narrative allure of his affair with his lover explains such a plethora of studies. There is, however, a deeper reason for this attraction. As with other pieces, such as spherical vaults, de l'Orme does not include in his treatise simpler variants of the trumpet squinch; he starts with a complicated case, involving asymmetrical springings, one flat and the other sloping, and a quite complex curved

[11] Sanabria (1984:1, 224–228); Evans (1988); Sanabria (1989); Evans (1995: 183–189); Potié (1996: 99–106); Trevisan (1998); Trevisan (2000); Fallacara (2003a); Lenz (2009).

front (Fig. 5.15). Forgetting Jousse or Derand, the Spanish sources, or even the corner arch explained by de l'Orme himself in the following section of his treatise, as we have seen in Sect. 5.1.2, some researchers have been dazzled by the brilliance of de l'Orme's solution and have not stressed that it is an accumulation of not-so-complex problems (see Sanabria 1984: I, 219, as an exception).

Thus, at risk of repeating what has been said by some of these researchers, I will try to explain once more de l'Orme's solution to the squinch of Anet, stressing its connections with the simpler variants we have seen in Sects. 5.1.2 and 5.2.1. The problem is basically that for the *trompe de Montpellier*, although the ascending right springing and the complex face add new difficulties. The key to the solution, as suggested by Potié's drawings (1996: 94–95) is the use of a generating arc, spanning the distance between the farthest ends of both springings, which are placed at different levels; similar solutions are also used by Chéreau (c. 1567–74: 105v-106r), Vandelvira (c. 1585: 11r-2v, 15v-16r) and Jousse (1642: 102–111), as we have seen. In any case, de l'Orme's explanation is quite unsystematic; in fact, after dealing with the whole process (1567: 92r-96r) he is afraid the reader will not understand him, so

Fig. 5.15 *Trompe d'Anet* (de l'Orme 1567: 89r)

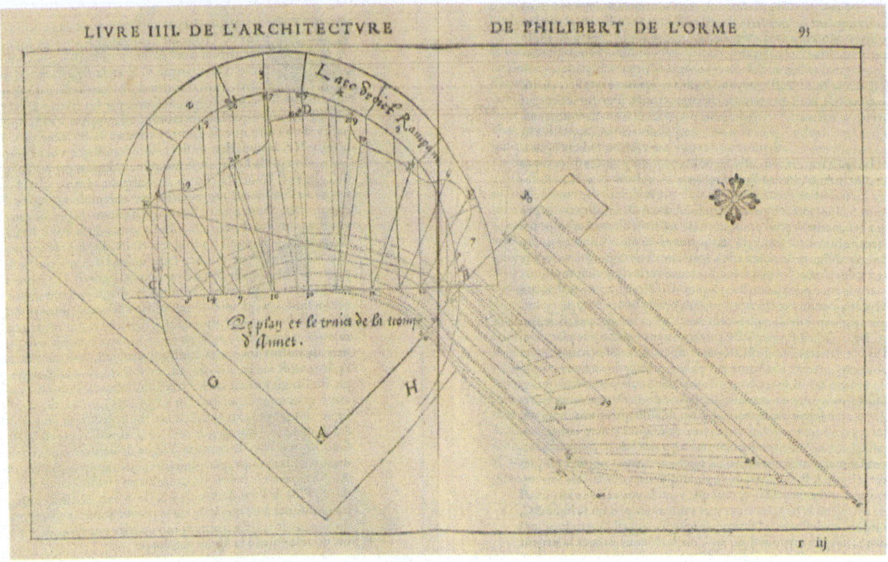

Fig. 5.16 *Trompe d'Anet*, main tracing (de l'Orme 1567: 92v-93r)

he starts anew, essentially repeating the same explanation with simpler figures (1567: 96r-99v);[12] such choice is extremely unusual in stereotomic literature.

De l'Orme (1567: 92v-93r) starts by drawing the plan of the squinch, including the springings and the complex, undulating face, intermingled with the generating arc, in typical stonecutter's fashion (Fig. 5.16). Given the slope of the right springer, the generating arch is constructed through an affine transformation. The heights of the corners of an ordinary round arch, which is not drawn, are transferred to perpendiculars starting from a slanting line that joins the ends of both springings. Next,[13] he constructs a diagram showing the intrados joints in true size and shape, as he did in the corner squinch. In particular, he transfers to the springings the distances between the apex and the intersections of the intrados joints with the generating arch, taken from the plan, in order to construct a separate diagram, as in the corner squinch. In the second explanation, for greater clarity, this diagram is presented as a separate drawing. He then transfers the heights of these intersections from the

[12] In particular, in the second explanation, he extracts the construction of intrados joint profiles from the main drawing, placing it on a separate drawing; also, instead of presenting the full sets of face, intrados and bed joint templates, he includes in the corresponding diagrams only those for the left section.

[13] At this moment, de l'Orme explains the first steps of the constructions of the face templates. However, to carry out this construction, he needs to know the height of the ends of the intrados joints, so he leaves aside the construction of face templates and constructs profiles of the intrados joints, in order to determine the height of their ends. Only when he knows these heights, does he come back to face templates. For the sake of clarity, instead of following such a convoluted explanation, I will explain first the construction of intrados joint profiles; after this, I will address the construction of face templates.

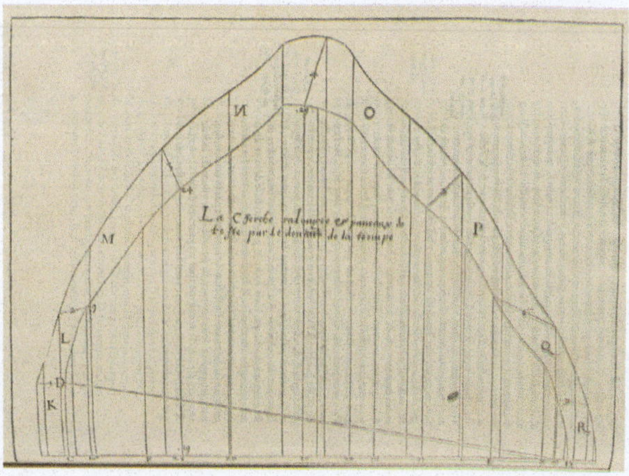

Fig. 5.17 *Trompe d'Anet*, face templates (de l'Orme 1567: 94v)

elevation to perpendiculars in the auxiliary scheme. In the next phase, he extends the intrados joints in this auxiliary diagram until they reach the complex face; this last operation is carried out taking into consideration the horizontal distance between the generating arc and the face, measured from the plan. Thus, the auxiliary diagram furnishes the length of the intrados joints and the height of their ends. In the next step, de l'Orme (1567: 94v), constructs face templates (Fig. 5.17), taking into account the heights of intrados joint ends, transferred from the auxiliary diagram, as well as the horizontal distances between these points, taken from the plan. In fact, he is applying the procedure he uses in the corner squinch a few pages later; the only difference is that these distances cannot be measured along a straight line, so de l'Orme measures the length of each segment of the jagged perimeter line. As in the corner squinch, the next step (de l'Orme 1567: 95v) is the construction of the intrados templates and bevel guidelines (Fig. 5.18), carried out by triangulation; he uses the length of intrados and extrados joints, computed in this case from the auxiliary diagram, and the length of the edges of the face templates.

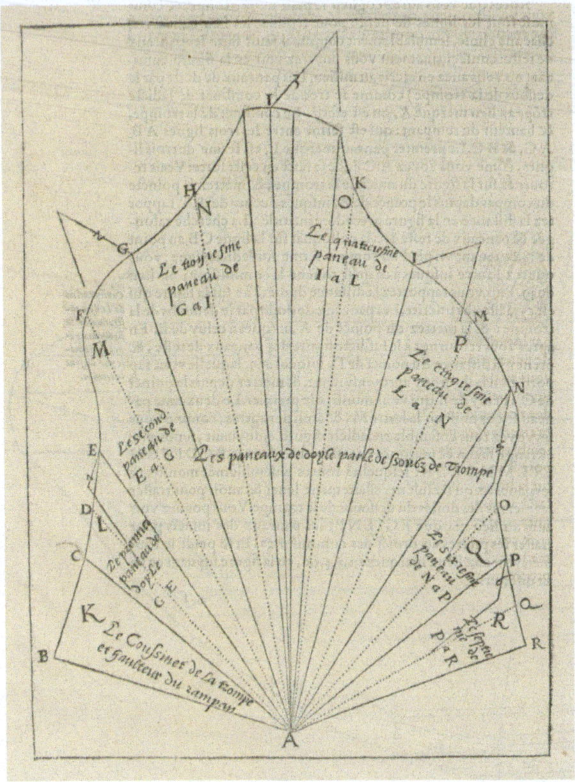

Fig. 5.18 *Trompe d'Anet*, intrados and bed joint templates (de l'Orme 1567: 95v)

Chapter 6
Arches

6.1 Splayed Arches, Gunports and Oculi

In this chapter, for the sake of clarity, I will include under the label *skew arches* (Fig. 1.12e) those starting from *parallel* springings oblique to the face planes. In contrast, I will classify as *splayed arches* (Fig. 6.1) those featuring *convergent* springings (Figs. 1.12f, 1.12g), including the ox horn, where one springing is orthogonal to the face while the other one is oblique to the faces and the opposite springing. While the intrados of skew arches is generally cylindrical, the intrados of splayed arches is conical, except for some complex cases. I will deal first with splayed arches since they include the simplest variants, in particular, the one where springings are symmetrical about the axis of the arch. Next, I will address skew arches in Sect. 6.2.

6.1.1 Symmetrical Splayed Arches

A diagram by Hand IV (Villard c. 1225: 20r, dr. 18-r) has been connected with this problem by Lalbat et al. (1989: 23–25); according to their interpretation, the drawing represents a double splayed arch, with different conical intrados surfaces in both faces. Such an interpretation is reasonable, since the marks in the middle of the opening converge approximately with the springings, as shown by Lalbat et al.; however, such a simple scheme does not offer more information about the dressing process.

Cristóbal de Rojas (1598: 99r) presents, as usual, a drawing without text, including *plantas*, that is, intrados templates, and *saltarreglas* or bevel guidelines (Fig. 6.2). Rojas notices that all face joints are equal, taking into account the rotational symmetry of this arch, so he uses the first one, at the springer, as a guideline to measure the angle between face and intrados joints. He also profits from rotational symmetry to construct a single intrados template. There are two possible interpretations of his diagram. He may be thinking about a development of a cone, as in contemporary

J. Calvo-López, *Stereotomy*, Mathematics and the Built Environment 4,
https://doi.org/10.1007/978-3-030-43218-8_6

Fig. 6.1 Splayed arch. Siracusa, Castello Maniace (Photograph by the author)

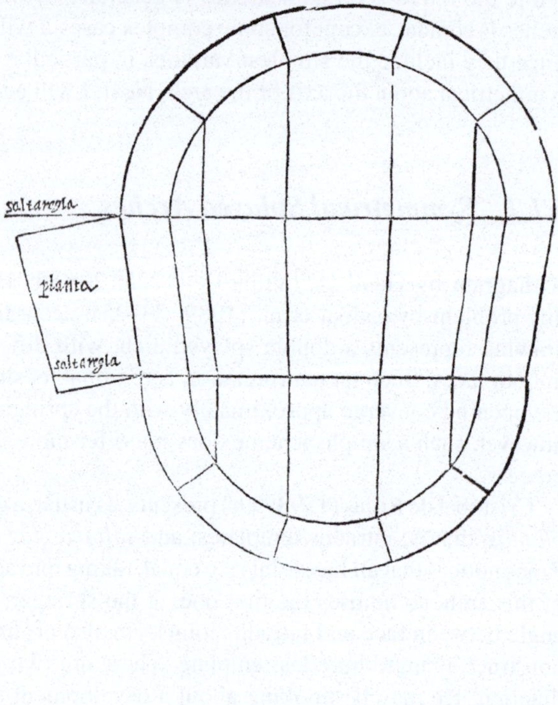

Fig. 6.2 Symmetrical splayed arch (Rojas 1598: 99r)

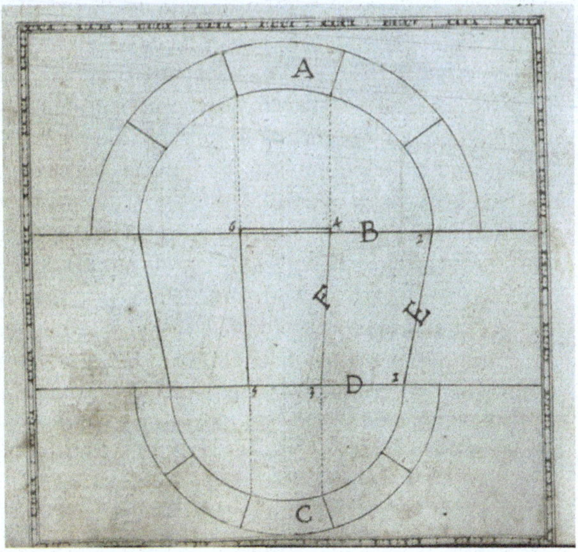

Fig. 6.3 Symmetrical splayed arch (Martínez de Aranda c. 1600: 33)

solutions for spherical vaults (see Sects. 9.1.1 and 12.4.3). However, the short sides of the template are straight; this suggests he is inscribing a polyhedral surface into the intrados of the cone. Also, if he were thinking about cone developments, it would be easier to place the template against the actual springing, rather than to lean it in the extrados of the arch, as Rojas does. The other interpretation is less problematic: Rojas may be using triangulations, as Vandelvira (c. 1585: 26v-27r, 27r-28v) does in the skew arch with circular faces, placing the template outside the plan for greater clarity. However, there is no hard evidence to confirm either of these hypotheses.

In contrast, Martínez de Aranda (c. 1600: 33–34) constructs an intrados template drawing an orthogonal to the voussoir edge in the larger face of the arch (Fig. 6.3), although he does not say this in so many words.[1] Next, he takes the length of the intrados joint from the plan, where it is represented in true size by the springer, relying again on rotational symmetry, and draws an arc whose radius equals this length with its centre at a back corner of the voussoir. Where this arc meets the orthogonal, he may place the front corner of the voussoir, while the front side is parallel to the faces. These results are consistent with nineteenth-century descriptive geometry: his procedure may be interpreted as a revolution around the chord of the keystone edge belonging to the smaller face. The back side of the quadrilateral formed by the four corners of the voussoir is placed at the axis of rotation and thus does not move. The front corners will move along orthogonals to this axis. All this does not necessarily mean that Martínez de Aranda was thinking in these abstract terms; I will come back

[1] He does not mention this line in the text explicitly. In the drawing, he includes two *plomos*, that is, lines connecting the vertical and horizontal projections of the relevant voussoir corners, akin to the projection lines of descriptive geometry. However, he draws only these lines for the corners used in the construction of the template; that is, he uses *plomos* exclusively to construct the template.

to this issue in Sect. 12.5. Of course, given the rotational symmetry of the piece, this template can be used for the rest of the voussoirs; as for the bed joint template, it can be taken directly from the keystone, as Rojas had done.

Jousse (1642: 82–83) addresses a similar problem by different means in the *trompe en cannonière*, that is, "trumpet squinch in a gunport", which is basically a trumpet squinch where the section near the apex has been removed, leaving a cone frustum as a result (Fig. 6.4). Of course, cutting away the end portion is essential if the piece has to be built in a casemate, but it can also be used as a splayed arch or even an ordinary squinch, completing it with a *trompillon* or single stone in the apex. Jousse easily solves the problem by constructing a face of a pyramid inscribed in the intrados of the piece. Taking into account that intrados joints converging to the intersection of both springings are equal in length, he draws an arc with its centre at the apex passing through the end of the springing. Face joints may be measured on both elevations, so he draws arcs with their lengths to locate the corners of the template, which can be used for all voussoirs, thanks to rotational symmetry.

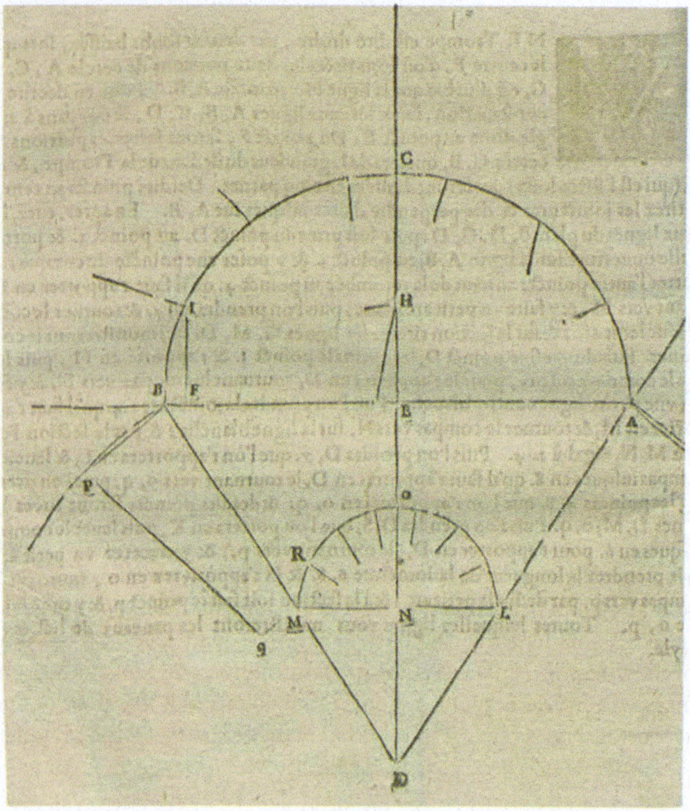

Fig. 6.4 Trumpet squinch in a gunport (Jousse 1642: 82)

6.1.2 *The Ox Horn*

The next step in complexity is the arch where a springing is orthogonal to the face planes, while the other one is oblique (Fig. 1.12g, 6.5); this piece is known for obvious reasons as *corne de bouef* or *corne de vache* (ox horn or cow horn). Most authors address the problem using bed joints orthogonal to the faces, so it is quite easy to dress the voussoirs by squaring. In contrast, in this case, intrados joints are neither parallel nor convergent, the intrados is a warped surface, and the use of templates is highly problematical.

Bed joints orthogonal to face planes. As usual, de l'Orme (1567: 69v-70v; see also Jousse 1642: 16–17) does not include the basic *corne de boeuf*, but rather a composite piece made up from two ox horns joined back-to-back (Fig. 6.6); the ensemble approximates a skew arch known as *biais passé*, as we will see. It features identical front and back faces and an intermediate, smaller circular groin. De l'Orme clearly explains the construction of the bevel guidelines used to control the dressing of the voussoirs. He measures in the elevation the apparent distance between the face and groin ends of each intrados joint and transfers it to the plan, in the vicinity of the springers. Then he forms a series of right triangles, placed at the springers, using as catheti these distances and half the thickness of the arch; their hypotenuses give the length of the intrados joints, while the angles to the face serve as bevel guidelines. Again, this operation is consistent with nineteenth-century descriptive geometry: it may be understood as a rotation around a line in point view passing through the centre

Fig. 6.5 Ox horn. Joigny, Parish church (Photograph by the author)

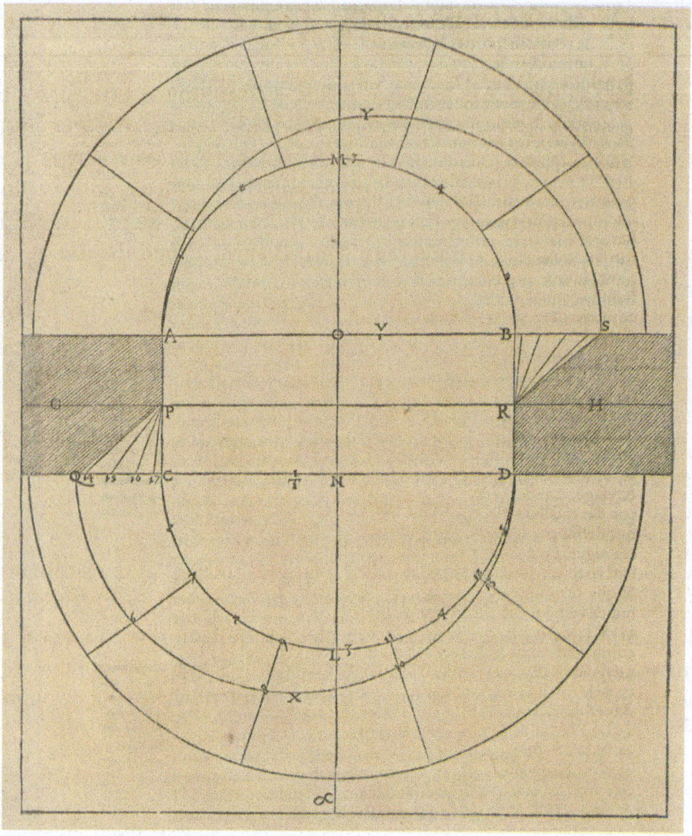

Fig. 6.6 Double ox horn (de l'Orme 1567: 70v)

of the groin. The groin ends of the intrados joints will describe circles along the groin, which is depicted in the plan as a straight line; the face ends will move along circles placed in the face, whose horizontal projections are straight lines overlapping the face. The radii of each of these circles exceed the radius of the groin by the apparent distances between the ends of the intrados joints, which are transferred to the plan. Thus, each of the bevel guidelines represents an intrados joint, rotated in order to bring it to a horizontal plane, allowing the angle between intrados and face joints to be measured directly. All this does not imply that de l'Orme was thinking in such abstract terms; I will come back to these issues in Sect. 12.5.

Jean Chéreau (c. 1567–74: 112r) includes a real ox horn, on the same sheet as a skew arch, so tightly intermingled that both pieces share the same plan. Although there is no explaining text, it seems clear that he applies de l'Orme's solution to a single ox horn. It is worthwhile to remark that Chereau's *panneaux de teste* (face templates) are really bevel guidelines, formed by a single line.

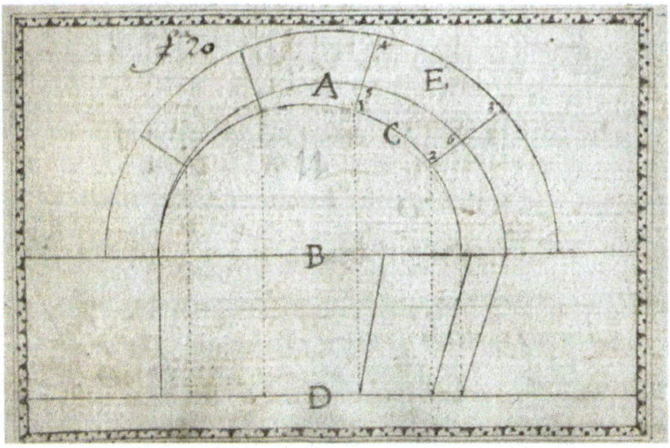

Fig. 6.7 Ox horn solved by squaring, with bevel guidelines (Martínez de Aranda c. 1600: 11)

These procedures are more explicit in the solution by Martínez de Aranda (c. 1600: 11–12). First, he presents a single ox horn rather than a pair (Fig. 6.7). Instead of placing bevel guidelines close to the springers, he draws them near their horizontal projections. He uses a line in point view, passing through the front end of each intrados joint, as an auxiliary line. Next, he draws the intrados joint placing its front end at the intersection of the auxiliary line with the face of the arch. In order to locate the back end, he measures the apparent distance between both intrados joints ends in vertical projection and transfers it to the back face, starting from its intersection with the auxiliary line. Again, this can be understood as a rotation around the auxiliary line, although Martínez de Aranda probably did not think in these terms. Moreover, such a solution has no real practical advantages. Compared with the solution of de l'Orme, it requires tracing projection lines; further, when dressing the voussoirs, the mason would have to walk on all fours to take the angles of the guidelines with the bevel. Martínez de Aranda probably put it forward for didactic reasons. He adds some short but interesting remarks about the dressing process:

> I suppose you want to dress the voussoir *E* by squaring; you should dress it first as a block with the shape *1–2–3–4* and the thickness shown in the plan; next you will take off a wedge *1–2–5–6* in the side of the smaller arch *C*, so the intrados joint of the lower bed will follow the bevel guideline *F*, while the upper one will follow *G*; all the voussoirs of the arch should be dressed in the same way.[2]

That is, the mason should dress an ordinary voussoir, taking its shape from the tracing; next, he will score the bevel guidelines on both bed joints and take a wedge below both marks in order to shape the intrados. In this way, the "great waste of

[2]Martínez de Aranda (c. 1600: 11–12) … *supongo que quieres robar el bolsor E lo labrarás primero de cuadrado con la forma que parece entre los cuatro puntos 1 2 3 4 que tenga de grueso lo que tuviere de ancho la planta y después la robarás por la testa que mirare al arco pequeño C con el robo que parece entre los números 1 2 5 6 que venga a quedar la cara por el lecho bajo con las saltarreglas que causare en los lados de la planta la línea F y venga a quedar la cara por el lecho alto con las saltarreglas que causare en ambos lados de la planta la línea G y de esta manera se han de robar todas las demás piezas de este dicho arco.*

stone" associated with the squaring method is reduced to a minimum, thanks to the orthogonal layout of the bed joints.

However, Martínez de Aranda (c. 1600: 12–14) endeavours to dress the voussoirs by templates as an alternative solution, despite the warped nature of the intrados (Fig. 6.8). He takes it for granted that a voussoir corner will move along an orthogonal to an intrados joint, so he places a corner of the upper intrados joint at this orthogonal, measuring its distance to the lower corner in the elevation. Next, he locates the fourth corner of the intrados quadrilateral using its distances to other corners. Such a construction is quite problematical; since the intrados joints are neither parallel nor convergent, the intrados quadrilateral is non-planar. Thus, if we understand this operation as a rotation, when the third corner of the quadrilateral reaches the horizontal plane, the fourth one has not reached it yet, and the template should be folded along its diagonal in order to be completely brought to the horizontal plane. As a result, such template depicts exactly the four sides and one diagonal of the intrados quadrilateral; however, it misrepresents the other diagonal and two corner angles. It can be used, however, as a folding template, like those used by Jousse (1642: 156–157) and de la Rue (1728: 44–46) for groin vaults; we will come back to this issue in Sect. 8.2.2.

Conical intrados. Alonso de Vandelvira (c. 1585: 26r-27v; see also Palacios [1990] 2003: 96–101) uses an ingenious approach (Fig. 6.9) to arrive at a correct solution to the problem. The key to his method lies in the division of the face arches. De l'Orme, Martínez de Aranda and Jousse divide *one* of these arches into equal parts, drawing lines to its centre; the intersections of these lines with the other face arch provide the division points for the second face arch and, indirectly, the layout

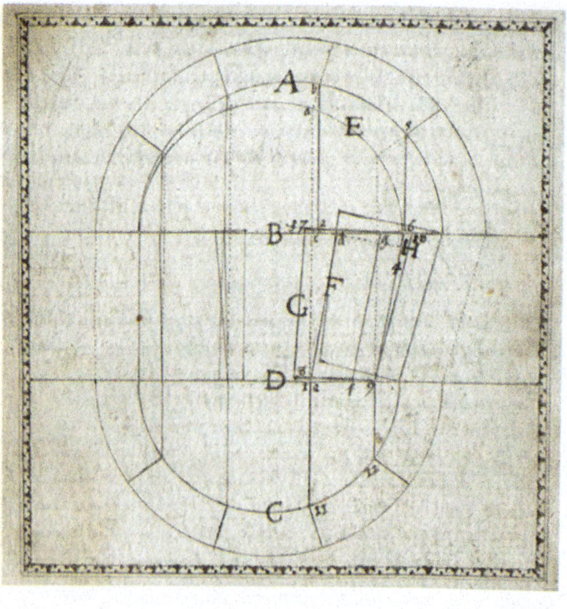

Fig. 6.8 Ox horn solved by templates (Martínez de Aranda c. 1600: 13)

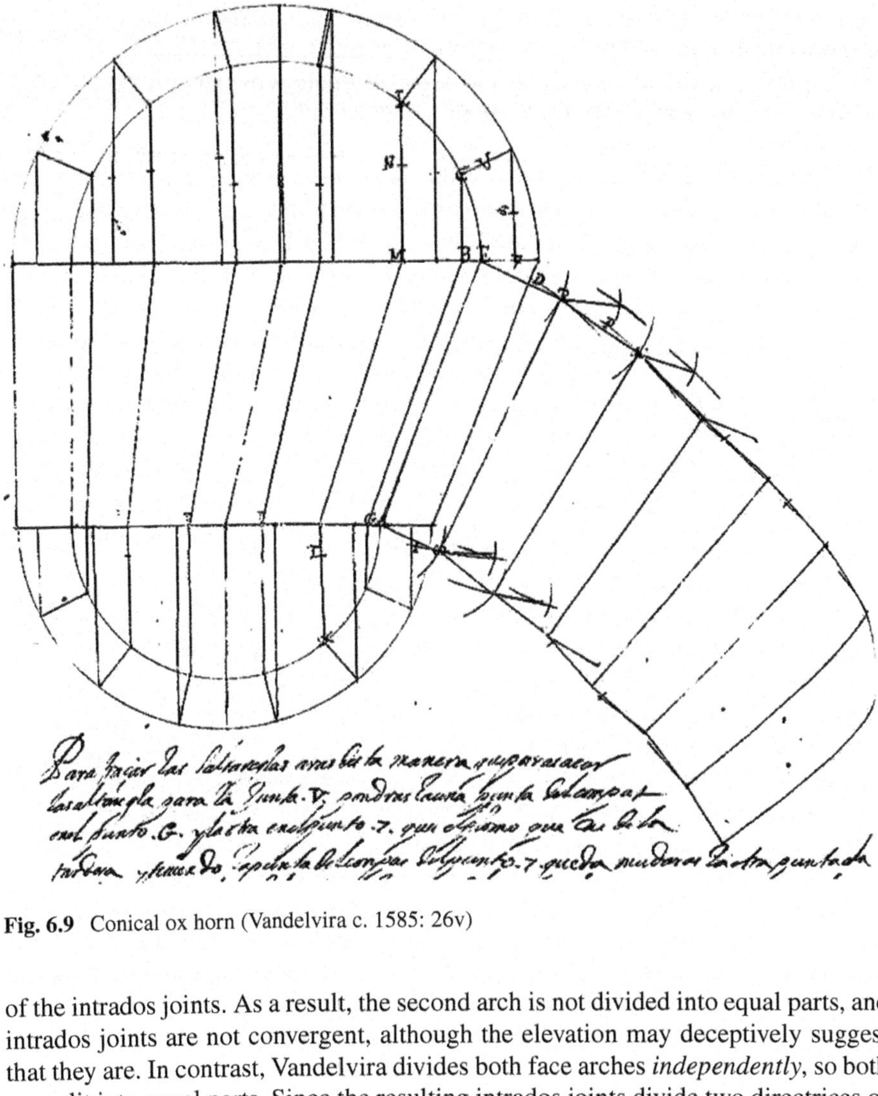

Fig. 6.9 Conical ox horn (Vandelvira c. 1585: 26v)

of the intrados joints. As a result, the second arch is not divided into equal parts, and intrados joints are not convergent, although the elevation may deceptively suggest that they are. In contrast, Vandelvira divides both face arches *independently*, so both are split into equal parts. Since the resulting intrados joints divide two directrices of a cone in equal parts, they are generatrices of the cone. Thus, the intrados surface of this arch is developable; in fact, Vandelvira performs an approximate development, based on a polyhedral surface (c. 1585: 27v-28r).

After drawing the plan and the elevation, he addresses the problem by triangulation, as he did in trumpet squinches. In order to draw the first intrados template, he starts from the springing, which plays the role of first intrados joint and is depicted in true shape in the plan. Next, he computes the length of the diagonals of the intrados template forming a right triangle, using as catheti their horizontal projections and the difference in heights between their ends; the hypotenuse furnishes the length of the diagonal. In typical stonecutting fashion, this operation is performed without actually drawing the projection of the diagonal or its true length, in order to save work in tracing. Next, he traces a circular arc whose radius equals the length of the diagonal, starting from the front end of the springing, and another arc with the length of the back face side of the template, which may be taken from the elevation; the back end of the upper intrados joint may be placed at the intersection of both arcs. Repeating the operation for the other diagonal, he can place the front end of the upper intrados joint and close the intrados template. Then, he repeats the same construction for the second intrados template; the upper intrados joint of the first template is reused as the lower intrados joint of the second template.

The same procedure is used by Vandelvira in order to construct bed joint templates, computing the length of both diagonals of each bed joint and determining the position of both ends of the extrados joint. However, he does not draw the full bed joint template, but only the segments representing face joints, called *saltarreglas* or bevel guidelines. This implies that the segments are transferred to the stone with a bevel, in order to measure the angle between the intrados and face joints. However, in this case, the bevel guideline stands for the face joint rather than the intrados joint, in contrast to Martínez de Aranda (c. 1600: 11–12, 15–16).

Although tiresome, recursive, and prone to the accumulation of errors, Vandelvira's method is exact and suitable to dressing with templates. However, other stonecutting authors took different routes. For once, Derand (1642: 124–125; see also de la Rue 1728: 27) includes only a short explanation of the double ox horn (Fig. 6.10), advising the reader to use the same method he had put forward earlier for a skew arch. Essentially, he follows the procedure of de l'Orme, Chéreau and Jousse (1642: 16–17), scoring bevel guidelines directly on the stones being dressed, although the guidelines are actually shown in the drawing, perhaps to make it clear that they should start from the groin in the middle plane of the arch.

As usual, Frézier takes exception to the warped surfaces in the standard solution; he stresses that "it is quite difficult to give a good reason for the irregularity of this construction; the only one, and a feeble one in any case, is that voussoirs may be dressed easily by squaring".[3] Thus, he puts forward a new solution (Frézier 1737–39: II, 269–271), remarking that it is solved using templates (Fig. 6.11). The key to his innovative solution is using an oblique cone as an intrados surface so that he may use templates; however, he will cut the piece by planes in edge view, in order to use the square when dressing bed joints. Thus, he is cutting a cone by planes that do not pass through the apex, so intrados joints are not straight lines, but parabolas.

First, he draws the plan of the piece and both face arches, marking the line that joins both centres, which is, of course, oblique to the faces, as well as an orthogonal to the faces drawn through the centre of the smaller face. Next, he divides the segment between the centres of both faces in several small portions; the division points provide the centres of several cross-sections of the piece. Then, he divides the narrower face

[3]Frézier (1737–39: II, 267–269): *On seroit fort en peine de rendre une bonne raison de l'irregularité de cette construction; la seule qu'on peut donner, & qui n'est d'aucune considération, est la facilité d'exécuter ce trait pour la voie de l'equarrissement.*

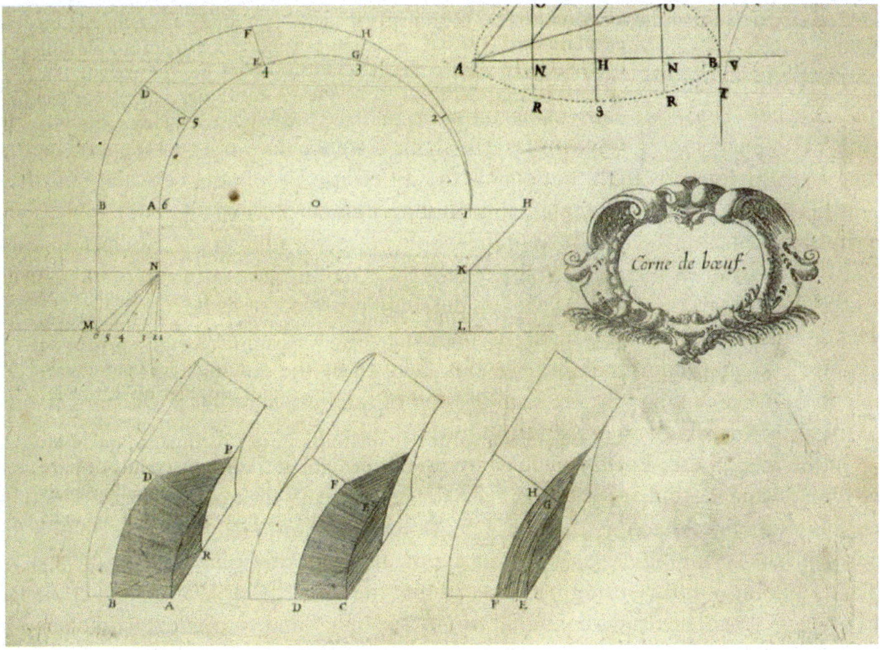

Fig. 6.10 Double ox horn (Derand [1643] 1743: pl. 60)

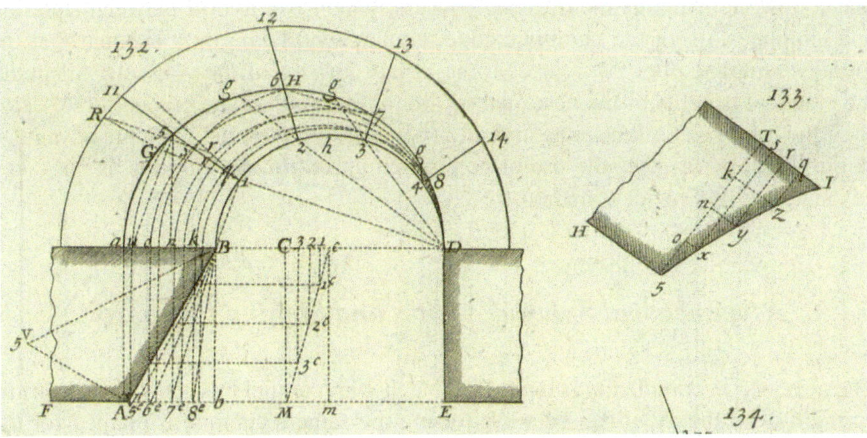

Fig. 6.11 Conical ox horn. (Frézier [1737–1739] 1754–1769: pl. 49)

edge in even portions. In the next step, he draws radii of the small face arch and extends them to the larger arch and beyond. This operation divides the piece into voussoirs, but since the projections of the arch edges are not concentric, the larger face is divided into uneven portions; Frézier argues that this irregularity will be less visible in the broader face.

Next, he constructs bed joint templates, starting by transferring the apparent distance between both ends of an intrados joint, *1* and *5*, to a separate diagram, where it appears as *1-T* (Fig. 6.11, right). This distance does not represent the actual length of the intrados joint, but equals the difference in lengths between both face joints. He then draws a perpendicular to this segment through its end, *T*; next, he transfers to this line the thickness of the whole piece, locating point *5* in the template. Now, he must draw a parabolic arc standing for the intrados joint, joining points *1* and *5*. He constructs it by points, using the set of cross-sections he has drawn before. The distances of the cutting planes to the faces give a coordinate for each point, *Tk*, *Tn*, etc. The other coordinate, *kz*, *ny*, etc., is given by the distance of each point to a line in point view *1* passing through the end of the intrados joint in the narrow face of the arch, which may be taken from the elevation.

In the next phase, Frézier endeavours to prepare an intrados template. However, he remarks that it is impossible to make a flat template for the quadrilateral defined by all four corners of the voussoir, *a-B-1-5*. Although the intrados surface is an oblique cone and the springing *A-B* is a generatrix, the first intrados joint *1-5* is not. Thus, he places a point u in the springing plane so that the quadrilateral *B-1-5-u* is coplanar. In order to do so, he draws a parallel to *1-B* through *5* and computes its intersection with the springing plane both in elevation and plan; next, he computes the length of a diagonal of this quadrilateral, *B-5*; using this diagonal and the four edges of the quadrilateral, he constructs the template for this auxiliary figure. In the sections that follow, he explains the dressing process, which is a hybrid between squaring and templates, using the parabolic edge template and taking into account that the intrados template does not represent the actual intrados of the voussoir. All in all, the practical benefits of the new method are far from clear. The crucial issue seems to be the use of an oblique cone instead of a warped surface, but the main advantage of developable surfaces, the use of precise templates for the intrados, is not taken advantage of by Frézier's method.

6.1.3 Asymmetrical Splayed Arches and Sloping Gunports

The next step in complexity is the splayed arch with parallel faces and asymmetrical springings. In theory, it can be solved using the same technique explained for the ox horn. However, Martínez de Aranda (c. 1600: 36–38) tries to solve it using full templates (Fig. 6.12). He puts forward a concise explanation, referring the reader to another variant, the splayed arch with an oblique face. However, there are significant differences between both pieces. In the splayed arch with parallel faces, the author divides both face arches into even portions; thus, the intrados is a conical surface, as in Vandelvira. In contrast, in the arch with an oblique face (Martínez de Aranda c. 1600: 34–36), he draws a circular cross-section and projects it orthogonally onto the face; as a result, intrados joints are neither parallel nor convergent, and the intrados

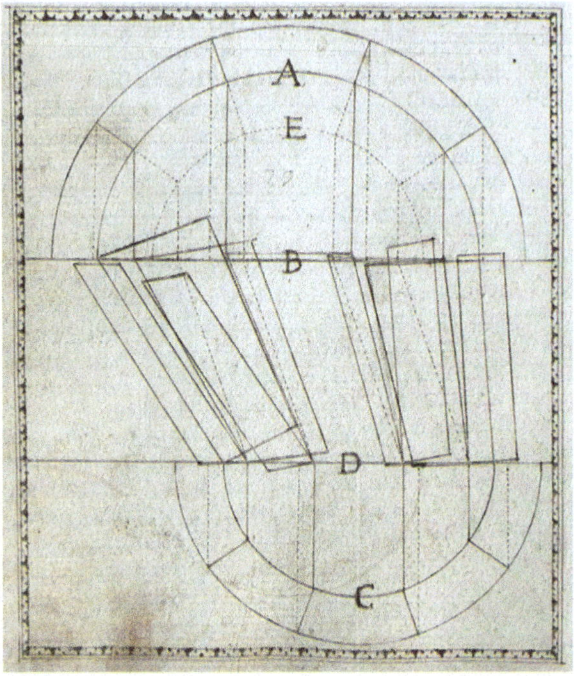

Fig. 6.12 Splayed arch with an oblique face (Martínez de Aranda c. 1600: 37)

is a warped surface. Further, in the arch with parallel faces, Martínez de Aranda applies a remarkable oblique projection: although the projection plane is vertical, the projectors are horizontal but oblique to the projection plane. Despite that, the author tries to apply the same technique to both arches, and his explanations are far from clear (Calvo 2000a: II, 113–121).

Jousse (1642: 83–84) addresses the same problem in the *trompe biaise en cannonière*, (oblique trumpet squinch for a gunport). In this case, there is no rotational symmetry, and thus the mason should construct all intrados templates separately, developing a pyramid frustum. Jousse tries to do this by triangulation, as in the trumpet squinches of De l'Orme or Vandelvira, but his explanation is somewhat hasty. For example, he determines the length of the first intrados joint correctly, but does not explain how to cut a portion of this line to materialise a frustum; the reader may surmise that he should apply the same technique to the section of the intrados joint from the apex to the smaller face. The explanation of the construction of bevel guidelines is also cryptic.

Sloping gunports. Vandelvira (c. 1585: 36r-39r; see Palacios [1990] 2003: 122–131) deals with similar issues, although he introduces two interesting details. First, in most cases springings slope downward (Fig. 6.13), to direct arrows or gunfire against an attacker; this excludes rotational symmetry even in the simplest case, the symmetrical one. Second, in some particular examples, the jambs and intrados of the openings are stepped, rather than planar or conical; in Vandelvira's own words "The dentils … are useful in fortification gunports, as a defence against arrows and

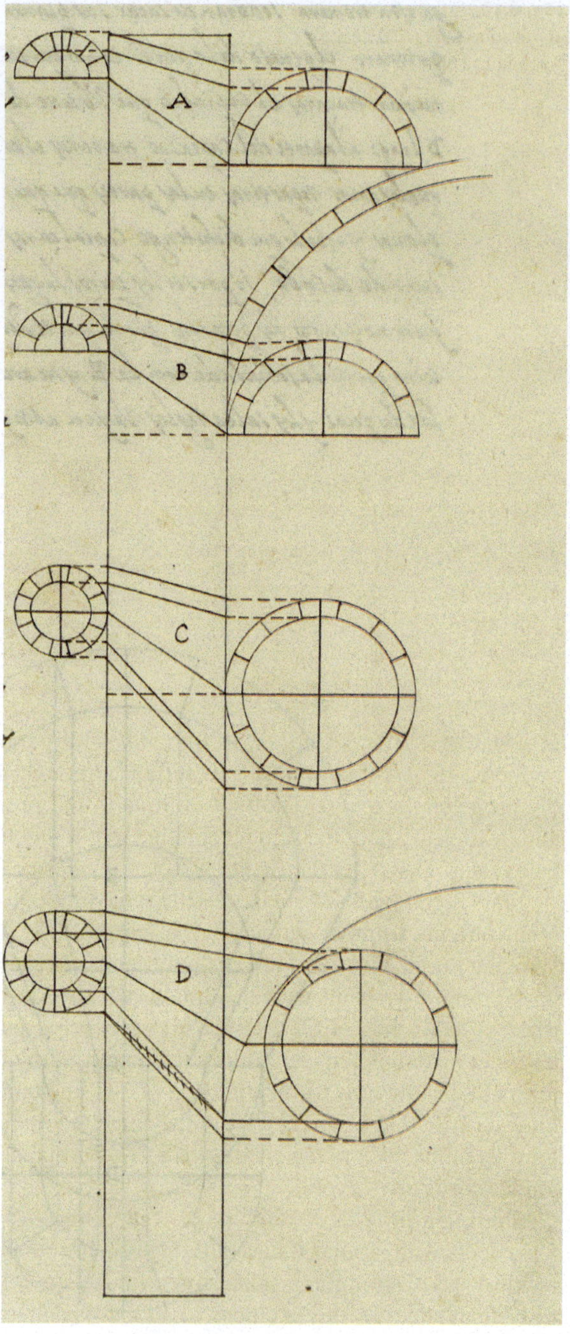

Fig. 6.13 Sloping gunports ([Vandelvira c. 1585: 36r] Vandelvira and Goiti 1646: 64)

Fig. 6.14 Skew oculus. Rome, San Carlo a Catinari (Photograph by the author)

arquebus fire from the enemy, since they are angled and not straight like those used to provide light".[4] He solves them applying the same triangulation techniques he had used in trumpet squinches and the ox horn, that is, computing the length of intrados joints and diagonals forming right triangles and constructing quadrilaterals for intrados templates or bevel guidelines for the bed joints. Faces are parallel, except in the cases of gunports opened in vaults or round walls, and thus no difficulties with warped surfaces arise.

6.1.3.1 Oculi

Most of Vandelvira's gunports feature semicircular fronts, like those he probably built in the castle of Sabiote. However, he mentions that they may be shaped as segmental arches or full circles (compare with Fig. 6.14) and includes a circular porthole opened in a barrel vault (Vandelvira c. 1585: 36r, 39r). Since both solutions feature sloping axes, the result would have been similar to a well-known oblique oculus in Seville Cathedral (Sakarovitch 1998: 148). However, he does not give detailed instructions for setting out or dressing these pieces; we may surmise that he would have solved them by triangulation, like those with semicircular openings.

---

[4]Vandelvira (c. 1585: 38r): *Los dentellones ... son buenos para las troneras de las fortalezas, para defensa de los tiros de flechas y arcabucería de los enemigos, por hacer aquellos ángulos y no en la línea recta como las que sirven para luces. Transcription taken from Vandelvira and Barbé 1977.*

Gelabert (1653: 48v-49r) includes a special oculus with a V-shaped cross-section (Fig. 6.15); that is, the front and back faces are equal, at least in radius, but the middle section is narrower. It resembles, and in fact, it should be built, as two splayed arches extended to fulfil a whole circle and joined by their narrower faces. Given the rotational symmetry of the piece, two templates, one for the bed joint and another one for the intrados, are enough. The bed template, formed by two symmetrical pieces, is an input of the process and can be traced directly. For the intrados template, he attempts a cone development, but the process is problematical. He does not notice that this line is not frontal and is thus shortened by projection, so he mistakes the vertical projection of a generatrix for its real length. However, this error does not have practical consequences, since it affects only the radius of the development of the directrices, which are in any case simplified to linear segments.

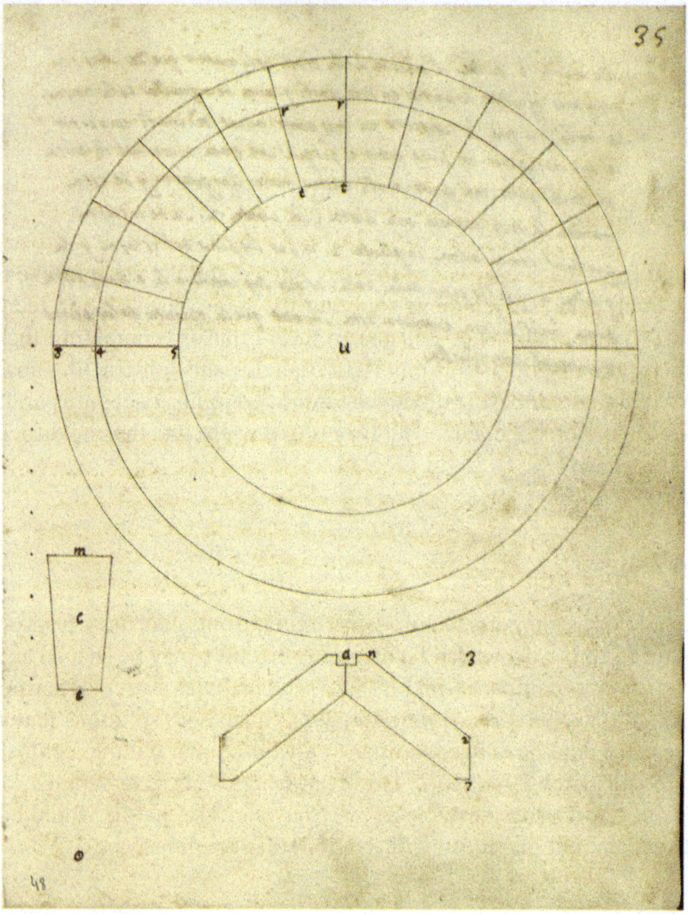

Fig. 6.15 Oculus (Gelabert 1653: 35r)

Gelabert describes the dressing process step by step. First, the mason should carve a flat face, scoring the intrados template on it. Next, he should dress the face, measuring the angle between the intrados and face joints in the bed joint template and transferring it with the bevel. This involves a slight error (Gelabert/Rabasa [1653] 2011: 122) since the angle between the flat intrados at this stage and the face does not equal the angle between the intrados and face joints; this shows that masons did not intuitively grasp the notion of a dihedral angle. Next, he should score face joints on the face using an arch square. This enables him to dress bed joints and score on their surface the outline of their templates. In the next step, the mason should carve the intrados using a pair of templets for the ends and a ruler to connect them; finally, he must carve the shorter end, where both halves of the piece meet; although this last phase is critical, Gelabert says nothing about it.

6.2 Skew Arches

6.2.1 Bed Joints Parallel to the Springings

Trapecial plans. The simplest skew arches are those with a right trapecial plan. They arise when a window or doorway is opened in a tapering wall; thus, both springings are parallel, while face planes are convergent. They should not be confused with the ox horn, which is built starting from convergent springings.

This variant is explained by Martínez de Aranda (c. 1600: 6–8) and in the manuscript attributed to Pedro de Alviz (c. 1544: 9r). After drawing the plan, with parallel springings, a face plane orthogonal to the springings and another face oblique to the jambs, Martínez de Aranda draws the elevation of the square face in true size and shape, as a round arch (Fig. 6.16). Next, he constructs an auxiliary view representing the oblique face also in true shape, using the horizontal projections of the intrados joints, drawn starting from the original elevation. Then, he draws projection lines for the new elevation, passing through the intersections of these joints with the oblique face; after this, he transfers the height of each intrados joint to each projection line. This operation gives the lower corners of each voussoir in the arch, and thus the outline of the oblique face arch. This line should be an ellipse since it corresponds to the intersection of the intrados cylinder with a plane oblique to its axis. However, taking into account an introductory section of his manuscript, we may surmise that Martínez de Aranda (c. 1600: 2; see also Sect. 3.1.2) joins these points in groups of three with the compass; that is, he substitutes a number of circular arcs for the semiellipse.

In the next step, he sets out to construct intrados templates. The width of these templates, equalling the distance between two consecutive intrados joints, may be measured in the elevation of the square face since the joints are lines in point view. This allows the mason to mark the back end of the intrados joint on the face plane and draw the upper intrados joint as a parallel to the projection of the lower one. In

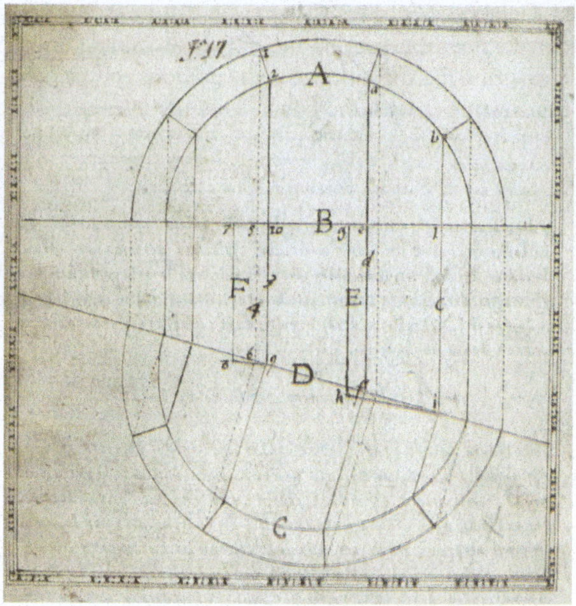

Fig. 6.16 Skew arch with a right trapezium plan (Martínez de Aranda c. 1600: 7)

order to place the front end of the intrados joint, which belongs to the oblique face, he measures the length of the upper intrados joint in projection and transfers it to the template; this furnishes the front end of this joint, allowing the mason to close the intrados template. As we have seen in Sect. 3.1.3, this method is exact, since it amounts to a rotation of the planar quadrilateral joining all four voussoir corners around the lower intrados joint until the other three sides reach the horizontal plane of the lower joint; thus, the template depicts the intrados joint in true size. However, nothing suggests that Martínez de Aranda was thinking in such abstract terms; this is why I have avoided explaining this procedure as a rabatment.

After constructing the intrados templates, Martínez de Aranda instructs the reader to draw bed joint templates by the same method. Alviz does not draw full bed joint templates, but rather the sides of the templates belonging to the faces, known as *saltarreglas* (bevels): this makes it clear that they were constructed in order to measure the angle between the intrados and the face joint. Moreover, there is an important difference between Martínez de Aranda's intrados and bed joint templates: while the latter stand for the finished joint, intrados templates represent a face of a polyhedral surface inscribed in the intrados, rather than the actual intrados surface, which is of course cylindrical. Thus, the mason should first dress a close approximation to the final form of the voussoir, including planar faces and bed joints, as well as a planar surface with the shape of the intrados template; next, he should hollow the intrados face in order to materialise the cylinder. Such an operation was surely a matter of course for sixteenth-century masons, so Martínez de Aranda does not mention it, except in a passing remark about a complex piece, a splayed arch opened in a curved wall, stating that "these intrados templates are to be placed on the faces of voussoir, dressed by boning with the straightedge; then, the voussoirs should be hollowed with the templets of the face arches, placing each one in the corresponding face".[5]

Martínez de Aranda (c. 1600: 8–9) also explains a variation of this piece with two symmetrical, oblique faces and parallel springings. The template-construction procedure is essentially the same one, but there is a significant difference. In the preceding example, he used the elevation for the square face, which amounts to a cross-section of the arch, to measure the distance between two consecutive intrados joints, which are lines in point view and thus orthogonal to the plane of the elevation. However, in this case, there is no square face, so Martínez de Aranda starts by drawing a *juzgo* (auxiliary scheme) with the shape of the cross-section of the arch, although he does not mention anything suggesting the concept. He then uses this auxiliary diagram to construct the actual elevation of both face arches in true shape, as well as the bed joint and intrados templates, as in the arch with a square face.

Frézier's (1737–39: II, 122–133) approach to this problem clearly shows the effects of the application of abstract geometry to masons' empirical practices. After drawing the plan, he takes it for granted that the stonecutter will begin drawing the oblique face arch, not the square face or the cross-section; he may choose a semicircle or a half-ellipse. Next, he should construct the cross-section or *cintre primitif*, (primitive arc). He remarks that if the face arch is a surbased ellipse "it may happen by chance that the cross-section is circular";[6] of course, this cannot happen if the outer arch is a raised ellipse. In any case, he states that it is advisable to draw the cross-section directly as an ellipse, rather than by points, as other theorists do. Next, he constructs bed joint templates and *panneaux de doëlle plate* (flat intrados templates); the name stresses that they are rigid templates, to be placed on a flat face before hollowing the intrados. Both are drawn by measuring the intrados joints, as in Martínez de Aranda, although they are all grouped in a separate diagram. However, he explains that such grouping, which master masons call *developpement*, is a mix of two different kinds of surfaces, and the only reason for such assembly is to show these surfaces clearly, so he will not use it frequently in the rest of his treatise. He goes on to a lengthy discussion of all geometric aspects of the piece, as well as the different methods that may be used when dressing it, remarking for example that there are three different carving methods, although it is advisable to start by carving the intrados. He also takes strong exception to the usual tracing methods for these pieces, where face joints are traced as radii of a half-ellipse, while they should be drawn as orthogonals to the tangents in voussoir corners.

Rhomboidal plans with elliptical faces. De l'Orme (1567: 71r-72r) gives an apparently clear explanation of a rhomboidal-plan skew arch (Figs. 6.17 and 6.18). First, he draws the cross-section of the piece, in the shape of a round arch. It does not lie on the surface of the wall, but rather on a virtual plane orthogonal to the

[5]Martínez de Aranda (c. 1600: 105; see also Sect. 3.2.1): *se entiende que estas dichas plantas por caras se han de plantar teniendo labradas las caras de las piezas a regla y borneo y después se han de afondar las dobelas en las dichas caras con las circunferencias de los arcos plantando cada una por la testa que le conviniere.*

[6]Frézier (1737–39: II, 123) … *il peut arriver par hazard que le ceintre de l'Arc-Droit devienne circulaire.*

Fig. 6.17 Skew arch. Syracuse Cathedral (Photograph by the author)

springings; in other words, it plays the same role as Martínez de Aranda's *juzgo*. He also draws the plan of the arch; its faces are mutually parallel, although oblique to the springings. Next, the mason should measure in the elevation the width of a bed joint, that is, the distance between the intrados and extrados joints, drawing a parallel to the intrados joint at this distance; this parallel will stand as the extrados edge of the true-shape template of the bed joint.[7] However, he has not yet located the ends of the extrados joint. To do so, he measures the distance from the springing line in the elevation, which is used as a reference in the plan, to the projection of voussoir corner *15*, transferring it to the true-shape template of the bed joint in *13*; repeating the operation with the opposite voussoir corner, he may close the bed joint template. This procedure is essentially the same as that used by Martínez de Aranda for both trapezial arches, particularly, the second one, since he cannot use the elevation of a square face to measure the distance between the intrados and extrados joints and therefore must use an auxiliary cross-section.

[7]The interpretation of this passage is tricky. In de l'Orme's (1567: 71v) words *vous prendrez la largeur des ioincts, comme de I à R, & la transporterez de 11 iusques à 13, faisant deux lignes perpendiculairement sur celle de A B*" (you should take the length of joints, such as *I* to *R*, & you should transfer it from *11* to *13*, drawing two orthogonals to *A-B*). Reading De l'Orme literally, it seems that he makes the width of the bed joint equal with the distance between voussoir corners 11 and 13. However, this does not seem to be the case, since the upper intrados joint is not drawn at this stage, so he is not able to locate the upper voussoir corners. Thus, it seems more natural to surmise that he makes the distance between intrados and extrados joints equal to with the width of the bed joint in the elevation.

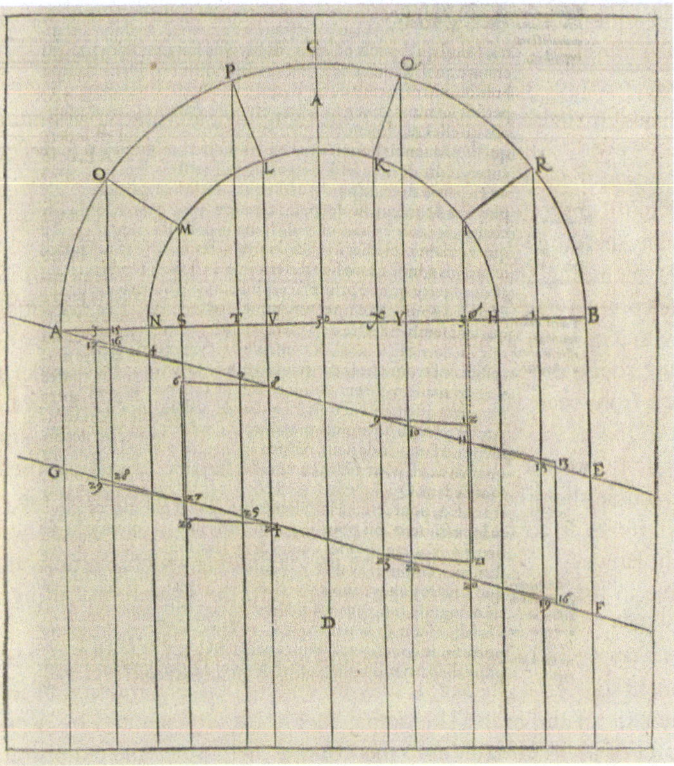

Fig. 6.18 Skew arch with joints parallel to the springings (de l'Orme 1567: 72r)

Startingly, de l'Orme makes no mention of intrados templates,[8] while he explains them thoroughly in other pieces (for example, 1567: 74v-77r), as we will see later. This absence seems to relate to a passing remark stating that "I take it for granted that you have squared the voussoirs and intradoses of your arch following the template that you should prepare *IR*, *HB*, according to the thickness of the wall, including any projection. This template should be used for all five voussoirs".[9] That is, before

[8]It may be argued that de l'Orme is making a deliberate ellipsis, since he says at the beginning of this section (de l'Orme 1567: 71r) *Il suffit, à ce qu'il me semble, d'en montrer seulement les principes et méthode, pour autant que ceux qui en après voudront prendre peine, en trouveront à tous propos, selon les oeuvres qu'ils auront à faire* (It is sufficient, in my opinion, to explain just the principles and the method, since those that will take the pains with find [the adequate solution] for the elements they need to execute). Transcription taken from http://architectura.cesr.univ-tours.fr; translation by the author. However, he makes some problematical remarks about the dressing method and no mention to intrados templates, as we shall see.

[9](De l'Orme 1567: 71v): *Je présuppose que vous avez déjà équarri les pièces et doiles de votre arceau, suivant le panneau qu'il faut lever IR, HB, le tout selon l'épaisseur de votre muraille, compris son avancement. Ce panneau servira pour toutes les cinq pièces de voussure.* Transcription taken from http://architectura.cesr.univ-tours.fr; translation by the author.

using the bed joint template, the mason should have cut a voussoir for a round arch, using a face template of the voussoir, taking into account the thickness of the wall and leaving space for any projecting element, such as mouldings. It can be implied that, once such a voussoir is cut, bed joint templates are placed in both sides of the voussoir; however, taking into account the obliquity of the arch, it is not easy to place bed templates at their relative positions.[10]

Quite significantly, Jean Chéreau (c. 1567–74: 113r), who elsewhere reproduces literally many drawings from de l'Orme, presents a different solution for this problem (Fig. 6.19), including simplified templates both for the bed joints and the intrados, with no accompanying text. However, Jousse (1642: 10–11) includes a similar diagram with a concise written explanation. After drawing the plan with an oblique face and the round cross-section of the arch, Jousse advises the reader to measure the distance from each voussoir corner to a vertical reference plane in the plan and transfer it to the nearest springing. Up to this stage, the operation is similar to the method of de l'Orme or Martínez de Aranda, although he brings the measure to the springers. Jousse then instructs the reader to trace a line by placing the ruler on the transferred voussoir corner and the intersection of the oblique face with the axis of the arch. However, he does not draw this segment, which overlaps the arch, but rather a line in the opposite direction, going into the springers, calling it the "bed joint template".

All this may seem quite striking at first glance. However, the procedure makes much geometrical sense; in fact, it may be understood as a rotation of the face joint around the axis of the arch. The actual face joint is generated by a projection of the face joint of a virtual arch onto the oblique face. Since the extension of the bed face of the virtual arch passes through the centre of the arch, its projection in the actual face should intersect the axis of the arch. Then, if the face joint is rotated around the axis, the intersection of its extension and the axis will not move in the revolution. In contrast, the voussoir corner will rotate along a circumference placed on a vertical plane perpendicular to the axis, which is projected horizontally as a straight line orthogonal to the axis. Since the virtual arch is circular, the distance of the voussoir corner to the axis equals half the span of the arch; as a result, the rotated voussoir corner is placed at the springing. The line that joins the rotated voussoir corner and the axis is a prolongation of the actual face joint, which can be drawn by extending this line. However, the face joint in the virtual arch does not represent the

---

[10]Further, at the end of this section (de l'Orme 1567: 72r), he adds *Mais pour couper le devant des pierres pour le faire biais, il se prendra après la ligne AB, et celle de AE, comme j'ai dit, et le pouvez voir par la figure presente* (However, before cutting the stones to make them inclined, you will take [the bias angle] from line *AB*, and line *AE*, as I have said, and you can see in the present figure). Transcription is taken from http:// architectura.cesr.univ-tours.fr; translation by the author. It is difficult to perform this operation *after* having applied the face template, since the mason cannot use a horizontal plane as a surface of operation in order to score the angle between the oblique face and a plane orthogonal to the springers. In fact, Jousse or Derand use the intrados template mainly to control the relative positions of the bed templates, as we will see. If the mason performs this operation *before* applying the face template, the result will be clearly inexact, because the face template is based on a virtual arch placed on a plane orthogonal to the springers, and not in the actual face, which appears when applying the bias angle between *AB* and *AE*.

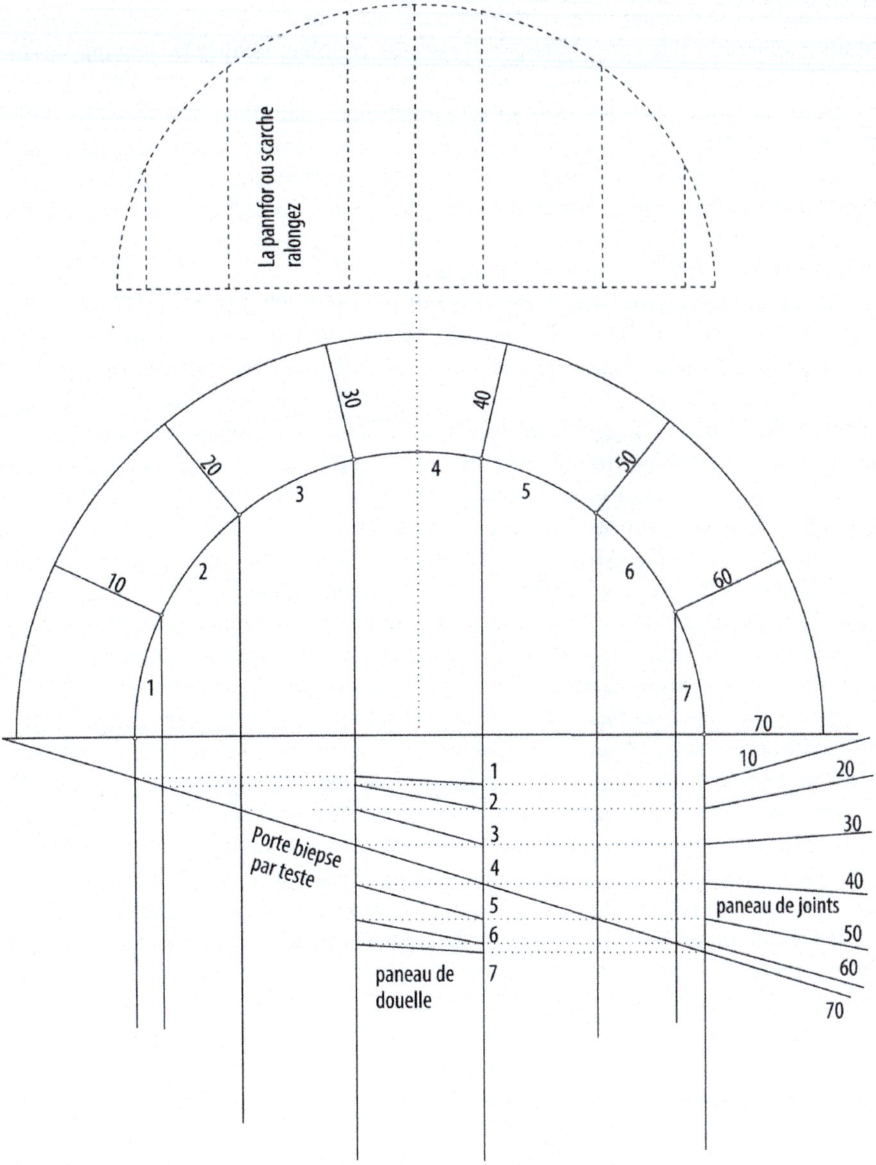

Fig. 6.19 Skew arch with rhomboidal plan and circular cross-section (Chéreau 1567–1574: 113r, redrawn by author)

true length of the actual face joint, which increases as a result of projection onto an oblique plane. In this case, Jousse eschews the problem simply by leaving the face joint undetermined; Chéreau extends it up to the edge of the sheet. This hints that both are planning to use it not as an actual template, but rather as a bevel guideline, that is, a measure of the angle between the face and intrados joints (see Sect. 3.2.5). Admittedly, all these explanations are anachronistic: they are useful to prove that the methods of Renaissance masons are consistent with nineteenth-century descriptive geometry, but nothing suggests that Chéreau or Jousse thought in these terms; we will come back to these issues in Chap. 5.

As for intrados templates, both Chéreau (c. 1567–74: 113r) and Jousse (1642: 10–11) use similar methods. Taking into account that the cross-section is divided evenly, and all intrados faces are equal in width, they may be placed with coincident intrados joints. Chéreau groups them at the keystone of the arch, while Jousse draws a separate diagram, starting with an auxiliary line and two orthogonals whose distance equals the width of the intrados of the voussoirs. Next, he transfers the distance of the voussoir corners to the diagram, starting from the auxiliary line; again, this leads to the same result as the distance transfers used by Martínez de Aranda in the trapecial-plan arch. Also, these templates include only one edge and two lines with undetermined length, in contrast to four well-defined edges in Martínez de Aranda; thus, it seems again that they were used as bevel guidelines. Finally, Jousse constructs the elliptical face arch in an auxiliary view, raising perpendiculars from one of the faces and transferring there the heights of the voussoir corners. This is, in essence, the method Martínez de Aranda had applied for the right trapecial plan arch, although Jousse also uses the midpoint of each voussoir, for greater precision. Chéreau performs the same operation in an independent diagram.

Spanish manuscripts such as Vandelvira (c. 1585: 19v) and Martínez de Aranda (c. 1600: 9–11) solve the rhomboidal-plan arch using a method that resembles, at first glance, that of De l'Orme (Figs. 6.20 and 6.21); however, they include both intrados and bed joint templates. As in the trapezial-plan arch or Chéreau's solution for the rhomboidal-plan arch, they represent the short sides of the intrados templates as straight lines. That is, templates stand for the development of a polyhedral surface inscribed in the intrados of the arch, rather than the actual intrados surface; in contrast to Chéreau or Jousse, these templates feature all four edges, although Vandelvira indirectly admits the possibility of using only bevel guidelines. Perhaps the absence of intrados templates in de l'Orme's treatise stems from a reluctance to use this system, since he seems to use cylindrical developments in arches opened in curved walls. Moreover, Vandelvira introduces an interesting variation: rather than using distances to a reference line, he transfers the position of the lower voussoir corners to the upper ones using orthogonals to the lower intrados joint. Although his written explanation is utterly befuddling,[11] these orthogonals will reappear in Martínez de Aranda's ingenious solution to the skew arch with circular faces, as we will see next.

[11]The sentence in Vandelvira (c. 1585: 19v) *comienzan de sus plomos y acaban conforme el plomo de adelante toca en el arco llevado a trainel de la línea plana con aquella línea de puntos hasta que encuentre con el altura … de una dovela del arco que hubieres echado desde el plomo hacia el medio*" (starting from their projection lines and ending where the front projection line meets the

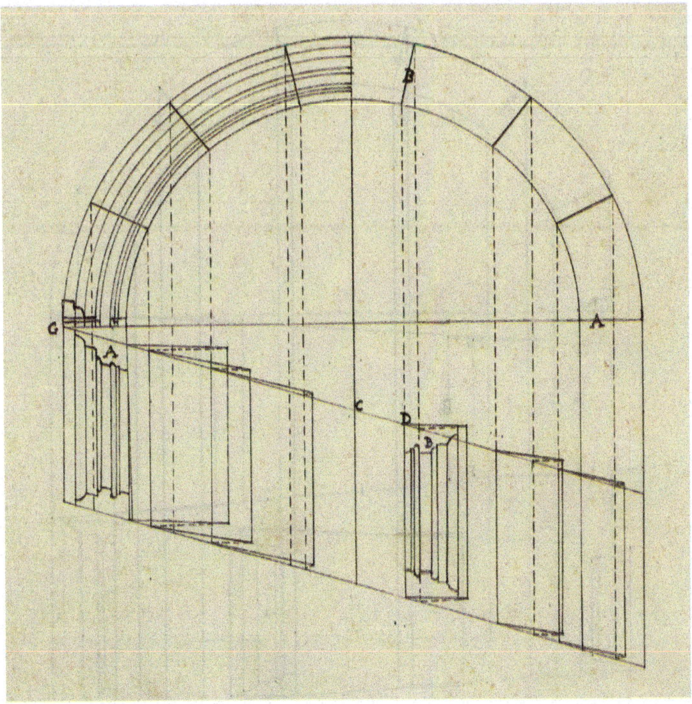

Fig. 6.20 Skew arch with circular cross-section ([Vandelvira c. 1585: 19v] Vandelvira/Goiti 1646: 29)

In any case, the approach of Chéreau, Jousse and the Spanish writers to skew arches is quite effective: it solves the problems created by the lack of intrados templates in de l'Orme's method. The mason can start by dressing an ordinary voussoir for a round arch, using the front face as the operating surface, and dressing the bed joints. However, instead of dressing a cylindrical intrados surface, he should prepare a planar surface and score the intrados template on it; the template represents exactly the planar quadrilateral formed by the four corners of the intrados surface of the voussoir since both intrados joints are parallel to the springers. Once these corners are located, the mason can place the bed joint templates, which furnish the edges of the face of the voussoir, allowing him to dress the oblique face. In order to achieve greater precision, Martínez de Aranda (c. 1600: 9–11) constructs the elevation of the face arch in true size and shape as an auxiliary view, as he had done in the trapezial-plan arches. Again, this arch should be an ellipse, since it is the intersection of a cylinder whose cross-section is a circle, as shown in the *juzgo*, but we may assume that Martínez de Aranda draws it as a series of circular arcs.

---

arch brought in parallel to the level [i.e., springing] line with this point line until it meets the height of a voussoir of the arch drawn from the projection line to the middle point) may be connected with this operation, but its meaning is far from clear.

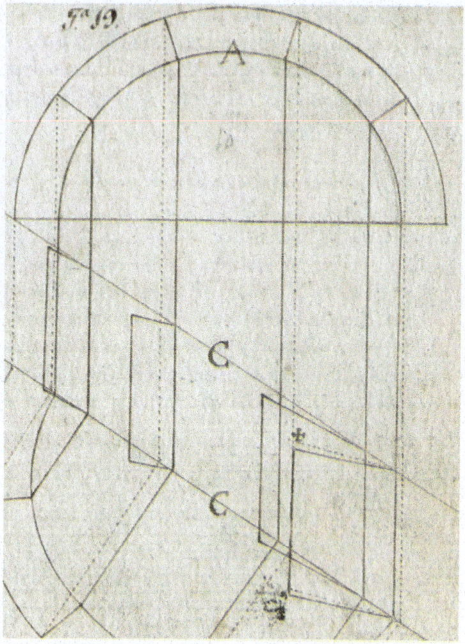

Fig. 6.21 Skew arch with a rhomboidal plan, detail (Martínez de Aranda c. 1600: 10)

Derand (1643: 153–161) offers a different solution, using a simplified development of the intrados cylinder. First, he starts by making it clear that the cross-section of a round cylinder is a half circle. If the cylinder is cut obliquely, the section will be an *ovale*; for him, this term is synonymous with an *ellipse* (see for example Derand 1643: 294–296). Next, he explains in great detail the tracing and dressing procedure. The mason should start drawing the plan of the oblique arch, as well as the cross-section, with two concentric circles and several face joints, dividing the arch into several voussoirs, of equal length if possible. In the next phase, "each of these divisions of the intrados will be divided into two parts, or more, if the size of the voussoirs demands it";[12] next, projection lines should connect the elevation with the plan. At this point, he introduces a distinctive trait of his method. First, the mason is to extend the "interior arc"—that is, the portion of the face edge that belongs to each voussoir—taking two or more chords and bringing them to a "directing line", and marking them one after the other; the combined length of the chords will give the full width of each intrados template. In other words, while Jousse, Vandelvira and Martínez de Aranda make the width of the intrados template equal to the distance between intrados joints, Derand attempts an approximate development of the intrados surface.

Next, Derand places the corners of the voussoirs in the developed intrados joints by transferring their distances to the directing line from the plan to the development.

[12]Derand (1643:155): *Chacune de ces diuisions dans la doüelle ou cherche interieure, se partagera derechef en deux parties, ou bien mesme en dauantage, si la grandeur des voulsoirs l'exige ansi.*

Thus, the face side of each template is defined by three (or more) points; in his example, Derand uses three points and joins them with a curve, approximating the theoretical development of a cylindrical surface cut by an oblique plane. This operation gives the back sides of all intrados templates. In order to draw the front sides, Derand advises the reader to transfer the length of the intrados joints; of course, if both faces are parallel, the same intrados joint will do, whereas if the thickness of the arch is variable, the mason must transfer all intrados joints, as Derand duly remarks. In other words, the rhomboid, the right trapezium and the symmetrical trapezium plans are treated as particular cases of the generalised trapezium.

It may seem startling that Derand, who had been teaching mathematics at the prestigious Jesuit college of La Flêche, resorts to a simplification instead of using π. We should take into account, however, that in the seventeenth century, the use of decimal notation for numbers was not widespread; that the usual metrological system of the period involved feet, spans, inches and fingers; that applying π to the radius of the arch of say, six feet, would have led to an approximate measure expressed in feet and inches; that dividing this measure into five, seven or nine parts would have made the problem even less manageable, since most masons did not know how to divide a quantity. All in all, the errors brought about by a numerical approach to the issue would have been greater than those involved by this graphical procedure.

The next step is the construction of the bed joint templates. Derand measures their width in the elevation and draws a parallel to the intrados joint in the developed templates; next, he measures in plan the distance of the upper end of the face joint to the directing line and transfers it to the template. If the mason does not wish to measure distances to the direction line, as Jousse does, then he can use orthogonals to the intrados joints, as Vandelvira does. Derand explains both procedures exhaustively, although, in essence, both are a variation of de l'Orme's method.

After this detailed exposition of the construction of the templates, Derand starts another lengthy account of their use; such careful explanations hint that this is an essential point in the theory of arch construction since the author does not go so far with other arches. First, the mason should select a block capable of containing each voussoir, and then carve with the arch square the outline of the face of each piece, dressing the intrados and both bed joints. Next, he should apply the intrados and bed joint templates to the appropriate surface; however, it is important to notice that Derand uses *coucher dans son creux,* (lying in its hollow), for intrados templates, while he talks about *appliquer* (applying), for bed joint templates; the distinction between the flexible intrados templates, snugly fit into the concave surface of the intrados, and the bed joint templates, just laid on flat surfaces, is quite clear. In any case, the primary use of these templates is to control the orientation of the skew face; after checking carefully that the intrados edges of both the intrados and bed templates match along the actual intrados joint, the mason should mark the face sides of all three templates and use them as marginal drafts to carve the front and back faces.

To construct bed and intrados templates just to control the angle of the face of the arch may seem excessive; this is why Derand (1643: 160) offers an alternative method using bed joints passing through the intersection of the oblique face and the axis of the arch, as do Chéreau and Jousse. He also draws several lines in the keystone, resembling the simplified intrados templates of these authors, although, for once, he does not explain their construction.

De la Rue (1728: 16–17) presents another sophisticated variation of Chéreau's and Jousse's method. He constructs intrados templates grouped at the keystone, as well as bevel guidelines passing through the intersection of the face plane and the arch axis, and a true-shape depiction of the face arch (Fig. 6.22). The result is a remarkably economical method, which justifies the name given to this particular stonecutting procedure, *biais par abregé*, (abbreviated skew arch).

Frézier (1737–39: II, 133–134) explains the same construction, citing Derand as a source; this is somewhat striking, since Derand includes the procedure in his drawing, but does not explain intrados templates in writing. Nevertheless, Frézier objects to

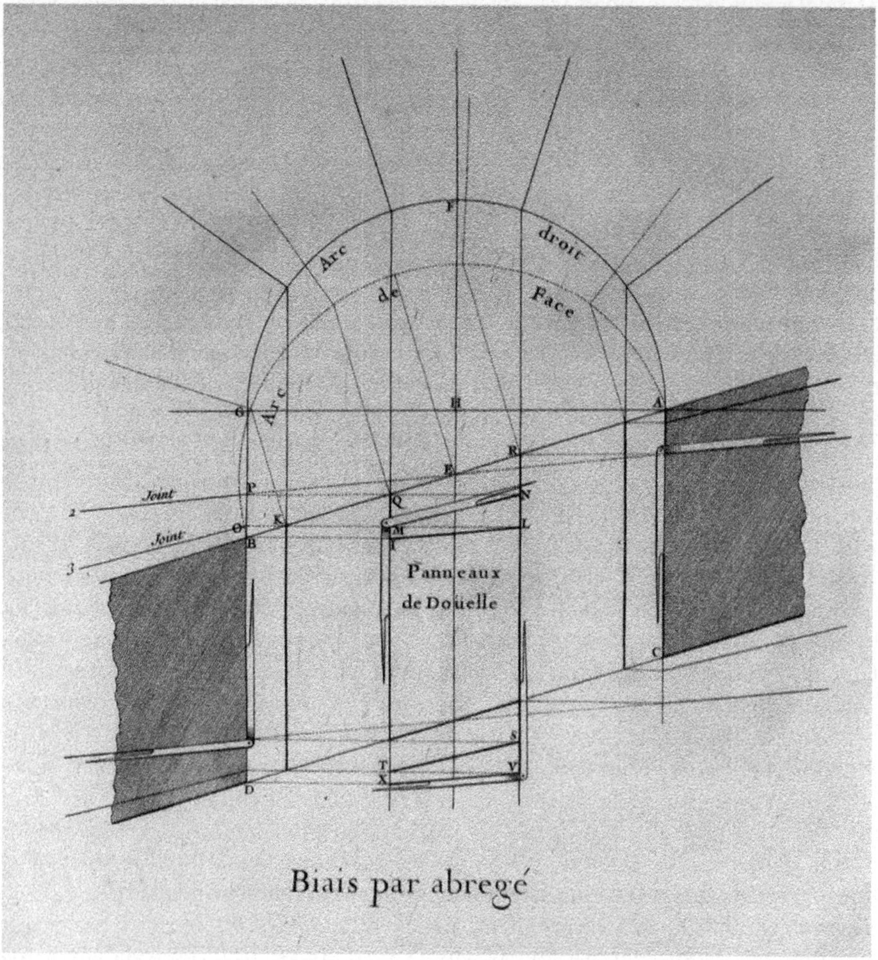

Fig. 6.22 Skew arch with rhomboidal plan and circular cross-section (de la Rue 1728: pl. 10)

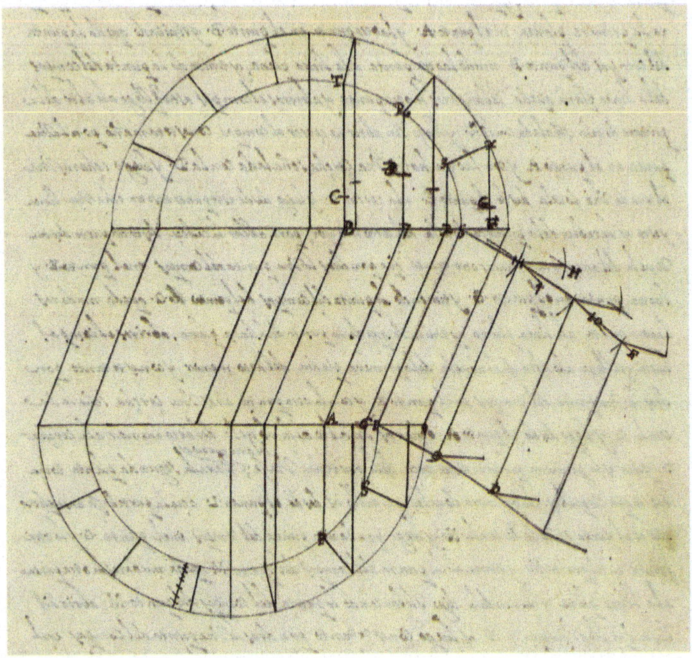

Fig. 6.23 Skew arch with rhomboidal plan and circular faces (Vandelvira c. 1585: 27v)

this method, explaining that the face arches are half-ellipses, and the extensions of the face joints in the drawings of Derand and de la Rue pass through the centre of the half-ellipse, rather than being orthogonal to the tangents to the ellipse at each point. Furthermore, he remarks that neither author has noticed this error, perhaps because they consider this method to be "a worker's practice, misshapen and irregular in this respect".[13]

Circular faces. As we have seen, de l'Orme and other writers approached skew arches projecting a circular cross-section orthogonally onto oblique vertical planes; this amounts to generating a circular cylinder and finding its intersection with a plane at an angle to the cylinder axis; of course, this operation results in *surbased* elliptical intersections. Alternatively, the circular arch may be placed at the wall plane and projected obliquely to this plane, giving as a result an elliptical cylinder, whose cross-section will be a *raised* ellipse (Sanabria 1984: III, ill. 92; Rabasa 1994: 147, Fig. 1).

Vandelvira (c. 1585: 27v-28r; see also Palacios [1990] 2003: 102–105) presents a remarkable attempt to construct the templates of a skew arch with round faces and parallel intrados joints (Fig. 6.23), applying a variant of the method used in the ox horn (see Sect. 4.3.1). He takes the length of the chords of voussoir edges from the elevation; next, he measures the diagonals of the intrados quadrilaterals, forming a

[13]Frézier (1737–39: II, 134): ... *une pratique d'Ouvrier difforme, & peu réguliere en ce point.*

right triangle with the horizontal projection of each diagonal and the difference in heights between its ends. Next, he forms a triangle with the chord, the diagonal and the springing; this allows him to place one end of the first intrados joint. He then places the other end, forming another triangle with the opposite diagonal and the chord so that he can close the intrados template for the first voussoir. It is worthwhile to remark that this complex procedure provides the shape of the intrados template *and* the length of the upper intrados joint. Although nothing prevents Vandelvira from measuring the length of intrados joints in the plan, he places the second end of the intrados joint using the diagonals. The simplest explanation for this unusual choice, which leads to a recursive procedure, is that Vandelvira was extrapolating the solution for his conical ox horn, where intrados joints are not horizontal (see Sect. 6.1.2), and he did not notice that he could take the length of intrados joints from the plan, since in this case they are horizontal.

Vandelvira's method shows a remarkable empirical command of many concepts in spatial geometry that had not entered the realm of learned science at this period. For example, he constructs only half the templates of the arch, understanding implicitly that the opposite ones can be obtained by a 180° rotation; thus, the left-side templates can be reused for the right half of the arch by simply turning them around. However, it is also tiresome and leads to the accumulation of errors. An alternative solution was anticipated by a drawing without text by Cristóbal de Rojas (1598: 99v, lower half; see Fig. 2.12) and fully explained by Martínez de Aranda (c. 1600: 16–17). He starts by tracing the plan and the elevation of a round arch, placed in the actual surface of the wall (Fig. 6.24). A first difficulty arises when drawing the horizontal projection of the intrados joints. He cannot draw them as orthogonals to the plane of the round arch, so he needs to use an auxiliary orthogonal to the springings, called the *juzgo*; we should recall that, in full-scale tracings, it is easier to draw a perpendicular than a parallel. Next, he follows broadly the method he had used in the trapecial-plan and rhomboidal plan with elliptical-face arches. However, here the task is not so simple: in this case, the intrados joint is not a line in point view. Thus, Martínez de Aranda

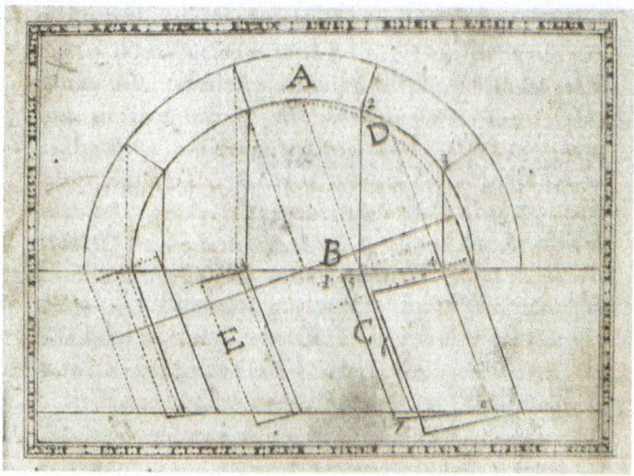

Fig. 6.24 Skew arch with rhomboidal plan and circular faces (Martínez de Aranda c. 1600: 16)

cannot take *the distance between two intrados joints* from the elevation, since an orthogonal to both intrados joints is not parallel to the projection plane. However, he can measure *the distance between the ends of both intrados joints* in the elevation, since these ends lie on the face arch, which is placed on the projection plane. In order to construct the intrados template, Martínez de Aranda draws an orthogonal to the lower intrados joint, passing through the horizontal projection of the end of upper intrados joint; however, he explains the operation as the construction of a parallel to the *juzgo*, using the gauge. This is somewhat surprising to the modern reader. It is not easy to say whether he is still thinking about distance transfers measured from a reference line, or if he is thinking about a perpendicular but recommends using the gauge to avoid the errors introduced by the use of the square to draw long orthogonals. In any case, in the next stage he draws an arc whose radius equals the distance between the ends of both intrados joints, with its centre at the end of the lower intrados joint; where this arc meets the orthogonal to the intrados joint, he places a corner of the intrados template representing an end of the upper joint.

Again, all these operations are consistent with nineteenth-century descriptive geometry. They may be understood as a revolution around the lower intrados joint, where the end of the upper joint moves on a circle in a vertical plane, which is projected as an orthogonal to the lower joint. Of course, Martínez de Aranda, a man "tied to material"[14] by his own confession, did not think in such abstract terms. In any case, the upper intrados joint can be drawn by starting from the known end as a perpendicular to the *juzgo*, with the same length as its projection; this furnishes the fourth corner of the intrados template and allows the author to close the intrados template. The drawing hints that the same method should be used for all the voussoirs of one side of the arch; this implies that the templates may be used for the opposite side by reversing them and that the horizontal projection of the keystone may be used as a template. Next, he uses the same procedure to construct bed joint templates. Of course, all this implies that Martínez de Aranda does not develop the actual intrados surface, but rather a polyhedral surface inscribed in the intrados of the arch.

A most interesting passage is the construction of the cross-section of the arch. As we have seen when dealing with the skew arch with elliptical faces, Martínez de Aranda (c. 1600: 9–11; see also de l'Orme 1567: 71–72; Vandelvira c. 1585: 19v) starts with the cross-section, which is projected orthogonally onto the oblique face planes; thus, the author must draw separately a true-shape depiction of the face of the arch. In the arch with circular faces, the situation is exactly the opposite. Martínez de Aranda starts with the actual face arch, so he does not need to construct its representation in true shape. However, he needs to construct the cross-section of the arch, that is, the intersection of an elliptical cylinder with a plane orthogonal to its axis, as an auxiliary view. He starts from an orthogonal to the springings, the *juzgo* I have mentioned in the preceding paragraphs. Next, he transfers the heights of voussoir corners taken from the elevation to the prolongation of the intrados joints, starting from the *juzgo*; in other words, the extensions of the joints play the role of

[14]Martínez de Aranda (c. 1600: preface, n. p. [i]): *por haber de estar los artífices continuamente asidos a la materia.*

the projection lines of descriptive geometry. The result is a *raised* elliptical arch, in contrast to the *surbased* elliptical face of the preceding variant. It is easy to understand that the span in the face arch is given by a line that bisects the springings obliquely, so it is larger than the span measured in the cross-section, given by an orthogonal to the springings, while the rise both in the face and the cross-section is given by half the span at the face (Rabasa 1994: 148).

To an architect or engineer trained in descriptive geometry methods, such a solution may seem pretty straightforward. However, masons arrived at it through a long and difficult research process. The oldest solutions involve bed joints that are not parallel to the springings, but rather orthogonal to the face planes; we will see them later on. As we have seen, Vandelvira's solution is quite ingenious, but also cumbersome and prone to errors; surely, if he had known the solution of Martínez de Aranda (c. 1600: 16–17), he would have included it in his manuscript. All this shows the fast evolution of stonecutting methods in the sixteenth century, fostered by the introduction, in the Renaissance, of a large catalogue of new or recovered architectural elements, which brought about the empirical use of a whole set of new concepts in spatial geometry.

In any case, the solution of Rojas and Martínez de Aranda did not enjoy much success. It was repeated by Alonso de Guardia (c. 1600: 80v-81r) and Juan de Portor y Castro (1708: 2v), but French treatises followed another line of thought. Derand (1643: 164–166; see also de la Rue 1728: 16, pl. 9) solves the problem by squaring (Fig. 6.25, compare with 6.26). After drawing the elevation and the plan of the arch, explaining that it is not necessary to draw all intrados joints since they are parallel and form the same angle with the face plane, he draws enclosing rectangles in the elevation. Afterwards, he explains the dressing procedure: the mason should carve a stone "as if he wanted to make a jamb of the arch";[15] that is, the plan should be a rhomboid with two internal angles equalling those between the face and the springings. After that, the mason should mark on the face of the block two corners of the voussoir, measuring the projected distance and the difference in heights between the ends of the intrados joints, and placing the arch square in order to score the section of the face edge belonging to the voussoir and the upper bed joint. Since the edges of the block follow the angle of the intrados joints, he can easily mark the position of the arch square in the back face and cut a wedge under the intrados of the voussoir, and another one over the bed joint. At this moment, the intrados, both faces and both bed joints of the first voussoir are already dressed, since the first bed joint, that is, the springing, is horizontal and was shaped with the rhomboid block. If necessary, the mason can draw the extrados with a templet or a compass, in order to finish this first voussoir; except for that, no template is necessary. The dressing procedure for the second voussoir is fairly similar, except that none of the bed joints coincides with the sides of the original rhomboid block and thus it must be carved

[15]Derand (1643: 164): *comme si on vouloit faire vn pied droit de la porte biaise.* This comparison is not really precise since a stone for the jamb of the skew arch would usually adopt the shape of a trapezium, with a skew side to form the jamb and a square one to adapt to the ordinary blocks of the wall. Perhaps Derand refers to the springer rather than the jamb since in the drawing he uses the word "piedroit" for the springer.

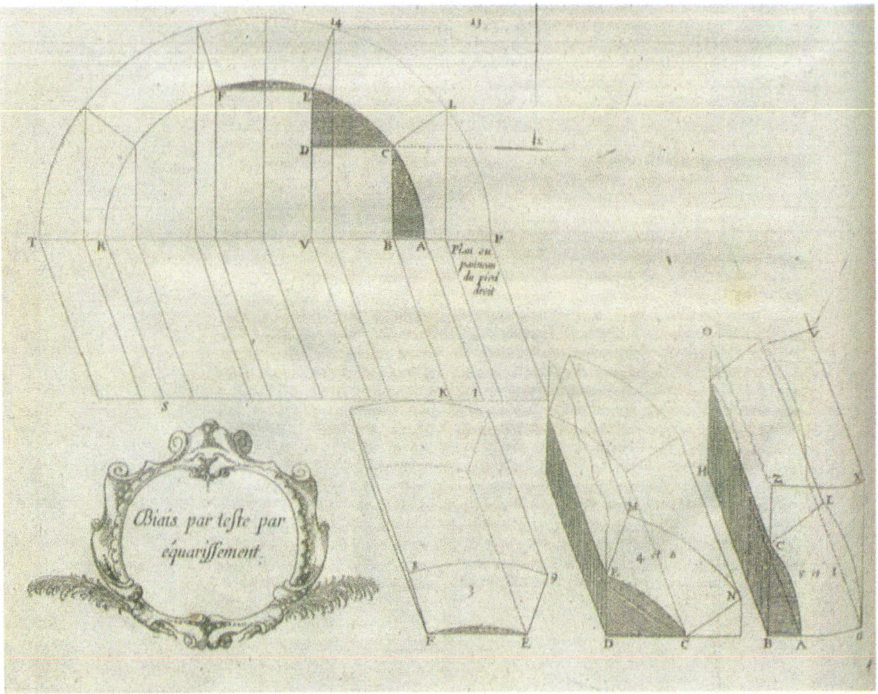

Fig. 6.25 Skew arch dressed by squaring (Derand [1643] 1743: pl. 19)

in order to arrive at the final form. A remarkable detail stresses the nature of this oblique method: in addition to the actual stonecutting tracing, Derand includes an axonometric drawing, with didactic purposes. Although this drawing may resemble at first sight an ordinary cavalier perspective, on close inspection it is a transoblique or "Hejduk" perspective, as remarked by Rabasa (1994: 148; see also Alonso et al. 2011: 660).

6.2.2 *Bed Joints Orthogonal to Face Planes*

The *biais passé*. Still another approach to the problem of the skew arch is possible. Instead of setting the planes of the bed joints parallel to the springings, the mason can choose to make them orthogonal to the faces; the solution is known in French as *biais passé*, literally "passed bias". More precisely, in the preceding solutions, both the intrados joints and the bed joint surfaces are parallel both to the axis of the arch and the springings. Moreover, it may be seen in the elevations that joints, both for the front and back faces, converge in the centres of their face arches; thus, all bed joints are convergent in an axis parallel to the springings. In contrast, in the *biais passé* bed joints converge either in two lines in point view, the axes of the face arches

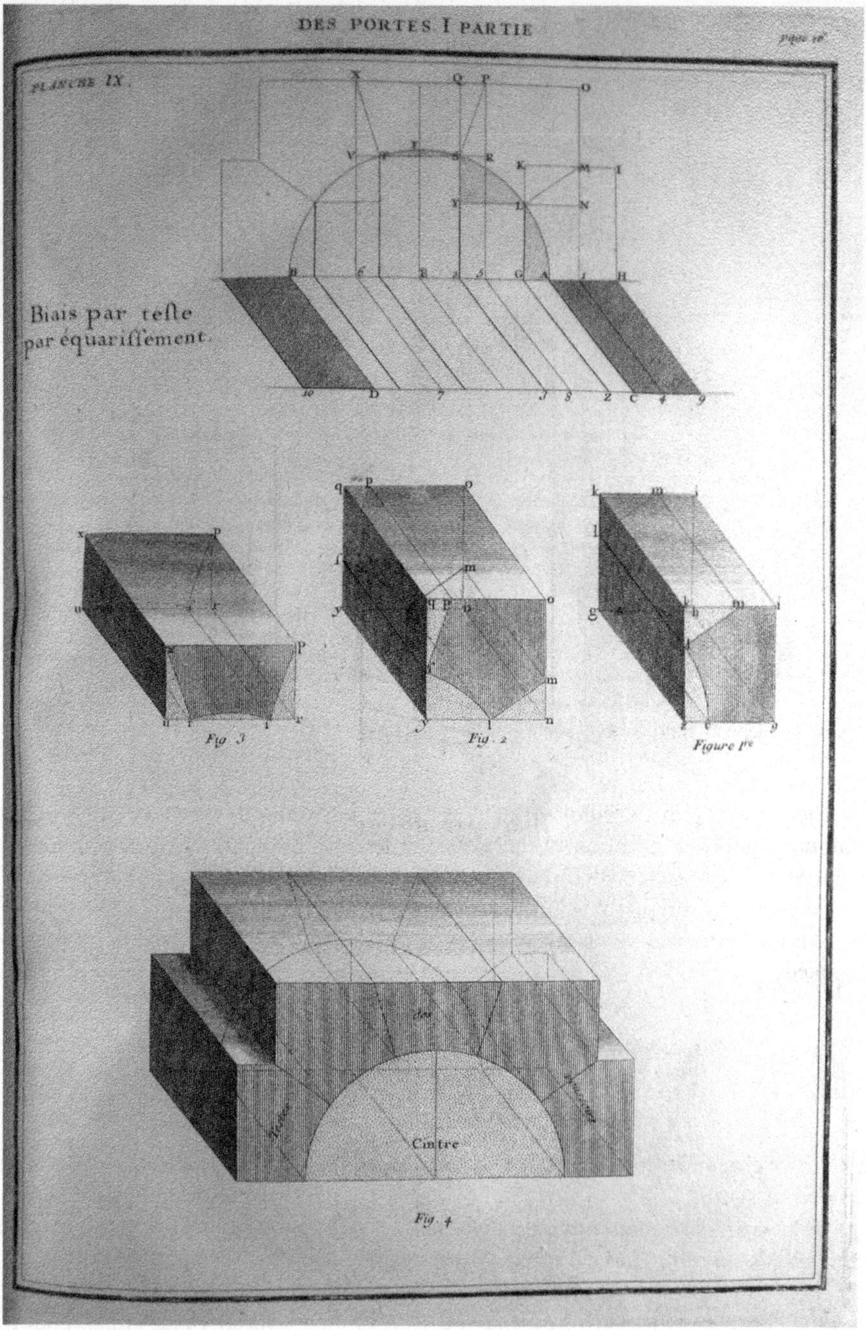

Fig. 6.26 Skew arch dressed by squaring (de la Rue 1728: pl. 9)

Fig. 6.27 Left, arch on a curved wall; right, skew arch. (Villard c. 1225: 20v)

(Martínez de Aranda c. 1600: 15–16; De la Rue 1728: pl. 17, bottom) or in a single axis placed halfway between them (Jousse 1642: 14–15; Derand 1643: 122–124): In both cases, the single or double axes are *orthogonal to the faces*, but not *parallel to the springers*.

Such an approach has remarkable geometrical consequences. The intersections of the planes of the bed joints with the faces are points placed at different heights; thus, the intrados joints are neither horizontal nor parallel. They are not convergent, either; as a result, the intrados surface is not a cylinder, but rather a non-developable surface (Rabasa 1994; see also Lawrence 2011). In theory, this excludes the use of intrados templates for this surface. Martínez de Aranda (c. 1600: 15–16) tries to use rigid templates, but the four corners of the intrados face are not coplanar, and his explanations are far from clear. As for flexible templates, nineteenth-century writers made it clear that they are unusable in this case since the intrados surface is a classic example of a warped surface. In contrast, dressing the voussoirs by squaring is a sensible choice. Since bed joints are orthogonal to the faces, once the mason has marked the outline of the face template on the operating surface, they can be dressed easily with the square. As a consequence of the orthogonality between faces and bed joints, the volume that encloses the voussoir is rather compact, reducing the "great loss of stone" mentioned by de l'Orme to reasonable limits.

As we have seen in Sect. 2.1.1, the sketchbook of Villard de Honnecourt includes a diagram for a skew arch with parallel springings, attributed to a different draughtsperson know as Hand IV (Fig. 6.27, right).[16] Since the scheme does not include an elevation, it is not easy to interpret. Branner (1957) identified the problem and attempted a first explanation of the procedure that Hand IV had in mind; however, he seems to be mixing two different solutions to the problem. Lalbat et al. (1987)

[16]The draughtsperson is also known as ms 2 or Magister II (Villard/Barnes [c. 1225] 2009: 20r, dr. 9; see also pages 130, 133, 13 and Villard/Hahnloser [c. 1225] 1972: 194–200).

have posited that it is substantially identical to the *biais passé* of Mathurin Jousse (1642: 14–15), although, in a later study, they add other hypotheses.[17]

Lalbat's first interpretation is quite daring. As we will see, Jousse solves the problem using several lines representing intrados joints and acting as bevel guidelines. In the scheme of Villard's sketchbook, there are no such lines, although some of the marks, grouped near the imposts, may represent the ends of the intrados joints. Other notches, spanning the full aperture of the arch, may stand for the horizontal projections of the corners of the voussoirs. However, in order to compute the projections of the voussoirs, not to speak of the intrados joints, the mason needs an elevation of the arch, which is missing. This may be explained if we consider Hand IV's schemes as aide-memoires rather than complete reproductions of actual tracings, as in later treatises and manuscripts. Moreover, other explanations of the drawing are less convincing, so Lalbat's hypothesis may not be so far off the mark. However, there is no evidence of the use of the bevel among medieval masons. It has been implied that they measured angles by means of gradients (Bechmann [1991] 1993: 180–181) and, in fact, the third group of marks in the diagram, placed near a square, may be used for this purpose. As we will see, Derand (1643: 122–124) actually draws the lines that represent intrados joints directly on the stone. This suggests that Hand IV may have used the same technique; instead of using a bevel, he may have transferred each notch to the relevant bed joint of a voussoir; then, he may have used a scriber to join the mark to the opposite corner of the voussoir, or else he may have carved a marginal draft directly with the chisel.

De l'Orme's solution to this problem (1567: 67v-69r; see also Rabasa 1994: 147–148) is somewhat puzzling since it includes a general drawing showing *two* different face joints for each bed joint (Fig. 6.28), as in the solutions with intrados joints parallel to the springings.[18] However, a separate axonometric scheme of an individual voussoir features the typical sloping intrados joints of the *biais passé*, while the text explains that the hatched area of the axonometric drawing is to be cut away, as Martínez de Aranda will also explain. However, de l'Orme makes no mention of the guidelines used by Martínez de Aranda or Jousse; this is striking since De L'Orme uses these lines in the double ox horn (see Sect. 6.1.2).

Martínez de Aranda (c. 1600: 15–16; see also 12–14) presents two solutions for this arch.[19] In the first one (Fig. 6.29), he tries to apply a derivative of the method he uses successfully for the arches with parallel intrados joints, using orthogonals to the projection of the joints. However, the problem is not so simple, since in this case, the joints are not parallel. Thus, he places a corner of the template along an orthogonal to the intrados joint, as he did in the arch with bed joints parallel to the springings and circular face. Next, he locates the fourth corner using its distance to other corners. The

[17]Lalbat et al. (1989). See also Villard/WG/Bucher (1979: 120); Sanabria (1984: I, 38); Bechmann ([1991] 1993: 169–175); Villard/Barnes ([c. 1225] 2009: 133).

[18]In the *biais passé* both face joints are of course different, but they overlap in the elevation since they are placed in a plane that is orthogonal to the faces, which act as the projection plane of the elevation; in other words, the bed joint is a shape in edge view (see Sect. 1.5).

[19]In both cases, the text is quite succinct, since Martínez de Aranda refers the reader to the ox horn (Martínez de Aranda c. 1600: 11–14).

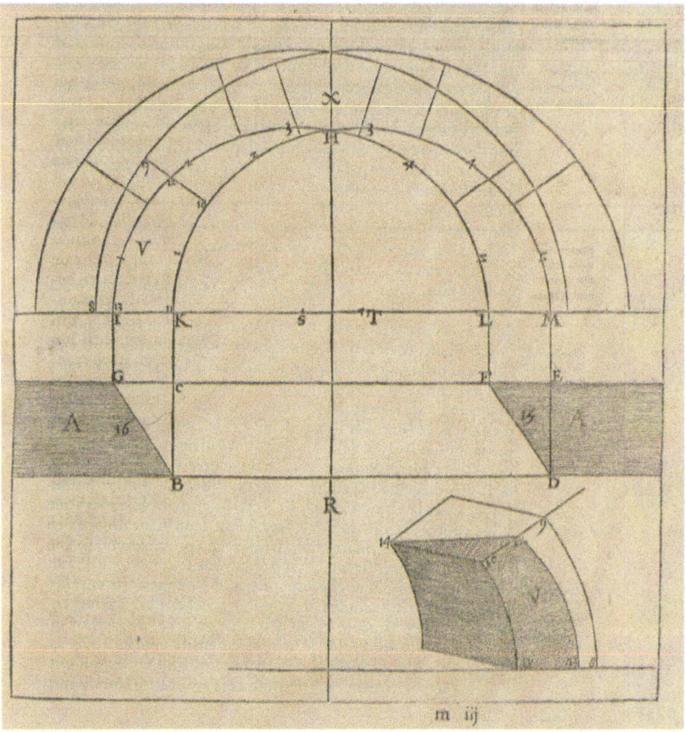

Fig. 6.28 Skew arch with rhomboidal plan and circular faces (de l'Orme 1567: 69r)

operation is inexact, since in this case the quadrilateral formed by the four intrados corners of a voussoir is not planar. As a result, the template represents the four sides, one diagonal and two angles of the quadrilateral accurately, but the procedure used by Martínez de Aranda distorts the other diagonal and two angles of the template.

In contrast, the second solution presented by this author (Fig. 6.30) is rather clear and practical. After drawing the plan and the elevation of the arch, he draws a projection line from the lower corner of each voussoir. Since bed joints are orthogonal to the vertical projection plane, all edges of the bed joint are projected as a single segment; in other words, the intrados and face joints are aligned. Thus, Martínez de Aranda measures the apparent distance between the two ends of the intrados joint in the elevation and transfers it to the plan, displacing the upper end of the joint by that amount. Once again, the operation is consistent with nineteenth-century descriptive geometry methods, although nothing suggests that Martínez de Aranda was thinking in these terms. It can be understood as a revolution around a line in point view; the lower end of the intrados joint, placed on the rotation axis, does not move, while the upper end travels along with a circle in a vertical plane, which is projected horizontally as a segment orthogonal to the rotation axis, overlapping the face plane. Once he has located the new position of the upper end, Martínez de Aranda can draw

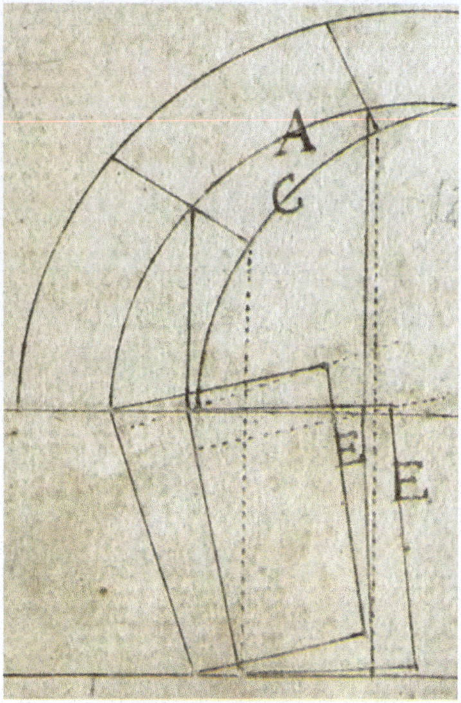

Fig. 6.29 Skew arch with bed joints orthogonal to the faces, solved by templates, detail (Martínez de Aranda c. 1600: 15, left)

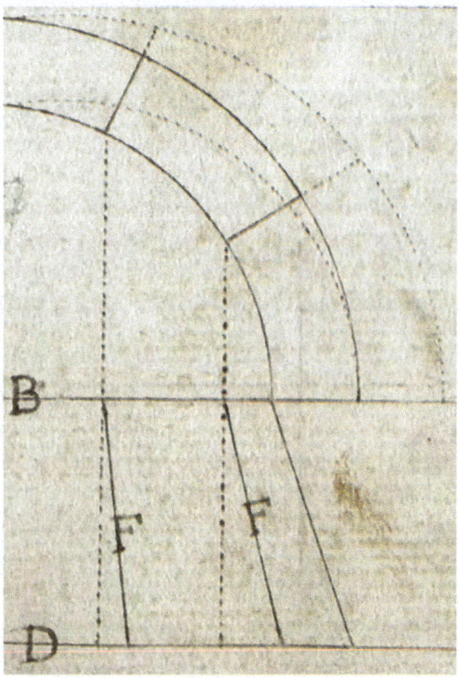

Fig. 6.30 Skew arch with bed joints orthogonal to the faces, solved by squaring and bevel guidelines, detail (Martínez de Aranda, c. 1600: 15, right)

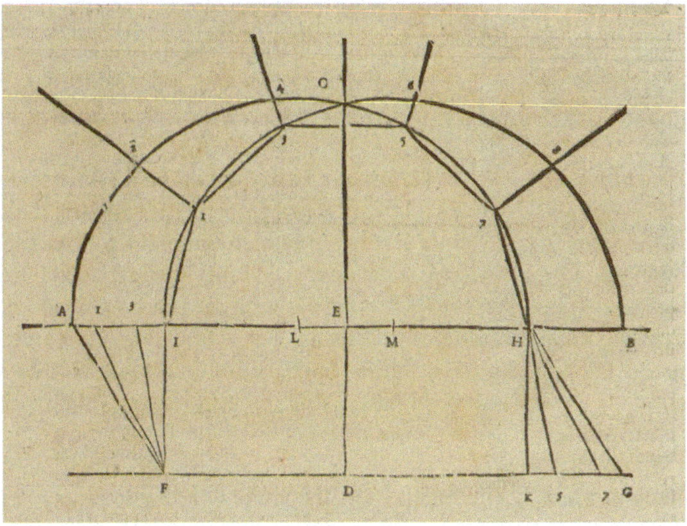

Fig. 6.31 Skew arch with bed joints orthogonal to the faces (Jousse 1642: 14)

the transformed intrados joint. He adds a puzzling remark: "lines *F* should be used as bevels of the joints after squaring them".[20] However, the author refers the reader to his solution for the ox horn (Martínez de Aranda c. 1600; 11–12); comparing the two, it seems clear that the mason should dress an initial block in the shape of a wedge, score the bevel guidelines in both bed joints, and cut away the surplus below the bevel guidelines, as suggested by De l'Orme's hatching.

Jousse (1642: 14–15; see also Guardia c. 1600: 70v) uses mainly the same method, although he groups the guidelines near the springers (Fig. 6.31), as in de l'Orme's double ox horn (see Sect. 6.1.2). In fact, the axis of the arch plays the role of Martínez de Aranda's line in point view, that is, an axis of rotation. Thus, the mason should transfer the apparent distances between the upper and lower corners of each intrados joint starting from an orthogonal to the springings; then, he can draw lines furnishing the length of the intrados joints and its angles with both face joints. This procedure allows him to trace as few lines as possible and reduce the need to crawl on all fours around a full-scale tracing; in fact, he dispenses with projecting lines. An interesting point is the division of the arch in voussoirs. Jousse instructs the reader to divide the arch into five even parts, but the front face lies under the back one at the left half of the arch and vice versa. Thus, the mason should divide a pointed arch into equal parts. Classical geometry methods do not offer a solution for this problem; it was performed quite probably by trial and error (see Sect. 3.1.2).

As we have seen in Sect. 3.2.5, Derand (1643: 122–124) repeats similar operations; however, rather than drawing the guidelines on a full-scale tracing and transferring

[20]Martínez de Aranda (c. 1600: 16): *las líneas F sirven para las saltarreglas de las juntas después de robadas.*

them with the bevel, he traces them directly on the stone. That is, he dresses a wedge-shaped enclosing solid, carving the bed joints carefully; next, he measures in the tracing the apparent distance between the lower and upper ends of each intrados joint and marks them on the face joint; then, he cuts away the portion of stone under this line.

Frézier's solution: joints orthogonal to the faces *and* cylindrical intrados. Frézier (1737–39: II, 137–142; see also Rabasa 1994: 150) explains the traditional solution to the *biais passé*. However, he is concerned about the warped intrados surface, which offends his pre-Enlightenment sensibility, so he tries to find a solution that excludes "irregular" surfaces. Of course, the piece can be solved with a skew, elliptical cylinder and bed joints set on planes perpendicular to the arch faces and thus oblique to the cylinder axis, but in this case the intersections of such planes with the intrados cylinder will be elliptical (Fig. 6.32); Frézier explains an elaborate geometrical construction, anticipating nineteenth-century descriptive geometry, to draw such joints.

Thick skew arches. If the thickness of the arch exceeds the practical maximum for a single voussoir, the solutions with intrados joints parallel to the springings may be elongated indefinitely, using courses and dividing them into voussoirs, that is, converting the skew arch in a skew barrel vault. However, this solution is impractical in the *biais passé*, in particular for a combination of strong obliquity and considerable depth; in the extreme instances, the elevations of the two face arches may not overlap at all. In the nineteenth century, the advent of railroad bridges led to new solutions to this age-old problem, as we will see in Sect. 8.1.2.

6.3 Corner Arches

Corner arches are built at the junction of two walls. At first sight, such an idea is preposterous: to place it at this particular point weakens the union between both walls, a critical point in masonry construction, not to mention the geometrical complexities brought about by the piece. However, some Spanish cities, in particular Trujillo and Ciudad Rodrigo, are packed with corner windows (García Baño 2019). At first sight, they may be fostered by the desire to watch two intersecting streets at the same time, either to hold a position in Late Middle Age urban wars and riots or simply for gossip. However, the tendency to embellish these windows with elaborate heraldic compositions, in particular in Trujillo (Fig. 6.33), suggests that these stereotomic stunts are simply part of a boastful display of wealth and power.

The oldest explanation of the tracing for such arches seems to be included in the manuscript attributed to Pedro de Alviz (c. 1544: 11r; see also García Baño 2017: 210–262). The face arches are quarters of a circle (Fig. 6.34), as in skew arches with circular faces, while both springings are parallel to the bisector plane of the two walls, that is, the symmetry plane of the whole ensemble. As in skew arches with circular faces, the cross-section is a raised elliptical arch; it is carefully drawn by the author, including face joints and extrados. This shows that the face joints in the

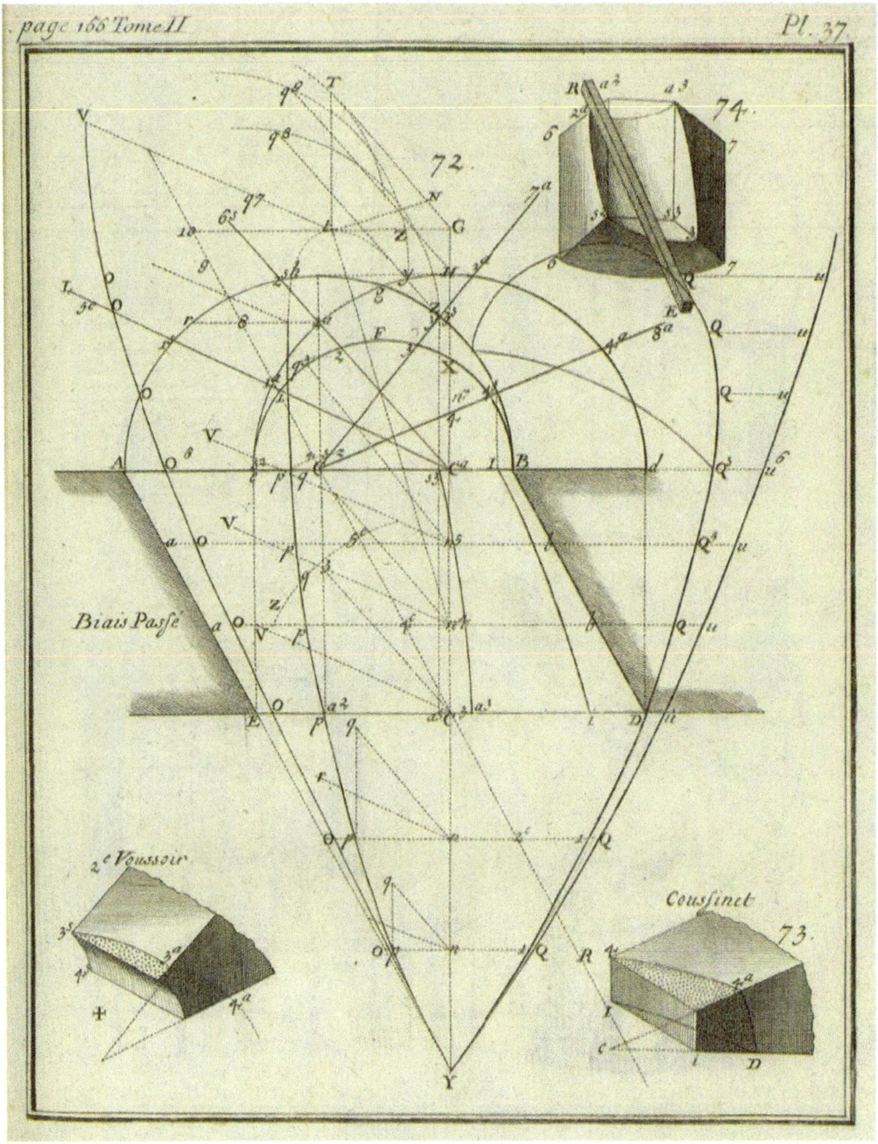

Fig. 6.32 Skew arch with eliptical intrados joints (Frézier [1737–1739] 1754–1769: pl. 37)

cross-section are not equal in length. The author of the manuscript does not quite understand this detail. Therefore, when he tries to construct the intrados, he takes the distance between two intrados joints from the actual faces of the arch. This method is geometrically incorrect, since the intrados joints are not orthogonal to the face plane, and thus the distance between intrados joints is misrepresented. Together with

Fig. 6.33 Corner arch. Trujillo, Palace of Hernando Pizarro, Marquess of the Conquest (Photograph by the author)

several drawing imprecisions, this leads to noticeable errors in the construction of bed joints and intrados templates, as García Baño (2017: 210–220) has shown.

A different solution was put forward by de l'Orme (1567: 71r-72v). As in the skew arch with elliptical faces, a cross-section is set on a plane orthogonal to the bisector plane of the faces and then cast onto the walls using projectors parallel to this symmetry plane (Fig. 6.35). The result of this operation is a pair of quarter-ellipse face arches, although de l'Orme does not represent them explicitly. As in the skew arch, he constructs bed joint templates by drawing a parallel to the lower intrados joint at a distance equal to the thickness of the arch, which stands for the transformed extrados joint. Next, he brings the distance of the back end of the extrados joint to the springings line, which is reused as a vertical reference plane, to the transformed

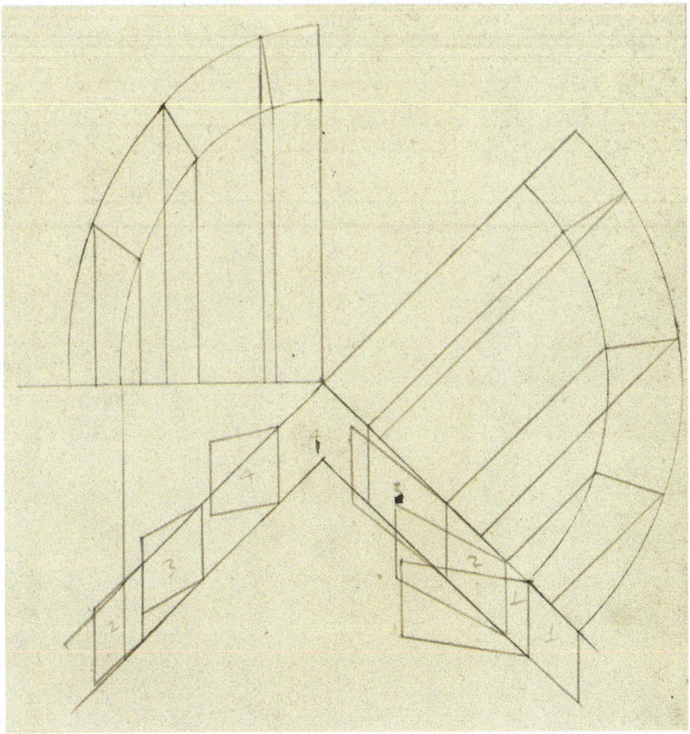

Fig. 6.34 Corner arch (Alviz, attr. c. 1544: 11r)

extrados joint; repeating this operation for the front end, he can close the bed joint template. Although he does not represent intrados templates in his drawing, he states in the text (1567: 73v) that the mason may use them.

The presence of this element in de l'Orme's treatise is a bit striking since it is quite scarce north of the Pyrenees and the Alps (Pérouse [1982a] 2001: 284–316; Calvo 2015b; Galletti 2017: 155). In any case, his solution was widely accepted by later treatises and manuscripts, both French and Spanish. Vandelvira (c. 1585: 20v) reproduces it, including intrados templates; Martínez de Aranda (c. 1600: 71–73) adds the construction of the surbased quarter-ellipse face arches, joined as always with arcs of a circle joining three points. Jousse (1642: 20–23) follows the same lines, although he groups the intrados templates in a continuous series and reduces the bed joint templates to a single line, stacking them on the springers.

Derand (1643: 167–162) also uses a chain of intrados templates, treating these shapes as flexible rather than rigid templates (Fig. 6.36). In geometrical terms, this involves a development of the cylindrical intrados surface, rather than a polyhedral surface. Of course, a correct development of the cylinder requires the use of π. Quite probably, Derand was aware of this notion but did not trust that masons would grasp it,

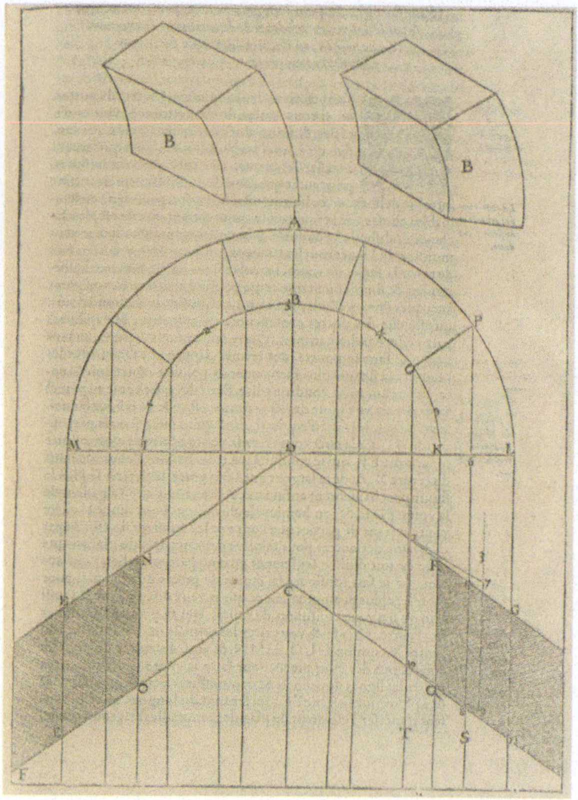

Fig. 6.35 Corner arch (de l'Orme 1567: 74r)

so he performed a simplified development using a single intermediate point; we will come back to these issues when dealing with arches on curved walls on Sect. 6.4.1.

Frézier (1737–1739: II, 135–137) discusses the choice between circular cross-section and circular faces; he prefers the first, particularly when the angles between the faces and the axis of the arch are different. He uses rigid templates, so he basically returns to de l'Orme's solution. However, what is more interesting is his remark that "the corner arch is a stonecutting design which is rarely executed, and a good architect should avoid it …".[21] Paradoxically, in the longest stonecutting treatise ever published, Neoclassical taste begins to raise objections to the colourful repertoire of the craft.

[21] Frézier (1737–1739: II, 137): *La Porte sur le Coin est un des Traits de la Coupe des pierres qu'on execute rarement, & qu'un bon Architecte sçait éviter, …*

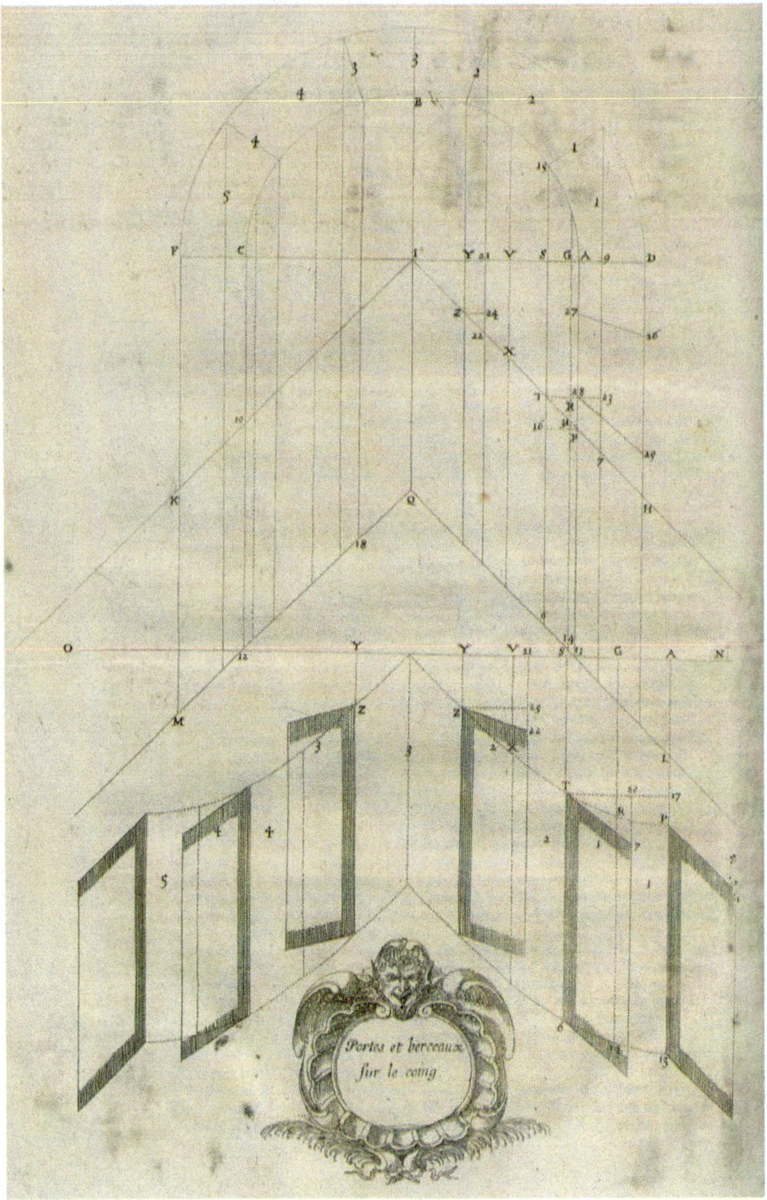

Fig. 6.36 Corner arch. (Derand [1643] 1743: pl. 81)

6.4 Arches in Round Walls

Arches on curved walls (Fig. 6.37; see Pevsner 1958) derive probably from two medieval sources. On the one hand, they are a typical Romanesque element, windows opened on apses, that resurfaced in the Renaissance and the Baroque periods; surprisingly, Pérouse ([1982a] 2001: 181) does not mention them when listing the Early Medieval sources of classical stereotomy. On the other hand, the French and Spanish

Fig. 6.37 Arch in a round wall. Turin, San Lorenzo (Photograph by the author)

names of these pieces, *porte en tour ronde*, *arco en torre cavada* and other variants (respectively, “door in a round tower” and “arch in a hollowed tower”), suggest connections with the cylindrical towers of medieval fortification. These elements may have a convex or concave front and a flat back; however, they usually present convex and concave surfaces on opposite faces. Springings may be parallel, although on many occasions, from Romanesque apse windows to Renaissance rotundas or drums, they are convergent. All this fosters a wide range of combinations: Martínez de Aranda (c. 1600: 19–33, 38–39, 103–109) presents no fewer than fourteen different variants of this kind of arch.

6.4.1 Parallel Springings

Polyhedral surface. Renaissance masons address the problem using different methods. Most of them use flexible templates which allow direct control of the curvature of the faces. However, a simpler approach is possible: to construct a rigid template standing for a face of a polyhedral surface inscribed on the intrados of the arch, projecting a point of the curved face edge on the template. Martínez de Aranda (c. 1600: 19–21) follows this path: he draws the plan of a curved wall and the cross-section of the arch, which is cast onto the curved surface (Fig. 6.38). Next, he constructs the intrados and bed joint templates, taking their width from the elevation and measuring distances to a reference line, as in the skew arch with elliptical faces. This enables him to dress the voussoirs in the same way but does not provide a method for the control of the curvature of the faces. To solve this problem, Martínez de Aranda uses an intermediate point in the face edge, measuring the distance from this point to the chord of the voussoir in horizontal projection and transferring it to the template. He can then construct an arc passing through this point and the corners

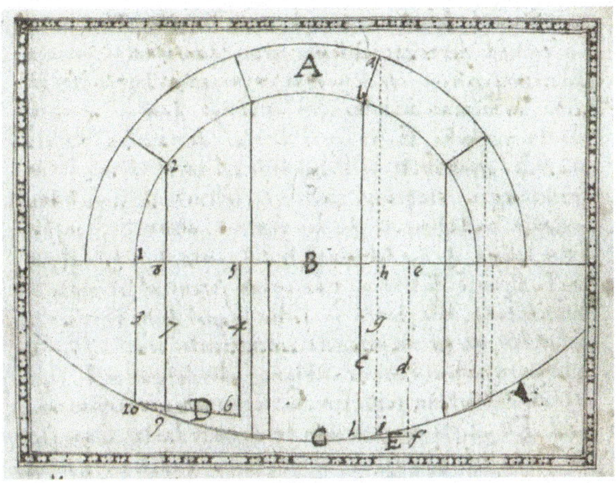

Fig. 6.38 Arch in a round wall (Martínez de Aranda c. 1600: 20)

of the voussoir, which furnishes the curved face edge of the template. Next, he repeats the same operation for the bed joint templates.

All this involves several simplifications. First, the midpoint of the voussoir edge is not on the plane of the chord and thus should not be included in a revolution. However, the procedure may be consistent with descriptive geometry, if we accept that the point which is rotated is the orthogonal projection of the voussoir edge onto the plane passing through both intrados joints. Even in this case, when carving the face, the mason should use a square leaning on this plane; this implies that the direction of the generatrices of the face cylinder is different in each voussoir. This result diverges slightly from the usual solution, where all voussoirs fit into a single cylindrical surface. Quite probably Martínez de Aranda found this solution through trial and error, or learned it from another master, and tested it through models, as he stressed in the introduction of his manuscript (c. 1600: n. p. [iii]) and did not notice the slight difference from the canonical solution involving a single cylinder. Moreover, bed joint templates do not pose this problem, since all points are placed on the same plane; probably, Martínez de Aranda extrapolated the solution for the bed joint templates to the intrados templates.

A remarkable variant is used to solve the arch opened in a round wall when the intrados joints are not orthogonal to the flat back face. In this case, it is not easy to measure distances to a reference line, so Martínez de Aranda (c. 1600: 23–24) draws orthogonals to the intrados joints to determine the position of voussoir corners in the template, as a nineteenth-century engineer would do when using a rabatment. However, it seems that he does not quite grasp the idea of the perpendicular since he does not mention it in the text and refers to the general case.

Another interesting detail in Martínez de Aranda is that he devotes an introductory section (c. 1600: 4–5) to a development of the face of an arch, presented as a set of face templates or *cimbria*. In order to do so, he takes the horizontal distances between voussoir corners from the plan and carries them to the base line of the development; next, he takes the heights of the corners from the elevation to complete the development.

Cylinder development. In any case, Martínez de Aranda stands alone in this approach based on a polyhedral surface. Most stonecutting writers try to solve this problem using flexible templates. De l'Orme (1567: 74v-77v) starts by drawing the plan of the curved walls and a round cross-section projected onto the round walls (Fig. 6.39; see also 3.30 for the full drawing). Next, he constructs the bed joint and intrados templates separately, to avoid confusion, although the method is essentially the same one that Martínez de Aranda will use later. However, when constructing extrados templates, there is a conceptually significant difference: in the drawing, the distance between intrados joints is slightly larger in the template than in the elevation. This is explained by a somewhat ambiguous passage in the text: "It remains for me to explain how to make the other intrados and extrados templates are to be constructed, for the practice of which we will begin with that of the one above. So take the length of the three points *OSN*, and draw separately three parallel lines of the same length,

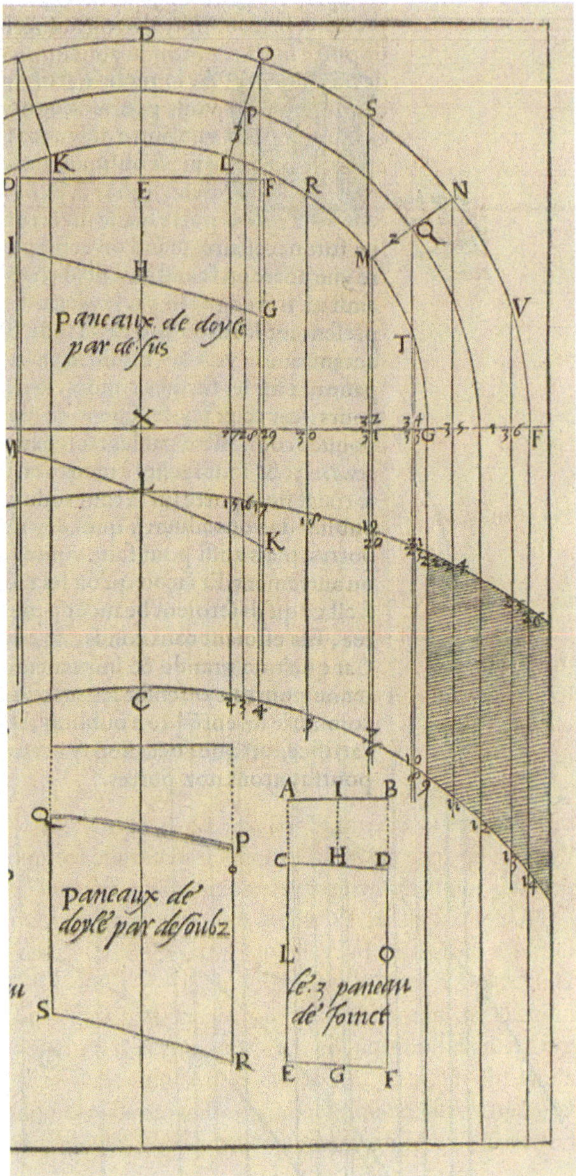

Fig. 6.39 Arch in a round wall, detail (de l'Orme 1567: 77r)

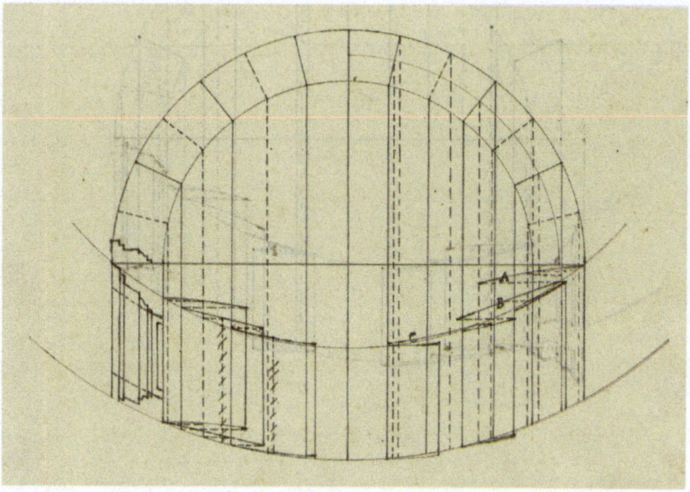

Fig. 6.40 Arch in a round wall ([Vandelvira c. 1585: 22r] Vandelvira/Goiti 1646: 34)

as you see in *DEF*".[22] At first sight, the sentence can be understood as meaning that the width of the intrados template equals the distance between *O* and *N*, but in this case, why does de l'Orme mention *S*? It seems more probable that the distance between *D* and *E* equals that between *O* and *S*, while *E–F* equals *S–N*. Thus, the width of the template does not equal the chord; it is instead a crude development of the face edge, making it equal it to the combined width of two chords. In fact, two centuries later, Guarini (1737: 197) and Frézier (1737–39: III, 43–44) will censure this practice (see Sects. 2.4.2 and 12.2). In any case, the issue is far from clear in de l'Orme; he does not seem to apply it to intrados templates since the distance between intrados joints in the template equals their distance in the elevation.

It seems that de l'Orme's method for extrados templates was also applied to intrados templates by Vandelvira (c. 1585: 21v; see also 22r, 22v, Palacios [1990] 2003: 80–83, and Wendland et al. 2015: 1788–1790) (Fig. 6.40). Like de l'Orme, he uses an intermediate point in each bed joint and each intrados face; following the Spanish tradition, he does not construct extrados templates. He is quite laconic about his method: "Since the arch is curved, you should use the midpoints of the voussoirs with their projections so that intrados templates and bevel guidelines should be curved".[23] At first glance, it is not clear whether the distance between the intrados

[22]De l'Orme (1567: 75v): *Reste maintenant d'entendre comme il faut faire les autres panneaux de doile, pour la pratique desquels nous commencerons à celui de dessus. Vous prendrez donc la largeur des trois points OSN, et en tirerez à part trois lignes de même largeur, qui seront parallèles, comme vous les voyez marquées DEF, et perpendiculaires, ainsi qu'il se voit au lieu écrit, panneaux de doile par le dessus.* Transcription is taken from http:// architectura.cesr.univ-tours.fr.

[23]Vandelvira (c. 1585: 21v): *Por ser en cercha es menester echar los medios de las dovelas con sus plomos para que vayan adulcidas las plantas y saltareglas.* Transcription is taken from Vandelvira and Barbé 1977.

joints in the template is larger than their actual distance in the elevation. However, in the variant featuring convex and concave faces he addresses the reader in these terms:

> You will ask why, in an arch with a round and a concave face, the templates for the hollow face are convex, while those for the round side are concave? You should know that the sharp rise of the arch causes this effect, as in the hollow-face trumpet squinches; if you want to prove this, prepare a model of this arch by squaring, as I will explain later, and then apply these templates, and you will have the proof that the method is correct.[24]

This passage hints that Vandelvira is using the same simplified-development method used by de l'Orme since the reversal of curvatures he is talking about does not occur with Martínez de Aranda's method. At the same time, this makes clear that knowledge of this method was not widespread in the sixteenth century since he anticipates that the result would befuddle the reader. Moreover, it must be remarked that neither de l'Orme nor Vandelvira explain the concept clearly, and de l'Orme uses it only for extrados templates and Vandelvira only for intrados ones. Another interesting point is that both De l'Orme and Vandelvira use midpoints for the bed joint templates, but their geometrical significance is entirely different since bed joints are flat. In bed joints templates, the midpoint is used only to compute the curvature of the face joint or voussoir edge, not to develop a cylindrical surface.

Further clues are provided by Jousse (1642: 46–47, 52–53), who uses several points along the face edge, placed at the corners and midpoints of the voussoirs, advising the reader to measure their distances to the line of the springings (Fig. 6.41). No doubt, the use of the springings as a reference stems from a desire to avoid drawing an additional line, but this practice will lead later on to the birth of the concept of the folding line. In any case, these horizontal distances, transferred to a new auxiliary line, give the corners of the templates. Since Jousse uses both voussoir corners and midpoints, it is safe to assume that template widths are larger than the distances between intrados joints and he is trying to construct an approximate development of the cylindrical intrados surface.

These issues are finally clear in Derand (1643: 172–175; see also de la Rue 1728: 8–9, 20–22). To begin with, he has used approximate cylindrical developments before in skew and corner arches (Fig. 6.36), as we have seen, so he expects his reader to be familiar with the concept. Notwithstanding that, he explains that "You should draw under the plan, or in any other place, the directing line *V-T*, on which you should extend the intrados *F–E–D*, transferring *D–18* to *T–2*, and *18–20* to *2–3* … and so on".[25] Since *D* and *20* are voussoir corners and *18* a midpoint, his intentions are

---

[24]Vandelvira (c. 1585: 22r): *Dirás ahora cómo, siendo el arco torre cavado y torre redonda, las plantas van al contrario que las primeras, van redondas a la parte del torre cabo y a la parte del torre redondo van cavadas; a lo cual has de saber que el mucho capialzo que las primeras capialzan les hace hacer este efecto, como parece en las pechinas torre cavada y si lo quisieres probar contrahaz un arco de éstos por robos, como te enseñaré adelante, y luego planta estas plantas y harás la prueba ser éstas ciertas*. Transcription is taken from Vandelvira and Barbé 1977.

[25]Derand (1643: 174): *Vous tirerez au dessous du plan, ou ailleurs, la ligne de direction VT, sur laquelle vous étendrez la doüele F E D, portant D 18 sur T2, & 18, 20 sur 2, 3 … ponctuée size au delà de 25, & ainsi du reste.*

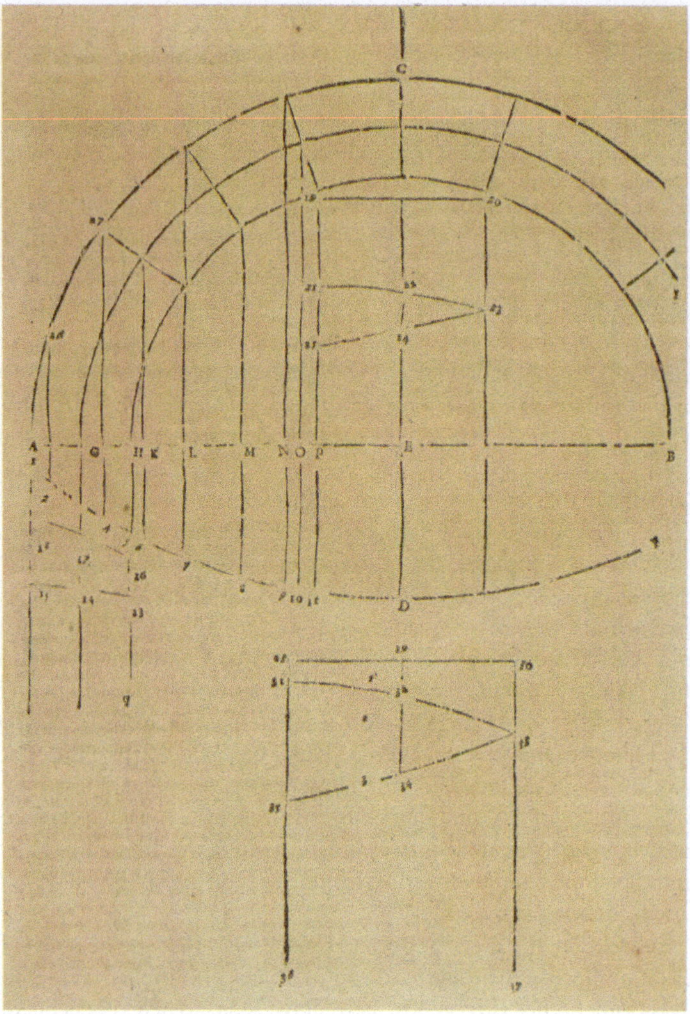

Fig. 6.41 Arch in a concave wall (Jousse 1642: 52)

clear: he means to perform an approximate development of the cylindrical surface by dividing the portion of the face edge belonging to each voussoir in two segments (Fig. 6.42, detail in Fig. 6.43). His approach is not as crude as it may seem since he has told the reader previously (Derand 1643: 155–156) that he may use more than two segments if necessary. He also explains how to dress the voussoirs by squaring, starting with the face cylinders. Frézier (1737–39: III, 45–47), explains the squaring method as the main alternative; this choice may stem from his reluctance to use simplified cylindrical developments.

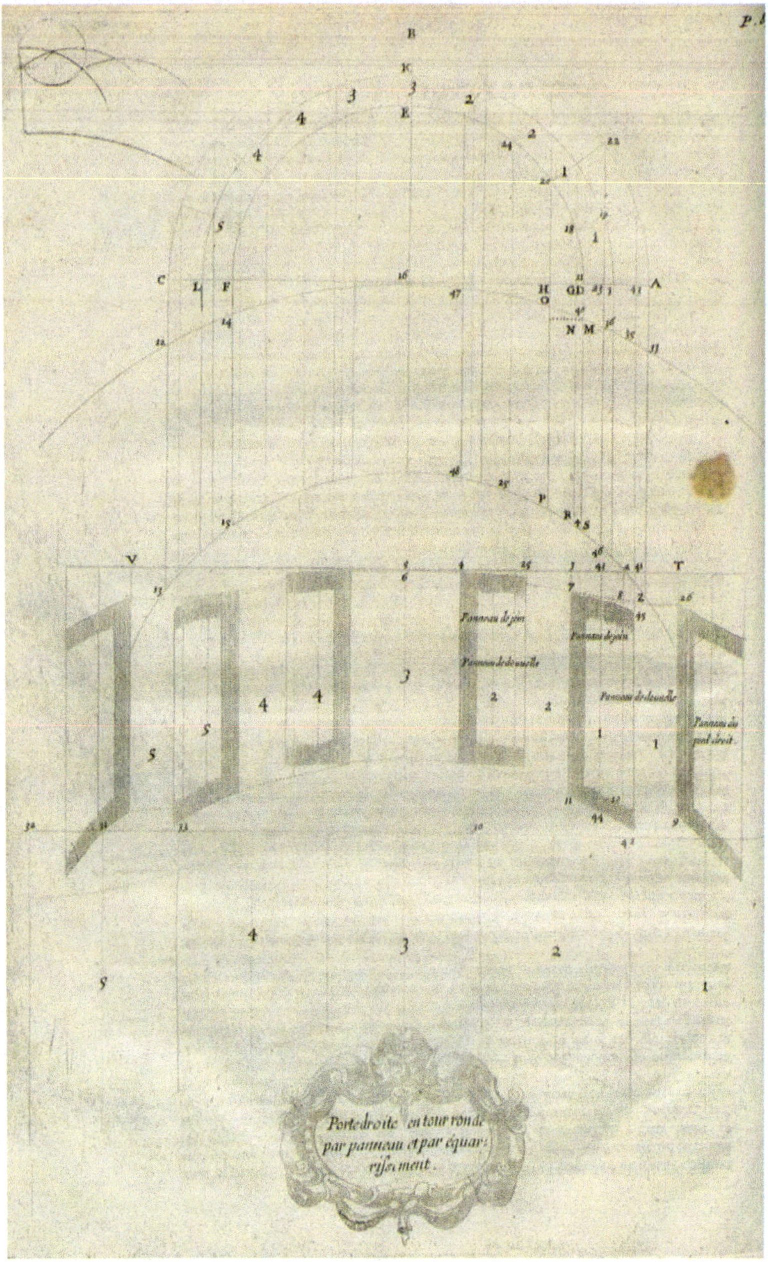

Fig. 6.42 Arch in a convex and concave wall (Derand [1643] 1743: pl. 83)

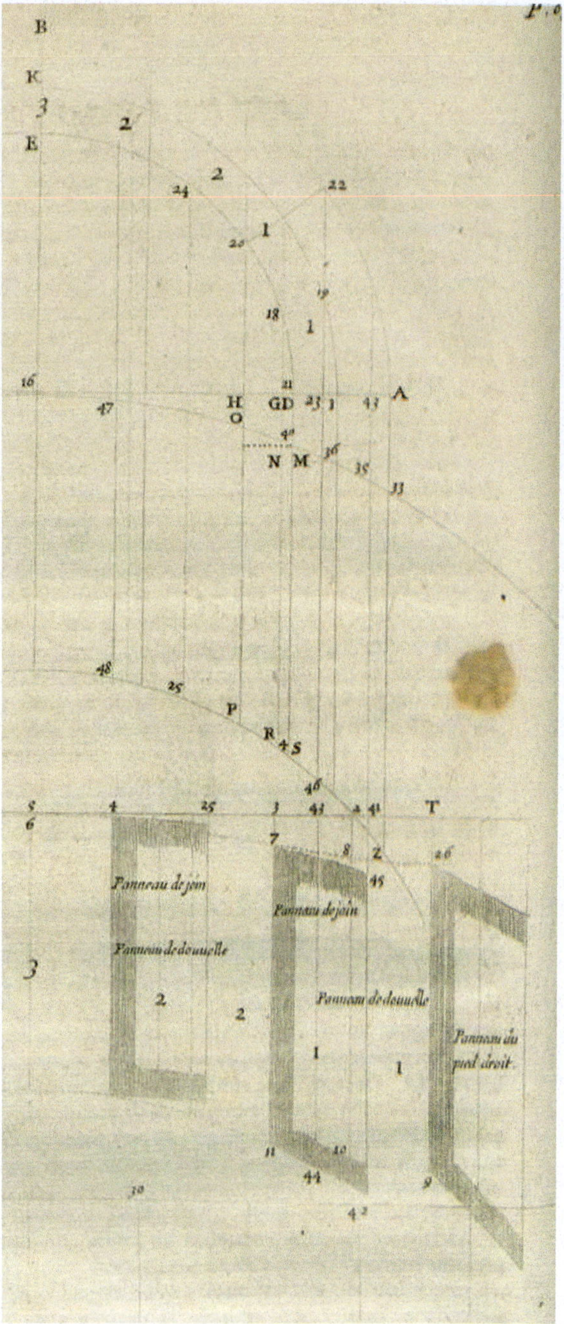

Fig. 6.43 Arch in a convex and concave wall, detail (Derand [1643] 1743: pl. 83)

Adjusting the width of the face arch. An unsolved problem remains. Should the generating arch have a face joint placed exactly at its highest point,[26] this joint would be parallel to a generatrix of the cylindrical surface of the wall and thus keep its size when projected onto the cylindrical surface. In contrast, a face joint near the springings would undergo a noticeable lengthening. Thus, the real thickness of an such arch (as opposed to the thickness of the projected generating arch, which remains constant) would increase gradually from the keystone to the springings. Different solutions have been put forward to avoid this effect. Martínez de Aranda addresses the problem directly: "… if you wish to make face joints equal, you should mark the thickness that you wish to have the arch in the face sides of all bed joint templates and cut these joints through this point".[27]

Derand's (1643: 190–192) approach to the problem is subtler. Vandelvira (c. 1585; 11r-12v) and Martínez de Aranda (c. 1600: 4–5) had used approximate cylindrical developments in order to obtain face templates in other elements. Derand (Fig. 6.44) reverses this process, starting from "the front opening of the arch, extended and, so to speak, developed"[28] to arrive at a transformed orthogonal projection. First, he draws a round arch; next, he measures horizontal distances between voussoir corners from this virtual arch and transfers them to the curved surface of the wall. He adds ironically that this can be done "transferring several compass apertures, to make the operation more precise and satisfy the most scrupulous minds".[29] Measuring voussoir corner heights in the development, he constructs an orthographic projection of the actual arch, which is, of course, slightly irregular and raised, although the intrados is a generalised cylinder. Once this is done, nothing prevents Derand from applying the general procedure for the templates of the arches on curved walls to this particular element.

6.4.2 *Convergent Springings*

All preceding examples are solved with parallel springings. However, in arches opened in rotundas or drums, it is quite frequent to set out the springings of the arch passing through the centre of the plan of the wall; in other words, to build an arch that is, at the same time, splayed and opened in a round wall.

---

[26]This is, of course, highly unusual in Early Modern arches, since the Classical tradition prescribes the use of a keystone, and thus face joints are placed at both sides of the keystone, not exactly at the axis of symmetry. This possibility is mentioned here only to explain geometrical issues as clearly as possible.

[27]Martínez de Aranda (c. 1600: 21): *y si quisieres que las juntas del dicho arco extendido sean todas iguales irás echando el grueso que le quisieres dar al arco en todas las testas de las plantas por lechos y cortando las dichas plantas por aquel lugar.*

[28]Derand (1643:190): *l'ouuerture de la porte par le deuant étenduë & comme déuelopée.*

[29]Derand (1643:190–192) … *ou bien par transports faits par diuerses ouuertures de compas, pour rendre l'operation plus precise, & satisfaire aux esprits plus scrupuleux.*

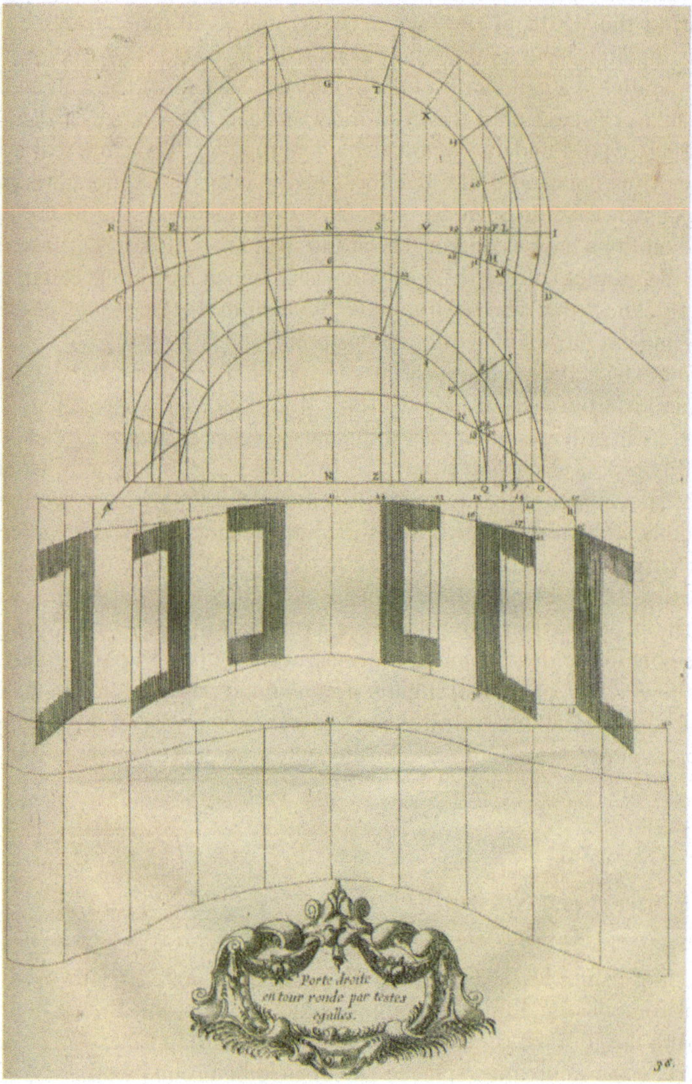

Fig. 6.44 Arch in a round wall, with equal faces. Note that the generating arch is not projected onto the wall (Derand [1643] 1743: pl. 90)

The diagram by Hand IV. An elemental diagram (Fig. 6.27, left) drawn by Hand IV in the sketchbook of Villard de Honnecourt (Villard c. 1225: 20r, drawing 8-h) seems related to this problem. The caption *Par chu tail um vosure des tor de machonerie roonde* (how to cut a voussoir for a tower of round masonry)[30] makes

[30]Translation from the author. Barnes (Villard/Barnes 2009:133) gives a different translation since he thinks that this drawing and the preceding one are mismatched, which is debatable in my opinion.

it clear that it represents the plan of curved wall with an opening, as stressed by most scholars (Branner 1957: 62–65; Bechmann [1991] 1993: 175–180; Lalbat et al. 1989: 26–30). A templet is laid with its straight edge tangent to the wall; since a straightedge could be used for the same purpose, Hand IV probably is suggesting its use in the construction of the wall. Some notches, similar to those on the drawing of a skew arch (see Sect. 6.2.2) hint that the mason should measure distances from the templet to the masonry. From this point on, Branner, Bechmann and Lalbat have put forward hypotheses about the dressing process, based mainly on squaring or hybrid methods. Barnes put forward a completely different, rather debatable interpretation: captions for this diagram and the preceding one would be switched, so the relevant caption would be *Par chu fait om cheir deus pires a un point si lons ne seront* (How to connect two stones at a common point if they are not far apart). Thus, the diagram would represent a round arch before the placement of the keystone (Villard/Barnes [c. 1225] 2009: 133). The problem with this interpretation, in addition to the awkward detail of the switched captions, is that it does not explain the presence of the templet and the marks.

Early Modern solutions. Martínez de Aranda (c. 1600: 38–39; see other variants in 103–108) offers a solution to this problem (Fig. 6.45). As he does in ordinary arches in curved walls, he constructs two virtual round arches whose diameter equals the span at both faces and projects them *orthogonally* onto the respective face cylinders. This operation poses a subtle problem. Had he projected the virtual arches *centrally*, a single arch would have been enough, and the resulting intrados would be

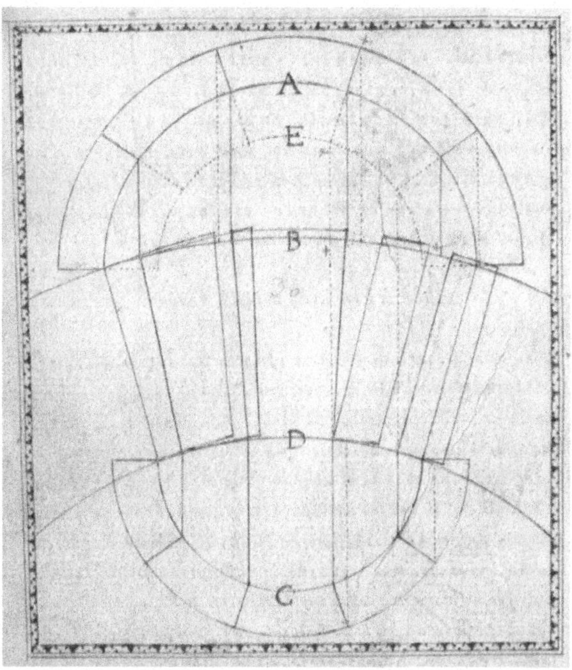

Fig. 6.45 Arch in a round wall with convergent springings (Martínez de Aranda c. 1600: 39)

conical and thus developable. In contrast, Martínez de Aranda's procedure leads to a slightly warped intrados. Further, he explains that the intrados templates should be constructed using the procedure explained in a preceding section for splayed arches with asymmetrical springings (c. 1600: 34–36), that is, using orthogonals to the projection of the intrados joints. Moreover, the curvature of the face edges should be controlled applying the method he had explained for ordinary arches in curved walls (c. 1600: 24–26). All this leads to small imprecisions in template construction.

Derand (1643: 188–190) addresses the problem by squaring (Fig. 6.46). Moreover, he chooses to use radial, horizontal intrados joints. In the main variant, he starts by drawing the plan of the arch and choosing a vertical plane orthogonal to the axis of the

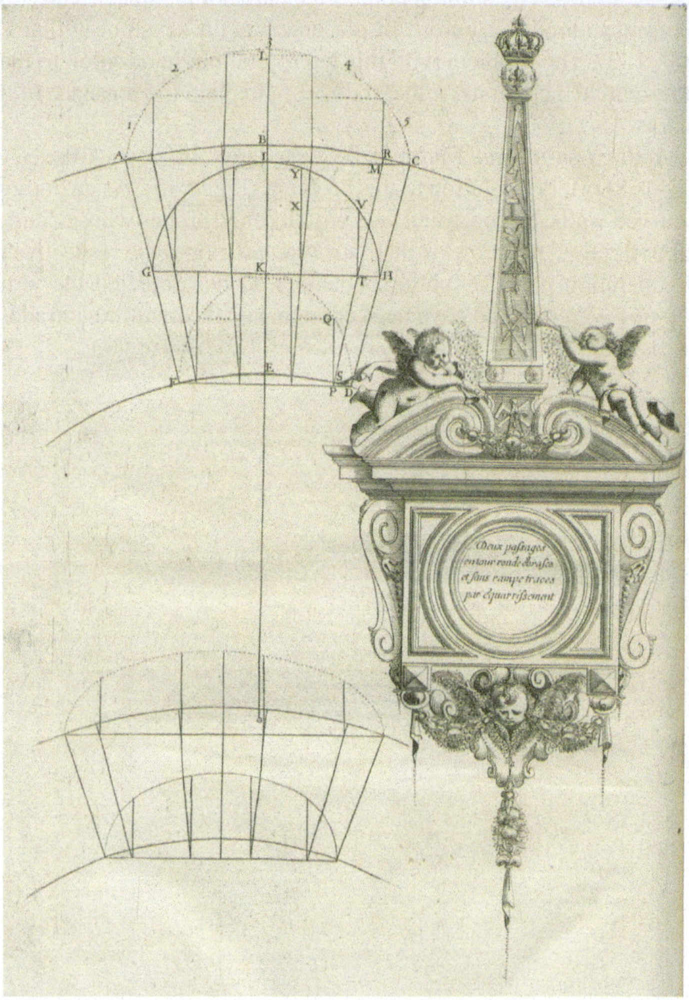

Fig. 6.46 Arch in a round wall with convergent springings (Derand [1643] 1743: pl. 89)

arch. Next, he draws the cross-section through this plane as a round arch; he divides it into equal parts and draws vertical lines from the division points. These vertical lines serve two purposes: first, they allow drawing the horizontal projections of intrados joints; second, since the intrados joints are horizontal, the verticals may be used to measure the height of each voussoir corner. Using these heights, the mason may construct the face arches or, to be precise, the cross-sections through vertical planes passing through the ends of both springings. The result of this operation may come as a surprise: since the keystones of all three sections are at the same level, the broader face is *surbased*, since its span is larger than the one of the middle section, while the narrower section is *raised* for the same reason. Moreover, all intrados generatrices meet a single vertical line, passing through the intersection of both springings; thus, the surface is a ruled, warped one, known as a *conoid*. As for the dressing phase, Derand simply instructs the reader to carve the voussoirs in the usual way, using the intrados joints as references, considering that the real faces go beyond the section plane in the convex face; the heights of the corners should be transferred in the usual way. He also remarks that the small curvatures in these elements do not justify the adjustment in the face arches we have seen in the preceding example.

6.5 Arches in Battered Walls

Arches opened in *battered* (sloping) walls are associated with Early Modern fortification, at least in theory. In contrast to vertical walls in medieval castles, designed to withstand the attack of siege machines, the emergence of gunpowder and artillery in the fifteenth and sixteenth centuries led to the use of battered walls in Renaissance forts. Cannonballs hit vertical walls frontally, causing considerable damage, while they tend to rebound on sloping curtains. However, Renaissance military engineers preferred materials that responded to the impact of projectiles by crushing or compacting (as rammed earth or brick) rather than by breaking (as stone does), and they were not prone to pierce openings in curtains. All these reasons made the use of arches on sloping walls infrequent; however, a few can be found in eighteenth-century fortifications in Cádiz, Cartagena, Palma de Mallorca and other Spanish cities (Fig. 6.47).

As for geometrical solutions to the problem, de l'Orme (1567: 78v-80r) presents a solution for an arch opened in sloping, curved wall. However, as usual in his treatise, he does not bother to explain the simple arch on a battered wall. Vandelvira (c. 1585: 23v) does, although his solution (Fig. 6.48) is unusual for several reasons. He starts by drawing the cross-section of the arch and *two* sections of the slanted wall. Next, he transfers the heights of the intrados joints from the section of the arch to both sections of the wall. Then, he focuses on one of the sections of the wall, measuring horizontal distances from the intersections of the intrados joints with the wall surface from a vertical plane passing through the start of the slope. Next, he draws orthogonals from the voussoir corners in the section of the arch, using them as projection lines; he then

Fig. 6.47 Arch in a battered wall. Cádiz, Muralla de San Carlos (Photograph by the author)

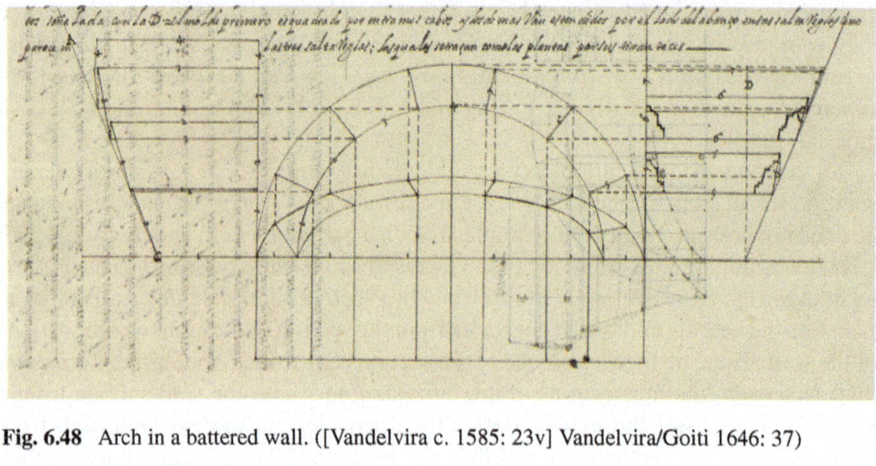

Fig. 6.48 Arch in a battered wall. ([Vandelvira c. 1585: 23v] Vandelvira/Goiti 1646: 37)

transfers the distances from the corners to the start of the slope to the projection lines. This operation furnishes the plan of the arch face.

Of course, this is quite unusual for a modern architect or engineer. Generally speaking, the horizontal projections of arch faces are overlaid on a single straight line or, in the case of arches opened in curved walls, on a single curve with the shape of the plan of the wall. However, in arches opened in sloping walls or vaults, the horizontal projection of the arch face shows clearly all voussoir edges and corners. Thus, Vandelvira draws this horizontal projection carefully, stating clearly that he does so in order to control the placement of the voussoirs using a plumb line, as we have seen in Sect. 3.3.4. Another interesting detail of Vandelvira's method is that he constructs the intrados and bed joint templates *in two cross-sections* and not in the plan, as we have seen up to this point. Translating his method to descriptive geometry concepts, he rotates the intrados and bed joint templates until they reach a vertical, not a horizontal plane. In any case, he takes the distances between two intrados joints or between the intrados and extrados from the virtual arch that has generated the element, as he did in his solutions for skew or corner arches. Quite probably, Vandelvira performs these "revolutions" in the cross-section, rather than in the plan, to avoid intermingling the templates and the horizontal projection of the arch face, which occupies the central area of the drawing. This concern for clarity also justifies the use of two wall sections, one for the intrados and the other for the bed joint templates. In a separate drawing, Vandelvira (c. 1585: 24r)[31] includes a true-shape depiction of the face arch or *cimbria*. He simply takes the horizontal distance between the corners of the voussoirs from the section of the arch and the distance from each corner to the start of the slope from the section of the wall. Of course, the result is a raised half-ellipse, resulting from the intersection of the cylindrical intrados with the sloping plane of the wall surface.

The more conventional solution by Martínez de Aranda (c. 1600: 53–54) confirms these ideas (Fig. 6.49). Like Vandelvira, he starts by drawing the cross-sections of the arch and the wall; next, he constructs intrados and bed joint templates in plan, as in other cases, taking the distance between two consecutive intrados joints, or between the intrados and extrados, from the section of the arch, and the positions of the corners of the voussoirs from the section of the wall. In contrast to other solutions of his, where he endeavours to offer the most complete procedure, in this case he does not include the horizontal projection of the face of the arch or its *cimbria* (development in true shape). In addition to the standard arch in a sloping wall, he offers no fewer than five other variants: two different skew arches opened in an ordinary sloping wall, and three solutions for an inverted or forward-leaning sloping wall, a straight one and two skew ones. At first sight, the forward-leaning solution makes no constructive sense; however, it can be used for the backside of the stone facing of an arch opened in a rammed earth curtain; there are extant examples in Cádiz, where Martínez de Aranda stayed for some years (Antón Solé 1975; Falcón 1994), although they date

[31] According to the original numbering of the School of Architecture manuscript, given by Barbé 1977 in round parentheses, the *cimbria* of this arch is placed in the page following the main explanation.

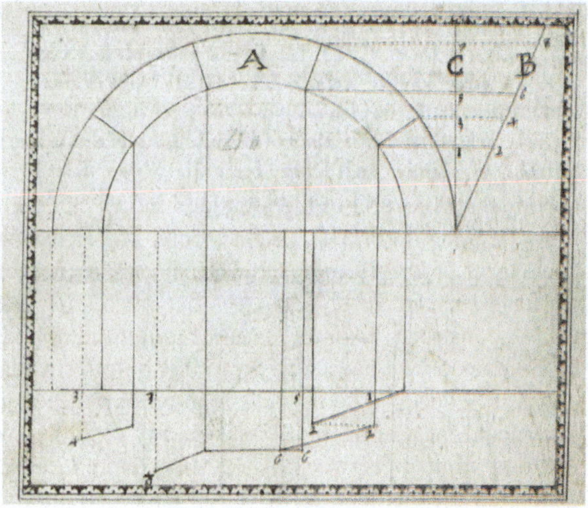

Fig. 6.49 Arch in a battered wall (Martínez de Aranda c. 1600: 54)

from later periods. In any case, it can be easily appreciated that all these variants use similar procedures, although explanations for some skew variants are not clear (see Calvo 2000a: II, 157–159, 164–165), and the most remarkable difference is that in skew variants all templates are drawn, while in symmetrical ones the templates for a side are reused for the other side of the arch.

An intermediate approach is followed by Jousse (1642: 26–27). After drawing the cross-sections of the arch and the wall, he constructs the horizontal projection of the arch face (Fig. 6.50). He then draws the intrados templates, taking their widths from the section of the arch and their lengths from the section of the wall; he groups them on a string of templates drawn separately, featuring a curved edge on the sloping side of the wall. The use of voussoir midpoints hints that he is trying to construct an approximate development of the cylindrical surface. Simplified templates of the bed joints are drawn using lines passing through the intersection of the axis of the arch with the start of the slope, as in skew and corner arches. As Vandelvira does, Jousse offers a *cintre*, the equivalent of the Spanish *cimbria*, that is, a true-size depiction of the arch face. He does not explain how the points representing voussoirs corners should be joined, but slight kinks hint that they were taken in groups of three points with the compass, as explained by Vandelvira (c. 1585: 18v) or Martínez de Aranda (c. 1600: 2).

De la Rue's solution (1728: 12–13) is quite clear (Fig. 6.51). He starts with a skew arch in a sloping wall, leaving out the straight variant, but it is evident that his solution may be applied to a straight arch as a particular case. The only difference is that he constructs all templates from both sides of the arch, as does Martínez de Aranda. Quite remarkably, he does not use midpoints, and draws two polygons, one for the intrados and the other for the extrados, in the cross-section. That is, in this case he is thinking about rigid templates and eschewing an approximate development of the cylindrical intrados surface. These templates are set apart in a continuous group,

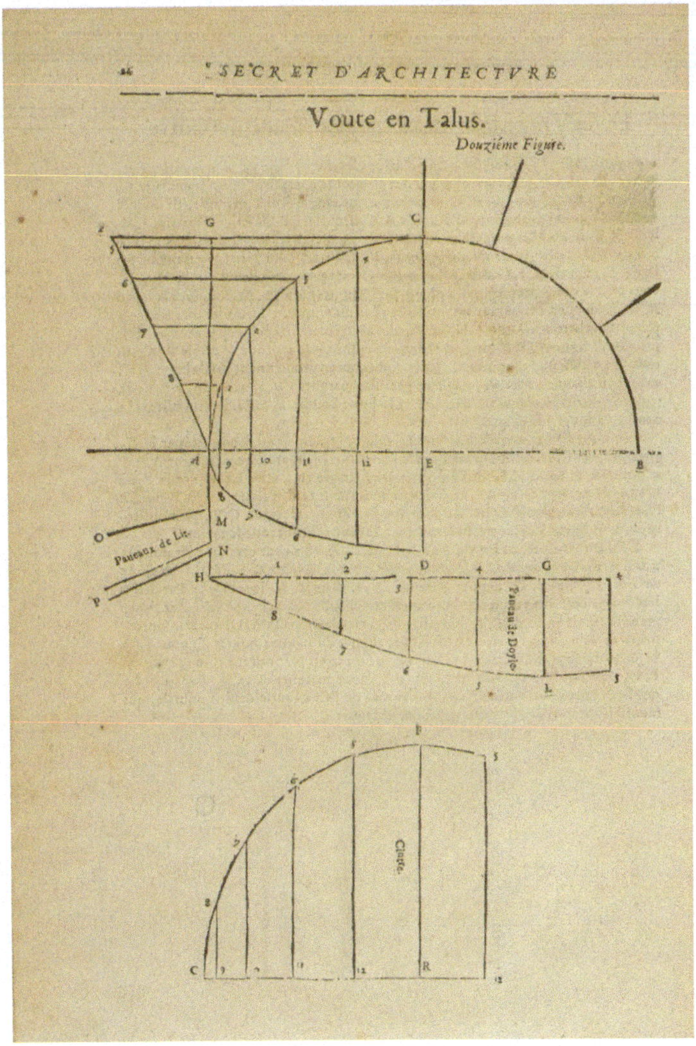

Fig. 6.50 Arch in a battered wall (Jousse 1642: 26)

as in Jousse or other solutions by Derand, in order to leave space for the horizontal projection of the arch face; the true-shape depiction of the arch face is not included.

Frézier (1737–39: II, 142–158) starts by explaining that "the skew arch without slope and the square arch with slope are really the same thing, rotated in different ways about their axes".[32] Building on this idea, probably taken from Desargues, he constructs the intrados templates measuring the width of each voussoir and the

[32]Frézier (1737–39: 142–143): *le berceau biais sans talud, & le berceau droit avec talud ne sont dans le fond que le même tourné différemment autour de son axe.*

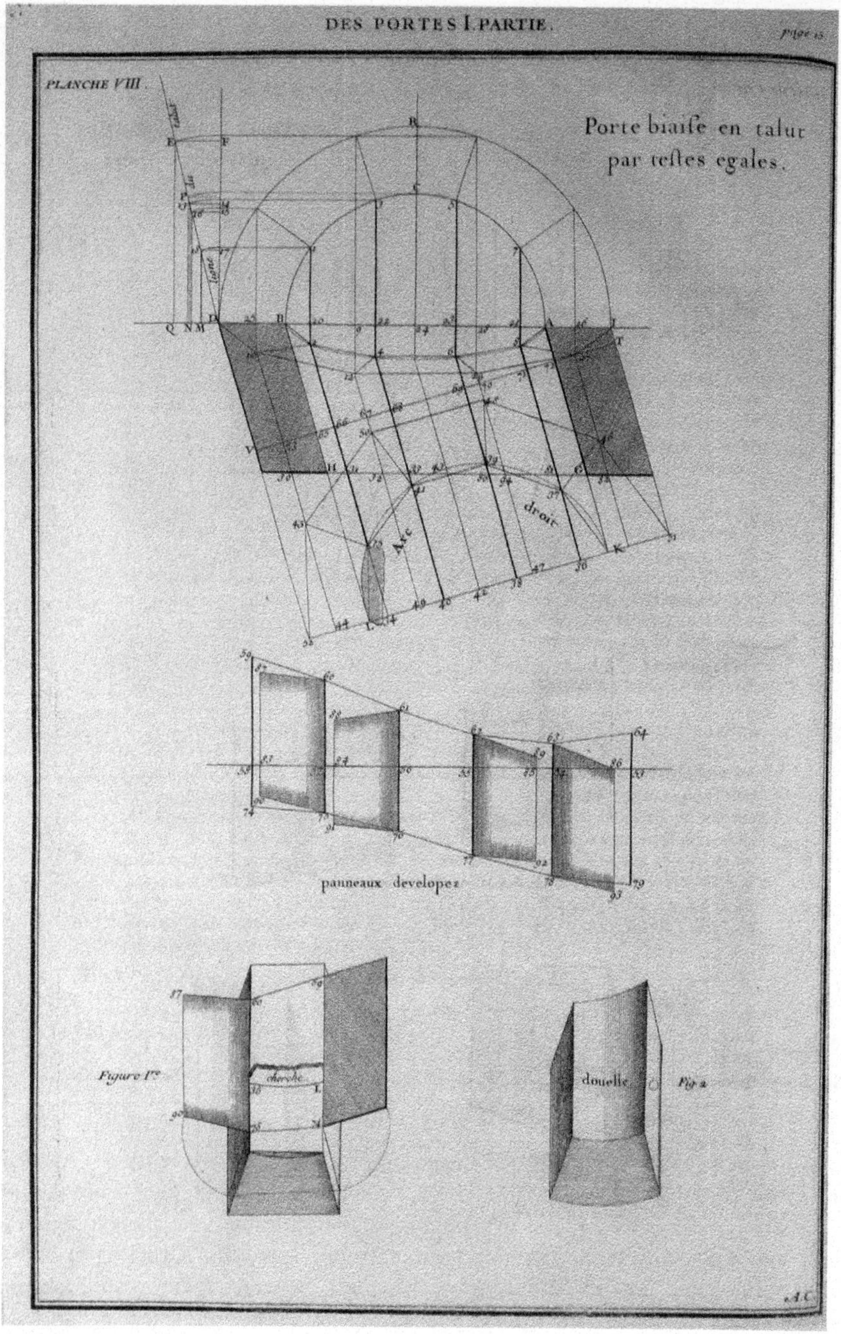

Fig. 6.51 Skew arch in a battered wall (de la Rue 1728: pl. 8)

distance of the voussoir corners to a reference plane, as done by most writers for the skew arch with elliptical faces; after all, the result of his procedure is not far from that of Vandelvira. What is most interesting in Frézier, however, is that he puts forward a single, systematic explanation that encompasses not only the square and skew arches on battered walls, but also the variants with round cross-sections and elliptical faces and those with circular faces and elliptical sections, using intrados joints to connect one with the other. As De la Rue, he treats cross-sections as polygons and uses rigid templates, probably out of distrust for approximate cylindrical developments.

6.6 Other Arches

All stonecutting manuscripts and treatises include a large number of combinations of the basic arch types, such as asymmetrical or splayed corner arches (Vandelvira c. 1585: 21r; Martínez de Aranda c. 1600: 74–75, 77–79), asymmetrical, splayed or skew arches in round walls (de l'Orme 1567: 77v-78r; Vandelvira c. 1585: 22v; Martínez de Aranda c. 1600: 21–24, 26–28, 30–33, 38–39, 103–111; Derand 1643: 176–177; de la Rue 1728: 22), skew arches in battered walls (Martínez de Aranda c. 1600: 54–57, 59–61; Jousse 1642: 18–19, 32–33; Derand 1643: 161–164; de la Rue 1728: 12–13; Frézier 1737–1739: II, 152–158), arches in the intersection of two battered walls (Jousse 1642: 28–31; de la Rue 1728: 18–19; Frézier 1737–1739: II, 159–161) or even asymmetrical arches in round battered walls (de l'Orme 1567: 78v-80r; Jousse 1642: 48–49; Derand 1643: 178–188, 193–198; de la Rue 1728: 24), corner arches involving a round wall (de l'Orme 1567: 80v-82r, 82v-84r), triangular-plan arches (de l'Orme 1567: 84v-85v), or arches with complex springing and longitudinal section outlines (Martínez de Aranda c. 1600: 46–48, 67–71, 90–103).

The list could go on and on, but most of these types are solved using the basic methods we have seen so far, so there is no need to explain them in detail. In contrast, Desargues (1640) and his follower Bosse (1643a) tried to present a radically new method for the combination of a skew, sloping arch or vault opened in a battered wall (Fig. 2.19) as a general solution or *manière universelle* for all stereotomic problems. Such a claim was a bit excessive since the method does not apply to staircases, arches opened in vaults or vaults themselves. In any case, since Desargues' method involves a sloping vault, I will deal with it in Sect. 8.5.

Chapter 7
Rere-Arches

Although arches and vaults are not used frequently in our times, anybody with the slightest interest in architecture is familiar with these terms. In contrast, many architects or historians educated in the last decades would be at pains to define or describe a *rear arch* or *rere-arch*, that is, an element designed to span an opening using a combination of arches, lintels, or both, placed at different heights; these expressions translate the French *arrière-voussure*.[1] This is a pity since rere-arches perform two important functions. First, they enlarge the area of the inside face of an opening, either door or window, in order to distribute evenly the light coming through the opening; second, on some occasions, they allow the designer to disassociate the exterior fenestration of a façade with the interior light outlet, achieving exterior regularity in difficult situations, in particular in staircases. As a result, rere-arches play a significant role in French stonecutting treatises of the seventeenth and eighteenth centuries, exemplified by the well-known *arrière-voussures* of Marseille, Montpellier, and Saint-Antoine, from the *Porte de Saint Antoine* in Paris.

I will classify this large repertoire in three broad groups. The first one, with lintels on both faces placed at different heights (Fig. 7.1), is explained only in Spanish sources, so I will refer to them as *capialzados*, a Spanish word of Catalan origin meaning "raised head".[2] The second group includes the most frequent kind of rere-arches, featuring a curved edge in one face and a straight one in the opposite front. The third one, with arches on both faces, are placed among rere-arches in

[1] The Merriam-Webster dictionary lists "rere-arch" as a variant spelling of "rear arch"; however, I will use "rere-arch" to avoid confusion with the back face of an ordinary arch.

[2] It may startle the reader to hear about rere-arches that do not actually include an arch, as those in the first group. The reason for this apparent contradiction is that the English word, in general use from the nineteenth century, stands for the French *arrière-voussure*, which is equivalent to the Spanish *capialzado*. Neither of these terms refer explicitly to arches; in fact, *arrière-voussure* means literally "rear-voussoir" while *capialzado* stands for "raised head". Thus, rather than set apart a particular category for pieces in the first group, I have included them with rere-arches, at the expense of a slight terminological incoherence.

J. Calvo-López, *Stereotomy*, Mathematics and the Built Environment 4,
https://doi.org/10.1007/978-3-030-43218-8_7

Fig. 7.1 Straight *capialzado*. Valencia, Dominican convent, now a military headquarters, refectory (Photograph by the author)

some sources, particularly French ones, although in Spanish manuscripts they are sometimes grouped with arches.

7.1 Lintels with Edges at Different Heights in Each Face

7.1.1 Planar Faces

Parallel faces. Alonso de Vandelvira (c. 1585: 44v; see also Palacios [1990] 2003: 138–143) includes the simplest variant of *capialzado* (Fig. 7.2). First, he draws the trapecial plan of the element, with a broader opening at the upper part of the sheet; we may assume that this side corresponds the interior. He also constructs the elevation, with the broader lintel raised above the shorter one, dividing it into voussoirs. Next, he draws projection lines from the division points until they reach the springing line; using these intersections, he may draw the horizontal projection of intrados joints.

In order to construct the intrados templates and bevel guidelines, he generally follows the triangulation procedure he used in the skew arch with circular faces (see Sects. 3.1.3 and 6.2.1). In this case, since face edges are placed at different levels, the impost is slanted. Thus, before constructing the intrados templates, Vandelvira draws a template for it, with a triangular shape, to solve the transition between the horizontal bed joints in the supporting wall and the slanting springing. The operation is performed simply by forming a right triangle with the length of the jamb and

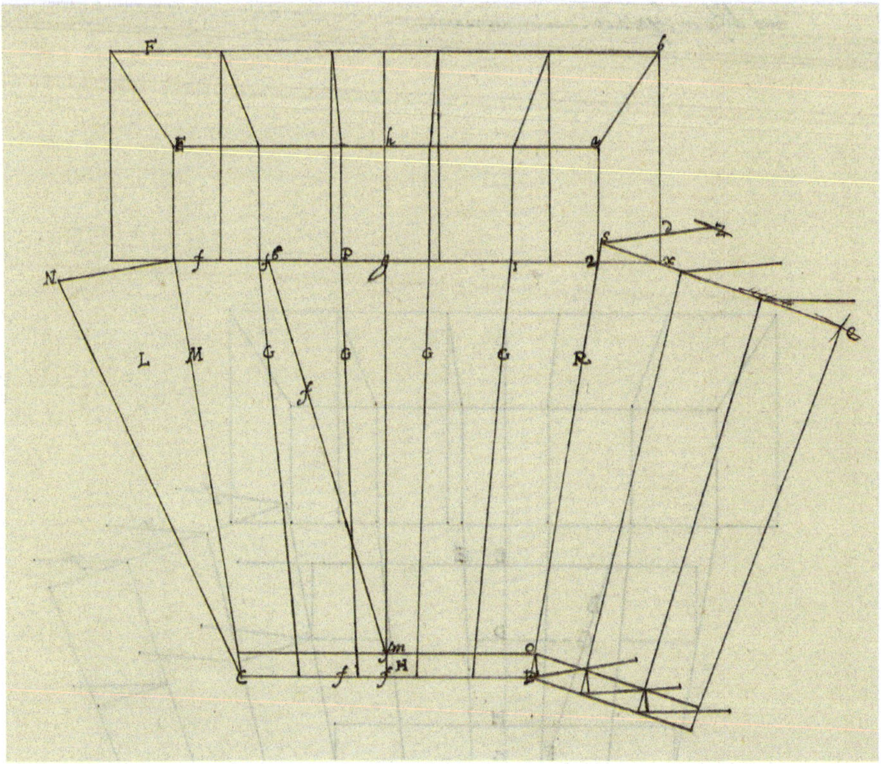

Fig. 7.2 Straight *capialzado* ([Vandelvira c. 1585: 44r] Vandelvira/Goiti 1646: 80)

the difference in heights between the face edges as catheti; the hypotenuse gives the length of the springing. Next, he may construct the first intrados template by triangulation, as he did in the skew arch with circular faces (Sect. 6.2.1). He starts with the length of the springing—that of the edge of the first voussoir, which he has just constructed—and the diagonals, computed using right triangles. Bevel guidelines are constructed using the lengths of the face joints, also taken from the elevation, and the diagonals of the bed joints, which are computed forming right triangles once more. Along with the shape of the first intrados template, this operation also gives the length of the first intrados joint; this enables the mason to begin the construction of the second intrados template, repeating the procedure as many times as necessary.

As in the skew arch with round faces, this method, although ingenious and elegant, is recursive and prone to accumulation of errors. Perhaps this led Martínez de Aranda (c. 1600: 116–118) to put forward an alternative method (Fig. 7.3) based on orthogonals to the projections of intrados joints and the upper face edge. First, he draws the plan and elevation of the piece, including a shallow section at the lower part, or *batiente*, in order to accommodate the window frame. Next, he draws an orthogonal to the horizontal projection of an intrados joint in order to draw a true-shape

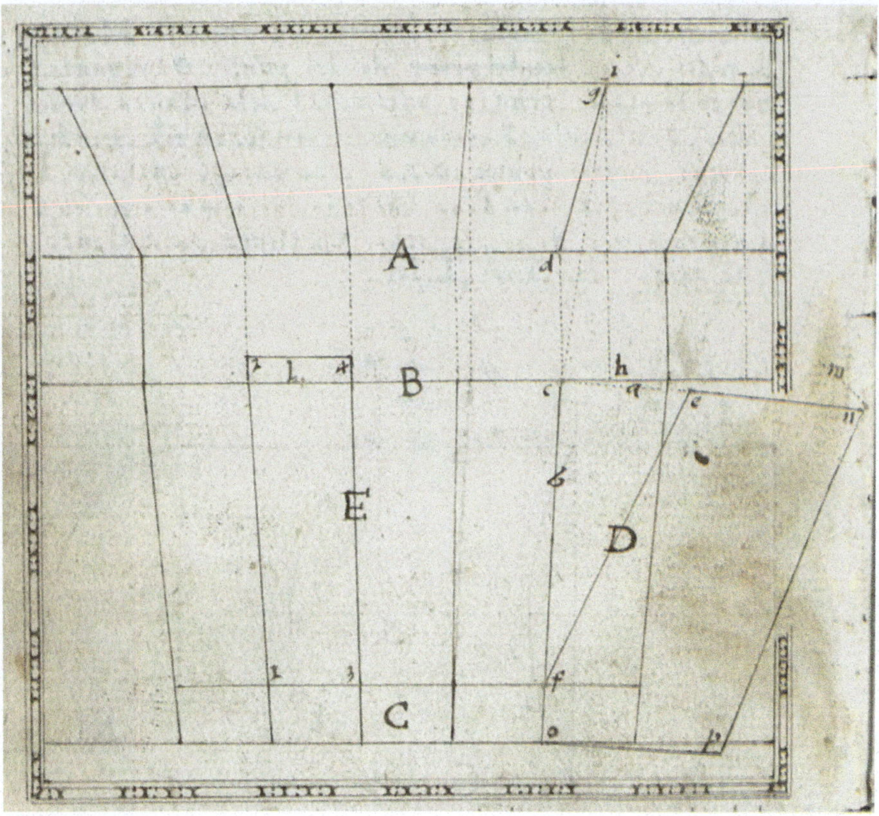

Fig. 7.3 Straight *capialzado* (Martínez de Aranda c. 1600: 117)

representation of the bed joint; bringing the height of the upper face edge to this perpendicular, he places a point representing the edge. Next, he tries to compute the length of the diagonal of the bed joint using its *frontal* projection, forming a right triangle with the length of this projection and that of the intrados joint; the hypotenuse will stand for the length of the diagonal.[3] Using this length and the measure of the face joint, which can be taken directly from the elevation, Martínez de Aranda finds the upper end of the face joint. Next, he constructs the rest of the bed joint template by triangulation. To our eyes, this operation is awkward, since the plane of the bed joint is not vertical and thus does not coincide with the plane passing through the

[3]There is a slight error in this operation. The length of the diagonal may be computed forming a right triangle with the frontal or vertical projection and the difference of distances to the vertical projection plane between its ends. Such operation is, in fact, a symmetrical reversal of the more usual construction based on the length of the horizontal projection and the difference of heights between its ends. Now, instead of measuring the difference in distances to the vertical projection plane, Martínez de Aranda takes the length of the intrados joint, which is slightly longer; the practical implications are quite small.

intrados joint and its horizontal projection. Thus, Martínez de Aranda rotates first the vertical projection plane of the intrados joint, and then the plane of the bed joint; this leads to a small error in the upper corner of the lower face.

In the next step, Martínez de Aranda departs entirely from Vandelvira's procedure, since he constructs the intrados template using projection lines as orthogonals to the edge between the intrados and the *batiente*, as he had done in the symmetrical splayed arch (see Sect. 6.1.1). The lower corners of the template, placed along this line, will not move; the upper edges will move along orthogonals to this line, which overlap projection lines. It is evident in the drawing that the upper edge of the template, being parallel to the frame strip, will keep its parallelism, although Martínez de Aranda does not mention this.

Both Vandelvira and Martínez de Aranda include several variants from this basic scheme. Vandelvira (c. 1585: 44v) addresses the problem of the rere-arch opened in a thick wall, where each course is to be divided into several voussoirs. As in large trumpet squinches (c. 1585: 8r, 10r; see Sect. 5.1.1), he divides intrados joints and constructs bevel guidelines for the division points; except for that, he uses the same procedure than in the basic *capialzado*. Both authors (Vandelvira c. 1585: 45r; Martínez de Aranda c. 1600: 121–123) deal separately with the case where the springings are not symmetrical. As always, this means that instead of drawing the templates for one half of the piece and reversing them for the other half, the mason should construct all of them; however, the rest of the steps of the setting out procedure are similar. Martínez de Aranda (c. 1600: 123–125, 126–128) includes two combinations of lintels and even actual arches grouped around a *capialzado*. He solves them using orthogonals to lines in point view; these operations are consistent with nineteenth-century descriptive geometry since in these cases bed joints are strictly planar.

Non-parallel faces. As we have seen, the convergence of springings, either symmetrical or asymmetrical, raises few practical problems. In contrast, when the faces are not parallel, and both lintels are horizontal, as usual, the issue is trickier. In this case, the intersections of the intrados with the faces are neither parallel to nor convergent with each other,[4] and the intrados surface is not developable; thus, the use of intrados templates, either rigid or flexible, is geometrically incorrect, and the voussoirs should be dressed by squaring, at least in theory.

Vandelvira eschews the issue entirely, so we must turn to Martínez de Aranda (c. 1600: 118–120). He uses basically the same method as in the *capialzado* with parallel faces, although he groups bed joint templates leaning on the axis of the piece. At first sight, this group of templates may resemble an orthographic projection. However, this is not the goal of Martínez de Aranda; he remarks that the templates are prepared

[4]In spatial geometry, two straight lines may be convergent with each other (when they intersect in a common point), parallel to each other (when they do not have a point in common but they lie in a single plane), or *skew* (when they do not lie on a single plane and thus do not intersect).

to place them on the bed joint surface,[5] so he needs a representation of the bed joint in true shape. The method is basically the same one used in the previous piece: he computes the length of the intrados joint forming a triangle with its horizontal projection and the difference of heights between both ends. Next, he determines the length of the diagonal of the bed joint and constructs a triangle with the lengths of the diagonal, the intrados joint and the face joint; this furnishes a schematic bed joint template. Although unconventional, this procedure is conceptually exact.[6] In contrast, he tries to construct the intrados template using a revolution around the line between the *batiente* and the intrados. Such an attempt is of course geometrically incorrect since the intrados surface passes through two skew lines, the lower edge of the upper face and the inner edge of the frame strip; therefore, it is non-developable. Strikingly, Martínez de Aranda is aware of this fact, since he remarks "after dressing, the intrados surface of each voussoir will be warped, while the faces should be planar, checking this with a straightedge and closing one eye".[7] Perhaps he does this consciously, as an approximation, since the template represents correctly three sides of the intrados surfaces and the errors on the fourth side and both diagonals are small.

In any case, in the next section Martínez de Aranda (1600: 120–121) offers an alternative solution (Fig. 7.4). He addresses the issue in a radical way: rather than using horizontal lintels set on converging vertical planes, which lead to warped surfaces, he uses a sloping lintel in the upper face; the result resembles some deconstructionist designs. In theory, this solution could lead to a planar intrados surface, although Martínez de Aranda does not explain a procedure to assure the convergence of the intersection of the intrados with both faces, which would guarantee the planarity of the intrados. Next, he uses a variant of the procedure he has applied in the previous example to construct the bed joint and intrados templates. First, he computes the length of the *first* intrados joint forming a triangle with its horizontal projection, transferred to the axis of the piece, and the difference in heights between its ends.

In contrast to the preceding example, the intradoses of all bed joint templates overlap. This is deliberate since he instructs the reader to "… draw a line to point *e*; it will be used as the intrados of all bed joint templates …".[8] Now, this is the correct

[5]He says literally *de esta manera sacarás las plantas por lecho de este dicho capialzado para plantarlas al justo*, that is, "in this way you will construct the bed joint templates in order to place them exactly". It should be remarked that in other passages, Martínez de Aranda uses templates based on orthogonal projection, but he says these orthographic templates are *para plantarlas de cuadrado*, and not *para plantarlas al justo*; this term is reserved for templates that represent voussoir surfaces in true size and shape.

[6]Martínez de Aranda neglects to apply triangulation to locate the upper corner of the lower face; this causes a slight error.

[7]Martínez de Aranda (c. 1600: 119–120): *las caras de las piezas de este dicho capialzado han de quedar después de labradas engauchidas y las testas de las dichas piezas han de quedar derechas a regla y borneo.*

[8]Martínez de Aranda (c. 1600: 120–121): … *a este dicho punto e tirarás la línea concurriente C que sirve de cara para todas las plantas por lecho de las dichas juntas.*

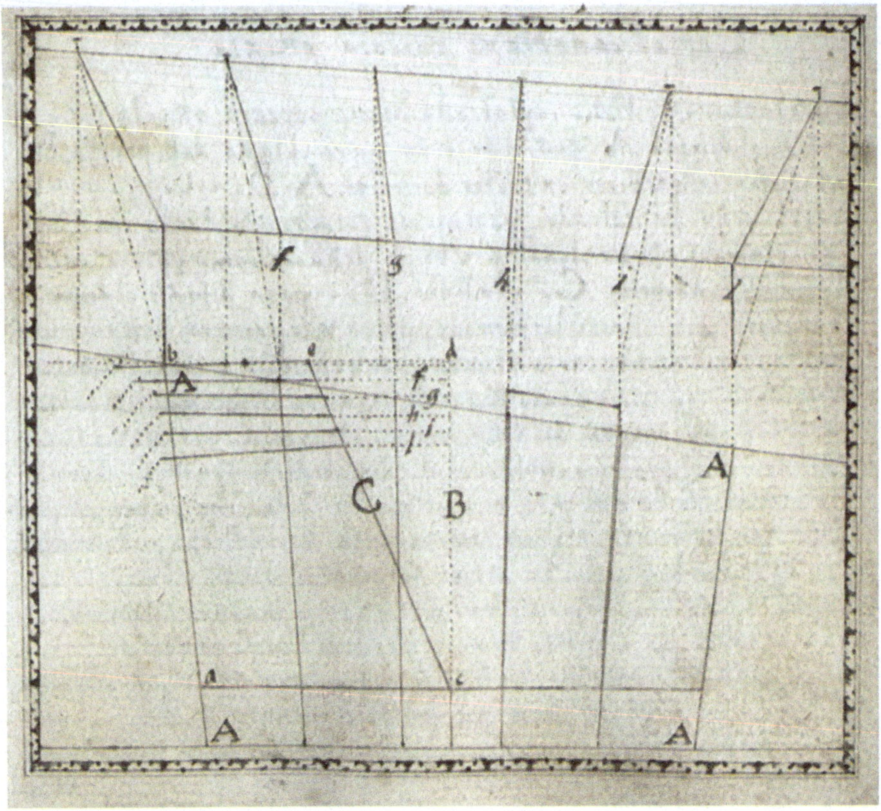

Fig. 7.4 Asymmetrical *capialzado* with sloping lintel (Martínez de Aranda c. 1600: 120)

result for an orthographic projection; if he had attempted to construct a true-shape representation, the slopes of the intrados joints would be different, since they are not parallel. Next, he constructs the upper corners of the upper face by triangulation, since they do not overlap in orthogonal projection; he does not close the template, which stands as a simple bevel guideline. All this strongly suggests that the template should be applied to an auxiliary surface parallel to the axis of the piece, used as an intermediate stage in the dressing process; after this stage, the mason should take off a wedge to materialise the convergence of the intrados joints. In the next step, Martínez de Aranda constructs the intrados templates using the same method he has applied in the preceding section; however, if the intrados surface is planar, this should lead to geometrically correct results.

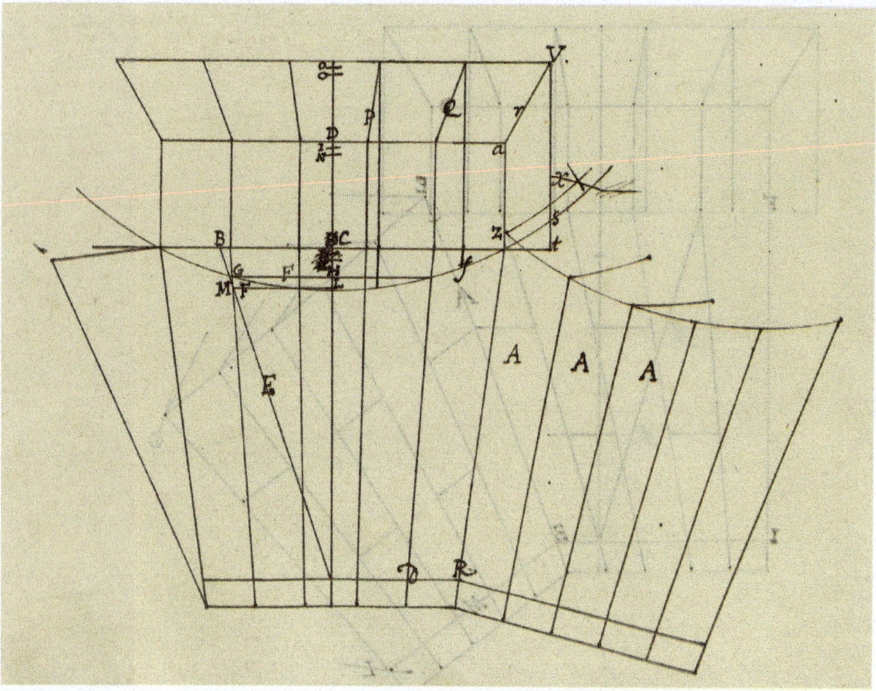

Fig. 7.5 *Capialzado* in a curved wall ([Vandelvira c. 1585: 45r] Vandelvira/Goiti 1646, 83)

7.1.2 Curved Faces

Both Vandelvira and Martínez de Aranda offer solutions for a *capialzado* between two lintels, one straight and the other one opened in a curved wall, either concave and convex. Vandelvira focuses on the concave case (Fig. 7.5). First, he prepares an auxiliary diagram leaning on the axis of the piece, as he does on other occasions. However, its purpose is entirely different. He makes a puzzling remark: "as the templates are shortened as a result of the concavity, they rise less".[9] This hints that Vandelvira is actually lowering the lintel in the central section, in order to preserve the planarity of the intrados surface; in fact, he marks these diminishing heights in the axis of the elevation, both for the intrados and the extrados, although this does not show in the elevation. Thus, the auxiliary diagram is used to compute the height of the upper ends of the intrados joints, taking into account their distances to the frame strip; in the next step, these heights are transferred to the elevation. From this moment on, he applies the procedure he used in the *capialzado* with planar faces, constructing the intrados templates and bevel guidelines by triangulation. Although Vandelvira does not say it explicitly, the length of the face edges of intrados templates seem to

[9]Vandelvira (c. 1585: 45v): *como se van acortando las plantas con el torre cabo van capialzando menos.*

be taken from the plan.[10] It is also essential to take into account that this diagram does not represent intrados joints in true size, so they are computed recursively, as on other occasions. All this is quite interesting since he tries at all costs to keep the intrados surface planar, in order to dress it with templates. In Sect. 7.2 we will see that in arch-and-lintel rere-arches, where this solution is impracticable, he resorts to dressing the voussoirs by squaring.

Martínez de Aranda (c. 1600: 128–133) offers two different solutions for the double-lintel *capialzado* opened in a convex wall. The first one seems to place voussoir corners in the central section at greater heights, adapting Vandelvira's method to a convex front; however, the text does not mention this detail. The templates share the same intrados joint. Martínez de Aranda states that these templates are *plantas al justo* (true-shape representations, see Sect. 3.2.1); as a result, the intrados joints share the same slope and the intrados surface is non-planar. Despite this detail, Martínez de Aranda constructs the intrados templates using orthogonals to the edge between the intrados and the *batiente.* An interesting detail of his solution is that he uses an intermediate point on the face joint in order to compute its curvature.

In his second solution, Martínez de Aranda places all voussoir corners at the same level; thus, the intersection between the intrados and the face is a curve placed on a horizontal plane, and the intrados is a slightly warped surface. Despite this, he uses the same procedure again; true-shape templates for bed joints include an intermediate point to compute the curvature of the face, while intrados templates are constructed using orthogonals to the intersection of the intrados proper and the *batiente*, which should lead to inexact results due to the warped nature of the intrados.

7.2 Rere-Arches with a Lintel and an Arch

7.2.1 Mainstream Solutions

This category includes the most usual rere-arch (Fig. 7.6), known as *arrière-voussure reglée* in French or *capialzado desquijado* in Spanish. A simple solution is offered by de l'Orme (1567: 64r-64v). As usual, he considers the basic type common knowledge; he presents a more complicated variant, opened on a thick wall, where the mason should divide each course into several voussoirs (Fig. 7.7). He does not elucidate the piece since he trusts the drawing to be self-explanatory. This is over-optimistic on de l'Orme's part; it is not easy to interpret the drawing and we can only guess that dressing is carried out mainly by squaring.

We must therefore turn to Vandelvira (c. 1585: 46r; see also Palacios [1990] 2003: 144–147), as on other occasions. First, he begins by explaining clearly that

[10] This involves a slight error, since the upper edge of the intrados is not horizontal, as we will see in the next paragraph. However, taking this width from the elevation would involve a greater error; and if Vandelvira meant to use a particular procedure to compute this dimension, he would have explained it.

Fig. 7.6 Simple rere-arch. Palma de Mallorca, Almudaina Palace (Photograph by the author)

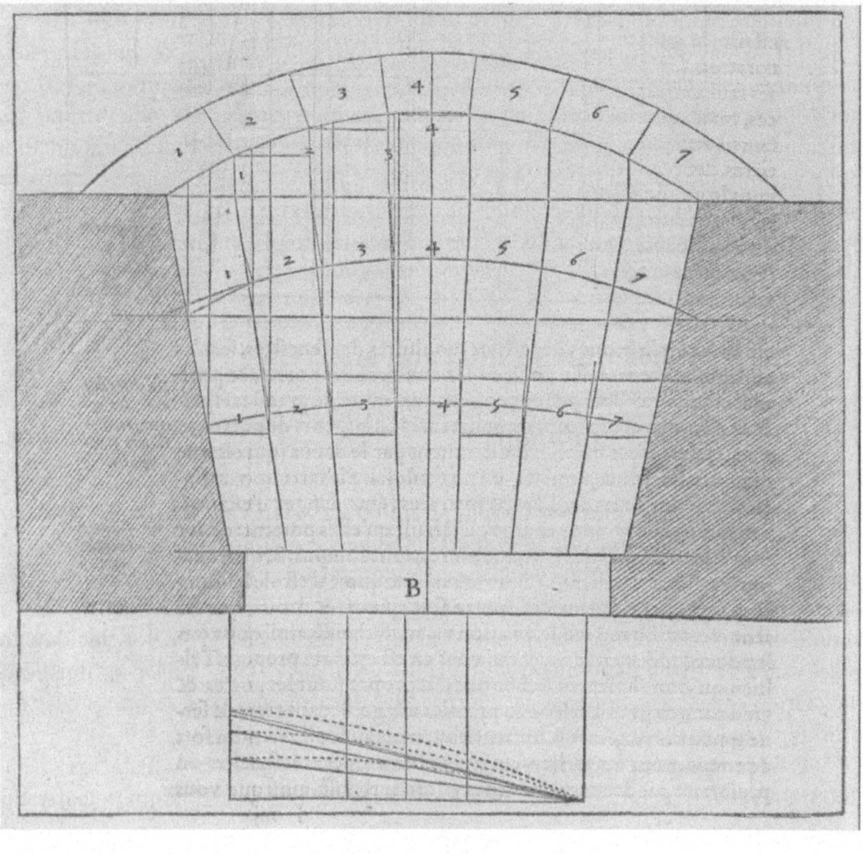

Fig. 7.7 Rere-arch in a thick wall (de l'Orme 1567: 64v)

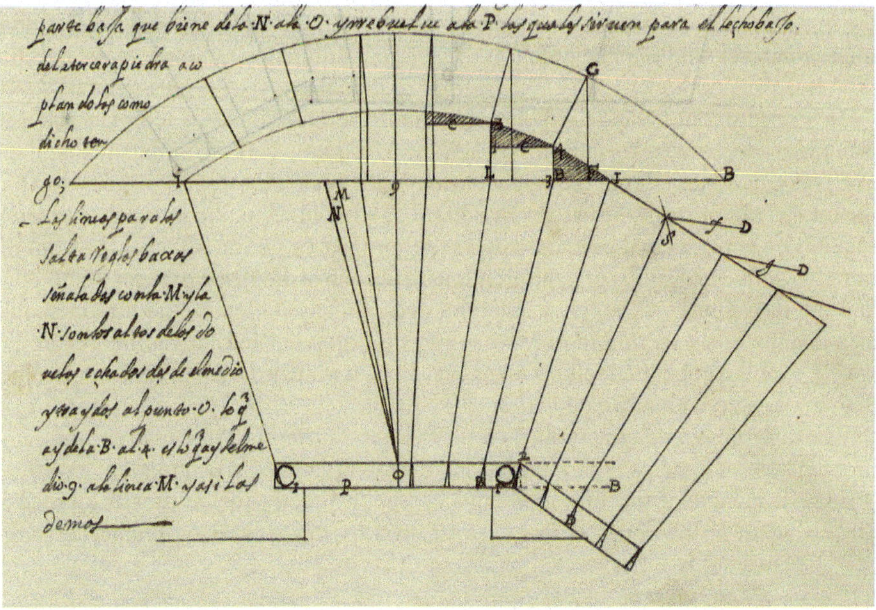

Fig. 7.8 Simple rere-arch ([Vandelvira c. 1585: 46r] Vandelvira/Goiti 1646: 84)

this rere-arch is to be dressed by squaring. He will construct some templates, but they are just auxiliary devices used as a preliminary step in determining the bevel guidelines. Next, he draws the plan of the element, with the arch or interior face slightly larger than the lintel or exterior one (Fig. 7.8). He then draws a segmental arc as elevation of the arc face, dividing the intrados of the arch into equal parts and drawing bed joints from the centre of the segmental arch; this design leads to a large springer, called *piedra del elegimiento.*[11] At the same time, he draws a schematic cross-section[12] including the intrados joints. Next, he constructs intrados templates as on other occasions, computing the length of their diagonals. Again, he computes the length of the intrados joints recursively; we should consider that the cross-section does not represent their true length. In this case, this procedure leads to gross errors, since the intrados joints are neither parallel nor convergent, and thus

[11] In other elements, for example in fol. 103v, Vandelvira uses *elegimiento* this term with the meaning of "foundation". The word may derive from the verb *elegir* (to choose), although it may also be related to a dialectal variant of the word *erección*, that is the act of erecting a construction. In any case, such a use seems to allude to the choice of a general plan for the building, implied in the act of laying out the foundations.

[12] "Cross-section" is used here and in the following sections in a loose sense, meaning "section through a plane orthogonal to one or both faces". In some rere-arches or *capialzados*, the distance between the two faces exceeds the one between springers and thus, in these particular cases, the section through a plane orthogonal to the faces should be classified as a longitudinal section. However, treatises and manuscripts do not generally differentiate this case, so it would be cumbersome and confusing to mention in the text cross- and longitudinal sections.

the intrados surface is non-developable. As a result, the length of the first intrados joint is inexact, and the construction of the second intrados template starts on false premises.

In any case, once Vandelvira has constructed the first intrados template, he uses it in order to compute the angle between intrados joints and face joints, using them as bevel guidelines. Using them in addition to the plan, elevation and cross-section, the mason can dress the voussoirs by squaring, without using the intrados templates. Vandelvira explains that he should dress the springer by taking off a wedge from the intrados of the voussoir, using the arch square to control the shape of the intrados edge and the face joints. Although Vandelvira does not say it clearly, the mason should also use the cross-section in order to control the slope of intrados joints before the wedge over the upper bed joint is taken out, since it is an orthogonal projection rather than a true-size-and-shape depiction.

Martínez de Aranda (c. 1600: 146–149) takes a different route to solve the same problem. He constructs both intrados and face templates by triangulation (Fig. 7.9), making it clear that the mason should use both directly in the dressing process, placing them in the voussoir faces, as stressed by the expression *plantar al justo*,

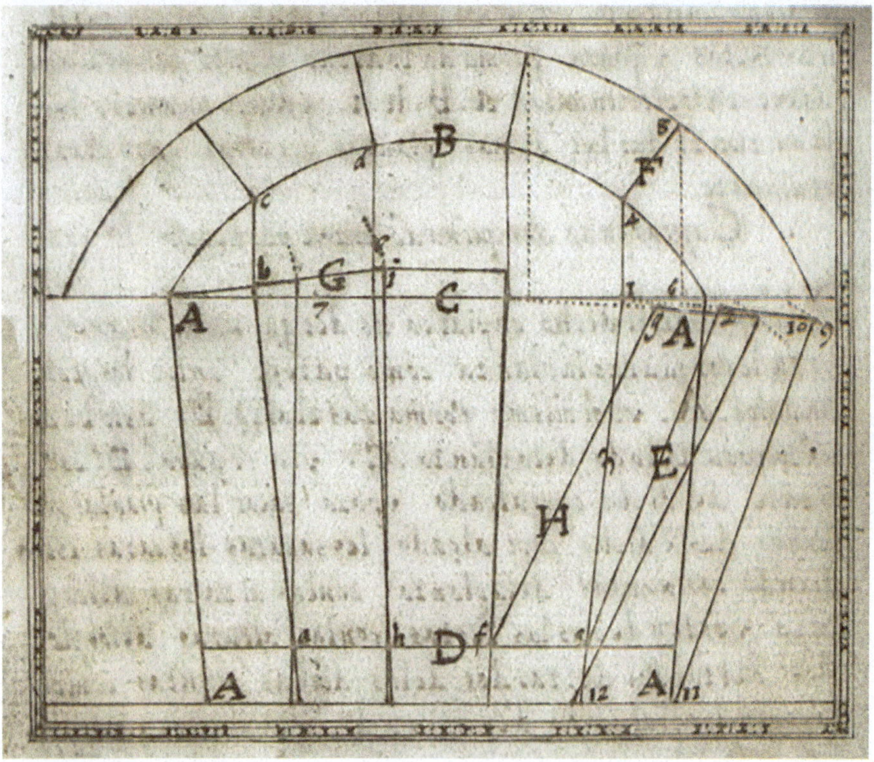

Fig. 7.9 Simple rere-arch (Martínez de Aranda c. 1600: 148)

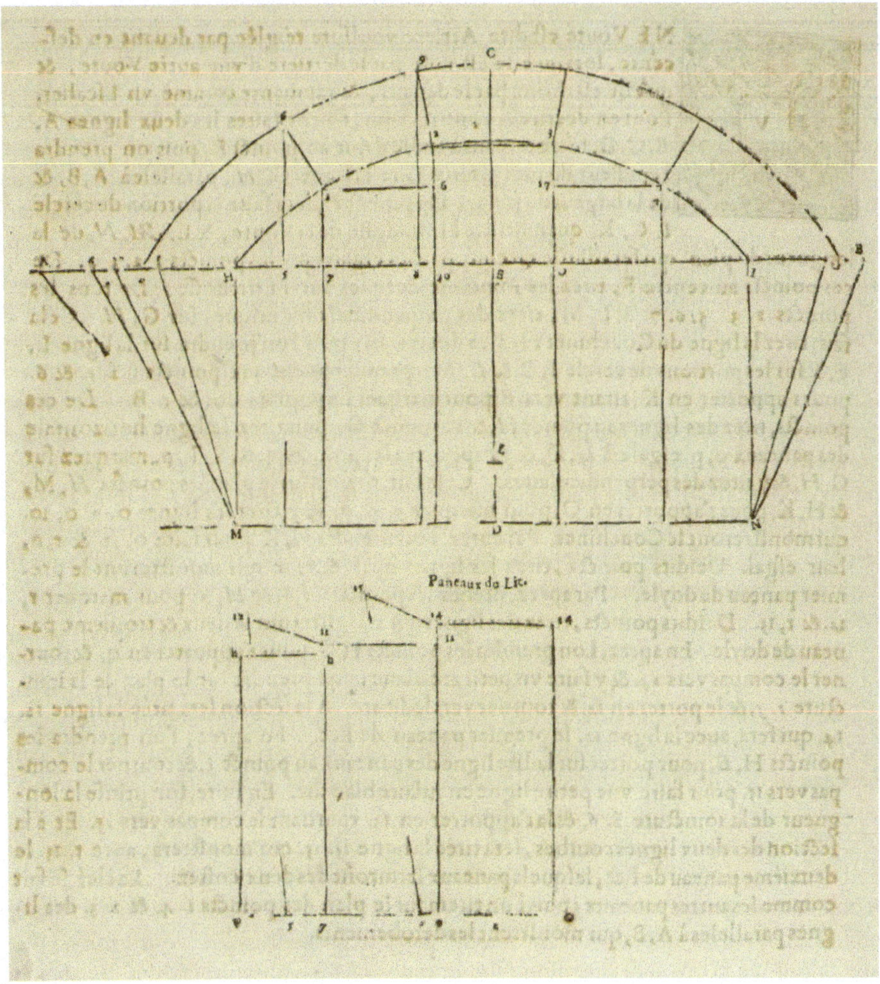

Fig. 7.10 Simple rere-arch (Jousse 1642: 214)

which means "to place exactly" or "to place snugly" (see Sect. 3.2.1).[13] Of course, this method is also inexact, at least for the intrados templates, since this surface is warped.

In contrast, Jousse (1642: 214–215; see also Guardia c. 1600: 71v, 72v) simplifies the problem by using parallel springings (Fig. 7.10), in contrast to Vandelvira and

[13]This is connected with *plantas al justo* or true-shape templates (see Sect. 3.2.1). In contrast, for templates that represent a face in orthographic projection, used to dress an auxiliary surface, Martínez de Aranda (for example c. 1600: 193, 196, 201, 236, 242, 245, 249) uses *plantar de cuadrado*.

Martínez de Aranda. After drawing the plan and the elevation, he constructs a cross-section standing for the slope of the intrados joints, leaning it on the springings. Next, he constructs auxiliary intrados templates, which do not represent the actual intrados surface; they stand for a planar surface that starts at the lower edge of the intrados surface (that is, the lintel side) and then ascends until it reaches the lower corner of the intrados of the voussoir at the arch side. Thus, this auxiliary planar surface passes through three of the corners of the voussoir, two in the lintel side and one in the arch side. However, due to the curvature of the arch face, the fourth corner of the intrados surface of the voussoir stands above the auxiliary surface. As a result, these templates may be used to dress a preliminary surface passing through three corners of the intrados; next, the mason should continue the dressing process in order to materialise the curvature of the intrados surface and the fourth corner of the voussoir. The geometrical construction of this template does not pose complex problems: it is drawn using orthogonals to the lintel edge: the distance of voussoir corners to the axis may be computed from the cross-section, while their width may be taken directly from the plan. Further, to dress the intrados and bed joint faces of the voussoir correctly, the mason needs face templates, which are taken easily from the elevation, and bevel guidelines, called *panneaux de lit* by Jousse. Although his explanations are not clear, it seems that he constructs them using the diagonal of the bed joint face.

Derand (1643: 136–139) takes further steps in the direction of the squaring method: he starts by drawing the plan, the elevation, and the cross-section but does not attempt to construct intrados or bed joint templates. Springings are convergent, but horizontal projections of intrados joints are parallel. Next, he explains the dressing process in detail: the mason should score on the largest face of the block the horizontal projection of the intrados joints. He should then measure the position of the upper corner in the voussoir face and transfer it to the stone, marking the face edge and the face joints. This operation can be performed with the arch square, but Derand recommends the use of a templet and a bevel, particularly for faces with basket handle arches. In the next step, the mason should carry over (he uses the word *trainer*, literally "drag") a horizontal line along the intrados of the voussoir. Derand finishes his explanations for the general voussoir here; however, it is implied that the mason should hollow out the intrados in the next phase and dress the bed joint, and in fact, these operations are mentioned for the springer.

Rere-arches on thick walls. In the preceding solutions, authors solve the problem with a single row of voussoirs spanning the distance between both imposts; each voussoir goes from the lintel side to the arch side. When the wall is too thick, it may be difficult to find stones large enough. In these cases, the rere-arch may be formed as a succession of courses going from the lintel to the arch, each of them divided into several single voussoirs. As we have seen, de l'Orme (1567: 64r-64v) includes a drawing of this solution without a real explanation, so we must start again with Vandelvira (c. 1585: 46v). He uses mainly the same method as in the single-voussoir piece. In addition, the mason should draw an arch between both faces of the wall to divide each course into voussoirs, compute the length of the portion of each intrados joint belonging to the first voussoir and carry it to the intrados template to divide it into two or more parts. He should also construct bevel guidelines for the division points, following the same procedure explained for the single-voussoir case. It is interesting to notice that Vandelvira draws a single division arch, and thus all joints are aligned, in contrast to usual practice. Thus, Vandelvira's explanation is purely didactic; he surely expected any experienced mason to manipulate the joints to avoid aligning them.

Jousse (1642: 216–217) and Derand (1643: 136–139) take the same route, repeating their general procedures to deal with the case of several voussoirs in each course; in fact, Derand does not present this case as a specific section of his treatise, but as a detail of the general procedure (Fig. 7.11). Even Martínez de Aranda (c. 1600: 152–154) eschews in this element his idiosyncratic use of true-shape templates and solves the problem by squaring, clearly showing the enclosing rectangles of each voussoir and following the mainstream procedure.

Other variants. Building on this basic variant, all authors address additional difficulties. Vandelvira (c. 1585: 47r) includes a skew rere-arch, with each course divided into voussoirs; the only real difference with the general solution is that the templates are not symmetrical, so he needs to construct all of them. He also includes other variants (Vandelvira c. 1585: 47v-49r) with the arch face opened in a concave wall, both symmetrical, asymmetrical or divided. All are solved using intermediate points for the intrados templates, including here and there developments for the arch face. Martínez de Aranda (c. 1600: 149–212), as usual, includes rere-arches with fronts on concave and convex walls, wall intersections and barrel vaults, as well as a large number of combinations of arches and rere-arches and even rere-arches with oval and annular plans. He tries to solve these variants by templates, but in the more complex ones he is forced to use hybrid methods making heavy use of auxiliary templates, as we have seen in Sect. 7.1. These templates represent an intermediate stage in the dressing process and should be placed orthogonally to the sides of the starting block (*plantar de cuadrado*) rather than on the actual faces of the voussoir (*plantar al justo*).

7.2.2 Rere-Arches with the Lintel Placed Above the Arch

In the preceding variants, the springings of the arch side are placed at the same level as the lintel, so the arch keystone lies over the straight edge. In these cases, the arch operates as an expansion of the lintel-side opening, complemented by the splaying of the jambs, softening the light coming from the exterior by distributing it over a larger surface. However, when the rere-arch is used to cast light into a basement room, the exterior lintel is placed above the springing or even the keystone of the arch. Martínez de Aranda (c. 1600: 161–165) offers two different solutions to this problem. In the first one, the springings are parallel, as in Jousse's solution for the basic problem; in the second one, he comes back to the usual splayed design. In any case, both solutions are based in the standard methods of this author: he tries to use full templates for the intrados and the bed joints, constructed by a combination of

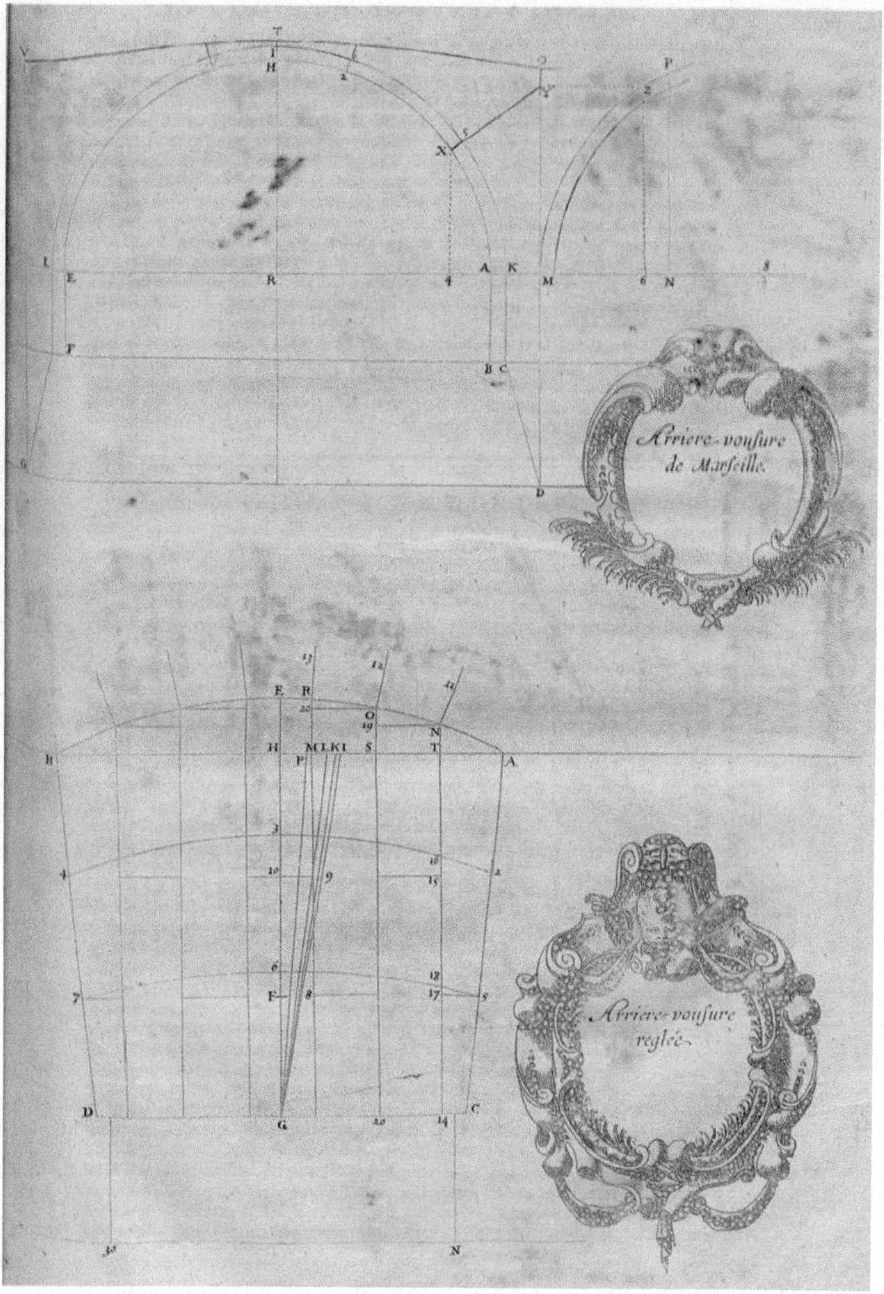

Fig. 7.11 Above, *arrière-voussure de Marseille*; below, rere-arch in a thick wall (Derand [1643] 1743: pl. 64)

orthogonals and triangulation; since the intrados is a warped surface in both cases, his solutions are inexact.

Jousse (1642: 212–213) includes a variant of his ordinary solution, which looks quite similar at first sight. However, on close inspection, the springings have a strong slant so the element resembles a sloping vault, and Jousse in fact associates it with stairs. In any case, he uses the same procedure he employed in the ordinary rere-arch, constructing auxiliary templates standing for an auxiliary planar face passing through three corners of the voussoir; later on, he dresses the curved intrados surface.

Frézier (1737–39: II, 476–484) included a refined version of Jousse's solution under the name of *arrière-voussure de Montpellier*, remarking that no stonecutting treatise had included it before and that it could be found only on Blanchard's (1729) carpentry treatise, as a variant of the rere-arch of Marseille. This hints that the rere-arch of Montpellier, although having a straight upper edge, shares some traits with those having two curved edges. In any case, Frézier explains that the bed joints can be arranged in three different ways, using the same intrados surface (Fig. 7.12). First, they may be set out as the voussoirs of the face arch, that is, as a fan of planes converging on the symmetry axis of the rere-arch. However, he eschews this first solution, arguing that it leads to exceedingly large voussoirs and acute angles in the straight face.

Fig. 7.12 *Arrière-voussure de Montpellier* (Frézier [1737–1739] 1754–1769: pl. 68)

He then explains the second solution: the intrados joints may be drawn as the intersection of the intrados surface with a set of vertical planes parallel to the symmetry axis. Frézier makes it clear that this applies only to *joints de doële* or intrados joints; of course, if bed joints were vertical, the piece would depend only on mortar adherence to be stable, a dangerous solution. Also, intrados joints may be placed at a sheaf of vertical planes passing through the intersection of the jamb planes; again, Frézier stresses that this applies to *joints de lit â la doële*, that is, the intersections of bed joints with the intrados, not actual bed joints. This second solution resembles Jousse's method since the horizontal projections of the intrados joints are parallel. However, in Frézier's solution, intrados joints are not straight lines as in Jousse but rather quarter-ellipses, drawn in a cross-section. This leads to a messy result, as Frézier himself remarks: the intrados and face joints cannot lie on the same plane, and thus must be manipulated, although this operation will not be noticed except from the extrados, which is usually hidden.

The third solution is neater: it admits planar bed joints, as usual. The mason should extend the springings in plan to find their intersection; next, he will divide the arch face into equal voussoirs, transfer the division points from the elevation to the plan and draw the horizontal projections of intrados joints passing through the meeting point of the springings. Once this is done, Frézier constructs quarter-ellipses standing for intrados joints, as in the preceding case. However, he warns the reader that he should not take as semi-axes the heights of each joint, but rather the distance between the round face and the straight one, taken along the frontal projection of the intrados joint, which should not be vertical. In the next step, he uses the chords of the quarter-ellipses to construct *panneaux de doële plate* (flat intrados templates), using orthogonals to the intrados edge. This term refers to templates that should not be applied to the finished intrados, but rather to an intermediate enclosing solid since their face edges are horizontal and their side edges are straight. Thus, they will fit the straight face but not the curved face or the intrados joints, which are half-ellipses.

Next, Frézier explains in detail the dressing process. The mason should carve a plane as a first approximation to the intrados, scoring on it the flat intrados templates, as well as the faces, marking the profile of the curved one. Next, he should dress the planar bed joints, leaning a straightedge on the face joints and the long edges of the flat intrados template. Then, he should apply the quarter-ellipses on the bed joints, using templets. At this stage, the four edges of the intrados surface are defined, and the mason may materialise it using templets.

7.2.3 Rere-Arches with a Double-Curvature Intrados

In contrast to the solutions of Martínez de Aranda and Jousse, the *Arrière-voussure de Montpellier* of Frézier is a double-curvature surface, since neither the directrix in the curved face nor the quarter-ellipse generatrices are straight lines. Thus, it can be seen as a particular case of rere-arches with non-ruled surfaces (Fig. 7.13). Before Frézier, several writers explained procedures leading to curved intrados joints,

Fig. 7.13 Rere-arch with double curvature. Valbonne, Chartreuse (Photograph by the author)

controlled with a templet rather than a straightedge. This choice does not seem to stem from practical considerations, but rather from aesthetical reasons, in imitation of a well-known example or even whimsy. In fact, such pieces are known in French as *arrière-voussure de Saint Antoine*, from the Gates of Saint Anthony in Paris, located next to the Bastille and demolished in 1778, one year before the storming of the fortress.

However, the oldest solution for this problem is not to be found in French sources, but in Martínez de Aranda (c. 1600: 165–166), who does not mention the Gates of Saint Anthony.[14] Rather than a full solution for a particular type of rere-arch, he explains a general procedure to hollow the intrados of a voussoir. Once the mason has constructed the intrados templates, he can divide each intrados joint into six parts and draw an orthogonal to the joint through its midpoint, marking on it the sixth part of the length of the joint. Next, he can draw an arc through the ends of the joint and the end of the orthogonal. This procedure may be applied to full bed joint templates, used by Martínez de Aranda on many other occasions; however, he illustrates it using simplified templates. The reader may wonder if there is a particular reason to measure exactly one-sixth of the length of the intrados joint. As far as I know, the procedure would be feasible with any other length, say one fourth or one eighth, but a shorter length may pass unnoticed, while a deeper hollowing would weaken the voussoir dangerously.

Derand (1643: 139–142) proposes a different solution, explained as an independent problem and not as a general procedure (Fig. 7.14). He deals first with the basic

[14]Spanish manuscripts include here and there references to French archetypes; in particular, Martínez de Aranda (c. 1600: 230) includes a *Vía de San Gil*, alluding to a well-known staircase in Saint-Gilles, in Languedoc; Vandelvira (c. 1585: 15v); mentions also the *Trompe de Montpellier*, or *Trompa de Mompeller*. See also Barbé (1977: 187–188), Pérouse ([1982a] 2001: 84, 125) and Aranda (2017).

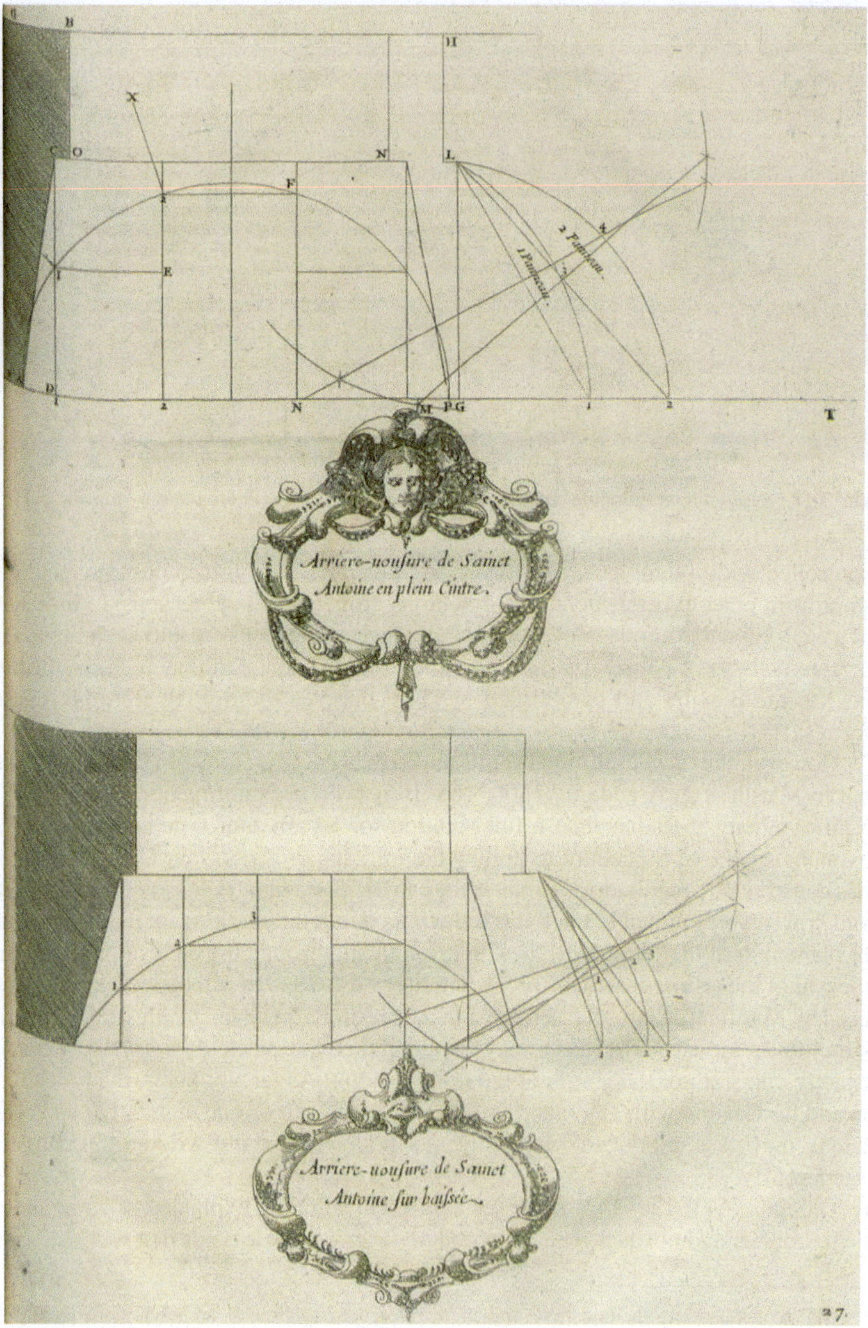

Fig. 7.14 *Arrière-voussure de Saint Antoine* (Derand [1643] 1743: pl. 66)

case, and then with the skew variant. Although the elevation is quite schematic, it is clear that the lower face is solved with a lintel, while the upper one is treated as an arch; that is, the solution is the opposite of Frézier's *Arrière-voussure de Montpellier*. In the square variant, he draws the plan of the piece, with convergent springings; however, the horizontal projections of the intrados joints are parallel. He warns the reader that the face arch can be either round or surbased, unless the thickness of the wall is less than half the span of the opening. In this case, the surbased solution should be used. Next, he draws a cross-section of the arch. As in modern multiview drawings, the construction of the cross-section starts from the plan; as a result, the verticals are parallel to the face of the arch. This may be a bit confusing but shows clearly the connection of the section with the plan. In order to trace the curvature of the intrados joints, he uses a sophisticated solution: he draws straight lines and orthogonals through their midpoints, as does Martínez de Aranda; next, he determines the centre of the intrados joints at the intersection of the orthogonals with the vertical plane of the round face. As a result, the radii at the end of these curves are vertical, so the tangents at the endpoints are horizontal. The cross-section also shows why Derand eschews the round arch in thin walls; the slope of the intrados joints would be excessive, and the centres may fall above the springings. In the next section of his treatise Derand (1643: 142–144) explains the construction of the skew variant, with a surbased face arch. The procedure is essentially the same, although in this case, he constructs a true-size representation of the face arch or *cintre*. The cross-section is based on this auxiliary view rather than the plan; of course, the bias of the element leads him to construct templates for all intrados joints.

De la Rue (1728: 31–33) takes as a literal model the side bays of the Gates of Saint Anthony, still extant in his time, with parallel springings (Fig. 7.15). He constructs a cross-section starting from the plan, like Derand, but he uses quarter-ellipses for the intrados joints, remarking that they are more graceful than Derand's circular arcs; of course, this solution allows him to use vertical tangents at the springings and horizontal ones at the endpoints. Frézier (1737–39: II, 489–501) follows basically de la Rue's solution, with parallel springings, giving two variants, the first one by squaring, like Derand and de la Rue, and the second one using flat intrados templates. Further, he gives an alternative solution with converging springings, remarking that the quarter-ellipses lie on sloping planes, and thus the half-axes of the quarter-ellipses should equal the frontal projection of the intrados joints, not their height, as in the Montpellier rere-arch.

7.3 Rere-Arches with Arches on Both Faces

Up to this moment, most rere-arches are used to provide an interior expansion of a rectangular window, either with another raised lintel, as in the Spanish *capialzados a regla* in Sect. 7.1.1, or with an interior arch, as in the rere-arches in Sect. 7.2.1. When an architect tries to build an arched opening in a building, a round arch in the interior face will lead to a cylindrical surface; when opening the window, its

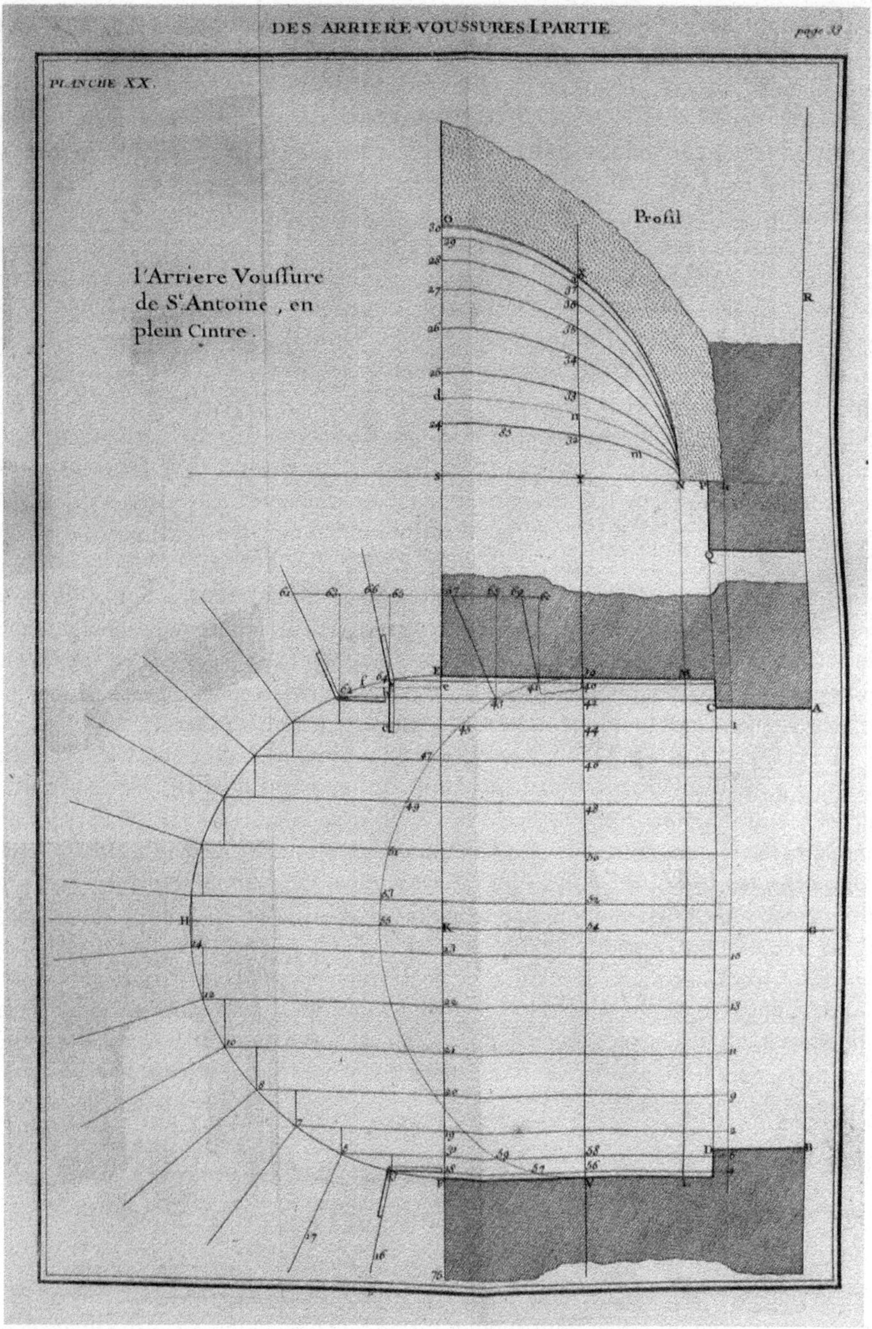

Fig. 7.15 *Arrière-voussure de Saint-Antoine* (de la Rue 1728: pl. 20)

Fig. 7.16 Rere-arch with arches in both faces. Barcelona, Convent dels Angels, now Biblioteca de Catalunya (Photograph by the author)

frame will bump against the cylinder (Rabasa 1996b: 32; Rabasa 2000: 278). A solution for this problem is to place an arch with raised springings and a larger radius of curvature, usually a segmental arch, on the inner face (Fig. 7.16). As in other problems, solutions evolve from crude solutions cobbled together in the sixteenth century to more sophisticated procedures in the seventeenth and eighteenth centuries, while in the nineteenth century, the problem is taken as an occasion to illustrate rather abstract mathematical concepts rather than practical construction issues.

Early solutions. De l'Orme (1567: 64) includes a drawing of a rere-arch with an exterior face in the shape of a round arch. The interior face is solved with a segmental arch with its springings placed well above the keystone of the semicircular exterior face so that the window can be opened without hitting the intrados surface (Fig. 7.17). No explanation is given since de l'Orme argues that

> it is easy to understand for those who know the practice and industry of the compass ... So, I will not make a long discourse since it is easy to construct the templates and dress the stones to execute the rere-arch, as you can see from the drawing.[15]

Perhaps de l'Orme is too optimistic, but at least a modern reader can notice that bed joints are represented by simple straight lines crossing the face, the intrados and

[15] De l'Orme (1567: 64) ... *elle sera facile de connaître à ceux qui ont commencement de la pratique, et industrie du compas ... Qui fait que je ne vous en ferai plus long discours, aussi qu'il est facile de pouvoir lever les panneaux, et faire couper les pierres pour mettre l'arrière-voussure en oeuvre, ainsi que vous le pourrez connaître par la figure ensuivant ...* Transcription is taken from http://architectura.cesr.univ-tours.fr; translation by the author.

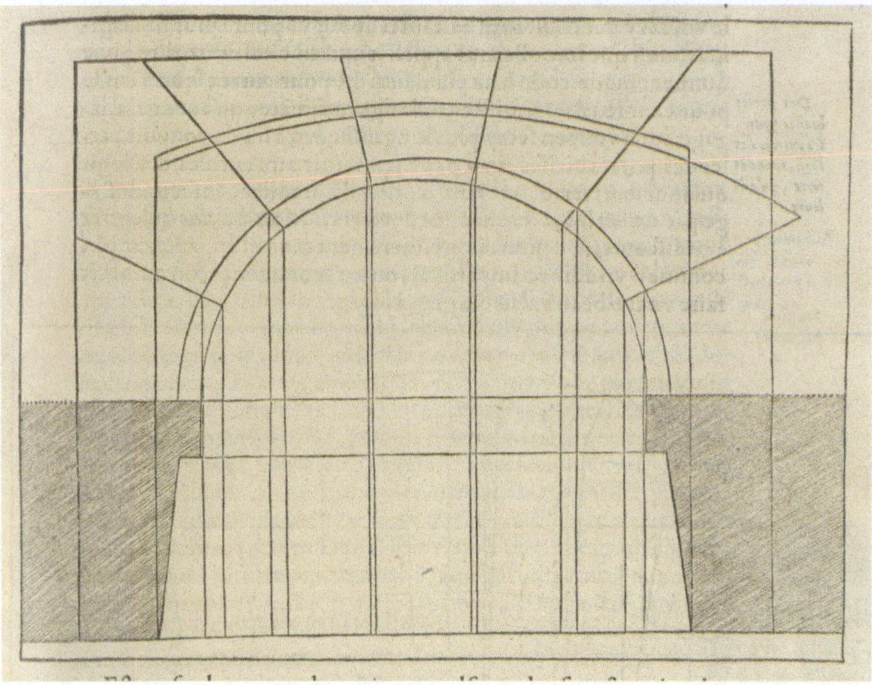

Fig. 7.17 Rere-arch with arches in both faces (de l'Orme 1567: 64r)

the strip holding the window-frame. This means that bed joints are set out on planes orthogonal to the face of the rere-arch, suggesting strongly that the piece is to be dressed by squaring, as in the *biais passé* solution for the skew arch (see Sect. 6.2.2); this prevents the "large waste of stone" brought about by the squaring method.

Two different solutions are presented by Martínez de Aranda (c. 1600: 81–85). In the first one, he constructs both face arches, round and segmental, with their keystones at the same level (Fig. 7.18). However, there is an important difference from de l'Orme's scheme: Martínez de Aranda divides the round arch into equal parts and draws verticals from the division points until they reach the segmental arch; from the intersection points, he draws the face joints of the segmental arch. As a result, the bed joints are not laid on planes orthogonal to the faces; dressing the voussoir by squaring is a complex process, dutifully explained by the author.

Perhaps it is for this reason that he also includes a problematic variant using templates. Martínez de Aranda tries to use orthogonals to lines in point view to construct the bed joint templates; however, since bed joints are not placed on planes orthogonal to face planes, the technique leads to inexact results. Next, he addresses the intrados templates, attempting a true-shape representation of a warped quadrilateral through a combination of triangulation and orthogonals to the intrados edge, as on other occasions, and the result is equally flawed.

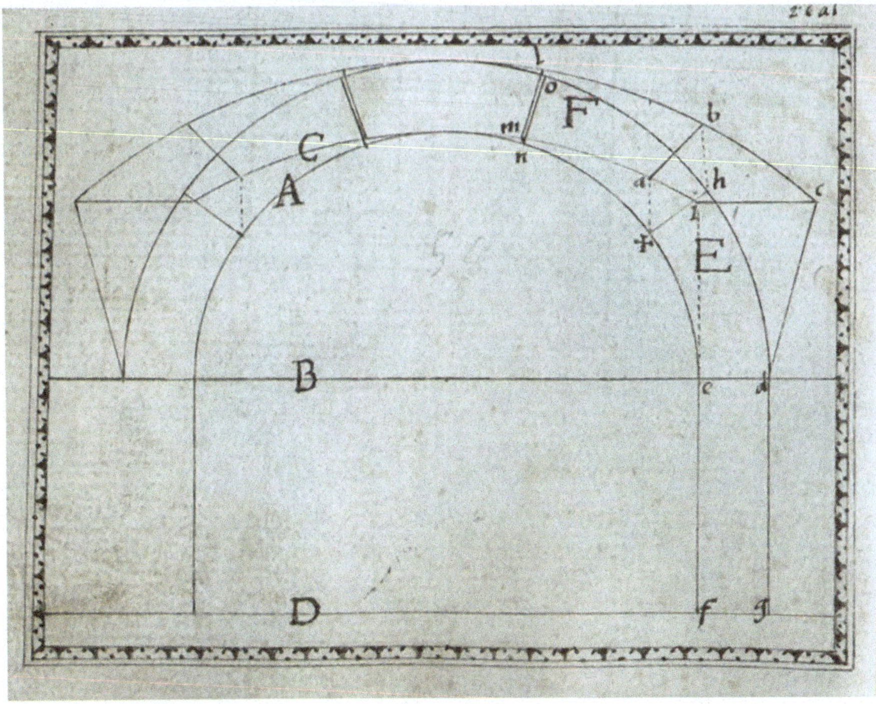

Fig. 7.18 Rere-arch with arches in both faces (Martínez de Aranda c. 1600: 81)

The *arrière-voussure de Marseille*. Although Jacques Gentillâtre includes two variants (c. 1620: 410r, 411r) of a rere-arch with one segmental edge and one round edge (Fig. 2.10), introducing the classic denomination of *arrière-voussure de Marseille*, he does not offer a systematic explanation, so we must begin with Jousse (1642: 208–209). He follows the essential traits of de L'Orme's solution (Fig. 7.17). The bed joints are set on planes in edge view; thus, it is easy to construct the bed joint templates using orthogonals to lines in point view. Further, Jousse mentions for the first time a typical feature of this element: the intersection between the jamb and the intrados surface. Since the springings of the round and the segmental arches are placed at different heights, the intersection of the jamb and the intrados surface must be a sloping line. However, neither de l'Orme nor Martínez de Aranda make it clear whether it is straight or curved, although a line in the first intrados template of Martínez de Aranda's second solution suggests it is straight.

In contrast, Jousse (Fig. 7.19) chooses to use curved intersections; this choice has practical advantages, giving more room for the window frame. First, he places the window frame at the middle of the thickness of the wall, so the rere-arch is divided into three sections: an outer strip, the window-frame support, and the rere-arch proper. Next, he draws a true-shape representation of the intersection of the jamb and the intrados surface, giving it the same radius as the round arch. He then prepares

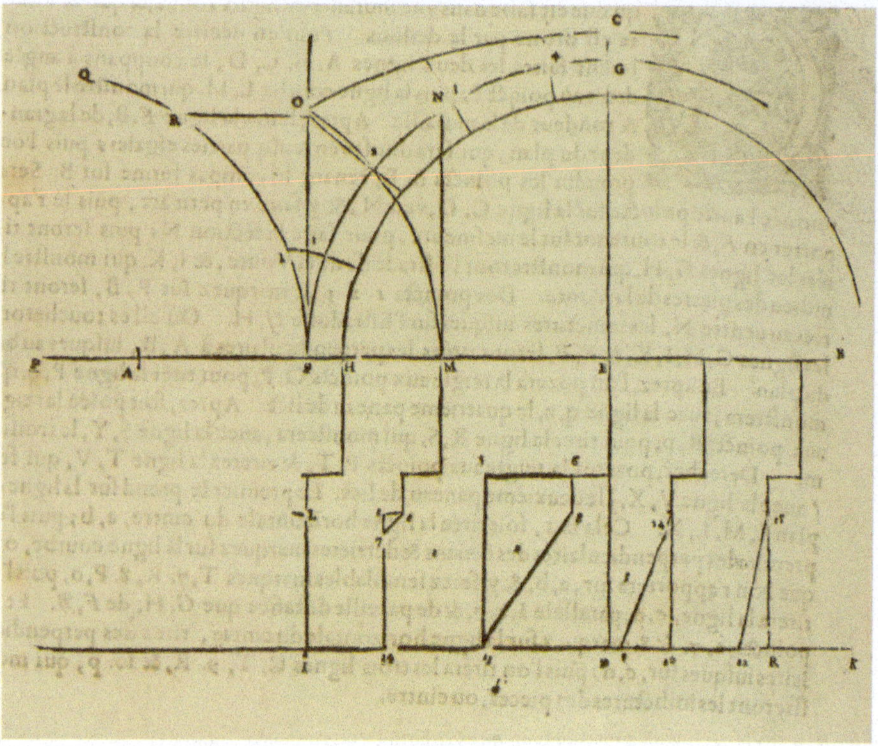

Fig. 7.19 *Arrière-voussoure de Marseille* (Jousse 1642: 208)

the bed joint templates taking into account the distances between the edges of the window-frame strip and the segmental face, taken from the elevation and transferred to orthogonals to the axis of the arch; of course, these operations amount to rotations around lines in point view.

Derand (1643: 134–136) uses a similar procedure: he first traces the intersection of the jamb and the intrados surface, using the same radius as in the round arch, and then the elevation of the segmental arch, taking care to make it consistent with the true-shape representation of the intersection. He adds an interesting detail: the round arch may be divided at will, but the mason may place a joint end at the intersection of the intrados and the jamb to simplify operations. He does not draw the bed joint templates, arguing that the springing may be used as the template for the first voussoir and that the following ones are identical.

De la Rue (1728: 28–39; see also Frézier 1737–1739: II, 440–443) does not agree. He starts with the intersection with the jamb, following Derand's procedure, first drawing its true-shape representation, computing from it the height of the springers of the segmental arch, bringing it to the elevation, and using it to draw the segmental arch. As for the bed joint templates, he states that "you should now construct the bed joint templates; the most complex are those that cross a kind of fork that is formed by the edge of the window strip and that of the intersection made in the jambs in order to allow the opening of the windows”[16] and thus he returns to Jousse's procedure, constructing the templates by using orthogonals to the axis of the arch (Fig. 7.20). He describes in detail the dressing process, which the mason should start by carving a block with the profile of the face template. Next, the mason should mark the bed joint templates on the sides on this block; he will then carve the different sections of the intrados, using arch squares for the outer round section and the frame strip and a straightedge for the warped inner surface.

Hachette (1822: 81–82, 283–286) explains the element in descriptive geometry terms, stressing that it should fulfil several conditions. First, the generatrices of the ruled intrados surface should intersect both the round and the segmental edges and the axis of the round arch. The bed joints should belong to a sheaf of planes passing through the axis; as a result, the intrados joints are generatrices of the ruled surface. Now, the segmental arch intersects the jamb plane at a point; the plane passing through this point and the axis belongs to this sheaf of planes and thus its intersection with the intrados is a straight generatrix. At the other side of this generatrix, the intrados is formed by a different surface which connects the round arch with the jamb rather than the segmental arch. In other words, the intrados is formed by three different surfaces: a central one passing through the round and the segmental edges and two lateral ones connecting the round arch with the jambs (Fig. 7.21). In order to assure the smoothest finish, the central surface should share the same tangent plane with the side ones along the common generatrices. Thus, Hachette finds a perfect opportunity to illustrate a descriptive geometry theorem stating that if two ruled surfaces intersecting at a straight line share the same tangent plane in three points, they also share it in all points of the common line (Hachette 1822: 84); he therefore proves in detail that both surfaces share tangent planes at the intersection of the common generatrix with the axis, the round arch and the segmental arch. As Sakarovitch (1992a: 534–536) and Rabasa (1996a: 32) and have shown, such a mathematical arsenal seems excessive for the practical needs of stonecutting: it is impossible to distinguish rere-arches executed following Hachette's procedure from their eighteenth-century predecessors.

[16]De la Rue (1728: 28): *Il s'agit présentement de trouver les panneaux de joints, dont les plus composés sont ceux qui coupent une espece de fourche qui se trouve formée, tant par la naissance de l'arête du derriere de la feillure que par celle du ceintre qu'on pratique dans l'ebrasement, exprès pour faciliter l'ouverture des ventaux.*

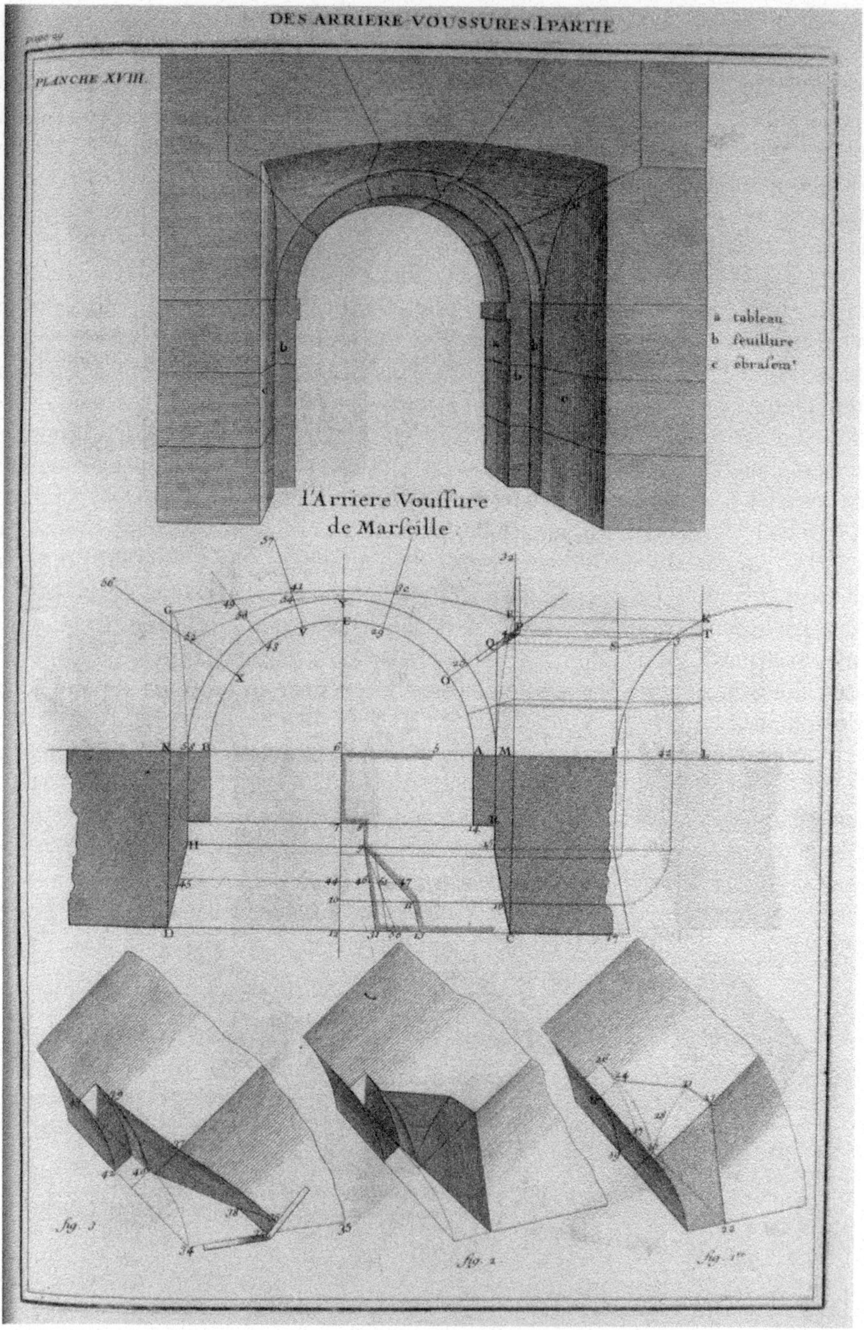

Fig. 7.20 *Arrière-voussoure de Marseille* (de la Rue 1728: pl. 18)

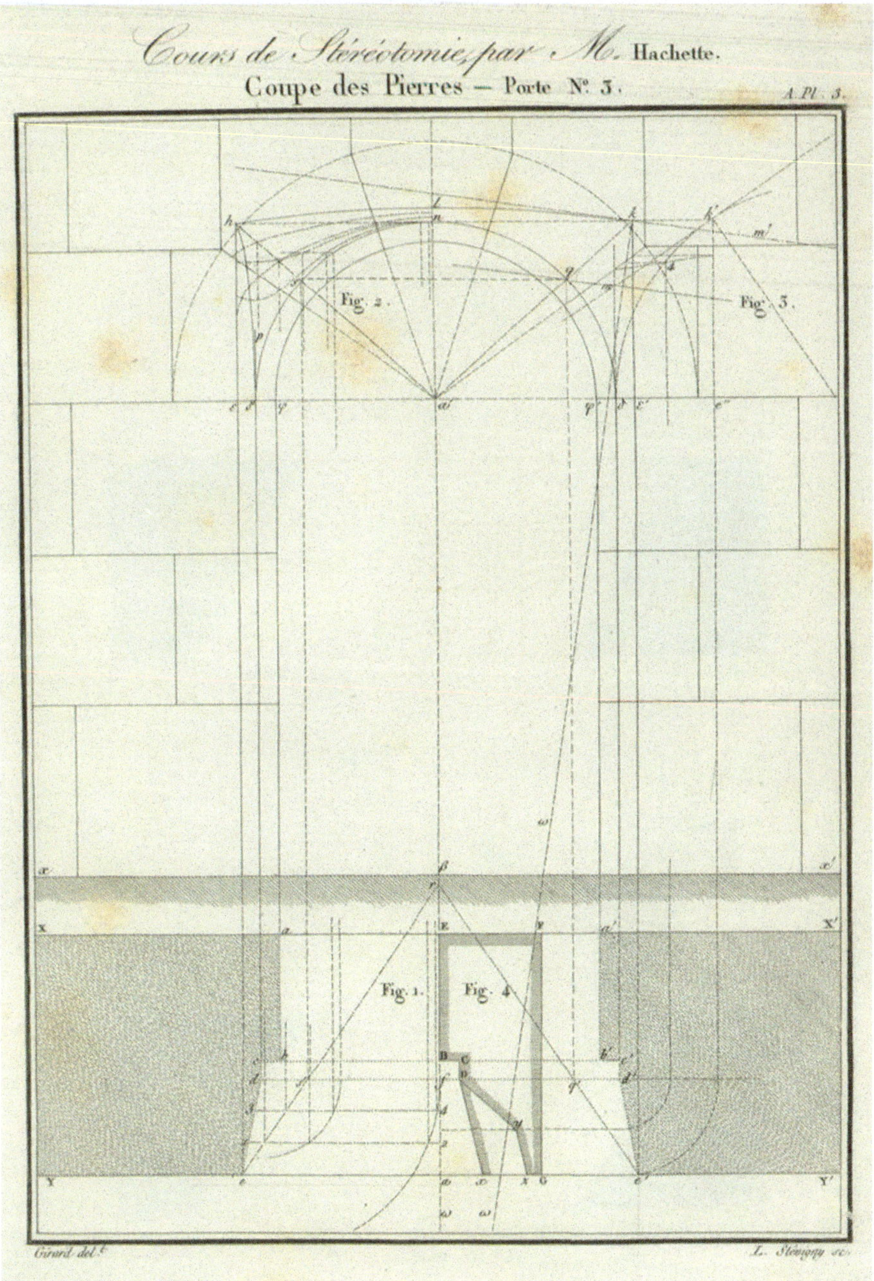

Fig. 7.21 *Arrière-voussoure de Marseille* (Hachette [1822] 1828, *Cours de Stéréotomie* ..., porte 3)

Chapter 8
Cylindrical Vaults

8.1 Barrel Vaults

8.1.1 Simple Barrel Vaults

In its simplest form (Fig. 1.14a), a barrel vault may be understood as an extended round arch, whose depth exceeds the maximum length for a single voussoir. Thus, the vault is divided into courses using bed joints parallel to the springings; each course is then divided into voussoirs by a secondary set of joints known as *transversal* or *side joints*. Intrados joints—that is, the visible edges of bed joints—are usually set out as continuous lines going from one end of the vault to the opposite one. In contrast, side joints are usually laid in staggered fashion; that is, they are discontinuous at bed joints to prevent the vault from working as a succession of independent arches and developing transversal cracks. Other precautions must be taken to assure a correct mechanical performance of the vault: it is advisable to place some dead load, usually in the form of rubble, over the initial sections of the vault, in the vicinity of the springers, to increase vertical forces and compensate thrust.

Although the most usual form of barrel vault is based on the round arch, other types of simple arches, such as segmental, basket handle or pointed arches, can be used as generating shapes. In all these cases, the dressing procedure is the same as in a round arch; the cross-section of the voussoir is traced and controlled using an arch square from a flat surface that furnishes one of the side joints, while the intrados, extrados and bed joints are dressed with the help of a square leaning on the flat surface. Round, segmental and basket handle barrel vaults usually include a course at the top that acts as a keystone; in contrast, pointed barrel vaults are built with a joint at the apex, avoiding the use of V-shaped voussoirs, just as in pointed arches. True elliptical barrel vaults are not frequent since they would require a different arch square for each course, and the difference from a basket handle vault is usually not noticeable.

None of the usual treatises, from de l'Orme to de la Rue, deal specifically with this kind of vault. Derand (1643: 18; see also de la Rue 1728: 8–9) mentions *arches, portes*

J. Calvo-López, *Stereotomy*, Mathematics and the Built Environment 4,
https://doi.org/10.1007/978-3-030-43218-8_8

and *berceaux*, (arches, doors and barrel vaults) but he explains all them together and does not address the issue of the continuity of side joints. In contrast, Frézier (1737–39: II, 83–87) although grouping both arches and barrel vaults under the heading *berceaux*, includes a theoretical explanation of the subject. He remarks that barrel vaults are usually built using simple curves, such as the circle and the ellipse, as directrices; however other shapes, such as the catenary, the cycloid or Cassini's oval, offer mechanical advantages. He adds that in elliptical vaults, the bed joints should be traced orthogonally to the tangent of the cross-section, rather than pointing to the centre of the ellipse.

8.1.2 Skew Barrel Vaults: Orthogonal and Helicoidal Bonding

In theory, the solutions for skew arches we have seen in Sect. 6.2 can be applied to barrel vaults with oblique faces. When using intrados joints parallel to the springings, with either elliptical or circular face arches, the obliquity of the element affects only the first and last voussoirs in each course, while the rest can be dressed as ordinary round voussoirs. However, the combination of an acute angle between the faces and the axis of the vault and a long axis usually makes the *biais passé* solution impracticable.

In the nineteenth century, with the advent of railroads, traditional solutions for skew vaults proved insufficient (Rabasa 1996b: 38–41; Sakarovitch 1992a: 539–540; Sakarovitch 1998: 313–319). Up to this moment, when roads crossed a river or gorge, they bent in order to intersect the waterway at a right angle. This solution is useful in thoroughfares for pedestrians, animals or carts; however, it is not advisable in railroad bridges. A sharp bend demands a sudden reduction in speed to avoid the danger of derailing;[1] afterwards, the train will recover slowly due to the upward slope of the river bank. Railroad bridges were thus designed to cross rivers without bending, and the old problem of skew arches was again at the forefront of construction technology. However, the scale had changed, and spans were not measured in feet, but rather in scores of meters. As a result, an old issue resurfaced: Martínez de Aranda had remarked that in skew arches with joints parallel to the springings the voussoirs may slip.[2] This behaviour is caused by the fact that the resultant of compressive stresses is oblique to bed joints, and then the thrust of the arch is cast "into the void". Also, elliptical face arches may be remarkably surbased, while circular face arches will lead to a remarkably raised cross-section. On some occasions, the problem can be addressed by using bed joints orthogonal to the faces. However, the combination

[1]Sadly, the issue is not merely theoretical. It appears that a combination of excessive speed and a bend caused the derailings in Angrois, near Santiago de Compostela in 2013 and Dupont, near Seattle, in 2017, the latter when circulating over a bridge.

[2]Martínez de Aranda (c. 1600: 11) … *si este dicho arco por tener tanto viaje hiciere algún género de garrote* … (if this arch assumes the shape of a cane as a result of being extremely skew).

of breadth and sharp obliquity makes this solution impracticable; in extreme cases, the vertical projections of the two face arches do not overlap.

French engineers, applying the theoretical tenets of rational mechanics and descriptive geometry, endeavoured to find a solution (Fig. 8.1, left), known as *orthogonal bonding*, fulfilling two conditions: first, intrados joints should be perpendicular to any frontal section of the intrados surface to nullify "thrust into the void"; second, bed joint surfaces should be ruled surfaces generated by normals to the intrados surface, as in Monge's ellipsoidal vault (see de la Gournerie 1855: 29, and Sect. 2.4.3). The angle between intrados joints and frontal sections is preserved in a development of the intrados surface, and thus intrados joints should be perpendicular to the sinusoids that stand for frontal sections in the development. Although this simplifies operations, to construct such ideal surfaces graphically was impracticable. Thus, several simplifications were proposed, although none was satisfactory: in some cases, intrados joints, starting from even divisions in the front face, do not reach the corresponding points in the back face, and some courses are even cut at the middle.

The final simplification was found in Britain, a country that distrusted descriptive geometry, a product of Napoleonic France (see Lawrence 2003). Simply put, intrados joints were traced as helixes (Fig. 8.1, right), and thus depicted as straight lines in the intrados development; joints between voussoirs in the same courses were represented by another set of straight lines, orthogonal to the intrados joints (Nicholson 1839: 3–20; see also de la Gournerie 1855: 32–36). Such a solution did not nullify the problem of "thrust into the void", but reduced it to reasonable limits so it could be withstood by friction between voussoirs or the adherence of mortar. This episode showed that descriptive geometry was not indispensable to solve all spatial problems

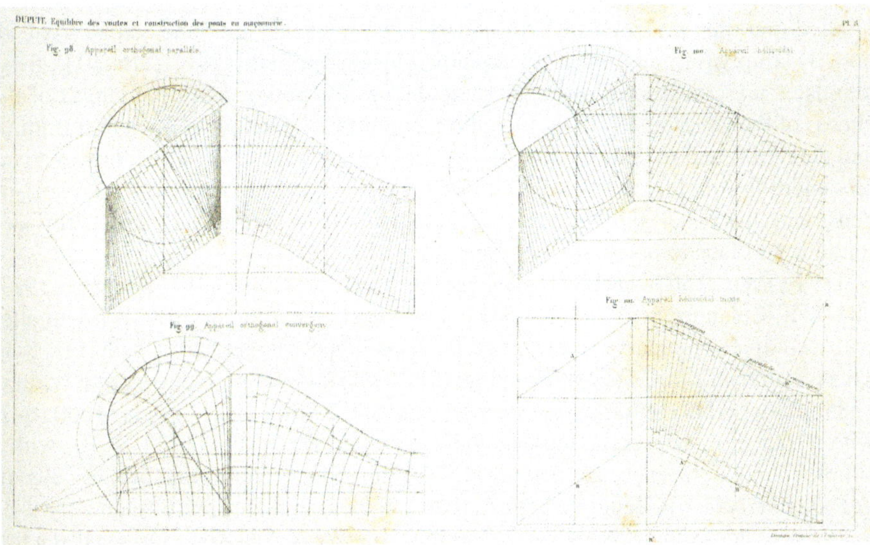

Fig. 8.1 Skew vaults. Left, orthogonal bonding; right, helicoidal bonding (Dupuit 1870: pl. 8)

raised by the Industrial Revolution (for alternative systems, see Lawrence 2003). In any case, the problem exited the technological scene as quickly as it had entered it: starting in the last decades of the nineteenth century on, stone bridges were pushed into obsolescence by iron and concrete.

8.1.3 Sloping Barrel Vaults

Probably the oldest appearance of a sloping barrel vault in stonecutting literature is the *embocinado de escalera*, literally a "splayed piece for a staircase", in MS 12.686 in the National Library of Spain in Madrid (Alviz c. 1544: 13; see also García Baño and Calvo 2015), probably a copy of an earlier manuscript in the entourage of Pedro de Alviz. It is a very short piece, intended either for a staircase crossing a wall orthogonally or a few steps set in the thickness of the wall (Fig. 8.2). In any case, the problem can be solved with a simple row of voussoirs; rather than a sloping vault, it is a slanted arch; in fact, Alviz avoids the word "vault" and even the traditional Spanish term, *decenda*, from the French *descente*, meaning "descent". The drawing carries no explanatory text, but the construction method has been reconstructed by García Baño (2017: 292–307). The author seems to start by drawing an elevation of the arch, but his graphic conventions may startle a modern reader. The extrados of the rear face of the arch is dashed, in accordance with present-day drawing conventions, since it is hidden; in contrast, the radial joints in this face, as well as the extrados joints, which are also hidden, are depicted in full lines. In any case, Alviz draws the plan easily with the aid of projection lines starting from voussoir corners. Next, the mason draws four templates for the right-hand voussoirs and the keystone. It is implied that the two sides are symmetrical, and the same template can be used for both by simply turning it over. As explained by García Baño (2017: 296–303), these templates are constructed through triangulation: the author takes the length of the chords of the voussoirs from the elevation, computes the length of the intrados joints and diagonals and repeats the procedure until he reaches the keystone; however, his procedure is simpler than the one used by Vandelvira (c. 1585: 28v-29v). Next, he constructs the bevel guidelines (the segments used to measure the angle between intrados and face joints), also by triangulation.

De l'Orme (1567: 58r-59v; see also Sanabria 1984: 208–212; Sanabria 1987: 284–288; Calandriello 2019: 64–66) opens the stonecutting section of his treatise with a complex element: a vault whose lower end abuts a larger barrel vault (Fig. 8.3). This is rather striking, since other authors start with simpler problems: Vandelvira (c. 1585: 6v-7r) with a symmetrical trumpet squinch; Martínez de Aranda (1600: 6–7) with a trapecial skew arch; Jousse (1642: 4–5) and De la Rue (1728: 9–10) with a simple round arch. There are two reasons for this. The French name of these pieces, *descente de cave* (descent to a cellar), makes it clear that they were frequently used to access underground spaces, often covered with vaults. Since de l'Orme talks a few pages later about a gentleman who wishes to restructure a house, it makes some sense to start his explanations with a vault going down to a cellar, then to address arches

Fig. 8.2 Short sloping vault for a staircase (Alviz, attr., c. 1544: 13)

opened in walls, vaults and finally staircases. Further, Galletti (2017b: 158, 162) has recently pointed out that de l'Orme seeks pieces that group a remarkable number of geometrical problems. This justifies the inclusion of the large barrel vault at the end of the sloping one in the drawing, although de l'Orme does not say a word about it the larger barrel vault in his lengthy explanation, which is limited to the problem of the simple sloping vault.

He starts by drawing a line that represents the slope of the vault, the cross-section of the larger vault, and a section of the sloping vault, in the shape of a round arch,

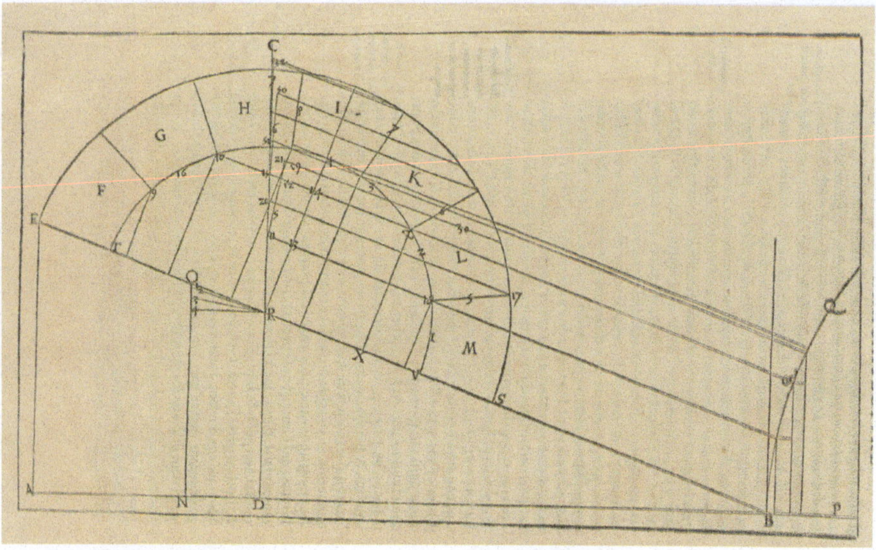

Fig. 8.3 Sloping barrel vault ending in a lunette (de l'Orme 1567: 59v)

divided into seven equal voussoirs. It is not clear whether this section is a cross-section through a plane orthogonal to the slanted axis of the vault or a section through a vertical plane; as a result, the correctness of the general procedure cannot be verified. Next, he draws intrados joints starting from voussoir corners taken from the section. In the next step, de l'Orme constructs the intrados templates through an apparently convoluted method. First, all templates will have the same width; thus he draws a separate diagram based on two parallel lines, the distance between which is equal to the length of the voussoir chord. Next, he measures the distances of two voussoir corners, namely *11* and *15*, to a plane perpendicular to the axis of the vault. It seems, however, that he is not interested in the absolute positions of voussoir corners, but rather in their relative positions. In order to measure them, he measures *11–13* in the first intrados joint and transfers this distance to the one passing through *14*; he obtains as a result that the difference in lengths between these intrados joints equals the segment *12–15*. He transfers this distance to the auxiliary diagram starting from point *4*, set in an orthogonal starting from point *R*, obtaining the point *2* as a result. Thus, *4–2* will equal the excess length of the intrados joint passing through *14* over the one passing through *13*, and the quadrilateral *D–R–2–N*, will stand for a three-sided template of the intrados of the voussoir included between both intrados joints. As on other occasions, this template is meant to be used before hollowing the voussoir intrados, so it represents a face of a polyhedral surface inscribed in the intrados of the vault.

This procedure is used for the ordinary voussoirs of the vault; templates for the uppermost one and the springings are constructed using different methods. Since the upper one is symmetrical about the vertical plane passing through the axis of the vault, both intrados joints should have equal lengths, and thus the end of the template, *R–4*, should be perpendicular to *R-D*. As for the springer, de l'Orme seems to be thinking that triangles *11–13–R* and *R-O–4* are equal, and thus *O–4* equals *11–13*. In fact, both triangles are similar, but not equal, since the projected distance between the intrados joint that passes through *18* and the springing is shorter than the real distance, that is, the chord of the voussoirs in any course; thus, de l'Orme's construction involves a slight error.

Next, de l'Orme constructs the bed joint templates through rotation around the intrados joints; in contrast to usual practice, he does not start from the plan, which is simply not drawn, but rather from the elevation; in modern parlance, he rabats the templates until they reach a vertical plane. In order to do so, he first draws a parallel to an intrados line, set at a distance equal to the width of the bed joint, taken from the round arch. Next, he draws a perpendicular from the projected end of the intrados joint, for example, *20*, until it meets the other intrados joint at *21*;[3] such a procedure is conceptually similar to the one used by Vandelvira (c. 1585: 23r) in cylindrical lunettes, as we will see in Sect. 8.4.2.

When dealing with the sloping vault, Vandelvira (c. 1585: 28v-29v; see also Palacios [1990] 2003: 110–113) first includes a simple one; it is longer than the one drawn in MS 12.686, but can still be solved with a single row of voussoirs (Fig. 8.4). He solves the problem by triangulation, as did Alviz, although his method is more complex. He adds some interesting remarks. First, "the sloping barrel vault and the skew arch with circular faces are traced in the same way regarding templates and bevel guidelines; however, in the sloping vault, you should mark the difference in heights between both faces of the vault as shown from A to B"; [4] in other words, he grasps that the two problems share similar traits, fifty years before Desargues (see Tamboréro 2008: 52–58). Another significant detail is that the vault should rest on a triangular-shaped block placed in the wall, whose length should equal the horizontal projection of the vault, while the other cathetus should equal the difference in heights between both ends of the block. After this, Vandelvira enters into a lengthy explanation of the triangulation procedure for constructing the templates and bevel guidelines of the element; as announced, the method is basically the same one used

[3] At first sight, *20–21* does not seem perpendicular to *17–20*. This impression comes from two facts: the perpendicular is not drawn, and *20*, which is given by the intersection of *CR* and the extrados joint passing through *17* seems to be placed on the projection line of another voussoir corner. However, there is no reason for this, so *20* may be slightly off the projection line, exactly as much as *21*. In any case, the text is clear (de l'Orme 1567: 58v-59) … *lequel point de 20 vous porterez perpendiculairement sur la ligne 19, au point de 21, et de ce point là de 21, vous tirerez une ligne jusques au point de 11, qui montre justement comme doit être le panneau de joint pour tracer au droit de la commissure, 5* (… you will transfer this point 20 orthogonally to line 19, to point 21, and from this point 21 you will draw a line to point 11, which shows the shape of the bed joint template for the joint 5). Transcription is taken from http://architectura.cesr.univ-tours.fr; translation by the author.

[4] Vandelvira (c. 1585: 29r): *La decenda de cava y la traza pasada viaje contra viaje se trazan de una manera en cuanto al capialzar de las plantas y saltareglas, sólo difieren en que la decenda de cava has de poner a la una parte lo que quieres que haga el arco de decenda que es lo que hay de la A. a la B.* Transcription is taken from Vandelvira and Barbé 1977. Translation by the author.

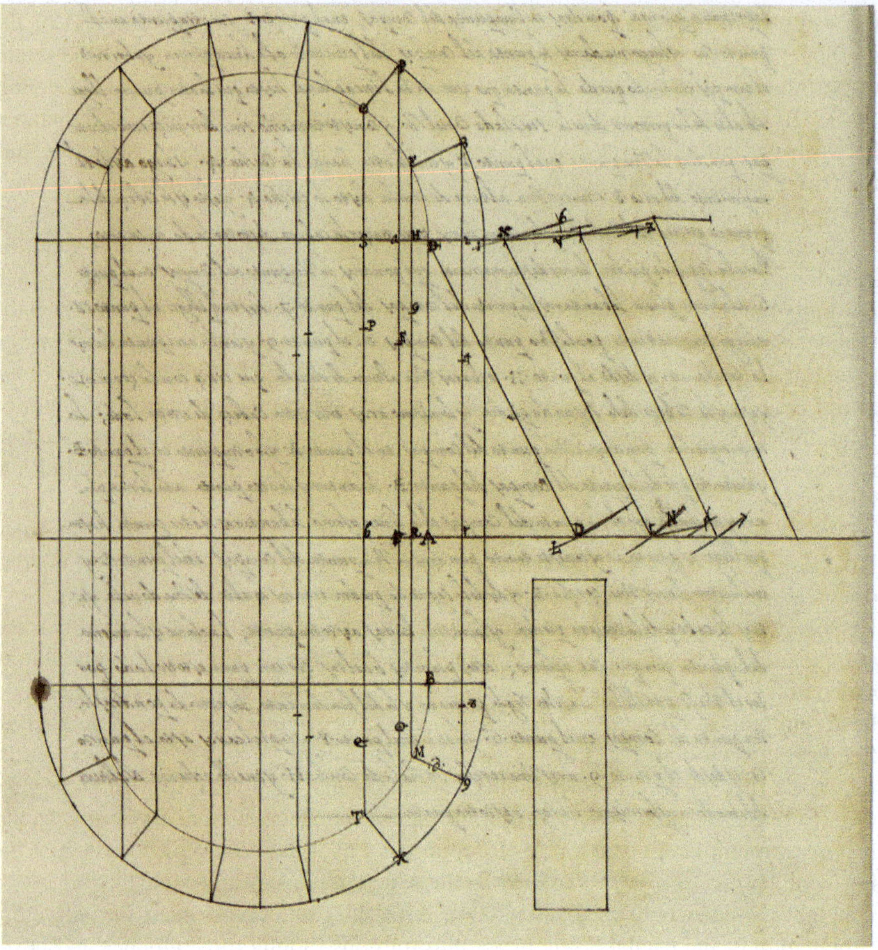

Fig. 8.4 Sloping vault (Vandelvira c. 1585: 28v)

in circular face skew arches. Finally, he makes it clear that since the element is not skewed, templates and guidelines from one side can be reused on the other one.

Jousse comes back to De l'Orme's solution: instead of triangulation, he constructs intrados templates measuring distances to a reference plane (Fig. 8.5). The heading of the first sloping vault in his book (Jousse 1642: 56–57) is *voute d'escalier ou descente de cave droite en demie circonference testes égalles*, which may be loosely translated as "Straight sloping vault for a staircase with a round arch and equal faces". Although the notation is not systematic, and the quality of the woodcut does not help, the essential traits of the procedure can be reconstructed in the light of the next problem in his book, a surbased sloping vault (Jousse 1642: 58–59), or the clearer explanation of Derand (1643: 22–29). Jousse starts by drawing a section through a vertical plane in the shape of a round arch and dividing it into five voussoirs; this departs from

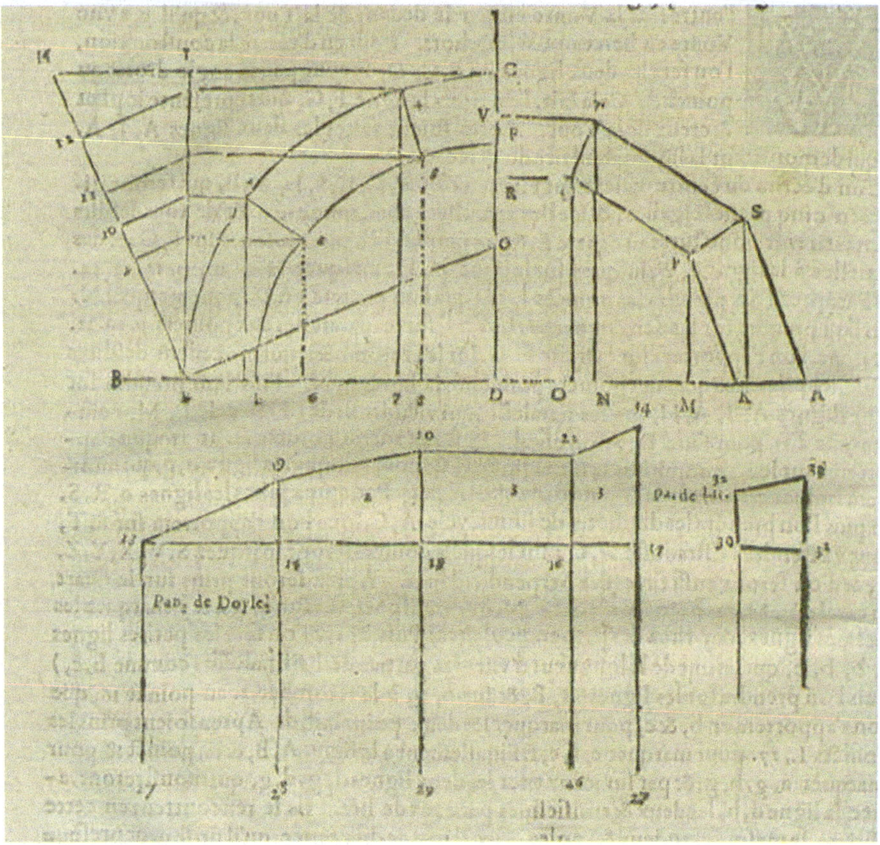

Fig. 8.5 Sloping vault (Jousse 1642: 56)

de l'Orme's apparent choice of a circular section through a plane orthogonal to the axis of the vault. In the next step, Jousse constructs the true cross-section using an auxiliary longitudinal section of the vault; he draws the sloping springing, the vertical plane and a line orthogonal to the springing, which stands for the plane of the cross-section. He then transfers the heights of voussoir corners from the round arch to the vertical plane using horizontal lines; next, he carries these points to the plane of the cross-section using lines parallel to the springing, which stand for the intrados and extrados joints. This operation furnishes the intersection of the intrados joints with the cross-section plane. Using the distances of these points to the springing plane[5] and the horizontal position of the voussoir corners, Jousse constructs the cross-section.

[5] In strictly geometrical terms, Jousse does not really measure the distances from the intersection points to the springing, but rather the length of the projections on a vertical plane passing through the axis of the vault of the segments measuring these distances. The lengths of these projections equal the distances from the intersection points to the sloping plane passing through the springers and the axis of the vault.

Since the intrados joints are parallel to the axis of the vault and orthogonal to the plane of the cross-section, the distances between intersections of intrados joints in the cross-section are shorter than the corresponding ones in the vertical section. In theory, the cross-section should be a surbased half-ellipse; however, Jousse does not bother to draw it and simply represents it as a succession of line segments. Therefore, the intrados and extrados surfaces of the vault are elliptical cylinders, with their longer axes joining the springings, and the shorter ones going from the axis of the vault to its top.

In a second phase, Jousse will construct a series of intrados and bed joint templates. First, he transfers the distances between voussoir corners in the cross-section to a straight line; next, he draws perpendiculars to the line through these points and transfers the distances between the intersections of each intrados joint with the vertical and cross-section planes to these perpendiculars. Joining these points, he obtains the edges of a series of intrados templates; in geometrical terms, he is constructing the development of an oblique section of an elliptical cylinder. The templates are left open at the other end; this suggests that Jousse is deliberately leaving this point undefined, and that the mason may extend the template as necessary since the following voussoirs can be treated as ordinary barrel vault voussoirs. As a final step, Jousse prepares the bed joint templates using the same method, measuring the width of the bed joint in the cross-section and taking the position of its corners along intrados and extrados joints from the longitudinal section, measuring the distance between the intersections of each intrados joint with the vertical and the cross-section planes. Remarkably, Jousse standardises the width of the bed joints. In the round arch, all bed joints have equal widths; however, when projecting them on the cross-section plane, their lengths are different. In spite of that, Jousse uses the same width for all bed joints, as proposed by Martínez de Aranda (c. 1600: 21; see Sect. 6.4.1) for arches opened in curved walls. Clearly, this is a conscious choice for Jousse, since he alludes to *testes égales* (equal faces), in both the headings of the drawing and the text.[6]

Following de l'Orme's example, François Derand (1643: 22–29) opens his treatise with a sloping vault intersecting a larger barrel. However, the other end of the barrel meets a vertical plane, and the author explains the construction of both ends of the templates dutifully. Basically, his method is the one used by Jousse, although his treatment is much better, since the text deals with every minute detail of the procedure, the notation is systematic and the copper engravings much are clearer. As in Jousse, the cross-section is drawn schematically, as a series of line segments, and the width of bed joints is unified.

---

[6] *Voute Descalier ou descente de Caue droitte en demie Circonference testes égalles* for the drawing (Jousse 1642: 56) and *De la descent de la Caue droite à son plain Cintre testes égalles* for the text (Jousse 1642: 58–57).

8.1.4 Skew Sloping Vaults and Other Variants

Vandelvira (c. 1585: 28v-31r, 32v, 35v) presents a skew sloping arch, a sloping arch opened in the intersection of two walls, a skew sloping arch with a large barrel in the low end, a sloping arch in a curved wall, and even a sloping arch with curved springings, resorting systematically to triangulation. Not to be left behind, Jousse (1642: 58–77) includes no fewer than ten variants, introducing surbased sloping vaults and *descentes* opened in counter-battered vaults. Derand (1643: 22–108) raises the bar with 22 variations; it is sufficient to say that the last one is skew, sloping, open in a battered wall on one end and a large barrel in the other, with equal faces; of course, this tour de force is solved by squaring. De la Rue (1728: 104–120) shows more restraint, with only nine variants, solved basically by the same methods.

Most of these variants are based on the skew sloping vault, often with a battered face. As stressed by Sakarovitch (1998: 157–170; 1999) different authors address this problem using different reference systems. De l'Orme (1567: 60r-62v) includes a skew vault abutting a large barrel on one end and a battered face on the other one. He starts by drawing the cross-section of the vault as a round arch and the plan of a skew wall; he adds an intermediate semicircle in the cross-section in order to place the midpoints of bed joints. He also draws two sections of the battered face and the barrel. Next, he draws a number of orthogonals to the plane of the cross-section, standing for the intrados and extrados joints, as well as projection lines for the midpoints of the joints. He constructs the bed joints in true shape taking their widths from the cross-section and transferring these distances to the intersections of the sloping vault with the battered face and the large barrel; this operation furnishes the short ends of the bed joint templates. Intermediate points are used at both the barrel and the battered ends, although in the latter they are not necessary. Thus, as remarked by Sakarovitch (1998: 157–159), he is trying to reduce the problem of the skew sloping vault to that of a barrel vault opened in a battered wall, tilting it until its axis reaches a horizontal position. However, the price of such simple approach is that the face arch, which is not drawn in De l'Orme's sheet, would result in a distorted shape (in fact, an apparently asymmetrical ellipse), resulting from the accumulation of slope, obliquity, and batter.

In contrast, Derand (1643: 48–53; see also Sakarovitch 1998: 376–379) uses a round face arch; as a result, he obtains an asymmetrical ellipse as a cross-section. He first explains the variant with a vertical front face, while the back one is opened on a larger barrel vault; however, his exposition can be easily extrapolated to the case of a battered face. Right at the start, he explains that the vertical section of a sloping round cylinder is a raised half ellipse (he uses the term *demi-ovale*); however, the section through a vertical oblique plane would be an asymmetrical half-ellipse, since the skew cutting plane would meet the springings at different heights. Since this result is not visually pleasing, this design may be trasformed in order to obtain a round face arch. In order to do this, the mason can start from the plan of the vault and the round face arch. Next, he can construct an auxiliary view, projecting the face arch onto a vertical plane passing through one of the springings of the vault;

since Derand is using as an example a vault abutting a larger barrel vault, he also constructs the projection of the resulting lunette on this auxiliary view. In order to do this, he needs to draw the springings, taking into account the difference in heights between its ends. He then sets out to construct the orthogonal cross-section of the vault, starting with a perpendicular to the springers, which are parallel to the axis of the vault. It is evident that the section plane meets the springings at different heights. The cross-section is drawn as a new auxiliary view, using as a folding line an orthogonal to the intrados joints in plan. Derand transfers to one of the springers the difference in heights between the springings; this allows him to draw a slanted line connecting both springers. He then brings the difference in heights between intrados joints and the line connecting both springers in the skew face to the line connecting both springers in the cross-section. In other words, he is reversing his theoretical explanation. In this practical application, the intersections of both springers with the vertical skew face are placed at the same height; however, to reach this height from the cross-section, the left springing must have ascended from a lower position, so the right springing is placed higher in the cross-section. Once he has drawn this asymmetrical cross-section, he can use the distances between voussoir corners to construct intrados and bed joint templates, as on other occasions.

8.2 Groin, Pavilion and L-Plan Vaults

Two intersecting barrel vaults with the same radius and axes set at 90° can cover a square space. The intersection of the two cylinders is given by two ellipses placed at vertical planes standing at an angle of 45° to both axes. Thus, the area is divided into four triangular quarters by the horizontal projections of the ellipses of intersection. Should both vaults span the entire area, the two cylinders would overlap, and each point in the area would be covered by two vaults. This is redundant, of course; thus, both cylinders are trimmed at the intersection curves, some portions of each vault are removed, and each point is covered by a single surface.

In particular, for each of the triangular quarters we may consider two different cylindrical portions. One of these portions has its axis and generatrices placed orthogonally to the base of the triangle, that is, one of the sides of the enclosing square. Thus, the intersection of this half-cylinder with the vertical plane stemming from the side of the square is a directrix of the cylinder, materialised as a round arch. The generatrices of this portion of cylinder go from the round arch to the elliptical intersections. The generatrices of the other cylinder are parallel to the sides of the square; in particular, one of them coincides with the side of the square, acting as a springing. Thus, this second portion is placed below the first one. The rest of the generatrices of this lower cylinder portion join two points, one in each elliptical intersection; as they go up, the segment of each generatrix between the ellipses is shorter.

If we choose to build the vault with the upper portions, that is, those whose generatrices are orthogonal to the sides of the square, the surface intersections will stand out when seen from below; they are called *groins*, and the resulting element is

Fig. 8.6 Groin vault. Ravenna, Mausoleum of Theodoric, lower chamber (Photograph by the author)

a *groin vault* (Fig. 1.14b). Generatrices can be extended beyond the square area, in order to cover a cross-shaped plan, as in a well-known example: the lower chamber of the Mausoleum of Theodoric in Ravenna (see Fig. 8.6). In any case, each quarter of the vault is divided by generatrices into a number of courses, which are, in turn, divided into voussoirs by directrices, just as in a barrel vault. In theory, courses can be cut at the elliptical intersections at an oblique angle, but this is not advisable, since a continuous joint at the intersection may cause cracks. In order to prevent this, the mason can use L-shaped pieces, connecting two courses in adjoining quarters of the vault at the same level.

In contrast, if we keep the lower portions, the result is entirely different. The portions cannot be extended outside the square area. The ellipses do not stand out as a groin but instead are set inside as a crease; the result is known as a *pavilion* or *cloister vault* (Fig. 1.14c). As in the groin vault, most voussoirs are like those on a common barrel vault; only the L-shaped pieces bridging the creases are special ones requiring a different carving method.

There is still another approach to the problem: if we divide the square into *two* triangles using *one* of the diagonals, we can use one of the cylinders for one of the triangles and the other one for the opposite triangle. In each triangle, we should use both the upper and the lower portions of a single cylinder. In two adjacent sides of the square, the edge of the vault will be given by a round arch, while in the other sides of the square, it will start with generatrices. As in the groin vault, the cylinders can be extended beyond the sides of the square, forming a vault with a plan in the shape of an L. This variant was used frequently to solve the junction of two orthogonal barrel vaults, in particular in Romanesque cloisters. As a result, the solution was originally called *arc de cloître* or *rincón de claustro* ("cloister arch" or "cloister corner"); however, the meaning of this term shifted in the eighteenth century to designate the four-lower-portions solution.[7] As far as I know, this variant has no particular name in English so I will call it "L-plan vault".

8.2.1 Square Pavilion Vaults

Alonso de Vandelvira (c. 1585: 79v-80r) includes an element called *Capilla cuadrada por arista*, which may be translated literally as "square groin vault". However, it is evident that the intrados joints in each quarter are parallel, rather than orthogonal, to the side of the square belonging to this quarter (Fig. 8.7; see also Fig. 1.11c); he is therefore explaining a pavilion, not a groin vault.[8] He draws a square plan and a circular cross-section. Next, he constructs an auxiliary view representing the profile of the elliptical intersections, raising perpendiculars to the diagonals through the intrados joints and the midpoints of each voussoir and transferring the heights taken from the cross-section to these points. He surrounds the voussoirs spanning the crease with enclosing squares not only in the cross-section but also in the plan; thus, it is clear that he means to carve the voussoirs by squaring. Furthermore, he explains the dressing procedure for the L-shaped crease voussoir in these terms:

> And to trace the first stone, it will be as wide the distance between the lower bed and the higher one as shown by that marked *B–B* and as high as that marked *C–C* and *B*. Once the stone is dressed in this way, you should take the triangle *E* from the side *D–D* and also from the *D* side on the bed joint the cantilevered portion *F* and from point *r* through point *g* you should transfer the arc *f*, so that the groin adopts the shape of arc *G*.[9]

In other words, Vandelvira starts with a square-base enclosing prism; next, he removes two wedges from the intrados, one for each quarter of the vault, as well

[7]In sixteenth-century Spanish, *rincón de claustro* was used for a reinforcement arch below the intersection of two barrel vaults with orthogonal axes (Martínez de Aranda c. 1600: 85–87). However, in subsequent centuries, its meaning shifted to "pavilion vault" or "cloister vault". In fact, the shift had begun in the late sixteenth century, since Vandelvira (c. 1585: 24r, 24v, 26r) uses it to describe a vaulted element, although not a pavilion vault, but rather the first starting sections of two intersecting barrel vaults. The same semantic shift also occurred in French; see Pérouse (1982: 112).

[8]Palacios ([1990] 2003: 250) includes a drawing of a groin vault before dealing with the *Capilla cuadrada por arista* and including Vandelvira's drawing (c. 1585: 80r) on page 251 of his book. However, it is evident that the continuous joints in his drawing are orthogonal to the nearest side of the square, while the unbroken joints in Vandelvira's drawing are parallel to this side.

[9]Vandelvira (c. 1585: 79v): *Y para trazar la primera piedra, será tan ancha por el lecho bajo y alto como demuestra la señalada con las B.B. y tan alta como la señalada con las C.C. y B. y hecha esta piedra de esta manera, quitarle has por la parte de las D.D. el triángulo E. y por la misma parte de las D. por el lecho bajo el avanzo F. y desde el punto r. al punto g. llevarás la cercha f. y así viene a hacer por el arista la cercha G.* Transcription by Vandelvira and Barbé 1977. Translation by the author.

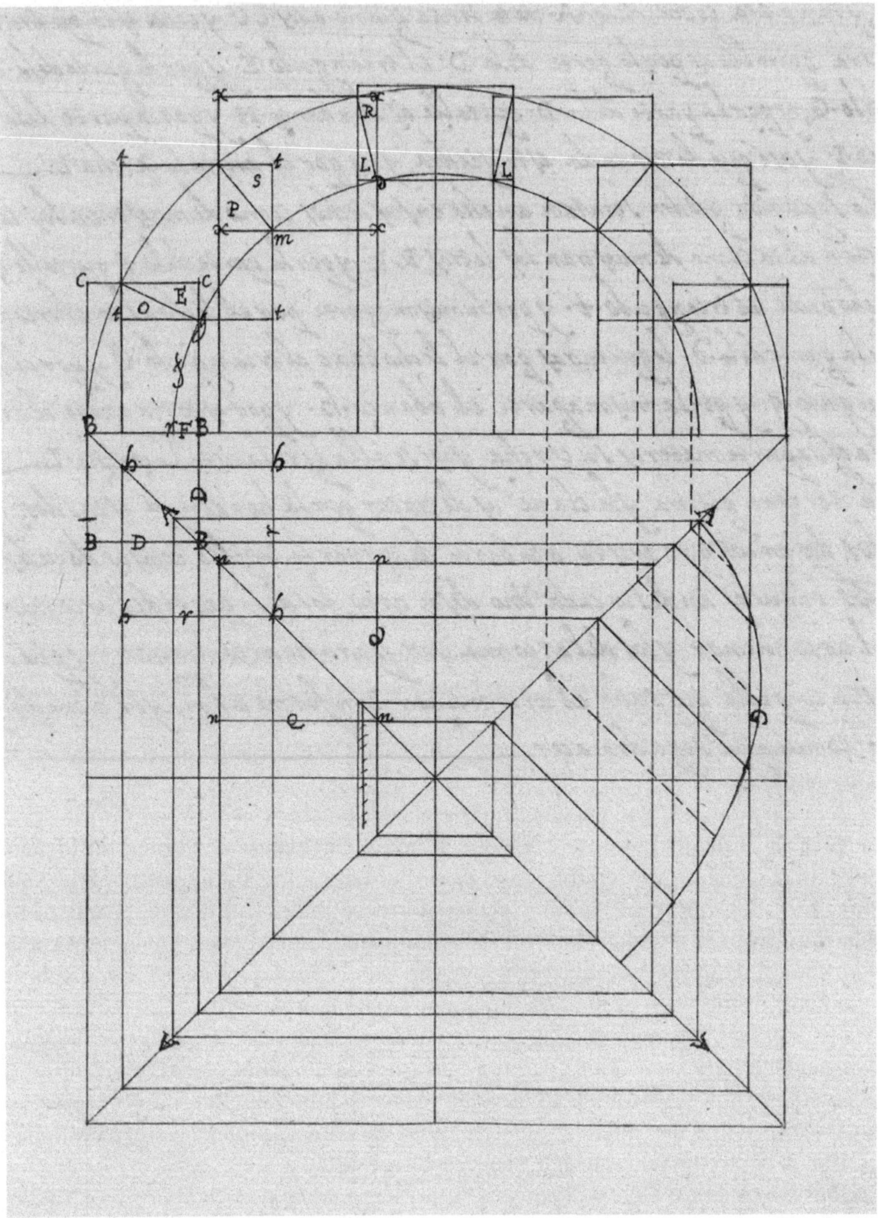

Fig. 8.7 Pavilion vault ([Vandelvira c. 1585: 80r] Vandelvira and Goiti 1646, 132)

as a wedge in the upper bed joint, using as a reference the cross-section. The shape of the groin should be controlled using its true-shape representation in the auxiliary view. The procedure for the second and successive voussoirs is more complex since the mason should take off wedges from the lower and upper bed joints, but it is conceptually identical.

Jousse (1642: 126–127) uses the same method, although the drawing is more schematic and does not show the extrados of the cross-sections, or the full enclosing rectangle (Fig. 8.8). He only includes two lines in the intrados of the voussoir to suggest that the voussoirs are dressed by squaring, as he remarks laconically in the text; there is no reference to templates, in contrast to his groin vault. Derand (1643: 336–338) explains both the square and the rectangular pavilion vault, although the drawing represents the rectangular variant; his method is basically the same as that of Vandelvira, although he stresses the use of templets with the curvature of the cross-section in order to control the dressing of the cylindrical surfaces.

8.2.2 *Square Groin Vaults*

The first appearance of this type of vault (Figs. 1.11b, 8.6) in stonecutting literature is probably a drawing in the manuscript of Alonso de Guardia (c. 1600: 73v) entitled *Capilla por arista* (groin vault), with no explanatory text. It includes the plan of the vault, four round arches at the perimeter, the generatrices and the axes of the piece; two lines are drawn under each course, suggesting the voussoirs are to be carved by squaring.

A much more detailed explanation is given by Jousse (1642: 156–157). He draws the plan and a single round arch; he divides the arch into voussoirs, brings the division to the plan using projection lines, and draws intrados joints for a quarter of the vault (Fig. 8.9). Starting from the intersections of these joints with the diagonals of the plan, he draws the intrados joints for the adjoining quarters of the vault at right angles with the joints in the first quarter. However, this is not really necessary, given the symmetry of the vault around the diagonals, and he draws only a few generatrices just to explain the problem to the reader. Next, he constructs the profile of the groin using an auxiliary view. This operation furnishes a number of points of the elliptical intersection; however, Jousse advises the reader to join them in groups of three; we must surmise that this operation is done with the compass; once again, an elliptical portion is approximated through a succession of circular arcs.

Jousse explains the dressing process using a remarkable method based on folding templates, although at the end of this section he mentions very briefly that the voussoirs can be also be carved by squaring. Jousse seems to be struggling with words, perhaps as a result of the innovative nature of his method, but the issue has been analysed in detail by Pérez de los Ríos and García (2009). As a first step, Jousse draws a line with the length of the first section of the cross-section. Next, he draws two orthogonal lines passing from the ends of the first line, so that one line will exceed the other one by the horizontal distance between the perimeter of the square and the

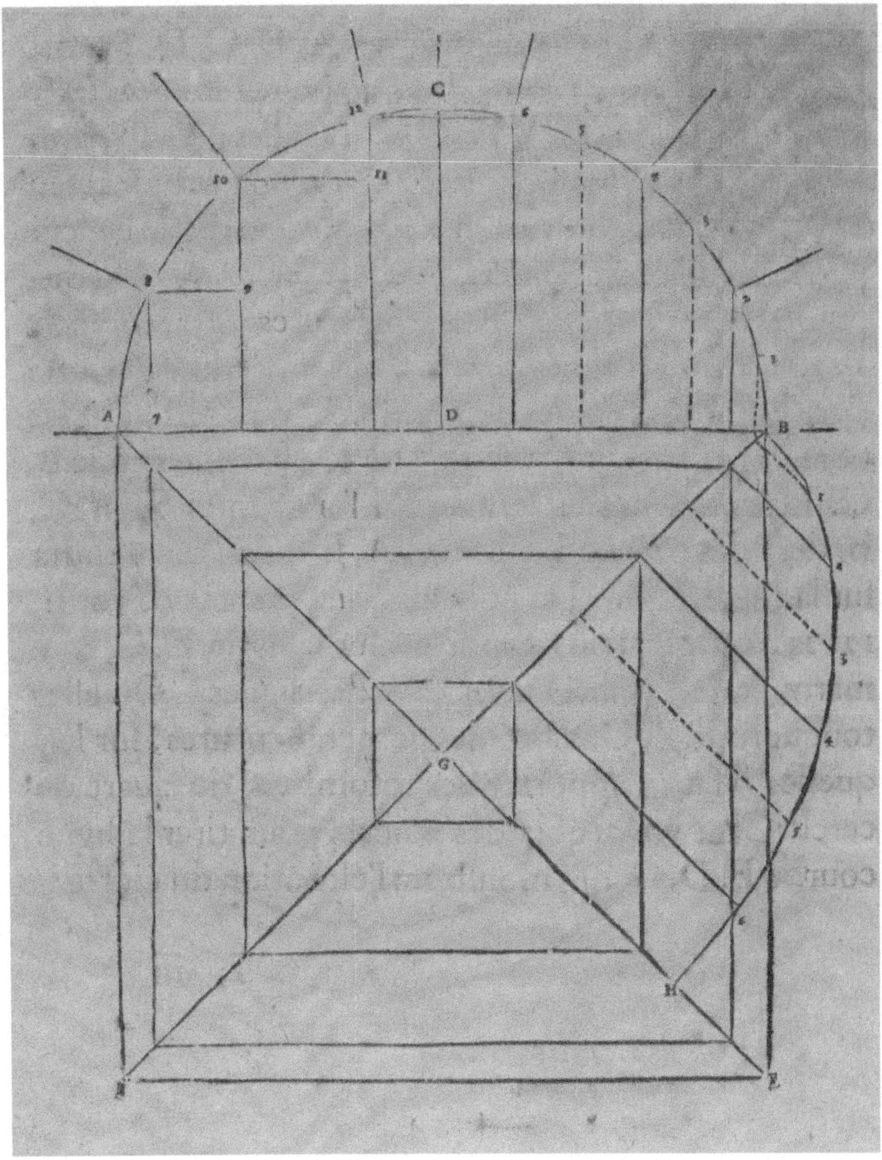

Fig. 8.8 Pavilion vault (Jousse 1642: 126)

diagonal, measured at the first intrados joint. This implies that Jousse is thinking about a cross-plan vault. Thus, the shorter horizontal line will represent the length of one arm of the cross, while the longer one will add the portion of generatrix between the side of the base square and the groin. Next, Jousse connects the ends of both orthogonal lines, obtaining a trapezium standing for an approximate development

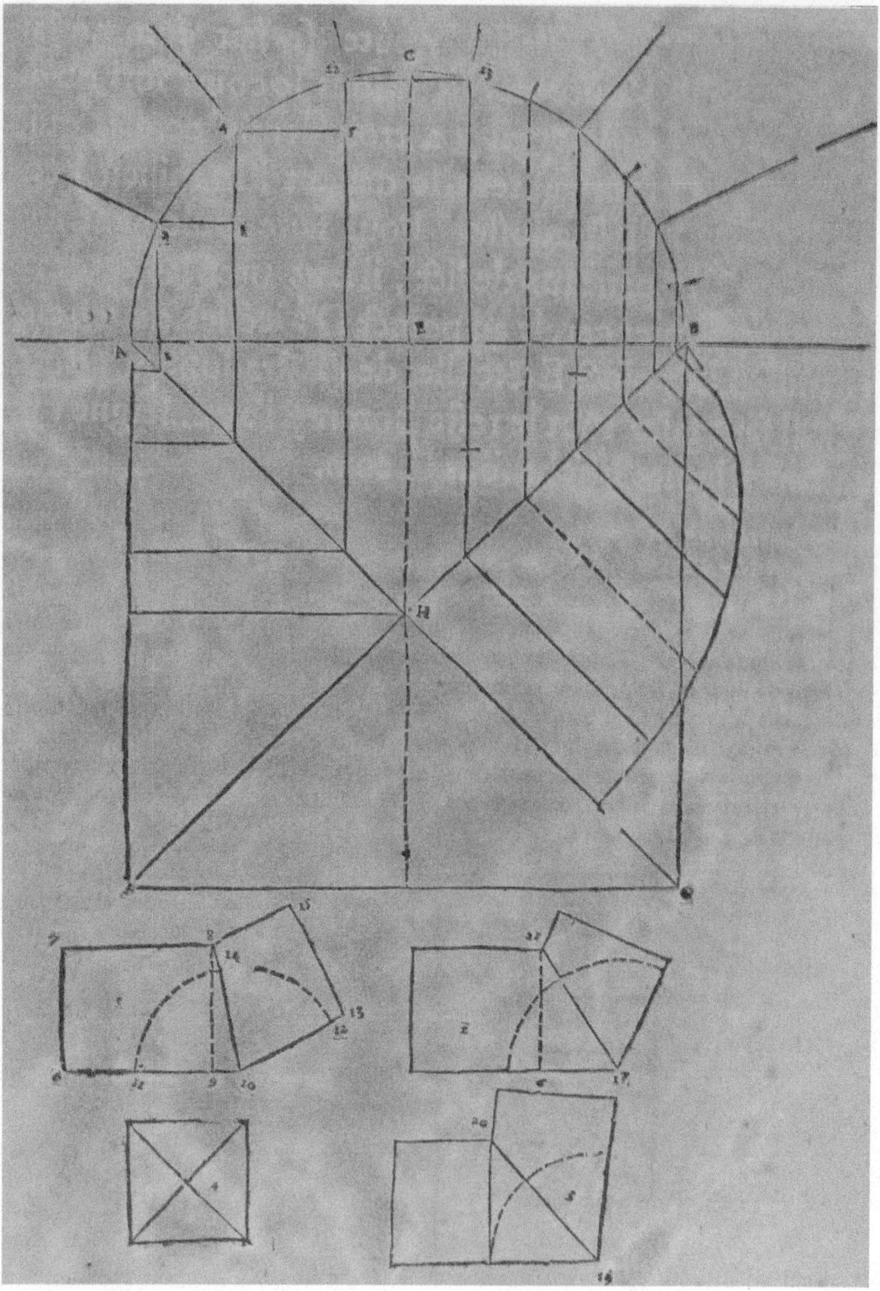

Fig. 8.9 Groin vault (Jousse 1642: 156)

of the course, or rather, the polyhedral surface inscribed in the intrados of the vault. In this development, the orthogonal lines stand for the generatrices; their distance equals that of the intrados joints, taken from the elevation, while their lengths may be taken from the plan.

Next, Jousse endeavours to construct a template for the course of the next quarter of the vault, which stands at a right angle with the one he has constructed, since both courses are symmetrical about the diagonal of the plan. He does not mention symmetry, but he describes in few words a simple and effective construction: “From this point *10* you will make the dashed arc *11 12*. Put the compass point in 11 and on the end of this arc in *14* in order to make *14 12* equal to this distance, then trace the line *10 13* and *8 15*, parallel to *10 13*”[10] In this way, he constructs a template for both courses, as posited by Pérez de los Ríos and García (2009).[11]

As remarked by these authors, Jousse does not say a word about the actual dressing procedure, probably taking it for granted that any mason would understand it. Pérez de los Ríos and García surmise that the two trapeziums of the template are used jointly as a folding template.[12] Since nothing suggests that Jousse knew the notion of dihedral angle, these authors assume that the mason should fold the template until the intrados joints of the two quarters form an angle of 90°. Such a hypothesis is quite brilliant; however, a simpler alternative interpretation, supported by Martínez de Aranda’s explanation for the L-plan vault, which we will see further on, may be put forward. The mason should dress the voussoir by squaring, starting from the side joints, taken from the elevation, and materialising the generatrices with the aid of a square. The groin will appear automatically at the intersection of the two cylinders; however, in order to improve the execution of the groin, the mason may use the folding template, applied to the intrados of the voussoir; the groin should coincide with the crease in the template.

In any case, Jousse’s method is somewhat cumbersome, and other writers follow different routes. Derand (1643: 329–335) eschews the simple groin vault and starts with the rectangular-plan groin vault, making it clear that the square-plan variant should be treated as a particular case. However, for the sake of clarity and consistency, here I will deal with the square-plan vault, coming back to the rectangular one in a few pages.

---

[10] Jousse (1642: 157): *Duquel poinct 10 sera faite la portion de cercle poncté 11 12. Portez le compas en 11 & sur ladite portion de cercle en 14 pour faire 14 & 12 son égal, pour tirer la ligne 10 13 et 8 15 parallele a 10 13.*

[11] Jousse does not mention the use of a folding template explicitly and, at first sight, his diagram can be interpreted either as a folding template or as a set of two independent templates, one for each quarter of the vault. However, if Jousse meant to use two different templates for both portions of the vault, he could it have used the same template turning it around, as in many other symmetrical vaults. See for example Jousse (1642: 20, 30, 36, 38).

[12] Jousse does not mention explicitly folding templates. However, he draws two symmetrical trapeziums joined by their oblique sides. These are not, in all evidence, independent templates, since many stonecutting theorists advise readers to reuse templates for symmetrical shapes; In contrast, Jousse explains in detail how to construct the second trapezium starting from the first one. Thus, the templates should be used together on two intersecting cylinders, so it may be surmised that they are to be folded, as shown graphically by de la Rue (1728: pl. 24 bis).

Derand's setting out method is identical to that of Jousse for all practical purposes. However, he dresses the voussoirs by squaring, like Guardia; the procedure has points in common with the one used by Vandelvira for the pavilion vault. It is implied that most voussoirs are simple barrel vault pieces, and thus Derand explains only the dressing of the ones that cross the groin, forming an *enfourchement*, which means literally "forking", probably because all *enfourchements* stem from the keystone or central voussoir. For the first voussoir in the forking, the mason should start preparing a block with a horizontal bed and two orthogonal vertical faces. Next, he should measure in the tracing the height of the voussoir as well as the horizontal projection of the cantilevered portion, at both sides of the diagonal, transferring these measures to a block. Then he should use arch squares to draw the cross-sections on the faces of the block. This allows him to dress the intrados of the voussoir as two cylindrical portions; since the generatrices are orthogonal to the faces of the block, they may be controlled with an ordinary mason's square. In contrast to the pavilion vault, the mason may happily carve one of the cylindrical portions and then go on with the next one, since the groin stands out; the surfaces will intersect forming the groin "by a kind of chance", in Frézier's (1737–39: II, 13–14) words. The vertical faces of the block, already dressed, will provide the side joints, while the bed joints are to be carved with the straight arm of the arch square. As with the pavilion vault, the second voussoir is more complex, since it includes two bed joints, but it may be executed using the same method.

De la Rue (1728: 44–46; see Bortot and Calvo 2019) also starts with the rectangular-plan groin vault, remarking that the solution includes the square-plan vault. He uses Jousse's folding templates, but the dressing method is quite different: it involves the computation of the dihedral angle between both templates (Fig. 8.10, details in Figs. 8.11 and 8.12). This procedure is highly abstract, at least in comparison to the empirical approach used by most preceding treatises, except for Desargues, and prefigures the descriptive geometry that will arise at the end of the century.

Each of the planes forming the dihedral angle passes through the upper and lower intrados joints of a branch of a "forking" voussoir; the intersection of the two planes is the chord of the groin portion belonging to this specific voussoir. De la Rue measures the angle using the intersection of the two planes with an auxiliary plane orthogonal to the chord. The operation is performed in a cross-section through the plane of the groin, which shows the chord *N–O* in true shape; since the auxiliary plane is orthogonal to the chord, it is shown in edge view (De la Rue 1728: pl. 24, upper section; see Fig. 8.11). This simplifies the construction to a great extent, but visually it is not exactly intuitive. The procedure is explained in nineteenth-century descriptive geometry manuals, but de la Rue's application to this particular case is quite ingenious. First, he draws the auxiliary plane in the elevation as an orthogonal to the chord *R–S* placed at will. Since intrados joints are horizontal and coplanar, the author can determine easily their intersections with the auxiliary plane: the auxiliary plane intersects the horizontal plane of the intrados joints at a horizontal line shown in the cross-section as a line *S* in point view. This line is represented in plan as *S–Y*; *Z* and *Y* give its intersections with the intrados joints. Up to this moment, de la Rue knows three points of the intersections of the auxiliary plane with the planes

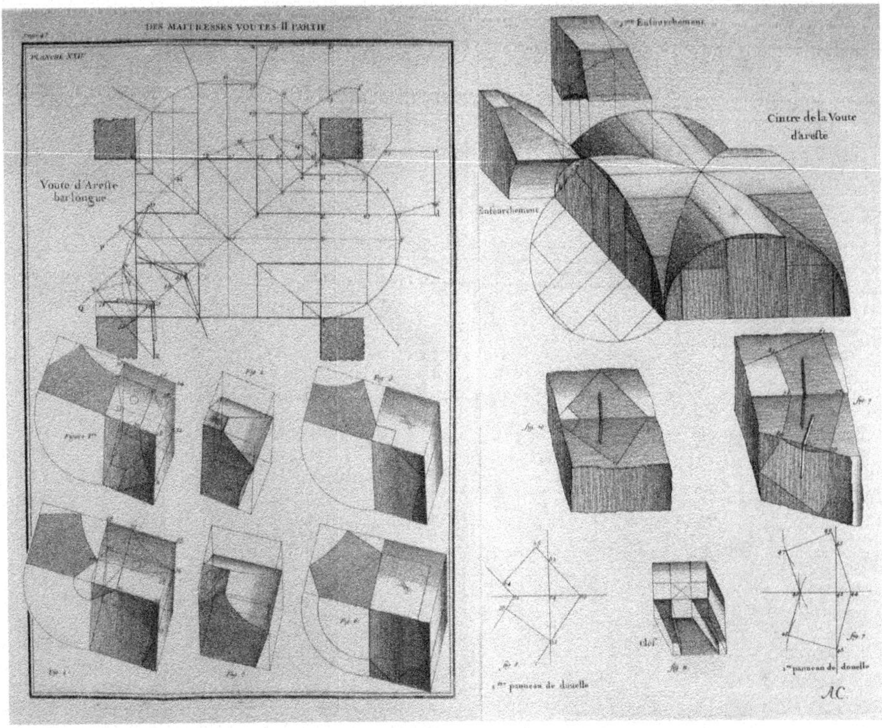

Fig. 8.10 Groin vault (de la Rue 1728: pl. 24, 24 bis)

of the dihedral angle, placed at the chord *R* and the intrados joints, *Z* and *Y*. The angle *Z–R–Y* stands for the dihedral angle; however, de la Rue needs to depict it in true shape. To do this, he can rotate the triangle *Z–R–Y* around the horizontal line connecting the intersections of the intrados joints and the auxiliary plane, in order to bring *R* to the horizontal plane passing through the intrados joints. *Z* and *Y* belong to the horizontal line, which acts as the axis of a rabatment, so they will not move in the revolution. As for *R*, its position after rotation may be determined by drawing an arc with its centre at *S* until it reaches the horizontal plane *P–O*; next, this unnamed point is transferred to the plan to locate point 41; *Z-41–Y* will provide a measure of the dihedral angle.

In addition to furnishing the dihedral angle, this triangle is reused (1728: pl. 24bis, lower section; see Fig. 8.12) in a new method for the construction of Jousse's folding templates. Since the auxiliary plane is orthogonal to the chord, the intersections of the faces of the polyhedral surface with the auxiliary plane *Z-R* and *R-Y* also be perpendicular to the chord. Thus, de la Rue will rotate the faces of the polyhedral surface around the chord, as Jousse did. However, instead of drawing first one face, then the other, he will start drawing two lines, *50–51*, standing for the chord and *53–54–52*, representing the intersection of the polyhedral faces with the auxiliary plane. He can measure in the rabatment the length of the intersections of the auxiliary

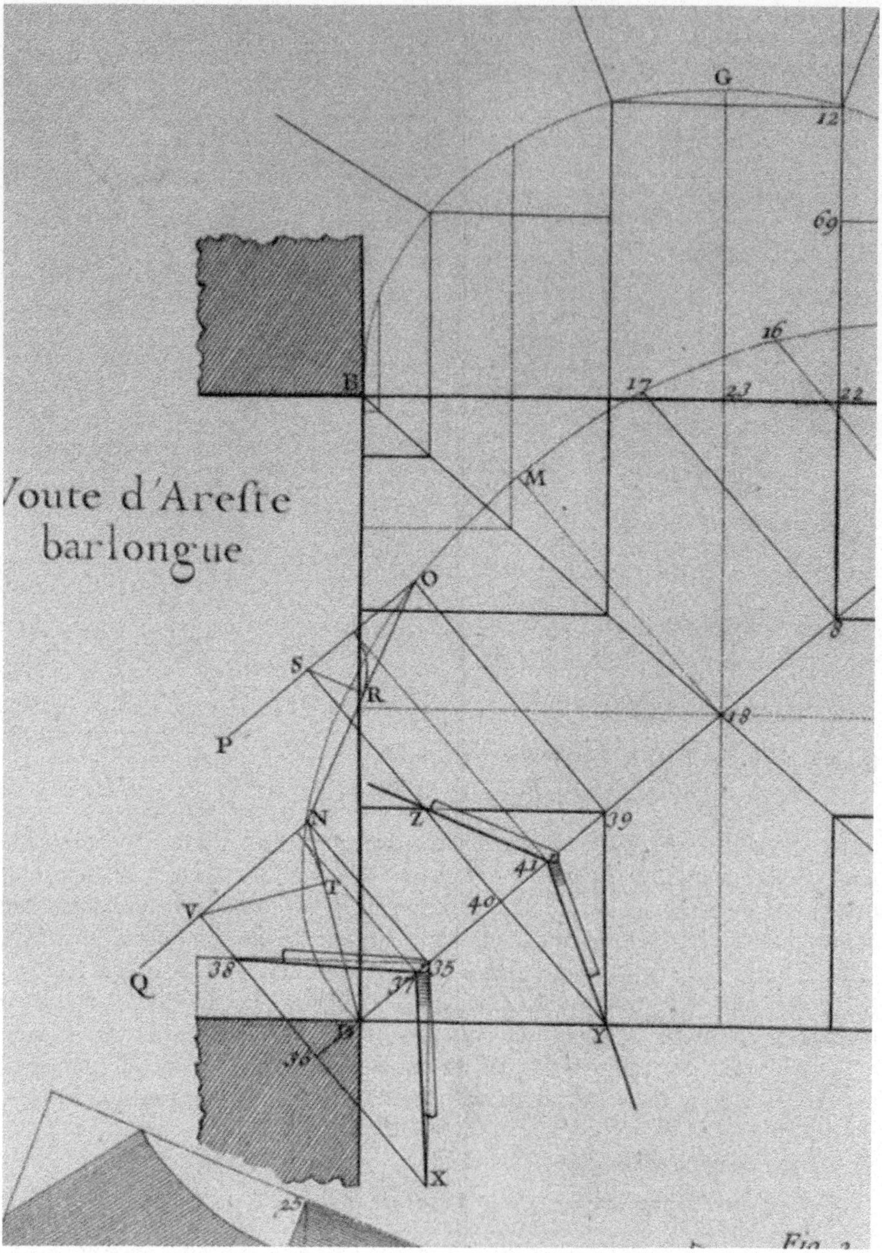

Fig. 8.11 Groin vault, determining the magnitude of the dihedral angle, detail (de la Rue 1728: pl. 24)

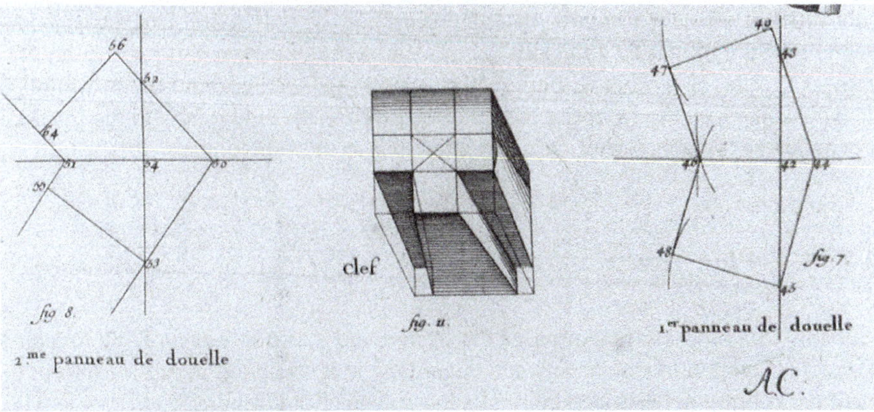

Fig. 8.12 Groin vault, construction of the folding templates, detail (de la Rue 1728: pl. 24 bis)

plane and the planes of the polyhedral surface *Z–41* and *41–Y* and bring them to the orthogonal to the chord, locating two points, *53* and *52*, in the upper intrados joints. This operation amounts to flattening the dihedral; accordingly, the triangle is transformed in two collinear segments. By joining points *50* and *52*, the mason may determine the direction of one of the intrados joints; measuring its length in the plan, he may place point 56, standing from the end of the intrados joints. Next, he can repeat the operation for the other intrados joint, if necessary;[13] he will end with the full intrados joints and three corners of the voussoir. He can then draw orthogonals to the intrados joints, *56–64* and *53–55*, standing for the side joints; drawing orthogonals to these joints from *51*, he may draw the *lower* intrados joints, closing the folding template.

After dealing with the construction of the angles and templates, de la Rue explains in detail the dressing procedure. First, the mason should materialise a flat face; next, he will dress a second flat face so that the angle between both planes equals the dihedral angle he has computed in the first phase of the tracing procedure. Of course, the two planes stand for the faces of the polyhedral inscribed in the intrados of the vault. The mason can then apply a folding template on both faces; the crease in the template should coincide with the chord of the groin. Each portion at both sides of the crease should be laid on one of the planar faces; this will allow the mason to score the outline of the template on both faces. Next, the mason can dress the side joints dividing the "forking" voussoir from the ordinary barrel-vault voussoirs. These sides are orthogonal to the intrados joints, and thus perpendicular to the polyhedral faces since they pass through intrados joints. Thus, the dressing of the side joints can be

[13]De la Rue has placed the auxiliary plane passing through one of the corners of the voussoir, namely *Y*. This detail simplifies construction, avoiding the location of a new point opposite point 56. In the square-plan case the mason may force the auxiliary plane to pass through both ends of the upper intrados joints, relying on symmetry about the diagonals; in this case, point 56 would coincide with 52, leading to further simplification.

controlled easily with a square, placing one arm on the polyhedral face and the other one in the side joint. Once the side joint is dressed, the mason can mark a template with its profile on it. This will allow him to dress the intrados and the bed joints of the voussoir, since the latter are planes orthogonal to the side joint and the former is a cylinder whose generatrices are perpendicular to the side joints.

8.2.3 L-Plan Vaults

Probably, the oldest explanation of the design and execution of an L-plan vault is a drawing, without a title or any accompanying text, included in MS 12.686 in the National Library in Madrid, a copy of a lost manuscript attributed to Pedro de Alviz (c. 1544: 25; see also García Baño 2015; García Baño 2017: 468–481, 735–767). The author draws a rectangular plan and a cross-section in the shape of a round arch; all voussoirs are surrounded by an enclosing rectangle, hinting that the piece is to be dressed by squaring. He also includes a true-shape representation of the elliptical intersections, obtained through an auxiliary view. As in other cases, we will see the implications of the rectangular plan under a separate heading.

Vandelvira (c. 1585: 25r) offers a solution based on the same method used in his pavilion vault. He draws a square plan, although the method can be applied to two intersecting, orthogonal barrel vaults of arbitrary length (Fig. 8.13); next, he draws the cross-section and advises the reader to start from an enclosing prism, taking off two wedges under the intrados and another pair of wedges over the bed joint. It is worthwhile to remark that although Vandelvira calls the element "cloister corner", he places it among arches and actually talks about "this arch". This shows that the word "cloister", in this context, referred originally to an L-shaped vault, and only in the seventeenth century was associated to a pavilion vault (see Pérouse [1982a] 2001: 111–112); of course, L-shaped vaults were useful at the intersection of two barrel vaults in Romanesque cloisters.

In contrast, Martínez de Aranda (c. 1600: 85–87) treats this element as a real arch (Fig. 8.14). He uses two orthogonal cylinders, as Vandelvira. However, he cuts them by two vertical planes parallel to the intersection of both cylinders so that the piece can be used as a reinforcement below the junction of two vaults in a cloister. As we have seen in Sect. 3.2.3, he dresses the voussoirs by squaring, starting with a complex prism in the shape of the plan of the voussoir. He then cuts four wedges from each side to materialise the bed joints, the intrados and the extrados. He stresses that a crease appears around the intersection of both cylinders in the outer half of the arch, as in pavilion vaults, while the other half resembles a groin vault.

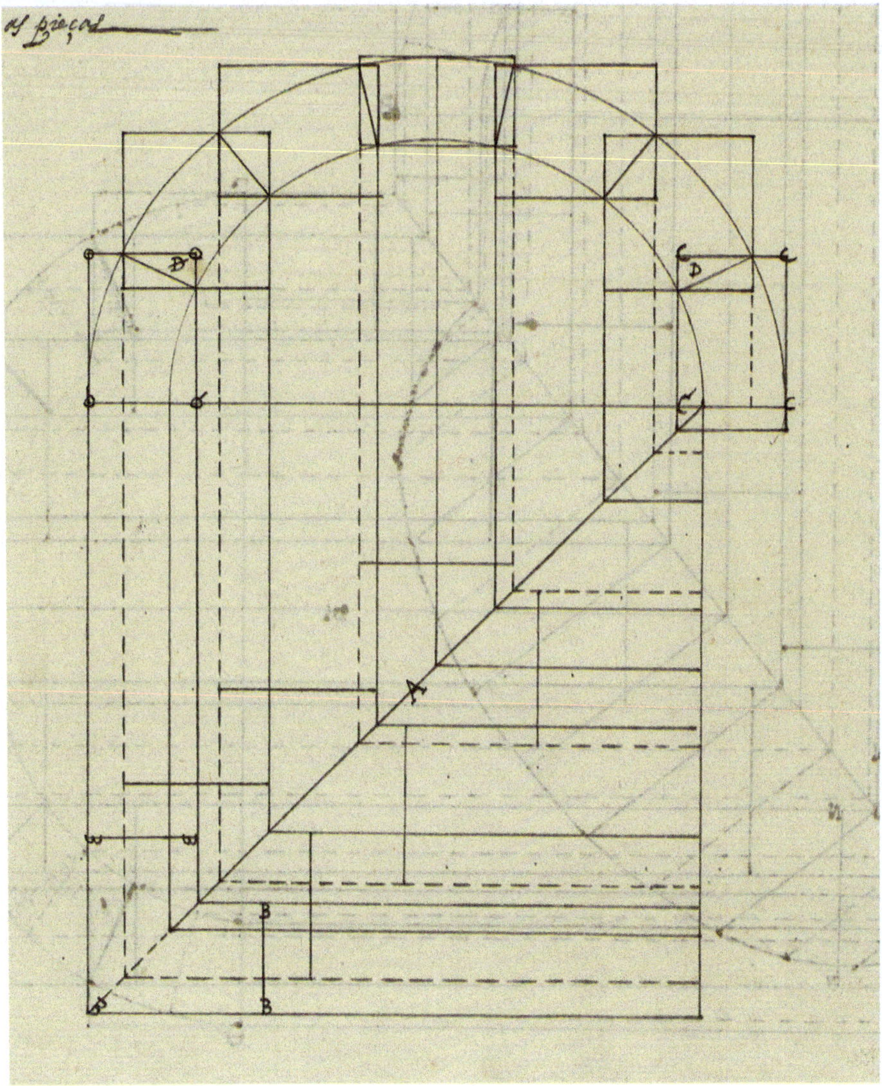

Fig. 8.13 L-Shaped vault ([Vandelvira c. 1585: 25r] Vandelvira/Goiti 1646: 40)

8.2.4 Rectangular-Plan Groin, Pavilion and L-Shaped Vaults

If we attempt to cover a rectangular area with a groin vault, the cylinders will have different radii and heights, and their intersections will not be a pair of ellipses, but rather two non-planar curves: such a solution is called *cylindrical lunette*. We will deal with this problem further on (see Sect. 8.4.2), but there is an alternative approach: the mason or architect may manipulate the height of one or both cylinders so that their

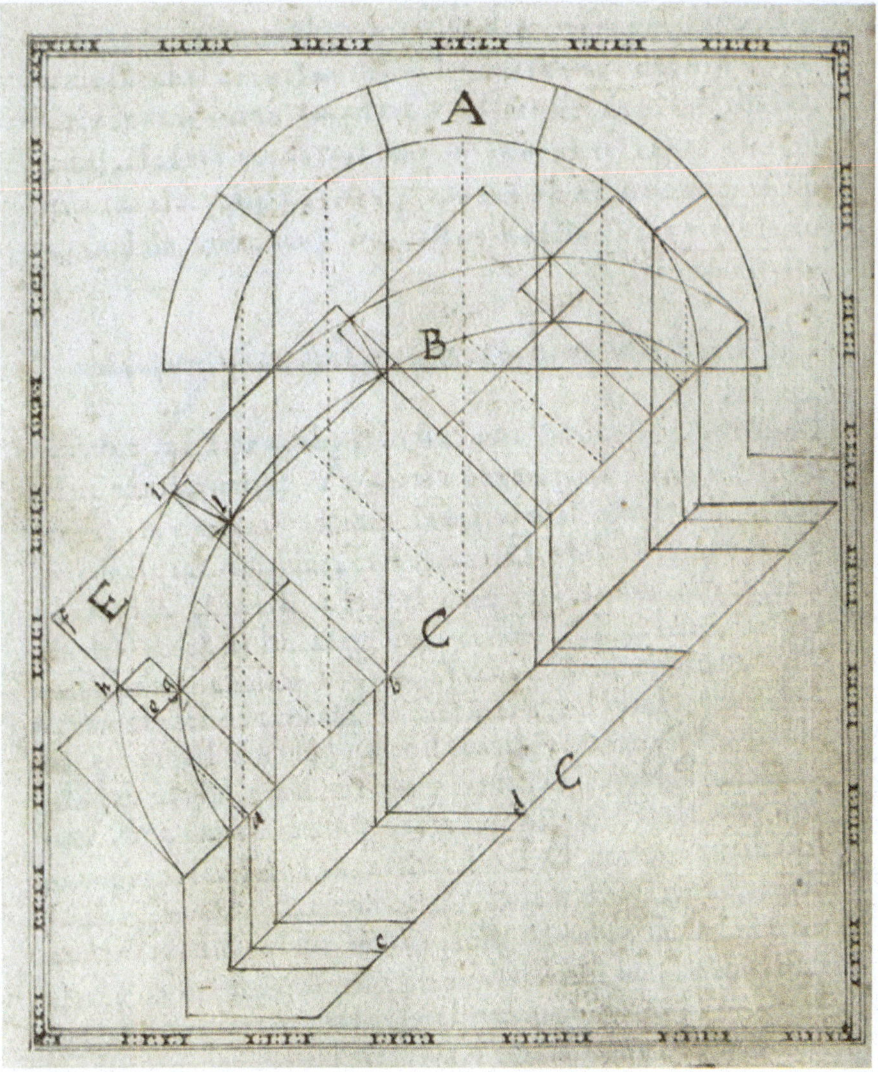

Fig. 8.14 Reinforcement arch for an L-shaped vault (Martínez de Aranda c. 1600: 86)

upper generatrices are placed at the same height. In other words, elliptical cylinders should be substituted for one or both of the circular cylinders.

This alternative appears in the early manuscript attributed to Alviz (c. 1544: 23; see also 21 for a simple L-shaped vault). In its simplest form, a drawing without title or text shows a rectangular-plan groin vault. Both elevations are drawn; while the shorter one is a round arch, the larger one is a surbased elliptical arch (García Baño 2017: 494–503). The drawing also includes the intrados joints in plan, a true-shape representation of a groin and full enclosing rectangles, suggesting that the voussoirs that cross the groin are to be carved by squaring.

Although there is no explanatory text, the elliptical arch was probably traced using intrados joints. In a first step, the mason can draw projection lines from the corners of the voussoirs of the round arch and extend them until they meet the groins in plan; this will furnish the horizontal projection of the longitudinal intrados joints. Next, the transversal intrados joints can be drawn at right angles with the longitudinal ones, starting from their intersections with the groin. Such lines are also the projection lines of the corners of the voussoirs of the elliptical arch. Thus, the side elevation can be drawn by transferring the heights of the voussoir corners from the front elevation to the corresponding projection lines, since intrados joints are horizontal and therefore keep the same level from one elevation to the other, passing through their intersection with the groin. This is an efficient and straightforward construction, but Alviz makes a mistake when drawing the extrados joints. He presumes that the width of bed joints placed at the same level will be equal; this is true in square-plan groin vaults, where the elevations are identical round arches, but does not hold for the elliptical arches of rectangular-plan groin vaults.

In addition to these simple groin and L-shaped vaults, Alviz (c. 1544: 22; see also García Baño 2017: 468–475, 482–492) presents a remarkable tour de force (Fig. 8.15): an oblique L-plan vault with different spans and elliptical sections. He starts with a regular barrel vault and cuts it by an oblique plane, obtaining an elliptical intersection by transferring the heights of the corners of the voussoirs. Next, he builds an elliptical-section barrel vault starting from the intersection, using the height transferring method again in order to construct the biased end of this second vault. In both cases, he repeats the mistakes in the construction of the extrados; he is not consistent about the construction of ellipses, using ovals in some sections.

Jousse (1642: 158–159) includes a rectangular-plan groin vault (Fig. 8.16), together with a laconic text saying that everything is to be solved as in the square-plan groin vault. The tracing procedure is the one used by Alviz, including the mistake when constructing the extrados; Jousse's main contribution is the addition of folding templates, as in the square groin vault. De la Rue (1728: 44–46) uses the same method as in the square-plan vault; in fact, he does not bother to explain the square piece, considering it a particular case of the rectangular one.

In contrast, Derand (1643: 330–335, 344–348) reverses the problem. He does not lower the widest barrel; instead, he pulls up the narrow one (Fig. 8.17), using a raised elliptical vault, as he did in the Parisian church of Saint-Paul-Saint-Louis (Fig. 8.18). In other words, the quarter in front on the short side of the area is covered by a portion of a raised elliptical cylinder, while the one near the long side is spanned by a round one. The author starts by drawing the plan and the circular cross-section; next, he divides the section into even portions and carries them to the plan, drawing the intrados joints for one quarter and marking their intersections with the diagonals. He then draws the intrados joints for the opposite quarter as orthogonals starting from the diagonals; he extends them beyond the side of the area, using them to construct the elliptical cross-section, transferring the height of each intrados joint, since all of them should be horizontal. He uses the same technique to draw the profile of the

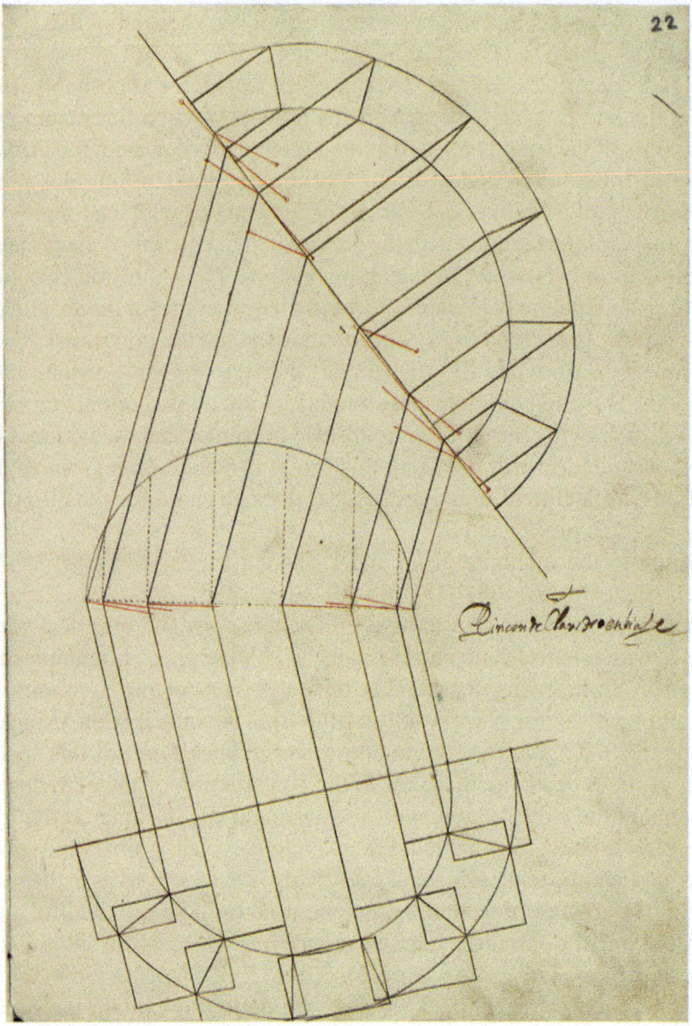

Fig. 8.15 L-shaped vault (Alviz, attr. c. 1544: 22r)

groin, raising perpendiculars from the intersections of the intrados joints with the diagonals and carrying the heights of the joints; in any case, he explains that this profile is essential when building this vault in brick or rubble, but not as much when using ashlar.

Fig. 8.16 Rectangular-plan groin vault (Jousse 1642: 158)

8.2.5 Other Groin and Pavilion Vaults

Stonecutting authors also explain other groin and pavilion vaults featuring a wide variety of plans. Vandelvira (c. 1585: 102v-103r, 104r-107r) specialises in octagonal pavilion vaults; since he divides courses radially and this approach poses different

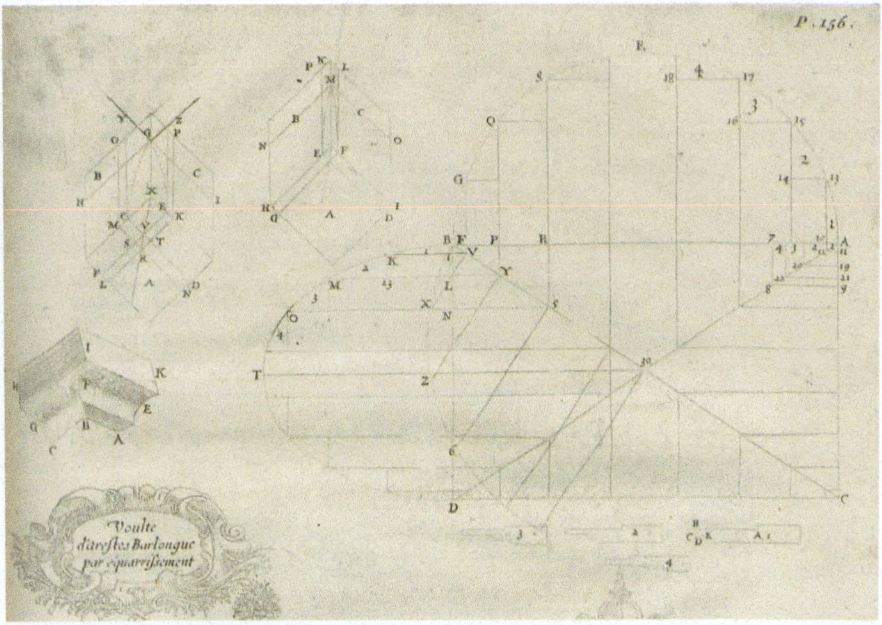

Fig. 8.17 Rectangular-plan groin vault (Derand [1643] 1743: pl. 156)

problems, we will deal with this problem in Sect. 8.3.[14] Jousse (1642: 166–160–163, 166–167, 172–173) includes groin vaults with pentagonal, triangular, and rhomboidal plans, as well as a hexagonal pavilion vault. In these cases, he follows the usual procedure, constructing true-shape representations of groins, creases, as well as an elliptical side elevation in the skew vault, and dressing the voussoirs by squaring; he makes no mention of his trademark folding templates.

Derand (1643: 336–354), includes groin vaults in the shape of rhomboids, hexagons, and equilateral, right and irregular triangles, "or any other figure, regular or irregular, as you wish".[15] An interesting addition is that of surbased or raised vaults. In the examples we have seen so far, operations started with a round arch; groins, creases or side elevations in rectangular or skew vaults adopt the shape of ellipses, constructed using intrados joints and joined with circular arcs. Generalising this procedure in raised or surbased vaults, Derand (1643: 338–340, 342–344, 351–354) starts with semiellipses as cross-sections and constructs groins, creases or side elevations by the same method, obtaining semiellipses with different proportions as a result. Although he is the first one to explain this issue systematically, the idea of constructing an elliptical section as a transformation of another one seems to have been used empirically by masons since at least the mid-sixteenth century, as shown

[14]Further, Vandelvira includes some spherical intrados vaults that may resemble pavilion vaults, either in name or form (c. 1585: 103v, 107v-108r, 126r).

[15]Derand (1643: 342): … *ou telle autre figure reguliere our irreguliere, que l'on voudra.*

Fig. 8.18 Rectangular-plan groin vault. Paris, church of Saint-Paul Saint-Louis (Photograph by the author)

by a drawing in the manuscript attributed to Pedro de Alviz (c. 1544: 22), as we have seen.

De la Rue (1728: 47) shows more restraint; he includes only skew and pentagonal groin vaults, remarking that the angles between intrados joints in the skew vault should be measured with the bevel, rather than the square. He adds laconically that the stones along the groins should be dressed using the same method as in the ordinary groin vault, that is, the sophisticated procedure involving a computation of the dihedral angle between the faces of the inscribed polyhedral. Later on (1728:

54–56), he includes a quite interesting element, the *Voûte d'arête en tour ronde*, literally "groin vault in a round tower". Actually, this is an annular vault penetrated by radial sections with conical intrados; its similarity with an ordinary groin vault is topological, rather than geometrical, so I will analyse it in Sect. 4.6.6.

8.3 Octagonal Pavilion Vaults

In theory, the scheme of a groin or pavilion vault can be doubled in order to cover an octogonal plan: four intersecting cylinders may be laid with their axes set at 45° degrees from one another; taking eight portions, either the upper or lower ones, we can build a groin or pavilion vault, respectively. Octagonal groin vaults are not frequent in traditional stonecutting, although there are interesting examples built in other materials during the twentieth century. In contrast, octagonal pavilion vaults are more frequent, from Roman examples such as the well-known hall in Nero's Domus Aurea in Rome to the Romanesque ceiling of the Florence Baptistery. However, these examples are built in concrete or brick; in sixteenth-century Spain, this architectural type was translated to stone, as were other Renaissance constructive solutions. In square-plan pavilion vaults, side joints are orthogonal to intrados joints, as we have seen; as a result, most voussoirs are ordinary barrel vault voussoirs, and the specific stonecutting problem is limited to the stones placed along the groin or crease. In contrast, in octagonal pavilion vaults, side joints are usually laid out radially; this justifies a specific section for these vaults.

Vandelvira (c. 1585: 102v-108r) presents a number of different solutions for *ochavos*, a Spanish word meaning "eighth", used for octagonal vaults, portions of them, or elements fulfilling their role. Some of them are full-fledged vaults, used here and there in sacristies to span octagonal areas, either regular or irregular. Others are actually half-octagons, used frequently in the chancels of parish churches. This problem can also be addressed with the *Ochavo de La Guardia*, a spherical-intrados vault built using a network of ribs; we will see this variant in Sect. 10.2.2.

Vandelvira (c. 1585: 102v-103r; see also Palacios [1990] 2003: 326–329) explains the problem in detail for the *Cabecera ochavada*, that is, a half-octagon suitable for the sanctuary of small churches (Fig. 8.19). He starts by drawing the plan, in the shape of a half-octagon, although he makes it clear that the mason should "trace the plan as befits the building".[16] In fact, in both copies of his manuscript, the central side is clearly larger than the lateral sides, which are approximately symmetrical; the final sections of the plan are approximately equal to half the lateral sides. It is not clear whether Vandelvira is using deliberately an irregular, although symmetrical, semi-octagon in order to enhance the central section, which usually houses the main section of the altarpiece, including the tabernacle. This design option is quite appropriate in liturgical terms, but it poses a geometrical problem, as we will see later on.

---

[16]Vandelvira (c. 1585: 102v): ... *trazada la planta como más convenga al edificio* ....

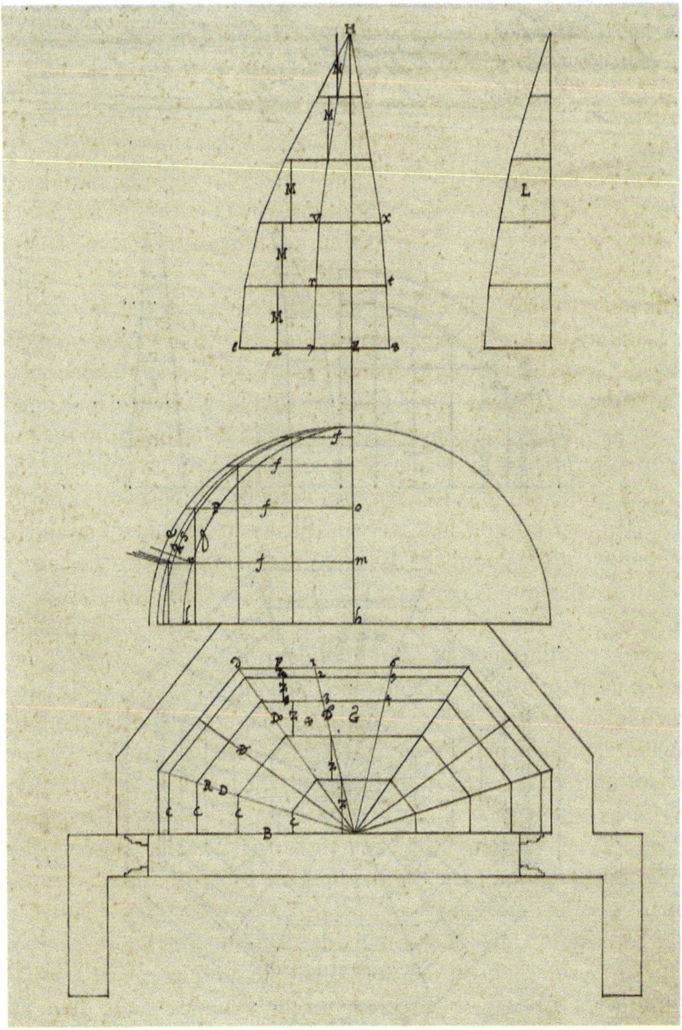

Fig. 8.19 Half-octagon pavilion vault ([Vandelvira c. 1585: 102v] Vandelvira/Goiti 1646: 164)

Vandelvira also draws lines from the midpoint of the line joining both ends of the half-octagon to its vertexes; such lines stand for the creases between cylindrical portions. Next, he draws a cross-section through the front plane of the element, that is, the vertical plane passing through the centre and the end of the perimeter. The cross-section is in the shape of a quadrant; since the front plane is orthogonal to the generatrices of the end sections, the final severies are round cylinders. He then divides the cross-section into equal parts, furnishing the position of the intrados joints, which are drawn in plan as parallels to the perimeter, changing direction at the creases. Next, he divides each side of the perimeter into two or three parts, again drawing lines to the centre of the figure, which stand for the side joints of the voussoirs. As a result of this construction method, the horizontal projections of the creases and the vertical planes passing through the side joints are different. Vandelvira is aware of this fact and constructs the cross-sections painstakingly by each of these planes, taking the heights of the intersections of the creases or side joints and the intrados joints from the quadrant cross-section and the horizontal distances between intrados joints from the plan. As a result, the intermediate portions, which are symmetrical with the end ones, are portions of a right cylinder; in contrast, the central severy is part of a raised cylinder.

All this is no more than a preparatory operation in order to construct the intrados templates, standing for the faces of a polyhedral surface inscribed on the intrados of the vault, as Jousse or de la Rue will do later. However, Vandelvira does not attempt to use folding templates; instead of grouping two templates in the same course on both sides of the crease, he prepares a series of templates starting from the lower course and going up to the top of the vault. Such templates can be constructed using two alternative procedures. In the simplest one, suitable for voussoirs placed along the symmetry axis of each portion, he starts drawing a perpendicular line and transferring onto it the lengths of the chords of the cross-section of the first portion, that is, the distances between the intrados joints in this portion. At this point, there seems to be a contradiction between the drawing and the text in both copies of the manuscript. This procedure would be quite appropriate if the sides of the octagon were equal; however, it is inexact in the case of an irregular octagon. This problem could easily be solved by constructing one more section in the elevation, corresponding to the symmetry axis of the central portion, and transferring measurements from this section; however, Vandelvira seems to be unaware of this problem. In any case, since the perpendicular line acts as the axis of symmetry of the portion, it is quite easy to transfer the lengths of the portions of each intrados joint belonging to a particular voussoir from the plan. By repeating this operation, Vandelvira obtains a series of intrados templates of the central voussoirs of the portion.

It would be easy to carry on for the side voussoirs, taking measurements from the plan and bringing them to the horizontal stemming from the axis. However, Vandelvira draws a new set of directrices of the cylinder, using them as reference points for transfers of measurements; this operation does not really seem necessary, maybe Vandelvira uses the directrices to control the parallelism of the horizontals. Even so, he does not seem to be satisfied with the precision of the operations; he puts forward an alternative solution, *capialzando en cruz*, that is, using the diagonals of the voussoir, as he had explained in oval vaults (Vandelvira c. 1585: 73v-74r). To the eyes of a modern architect or engineer, this sounds unnecessary, but we should not forget that Renaissance masons were not drawing on a computer, nor even on tracing paper, but rather on floors and walls.

When dealing with the *ochavo igual por dovelas*, or full octagonal vault, (Figs. 8.20, 8.21) or, Vandelvira (c. 1585: 104r; see also Palacios [1990] 2003: 336–339) is quite concise, referring the reader to the half-octagon (Fig. 8.19). However, close inspection reveals a striking detail: he starts with a section through the plane of the side joints, giving it the shape of a quadrant and dividing it into seven courses.

Fig. 8.20 Octagonal pavilion vault. Tripoli: Arch of Marcus Aurelius (Photograph by the author)

Next, he draws the intrados joints as in the half-octagon, and then constructs the section through the creases. Thus, all severies are portions of raised elliptical cylinders, although the cross-section is neither drawn nor mentioned. As for templates, he states explicitly that they should be constructed *capialzando en cruz*, that is, computing the length of diagonals; this explains the absence of the cross-section.

8.4 Lunettes and Lunette Vaults

8.4.1 Pointed Lunettes

A room covered with a barrel vault can, of course, be lit by windows placed under the springings of the vault. However, in many cases, openings are placed over the springing to direct natural light to the vault surface in order to provide zenithal illumination. There are several ways to solve this geometrical problem. The simplest one is to place a round arch on the vertical plane passing through the springings of the vault, cutting the vault by two oblique vertical planes, obtaining elliptical arches as a result[17] and to set out a ruled surface passing through the round arch and the

[17]In practice, these elliptical arcs were materialised as a succession of circular arcs, as in other elements such as skew arches. When opening lunettes in brick vaults, an alternative technique is to construct the full vault; then a rope can be tightened on the vault surface, forming a helix; next, the

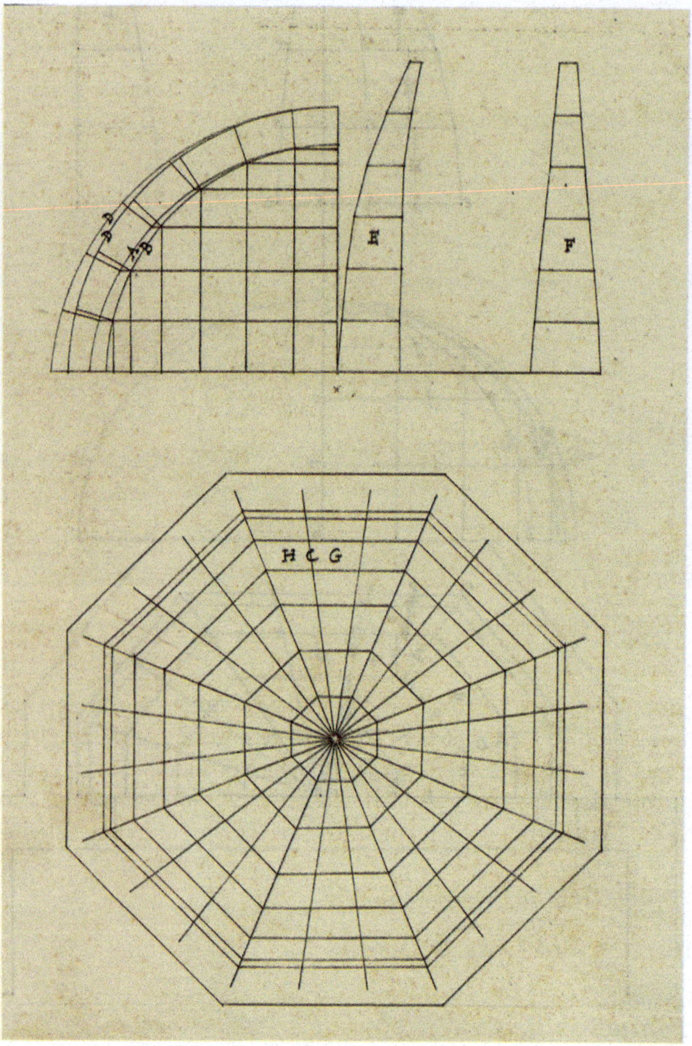

Fig. 8.21 Octagonal pavilion vault ([Vandelvira c. 1585: 104r] Vandelvira / Goiti 1646: 165)

elliptical arches (Fig. 8.22); the resulting constructive element is known as *lunette*. However, in order to place the meeting point of the vertical planes and the vault surface at the same level as the apex of the round arch, the lunette must be quite shallow. A widely used solution (San Nicolás 1639: 103v; see also Calvo 2000b) uses cutting planes at 45° degrees with the springing of the vault and round arches whose diameter is half the radius of the vault. In that case, the meeting point of the

vault portion between two symmetrical helixes can be demolished in order to open the lunette. Of course, in order to avoid such waste of material, the helixes can be drawn on formwork.

Fig. 8.22 Pointed lunette. The Escorial, Palace Courtyard (Photograph by the author)

elliptical arches will be placed much higher than the apex of the round arch; this solution is precisely described in the documents of the Escorial as *luneta apuntada y capialzada* (pointed and raised lunette). As a result, intrados joints in the lunette are usually neither horizontal nor parallel.[18]

This solution is employed frequently in Italian Quattrocento architecture, in such well-known buildings as the Ducal Palace at Urbino, the Strozzi Palace in Florence or the Sistine Chapel in Rome. Usually, it is executed in brick, although the Sistine vault is materialised in a typically Roman kind of concrete, painted in the sixteenth century with the well-known frescoes. The first examples in Spain, such as those at the palace at La Calahorra, are also built in brick; however, in the sixteenth century, most Spanish clients expected important works to be executed in stone. One of the first examples, in the crypt of the chapel of the palace of Emperor Charles V in the Alhambra of Granada, uses pointed arches instead of round arches; this allowed the builders to use the same radius in the vault and the pointed arch. The result is quite neat: intrados joints are parallel and horizontal, and they meet the intrados joints of the vault at the elliptical groins (Salcedo and Calvo 2015; see also Salcedo 2017: 277–292, 323). However, the use of pointed arches in such a canonical Renaissance building may have appeared heretical; this ingenious solution was not repeated in other pieces in the palace (Salcedo 2017: 307–323) or the hundreds of lunettes in the

[18]In theory, horizontal joints can be traced, joining points in the round arch with points in the elliptical arcs placed at the same level. However, such a solution would not lead to parallel joints and, what is worse, would leave an awkward-looking triangle at the apex of the arch. I do not know of any executed examples of this theoretical solution.

Escorial complex (López-Mozo 2005; López-Mozo 2009: 293–309), most of which were solved as pointed and raised lunettes. In contrast, lunettes are all but non-existent in sixteenth-century French architecture; when they appear in the following century, they assume the form of cylindrical lunettes (Pérouse 1982a: 114, 208).[19]

Paradoxically, pointed lunettes in stone, built almost exclusively in Spain, were explained first by French treatises, such as Jousse (1642: 168–169). The mason should start drawing the plan of the main vault and its cross-section, a round arch with evenly spaced voussoirs, carrying the projection lines of the voussoir corners into the plan to represent intrados joints (Fig. 8.23). Next, he will draw two oblique lines in plan, standing for the projections of the intersections of the vault and the lunette. From the intersections of the intrados joints with these oblique lines, he should draw lines at right angles, standing for the intrados joints of the lunette. Then, the mason should construct a cross-section of the lunette, transferring the heights of voussoir corners taken from the cross-section of the main vault to the horizontal projections of the intrados joints, starting from the springing of vault, which is used as a folding line. The important point is that the heights are taken from the cross-section of the main vault. This implies that the intrados joints of the lunette are horizontal; as a result, the cross-section of the lunette assumes the shape of a pointed arch, exactly as in the crypt of the palace of Charles V. After this, Jousse constructs a true-shape representation of the groin, drawing perpendiculars to its horizontal projection and transferring once again the heights of the intrados joints. He does not mention the use of templates, so we must presume that the dressing process is carried out by squaring.

As for Spanish literature, a schematic drawing in the manuscript of Alonso de Guardia (c. 1600: 76v; see also Calvo 2015a: 450–451) seems to resemble a pointed lunette vault, but it seems unfinished and does not carry an accompanying text, so it does not count as an explanation. Pointed lunettes are mentioned by Fray Laurencio de San Nicolás (1639: 104r) (Fig. 8.24), but he directs the reader to a previous chapter on groin vaults. The solution cannot be exactly the same, since the intrados joints in groin vaults are horizontal, while San Nicolás's solution for lunettes leads to sloping intrados joints. In fact, he seems to be more interested in lunette vaults built in brick. In order to find an explanation of ashlar lunette vaults in Spanish texts, we must analyse the late manuscript of Juan de Portor y Castro (1708: 47r-47v; see also Calvo 2000b), who explains not only the ordinary pointed lunette but also a skew variant (Fig. 8.25). He draws the plan and elevation of the lunette, dividing it into voussoirs using longitudinal and transversal joints. Next, he draws a cross-section of the main vault; he then draws the intrados joints of this vault starting from the intersections of the lunette joints with the groin, in order to connect the courses in

[19]Pérouse (1982: 208) remarks that *A l'avantage des Français, on peut seulement dire que la lunette espagnole garde pendant tout le XVIe siècle un tracé maigre et aigu qui n'est en rien comparable au tracé ample et souple des lunettes de Le Mercier* (The only argument in defence of the French is that Spanish lunettes keep during the entire sixteenth century a lean and pointed shape, that cannot be compared with the wide and smooth outline of the Lemercier lunettes). This is generally true; however, there are cylindrical lunettes in the crypt and the sacristy of Jaén Cathedral and other Spanish examples in the sixteenth century, built almost a century before the "wide and smooth" Lemercier lunettes.

Fig. 8.23 Vault with pointed lunette (Jousse 1642: 168)

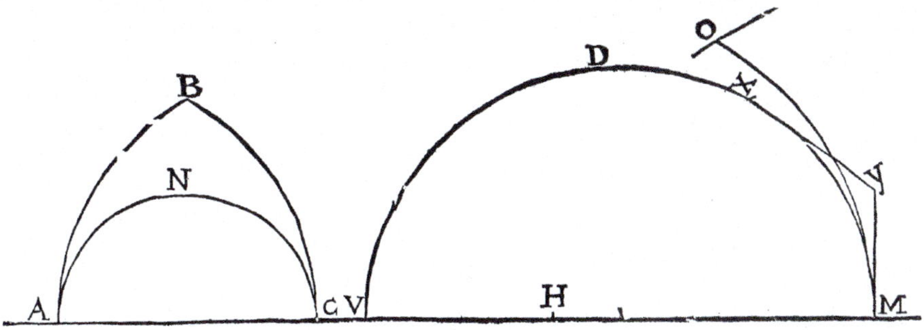

Fig. 8.24 Lunette vault (San Nicolás 1639: 104)

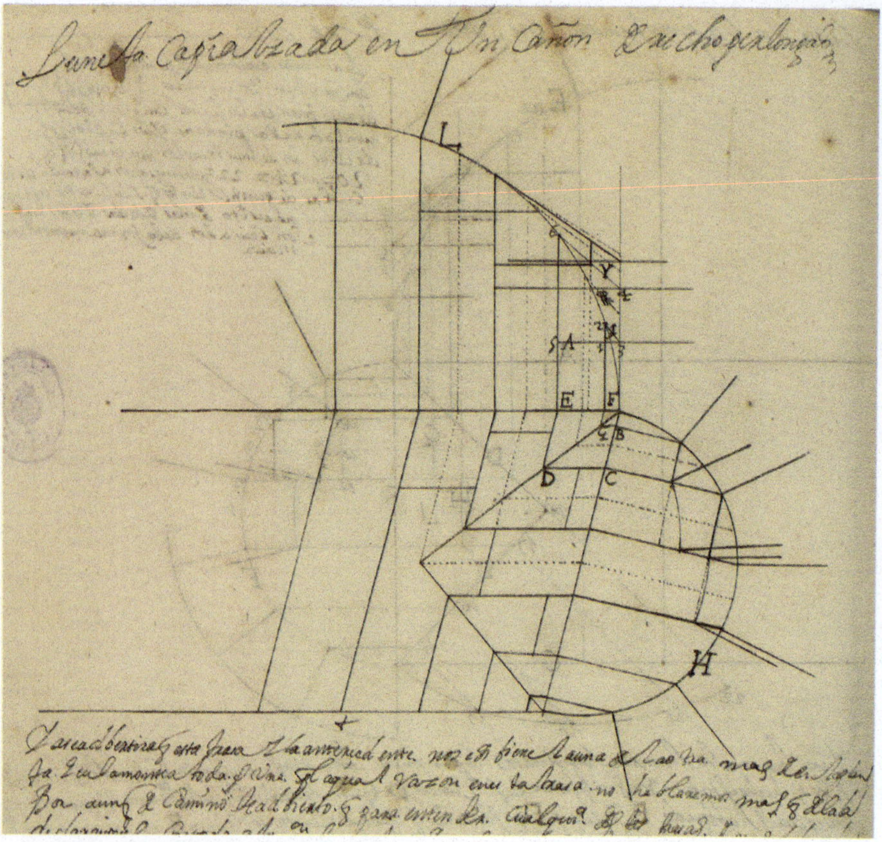

Fig. 8.25 Skew pointed lunette (Portor 1708: 47v)

the lunette and the main vault, avoiding a continuous joint at the groin, as Jousse had done. The intrados joints of the main vault are treated as projection lines, extending them until they reach the cross-section. In the next step, Portor constructs the profile of the intrados joints of the lunette, showing clearly their slant. All this gives him enough information to dress the voussoirs by squaring.

8.4.2 Cylindrical Lunettes

Cylindrical lunettes as independent elements. Lunettes can be set out using a completely different strategy. Instead of starting with two comparatively simple curves—a round arch and two elliptical arcs—and using a complex surface passing through them, the problem can be addressed starting with a simple surface for the lunette—a portion of a cylinder—and then determining its intersection with the larger

Fig. 8.26 Cylindrical lunette. Jaén Cathedral, crypt (Photograph by the author)

cylinder of the vault (Fig. 8.26). Of course, the radius of the lunette must be shorter than the radius of the main vault; otherwise, the result would be a groin vault. The intersection of two cylinders of different radii is a double-curved, fourth-degree line; that is, simple surfaces lead to a complex intersection.[20]

This solution, used here and there in the Romanesque period, reappears almost at the same time in brick in the Villa Pisani in Montagnana by Andrea Palladio and in ashlar in the crypt of Jaén Cathedral, built by Andrés de Vandelvira, the father of Alonso, which may be dated to 1553–1555 and 1555–1560, respectively (Puppi 1973: 52, 131–132; Chueca 1971: 156). However, when executing the piece in hewn stone, there are two different approches to the problem. In the first one, called *pénétration extradosée* by Pérouse ([1982a] 2001: 111, 114) and used by Vandelvira in Jaén, the lunette penetrates the larger barrel vault entirely. The face of the lunette appears in the intrados of the main vault. Although it adapts its curvature to the main piece, it is still clearly independent; it is placed between the arris at the intersection of the lunette and the main vault and a joint separating the lunette face from the intrados of the vault.

Paradoxically, this solution was explained in sixteenth-century Spanish manuscripts, but it was scarcely used until it came into fashion in France in the

[20]In strictly geometrical terms, the intersection is split into two independent closed curves. In actual building practice, these curves are usually open, since only the upper part of a cylinder is used in the lunette. In symmetrical designs, both curves are materialised, generating a pair of symmetrical lunettes at both sides of the main vault, although nothing prevents the use of a single lunette without its counterpart.

next century, as we will see later on. As in other cases, de l'Orme explains the solution to the problem of a sloping barrel vault penetrating a larger one but does not bother to include the simpler case of a straight, horizontal arch opened in a barrel vault. Thus, we must turn to Vandelvira (c. 1585: 23r; see also Palacios [1990] 2003: 88-91) in search of an early solution to this problem. He uses the same technique we have seen in the arch in a battered wall (see Sect. 6.5): after drawing the generating arch of the lunette and *two* cross-sections of the vault, he constructs the horizontal projection of the face of the lunette (Fig. 8.27). He again explains that this projection or *cimbria* is to be used to control the placement of the voussoirs. Next, he constructs the intrados and bed joint templates in the cross-sections, rather than placing them on the plan, as usual. Except for that, his method is the ordinary one: he takes the width of the templates from the generating arch and the position of voussoir corners by drawing verticals from their projections. The only substantial variation from the procedure used in the arch on a sloping wall is that he uses the midpoints of the voussoirs and face joints to compute the curvature of the face sides of both the intrados and the bed joint templates.

Again, a different solution is offered by Martínez de Aranda (c. 1600: 48–50). He does not include the horizontal projection of the face, so he is free to construct the intrados and bed joint templates in the plan, using orthogonals to intrados joints (Fig. 8.28). He uses the midpoints of voussoirs and bed joints to compute the curvature of the face sides of the templates; however, is clear that the width of the intrados template equals the distance between voussoir corners and he is not trying to construct a development of the intrados surface.

As usual, Martínez de Aranda includes a solution to the straight lunette (that is, the one whose axis is orthogonal to the springing of the vault) and two different solutions for a skew lunette. In the first one, the generating arch of the lunette is laid out in a plane orthogonal to the axis of the lunette, although oblique to the springings of the main vault. (Martínez de Aranda c. 1600: 50–51) For other operations, he follows the method he used in the straight version. This leads to a curious result. The

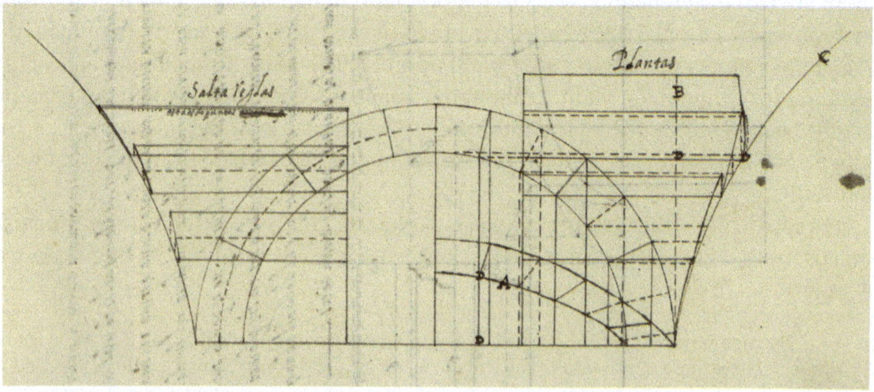

Fig. 8.27 Cylindrical lunette (Vandelvira c. 1585: 23r)

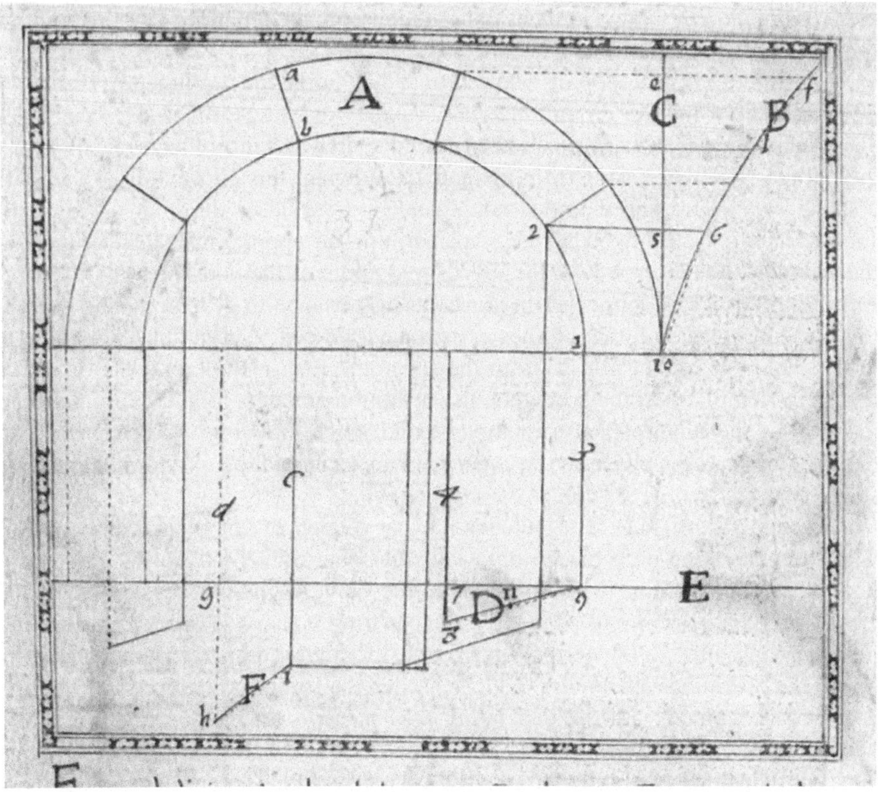

Fig. 8.28 Cylindrical lunette (Martínez de Aranda c. 1600: 49)

section of the main vault by a vertical plane passing through the axis of the lunette is semicircular; as a result, the cross section of the large vault should be a raised ellipse, although it is not clear whether Martínez de Aranda is aware of this fact, since he does not draw the the cross section. As for the intersection of the lunette with the main vault, it is a non-planar curve, as in the standard lunette; however, due to the oblique intersection of the lunette with the main vault, its span is larger than twice the rise, as in surbased arches.

In (Martínez de Aranda c. 1600: 51–53) the author seems to try to correct the “surbased arch” effect. He places the generating arch of the lunette at a vertical plane parallel to the springing of the main vault, using the circular cross section of the vault to measure distances from voussoir corners to the vertical plane that passes through the springings; he explains clearly, both in the drawing and the text, that *these distances should be carried orthogonally to the springings of the main vault and not along the intrados joints* in order to place the horizontal projections of voussoir corners. Thus, in this case, the cross section of the main vault is semicircular; however, the cross section of the lunette will be a raised ellipse. Although Martínez de Aranda draws a *juzgo* orthogonal to the axis of the lunette, it is not used to construct the cross section of the lunette, but rather to measure the length of intrados joints in order to construct templates.

As in the arch in a sloping wall, Jousse's solution for the lunette (1642: 38–39) stands on a middle ground between Vandelvira and Martínez de Aranda. He constructs the horizontal projection of the face arch or *cintre*; however, he places the intrados templates aside from the general drawing; this allows him to construct a string of intrados and extrados templates (Fig. 8.29). As on other occasions in Jousse, it is not clear whether he is performing an approximate development of the intrados surface; however, the use of midpoints of the voussoirs and the word *éstenduë*,[21] which is connected in stonecutting parlance to developments, point in this way. Before doing that, he has constructed the bed joint templates measuring the distances of corner voussoirs to the springing line in the cross-section. Since he groups templates at the springers, the operation amounts to a revolution around the axis of the lunette. These methods avoid much crawling on all fours in the template-preparation phase, at the expense of an accumulation of lines in the tracing stage.

Jousse (1642: 40–41) also includes a skew version of this element (Fig. 8.30), with the generating arch placed on a vertical plane orthogonal to the axis of the lunette, and thus oblique to the springing of the vault. As in Martínez de Aranda, the cross-section allows him to measure distances from voussoir corners to the vertical plane passing through the springing and transfer them to intrados joints; again, the rest of the operations are similar to those used in the straight version. Wisely, Jousse eschews the variant with the generating arch placed over the springing, avoiding Martínez de Aranda's pitfall.

Cylindrical lunettes integrated with a barrel vault. There is another approach to the problem of the cylindrical lunette, called *pénétration fileé* by Pérouse ([1982a] 2001: 111, 114). In this case, there is no separation between the face of the arch and the intrados of the vault, and the apparent bed joints of the vault meet those of the lunette at the intersection of both intrados surfaces. In other words, in the preceding solution the arch retains its independence, while in this one, the lunette and the vault amalgamate into a single entity. The key to this tight integration of both elements is the coordination of intrados joints at the intersection of both surfaces. As we have seen, in the Granada crypt and in Jousse's solution this coordination arises automatically, since the radii in the lunette and the vault are identical and the intersection between both surfaces is set on a vertical plane. Thus, intrados joints in the two surfaces are symmetrical about the intersection plane and meet at the common intersection (see Salcedo and Calvo 2015); of course, in a groin or pavilion vault on a square plan, intrados joints are symmetrical and meet at the groin.

However, in the cylindrical lunette, the joints in the two surfaces cannot be symmetrical, since they are placed on cylinders with different radii. Derand (1643:

[21] The full sentence is: *Cela fait, soit tirée la ligne 6.q & sur icelle r'apporter l'estenduë du cercle F, E, D, qu'il faut prendre sur les poincts marquez 1.2.5.6.E que signerez des mesmes marques* (Jousse 1642: 39) (Once this is done, you should draw the line 6.q and the extension of the circle F, E, D should be transferred on it, taking the points 1.2.5.6.E, which are marked with the same signs [in the extension]).

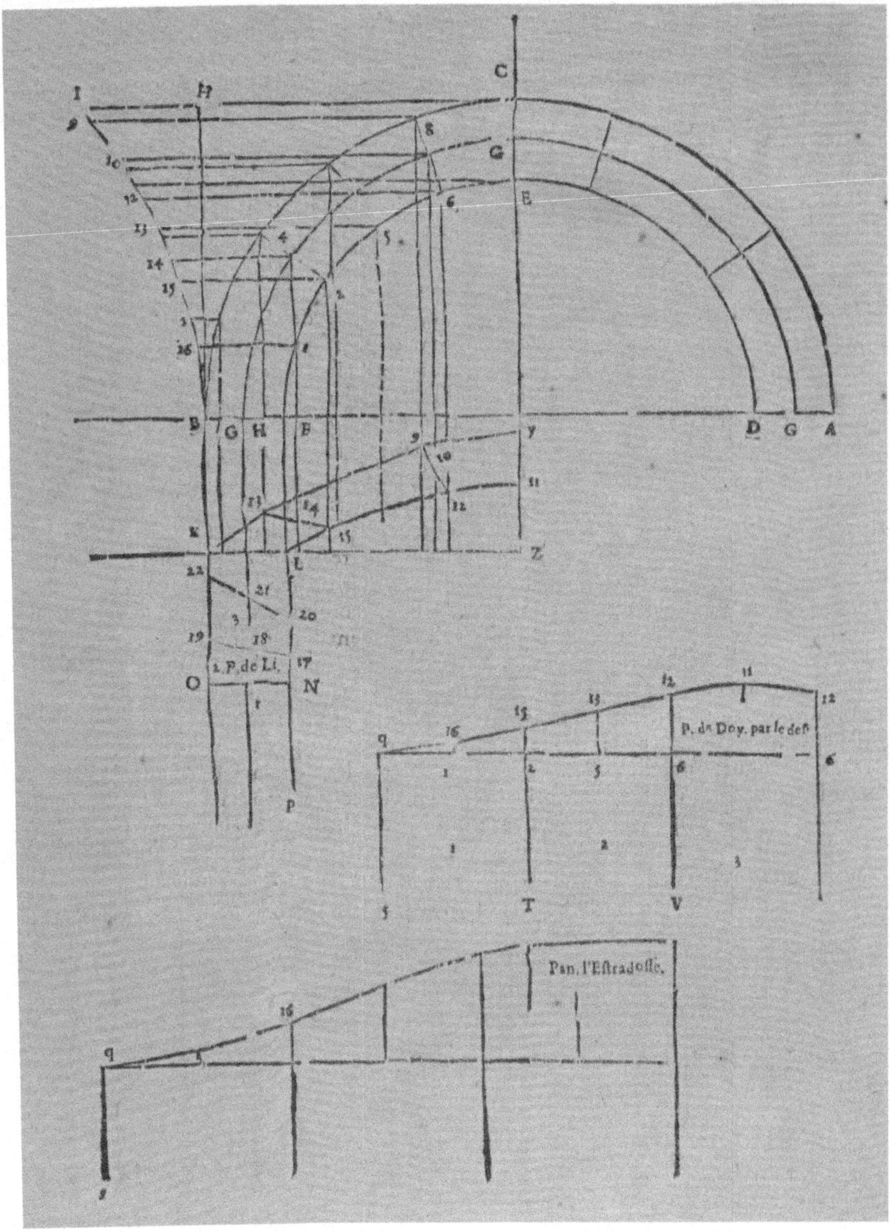

Fig. 8.29 Cylindrical lunette (Jousse 1642: 38)

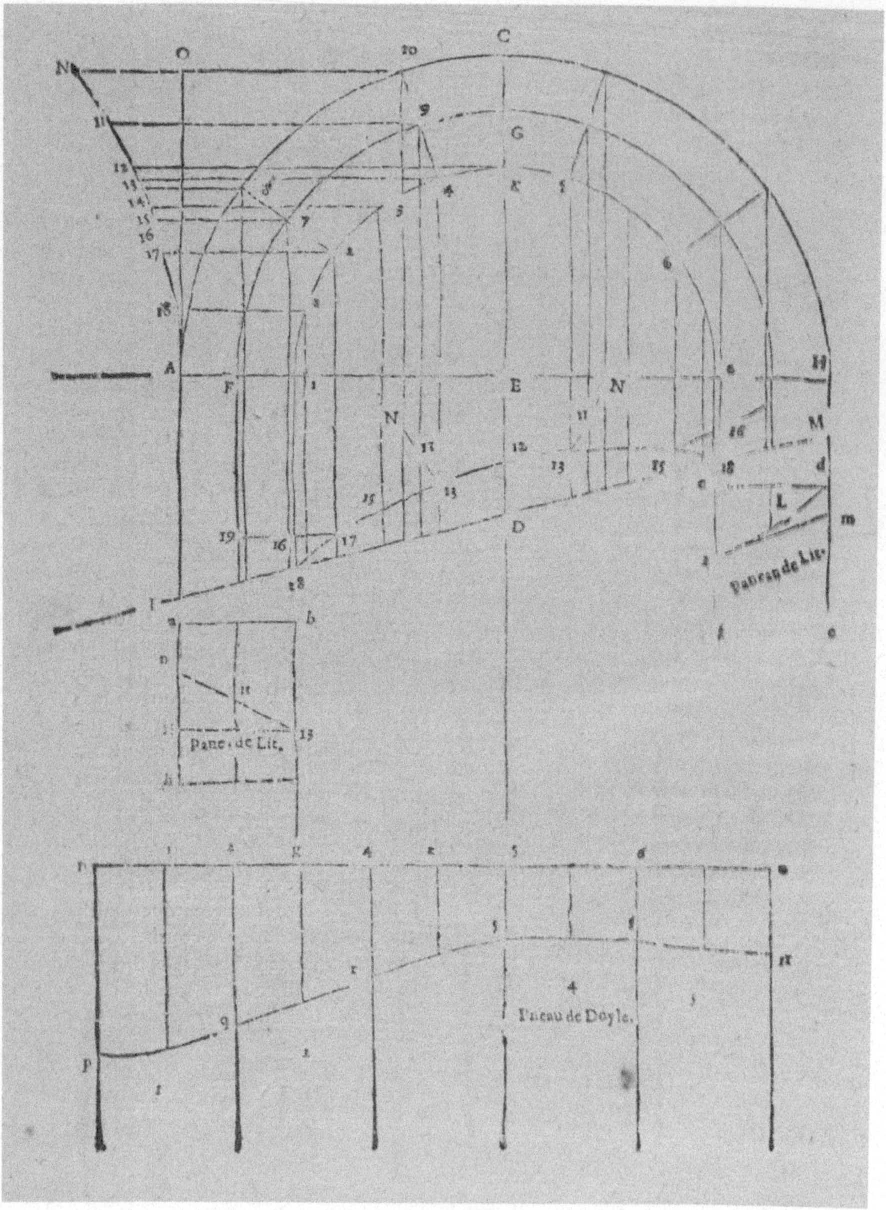

Fig. 8.30 Skew cylindrical lunette (Jousse 1642: 40)

344–345)[22] includes a short text on the issue, together with a schematic drawing, in contrast with his usually verbose explanations. He draws the cross-section of the lunette intrados as a round arch and divides it into evenly spaced voussoirs. Next, he draws the section of the main vault, or rather a portion of it, and draws the horizontal intrados joints starting from the section of the lunette up to where they meet the section of the main vault; next, he measures the distances from the intersection points to the vertical plane passing through the springing. Transferring these distances to a schematic plan, he can draw the horizontal projection of the intersection between the main vault and the lunette. And that is about all; no further constructions are included in the drawing, and Derand does not bother to mention any kind of template. Since this construction furnishes only the horizontal projection of the intersection line, it can be used both for independent and integrated lunettes. However, the text gives some clues, since it refers the reader to the groin vault and uses the same terms, such as *arête* for the groin or *enfourchements* for the voussoirs placed along it; all this suggests that Derand is thinking about the integrated lunette.

Frézier (1737–39: III, 36–41) is much more explicit (Fig. 8.31). He approaches the problem simply by manipulating the spacing of the intrados joints in the main vault. The choice of the skew lunette as a general case obscures this issue; the author does not explain the orthogonal lunette, arguing that the skew one encompasses the right one as a particular case. Admittedly, the same method can be applied to both variants, although it is much easier to understand it in the orthogonal case. After drawing the plan and the sections of the main vault and the lunette, Frézier divides the cross-section of the lunette into five voussoirs. Next, he measures in the section the heights of the division points, which represent the intrados joints of the lunette, and transfers them to the cross-section of the main vault; this operation provides the height of the intrados joints of the main vault. He can draw the intrados joints easily in plan, both for the lunette and the main vault, starting from the cross-sections. The intersection of the first intrados joint of the lunette and that of the main vault, which are at the same level and thus intersect, will provide a point of the groin between the lunette and the main vault. Repeating this procedure for other joints, he will locate a number of points of the groin; joining them, he may draw its horizontal projection. Thus, in a single operation, Frézier has established the position of the intrados joints and the horizontal projection of the groin. The procedure is quite ingenious, but the intrados joints in the main vault are unevenly spaced. In any case, since the upper generatrix of the lunette is placed below the upper section of the main vault, this method does not give the position of the upper generatrices of the vault, and thus Frézier is free to distribute them evenly, in contrast to the lower ones.[23]

---

[22]A numbering error in the first edition of Derand's treatise affects these pages. The lunette is explained in the first page numbered as 344, which is correctly paginated, and drawn on the following page, which is left unnumbered as in all plates in the volume. The second page numbered as 344 should have been paginated as 346 and is not connected with the lunette.

[23]Of course, Frézier's procedure can be reversed, distributing the joints evenly in the main vault and adapting the joints in the lunette to place them at coincident heights. However, he chooses to manipulate the heights of the joints in the vault; he seems to think that the larger scale of the vault will allow this operation to go unnoticed.

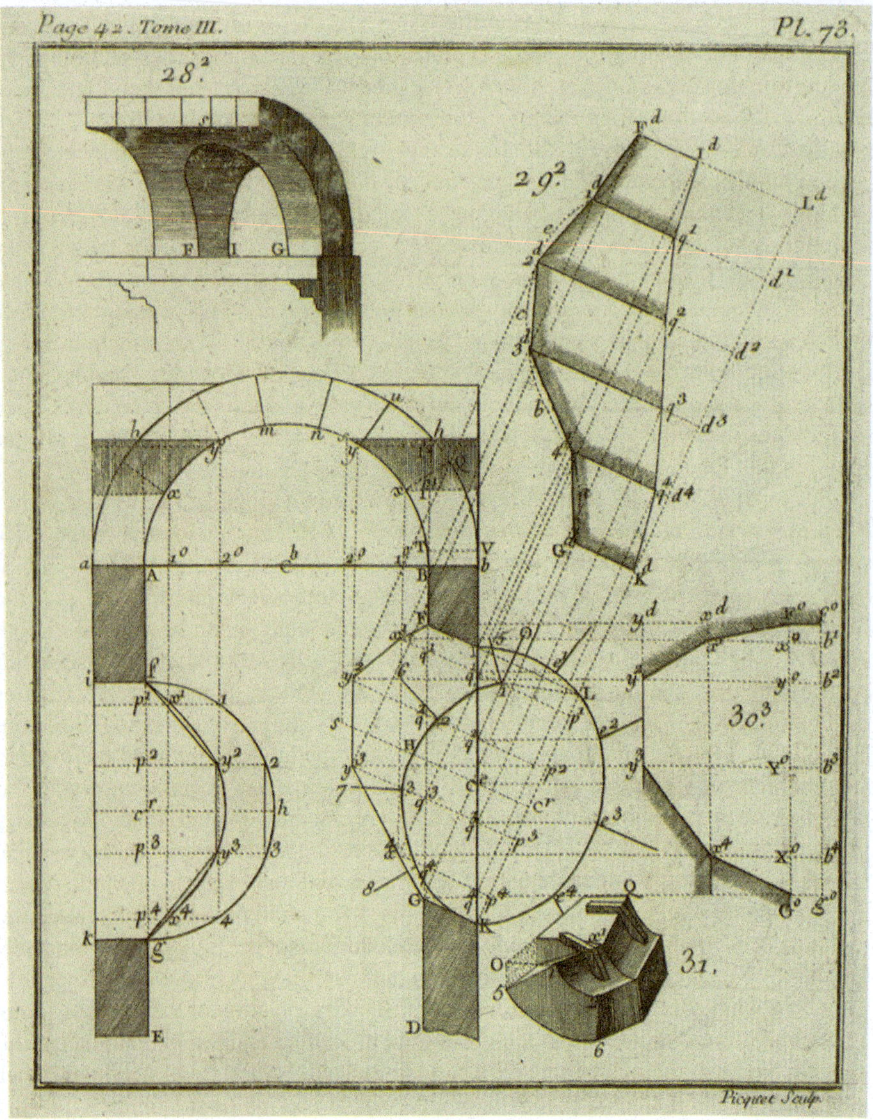

Fig. 8.31 Cylindrical lunettes (Frézier [1737–1739] 1754–1769: II, pl. 73)

The same procedure can be applied to a skew lunette whose axis is oblique to the one of the vault. As in Jousse or the first variant of Martínez de Aranda, Frézier uses a cross-section of the lunette placed on a vertical plane orthogonal to the lunette axis. He measures the height of the intrados joints in the lunette cross-section, transfers them to the main vault, draws the horizontal projection of the vault joints and finds their intersection with the lunette joints. Although the lunette joints are symmetrical about the lunette axis, these intersections result in an asymmetrical groin, as a result of the skewness of the piece. And of course, as in the orthogonal solution, the lower intrados joints are slightly uneven; thus, the *pénetration filée*, the canonical solution for Pérouse, is the result of fiddling with the position of the joints of the main vault.

8.4.3 Sloping Vaults Abutting on Another Barrel Vault

As we have seen, in his drawing of the straight sloping vault, de l'Orme draws the section of a large barrel vault at the lower end; however, he omits any reference to this element in the text, perhaps because the problem can be solved using the same techniques employed at the other, vertical end. This is evident in Derand's (1643: 22–29) solution to the same piece. Since the axis of the larger barrel is orthogonal to the symmetry plane of the sloping vault, it is represented in the longitudinal section through a directrix of the cylinder, a simple circular arc. Thus, it is quite easy to determine the intersections of the intrados joints with the larger barrel and transfer their distances to the plane of the cross-section, just as Derand did with the intersections with the plane of the cross-section.

In contrast, Vandelvira (c. 1585: 34v-35r) sticks to triangulation; he draws no fewer than the plan of the vault and the face arch in the large barrel side, the cross-section, the longitudinal section, the section through a vertical plane and the elevation of the face arch, plus the intrados templates and the bevel guidelines.

8.5 Desargues's "Universal Method"

Desargues's approach to the problem of skew sloping vaults, looking for a general solution for cylindrical vaults, is quite complex. It involves four planes, four axes and several angles (Desargues 1640: 2–3; Bosse and Desargues 1643a: pl. 8–11;[24] Schneider 1983: 72–73; Sakarovitch 2010: 127–128; Boscaro 2016: 52–64). The key elements of the problem are: the *essieu* or axis of the vault,[25] marked with oval ends; the *plan de chemin* or route plane, literally the plane of the floor of the passageway under the vault, which is parallel to the springings plane; and the *plan de face*, or face plane, that is, the plane of the skew and/or battered wall. Desargues solves it by starting from three given angles: the *angle de talus* τ or batter angle, that is, the one between the face plane and a horizontal plane; the *angle du biais* α or bias angle, the one between the horizontal projection of the axis of the vault and a horizontal line in

[24]Pages in Bosse and Desargues (1643a) are not numbered. However, each section of the text begins with a heading referring to a plate; thus, in the text of this book, "Bosse and Desargues 1643a, plate x" may refer either to Plate x itself or its accompanying text.

[25]Desargues uses the word *essieu*, as in the axis joining two wheels in a vehicle. Although Schneider (1983: 72) translates it as "axle", I think it is clearer to render the word as "axis", for consistency with other examples by different authors.

the face plane, and the *angle du pente du chemin* γ or slope angle, the one between the axis of the vault and a horizontal plane

In addition, Desargues uses a number of auxiliary axes and planes, starting with the *plan droit a l'essieu*, which is orthogonal to the axis and passes through its intersection with the face plane, and the *plan sous-essier*, which is orthogonal to the face plane and also passes through the vault axis; thus, it is orthogonal to the preceding plane. Next, the *sous-essieu*, or sub-axis, marked with convex ends, is the intersection of the face plane and the *plan sous-essier*; it may be understood as an orthogonal projection of the axis on the face plane. The *contre-essieu*, or counter-axis, is the intersection of the plane orthogonal to the axis and the *plan sous-essier*, that is, the plane which projects the axis on the face; it is identified with concave ends. Thus, the axis, the sub-axis and the counter-axis all belong to the *plan sous-essier*. In contrast, the *traversieu*, drawn with square ends, is the intersection of the face plane and the *plan droit a l'essieu*. It is orthogonal to the axis and belongs to the face plane. In the general case, it does not belong to the *plan sous-essier*.

The first and crucial step in Desargues' method (1640: 3, pl. 2.1; Bosse and Desargues 1643a: pl. 12–24; Schneider 1983: 78–84; Sakarovitch 2010: 128–129; Boscaro 2016: 67–71) is the determination of the angles between axes (Fig. 8.32). He tackles the problem mainly through triangulation, although his method prefigures one of the standard nineteenth-century procedures for changes of horizontal projection planes, as remarked by de la Gournerie (1860: vii; 1874: 154). He considers another pair of vertical auxiliary planes, one passing through the axis and another one orthogonal to the face, called *plan droit au face et au niveau* (plane perpendicular to the face and level planes), as well as five triangles. *A–B–N* lies on a horizontal plane passing through the intersection of the *essieu* and the face plane; *A–K–N* is placed on the vertical plane passing through the axis; *B–K–N* and *B–K–H* are placed on the vertical plane orthogonal to the face; and *A–H–K* lies on the *plan sous-essier*.

In order to compute the angles between axes, Desargues starts by drawing a horizontal line standing for the intersection between the face plane and a horizontal plane passing through both springings. Next, he marks on this horizontal line the intersection of the axis of the vault and the face plane, *A*, and another point, *B*, placed at will on the auxiliary *plan droit au face et au niveau*. Next, he draws a line at an angle α, (that is, the bias angle) with *A–B*. This line meets an orthogonal to *A–B* drawn through *B* at point *N*. Next, he draws the line *B–G* at an angle τ (the slope angle) to *B–N* and then an additional line *B–K* set at an angle γ (the slope angle of the passageway) with *B–G*. Next, he transfers *N* from *B–N* to *B–G* in order to mark point *N*. Then, he draws an orthogonal to *B–N* until it meets *B–K* at point *K*, in order to construct the triangle *B–K–N*. Next, he projects *K* on *B–N* in order to get the point *H*, which is the projection of *K* on the face plane. *A–H* stands for the sub–axis. Next, he draws an orthogonal to *A–H* through *H*, and he transfers *H–K* to this line in order to locate point *K*. *A–K* is the axis of the vault, rotated so that it lies on the face plane. The orthogonal to the axis drawn by *A* is the *contre-essieu* or counter-axis, while a perpendicular to the sub-axis passing through *A* is the *traversieu* or cross-axis. Bosse, the first interpreter of Desargues, dutifully explains that neither of them is strictly

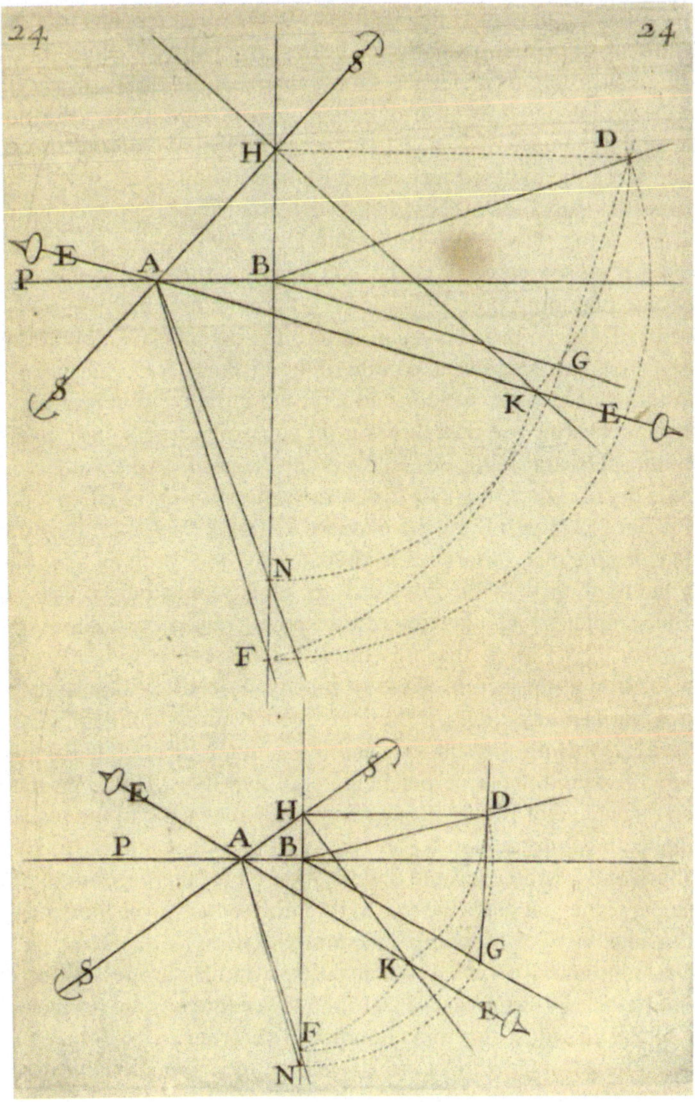

Fig. 8.32 Computing angles for a skew sloping vault in a battered wall (Bosse and Desargues1643: pl. 24)

necessary, although they are useful in order to draw orthogonals to the *essieu* and *sous-essieu*, as he does later on (Bosse and Desargues 1643a: plates 26, 28 and 29).

Of course, such a complex construction may be flawed in practice, either by a conceptual error on the part of the mason or simply by an accumulation of graphical errors. Thus Bosse and Desargues (1643a: pl. 24) furnish a specific procedure to check its exactitude; this justifies the phrase *trait a preuves*. The mason should draw an arch with its centre in B and radius equal to *B–D* until it reaches line *B–N* at point *F*; he should then construct another arc starting at *F* with its centre in *A*, which, if the preceding constructions are exact, should intersect the *essieu* exactly at point *K*. Moreover, the mason can draw a straight line connecting *A* and *F*; the angle *P–A–F* (that is, the supplementary of *B–A–F*), should equal the *biveau de la nivelée en face* (the angle between the battered face and the horizontal plane).

Desargues then sets out to construct to construct the cross-section of the vault (1640: 3–4, pl. 3, Fig. 1; see also Bosse and Desargues 1643a: pl. 29, 30; Boscaro 2016: 76–84). For each voussoir corner—for example, *R*–Desargues projects it on the *sous-essieu*, obtaining point *9* (following the notation in Bosse and Desargues 1643a: pl. 30). Then, he projects again *9* onto the counter-axis, obtaining *b*. Next, he extends the projector and transfers there the distance *R–9*, starting from *b*; this allows him to locate point *r*, which stands for the intrados joint starting in *R* in the cross-section. Repeating the operation for all voussoir corners, both in the intrados and the extrados, Desargues locates all the points he needs to draw the cross-section. A remarkable detail is that he draws the cross-section as a crooked line, rather than a curve, as Frézier (1737–39: II, pl. 36–67, 38–79, 48–127, 48–129, for example) will do later. This means that Desargues is choosing the face arch as a regular one and eschewing the regularity of the cross-section; it also hints that he uses rigid rather than flexible templates. Another interesting point, raised by Sakarovitch (1994b: 353–354; 2010: 129–130) is that this operation has some traits in common with the change of frontal projection plane in nineteenth-century descriptive geometry (that is, in American terminology, the construction of an auxiliary view; see Paré et al. 1996: 22:25). Both the sub-axis and the counter-axis act as folding lines, so distances are measured about perpendiculars to both lines. However, the operation is not strictly a change of frontal projection plane, since none of the planes involved (the face plane, the *plan sous-essier* and the plane orthogonal to the axis) is actually frontal or horizontal. Moreover, the understanding of the procedure is obscured by the fact that the plane that plays the role of the "horizontal" projection plane is shown as a plane in edge view overlapping the counter-axis.

Next, Desargues (1640: 4, pl. 4; see also Bosse and Desargues 1643a: pl. 28, 31–36; Schneider 1983: 84-86; Boscaro 2016: 85–87) addresses the problem of finding the shape of the intrados and joint templates. In order to construct the intrados template for a voussoir, he draws orthogonals to the sub-axis from the front ends of the intrados joints of the voussoir, for example, *O–V*; these perpendiculars meet the *sous-essieu* at *4* and *7* (following the notation in Bosse and Desargues1643a: pl. 36). Then Desargues draws perpendiculars to the axis from these points, until they meet the *essieu* at *d* and *z*. Next, he draws an arc with the centre in one of the points in the axis, let us say d, and radius equal to the distance between the ends of the intrados joints, *O–V*, until it reaches the other perpendicular at *u*. The angle *z–d–u* will furnish the angle between the chord of the voussoir and the intrados joints, that is, the shape of the end of the intrados template.[26]

[26]Of course, if the mason had drawn the arc from the opposite corner *z*, he would have obtained a symmetrical figure, which may be used turning the template around.

The same method may be used to compute the joint templates: the mason should draw orthogonals to the sub-axis from the ends of the face joints, for example, *V* and *R* (notation taken from Bosse and Desargues 1643a: pl. 35). From the intersection points, *7* and *9*, he constructs perpendiculars to the axis, which intersect it at *z* and *b*. Then he draws an arc with centre at z and radius equal to *V–R*; it will intersect the orthogonal passing through *b* at *r*; the angle *A–z–r* is the one between the intrados and face joints, which plays the role of bed joint template, exactly as do the bevel guidelines of Vandelvira (c. 1585: 32v) or Jousse (1642: 32).

From de la Gournerie (1860: vi–viii) to Boscaro (2015: 51, 65), all scholars dealing with Desargues stress the radical shift he brought about in stonecutting procedures. Admittedly, he detached the reference planes from verticals and horizontals, replacing them with a plane orthogonal to the axis and the *plan sous-essier*, that is, the projecting plane of the axis on the face plane. However, he used traditional resources such as triangulation, bevel guidelines and rigid templates. Another significant detail has passed unnoticed, as far as I know. When starting to construct the cross-section of the vault, Desargues supposes that the construction lines in the preceding steps are suppressed, leaving only the results[27]. This may just be a rhetorical figure, but Bosse takes it literally, stating clearly that the mason should erase unnecessary lines.[28] This can be done when drawing in paper, but it is unpractical in large-size tracings. This makes clear that neither Desargues nor Bosse were acquainted with the practical requirements of the stonecutter's craft; as a result, their procedures involve a large number of lines, at least in comparison with de l'Orme. Perhaps the main reason for the rejection of Desargues's procedures lies here; after all, the change of reference planes is not so strange to masons, who were familiar with stone rotations during the dressing process.

[27]Desargues (1640: 3): *En conceuant pour vne espece de commodité, que les droictes desormais inutiles Ak, BK, Hk, HK, sont disparües* (supposing that, for the sake of convenience, lines Ak, BK, Hk, HK, which are now useless, have disappeared).

[28]Desargues and Bosse (1643a: comments to pl. 26) *Et puis que vous n'auez affaire que de ces deux lignes de* sous-essieu & d'essieu *& que les autres lignes que vous auez tirées ne vous seruiront plus de rien, effacez les comme en la figure* d'enhaut *vous voyez qu'elles commençent à estre* effaceés, *& qu'en la figure* d'embas *elles sont tout à fait* effacées.

Chapter 9
Spherical, Oval and Annular Vaults

9.1 Spherical Vaults Divided into Horizontal Courses

9.1.1 Hemispherical Domes

Hemispherical domes (Figs. 1.14f, 9.1) are one of the most distinctive features of Renaissance architecture; in the words of Alonso de Vandelvira (c. 1585: 60v) they are *principio y dechado de todas las capillas romanas*, (starting point and paragon of all Roman vaulting). This phrase stresses simultaneously their aesthetical significance and the fact that stonecutting methods devised for domes are also applied to quarter-sphere, sail, oval, surbased and annular vaults.

These standard methods, based on cone developments, were already in use around 1543; a full-scale tracing for a small dome built over a spiral staircase was found by Ruiz de la Rosa and Rodríguez (2002) in the flat rooftops of Seville Cathedral. De l'Orme (1567: 111v-112v) includes a section on "vaults in spherical form", also using cone developments, but he is actually talking about a sail vault. Thus the first clear, written explanation of a strictly hemispherical vault is included in Vandelvira's manuscript (c. 1585: 60v-61v). As usual in stonecutting treatises and manuscripts, he starts with a prescriptive exposition, giving directions to the mason on how to prepare a full-scale tracing (Fig. 9.2). First, he should start drawing the plan of the vault in the shape of two concentric circles (Palacios [1990] 2003: 189–195; Rabasa 1996: 429; Rabasa 2000: 170–174). He then reuses this plan as a cross-section; that is, he understands at least empirically that any section of the sphere through a plane passing through its centre is a great circle and thus has the same radius as the sphere. Next, he divides the upper half of the section into an uneven number of portions, standing for courses; the middle one stands for the keystone. The division points will furnish, of course, the position of the intrados joints. The mason should also draw the axis of the vault, that is, a long auxiliary vertical line passing through the centre of the sphere. He should then construct a set of sloping lines connecting each intrados joint with the next one, extending them until they reach the axis of the vault. In the next step, Vandelvira performs, for each course, a development of the cone

J. Calvo-López, *Stereotomy*, Mathematics and the Built Environment 4,
https://doi.org/10.1007/978-3-030-43218-8_9

Fig. 9.1 Hemispherical dome. Escorial, main church, crossing (Photograph by the author)

generated by the rotation of the sloping line around the axis of the vault, although he does not explain in this section the rationale of this method (in fact, he never uses the word "cone" in his entire manuscript). However, he performs the operation in a most elegant manner, simply drawing two circular arcs, with their centres in the intersection of the generatrix and the axis, passing through both intrados joints. This procedure leads to a simplified, three-side intrados template, where the circular arcs stand for the upper and lower edges of the template, while the part of the generatrix between both divisions in the section provides one of the sides of the template.

In order to locate the other side of the template, a modern engineer, architect or well-trained mason would measure the angle between the end radii of the voussoir, compute the length of the intrados joints using π and recompute the angle between generatrices in the development, which is of course different from the angle between the radii of the voussoir. Vandelvira does nothing of the sort; in fact, he does not even draw voussoir radii or side joints. Felipe Lázaro de Goiti, the author of the seventeenth-century copy in the National Library of Spain added "you will close the arcs as you wish"[1]; that is, the mason may draw a second generatrix at will to close the template (Rabasa 2000: 172). Such an easy-going attitude may be justified by the condition of the stone arriving from the quarry to sixteenth- and seventeenth-century worksites, with irregular dimensions according to the geological properties

[1] Vandelvira and Goiti (1646:118): … *las cuales dos cerchas cerrarás por do quisieres que miren al punto G* … Transcription taken from Vandelvira and Barbé 1977.

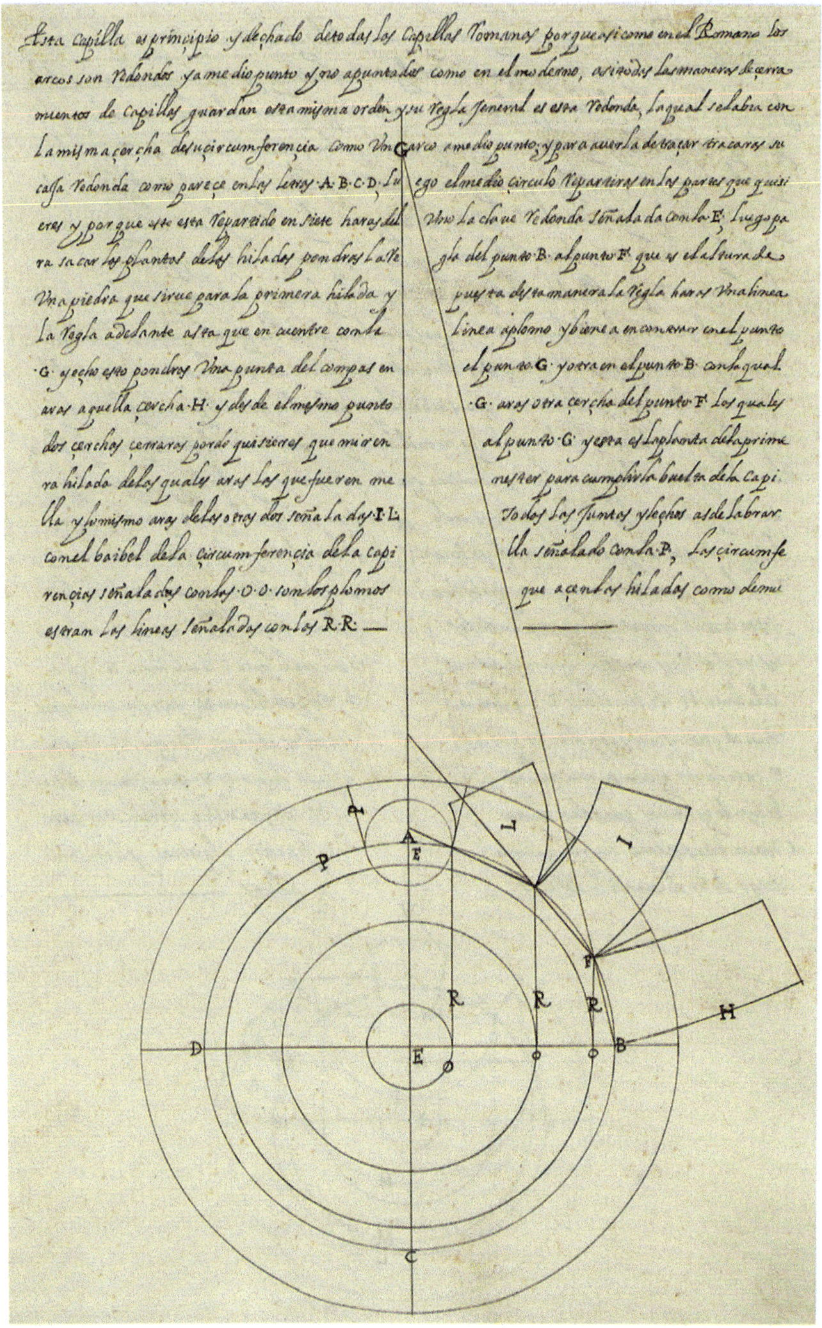

Fig. 9.2 Hemispherical dome ([Vandelvira c. 1585: 61r] Vandelvira /Goiti 1646: 118)

of the rock. Thus, when a particular element was to be covered with stucco or other veneer—as, for example, in the main dome of the Escorial church—it was usually built with voussoirs with different lengths[2]; in contrast, when a dome was to be left uncovered, and especially when it carried a decoration that had to be coordinated with the division into voussoirs, it was essential to control the length of voussoirs; I will come back to this problem when dealing with quarter-sphere vaults.

Vandelvira is afraid that his reader will be at pains to understand this method, which is radically different from medieval practice and other Renaissance procedures; he remarks that "some curious people have asked me how it is possible that templets [with different radii] can be put together without leaving space between them".[3] As a result, after these operative instructions in typical stonecutting fashion, he embarks on an additional explanation (Fig. 9.3), which is quite unusual in stonecutting literature before Frézier. He compares the courses in the vault with a *pechina cuadrada a regla*, that is, a symmetrical trumpet squinch; a splayed arch would have provided a more precise comparison, but Vandelvira does not include this piece in his manuscript. Then he performs conical developments for the squinch or splayed arch; in fact, he uses the same technique employed by Martínez de Aranda (c. 1600: 66–67) for a splayed arch with a spherical intrados. Vandelvira repeats the operation for a number of splayed arches, each one ensconced on the preceding one, stressing that "what stands for bed joints in the arch will be side joints for the vault".[4] In any case, his explanation is purely empirical and rather unconclusive; it seems that Vandelvira himself does not understand that, generally, the radius of a curve in a development is not equal to the real radius of the curve.

In neither the prescriptive section nor the comparison with the splayed arch does Vandelvira provide much information about the dressing procedure. He remarks that "all bed joints and side joints should be dressed with the curve of the circumference of the vault marked as p.P"[5] and instructs the mason to "dress this stone with this template and use its bevel on all sides and the curve of the foundation on the upper and lower bed joint".[6] As remarked by Palacios ([1990] 2003: 193) this means that, in addition to the templates he has dutifully explained, the mason should use an arch square both on the bed joints, which are shaped as cones, and the side joints, which are meridional planes. Rabasa has compared Vandelvira's explanations with the ones given by Guardia for the sail vault:

[2]This dome was intended to be rendered in stucco, but it was finally left with apparent stone; this justifies the irregular layout of side joints.

[3]Vandelvira (c. 1585: 61v): *he sido de algunos curiosos preguntado que como siendo diferentes las cerchas de los lechos pueden acudir en obra sin que hagan mala consonancia.* Transcription taken from Vandelvira and Barbé 1977. See also Palacios [1990] 2003: 192; Rabasa 2000: 169.

[4]Vandelvira (c. 1585: 61v): *lo que sirve de lechos para arco sirva de juntas para capilla.* Transcription taken from Vandelvira and Barbé 1977.

[5]Vandelvira (c. 1585: 60v): *Todas las juntas y lechos has de labrar con el baivel de la circunferencia de la capilla señalada con la p.P.* Transcription taken from Vandelvira and Barbé 1977.

[6]Vandelvira (c. 1585: 61v): *labra esta piedra con esta planta y métele por el lecho bajo y alto la cercha del fundamento y su baivel por todos cabos.* Transcription taken from Vandelvira and Barbé 1977.

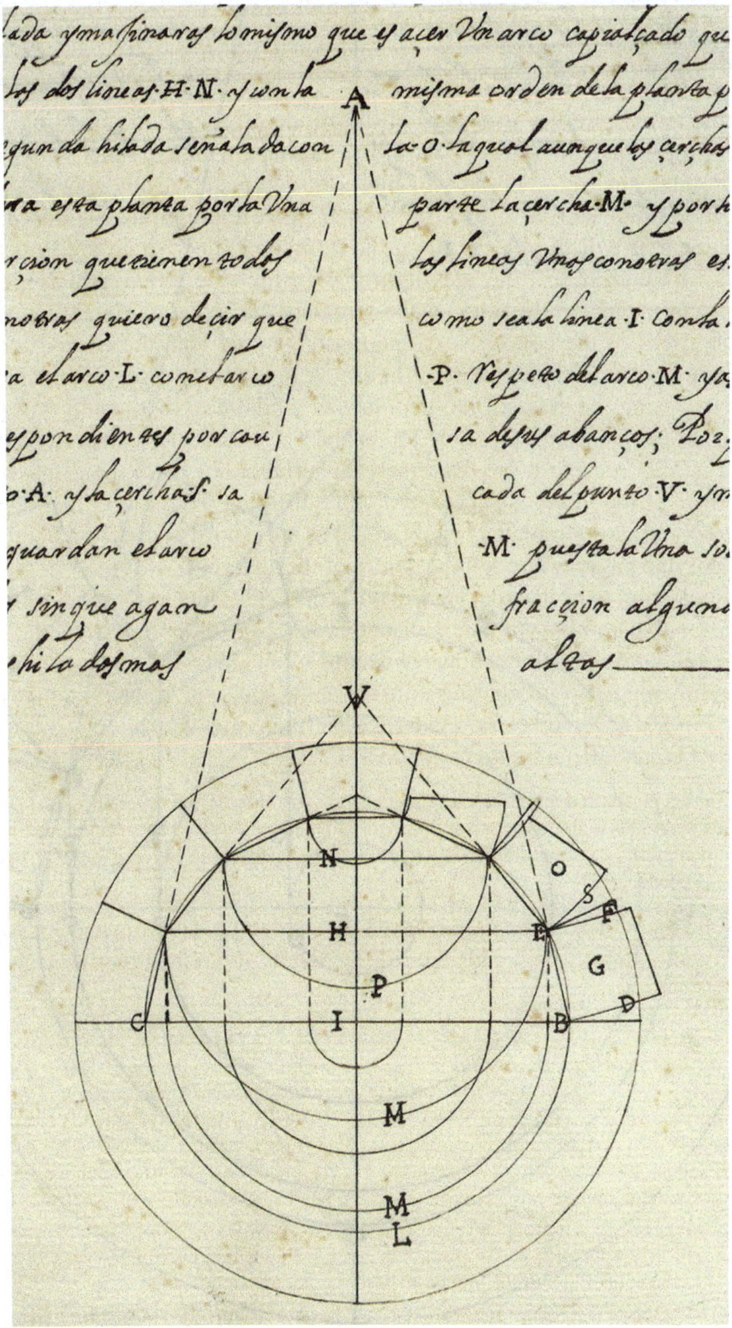

Fig. 9.3 Explanation of the hemispherical dome ([Vandelvira c. 1585: 61v] Vandelvira/Goiti 1646: 119)

> in order to dress the pieces of the pendentive you should dress their intrados with the radius of the sphere marking the templates on them and cutting them as shown in figure F, dressing the bed and side joints with the straight branch of the arch square for this vault[7]

That is, to execute the voussoirs of the pendentives, the mason should dress a portion of the spherical surface, controlling its shape with a templet with the radius of the intrados sphere. Then he will place the template on this spherical surface, scoring its outline with a scriber. Of course, since the template is based in cone developments, it will fit exactly along the intrados joints, which are directrices of the cone, although it will deviate slightly from the side joints, which are arcs of spherical meridians, while sides of the template are straight, since they stand for the generatrices of the cone.[8] Next, the mason should lean the curved arm of the arch square on this surface; the straight arm will adopt the direction of the radii of the sphere. Using the sides of the template as guidelines, these radii will generate both meridional planes used as side joints and a new series of cones furnishing bed joints. It is important to notice that these material cones, with their apexes placed at the centre of the intrados sphere, are quite different from the virtual cones used to construct intrados templates, whose vertexes are placed at different heights along the vertical line passing through the centre of the sphere.

It is interesting to notice that Guardia (c. 1600: 69v) includes on a different sheet two additional diagrams, without explanatory text; one of them repeats the template construction, while the other shows two sides of an enclosing rectangle under each voussoir; this suggests Guardia was thinking about squaring as an alternative dressing method (Fig. 9.4). Derand (1643: 354–357) mentions the same alternatives, although he clearly favours the template method:

> Now, in this piece, in contrast to other precedents, it is shorter and more efficient … to use intrados templates to dress the voussoirs, rather than to cut them by squaring … for the sake of completeness … I should make it clear how those that wish to execute them by squaring should proceed.[9]

Thus, he recommends the method used by Vandelvira, Guardia and Jousse (1642: 122–123), although Derand adds interesting details about the dressing technique:

---

[7]Guardia (c. 1600: 87v): … *para labrar las piezas desta dicha pechina les labrarás las caras con la vuelta de horno cortándolas con las formas que tuvieren sus plantas por caras y enjutándolas conforme parece en la figura señalada con la F labrando los lechos y juntas con la tirantez que tuviere el baivel de la dicha vuelta de horno.* In this passage, Guardia refers explicitly to a pendentive supporting a hemispherical vault; however, it is clear from a comparison with the hemispherical vault itself and another dome on fol. 69v that the tracing procedure is identical and, we may surmise, the same dressing technique can be used both in the pendentive and the hemispherical vaults.

[8]Rabasa (2003: 1681) has reproduced the use of conical templates on spherical surfaces in practical courses taught in the Centro de los Oficios in León, finding that even if the mason presses the template onto the spherical surface, the issue has no practical significance.

[9]Derand (1643: 356–357): *Encor bien qu'il arriue en ce trait, & en l'vsage d'iceluy, tout le contraire de ce qui s'est remarqué en plusiers des precedens; sçauoir est, qu'il est plus court & plus vtile … de se seruir des paneaux de doüele pour tracer les voulsoirs, que de les faire par équarrissement: … pour ne laisser rien à dire … ie dois declarer en suite de ce que dessus, comme ceux qui desireront tracer les voûtes de four par équarrissement, y deuront proceder.*

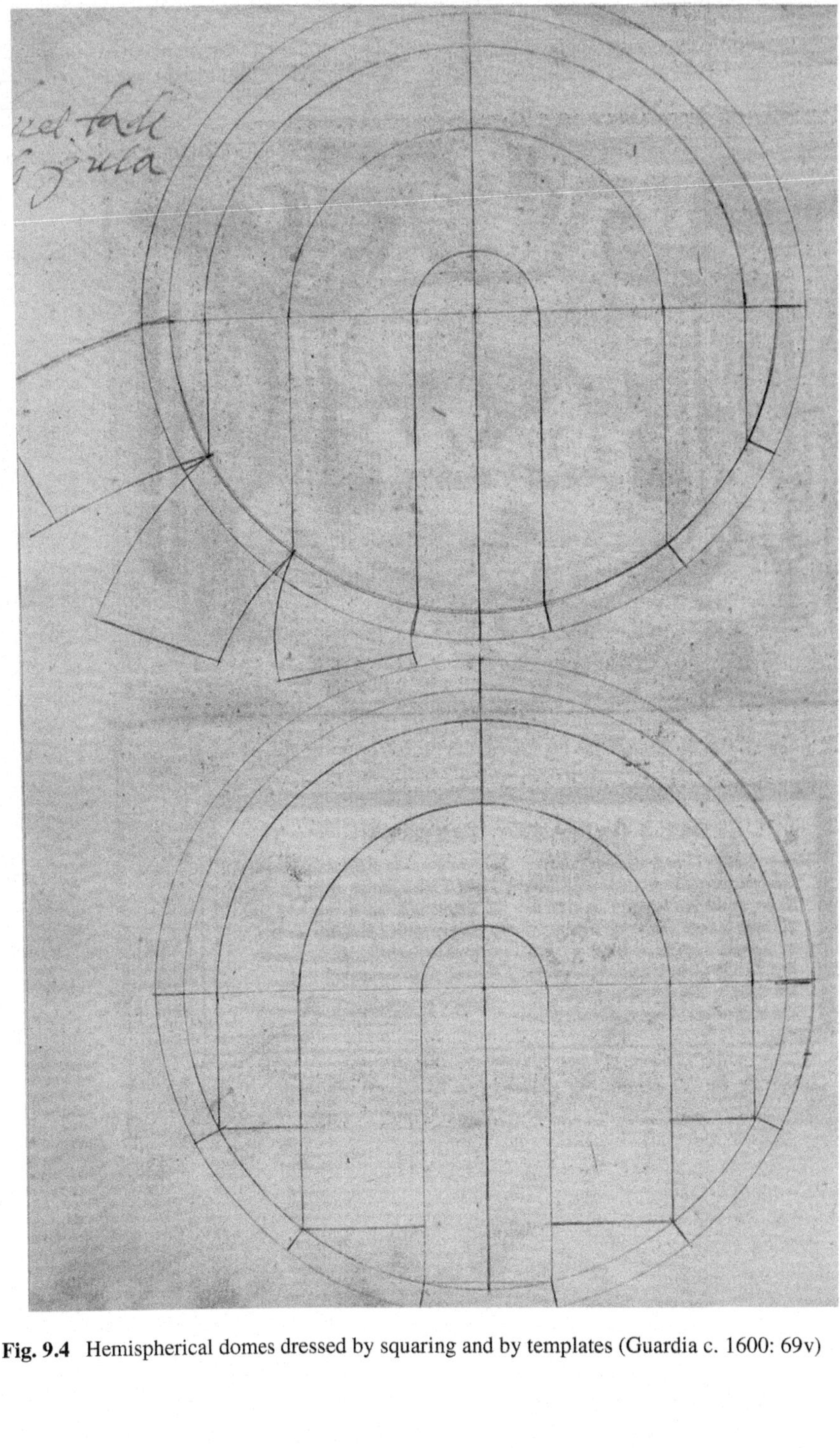

Fig. 9.4 Hemispherical domes dressed by squaring and by templates (Guardia c. 1600: 69v)

> To dress the stones, you will use the templates and tracing that we have prepared. First, it is necessary to hollow the surface of the stone with the templet with the radius of the intrados of the vault; this templet should turn in all directions inside the hollowed surface. In order to perform this task easily, you can mark diagonal lines on the surface you want to hollow, from one corner to the opposite one, making along them drafts with the curvature of the templet; and then take off the portion of stone between these drafts with the same curvature, applying the templet now and then, in order to make your work more dependable and free from errors.[10]

Once the spherical surface is dressed, the mason should place the template on it and score its outline, in order to materialise bed and side joints using the arch square, as Rabasa (1996a) pointed out for Vandelvira and Guardia; in Derand's words:

> Second, on the hollowed surface, you should place the template for the course of the voussoir you are dressing; and after marking its sides on the concave surface, you will trace the edges of the bed and side joints with the arch square of the radius of the sphere, and you will cut the stone along the joints of the sides and the beds with the straight arm of the arch square.[11]

Thus, the process involves four phases: first, the mason should dress a spherical surface controlling it with the templet, since the sphere has the same curvature in all directions; second, he should apply the flexible template on the spherical surface, marking its outline with a scriber to define the apparent bed and side joints; third, he should open marginal drafts at the intersection of bed and side joints, representing their internal edges, controlling their direction with the straight arm of the arch square; fourth, he should actually dress the bed and side joints starting from these marginal drafts (Fig. 9.5). This procedure clearly implies that the templates are

---

[10]Derand (1643: 356): *Or pour façonner la pierre, on se seruira & des paneaux & du trait que nous venons de faire. Et premierement il faut creuser le parement de la pierre choisie auec la cherche du plein circle de la voûte de four; en sorte que cette cherche puisse tourner de tous costez dans le creux qui y sera fait. Pour à quoy paruenir plus facilement, il faut tirer sur le parement qu'on veut creuser des lignes diagonales d'angle en angle, faisant sur icelles des cizelures ou entailles suiuant la curuité de la dite cherche; & puis abatre ce qui se trouuera entre ces entailles suiuant la curuité d'icelles, appliquant de temps en temps sur votre ouurage la susdite cherche, afin de rendre vostre trauail plus asseuré, & exempt de fautes.*

[11]Derand (1643: 356): *En second lieu, sur le parement creusé, comme dit est, se couchera le paneau de l'assise, à laquelle appartient le voulsoir que vous auez en main; & ayant repairé ses costez dans la doüele creuse de vostre voulsoir, vous en tracerez le ioints des bouts & les lits auec le buueau du plein cintre de ladite voûte de four C A D, lesquels lits & ioncts se couperont suiuant le bras dudit buueau.*

The phrase *tracerez les ionts … auec le buueau du plein cintre de la dite voûte* is incomprehensible at first sight. To draw intrados joints, that is, the edges of the bed and side joints, with an arch square is rather cumbersome and also inexact, since the edges of side joints are great circles and thus their radii are equal to that of the sphere, but the edges of the bed joints are small circles. Moreover, the operation of marking these edges is described as a previous phase: *ayant repairé ses costez dans la doüele creuse de vostre voulsoir, vous en tracerez les ionts …* Thus, the phrase *tracerez le ioints … auec le buueau du plein cintre de ladite voûte* must refer to a second phase: the mason starts by dressing the side and bed joints, carving drafts that follow the direction of their *internal* edges, which are segments with the direction of the normals to the spherical intrados surface and thus materialised by the straight arm of the arch square; in a third phase, the mason should dress the full side and bed joints starting from these marginal drafts, using again the straight arm of the arch square, as stated in the phrase *lesquels lits & ioncts se couperont suiuant le bras dudit buueau.*

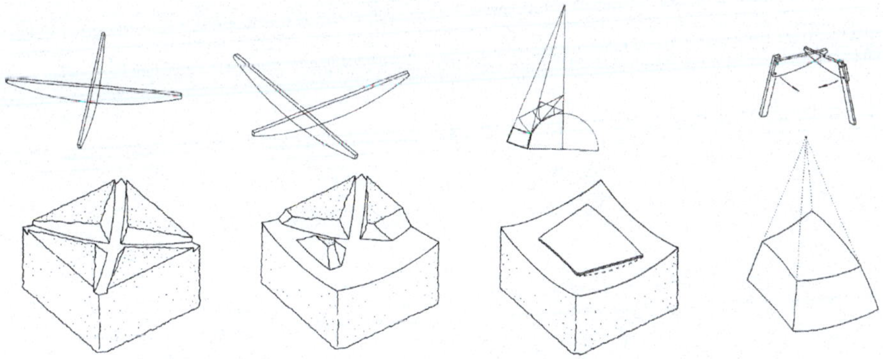

Fig. 9.5 Dressing a voussoir for a hemispherical dome by templates (Drawing by Enrique Rabasa)

flexible. However, all authors take this for granted; only Gelabert (1653: 50v) remarks that the template should be made of cardboard or another flexible material, as we have seen in Sect. 3.2.6.

Thus, from the start of modern stonecutting, from the Seville tracing and de l'Orme's sail vaults, there is a basic consensus: all authors accept that the canonical method for spherical vaults is based on cone developments. De la Rue (1728: 50–52) broke this agreement suddenly (Fig. 9.6). First, he considers squaring as the standard solution to the problem. After drawing a cross-section and plan of the vault, including the bed and side joints, the mason should dress a block with the plan and the height of a voussoir, cutting its sides orthogonally; this operation furnishes vertical, meridional planes for side joints. Next, the mason should measure in the elevation the height of the upper bed joint and transfer it to the edges of the block. This operation will provide both edges of the bed joint, which should be joined with a flexible ruler, which recalls Alberti's *tabula gracilis* (see Sect. 4.2.2). Next, the lower corner of the voussoir should be marked on the meridional plane; at this point, the mason may place a side template, taken from the elevation, on the meridional plane; this enables him to take off four wedges: one for the intrados, two for the upper and lower joints, and another one for the extrados, if necessary. The intrados surface, of course, should be dressed using a templet with the radius of the intrados surface; as for the wedges over and under the bed joints, de la Rue remarks that the mason should draw two curves using a templet taken from the plan and dress them using the ruler; this suggests that the mason should generate the cone in bed joints through the movement of the ruler leaning in two curves; that is, the ruler acts as generatrix and the curves as directrices.

After explaining the squaring method, de la Rue makes clear the reason of his rejection of templates for this piece. In his opinion, de l'Orme, Jousse and Derand were wrong, since the chord of the voussoir should not be mistaken for its edge. However, in order to eschew the waste of stone brought about by the squaring method, he proposes an alternative procedure, based on the use of a compass on the three-dimensional surface of the intrados sphere. The mason should dress a spherical

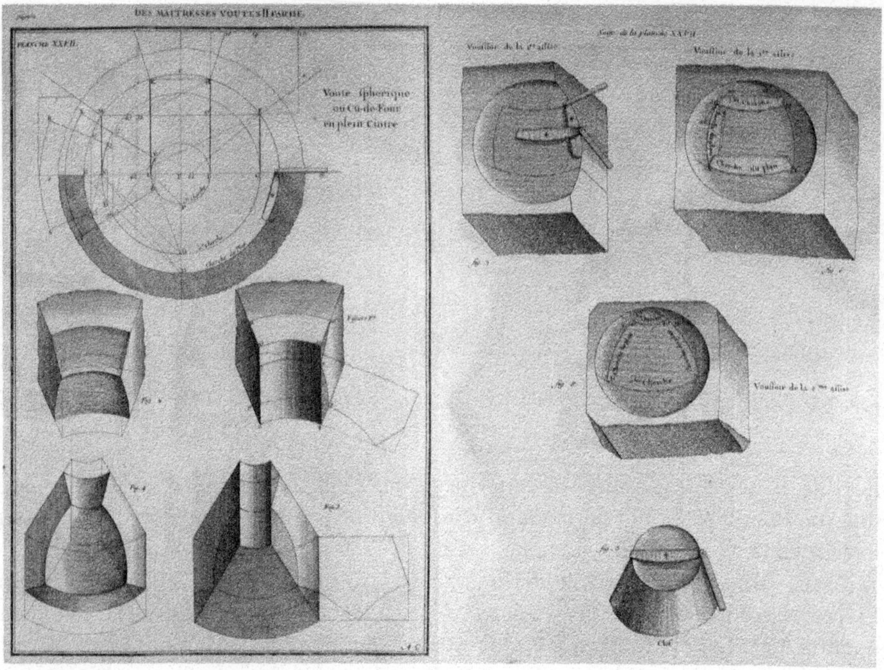

Fig. 9.6 Hemispherical domes solved by squaring and by "drawing in space" (de la Rue 1728: pl. 27, 27 bis)

surface using a templet with the radius of the intrados sphere, as usual; the result is called *écuelle* (literally, "bowl"). Next, he should mark the two lower corners of the voussoir intrados on the spherical surface, taking their distance from the plan. The mason should then measure the length of the diagonals of the intrados side of the voussoir, forming a right triangle; one of the catheti equals the difference in heights of opposite corners of the voussoir. Although the other cathetus is left unexplained, we may surmise that it stands for the horizontal distance between both corners and, in fact, some approximate measurements on the drawing support this idea. Thus, the hypotenuse equals the length of the diagonal; the mason may draw an arc on the stone surface with this radius. Next, he should mark on the intrados the length of the side of the inner face of the voussoir, taken easily from the cross-section of the vault. The intersection of both arcs gives a third corner of the intrados surface; repeating the procedure, the mason may locate the fourth corner. De la Rue dutifully remarks that the distance between the third and fourth corners should equal the length of the upper bed joint, taken from the plan. In the next step, the mason should connect all four corners with a templet with the radius of the intrados surface in order to score the edges of the voussoir. Of course, the correct use of the templet in this step is essential. Given two points in the spherical surface, there is one and only one great circle passing through both points; thus, the mason should carefully orient the template, placing it orthogonally to the intrados surface. After this, he can dress the bed and side joints using an arch square, as suggested by other writers.

With his usual encyclopaedic approach, Frézier (1737–39: II, 312–331) puts forward no fewer than four different methods to address this problem (Fig. 9.7): de la Rue's method of "drawing in space"; the standard method using flexible templates based on cone developments; the squaring method implied by Guardia and explained

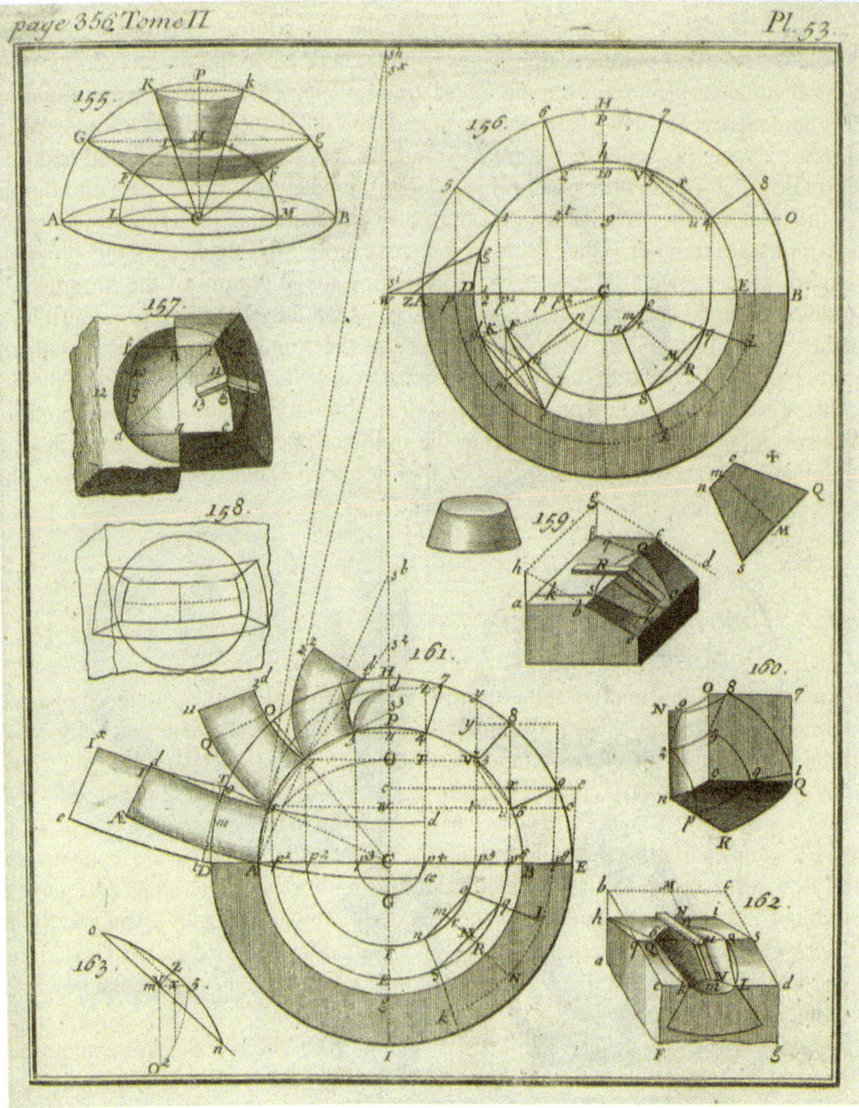

Fig. 9.7 Hemispherical dome (Frézier [1737-1739] 1754–1769: II, pl. 53)

by Derand; and a new method based on rigid templates, "reducing the sphere to a polyhedron".[12] In this case, symmetrical trapeziums are substituted for the conical portions of the standard method. The axis of each trapezium should be drawn in the cross-section, taking from the plan the distance of their ends to the midpoint of the corresponding intrados joints. Once this is done, the intrados template can be constructed easily starting from the axis and transferring the half-bases from the plan; the sides of the trapezium should equal the chords of the portions of the meridians belonging to each voussoir. Once the tracing is finished, the mason should dress a flat face and place the template on it. So far, the procedure seems straightforward; however, from this point on, things get complicated. Now, the mason should dress a provisional surface, representing a horizontal plane, taking into account the dihedral angle between such plane and the face of the polyhedron in the intrados of the voussoir. Once this is done, it is easy to mark on this surface a "horizontal bed joint template", that is, the horizontal projection of the bed joint, including a circular arc for the intrados joint. This allows the mason to carve the side joints easily; however, bed joints are another issue. Prior to their execution, the mason should dress the intrados with a special arch square, representing the curvature of the intrados and its angle with the horizontal auxiliary surface, which may be taken from meridional planes. Once this is done, the mason may use the standard arch square, with the angle between the intrados and the bed joints, to carve the conical surfaces of the bed joints. Thus, the procedure involves dressing an auxiliary surface and the preparation of a special arch square, in addition to the ordinary operations when using flexible templates; it is no wonder that the preceding writers had not bothered to follow this route.

9.1.2 Quarter-Sphere Vaults

On many occasions, quarter-sphere vaults, also called half-domes, are used to cover niches or apses (Fig. 9.8); in particular, such vaults, executed in carefully-hewn ashlars, were used in Armenia and Syria (Cuneo 1988: 98–101, 118, 143, 166–169, 175, 184, 194–195, 212–213, 234–237, 244–245, 254, 282–283, 358–359; Sakarovitch 1998: 105–107) from as early as the fifth to the seventeenth centuries. In these ancient examples, and in many later ones in Western Europe in the Romanesque and Renaissance periods, vaults are divided using horizontal parallels and vertical meridians, as in the standard hemispherical dome. In particular, in Spain and Sicily (Casaseca 1988: 176; Nobile 2013: 21), such half-domes are frequently treated like a big scallop shell, probably as a symbol of the Order of Saint James, quite powerful in Spain in this period.

Vandelvira (c. 1585: 66v-69r; see Palacios [1900] 2003: 208–219) explains no fewer than four solutions for quarter-sphere vaults, for both the horizontal- and vertical-course variants. All of them deal with special cases; he seems to think that the

[12] Frézier (1737–39: II, 325): *En réduisant la Sphère en Polyédre.*

Fig. 9.8 Quarter-sphere vault. Yererouk, basilica, chancel (Photograph by the author)

reader can solve the ordinary quarter-sphere vault by generalising the procedure for hemispherical vaults. First, he explains a niche opened in a cylindrical wall, solved with horizontal courses; as in the hemispherical dome, he prepares the intrados templates using auxiliary cones. However, he also explains how to construct the templets for face joints, computing the position of an intermediate point from the plan and constructing an arc passing through this point and the ends of the joint, as he did in arches opened in concave or convex surfaces (see Sect. 6.4). In the second variation, he addresses again the problem of a niche-within-a-niche, but in this case, the intrados is decorated with arrises and it is necessary to compute their curvatures.

Since the third variant uses vertical courses, I will deal with it in Sect. 9.2.1. In the fourth variation, Vandelvira returns to horizontal courses, constructing the templates by triangulation, since cones of revolution cannot be inscribed between oval sections of a surface; I will come back to the issue in Sect. 9.4.2.

9.1.3 Sail Vaults

Sail vaults (Fig. 9.9) are used to cover square, rectangular or, more generally, polygonal shapes with portions of a spherical surface; in this way, the mechanical advantages of vaults can be applied to layouts that fill the plane, as opposed to circles or ovals. Such vaults were used on some occasions in Antiquity, for example in the baths in Gerash (Choisy 1883: 88–89; López-Mozo 2009: 64; Hara et al. 2013) or the tomb of Ummidia Quadratilla in Cassino (Etlin 2012: 8; Piccinin and Natividad

Fig. 9.9 Sail vault with round courses. Cairo, Fatimid walls, Bab-el-Futuh (Photograph by the author)

2019); there are many examples in medieval Spain executed in more or less carefully dressed small-gauge masonry (Marías 1991: 73–74; Molina and Arévalo 2014: 167–170). However, the use of brick vaults with this shape in such well-known Florentine buildings as the Foundling Hospital or the aisles of San Lorenzo and Santo Spirito, fostered the construction of sail vaults in carefully hewn stone in Eastern Spain in the early sixteenth century (Natividad 2012a; Calvo et al. 2005a: 79–92, 197–210; Natividad 2017: I, 252–257); in the next decades, this solution spread to other regions of Spain and, to a lesser extent, France (Natividad 2017: I, 258–318; Pérouse 2000: 117, 344–345).

All of this poses the problem of the intersections of the intrados sphere with the vertical planes enclosing the area covered by the vault. Such planes may be materialised by either walls or transverse arches separating the vault from adjacent ones; in the first case, the junction of the vault and the wall is usually solved by an arch embedded in the wall. For simplicity, I will group these arches, together with the freestanding ones, under the cathegory of *perimetral arches*. The vault can be divided using different schemes: either with horizontal parallels and vertical meridians, as in the standard dome, or using two sets of vertical parallels dividing the vault in courses and the courses in voussoirs. The geometrical problems posed by such operations are complex, in particular when the plan is based on a rectangle or an irregular polygon; thus, I will deal first with square vaults and then with other variants.

Square-plan sail vaults with horizontal courses. When a spherical vault is built over a square area, usually the radius of the sphere equals half the diagonal of the area. Perimetral arches directrices are given by small circles of the sphere whose diameter equals the side of the square; given the symmetry of the figure, their keystones are placed at the same height, half the side of the square above the springers. Thus, a parallel of the intrados sphere connects all four keystones; its diameter equals the side of the square area covered by the vault and the diameter of the perimetral arches. This parallel divides the vault in an upper section or *cap*, placed over the keystones of these arches, and four lower portions called *pendentives*, each one placed between two perimetral arches and the parallel connecting their keystones. Courses over the cap describe full circles, and thus can be set out and dressed exactly as in the standard hemispherical dome. In contrast, courses in the pendentives meet perimetral arches; thus, the final voussoirs in each of these courses must be cut obliquely, following the shape of this arch. The usual choice of a radius that equals exactly half the side of the area for perimetral arches leads to a striking result: both perimetral arches meet at a point at the equatorial plane of the sphere, and the springing is reduced to a point, or a line if we consider the full thickness of the vault. This solution would lead to very sharp edges at a key point of the vault; usually this theoretical springer is fused with the springers of the perimetral arches.

De l'Orme does not address this problem: he starts with vaults with intrados joints laid in vertical planes parallel to the diagonal of the area, as we will see at the end of Sect. 9.2.2 (see Figs. 9.17 and 9.20). Vandelvira (c. 1585: 81v-82r; see Palacios [1990] 2003: 254–259 and Natividad 2017: I, 50–57)[13] starts by drawing the square plan of the vault, including the thickness of the perimetral arches. Next, he constructs a circle passing through the corners of the square, standing for the section of the sphere by a horizontal plane at the springing level (Fig. 9.10). This circle is reused as a cross-section of the vault. A quadrant is divided into six equal portions, representing courses: three portions stand for courses in the pendentives, two for courses in the cap, and the last one makes up the keystone, together with a symmetrical one in the next quadrant.[14] Vandelvira then draws generatrices, finds their intersections with the axis, and develops cone portions, as in the hemispherical dome.

However, in the templates for the pendentives, he needs to draw the junctions of the voussoirs with the perimetral arches. In theory, such curves are portions of hyperbolae, since they result from the intersection of vertical-axis cones with vertical planes that do not pass through the axis. However, Vandelvira uses a practical approximation. The first voussoir starts from the springing, which is in theory a point, and

[13]The sheet with the drawing of this piece in the manuscript of the School of Architecture is misplaced, probably as a result of the binding of the manuscript; at present, it is not placed immediately after the text, as usual in this manuscript. Thus, I am quoting both pages, text, and drawing, following the original numbering of the manuscript, transcribed in Vandelvira/Barbé (1977) in round brackets: this makes clear that the sheets were originally placed one after the other. In this particular case, 82r is included only in the second volume of Vandelvira/Barbé 1977 (the one with the facsimile), since it carries only a drawing, while the accompanying text is placed in fol. 81v (included both in the transcription in vol. I and the facsimile).

[14]This solution leads to a double-width keystone. However, dividing the quadrant or the full semicircle in an odd number of parts, as usually done in hemispherical vaults and other cases (see for example Vandelvira c. 1585: 61r, 70 r, 72r, 83r) would lead to courses placed partly in the pendentives and partly in the cap, with an untidy result.

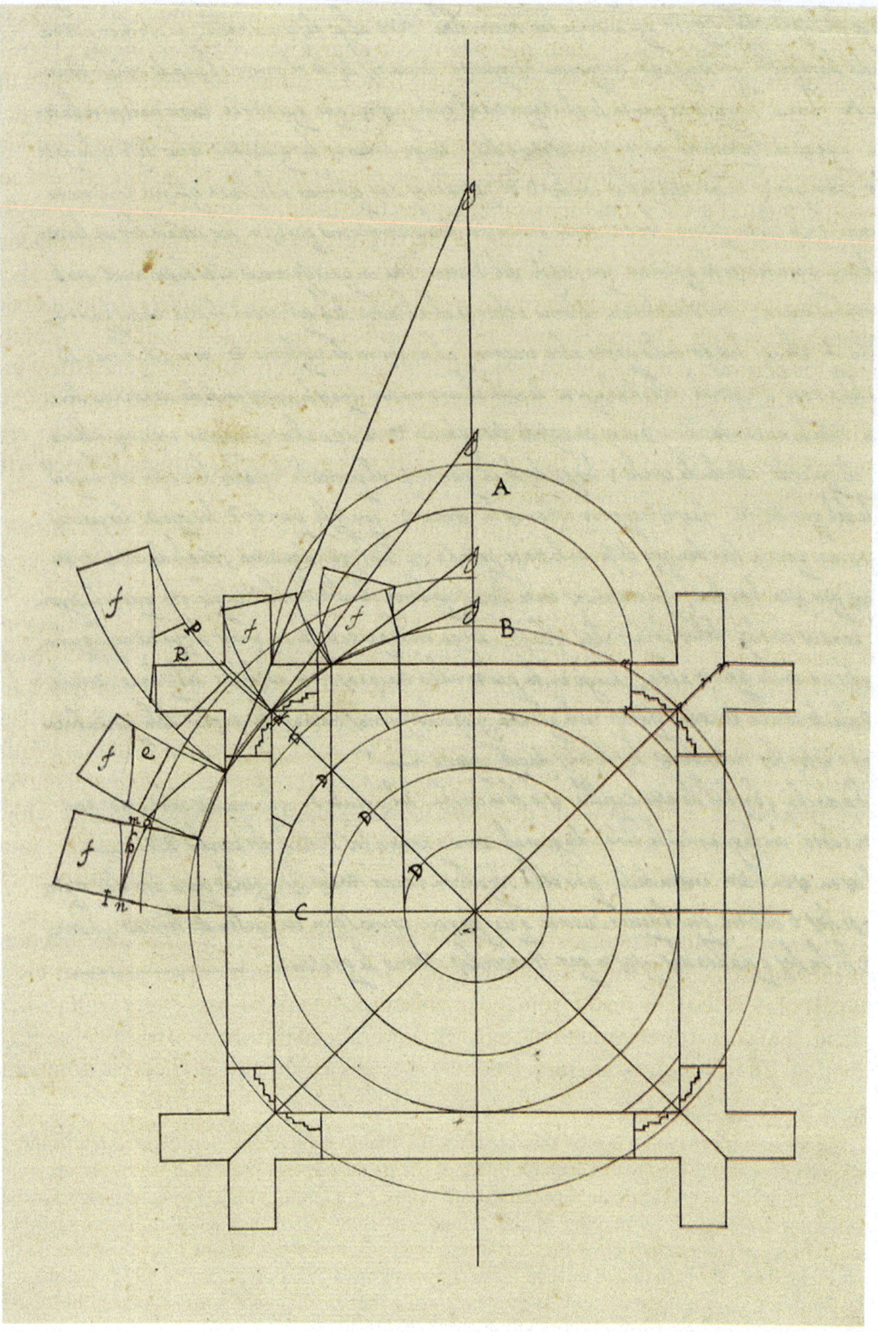

Fig. 9.10 Sail vault with round courses ([Vandelvira c. 1585: 82r] Vandelvira/Goiti 1646, 136)

goes up to the first bed joint; thus, its shape is triangular. However, Vandelvira notices that the intrados of this piece is a portion of a (virtual) regular voussoir, so he inscribes the triangular template in a regular template, such as those for a hemispherical dome. Next, he chooses an arbitrary point in the template and draws a generatrix, standing for the one placed over the diagonal of the area. He then measures the chord of a voussoir in the plan and transfers it to the upper template egde, at both sides of the generatrix. This operation involves a subtle error. Both the plan and the template represent the real length of the voussoir edge in true size. However, Vandelvira and other writers do not seem to be measuring the real length with a rope, but rather the distance between its ends with a compass or gauge; since the radii of the voussoir edge in the plan and the template are different, the results are not exactly equal. In any case, such error is negligible in practice, particularly for the first voussoir. In order to finish the triangular template, he must draw the side edges; for this task, he uses the curve for the joint passing through the arch keystones, which has the same radius as the perimetral arches. However, Natividad (2012b) has shown that Vandelvira's reasoning is subtler: in other pieces, such as rectangular-plan sail vaults, he uses the development of a horizontal-axis cone passing through the perimetral arch. Applying this method to the square-plan vault, he would get the same results from the development of the circle joining the keystones as a part of a vertical-axis cone and the development of the perimetral arch directrix as a part of a horizontal-axis cone.

The rest of the pendentive templates are prepared using the same technique, although of course the outline is a quadrilateral, with two sides standing for bed joints and the other two for intersections with the perimetral arches; all of them are curved. Thus, Vandelvira uses different developments in order to prepare the template: he uses a set of cones with vertical axes for the bed joints and two additional ones, with horizontal axes, for the side joints. Of course, the operation is geometrically inconsistent, since a single template cannot reproduce three different cones, but as shown by Natividad and Calvo (2013), the practical effects of these errors are small.

Another interesting detail in Vandelvira's procedure is an apparently cryptical remark: "Arch templates should follow the radius of the diagonal line, so the voussoirs that rest over them can find a suitable place to lean in, with the same shape that these voussoirs are cut; thus you will use the template marked as r.r. for all the voussoirs of the arches".[15] This means that the extrados of the perimetral arches should be dressed in the shape of a conical surface, and the angle between the generatrices of that surface and the face of the arch should be equal to the one between the diagonal of the plan and the horizontal projection of the arch face. Of course, since the extrados is a circular conical surface, these angles are equal for all voussoirs. Moreover, the rise of the arch equals half the side of the area; thus, the angle between the radius of the sphere at the keystone of the perimetral arch and the face equals 45°, which is also the angle between the diagonal and the face. Since the normal to the spherical

[15] Vandelvira (c. 1585: 81v): *Los moldes de los arcos se han de sacar con la tirantez que corta la línea diagonal por manera que las dovelas que cargan encima hallen adonde cargar, que correspondan con el mismo baivel que ellas van labradas y así será el molde el señalado con las r.r. y con éste se labrarán todas las piedras de los arcos.* Transcription taken from Vandelvira and Barbé 1977. Translation by the author.

surface is given by the radius, the generatrices of the conical surface coincide with the normals to the spherical one, and this result extends to any point of the transverse arch, since both the sphere and the arch feature rotational symmetry around the axis of the arch, which is a horizontal diameter of the sphere. As a result, the shape of the arch extrados furnishes an optimal support for the vault.

Despite the scarcity of such vaults in France, Jousse (1642: 138–139) explains the problem in the same terms as Vandelvira. He is less careful with course divisions,[16] although he places a bed joint along the parallel joining the keystones of the perimetral arches; he does not address the problem of the intersection with these arches. Derand (1643, 375–378) uses basically the same methods, although his explanations are clearer and more practical. To start with, he presents separate drawings for each template in the pendentives. He first draws the axis of each template, taking its width from the cross-section. Next, he draws the bed joint edges of the template, taking their radii from the cross-section, using implicitly the cone-development method. In any case, he explains that the first bed joint can be drawn as a straight line, unless it is appreciably long. He then closes the template by taking the curvature of the side joints from a true-size-and-shape representation of the perimetral arches. Although the rationale of the procedure is more straightforward than Vandelvira's one, the results for a square-plan vault are identical.

De la Rue (1728: 62–63) takes a different route. He comments that the bed joints in the first courses, (that is, those for the pendentive) should be dressed as horizontal surfaces, in contrast to the cones used by other writers. Taking advantage of this simplification, he dresses the voussoirs basically by squaring, although he uses a templet with the radius of the perimetral arches in order to control the curvature of the edges of the intrados surface of the voussoirs. About the courses in the cap, he refers the author to the spherical vaults with vertical courses (de la Rue 1728: 57–60; see also Sect. 9.2.3) where he applies a version of the "drawing in space" method (see Sect. 9.1.1).

Pendentive vaults. The cap in a sail vault is a spherical segment, whose rise is much smaller than a half-sphere. If the designer wants to increase its rise, he or she can use a *pendentive vault*, such as those used in the Old Sacristy of San Lorenzo or the Pazzi Chapel in Florence. In the sail vault, as we have seen, the spherical surface is cut by four vertical planes passing through the sides of a square, in order to adapt the spherical surface to a square plan. Since the springing line of the vault passes through the four corners of the square, at least ideally, the radius of the sphere is equal

[16]In the diagonal cross-section of a square sail vault, the length of the section corresponding to each pendentive is half the length of the cap. Vandelvira divides both the pendentive and half the cap into three parts; the resulting courses feature equal widths, although the diameter of the keystone doubles these widths. In contrast, Jousse divides the pendentive into three sections and the cap into three and a half; as a result, the width of the courses in the pendentive is slightly larger than the courses in the cap and the diameter of the keystone. As we will see further on, Vandelvira uses this latter scheme in sail vaults with vertical courses; in this case, the lower courses are not materialised and this solution poses no problem; this suggests that Jousse and the French theorists are thinking of the sail vault with round courses as a variation of the one with square courses, not the other way around. This may reflect the early appearance in France of square courses in the *Pendentif de Valence*, as opposed to the earliest sail vaults in Spain, which feature round courses.

to half the length of the diagonal of the square. In the pendentive vault, this sphere is cut again by the horizontal plane that passes through the keystones of the perimetral arches; the intersection of that plane with the sphere is a circle whose diameter equals the side of the square spanned by the vault. This circle divides the pendentives of a sail vault spanning this area from its cap. The cap, which is a spherical segment, is removed and replaced by a full hemisphere. The radius of this hemisphere is smaller than that of the cap in a sail vault, but its rise is higher.[17] If the builder wishes to attain a higher rise, he may place a cylindrical drum between the pendentives and the hemisphere as done in a plethora of examples after the Tempietto in San Pietro in Montorio and the first projects for Saint Peter's in Rome.

Thus, the pendentive vault is a combination of the pendentives of a sail vault and a hemispherical vault lying on them. Therefore, it is seldom explained in treatises and manuscripts as an independent problem. An interesting exception is Guardia (c. 1600: 87v; see also Calvo 2015a). The drawing and the explanations make it clear that the piece uses portions of two spheres, a larger one for the pendentives and a smaller one for the top dome, but both are controlled using cone developments, including also a diagram of the intersection of the pendentives with the perimetral arches.

9.2 Spherical Vaults Divided into Vertical Courses

9.2.1 *Quarter of Sphere Vaults*

When a quarter-sphere vault is divided into horizontal courses, the shell hinge (the intersection of the sphere with the axis of the vault), is placed at the top of the vault; however, on many occasions, the hinge is placed at the midpoint of the springing. This involves some interesting geometrical and mechanical problems. First, placing the hinge at springing level involves using a horizontal axis and vertical parallels and courses; meridional planes will revolve around the axis and, rather than being vertical, will adopt different slants. It is interesting to notice that Vandelvira (c. 1585: 68r) uses the word *clave* (keystone) for the single stone placed on the springer at its intersection with the horizontal axis of the vault; in this case, the keystone opens the vault, rather than closing it.

This layout has structural advantages. In a hemispherical dome divided into horizontal courses, each course adopts the shape of a ring and compression stresses are

[17] If the side of the spanned square area is *L*, the diameter of the spherical surface of a sail vault equals the diagonal of the square, that is $L\ x\ \sqrt{2}$, since the equatorial section of the sphere must pass through all corners of the square. The rise of the sail vault equals the radius of the sphere, that is, $L\ x\ \sqrt{2}/2$. The rise of the perimetral arches equals half the side of the square, that is, L/2. Thus, the rise of the cap equals $L\ x\ (\sqrt{2}\text{-}1)\ /2$. Now, in the pendentive vault, the rise of the sphere which replaces the cap also equals $L/2$ and is thus higher than the rise of the cap of the sail vault. Moreover, the total rise of the pendentive vault (without the drum) equals $2L/2 = L$, which is of course higher than $L\ x\ \sqrt{2}/2$.

compensated by the next voussoir. It is true that the lower courses are subject to tensile stresses and cracks may appear, but they do not generally imply a threat to the stability of the vault (Heyman 1995: 35, 42). In contrast, if a quarter-sphere vault is divided using horizontal parallels, the horizontal compressive forces are left uncompensated at both ends of each course, generating thrusts on the abutment, usually a round arch. It is therefore more effective to build the vault as a series of round arches, that is, dividing it in courses by means of vertical parallels; each of these courses is then divided into voussoirs using meridans rotating around a horizontal axis. It may be argued that such vertical courses may tend to work as independent arches, developing cracks between each other. Although apparent joints between courses are laid on vertical planes, internal joints are usually materialised as portions of cones with their apexes on the vault axis, just as joints in the standard hemispherical domes; this minimises the risk of cracks between courses.

Vandelvira's third solution for quarter-sphere vaults follows this scheme. Rather than actual quarter-sphere vaults, Vandelvira (c. 1585: 68r) presents a vault with a half-oval plan and a round face (Fig. 9.11). However, his explanations can be particularised easily for a semicircular plan. As we would expect, he uses auxiliary cones, but their apexes are placed at the axis of the vault, which is horizontal, rather than vertical; the fact that the plan is oval does not introduce any significant change from a round plan.

9.2.2 *Sail Vaults*

Square-plan sail vaults with vertical courses parallel to the perimetral arches. The solution we have seen in Sect. 9.1.3, based in horizontal parallels and vertical meridians, is used frequently in pendentive vaults in Western Europe. When applied to sail vaults, however, it causes a great waste of stone in the pendentives and is not aesthetically ideal. Around 1540, a variant arose at the same time in France and Spain, both in the *Pendentif de Valence*, the funerary monument of Canon Nicole Mistral, probably built by De l'Orme (Pérouse 2000: 117, 344–35) and in the sacristy of the Sacred Chapel of the Hospital of El Salvador, in Úbeda (Fig. 9.12), designed and built by Andrés de Vandelvira, the father of Alonso (Chueca 1971: 129–130; Galera 2000: 77–81). In this solution, the vault is ideally divided into four quarters by vertical planes passing through the diagonals of the area. In each of these quarters, the vault is split into courses by a set of vertical planes, parallel to the nearest side of the plan. However, the internal joints between courses are not vertical planes, but rather conical surfaces with their apexes placed at the centre of the sphere, generated by normals to the spherical surface. This solution avoids the waste of stone brought about by the use of round, horizontal courses. However, it raises a new problem, namely the junction of each course with another one set at right angles, in the next quarter of the vault. The most sophisticated solution involves V-shaped pieces jumping across diagonals. Another less elegant, more practical variant uses ordinary voussoirs crossing the diagonals with their tips; usually voussoirs from both quarters alternate in these slight

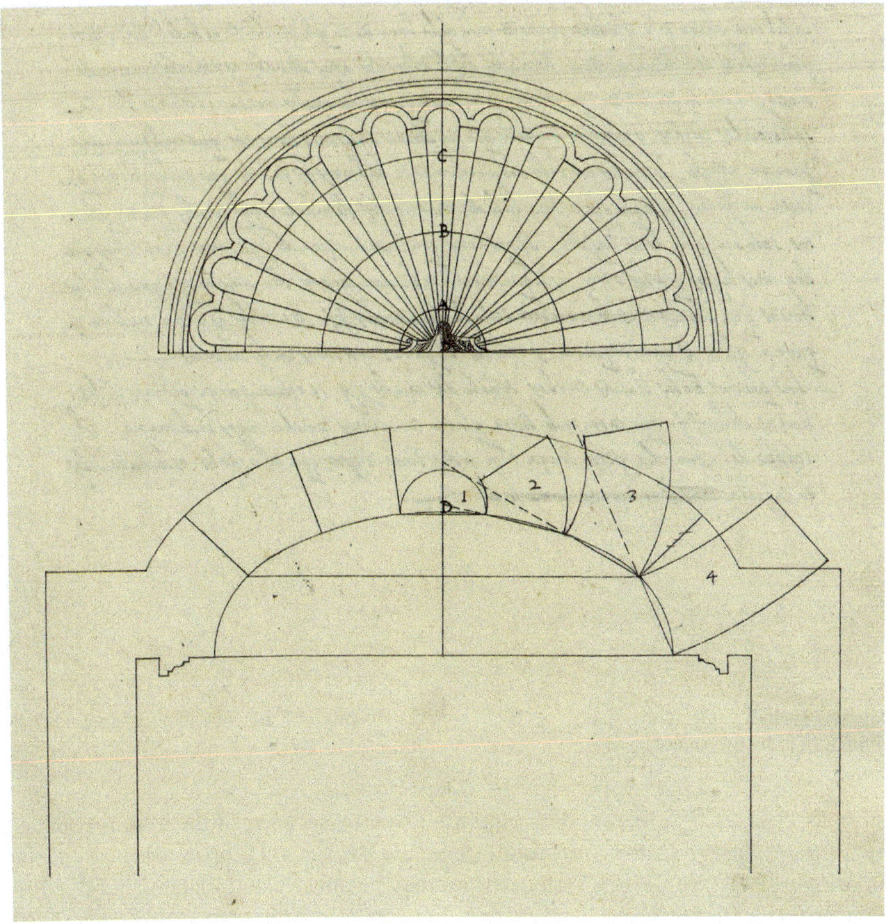

Fig. 9.11 Quarter-oval dome ([Vandelvira c. 1585: 68r] Vandelvira/Goiti 1646: 124)

penetrations (Fig. 9.13). In any case, all texts, including those closely connected with workplace practices, such as Alonso de Guardia (c. 1600: 65v, 67v, 84v), focus on the V-shaped solution. Both solutions involve templates based on cone developments, just like as in hemispherical domes; however, in these cases the directrices of these cones are vertical, and thus the axes are horizontal.

As we have seen, de l'Orme starts with a complex variant of this solution, explaining only sail vaults with vertical courses parallel to the diagonals of the area, so I will start again with Vandelvira (c. 1585: 83v-84r; see also Palacios [1990] 2003: 264–267; Potié 1996: 120; Natividad 2017: I, 60–64). After drawing the square plan and the cross-section along the axis of the area, he divides an eighth part of the circle, standing for half the cap, into three portions and a half (Fig. 9.14). The half part will furnish half the keystone; in this case, the diameter of the boss equals the width of

Fig. 9.12 Sail vault with square courses. Úbeda, Hospital of El Salvador, chapel sacristy (Photograph by the author)

ordinary courses.[18] Apparent bed joints are given by sections of the intrados surface by vertical planes, so they are small circles; however, they are projected in plan as straight lines (see Sect. 1.5); these lines may be constructed transferring the divisions from the section to the plan and drawing parallels to the perimetral arches. Next, Vandelvira applies the cone-development method to construct ordinary templates, using a horizontal axis passing through the centre of the area.

So far, so good, but now he must address the construction of the V-shaped templates for the voussoirs across the diagonals. Vandelvira's exposition seems cryptic at first sight, since he tries to explain a complex operation in a few lines; however, the first phase of the operation is sound and quite elegant. He reuses once again the circle surrounding the plan, in this case as a cross-section through the vertical plane passing through the diagonal of the area. However, in contrast to our usual graphic conventions, he employs the axis of the plan again as a folding line, instead of using the diagonal, as a twentieth-century architect or engineer would do;

[18]This solution is used by Jousse in the round-course sail vault. Vandelvira eschews it in that case, since it leads to different course widths in the pendentives and the cap. Instead, he uses it in the square-course vault, since in this case the lower portions of the cross-section correspond to four circular lunettes that are cut by the planes of the perimetral arches and left out the area covered by the vault.

Fig. 9.13 Sail vault with staggered voussoirs along the diagonals of the area. Navamorcuende, parish church (Photograph by the author)

in any case, he uses the right half of the enclosing semicircle as diagonal section, to avoid confusion with the ordinary templates placed on the left half. Next, he measures in the plan the distances between the points where the bed joints belonging to two different quarters meet, transferring them to the folding line of the diagonal section. He then projects these points onto the diagonal section; this operation provides their distances in true size. After this, he prepares the templates for corner voussoirs in a separate diagram, leaning on a common axis. First, he transfers the distance between the springing and the upper corner of the first template to this axis. Using the upper corner as a centre, he draws an arc whose radius equals the breadth of the first course, taken from the axial section. Next, he uses a templet with the curvature of the outer edge of the template for the first course, placing one end on the springing and adjusting it so that it touches the arc with the other end. He repeats the same procedure for other courses until he reaches the keystone. He then draws lines connecting bed joints orthogonally, standing for side joints; it is important to notice that these lines can be placed at will and bear almost no relation to the rest of the procedure. Although the operation is quite ingenious, it poses some problems. For

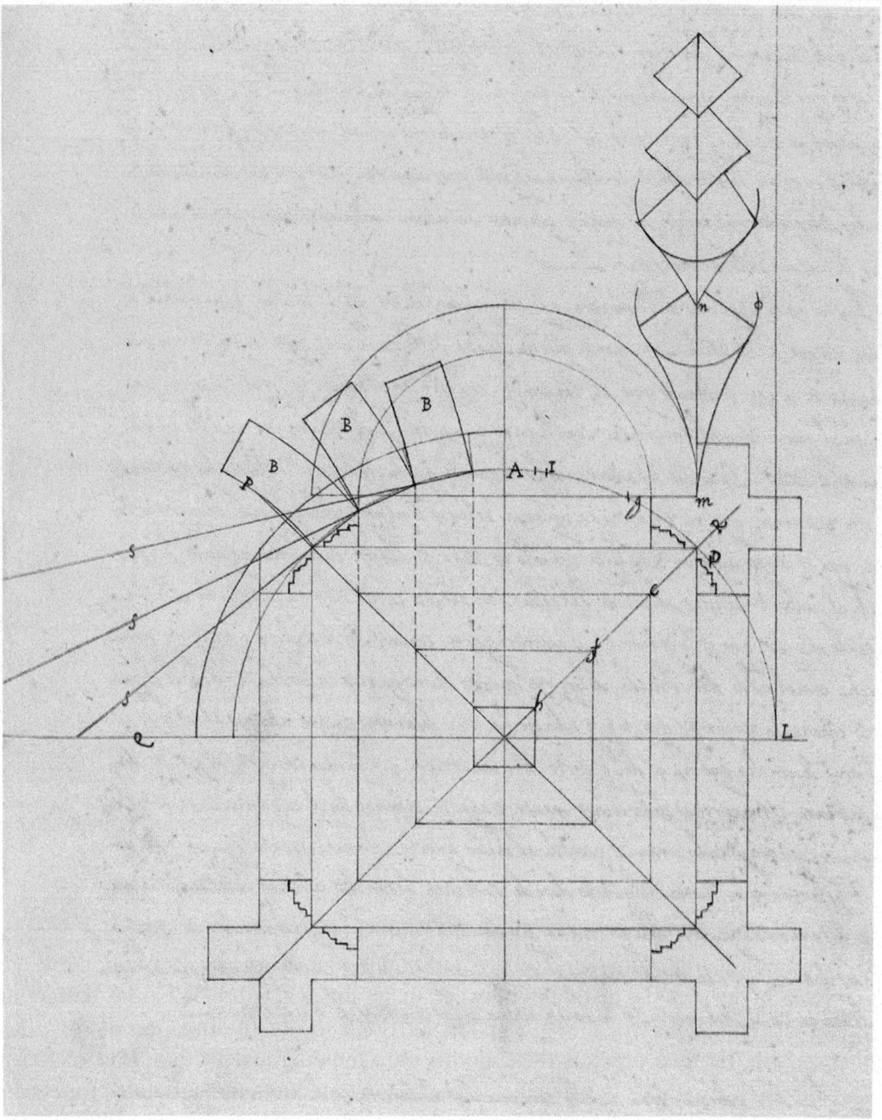

Fig. 9.14 Sail vault with square courses ([Vandelvira c. 1585: 84r] Vandelvira/Goiti 1646: 140)

each course, the author is mixing the developments of two different cones, one for each quarter of the vault. Further, Vandelvira himself (c. 1585: 61v) had remarked that the curvatures of the templates for voussoirs that meet at a given bed joint are different. In this case, he uses a single curve for each bed joint, the one for the lower edge of the uppermost template. Thus, for each voussoir overlapping the diagonal, he is combining no fewer than four cone developments: two for the branches of the given course, and a further pair borrowed from the next course.

At first sight, Jousse's drawing (1642: 132–133) seems to represent a vault with diagonal courses, but closer inspection reveals that he has rotated the whole diagram to fit the page, and courses are parallel to the sides of the area. In other aspects, he uses basically the same method as Vandelvira. However, to minimise the work of tracing, he draws all templates for the same course together, including ordinary voussoirs, with the centres of the arcs standing for bed joints at the intersections of generatrices with the axes of the vault. This furnishes another method for the control of the angle between bed joints, at the expense of placing the set of templates for the uppermost courses far from the rest of the tracing. In contrast with Vandelvira, different courses are placed separately; the edges of the template standing for the same bed joint feature different radii in different courses. Still, each template is a combination of developments of two different cones.

Neither Derand nor De la Rue include a solution for this problem; however, they address the hemispherical dome divided by vertical courses, which includes a square sail vault by vertical courses as its kernel. Derand (1643: 364–365) follows basically Jousse's solution, although he explains it more clearly (Fig. 9.15, below); as in other occasions, De la Rue takes exception against the cone development method and puts forward other solutions. I will deal with these issues in detail in Sect. 9.2.3.

Square-plan sail vaults with diagonal courses. Soon after the emergence of sail vaults with square courses, de l'Orme introduced a new variation: instead of laying out the courses in vertical planes parallel to the perimetral arches, he set them parallel to the diagonals of the area. Of course, the reason for this choice is purely aesthetic; it causes a great waste of stones in the corners of the vault. This fact probably discouraged the use of this solution for almost a century; as far as I know, they were used for the first time by Juan de Aranda Salazar, a nephew of Ginés Martínez de Aranda, in the Jaén Cathedral, starting in 1653 (Fig. 9.16).

De l'Orme (1567: 111v-113r; see also Potié 1996: 118–123; Carlevaris 2000: 88–91; Natividad 2017: I, 159–163) draws first a square plan and an enclosing circle (Fig. 9.17). Next, he uses this circle as a diagonal cross-section, dividing a quadrant in thirteen and a half parts.[19] At first sight, this choice may seem arbitrary. However, de l'Orme uses nine divisions for the lower portion and four and a half for the cap; adding the symmetrical half, the cap encompasses nine portions. As a result, the bed joint dividing the pendentives from the cap falls exactly on the ninth division, while the width of the keystone equals that of an ordinary course.[20] Next, de l'Orme

[19] De l'Orme's explanations do not make it clear whether he uses this diagonal section or an additional elevation of the perimetral arch in order to divide the pendentives into courses. However, Natividad (2017: I, 159, 162) has reproduced the drawing in a CAD program, reaching the conclusion that de l'Orme uses the diagonal section for dividing both the pendentives and the cap.

[20] The diagonal of the large square equals the diameter of the enclosing circle. Thus, the side of the large square and the diagonal of the small square equal $D/\sqrt{2}$, and thus the side of the small square equals $D/(\sqrt{2}x\sqrt{2}) = D/2$. The diameter of the cross-section can be divided into a section equalling the side of the small square, measuring D/2 and two portions which are coincident with the heights of the triangles between both squares; each of this portions measures $(D/2)/2 = D/4$; this gives the

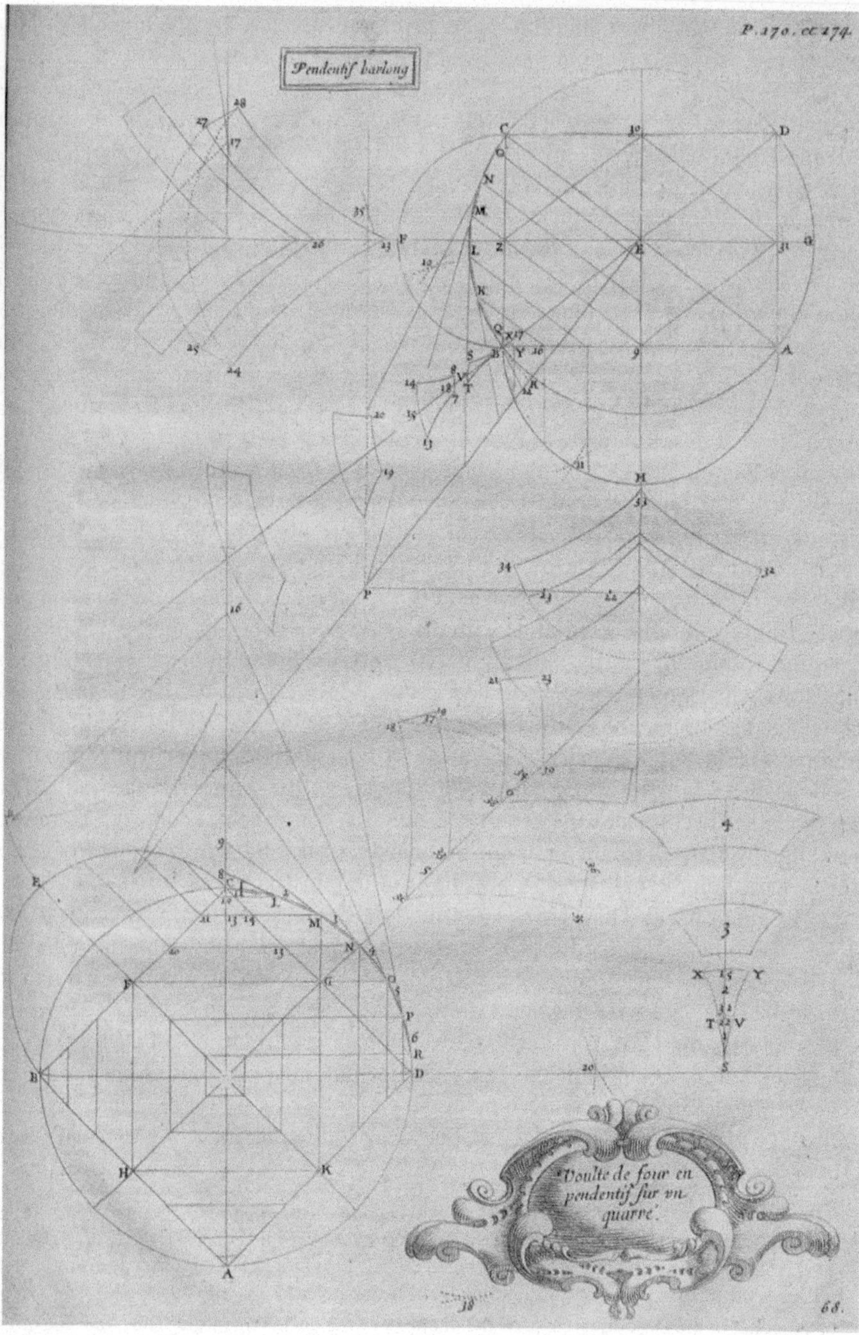

Fig. 9.15 Above, rectangular sail vault with diagonal courses. Below, hemispherical vault with square courses (Derand [1643] 1743: pl. 170, 174)

Fig. 9.16 Sail vault with diagonal courses. Jaén, cathedral, aisle (Photograph by the author)

projects these divisions onto the diameter of the cross-section; the intersection points allow the construction of the intrados joints in plan. The ensemble of the four joints dividing the cap from the pendentives forms a smaller, rotated square; its corners are placed at the keystones of the perimetral arches. The courses at the cap turn at the corners, and thus require V-shaped voussoirs for an elegant result, while the courses

height of the triangles. Now, a half-hexagon can be inscribed in the cross-section, dividing it into 3 parts. If we project a corner of the half-hexagon onto the diameter of the cross-section, the distance from this point to the springer may be computed taking into account that the angle between the side of the half-hexagon and the base of the cross-section equals (D/2) x cos 60º = D/4. Thus, the projection of the point that divides the cross-section into equal parts is coincident with the edge of the lesser square. Now, this point divides a quadrant of the cross-section into 2 portions, the lower one twice the upper one. Thus, dividing the quadrant into 18 and 9 parts will furnish a point standing exactly over the division between the lesser square and the triangle. However, this scheme would lead to very small voussoirs, so de l'Orme takes 2 parts for each voussoir, using 9 portions for the peripheral triangles and 4 ½ parts for each half of the cap.

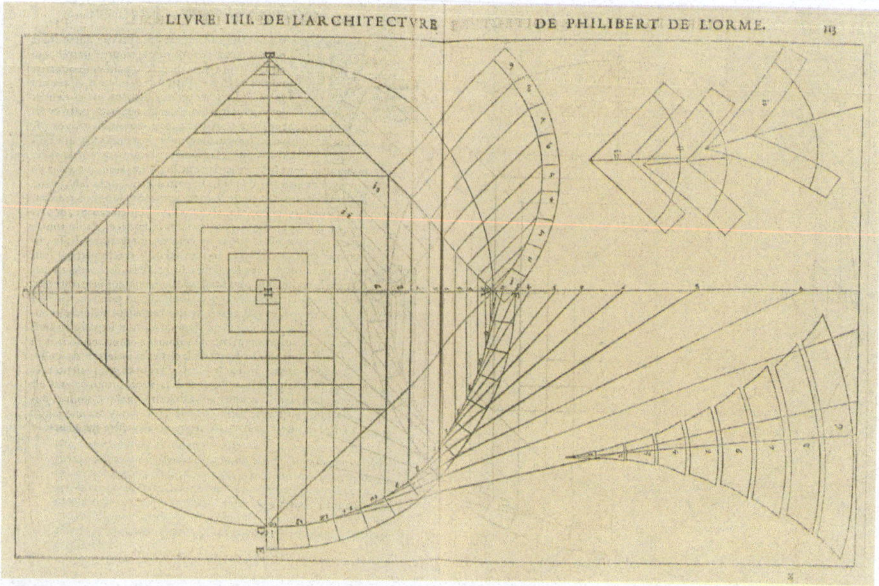

Fig. 9.17 Sail vault with diagonal courses (de l'Orme 1567: 112v-113r)

at the pendentives go neatly from one of the perimetral arches to the next one and can be solved with ordinary voussoirs, provided they are cut at the ends to fit the perimetral arches.

In the next phase, de l'Orme uses the cone-development method to construct the templates for all courses, both in the pendentives and the cap, although he does not mention the word "cone", as usual. He draws generatrices passing through two consecutive bed joints until they reach a horizontal axis. However, he does not construct the templates starting from the cross-section, as Vandelvira did; he places them separately. He explains that the curvature of these templates should equal the distance between these intersections and the axis of the vault; however, he says nothing about the length of the templates or the construction of the V-shaped templates, arguing that "it would be much more expedient to show the practice of these vaults visually by making [models of] them manually, than writing everything that is necessary to understand this practice … so I beg you to be satisfied with my figures and lines".[21]

[21]De l'Orme (1567: 112v): *Il serait beaucoup plus expédient de montrer à l'œil la pratique de telles voûtes pour les contrefaire manuellement, que vouloir entreprendre d'écrire tout ce qui serait nécessaire pour faire entendre la dite pratique... Pour ce est il que je vous prie de vous vouloir contenter, de ce que je vous en montrerai par figures et traits.* Transcription taken from http:// architectura.cesr.univ-tours.fr. Translation by the author. The meaning of *contrefaire* is not completely clear since, in stonecutting parlance this word has two meanings: "copy literally", as when reproducing a personal notebook with self-instruction purposes and "make a model". Thus de l'Orme may be suggesting the reader to reproduce the drawing in order to grasp its construction, or either to build a model to understand the rationale of this element. .

Natividad (2017: I, 162–163) has put forward an interesting hypothesis about the construction of the V shaped templates. First, de l'Orme draws two arcs with the curvatures of the templates, taken from the generatrices drawn in the cross-section. Next, he may have measured the distances between corners of the courses from an additional cross-section of the vault by a plane passing through the centre of the vault, parallel to the sides of the area.[22] This section is not explicitly drawn, but Natividad argues that he may have reused the diagonal cross-section as a section through the axis, marking four points on it; in fact, de l'Orme used this method in his triangular sail vault (1567: 116r-117). This operation may furnish the length of the axis of the template, that is, the line where both branches of the V meet. Next, it is easy to draw another pair of arches, since they are symmetrical about the axis. All this would explain why the templates and the axes are arranged apparently at random, in contrast to Vandelvira (c. 1585: 89v-90r; Palacios [1990] 2003: 278–281; Natividad 2017: I, 84–87).

Vandelvira uses a similar method (Fig. 9.18). For example, he divides the quadrant in seven and a half portions. This leaves five courses for the pendentives, while the cap includes two courses and the keystone; the inner square coincides exactly with the fifth division. An interesting addition is the inclusion of an elevation of a pendentive, using a projection plane parallel to the diagonal of the plan. As suggested by Natividad (2017: I, 84–87), this elevation is used to compute the length of the bed joints of the templates for the pendentives; as for the V-shaped templates for the cap, they are constructed using the same method used for the templates of the sail vault with square courses—that is, starting from a vertical axis, not from the edges of the templates.

Again, Jousse (1642: 140–141) uses basically the same method as de l'Orme and Vandelvira; however, he introduces an interesting innovation. For each template, he uses three parallels: both side joints and a virtual parallel between both, which furnishes a point of the side joint. Thus, he draws the side joint as an arc passing through three points, rather than taking it for granted that the template has the same curvature as the perimetral arch.

Although Derand (1643: ii r; see also Sect. 2.2.6) had direct knowledge of Jousse's book, he ignored this innovation and reverted to de l'Orme and Vandelvira's method: taking the curvature for the side joints of the lower voussoirs from the perimetral arches (Derand 1643: 370–375).[23] Also, his method of division seems a bit clumsy: he divides a perimetral arch into equal parts and transfers the divisions to the plan, drawing the bed joints of the pendentives. However, this method cannot be used in order to construct the bed joints of the cap, so he uses the plan instead; as a result,

---

[22]In addition to the diagonal cross-section, there is another quadrant divided into voussoirs in the drawing, but it represents an elevation of a perimetral arch; that is, a section of the vault through a vertical plane *parallel* to the horizontal axis. However, the section through a vertical plane *passing through* the axis is not explictly drawn.

[23]There is an error in the numbering of these pages; the actual numbering is 370, unnumbered plate, 378, 379, 380, 375; pages 378, 379 and 380 should be numbered 372, 373, 374. In fact, page numbers 378 and 380 reappear at their proper places.

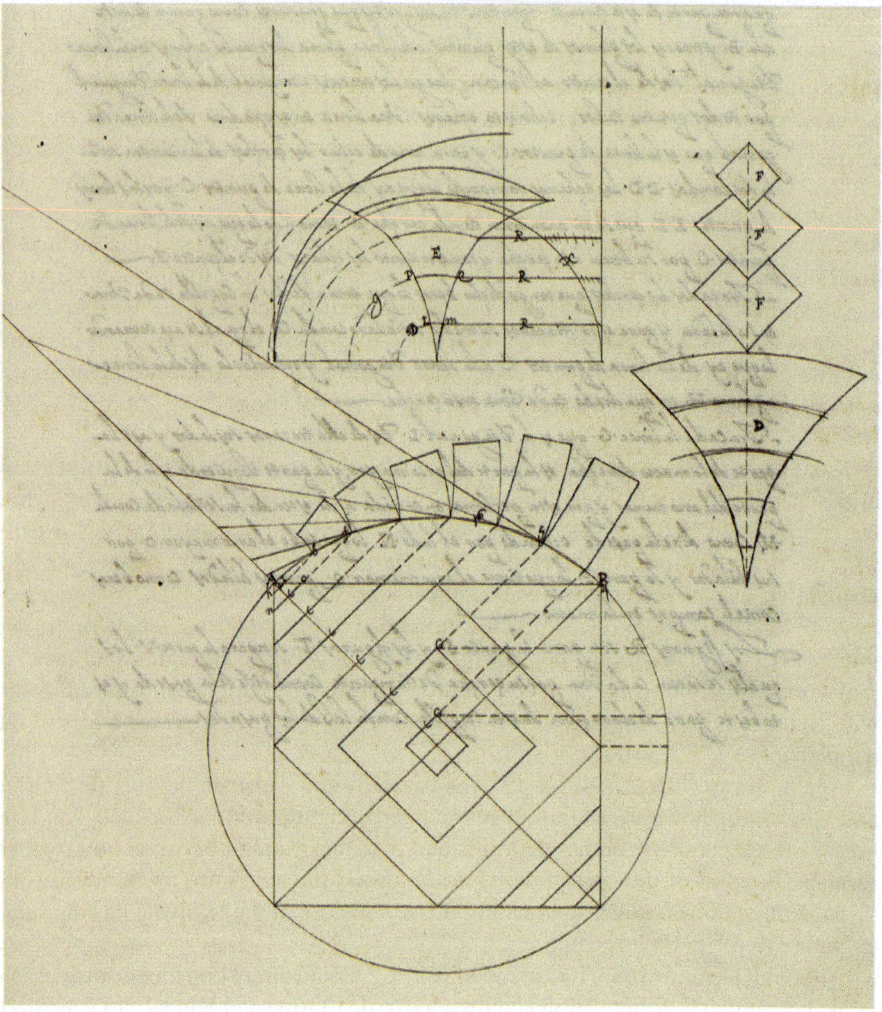

Fig. 9.18 Sail vault with diagonal courses ([Vandelvira c. 1585: 94r] Vandelvira/Goiti 1646: 152)

course widths are uneven. De la Rue (1728: 64, pl. 35) rejects templates here, as in other spherical vaults, so he dresses the voussoirs by squaring.

Rectangular-plan sail vaults. As with square-plan sail vaults, courses in rectangular-plan sail vaults can be round, parallel to the sides, or parallel to the diagonal. Vandelvira's (c. 1585: 82v-83r) solution for the rectangular vault with round courses uses the same techniques he employed for its square-plan counterpart, but it is quite clear from the start that new problems arise. First, perimetral arches have different diameters, and their keystones are placed at different levels; as a result, it is not possible in practice to adjust courses to make a particular bed joint pass through all four keystones.[24] In spite of this, Vandelvira prepares a set of ordinary templates leaning on the cross-section, developing cones with vertical axes, as he did in the round-course square-plan vault. Next, he draws another set of templates for the voussoirs that meet the perimetral arches. Natividad (2012b; 2017: I, 57–60) remarks an interesting detail in this phase: Vandelvira computes the curvature of the intersections of the pendentives with the perimetral arches developing horizontal-axis cones, with the perimetral axes acting as directrices. Such developments are used for the sides of the templates of the pendentives, as in the square-plan vault, but the result is quite different: the curvatures of both sides are different, since perimetral arches have different diameters; further, some courses, in this case the fourth one, intersect the larger perimetral arch, but not the smaller one. All this gives Vandelvira's result an untidy aspect; it is no wonder that other writers leave this problem aside.

In contrast, the rectangular-plan vault with vertical courses is explained by most stonecutting authors, starting again with Vandelvira (c. 1585: 84v-85r). He draws the rectangular plan with its diagonals, the perimetral arches and the enclosing circle (Fig. 9.19). Next, he divides a quadrant in a remarkable way. Of course, the diagonal intersecting the quadrant divides it in two portions of uneven length; Vandelvira divides each of these portions in four and a half parts. As a result, the widths of the courses in the quarters that meet at the diagonal are different, but intrados joints intersect neatly at the diagonal, in contrast to some built examples such as a vault in the parish of Navamorcuende, where V-shaped voussoirs are not used and a course in one quarter meets two courses in the other quarter. Going back to Vandelvira, since the courses in both quarters are different, he prepares two different groups of ordinary voussoirs, using two series of cones with their vertexes placed along both axes of symmetry of the plan. Of course, all this was unnecessary in the square-plan vault, since it is symmetrical about the diagonal of the plan. As for the V-shaped templates, Vandelvira solves the problem with a single set of templates, relying on the bilateral symmetry of diagonals about the vault axes. However, the vault is not symmetrical about the diagonals of the area, the templates themselves are asymmetrical, and Vandelvira must repeat constructions at both sides of the axis using different distances and radii. However, the voussoirs belonging to the same diagonal are identical at both sides of the vault, since they may be duplicated by a revolution of 180°, while those at the other diagonal are symmetrical to the first set around one of the axes of the vault, so they may be reused by turning them around the axis.

As for the rectangular vault with diagonal courses, the first explanation is offered by de l'Orme (1567: 114v-115r), as we would expect (Fig. 9.20). He follows the same route he took in the square-plan vault. However, in this case he makes an important mistake, as Natividad (2017: I, 167) has remarked. In order to employ the cone-development method, he uses a cross-section through a plane passing through one of the diagonals of the plan. This procedure is quite sound in the square vault, since

[24]Of course, in theory it is possible to draw a plan for a square vault with round courses and adjust the sides of the area so that they are tangent to different bed joints, arriving at a rectangle. But this solution has no practical advantages, since rectangular plans arise from constraints that lead architects or masons to use rectangles with predefined dimensions.

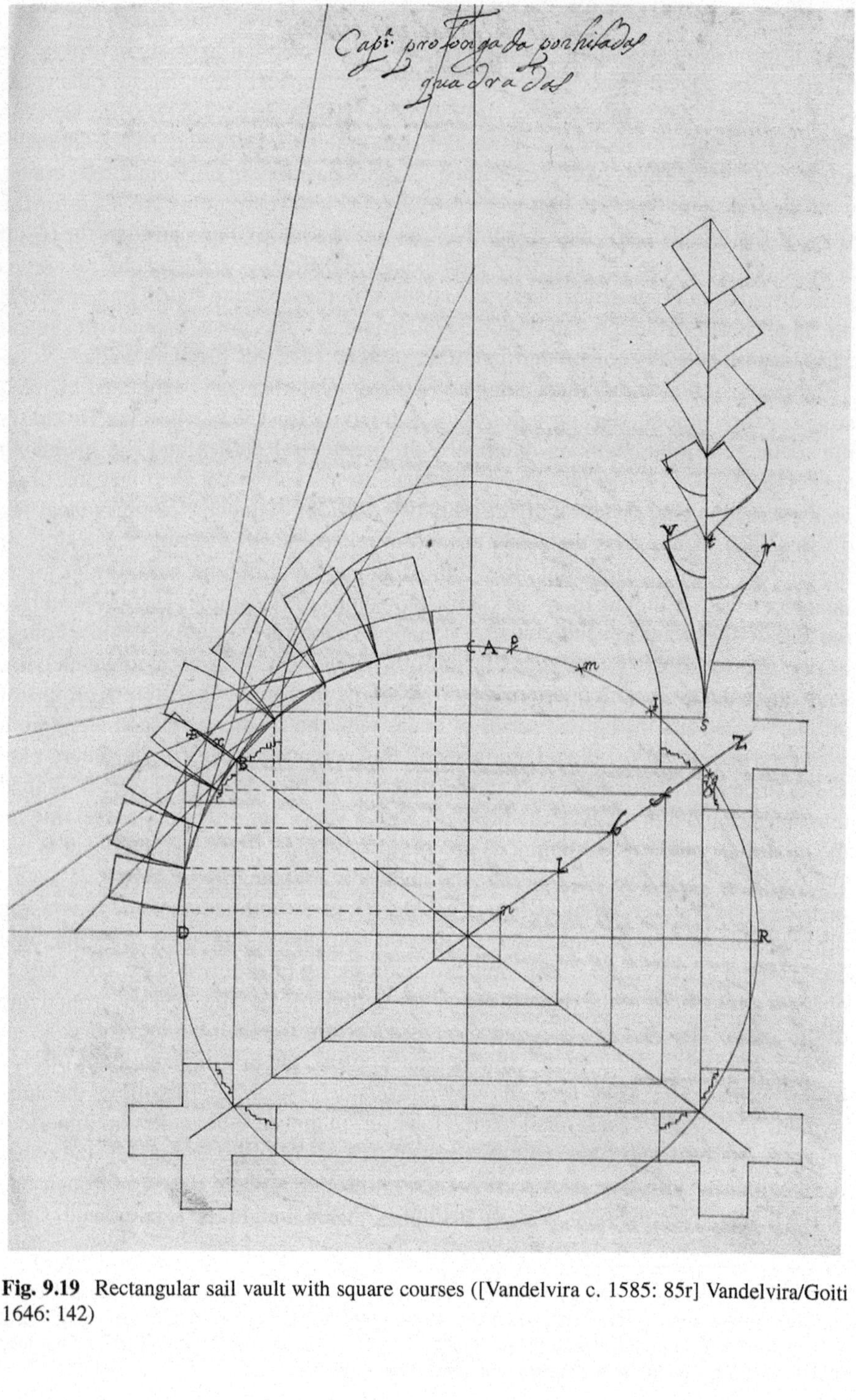

Fig. 9.19 Rectangular sail vault with square courses ([Vandelvira c. 1585: 85r] Vandelvira/Goiti 1646: 142)

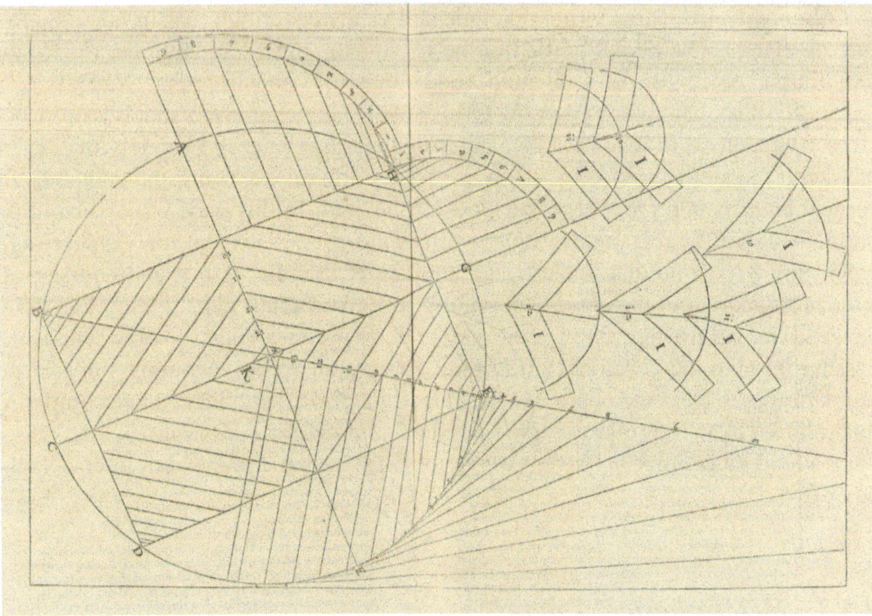

Fig. 9.20 Rectangular sail vault with diagonal courses (de l'Orme 1567: 114v-114v bis)

both diagonals are set at right angles; thus, one of the diagonals may be used as the axis for a set of cones whose directrices are parallel to the other diagonal. However, when applied to the rectangular-plan vault, the two diagonals are not orthogonal, and de l'Orme does not take this into account; he still uses a diagonal as the axis of a set of cones whose directrices are parallel to the other diagonal. In other words, he is trying to develop an oblique cone, but he is not aware of this fact and draws the edges of the templates as circular arcs, as in the square-plan vault, so his result is seriously flawed.

Vandelvira (c. 1585: 90v-91r; see Natividad 2017: I, 87–91) notices that something is wrong with de l'Orme's method, but his explanations are not completely clear. He draws an orthogonal to one of the diagonals and starts drawing a section through the vertical plane stemming from this diagonal; however, he does not use this section to construct templates, so his rationale is blurred. Jousse (1642: 142–143) also draws orthogonals to the diagonals, but he does not say a word about the whole procedure. In any case, Derand (1643: 378–381)[25] solves the problem correctly: he draws an orthogonal to one of the diagonals, uses it to construct the cross-section and the cones, and prepares the templates with the radii resulting from this construction.

[25]It should be noticed that the 1643 edition includes a misnumbered p. 378, two pages after p. 370; thus, the section about the rectangular vault with diagonal courses starts in the second p. 378, put in its proper place.

Other variants. Most writers include sail vaults covering areas in other shapes, such as triangles, both regular (de l'Orme 1567: 116r-117r; Vandelvira c. 1585: 87v-88r; Jousse 1642: 136–137; Derand 1643: 360–363) and irregular (Vandelvira c. 1585: 88v-89r), pentagons (Derand 1643: 368–370); octagons (Vandelvira c. 1585: 107v-108r; Jousse 1642: 144–145); and irregular quadrilaterals (Vandelvira c. 1585: 85v-86r). Generally speaking, these problems are solved using the methods for vertical courses laid parallel to the sides of the area. In this way, writers avoid the problems raised by rectangular vaults with round courses, except for Vandelvira (c. 1585: 86v-87r), who also includes an equilateral triangle with round courses. He divides a quadrant of the cross-section into seven and a half parts, giving five to the pendentives and two and a half to the cap. In this way, he manages to draw a bed joint passing through the keystones of all three perimetral arches, avoiding some of the problems posed by round courses in irregular figures. This is evident when comparing this solution to an irregular triangle solved using round courses (Vandelvira c. 1585: 88r), where a bed joint is tangent to one of the perimetral arches but intersects the other two.

9.2.3 Hemispherical Domes

Square courses. Jousse (1642: 130–131), Derand (1643: 364–368), De la Rue (1728: 57–61) and Frézier (1737–39: II, 331) present solutions for a hemispherical dome divided into courses by vertical-plane intrados joints. This seems an afterthought, derived from sail vaults; while vertical courses avoid awkward joints in sail vaults, this solution is not necessary in hemispherical domes, where vertical courses seem to be brought about by fancy. Following Derand's (1643: 364) explanations, which are the clearest, the solution involves four intrados joints forming a square and meeting the springings at the corners of the square. The kernel, that is, the portion inside this square, is identical to an ordinary sail vault, while the sections outside it take the shape of four peripheral half-caps divided by means of vertical intrados joints. He instructs the reader to draw a circle and divide a quadrant into five portions. However, this is a half-truth. Actually, he divides it into six parts; the length of the end ones is half the length of the middle ones; in other words, he divides each of the cross sections of the kernel and the cap into two and a half parts. He then draws bed joints for these divisions. The templates for the caps are constructed just like the templates for spherical vaults, although of course the axes of the cones are horizontal. The templates for the interior square are constructed using the same method used by Jousse both for hemispherical domes with vertical courses and sail vaults (1642: 130–31, 132–133), but there is an interesting variation. In order to avoid the templates for the uppermost courses being placed too far from the vault, outside the available floor or wall area, Derand advises the reader to draw a parallel to the diagonal of the square kernel and move the centre of the template edges along this diagonal. Such insistence in keeping the distance between the centre of the edges and its intersection constant confirms that Derand and Jousse use this factor to control the angle between template edges.

De la Rue (1728: 57–61, pl. 31–33) uses a completely different strategy. First, he takes exception to the use of conical developments; at the same time, he rules out the use of the squaring method, to avoid a large waste of material. Thus, he constructs the templates directly on the stone, as he did in ordinary hemispherical domes. In the peripheral caps, he resorts to the method he used in the spherical vault with horizontal courses (1728: 50–52, pl. 27 bis; see Sect. 9.1.1), constructing the templates in the spherical surface of the "bowl" by triangulation. However, for the V-shaped templates along the diagonals, things are not so simple. In order to construct the intrados template for the first-course voussoir, he scores on the "bowl" a line standing for the diagonal section of the vault, and he marks on it the distance between the upper, *37*, and lower, *25*, corners of the voussoir standing on the diagonal; the latter is the springer of the whole vault. Next, he endeavours to place the ends of the lower edges of the voussoir, drawing an orthogonal to the preceding line through a point, *28*, placed at the same level. Of course, he cannot use a square for this operation; instead, he uses two points on the line, *26* and *29*, and draws four arcs with the same radii; the intersections of the arches at each side of the line furnish two points of the orthogonal. Then, he transfers the distances between the edge ends from the plan to the "bowl", placing the voussoir corners at points *32* and *33*. Repeating the same operations for the upper corners, he places points *34* and *35*. In order to join these points, he notices that the arcs joining *25* and *33*, *37* and *36*, etcetera, are not great circles of the intrados surfaces; thus he constructs the lesser circles materialised by intrados joints as rabatments in the plan, in order to construct a series of templets to draw the edges of the voussoir on the "bowl"; the same method can be applied for the voussoirs of the second and successive courses. Moreover, De la Rue (1728: 61–63) remarks mercilessly that the method used by de L'Orme, Jousse and Derand for the construction of the V-shaped templates is wrong, devoting a full section to the issue and showing graphically that the sides of the template for the first course intersect one another; he admits that this problem is negligible in the second and following courses.

Other variants. Derand also explains other variants of these schemes; the outline dividing the inner section from the half-caps may be triangular, rectangular or pentagonal (Derand 1643: 360–363, 366–370); besides, the intrados joints may be laid *faisant le plan d'une voute d'arestes*, that is, following the plan of groin vault (Derand 1643: 386–390); if the vault with square courses follows the plan of a pavilion vault projected into a spherical surfaces, the plan of a groin vault leads to courses laid out in the shape of a central cross surrounded by L-shaped courses. While this solution has some precedents in the Escorial (Natividad 2017: I, 219–221), the triangular and pentagonal vaults are included by Guarini in *Euclides adauctus* (1671: 588–589; see also Guarini 1737: 246–28, pl. 12); Boetti (2006; see also Borin and Calvo 2019) has connected the interest of Guarini in these problems with the use of crossing arches in San Lorenzo and Santissima Sindone in Turin. Moreover, the use of spherical-intrados ribs in these churches guarantees continuity between ribs at the intersection, in contrast to the Islamic ribbed vaults frequently quoted as a precedent for Guarini's work, which use cylindrical intradoses leading to groins and creases in the intersections.

9.3 Other Division Schemes

9.3.1 Hemispherical Spiral Domes

Hemispherical domes can be divided using other schemes than parallels and meridians. A striking pattern, based on a spherical spiral drawn on the intrados surface, can be found in some built examples (Fig. 9.21). The result is paradoxical: although all stones belong in the same course (Rabasa 2003), their lengths can be adjusted to the size of the blocks coming from the quarry. Meridional joints separate each

Fig. 9.21 Spiral dome. Murcia Cathedral, antesacristy (Photograph by the author)

Fig. 9.22 Double-spiral vault. Jerez, San Juan de los Caballeros (Photograph by Pau Natividad)

voussoir from the next one. The earliest and largest example, dating probably from 1531 (Belda 1971: 222; Gutiérrez-Cortines 1987: 155–161; Vera 1993: 108–109; Rabasa 2003: 1686–1687; Calvo et al. 2005: 123–136) was built in the antesacristy of Murcia Cathedral; smaller ones are used to cover spiral staircases in Plasencia or León. A most interesting example, treated as a sail vault, is placed over the atrium of the church of San Juan de los Caballeros in Jerez de la Frontera, featuring a double spiral (Rabasa 2003: 1687; Natividad and Calvo 2014) (Fig. 9.22).

Despite these Spanish instances, the first explanation of the tracing procedure for such vaults (Fig. 9.23) is included in De l'Orme's treatise (1567: 119r-119v). He mentions that the vault can be used "to cover a round tower, or the ceiling of a spiral stair"; this suggests a connection with the Spanish precedents. He adds that it can be also placed "over a staircase that may be built in the shape of a pyramid"; he is probably thinking about a conical roof.[26] Other than that, he offers no real explanation of the tracing and dressing procedure in the text, so any interpretation must start from the drawing as an exclusive source.

As explained by Rabasa (2003: 1681–1683), de l'Orme's method raises intractable problems. After drawing a plan, reused as a cross-section as usual in ordinary domes, he constructs the spiral in plan, increasing its radius by the same amount in each revolution, in the fashion of an Archimedean spiral. In other words, the width of the course, measured in horizontal projection, remains constant. Next, he projects the spiral onto the cross-section. However, the slope of the hemispherical surface

[26]De l'Orme (1567: 119r): … *l'accomoder dessus une vis qu'on pourrait faire en forme de pyramide* ….

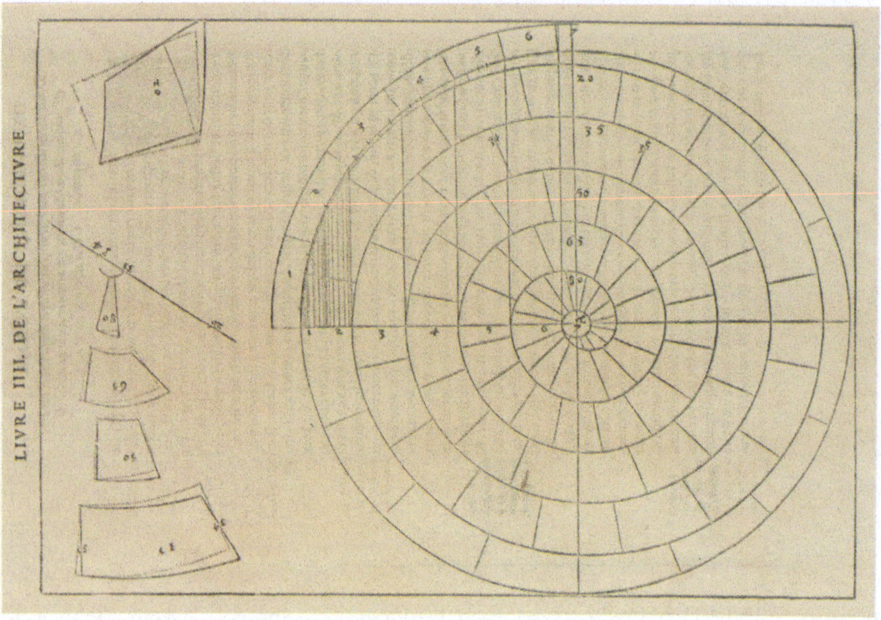

Fig. 9.23 Spiral dome (de l'Orme 1567: 119v)

is quite steep near the springer and almost horizontal close to the keystone. As a result, the real width of the course, measured along the cross-section, increases as it approaches the springing; the height of the first coil is excessive. De l'Orme is aware of this problem: he tries to solve it by dividing the first revolution in two, but this solution is not carried to the plan, and thus the cross-section does not match the horizontal projection. The drawing also includes five templates; they are clearly not sufficient, since each voussoir is different from the others, but we may accept that he is trying to present some examples, so that the reader may understand his method. However, this is not easy, since there is no real written explanation. The only clue to the problem is a set of additional templates, drawn in dashed lines over the main templates. The dashed shapes resemble templates for ordinary domes, based on cone developments; since the solid ones depart slightly from the dashed figures, we may surmise that they are obtained from the dashed ones by measuring in some way the displacements of voussoir corners from their theoretical position. It is no wonder that, when confronted with the task of building a spiral vault in the chapel of Anet Castle, de l'Orme avoided the problem, dividing the vault by parallels and meridians and treating two interwoven spirals as sculpted decoration (Potié 1996: 115–118)

In contrast, Vandelvira's explanation (c. 1585: 65v-66r), although not free from problems, is far more practical. Rather than drawing an Archimedean spiral in plan and projecting it onto the vault, as de l'Orme does, he starts from the elevation and projects it on the plan (Fig. 9.24). After drawing the outline of the plan and the cross-section, he divides a quadrant of the section into three and a half parts; each of these

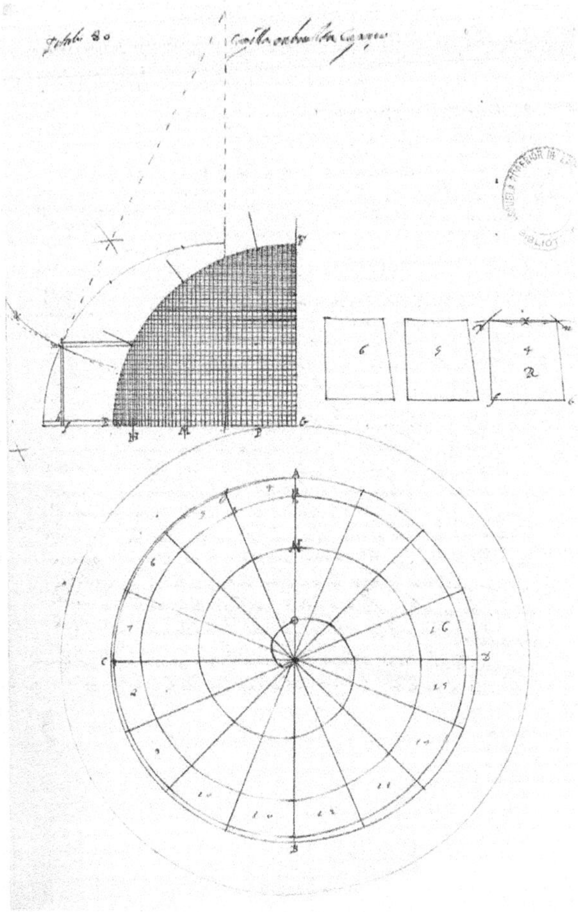

Fig. 9.24 Spiral dome (Vandelvira c. 1585: 66r)

parts will correspond to a coil of the spiral. Next, he divides each of these parts into sixteen small portions; he also divides the plan into sixteen parts using radii of the area covered by the vault. Thus, each of this divisions corresponds to one sixteenth of a revolution of the spiral, both in plan and in cross-section.[27] Then, he draws no fewer than fifty-six horizontal and vertical lines, one from each division point in the cross-section, standing for voussoir corners. The vertical lines will allow the mason to measure the horizontal distance from each corner to the centre of the vault. Placing each of these points along consecutive radii, Vandelvira draws a remarkable spiral. It is not an Archimedean spiral, of course, since the distance between points in the same radius in consecutive revolutions is not constant; nor is it the projection of a loxodromic spherical spiral, since it reaches the uppermost point of the sphere in a

[27]The half-section at the top is divided into eight parts and corresponds to eight wedge-shaped portions of the plan, so it generates a half-coil. It seems that Vandelvira is applying mechanically the division scheme of a vault with a keystone, although in this case there is no real keystone.

finite number of revolutions. However, it has an interesting property: the width of the course (that is, the actual distance, as opposed to the horizontal distance, of two points in the same meridional plane in consecutive coils) is constant. In other words, while de l'Orme conceives his spiral in the springing plane and projects it onto the vault, Vandelvira starts thinking about a spiral drawn on the surface of the vault and then projects it onto the plan (Rabasa 2003: 1683–1684).

Vandelvira gives a fairly detailed explanation of the template construction procedure. In contrast to standard solutions for spherical vaults, he uses triangulation, computing the length of all edges and diagonals of each individual voussoir by means of right triangles with the horizontal distance and the difference in heights of both ends of the segment as catheti. Of course, the horizontal distance is taken from the plan, while the difference in heights is measured using the horizontal lines that go from each voussoir corner to the axis of the vault. While in skew arches this procedure is correct, in this case it is rather problematical, since two corners of the voussoir belong to a meridional plane, while the other two stand on a different meridional plane. Each of these pairs are placed on a chord of a circular cross-section, but these chords intersect the axis of the vault at different points. As a result, the four corners of the voussoir do not lie in the same plane, and Vandelvira's method, which amounts to the construction of a flat template for a warped quadrilateral, miscalculates its upper edge. Further, Vandelvira tries to compute the curvature of the upper side using the cone-development method. However, instead of inscribing a cone in the intrados surface and taking the chord of the meridional section as a generatrix, he seems to be trying to use a tangent to the section (Rabasa 2003: 1683–1684).[28] In addition to these theoretical problems, Vandelvira's explanation does not solve practical difficulties: for example, all radii are aligned, giving continuous side joints as a result; this may cause the formation of cracks.[29]

[28]It may be argued that Vandelvira is drawing a small chord rather than a tangent, since he uses a separate scheme for this construction and extends two horizontal lines from the main drawing to the independent diagram; he also draws vertical lines from the same points. However, Vandelvira actually mentions orthogonals to the radii: *Las cerchas de los lechos se sacan echando una escuadría por la línea de la tirantez…* (the templets of the bed joints are drawn orthogonally to the radii). Further, the intersection points corresponding to the horizontal lines are drawn so closely that it is impractical to draw a line passing through them. Another interesting point is that Vandelvira draws only the first three templates, lying on the springing; the lower joints of these templates are drawn as straight lines, while the curved edges constructed by the cone-development method are applied only to the upper edges. The reason for these differences does not seem to lie in the fact that the lower joints are horizontal, while the upper joints ascend, since they belong to the spiral. In fact, Vandelvira uses round edges for the horizontal sides of the templates of the standard hemispherical vault. Rather, straight edges are paradoxically consistent in this case. The tangent solution leads to generatrices parallel to the axis of the vault; their intersections with the axis are placed at infinity, giving as a result straight edges over the springer. Of course, I am using the vocabulary of projective geometry for the sake of clarity, but I do not imply that Vandelvira used these nineteenth-century concepts.

[29]Perhaps Vandelvira took it for granted that the mason should manipulate these joints in order to avoid continuity, but this is not consistent with the text, which states that *cada planta capialza una parte de las diez y seis en que está repartida cada parte* (each template is raised in one portion of the sixteen parts in which each section is divided). Moreover, this solution fosters long voussoirs in

All in all, Vandelvira's solution is much closer to actual building practice than that of de l'Orme. Strikingly, later treatises and manuscripts include several variations of de l'Orme' scheme, ignoring Vandelvira's method. Chéreau (c. 1567–74: 118v) follows de l'Orme, as usual; Milliet ([1674] 1690: II, 681) worsens the issue, using four centres so that the horizontal width of the course widens as it approaches the springer. Tosca ([1707–15] 1727: 230–232, pl. 16) takes his cue from Milliet, but his solution is weird: he uses two centres, one placed at the actual centre of the vault and another one, set at an arbitrary distance from the first one, which acts as the centre of the first half coil. As a result, the spiral only increases its distance from the centre in half the arcs. When projected to the spherical surface, the result would be striking, since the joints would be horizontal in half of the sphere, while they would ascend sharply in the other end. Tosca's connection with French solutions was to be expected, but the fact that Portor (1708: 92v), the last example of the Spanish tradition, repeats Tosca's solution literally, is quite surprising.

9.3.2 *Pseudo-fan Vaults*

A striking variant is presented by de l'Orme (1567: 112v-113r). He draws the plan of a square vault resembling an English fan vault (see Sect. 10.1.6), but he projects it onto a spherical surface; the result is exactly the opposite of an actual fan vault (see Leedy 1980: 38, 41, 46, 56) since the intrados surface is concave rather than convex. As usual, the text gives no clue about the geometrical procedure; however, Natividad (2017: I, 163–167) has put forward a plausible hypothesis about de l'Orme's method. In this case, intrados joints are not small circles of the spherical surface, but warped curves resulting from the intersections of cylinders[30] with the intrados surface. De l'Orme seems to be aware of this fact; thus, he addresses the problem using distances between voussoir corners rather than cone developments. He seems to decide, somewhat arbitrarily, that the curvature radius of the sides of the templates of the pendentive equals the side of the area covered by the vault; using this curvature and the distances between voussoir corners, taken from an elevation of the perimetral arch and the cross-section, he constructs the templates, although there are some inconsistencies.

the first revolution and quite short ones near the top of the vault. In contrast, in the vaults actually built in Murcia, León, Plasencia and Jerez, all voussoirs feature comparable lengths. This suggests that in these examples two division schemes may have been used: first, an abstract scheme like that of Vandelvira, with a constant number of divisions for each coil; second, a different scheme with approximately equal voussoir lengths and, of course, a different number of voussoirs per coil. As shown in Calvo et al. (2005a: 134–136), this procedure would lead to results similar to the built examples in Spain.

[30] As in the spiral vault, Philibert is arbitrarily choosing a projection and then operating with its spatial result. The projectors of the quadrants in the plan generate a number of virtual cylinders, with the quadrants as directrices and the projectors as generatrices; the intersections of these cylinders with the spherical surface are warped curves.

9.4 Surbased and Oval Vaults

Surbased and raised vaults over a circular plan and vaults over elliptical or oval plans are similar from a geometrical standpoint since both involve an ellipsoidal intrados, or at least an oval approximation to this abstract geometrical concept. However, they raise different constructive problems, as we will see.

9.4.1 Surbased Vaults

Surbased circular vaults are used frequently when the available height is limited; they materialise, or at least approximate, oblate ellipsoids, since both horizontal axes are equal and the vertical one is shorter. In theory, the profile of a surbased vault may be elliptical, basket handle or even segmental. As we have seen in Sect. 3.1.2, Vandelvira (c. 1585: 18v) advises the reader to construct an ellipse by points when an arch or vault needs to be adjusted to a particular height. However, stonecutting writers favour the basket handle variant. This choice may be justified because the elliptical arch requires the use of different arch squares for each voussoir, in contrast to the basket handle profile, which requires no more than two arch squares. As for the segmental arch, it exerts a great amount of thrust in the supports. In all these cases, segmental, basket handle or elliptical, the vault surface may be generated by a revolution of the profile around a vertical axis starting from the centre of the plan.

In order to solve the problem, de l'Orme (1567: 117r-118v) constructs the plan and the cross-section, which is drawn clearly as a three-centre oval (Fig. 9.25). Of course, the plan cannot be reused as a cross-section, as in ordinary domes. Thus, only a half-plan is drawn, while the place of the other half is used for the section. Next, de l'Orme divides the section into voussoirs and uses the cone-development method to construct intrados templates; the oval section does not bring about any change in this procedure. In any case, de l'Orme admits grudgingly that some readers may wish to dress the stones by squaring, so he draws part of the enclosing solid, both in plan and section.

Vandelvira (c. 1585: 62r) uses basically the same tracing method (Fig. 9.26), minus the squaring part. However, he notes that the curvature of the intrados surface is not the same in all points and directions. Thus he instructs the reader to use a complex set of templets and arch squares. In particular, for the first course the lower edge of the intrados should be controlled using a templet with the radius of the springing; for the lower and front edges of the side joint, the mason should use an arch square with the same radius. For the upper edges of the intrados and the side joints, the mason should use a templet and an arch square with the radius of the first bed joint; and for meridional joints, he needs a templet with the radius of the lower section of the basket handle arch. Derand (1643: 358, 359) uses the same procedure again. At first sight, it is not easy to tell if the cross-section is oval or elliptical; however, he recommends the reader to use a pair of meridional arch squares, one for the lower

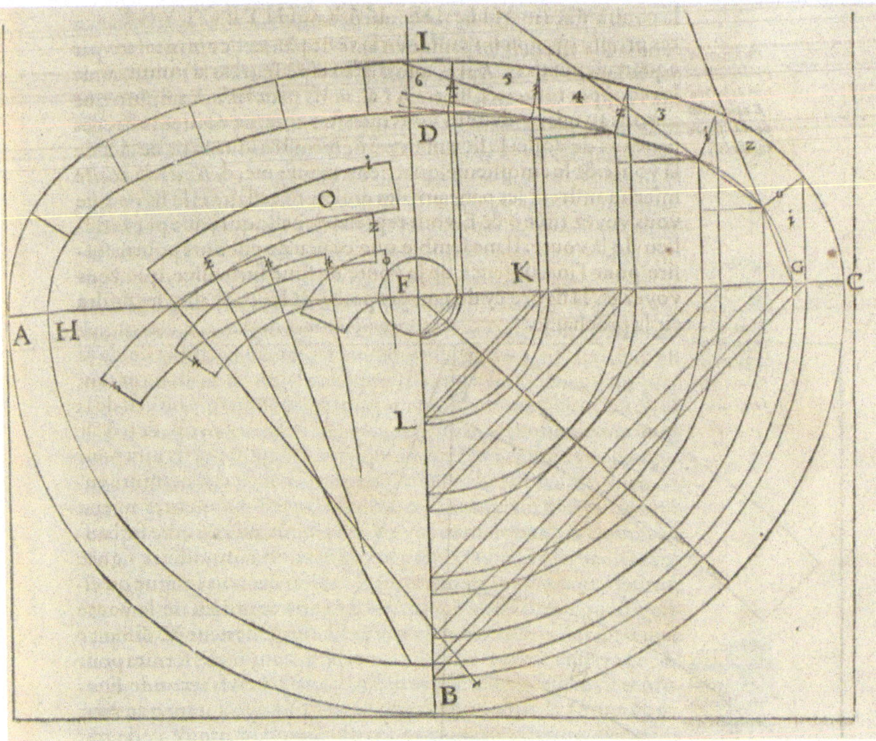

Fig. 9.25 Surbased circular vault (de l'Orme 1567: 118r)

part of the cross-section and other for the upper part, rather than Vandelvira's large set of instruments. This makes it clear that the section is a three-centre oval; an ellipse would have required an arch square for each course. Obviously, this is one of the practical advantages of ovals over true ellipses.

9.4.2 *Oval and Elliptical Vaults*

Oval-plan versus elliptical-plan vaults. So-called oval vaults are used either to respond to constraints in the plan or as a compromise between longitudinal and central plans, especially in the Late Renaissance and Baroque period.[31] As with arches, the builder can choose between an oval plan, employing Serlio's (1545: 17v-18r) ovals or a variant used by Vandelvira (c. 1585: 18r), and the construction of an ellipse by points (Serlio 1545: 13v-14r; Vandelvira c. 1585: 18v).

[31] The body of literature on this topic is too large to be cited here. See for example Wölfflin (1888); Lotz (1955); Lotz ([1974] 1995: 119–120); and, specifically for a construction in ashlar, Gentil (1996).

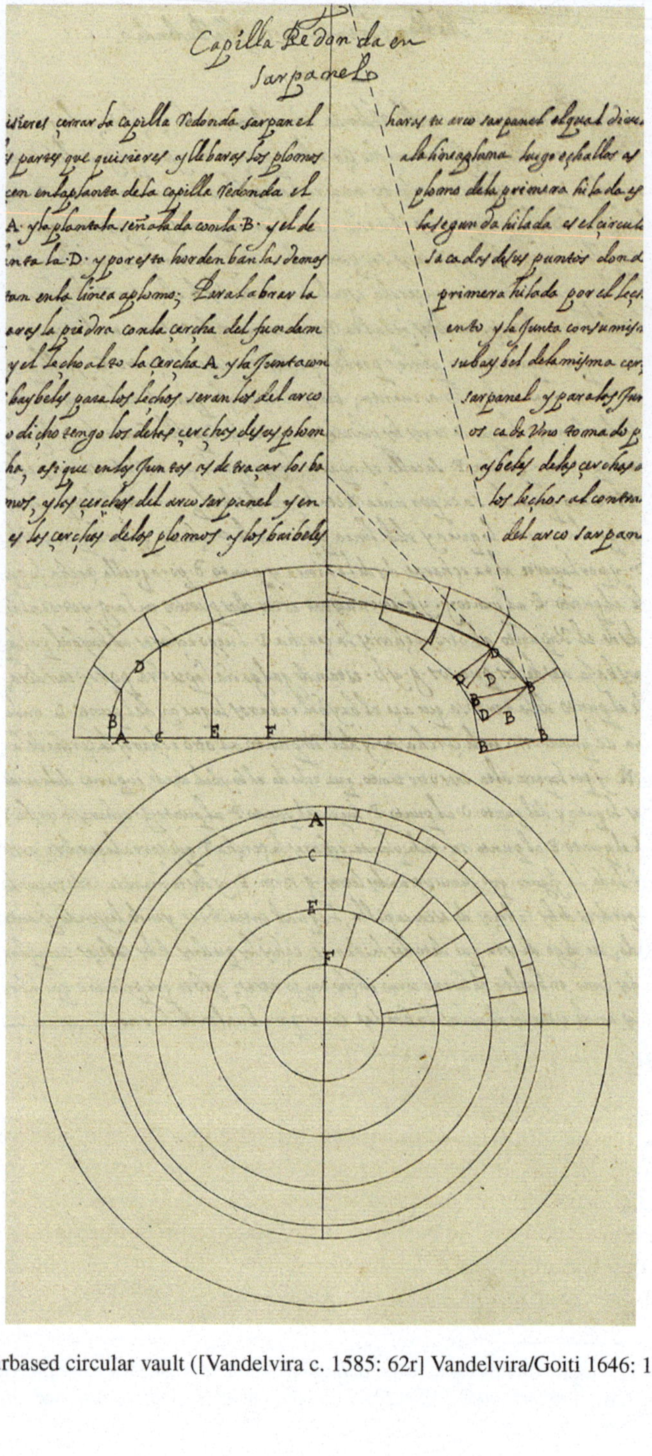

Fig. 9.26 Surbased circular vault ([Vandelvira c. 1585: 62r] Vandelvira/Goiti 1646: 120)

Fig. 9.27 Prolate oval vault. Seville Cathedral, chapter house (Photograph by the author)

Ellipsoidal and oval vaults pose different problems. Comparing the most usual elliptical solution, the prolate one, with an hemispherical dome, it is evident that ellipses replace circles in the intrados joints, that all meridians and all voussoirs in the same quarter of the vault are different, and that angles between parallels and meridians, which are right in the hemispherical dome, now are acute or obtuse (Fig. 9.27); acute ones may be easily damaged when dressing, moving or placing voussoirs.

Some of these difficulties can be avoided by using an oval plan for the vault, but new difficulties arise in this case. Since the outline of the plan is made up from four arches, the vault is an ensemble of two spherical portions at the ends of the larger axis and two toroidal ones in the middle section. In each of these portions, all the voussoirs in a course are equal, and the intersections of meridians and parallels are orthogonal, minimising the risk of broken corners. However, the junction of two sections over the longer axis of the plan poses new and challenging problems. First, the toroidal sections are not mutually tangent at the junction and thus a crease appears at the central section of the longer axis of the vault.[32] Also, the joints in the upper courses of the central portions are mutually intersecting circular arcs, and

[32]Strictly speaking, this happens when using the curvature of the centre portion of the oval for both the toroidal and the spherical portions in the vault. If the builder uses the curvature of the end portions of the oval for both portions of the vault, the plane tangents at the curves of the vesica piscis will be flat, and the simplest solution is to cover the vesica piscis with a flat surface; this solution is not usual. Finally, using different curvatures for the spherical and the toroidal surfaces of the vault will lead to noticeable discountiuities.

there are no courses corresponding to the end portions of the ovals at these levels. As a result, the uppermost portion of the vault takes the shape known as *vesica piscis* or *mandorla*, rather than oval joints. Further, in order to assure continuity between the spherical and toroidal portions, some voussoirs must cross the dividing line and straddle a spherical and a toroidal portion; these voussoirs will show a sharp change of curvature at the dividing lines (Rabasa 2009a: 62–66).

In both in oval and elliptical vaults, the builder has three design choices: first, he can make the height of the vault equal to the length of the longer axis of the plan, obtaining as a result an *oblate ellipsoid*. Geometrically, the result is akin to the surbased vault: two axes of the oblate ellipsoid are equal and larger than the remaining third. However, the constructive implications of this choice for an oval or elliptical vault are exactly the opposite: the rise equals the longer axis; this rise is usually considered excessive, and so the oblate oval or elliptical vault is rarely used in practice. Alternatively, the builder may make the height of the vault equal to the length of the lesser axis of the vault, obtaining as a result a *prolate ellipsoid*, that is, a figure recalling an American football or rugby ball; this solution is often used in actually built oval vaults. Both the prolate and the oblate ellipsoid are surfaces of revolution, since they can be generated by the rotation of an ellipse, and thus are called *ellipsoids of revolution* or sometimes *spheroids*. Finally, the builder can choose to use a different measure for the rise of the vault; in this case, all three axes would be different, and the result would be a *tri-axial, scalene* or *general ellipsoid*.

The reader may object to such undiscriminating use of the terms "oval" and "ellipsoidal". Early Modern writers mix both words (Duvernoy 2015: 450); as we have seen, Derand (1643: 294–296), who had taught mathematics at the Jesuit college in La Flêche, talks about an "oval compass, used to form ellipses".[33] Vandelvira (c. 1585: 18r-18v) is also quite eclectic, as we have seen; he proposes his idiosyncratic oval as a first choice, but he advises the reader to construct an ellipse by points when "you want to raise or lower the basket handle arch [or] build an oval vault according to the available area" as we have seen in Sect. 3.1.2. Actually, most oval vaults in his manuscript (c. 1585: 72r-78) are drawn as four-centre ovals; since he starts from a blank page, he is not constrained to use a particular proportion. In contrast, when the need arises, he uses Serlio's two-circle construction (Vandelvira c. 1585: 120v, 124v).[34] Despite these difficulties, both ellipsoidal and oval ashlar vaults were built here and there during the Early Modern period.[35] The use of these solutions, although not as efficient as the hemispherical dome, may still be justified by architectural intention; these elements are usually interpreted as a compromise between central and longitudinal vaults.

[33]Derand (1643: 294): *Du compas à ouale, ou pour former des ellipses.*

[34]The construction is used explicitly on fol. 120v; it is also used in fol. 124v, although the horizontal and vertical lines used to locate the points are drawn in drypoint and are thus difficult to locate. This detail suggests that Vandelvira is not using this construction just as a theoretical exercise but rather as a practical procedure (for a different opinion, see Huerta 2007: 230).

[35]Such vaults were used quite frequently in the Baroque period, of course, but they were usually built in brick.

Prolate oval vaults. The intrados of the surbased vaults of De l'Orme and Vandelvira approximates an oblate ellipsoid, since two axes are equal and larger than the third axis. In contrast, the most usual type of oval-plan vault approximates or materialises a prolate ellipsoid; that is, a surface with two equal axes, both smaller than the main axis. In other words, the rise of the vault equals the shorter axis of the plan. Vandelvira includes no fewer than three different solutions for this problem, not counting coffered oval vaults.

In the first one, the *capilla oval tercera*,[36] he divides the vault using parallels and meridians (Fig. 9.28). Since bed joints are oval or elliptical, the use of the cone-development method would be impractical, so Vandelvira (c. 1585: 73v-74r) relies on triangulations. He remarks that this solution is used when the mason wants to make the rise and the shorter axis of the plan equal; however, the construction is still consistent if both dimensions are different. First, he draws the plan as a four-centre oval; of course, he could have used Serlio's construction if necessary, to adapt the plan to any predetermined dimensions. Next, he draws a longitudinal section as a half-oval with the same axes. He then divides the plan radially into fourteen portions and the section into seven. This choice leads to a joint over the longer axis and a voussoir at the shorter axis. In the section, the choice of seven divisions leads to three courses and a keystone whose width equals that of a course. The reader may object that this procedure does not stagger the side joints between courses. However, as he has explained before when dealing with the horizontal-axis annular vault (Vandelvira c. 1585: 69v), the mason may manipulate the position of side joints if he wants to stagger the joints.

In the next phase, Vandelvira draws horizontal joints in the section and carefully constructs a true-shape representation of all meridional joints, transferring the horizontal distances between voussoir corners from the plan. Next, he constructs intrados templates through triangulation. He measures the lower edge of each voussoir in the plan and the length of side joints in the true-shape representations. He also computes the length of the diagonals, forming a right triangle with their horizontal projection, and the difference in height between voussoir corners, as done on other occasions (see Sects. 3.1.3, 5.5.1, 6.1.2, and 6.2.1). All this allows the construction of intrados templates; since the chords of the meridional joints meet at the vertical axis of the vault, the corners of each voussoir lie on the same plane and the construction is exact in theory. In order to dress the side and bed joints, the mason needs bevel guidelines. In this case, they are not orthogonal to the joints; Vandelvira draws them passing through the centre of the area, at springing level.

This ingenious and correct procedure does not seem to have been widely used. Derand (1643: 398–400) draws the lower and inner sides of an enclosing rectangle (Fig. 9.29, above); this suggests that he means to dress the stones by squaring, although he does not say that in so many words. In contrast, de la Rue clearly indicates that dressing should be conducted by squaring; it is worthwhile to remark that he presents this problem under the heading "Dome on an oval plan", but he starts

[36]This translates as "third oval vault"; the first two are oblate ellipsoids, both in plain and coffered versions; I will deal with them shortly.

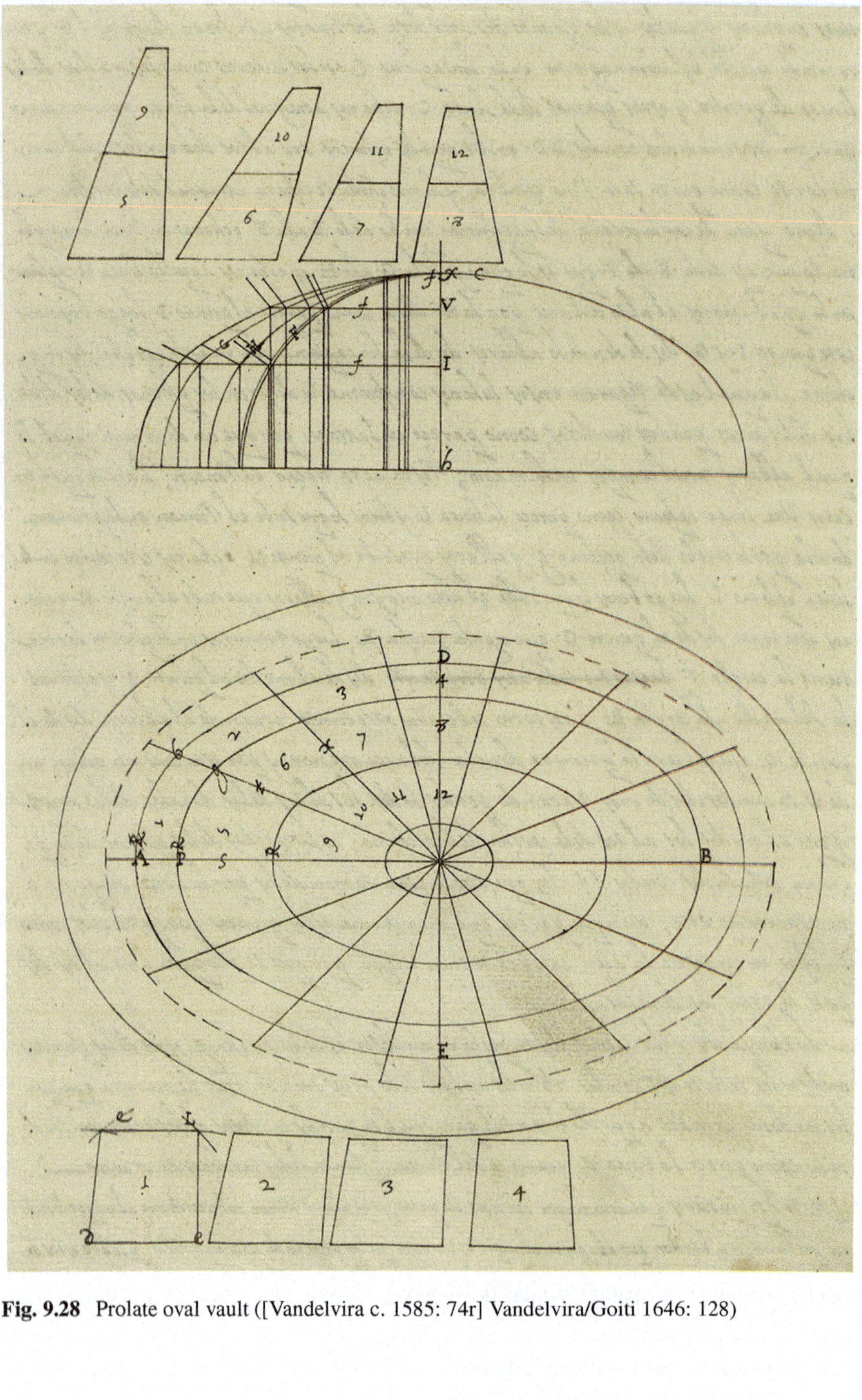

Fig. 9.28 Prolate oval vault ([Vandelvira c. 1585: 74r] Vandelvira/Goiti 1646: 128)

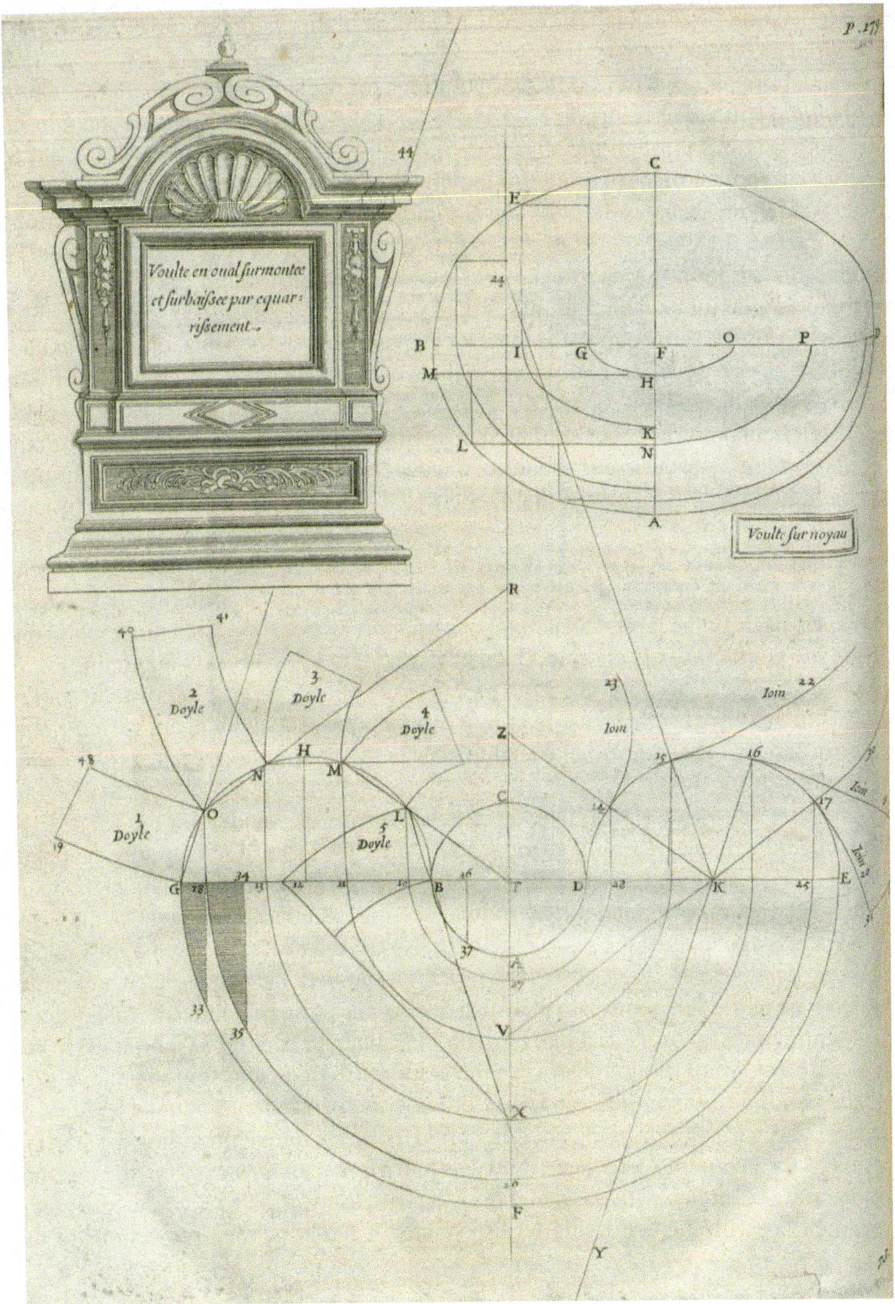

Fig. 9.29 Above, prolate oval vault. The scheme may be also read as a surmounted vault. Below, vertical axis annular vault (Derand [1643] 1743: pl. 179)

his explanation saying that "The correct execution of this vault is quite difficult, given its elliptical shape".[37]

Vandelvira presents two additional solutions for prolate oval vaults. In the *capilla oval quinta* (fifth oval vault),[38] (c. 1585: 76v-77r) he uses as bed joints a sheaf or fan of planes passing through the shorter axis of the plan; the side joints are given by a series of vertical planes parallel to the larger axis. Again, since the bed joints are oval, Vandelvira eschews the cone-development method and relies on triangulations. In contrast, in the *capilla oval sexta* (sixth oval vault), Vandelvira (c. 1585: 77v-78r) inverts this solution: he uses a sheaf of planes passing through the longer axis of the plan as side joints, and a set of vertical planes parallel to the shorter axis of the area, whose intersections with the intrados provide bed joints. Thus, the apparent bed joints are circles of different diameters, and Vandelvira can use a series of auxiliary cones passing through two consecutive circles, developing them as he had done in the hemispherical dome or other similar vaults. However, he also includes an additional set of rigid templates in order to control the decoration of the vault; although he does not say this explicitly, it seems that the templates in this second set are constructed through triangulation.

Oblate or raised oval vaults. As we have seen, an oval vault may be set out with its rise equalling half the larger axis of the oval or ellipse in the springing. In this case, two axes, namely the larger horizontal one and the vertical one, are equal, while one of the horizontal axes is shorter. Geometrically, this layout fulfils the definition of an oblate ellipsoid or spheroid, as do the surbased vaults we have seen at the start of this section. However, from a constructive standpoint, such vaults are *raised* rather than surbased vaults, since their height exceeds the shorter horizontal axis (Derand 1643: 398–400).

The first and most elaborate explanation is given by Vandelvira (c. 1585: 71v-72r), under the name of *capilla oval primera*, (first oval vault); remarkably, he opens his section on oval vaults with this unusual piece. To start, he draws the plan, again in the shape of an oval, and a longitudinal section, as a half-circle whose diameter equals the longer axis of the plan (Fig. 9.30). Next, he divides the section into an odd number of parts, in this case, seven; then, he draws radii of the half circle, which act as projections of the side joints. He then also divides the springing into several portions. The central ones are used as a pair of oculi placed at the end of the shorter axis. The two portions of the springing between the oculi are divided into seven parts; vertical planes passing through these divisions provide the bed joints. The intersections of these planes with the springing are transferred to the elevation, providing the starting points of the bed joints, which are drawn as half circles. In the next step, the meeting points of these circles with the radial divisions are transferred back to the plan; their intersections with the bed joints furnish points of the horizontal projections of the

[37] De la Rue (1728: 52): *Voute de four sur un plan ovale... Cette voûte, à cause de sa figure elliptique, est assez difficile à bien executer*.

[38] The "fourth oval vault" is a coffered version of the third one; I will deal with it in Sect. 10.2.4.

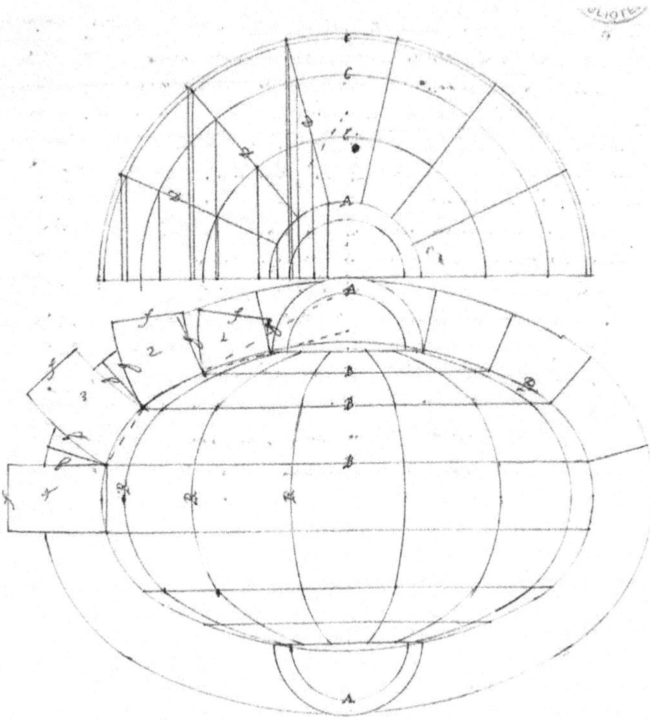

Fig. 9.30 Oblate raised oval vault (Vandelvira c. 1585: 72r)

side joints, which should be ellipses, at least in theory.[39] At this stage, Vandelvira has constructed all bed and side joints, both in plan and elevation, and may start constructing the templates by the cone-development method. Once again, Derand (1643: 398–400) eschews this elaborate procedure and seems to dress the voussoirs by squaring, simply rotating the scheme he has used for the prolate oval vault.

Oval vaults with non-planar joints. As we have seen in Sect. 2.4.3, Gaspard Monge (1796: 162–163; see also Hachette 1822: 288–290, 291–293, de la Gournerie 1855: 27–28; Sakarovitch 1992a: 536–539 and Sakarovitch 2009c) suggested in a scientific paper that the French National Assembly should be covered by a scalene ellipsoid, with bed joints passing through the lines of curvature of the intrados surface. In contrast to ellipsoids of revolution, where lines of curvature simply follow meridians and parallels, in the scalene ellipsoid lines of curvature go up and down (Fig. 9.31), as in Armenian pendentives and vaults (López-Mozo et al. 2013; Calvo et al. 2015c). Monge explained nothing about actual stonecutting methods, although we may surmise that the bed joints were to be dressed using flexible templates, to take

[39]Even when the springing is set out as an oval, the projections should be ellipses, since they are the horizontal projections of slanted half circles. However, Vandelvira draws them as a series of circular arcs taking points in groups of three, as usual.

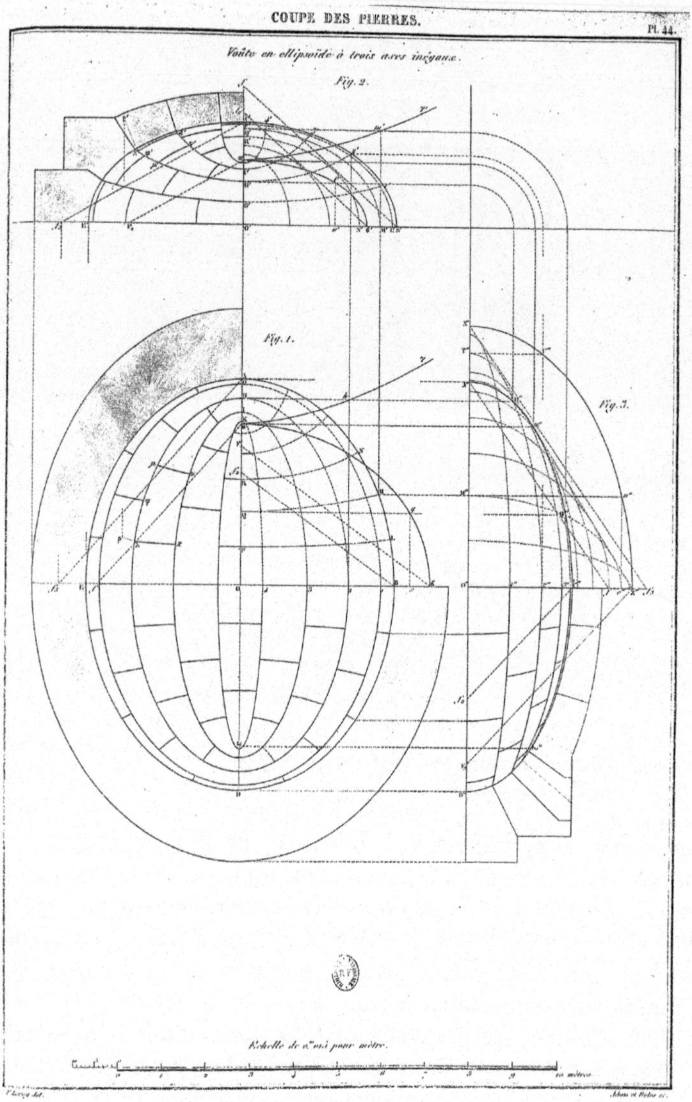

Fig. 9.31 Scalene oval vault (Leroy 1877: II, pl. 44)

advantage of their developable nature. The practical advantages of this solution, such as the orthogonality of joints and their developable nature, are far outweighed by the complexities of the setting out and dressing process; it is no wonder that no vault has ever been built following Monge's method, as far as I know (see also Sakarovitch 1998: 309–313; Rabasa 2009a: 66).

Strikingly, an oval vault with ascending and descending bed joints was built much before Monge; we can take it for granted that its builders knew nothing about ellipsoids or lines of curvature. The ceiling of the Nativity Chapel in Burgos Cathedral was erected by Martín de Bérriz and Martín de la Haya, starting probably in 1571 and finishing around 1580. As in Monge's method, intrados joints are not placed in horizontal planes, and joints between voussoirs in the same course do not follow meridians. As far as I know, no treatise, manuscript, full-scale tracing or bit of archival evidence explains the stonecutting method used in this isolated example. However, a survey conducted by 3D laser scanning has shown that bed joints approximate the shape of cones of revolution. This explains the rising and descending shape of intrados joints, since the intersection of a cone of revolution with an ellipsoid, or an oval surface approximating an ellipsoid, results in this shape. The voussoirs were probably dressed in place, using a bar attached to a fixed point at one end (Alonso et al. 2009).

9.5 Annular Vaults

Annular vaults are based on the torus, a surface of revolution generated by a circle rotating around an axis set in the same plane that does not pass through its centre. The full surface looks like a tyre or a lifebelt. Meridional sections—that is, those taken through planes passing through the axis—are given by two opposite positions of the generating circle during the revolution. Each section through a plane orthogonal to the axis is given by two circles of different diameters with their centres placed on the axis, except for two of these sections, corresponding to plane tangents to the generating circle that are orthogonal to the axis; these singular sections are given by single circles. A perpendicular to the axis can be drawn from the centre of the generating circle; its intersection with the axis is the centre of the torus. It is interesting to note that in the inner half of the torus, generated by the half of the circle closer to the axis, all points are hyperbolic; this means that seen from outside, the surface is concave in the direction of orthogonal sections and convex in the direction of meridional sections. In contrast, in the outer half of the torus, all points are elliptical, and all sections are convex.

This figure may be used in construction by taking the upper half of a vertical-axis torus; of course, the lower half is redundant, as in groin or pavilion vaults. In this case, the surface is cut by a horizontal plane passing through its centre; the resulting section, consisting of two concentric circumferences, provides the springings of the vault. Two concentric circles give other horizontal sections, that is, the paths of two points of the generating circle during the revolution; these sections stand for the intrados joints. Meridional sections are half-circles, furnishing the side joints. Seen from below, the outer half is concave in both directions, as with most vaults; In contrast, the inner half is convex in the direction of the bed joints and concave in the direction of the meridional sections. This constructive type has been used in many periods, from Romanesque crypts below deambulatories, as in the Abbey of Montmajour, near Arles, to Renaissance vaults such as the one in the portico of Villa Giulia in Rome, plastered and painted, or the ashlar vault around the courtyard of the palace of Charles V in Granada. In the seventeenth century, annular vaults were combined with transverse conical vaults, for example in the Grande Écurie in Versailles, built by Jules-Hardouin Mansart; the result is known in French as *voûte d'arete en tour ronde* (round groin vault) (Erlande-Brandenburg 1995: 224–225; Pérouse de Monclos [1982a] 2001: 142, 180, 181, 208; Lotz [1974] 1995: 111–112; Rosenthal 1985: 112–119; Salcedo and Calvo 2016; Senent 2016: vii–ix, 527–549).

Another architectural use of the torus is less frequent. The generating circle may rotate around a horizontal axis; in this case, meridional sections are generally slanted, although horizontal and vertical sections appear at the springings and the symmetry plane; sections through planes orthogonal to the axis are pairs of vertical circles. However, only the fourth part of the surface needs to be used to cover a space. The torus is cut through the horizontal plane passing through its centre, leaving a portion of the surface that covers an area in the shape of a rectangle capped by two semicircles. However, most points of this area lie below two different points on the surface. To avoid unnecessary expense, only the external, upper part of the surface is used. Meridional sections provide the bed joints, while vertical ones give the side joints; the intrados surface is concave in both directions. This solution was used in the inner chamber of the funerary chapel of Gil Rodríguez de Junterón in Murcia Cathedral, fostering several derivatives in the surrounding area. However, one of these, the niches in the presbytery of the church of Saint James in Orihuela, belongs to a widely used type. The vault is again cut by a plane perpendicular to the axis passing through its centre, and one half is discarded; that is, only one eighth of the toroidal surface is used. This figure is used frequently to cover a shallow niche as an alternative to the usual quarter-sphere niche, whose depth equals half its width (Gutiérrez-Cortines 1987: 161–170, 174–178; Calvo et al. 2005a: 151–170; Alonso et al. 2011a; Alonso et al. 2013).

9.5.1 Vertical-Axis Annular Vaults

A strange drawing is included in the final section of Hernán Ruiz's manuscript (c. 1560: 151v). Palacios ([1990] 2003: 169), Pinto (1998: 212–214) and Sanjurjo (2015: 182–183) interpret it as an auxiliary drawing for a spiral staircase, cantilevered from the wall. The main arguments for this stance are the small diameter of the central well and the presence of three pairs of parallel lines, as usual in simple staircases; however, such lines are not usual in vaulted staircases. There is an alternative reading, although it is not free from problems. The drawing may represent a (very small) vault based on a self-intersecting torus, that is, the surface generated by a circle rotating around an axis that intersects the circle but does not pass through its centre. Only the outer half is used, so the result may be a cantilevered annular vault; the enclosing rectangles around each voussoir suggests that they were dressed by squaring, while the pairs of parallel lines may correspond to mortises intended to secure each voussoir to the next one.

Vandelvira includes a detailed architectural drawing of a vertical axis annular vault (c. 1585: 111r) supported by a wall on the outer side and columns on the inner one, just like the vault in the courtyard of the palace of Charles V (Fig. 9.32; see also Fig. 1.14n). The text does not mention this archetype but refers the reader to the staircase of Saint Gilles (see Sect. 11.1.3); conversely, when dealing with this staircase, Vandelvira (c. 1585: 52v-53r) brings up the Granada vault. Moreover, the written explanation of the annular vault focuses in the concave-face arches between

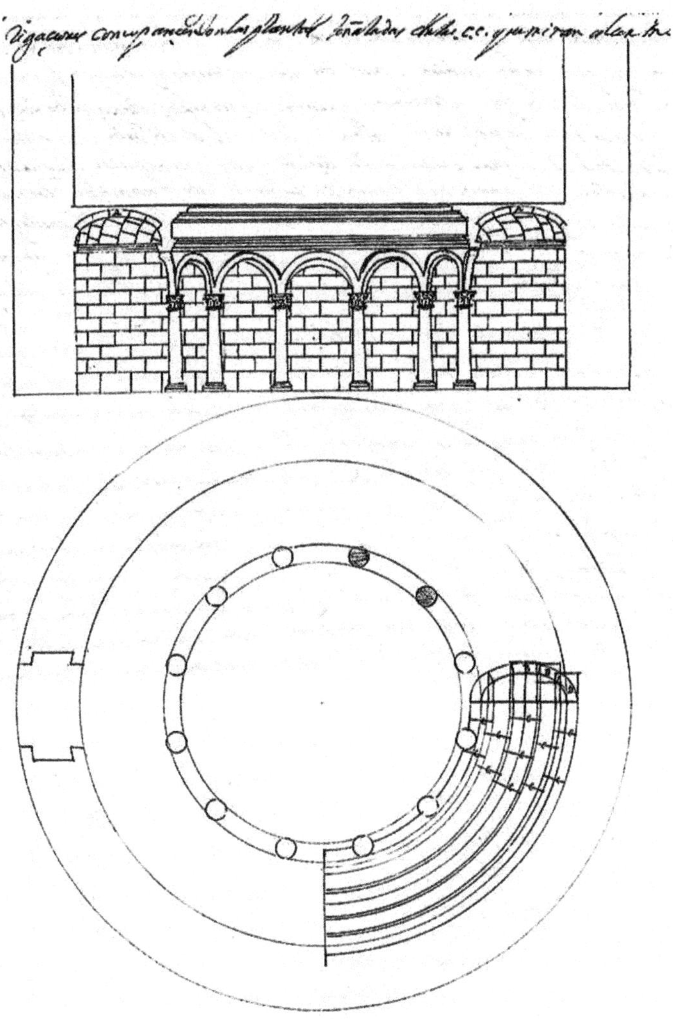

Fig. 9.32 Vertical axis annular vault (Vandelvira c. 1585: 111r)

the columns and the springer of the vault, while in the actual vault in Granada this role is played by a lintel. This insistence may be connected to some discussions and reports during the construction of the vault in the palace around 1576. Juan de Maeda, a master trained in the Spanish tradition, wanted to abut the vault in a number of arches resting on piers and columns, while the Italianate Luis Machucha opted for the column and lintel solution (Rosenthal 1985: 118; Rodríguez 2001: 441–445; Salcedo and Calvo 2016). Vandelvira seems to join this debate, siding with the masons and implying that Machucha did not know how to build arches on curved walls. The actual stonecutting solution for the vault gets lost in this controversy, although the drawing includes the usual enclosing rectangles, suggesting that voussoirs are dressed by stonecutting. Just before this section, Vandelvira includes another drawing of a cantilevered torus vault (c. 1585: 110v), without explanatory text or constructive details.

Guardia (c. 1600: 68v) includes two drawings of annular vaults with identical dimensions and proportions. The inner radius of the vault is quite small in comparison with the outer one. As in the Hernán Ruiz drawing, this may suggest that the drawings represent a spiral staircase like the *vis de Saint Gilles*; however, the drawings themselves present no trace of such use. They may represent, however, an actual annular vault designed to cover a staircase of this type, or it may be merely a didactic diagram, unconnected with a built vault. In any case, one of the drawings seems to be solved by squaring; although the usual enclosing rectangles are lacking in the cross-section, the detailed division of the plan by bed and side joints hints in this direction. Quite interestingly, the other variant uses the cone-development method. Since the bed joints are circular, this procedure may be used without problems in the tracing stage: cones may be inscribed in the interior of the intrados surface, passing through two consecutive bed joints. Of course, problems arise in the dressing stage, since the curvature of the surface is not the same in all directions, and it even bends in opposite directions in the inner half. Thus, many different arch squares must be used in its execution, including the singular inverted arch squares in a drawing by de l'Orme (1567: 56v).

The use of cone developments for this kind of vault is clearer in French treatises of the seventeenth century. Jousse (1642: 184–185) deals with an annular vault solved using intrados and bed joint templates (Fig. 9.33). After drawing the plan of half the vault and a cross-section including two semicircles, he constructs the intrados templates by drawing, as on other occasions, generatrices passing through two consecutive bed joints and developing the resulting cones. In the outer half of the vault, apexes are placed over the corresponding intrados joint, as in spherical vaults; in contrast, in the inner half they are placed *below* the joint or even below the impost. Since the toroidal surface does not have the same curvature in all directions, the dressing stage poses new problems; at least three different arch squares, one for meridional sections and two for the end parallels, must be used for each voussoir.

An interesting addition is the procedure used to construct the so-called "bed joint templates" shown on the right side of the drawing. The intrados and extrados edges of the same bed joint are circles located on different planes; however, their centres are joined by a vertical line passing through the centre of the vault. Thus, the bed

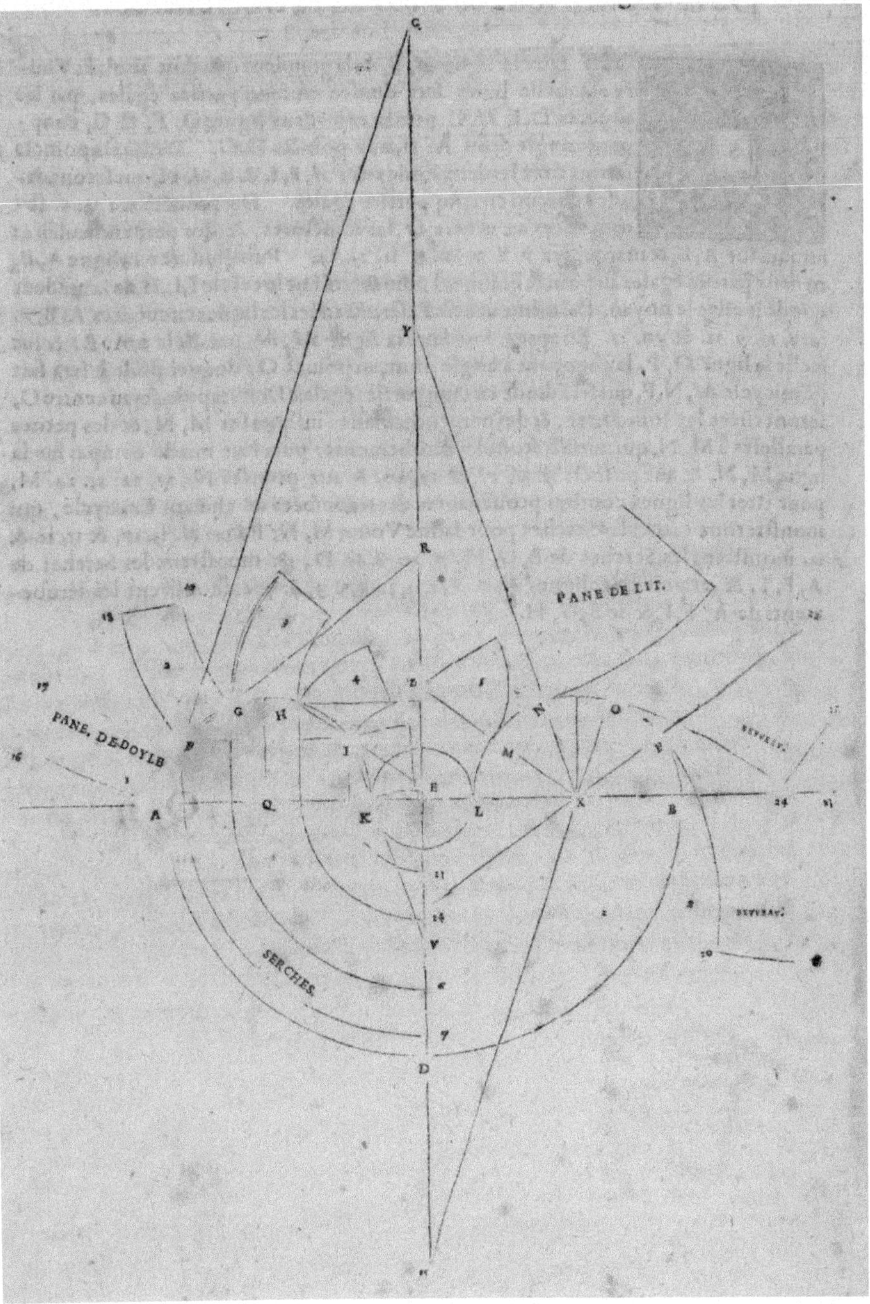

Fig. 9.33 Vertical axis annular vault (Jousse 1642: 184)

joint can be executed as a cone with its apex on this line. In contrast to intrados templates, the centres of the cones for the inner section of the vault are placed over the springing, while those for the outer half are located below the equatorial plane. After locating the cone apexes, Jousse draws the intrados edge of the bed joint template as an arc with its centre on the apex. However, he omits the extrados edge, so the template includes only the arc standing for the intrados side and *a single* generatrix representing the edge of the side joint. All this implies that the thickness of the voussoir is left undefined, and that the extrados is not dressed as carefully as the intrados.

In addition to this standard procedure, Jousse (1642: 182–183) includes a *voûte sur le noyau a trois entreés*, that is, a semicircular section of an annular vault with the fanciful addition of another branch, also in the form of an annular vault with the same width. This detail guarantees that the bed joints are placed at the same height in the main vault and the additional branch, so both pieces meet tidily, as in a groin vault. However, the groin, which should be a complex warped curve, is not drawn. Both vaults are dressed by squaring; nothing is said about voussoirs placed next to or across the groin.

Derand (1643: 395–398) presents a solution for the standard annular vault akin to that of Jousse, including both the intrados and bed joint templates. However, he explains the dressing procedure in detail. The mason should start with a block, marking on it the radius of the intrados edges of two consecutive bed joints. After having dressed the intrados surface according to these cues, he should apply on it the intrados template, in order to score the lower and side edges of the template; the upper one should be controlled with the arch square. In other words, Derand is applying a hybrid method, combining squaring and full templates; in fact, he states that "this must be done mixing squaring with the method of tracing stones by templates",[40] although later on he admits that "strictly speaking, you can dispense with templates; templates constructed following the circles that represent the intrados joints in the plan may be sufficient and substitute for them".[41] De la Rue (1728: 53, pl. 29), eschews the cone development method, as usual, going back to squaring; he remarks that "the templates shown by Father Derand in order to dress the voussoirs of these vaults are false … and they are useless since it is necessary to dress these voussoirs by squaring to hollow the intrados, before using templates".[42]

---

[40]Derand (1643: 396): *Cela se doit executer, meslant l'equarrissement auec la façon de tracer les pierres para paneaux..*

[41]Derand (1643: 398): *On pourra absolutement parlant, se passer en ces operations des paneaux; les cherches tirées des cercles, qui sur le trait designent le plan des assises, pouuans suffire & suppléer à leur defaut.*

[42]De la Rue (1728: 53): *Les panneaux que le Pere Derand donne pur tracer les voussoirs de cette voûte, sont faux … outre cela ils deviennent inutiles, pusqu'il faut couper les dits voussoirs par équarrissement pour fouiller les douelles, avant que de pouvoir se servir des panneaux …*

9.5.2 *Horizontal-Axis Annular Vaults*

Alonso de Vandelvira (c. 1585: 69v-70r) includes a horizontal-axis annular vault under the heading *bóveda de Murcia* (Fig. 9.34). As remarked by Chueca (1971: 340) and Gutiérrez-Cortines (1987: 167) its archetype is the vault in the inner chamber of the chapel of Junterón in Murcia Cathedral (Fig. 9.35). After drawing the plan in the shape of a rectangle capped by two semicircles, Vandelvira splits these half circles into an odd number of equal parts, using the division points to locate vertical planes parallel to the longer axis of the plan, which provide the side joints. In the next step, he constructs the intrados templates using cone developments: he draws generatrices until they reach the horizontal axis of the vault, using the intersection points as cone apexes. Next, he draws a longitudinal section in the shape of a semicircle and transfers the division points from the plan to the base of the section, drawing a number of semicircles, standing for the side joints. He then divides the outline of the section into an odd number of parts and draws radii of the semicircle, standing for the bed joints. These straight radii stand for the vertical projections of several generatrices of the vault, which are actually semicircular; however, their horizontal projections will be ellipses. In order to draw them, Vandelvira uses the intersections between the bed and side joints in the longitudinal section, transferring them to the springing line by means of verticals; then, distances to the axis of the vault are transferred again to the side joints in the plan using a gauge.[43] The intersection of each projecting line with the corresponding side joint in plan will furnish a point of the elliptical projection of the bed joint. However, Vandelvira constructs the projection using circular arcs, grouping the points into threes, as usual (Calvo et al. 2005a: 151–170; see also Sects. 3.1.2 and 3.1.3).

Next, Vandelvira gives some hints about the dressing process: the bed joints, depicted in the templates as straight lines, should be dressed with an arch square with the radius of the corresponding side joint, while the side joints, depicted as arcs, should be dressed with the radius of the semicircles in the plan. He also adds an interesting remark: "If you wish to stagger the joints you should make some stones larger than the others, and if you wish all the stones to be of equal length … you should draw them as shown in the drawing if you wish to put coffers in the voussoirs."[44]

---

[43] Actually, drypoint distance marks can be seen with the aid of a loupe in the copy of the manuscript in the library of the School of Architecture of Madrid. In contrast, these marks are not present in the copy of the National Library of Spain; this suggests the latter was traced from the former or a close copy. See Calvo et al. (2005a: 240–242, 243, 245) and Sect. 2.2.3.

[44] Vandelvira (c. 1585: 69v) *Si quisieres que vayan haciendo ligazones harás unas piedras más largas que las otras y si quisieres que vayan iguales … trazarlas has como en la traza parece y esto es para si quisieres echarles algunos artesones en las dovelas.* Transcription is taken from Vandelvira and Barbé 1977.

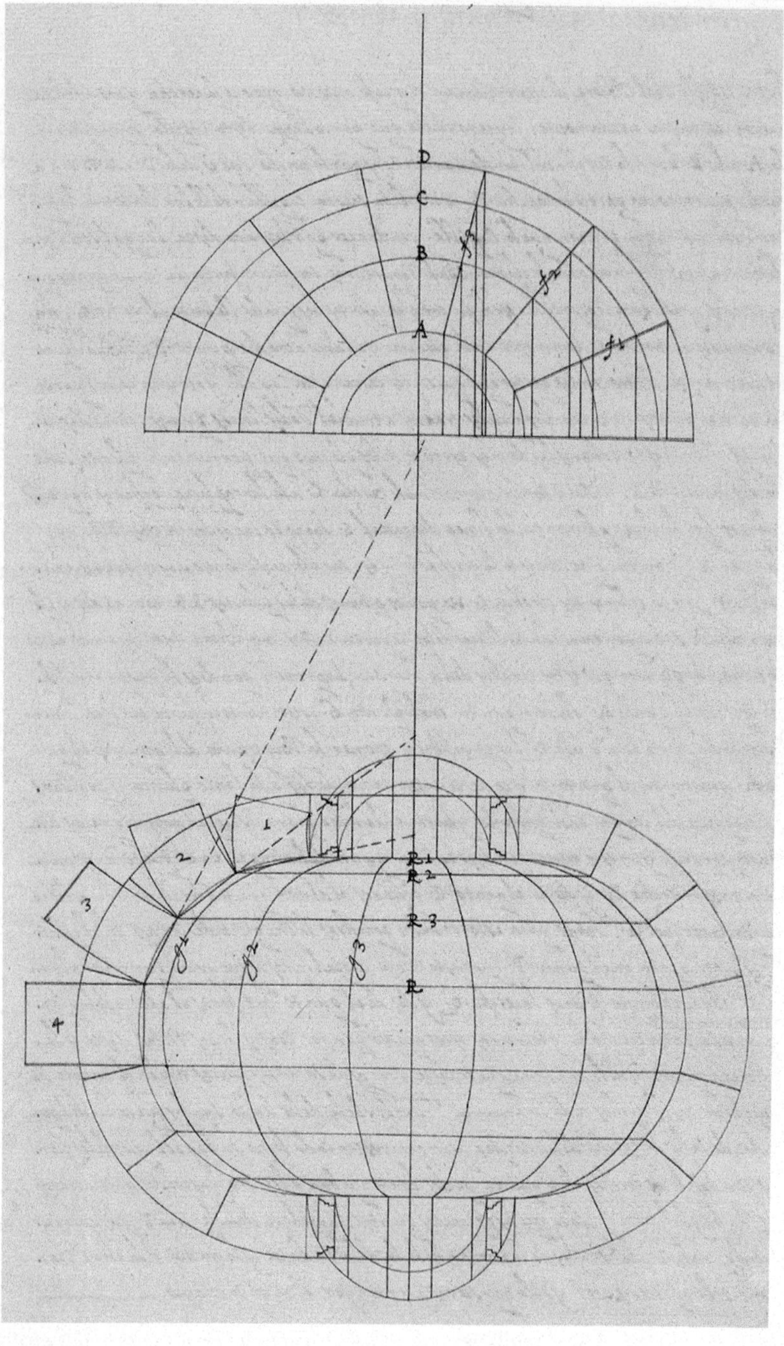

Fig. 9.34 Horizontal-axis toroidal vault or *bóveda de Murcia* ([Vandelvira c. 1585: 70r] Vandelvira/Goiti 1646: 126)

Fig. 9.35 Horizontal-axis toroidal vault. Murcia Cathedral, chapel of Junterón (Photograph by David Frutos)

9.5.3 Annular Groin Vaults

Derand (1643: 448–453) finishes his treatise with a sloping annular groin vault, which can be used for staircases; he must have in mind stairways of the *geometrical staircase* type, with a large well. He remarks that such vault may also be used for a non-sloping vault, nullifying its ascending movement. However, it will be easier to analyse De la Rue's (1728: 54–56, pl. 30; see also Senent 2016: 523–527, 536–542) version of this piece, since he explains in detail the non-slanting solution (Fig. 9.36). The vault is formed by a fair number of portions, each one set between four substantial piers. Each of these portions is formed by two different parts, an annular vault going around the whole ensemble, with round horizontal intrados joints and radial side joints; and

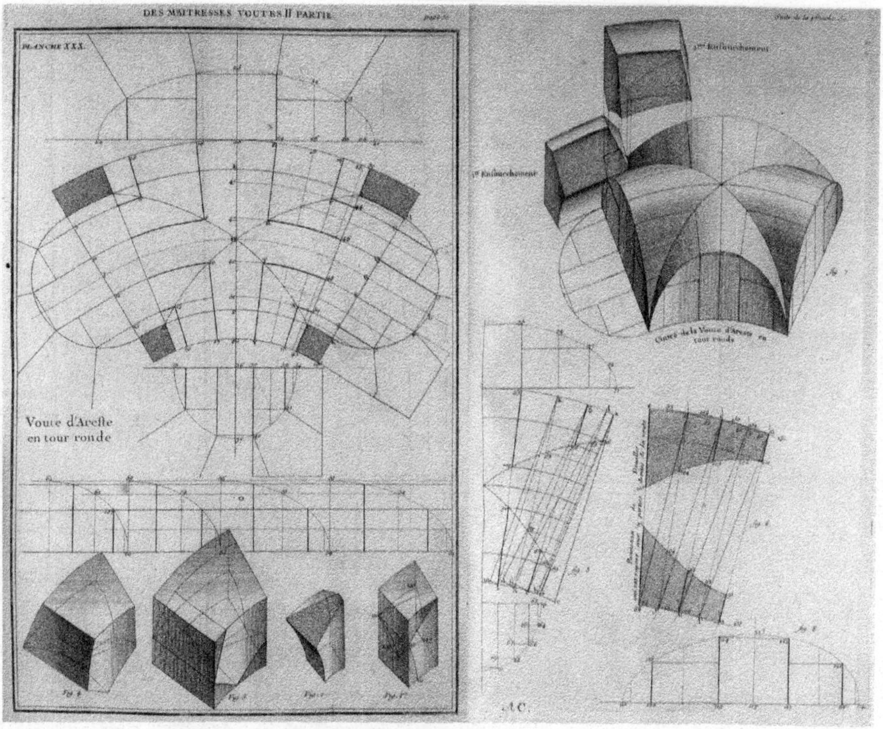

Fig. 9.36 Annular groin vault (de la Rue 1728: pl. 30, 30 bis)

a transverse vault, specific for each portion, in the shape of a conoid.[45] This means that the intrados joints of the transverse vaults are also horizontal, and they intersect a vertical axis passing through the centre of the overall plan. The annular vault and the conoidal ones intersect at curved groins going from one pier to the diagonally opposite one.

De la Rue starts by drawing a part of the annular vault between two radii of the overall plan. Next, he draws a semicircular cross-section through these radii, dividing it into equal parts; he transfers the division points to the plan, using them as a cue to draw the intrados joints of the annular vault, as well as a virtual line passing through the midpoint of the cross-section, representing the highest points in the annular vault. The highest point of both groins should be placed at the intersection of this line with the axis of the vault portion. However, this point is *not* aligned with the springings—that is, the pillar corners—so de la Rue draws the groins in plan as circular arcs passing through two opposite springings and the intersection of the axis and the uppermost line of the annular vault. Next, he draws the intrados joints of the transverse arches passing through the centre of the overall plan and the

[45]As we will see, de la Rue's solution deviates slightly from a true conoid, although the difference is minimal.

intersection of the intrados joints of the annular vault with the groins. In the next step, he draws the development of the front and back faces of the transverse vault, taking distances between voussoir corners from the plan and heights from the cross-section of the annular vault. Of course, this operation is justified because all intrados joints are horizontal and both the ones in the annular vault and the transverse one meet neatly at the groins. In the next phase, de la Rue applies the same procedure to the development of several circular sections of the transverse vault, obtained through the vertical cylinders passing through the intrados joints of the annular vault.

Each of these sections includes three parts. They begin with horizontal lines at the ends, belonging to the annular vault. However, these lines are not placed at the top of the vault, since no intrados joint passes through the intersection of the groins. Thus the part of the section crossing the axis of the transverse vaults must rise in order to reach the level of the junction of the groins and go down to reach the opposite joint of the annular vault. In any case, de la Rue is not interested in this complex section, just in the central part; he draws full sections of the conoid, furnishing the height of voussoir corners and templates for the side joints of the transverse vaults. Next, he endeavours to construct intrados templates for the transverse vault, but he is aware that the surface is non-developable; he divides it into a number of triangles, which act in a way similar to the folding templates he used for the ordinary groin vault. However, as explained by Senent (2016: 536) such templates are ancillary tools for a dressing process carried out by squaring.

Frézier (1737–39: III, 245–253) detected a subtle error in de la Rue's explanation: he draws the horizontal projections of the groins as circular arcs, positing that the intrados joints in the transverse vaults would meet at the centre of the ensemble. Both statements cannot be right at the same time: circular groins lead to intrados joints missing the centre (see Senent 2016: 529–533). Although the issue has no practical consequences, he inverted de la Rue's procedure, first drawing a true conoid and constructing the groin afterwards by points; as expected, they are not exactly circular in plan.

Chapter 10
Rib and Coffered Vaults

10.1 Rib Vaults

In all the vaults we have seen so far,—whether cylindrical, spherical, oval, or toroidal,—all voussoirs belong to a single constructive scheme. Admittedly, in some types, a few voussoirs present particular traits, such as the keystones, the springers or those at the diagonals in groin, pavilion or square-course sail vaults, but each voussoir must fit with those adjacent to it, both in the same course and in the courses behind and above it. Rib vaults are based on a different strategy. First, a network of linear elements, the *ribs*, is placed at the perimeter and along some internal lines over the area spanned by the vault. Second a *web*, formed of a number of *severies* is laid over the ribs to provide a continuous surface covering the entire area.

This approach has several advantages, particularly when resources are scarce. In the first phases of the introduction of this technique in Europe, in the late eleventh and the twelfth centuries, ribs are usually dressed carefully while the web or severies are materialised in rubble masonry, minimising carving effort. Later on, the quality of the severies gradually improved; by the sixteenth century, in many cases the dressing standard of the webs is comparable to the ribs or the spherical vaults of the period. Further, single-tier vaults executed in rubble masonry present a problem in the places where different surfaces meet, as exemplified in Romanesque groin vaults; the use of the ribs hides these problematic points (compare Figs. 10.1 and 10.2).

Also, this two-stage construction process allows a remarkable economy in false-work. In contrast to Romanesque barrel vault, which requires formwork for the entire surface, rib vaults can be executed using centring for the ribs; next, masons can use moveable planks or boards leaning on the ribs in order to place the severies, as seen in Sect. 3.3.2.[1] This issue was crucial at a moment when forests were approaching

[1] These planks or boards were named *cerces* by Viollet-le-Duc (1854-1868: IV, 105-108; see also Fitchen [1961] 1981: 99-102), implying that severies were double-curvature surfaces; however, recent surveys by Maira (2015: I, 166, 350, 442, 520) have shown that, generally speaking, the severies of French, English and Spanish sexpartite vaults approximate ruled surfaces, although

J. Calvo-López, *Stereotomy*, Mathematics and the Built Environment 4,
https://doi.org/10.1007/978-3-030-43218-8_10

Fig. 10.1 Groin vault without ribs. Church of Saint Vincent, Cardona (Photograph by the author)

depletion; this may be considered one of the relevant factors in the emergence of this process and, up to a certain extent, Gothic architecture (Bechmann [1981] 1996: 141–142).

In addition to these issues, much has been written about the mechanical behaviour of rib vaults. For the school of Viollet-le-Duc (1854–1868: IV, 1–126; for a summary, see Choisy 1899: II, 267–271), the role of ribs is essential: they convey stresses to four or six points at the springing of the vault, in contrast to the linear support line of the Romanesque barrel vault. From these points, loads are transferred to the foundations by piers, rather than walls; flying buttresses can also resist thrusts. Since walls are unnecessary, they can be replaced by windows, generating well-lit interiors, contrasting again with dim Romanesque churches. During World War I, the effect of shellfire on the cathedrals of Reims and Soissons, where some sections of the severies were left standing despite the loss of the ribs, cast some doubts about these theories, triggering a full-fledged attack on them by Pol Abraham (1934; see also Gilman 1920 and Coste 2003). However, more nuanced recent theories (Heyman 1995: 51–54) stress that while adding more strength, ribs are not strictly indispensable for the structural stability of such vaults. In fact, Romanesque groin vaults, executed in rubble masonry, are essentially rib vaults without ribs, and of course they concentrate

there are relevant examples of double-curvature surfaces in Sens, Paris and Bourges. In contrast, severies in Italian sexpartite vaults, usually executed in brick, are double-curvature surfaces.

Fig. 10.2 Ribs and severies. Sens Cathedral (Photograph by the author)

loads in the corners of the area. Also, in Late Gothic, exquisite ribless vaults were built in Valencia and central France in the fifteenth and sixteenth centuries, as we will see.

Another line of thought about Gothic architecture downplays structural explanations and focuses on aesthetical issues. Although this approach is one-sided and outdated, the merit of this school lies in its explanation of the Late Gothic proliferation of ribs. Tiercerons and ridge ribs still offer constructive advantages, since they divide severies into manageable dimensions. Other ribs, such as curved liernes, can hardly be explained by structural reasons, since a curved rib can withstand neither compression nor tension; in fact, in the more extreme versions of elaborate ribbed vaults in Germany, voussoirs are tied together with iron clamps or hung from the severies. The aesthetical theory and the desire to show the wealth of the client can explain this Late Gothic evolution.

While these contrasting theories have triggered fierce intellectual wars in the past, now a more eclectic stance is spreading; in Jacques Heyman's words (1995: 54):

> The rib, then, serves a structural purpose as a very necessary, but perhaps not finally essential, reinforcement for the groins; it enables vaulting compartments to be laid out more easily; it enables some constructional formwork to be dispensed with; and it covers ill-matching joints at the groins. As a bonus, the rib has been thought to be satisfying aesthetically, and all those functions may be thought as the 'function' of the rib.

10.1.1 Quadripartite Vaults

Leaving aside triangular vaults, which are employed for special purposes, the simplest kind of rib vault is the *quadripartite vault*, used to cover square or rectangular areas (Fig. 10.3). It includes four perimetral arches and two *diagonal ribs*. Perimetral arches may be *wall arches*, when a single rib vault spans a room; *transverse arches*, for example between adjacent vaults in the nave of a church or the central section of a nine-vault chapter house; or a mixture of both. Diagonal ribs are usually round arches. In square or rectangular areas, the radii and rises of both diagonal ribs are equal, and both ribs meet at their uppermost points; usually, this union is solved with a single keystone, in contrast to the apex of pointed or Tudor arches, usually solved in Medieval Gothic with two voussoirs.

Perimetral ribs may also be round arches. However, the length of the diagonal of a square or rectangle clearly exceeds the dimension of its sides; as a result, this

Fig. 10.3 Quadripartite vault. Amiens, cathedral (Photograph by the author)

solution leads to a domical form, known as *rampante redondo* or "round ridge" in Spanish; it was scarcely used until Late Gothic. To make the height of the perimetral and the diagonal ribs approximately equal, masons can use pointed arches at the sides of the area. On some occasions, particularly when transverse or wall arches equal the height of the keystone of the vault a horizontal rib, known as *ridge rib,* ties the bosses of the perimetral arches and the main keystone. This solution is common in England; later on, it was used in other countries, such as Spain, with the name of *rampante llano* or "level ridge" (Gómez Martínez 1998: 123–135). In other cases, perimetral arches are semicircular, their keystones are placed much below the main boss, and the web surface approximates a hemisphere. However, most quadripartite vaults stand between both extremes; pointed arches are used in the perimetral ribs, but their highest points are below the main keystone.

Very few tracings for quadripartite vaults, either at full scale or reproduced in treatises and manuscripts, have survived (see, for example, Taín 2006: 1892, for an eighteenth-century tracing for a quadripartite vault in the monastery of Montederramo). In addition to the practice of tracing on scaffoldings, the symmetry of the form may explain this absence. Perimetral arches do not interact with other elements except for columns and supported walls, while symmetry guarantees that diagonal ribs will meet at their common highest point. Next, severies can be placed spanning the triangles between perimetral arches and diagonal ribs. Thus, only the most schematic tracing, including the sides of the perimetral arches and the axes of the diagonal ribs, is needed to guide the placement of the struts and the centring that will support the ribs.

In any case, rib vaults raise complex problems at two particular points: the springers and the keystone. Over each corner of the vault, three ribs—namely two perimetral arches and a diagonal rib—interpenetrate. Given the usually complex profile of each of these elements, the intersection would include many small lines and curves. However, the issue was simplified by materialising bed joints as horizontal surfaces, at least until the level where the ribs depart from one another. This layout, known in French as *tas-de-charge* (plane of charge), has both geometrical and mechanical advantages. Structurally, it places the bed joints horizontally at a section where vertical stresses predominate. Geometrically, it simplifies the composition of complex profiles: the mason can simply dress a horizontal plane, mark on it the axes of the ribs, and apply the template for each rib, solving its interpenetrations directly on the stone (Willis [1842] 1910: 8–9; Rabasa 1996: 423–424).

The junction of the two diagonal ribs is usually solved with a cylindrical member, the *keystone*. Dressing this cylinder independently from the ribs would bring about mechanical and constructive problems: a vertical joint between cylinder and ribs would be dangerous; in addition, a concave surface at the end of the rib would be difficult to dress. To avoid these problems, four short fins with the curvature and the profile of the ribs are carved with the cylinder. To control the dressing of the keystone, masons usually dressed an upper horizontal face, known as *surface of operation*. Next, they marked a circle on it, standing for a directrix of the cylinder, and four axes meeting at the centre of the circle. The mason carved the fins and their bed joints with the aid of such axes, using also an arch square; a template

Fig. 10.4 Dressing a keystone for a rib vault at the Centro de los Oficios de León. From left to right and top to bottom, **a** Scoring a circle and rib axes in the surface of operation **b** Scoring the position of bed joints with a fixed bevel **c** Scoring the profile of the ribs on the bed joints **d** The finished keystone (Photographs by Agustín Castellanos)

with the section of the ribs was placed at the bed joints at the end of the fins, in order to assure the junction with ordinary rib voussoirs (Willis [1842] 1910: 24; Rabasa 1996a: 424–426) (Fig. 10.4). Despite all these precautions, small errors in the layout of the underpinning or the centring supporting the ribs can lead to an appreciable mismatch at the junction of ribs and keystones. Usually, a protrusion, known as the *boss*, is added to the lower face of the cylindrical nucleus, hiding the junctions between ribs and keystone. Bosses, placed at the centre of the vault and usually executed in good stone, provide an excellent opportunity to carve heraldic or religious images, so their primary function is frequently forgotten (Fig. 10.5). Only in quite carefully controlled constructions, such as the Castello Svevo in Catania or the Castell de Bellver in Majorca, builders dispense with bosses, showing boldly the junctions of the ribs in a cross-shaped piece.

In twelfth- and thirteenth-century Gothic architecture, the web is generally treated as an informal filling; it seems that the only means of formal control acting on the severies are the curved or straight planks used as light formwork. However, a specific method for web control is explained in minute detail by Josep Gelabert (1653: 151v–154v; see also comment in Gelabert/Rabasa [1653] 2011: 412–419). The master

Fig. 10.5 Boss with carved decoration protruding from the cylindrical nucleus of a detached keystone. The keystone belonged to a demolished vault in the church of Santa Catalina, Valencia, and was later reused in a wall; in the photograph it is inverted and leans in the operation surface in order to show the boss (Photograph by the author)

mason, with the help of two apprentices, should divide the length of a branch of a perimetral arch in equal-length segments, "starting from the middle", this hints that the division was carried out by trial and error and the end portions were adjusted empirically. Next, the apprentices should place a tight rope passing from the midpoint of the perimetral arch to the middle of the diagonal one; as a result, the rope would be set orthogonally to the plane of the perimetral rib, although Gelabert does not mention this, since he scarcely uses the notion of "plane". Then, the master should take the distance between the divisions in the perimetral arch and transfer it to the diagonal one, drawing on the extrados of a diagonal rib a segment with this length. This segment should be approximately parallel to the perimetral rib, and thus oblique to the diagonal one; at its end, a new division is marked on the diagonal rib. If the diagonal rib were a projection of the perimetral one on the vertical plane passing through the diagonal of the area, the lines joining the divisions in the perimetral and the diagonal ribs, and thus the joints in the severy, would be parallel. However, this solution is never used in medieval and Early Modern Gothic; otherwise, the directrix of the diagonal rib would be elliptical. Thus, the joints between consecutive stones in the severy deviate slightly from parallelism. Consequently, Gelabert advises the reader to dress the stones roughly with parallel edges; in a second step, the mason should adjust them carefully, following the detailed instructions given in the manuscript.

10.1.2 Sexpartite Vaults

Sexpartite vaults (Fig. 1.15b, 10.6) are associated with the initial phase of Gothic architecture, broadly encompassing the mid- and late-twelfth century. However, we should bear in mind that quadripartite vaults appear a few decades earlier, for example in Durham and in Rivolta d'Adda, near Milan, both built around 1100; that leaving aside exceptions like the atrium in Notre Dame in Dijon, sexpartite vaults are usually placed in the naves of important churches, and thus used concurrently with quadripartite vaults placed at the aisles; and that sexpartite vaults are used, although sparsely, at later periods, for example in the reconstruction of the choir at Beauvais Cathedral (Bony 1984: 7–10; Branner 1960a: 54–62)

These vaults include diagonal ribs, two main transverse arches and an additional or secondary transverse arch (Fig. 10.6); such layout creates three severies at each side of the secondary transverse arch, so there are six in the whole vault. Thus, a supporting element is needed at the midpoint of the sides of the vault, and two wall arches are placed at each side. The alternance of primary and secondary transverse ribs fosters in many occasions a binary *a-b-a-b* rhythm in nave elevations, for example in the cathedrals of Sens or Laon. As with many architectural rules, there are remarkable exceptions: although sexpartite vaults are used in the nave of Notre Dame in Paris, the internal elevations feature a basically unitary rhythm, although the binary rhythm reappears in the piers separating first and second aisles. In contrast, Noyon Cathedral,

Fig. 10.6 Sexpartite vault. Sens, Cathedral (Photograph by the author)

covered with quadripartite vaults, shows a binary rhythm in nave elevations and even primary and secondary transverse arches.

In contrast to quadripartite vaults, where diagonal ribs meet "automatically" at their highest points, in sexpartite vaults, some control device is needed to guarantee that the secondary transverse rib meets the diagonal ribs at the keystone. After measuring a large number of sexpartite vaults, Maira (2015: I, 154–158) has identified some strategies in order to solve this problem. First, the span of transverse ribs, being equal to the side of a square or rectangle, is shorter than its diagonal; thus, the use of pointed arches for transverse ribs is widespread, as in quadripartite vaults. In theory, the radii of transverse arches, both primary and secondary, may be adjusted, so the arch starts at the springers with a vertical tangent and rises to the exact height of the main keystone. However, this solution requires a complex tracing; a diagram in the sketchbook of Villard de Honnecourt by Magister II or Hand IV may be an attempt to solve this problem (Villard c. 1225: 21r, dr. 8-h); however, Maira's surveys show that this solution is not at all frequent.

Another approach addresses the problem using transverse ribs in the shape of pointed arches with the same radius as the semicircular diagonal ribs. Maira associates this school with Bourges Cathedral, although she mentions other significant earlier examples such as Laon Cathedral, as well as derivatives in Spain. This solution has practical advantages: it leads to the standardisation of voussoirs, formwork, and, when used, arch squares, as stressed by Maira (see also Palacios and Martín 2011: 538–540). In theory, for a span of 10 m, a rough average of early vaults, the keystones of transverse arches with the same radius as the diagonal ribs and vertical tangents at the springings should stand 31 cm below the keystones of diagonal ribs; thus secondary transverse ribs should be raised in order to meet diagonals at the main keystone. Significantly, Maira (2015: I, 155–158) has detected that many vaults in this school are slightly stilted, that is, they include short straight sections in the springers, around 30 cm; this is consistent with the theoretical calculation.

Another school, centred around the Paris area and associated with Nôtre-Dame (Maira 2015: I, 154), uses semicircular transverse arches; in this case, since the radius of the transverse rib is much shorter than the diagonal one—theoretically in a ratio of $\sqrt{2}$—, transverse arches are much more heavily stilted. In this case, theoretical computations result in 2.07 m for a 10 m span; adjusting this figure to the span of Nôtre Dame, 13.26 m, gives a difference in heights of 2.73 m, while Maira estimates the initial straight section at around 2.5 m. Of course, all these computations are rough approximations; going beyond this level is useless, since widths of constructive elements, execution tolerances and mechanical deformations make exact comparisons meaningless. In any case, these gross computations show that the geometrical problems posed by these vaults were solved mainly through stilted supports.

Another remarkable detail stressed by Maira (2015: I, 166) is that in the ribs of the earliest sexpartite vaults individual stones are rather short, so their intrados profile is not curved, but rather straight. Thus, the general curved outline of the rib is a result of wedge-shaped mortar joints. This is consistent with the almost non-existent evidence of the use of the arch square in the Gothic period, or at least in its early phases.

Fig. 10.7 Semipolygonal vault. Saint-Quentin, basilica, chancel (Photograph by the author)

10.1.3 Polygonal, Trapecial and Triangular Vaults

Polygonal and semipolygonal vaults. Polygonal vaults are used in regular spaces, such as towers or chapter houses. Typically, given a regular polygon, the meeting of ribs at a single keystone is guaranteed; thus, the problems raised by this kind of vaults are located at the same points as in quadripartite vaults: the springers and the keystone. Frequently, polygonal vaults are used in chancels, cutting them by a vertical plane passing through two corners of the polygon. This plane may be a symmetry plane of the polygon; in this case the vault is strictly *semipolygonal* (Fig. 1.15.d) and the keystone is placed on the cutting plane. However, on other occasions, Gothic architects used another design, placing opposite sides of the polygon at both ends of the chancel to smooth the junction with the nave (Fig. 10.7); these vaults are usually classified as semipolygonal, at the expense of a slight terminological incoherence.[2] As with polygonal vaults, all ribs are usually equal, except of course the transverse arch dividing the vault from the quadripartite or sexpartite vaults in the nave; usually, no rib connects the upper point of this transverse arch with other ribs in the vault. Thus, the intersection of ribs at the keystone is assured, and the main stonecutting problems are posed again by the springers and the keystone.

---

[2]In this solution, architects used even-sided polygons, with n edges, placing one side at the axis and ending the vault with full sides at each end; thus $(n + 1)/2$ sides of the polygon are materialised, and the chancel vault is slightly larger than a strictly semipolygonal one.

Fig. 10.8 Trapecial vault. Saint-Denis, Abbey, ambulatory (Photograph by the author)

A quick look at an Early Modern Gothic text such as Gelabert confirms this. The basic polygonal vault (Gelabert 1653: 137r) is an octagon from which three sides are cut to match the nave of a church. The remaining section includes six identical ribs, so Gelabert draws a single one standing for them all. In contrast, in other complex schemes (Gelabert 1653:138r), where the distances from the corners of the chancel to the main keystone are unequal, and thus the horizontal projections of ribs are different, he represents each kind of rib in the elevation in order to guarantee that all of them meet at the keystone.

Ambulatory vaults. The earliest solutions for trapecial areas in deambulatories—for example in the inner ambulatory in Saint-Denis Abbey (Fig. 10.8) and in Sens cathedral—use a rather sophisticated solution. If masons—or Abbot Suger, who boasted of having laid the foundations with arithmetical and geometrical instruments—had simply used the diagonals of the trapecial area, the inner severy would be quite small, and the outer one rather large. Thus, both in Saint-Denis and in Sens, the intersection of diagonal ribs is pushed outwards to enlarge the inner severy and reduce the outer one.[3] As a result the ribs are "crooked", that is, they are not placed at the same plane at both sides of the keystone. If the ribs were working alone, this would bring about mechanical problems, since a crooked element cannot withstand compressive stresses at the breaking point; however, the compressive stresses of the four half-ribs compensate each other at the keystone. In any case, this solution poses a geometrical problem. In order to use ribs with equal horizontal projections, as in semipolygonal vaults, masons should place the keystone at the intersection of the symmetry axis of the vault with the bisectrices of the lateral sides; in this solution,

[3]The simplest solution, placing each diagonal rib in a single vertical plane, so that the rib does not change direction at the keystone, is quite rare; however, it can be seen in some sections of the ambulatory in the Abbey of Saint-Germer-de-Fly.

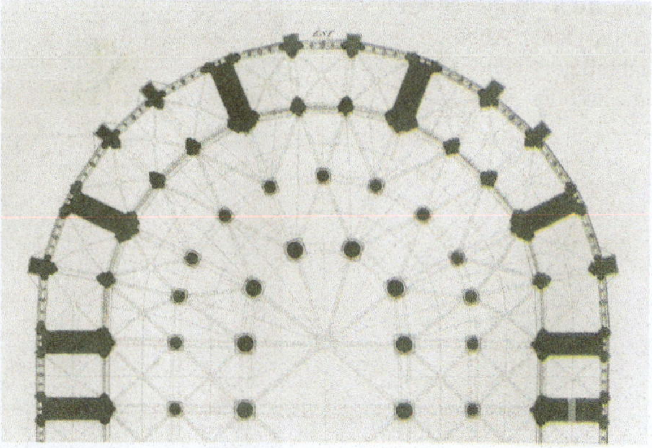

Fig. 10.9 Plan of the choir of Notre-Dame, Paris (Celtibère 1860: pl. 50-51)

the keystone is placed close to the outer wall. As a third alternative solution, the keystone may be placed approximately at the midpoint of the symmetry axis. Thus, the horizontal projections of the ribs coming from the inner springers are shorter than those arising from the outer wall; since both ribs had to reach the same height at the keystone, some adjustment is needed, either by raising the inner ribs at the springers or by using different radii.

The radial chapels in Saint-Denis raise a different problem. Their vaults are divided into five severies by an additional rib placed at the symmetry axis of the radial chapel. However, the radii of the three outer ribs seem to be approximately equal and larger than those in the inner part; that is, the outer part of the vault is treated as a simplified polygonal vault, with its centre at the keystone where five half-ribs meet.[4]

Although the solution used in Saint-Denis and Sens spread through most Gothic churches in the following centuries, bold masons attempted several experiments while seeking alternative layouts. For example, in the inner ambulatory of Nôtre-Dame (Figs. 10.9, 10.10), the vaulting is based in groups of three triangles; in each bay, the tips of the extreme triangles look inside, while the central one points outward. The solution is repeated in the outer ambulatory; here, each section is divided into five triangles, again alternating inward- and outward-pointing ones. Despite the geometrical cunning of this layout, it was not often repeated. The outer deambulatory of Le Mans Cathedral avoided trapeziums using a combination of rectangles and triangles. Such solution is augmented in Toledo Cathedral: the inner ambulatory is divided into rectangles and triangles, as in the outer one in Le Mans. However, the designer managed to make the bases of the triangles equal to the outer sides of the rectangles, so that the line that separates the two ambulatories is, or at least approximates, a regular polygon; he then divided again the outer ambulatory into rectangles abutting

[4]In contrast to the standard layout of radial chapels, those at Saint-Denis feature arches connecting each chapel with the adjacent ones. Thus, the ensemble can be described as a string of radial chapels, an outer ambulatory, or a combination of both. However, this issue is not central to the problem I am analysing in this paragraph, so I have described these elements as radial chapels for the sake of clarity.

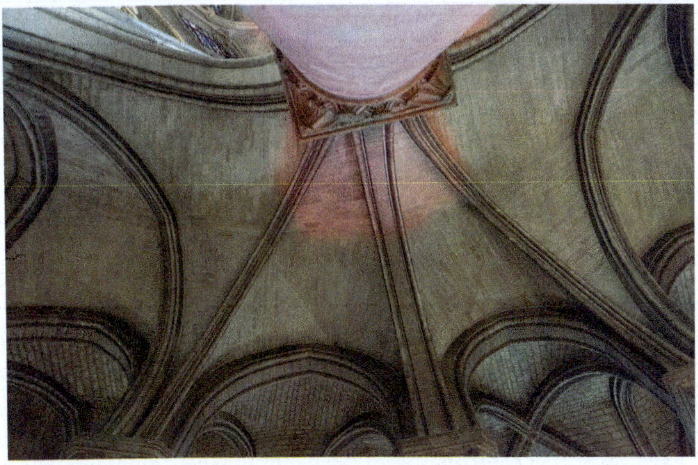

Fig. 10.10 Triangular vaults. Paris, Nôtre-Dame Cathedral, ambulatory (Photograph by the author)

the sides of the regular polygon and triangles inserted these between rectangles. The sight of flying buttresses in Le Mans is striking; if we could see those of Toledo without the obstruction of narrow city streets and later additions, we would appreciate a fork of two flying buttresses coming out from each edge of the chancel, each one arriving at an intermediate pier and being further divided into another pair of buttresses. No wonder a scholar as serious as Robert Bork (1997–98) chose the exclamation exclamation "Holy Toledo" as the title of a study on these issues.

Triangular vaults. However, the Le Mans-Toledo solution again poses the problem of the meeting of ribs in space. In Paris, this issue was absent, since triangles were small enough to be vaulted with a single severy. In contrast, in Le Mans and Toledo triangles are divided into three different severies by the ribs. Since the lengths of these ribs are approximately equal, they were probably built without the need of true-size elevations.

Another kind of triangle is often found in octagonal vaults in the fourteenth and fifteenth centuries in the Crown of Aragon. When octagons are circumscribed by squares or rectangles—for example in the chapel of the Christ of Lepanto in Barcelona Cathedral, the chapel of the Holy Grail in Valencia Cathedral or the Sala dei Baroni in Castel Nuovo in Naples—the space between the octagon and the enclosing square is covered by vaults in the shape of a right equilateral triangle (Rabasa et al. 2012). This area cannot be divided into three severies using ribs of the same length,[5] and thus a tracing, however schematic, is needed. Josep Gelabert (1653: 127r) presents a solution for this kind of vault, known in Catalan as *tercerol*.

[5]In order to divide a triangle in three smaller triangles so that the division lines are equal in length, all three lines must meet in the circumcentre of the triangle, that is, the point placed at equal distances from the corners of the triangle, which is coincident with the intersection of the orthogonal bisectors of the three sides of the triangle. In the case of the right equilateral triangle, this point lies on the hypotenuse, and thus it cannot be used to divide the triangle in three severies.

Unusually, he includes a drawing, showing the plan and the elevation of the vault, with no explanatory text. Next, he superimposes on the elevation true-size-and-shape depictions of the ribs, both the larger side ribs and the shorter central one; in this way, the mason can compute the radius of the ribs and dress the voussoirs with this curvature.

10.1.4 Star Vaults

Stellar, star or tierceron vaults appeared in the first decades of the thirteenth century, first in Lincoln Cathedral and soon after in French examples, such as the crossing of Amiens Cathedral (Fig. 1.15.e, 10.11). These vaults use additional ribs to divide the web into smaller portions, reducing the length of its courses and the overall surface of each severy. In each quarter of the vault, two new elements, the *tiercerons*, are placed between the diagonal ribs and the perimetral arches, starting from the springers. The two tiercerons meet at a secondary keystone; another rib, known as *lierne*, connects this boss with the central one. Later on, particularly in the fifteenth and sixteenth centuries, star vaults were enriched with still more ribs: short ribs, known also as liernes, circles around the main keystone or circular arcs starting from the uppermost points of the perimetral arches, known in Spanish as *pies de*

Fig. 10.11 Tierceron vault. Lincoln Cathedal, nave (Photograph by the author)

gallo (rooster's feet). However, in England, France and Spain, the tierceron was the fundamental element for these complex Late Gothic combinations. Thus, when writers began to include Gothic vaults in their treatises and manuscripts, from the sixteenth century on, the simple tierceron vault was often used as a didactic example to explain the geometrical complexities of rib vaults, in clear contrast to the scarcity of explanations about sexpartite, triangular or trapecial vaults.

The role of the diagonal rib. Rodrigo Gil de Hontañón (c. 1560: 25r) includes a section on rib vaults, together with a fairly detailed drawing (Fig. 10.12). To control the layout of the ribs the mason should prepare a full-scale tracing on a scaffold placed exactly under the vault in the making, as we have seen in Sect. 3.3.4. This tracing includes a plan of a tierceron vault with a circle around the main keystone, rooster's feet and other curved liernes connecting the circle and the rooster's feet. The drawing also includes an oblique elevation, showing in true shape the diagonal rib, which is strictly semicircular, as usual in medieval and Early Modern Gothic. This semicircle seems to be also used as a drawing aid for the construction of the plan: the tiercerons are drawn so that their extensions pass through the intersection of the vault axes with the semicircle. As a result, the tierceron follows the bisector of the angle between the wall arch and the diagonal rib. The drawing of a square rib vault in MS 12.686 of the National Library in Madrid (Alviz c. 1544; see also García

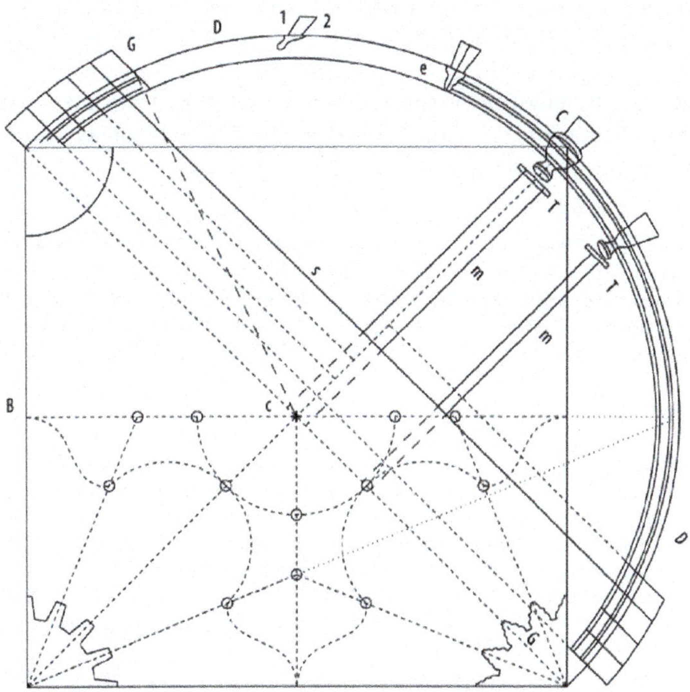

Fig. 10.12 Star vault (García 1681: 25r, after Gil de Hontañón c. 1560, redrawn by author)

Baño and Calvo 2015 and García Baño 2017: 612–621)[6] also includes portions of this semicircle, but no attempt is made to draw the elevation of the diagonal rib (Fig. 2.4); once again, the tiercerons pass through the intersection of the axes and the semicircle.

Rodrigo Gil recommends the mason to take the height of the main keystone from the semicircle, controlling it through the use of *mazas* or vertical struts supporting the centring and the vault during construction. Next, he adds that for

> … but for the struts of all the other [keystones] you should do this: having placed the main keystone at its proper height, you will draw a circle reaching from *B* to *C*, which is from the rooster's foot to the main keystone with the radius of the diagonal g; from that circle you should drop plumb bobs to the keystones that are drawn on the planks; this will furnish the length of each strut.[7]

Palacios (2009: 89–91) interprets this passage[8] by considering the surface of the vault to be hemispherical, arguing that Gil de Hontañón describes a hemispherical vault where diagonal ribs and straight liernes are great circles of the intrados surface, featuring the same curvature. Tiercerons and wall arches, not depicted in the elevation, would be small circles of this sphere. The formal control of the perimetral arches does not pose any problem; they can be drawn as round arches, with their radius equalling half the side of the enclosing square of the vault. Determining the radius of the tiercerons is not so simple. They were probably laid out by trial and error, as we will see in the next paragraphs, taking their horizontal projections from the plan, and the height of their upper ends (that is, the secondary keystones) from the diagonal rib.

Curvatures and keystone heights. The manuscript by Hernán Ruiz II (c. 1560: 46v) includes a drawing of a rib vault (Fig. 10.13), without any accompanying text. The drawing represents a square-plan vault with tiercerons, straight liernes

---

[6]The drawing of rib vaults in this manuscript are placed in unnumbered sheets at the end of the volume.

[7]Gil de Hontañón and García ([c. 1560] 1681: 25r-25v): *Mas para las mazas de todas las otras se hará así: puesta la clave mayor al alto que le toca, harás una cercha tan larga que alcance desde B a C, que es desde el pie de gallo a la clave mayor con la vuelta de la diagonal, y desde estas cerchas dejar caer plomos a las claves que están señaladas en los tablones, y aquello será el largo de cada maza…* Transcription by Cristina Rodicio García, taken from García et al. (1991) and modernised by the author. See also Sect. 3.3.4 for the transcription and interpretation of the preceding paragraph.

[8]Palacios states that the right half of the semicircle in the tracing represents the diagonal rib, while the left part stands for the lierne. Although there are a number of difficulties in this passage (see Calvo 2017, 29-32), the main point in Palacios' interpretation is unquestionable, since the phrase "you will draw an arc … from the cock's leg to the main keystone with the radius of the diagonal" states clearly that the curvature of the straight lierne equals that of the diagonal. While accepting this idea, Gómez Martínez argues that if perimetral arches are pointed, the straight lierne may have a salient point at the main keystone, in order to reach more height at the keystones of the wall arches. In my opinion, this is quite improbable. We should not forget the presence of a round rib joining eight secondary keystones, four in the diagonal ribs and four in the straight liernes. A smooth lierne, without salient point at the main keystone, with horizontal tangent at this point, and the same curvature than the diagonal ribs, guarantees that the eight secondary keystones along the round rib are placed at the same height.

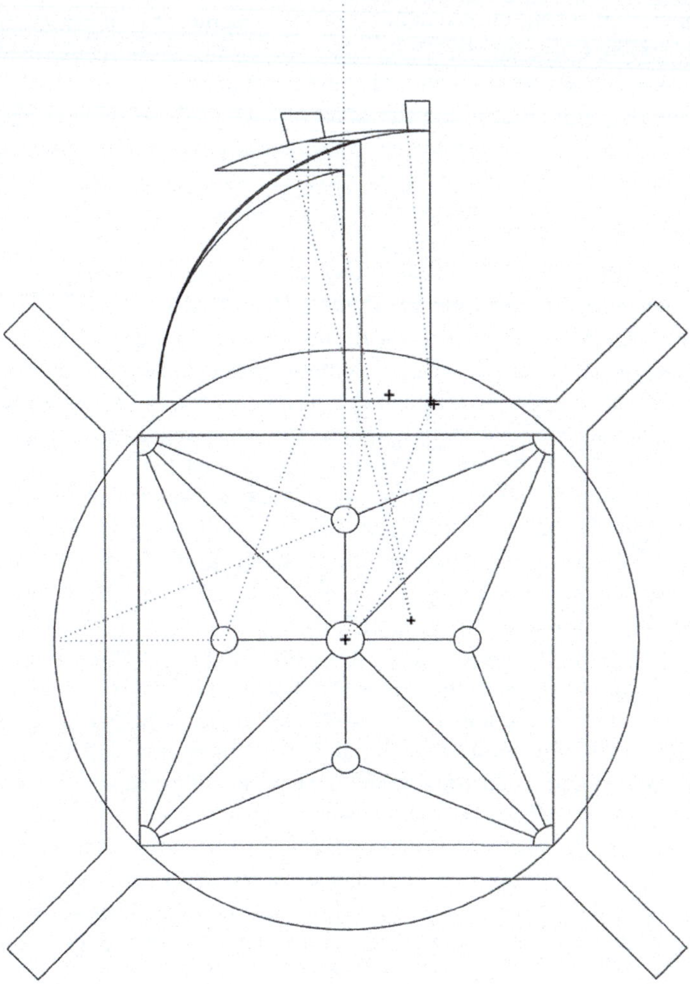

Fig. 10.13 Star vault (Hernán Ruiz II c. 1560: 46v, redrawn by author)

and four secondary keystones, without circles or rooster's feet. It includes elevations of the pointed perimetral arches, diagonal ribs, liernes and tiercerons; the tiercerons follow the bisector of the sides of the area and the diagonal rib. As Rabasa (1996a) has shown, rather than using a single orthographic projection, Ruiz uses a notable variety of geometrical resources to determine the curvature of the ribs, the height of the secondary keystones and even the slope of the beds of the voussoirs standing next to the primary and secondary keystones, in order to control their execution precisely. Speaking in descriptive geometry terms, perimetral arches are represented in their own vertical plane, liernes are projected orthogonally, and diagonal ribs and tiercerons are rotated in order to bring them to the projection plane. However, as we will see further on, masons did not seem to think in these terms; instead, they showed a strong penchant for the disarticulation of these elements (Rabasa 1996a; Rabasa and Calvo 2009b; Calvo 2017).

In any case, Hernán Ruiz's method furnishes a depiction of each rib in true shape, allowing the mason to control the curvature of each of these elements. Further, the drawing includes a horizontal line connecting the upper end of the tierceron and the starting point of the lierne. To our eyes, this line seems rather awkward: it connects two representations of the same point in the elevation. However, we should bear in mind the disarticulation of the tracing: although the lierne and the tierceron are set in different vertical planes, both are drawn in true size and shape, and Ruiz does not make any effort to construct an orthogonal projection of the whole ensemble (Rabasa 1996a). In any case, this remarkable graphic practice guarantees that the tierceron and the lierne meet at a single point in space, the secondary keystone. All this places a constraint on the shape of the tierceron; as Rabasa pointed out, Hernán Ruiz determined its centre through trial and error; the manuscript bears a number of compass marks, near the horizontal plane passing through the springings, attesting to his efforts. This implies that Ruiz tried to draw the tierceron with a vertical tangent at its lower end, but his notion of tangency was purely empirical. Otherwise, he would have started by placing the centre of the tierceron at the horizontal line passing through the springing.

The elevation shows clearly a pointed perimetral arch, while the lierne features a larger radius than the diagonal ribs, in contrast to Gil de Hontañón, who made both radii equal. In the plan, the lierne starts at the secondary keystone, while in the elevation it is extended until it reaches the keystone of the perimetral arch. This suggests that the prolongation of the lierne is an auxiliary line, and that the height of the perimetral arch sets a constraint both on the profile of the lierne and on the height of the secondary keystone. The tracing also includes two schematic diagrams of the primary and secondary keystones, representing the bed joints of the upper voussoir in the tierceron and the lower and upper beds in the lierne, in order to assure their correct match with both keystones, as well as the operating surface (the horizontal upper face of the keystone) in order to control the slope of the bed joint with the bevel (Willis [1842] 1910: 24; Rabasa 1996a) (Fig. 10.14).

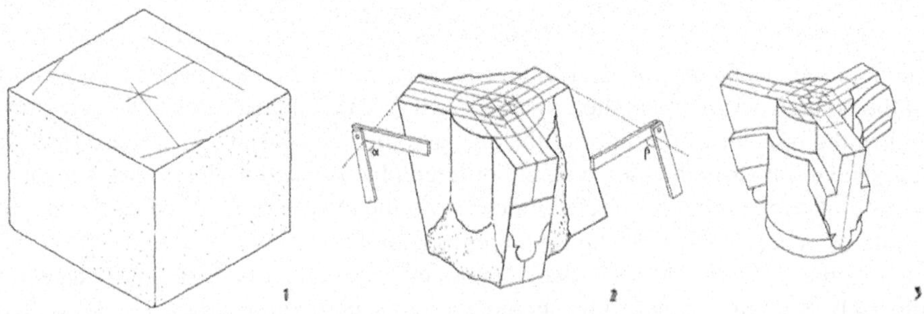

Fig. 10.14 Dressing the secondary keystones of a tierceron vault (Drawing by Enrique Rabasa)

De l'Orme (1567: 107r-108v) includes in his treatise a vault with five keystones. Although using different graphic resources (Fig. 10.15), the underlying geometrical concepts are similar to the method of Hernán Ruiz. Wall arches are slightly pointed and thus the intrados surface is not exactly spherical. The plan is drawn with the diagonal ribs parallel to the edges of the sheet, in contrast to all other writers on this subject. The disarticulation of the elevations is particularly evident, since the springers of the diagonal ribs in the elevation are placed over the central keystone in the plan, and vice versa. There are other significant differences with Hernán Ruiz: de

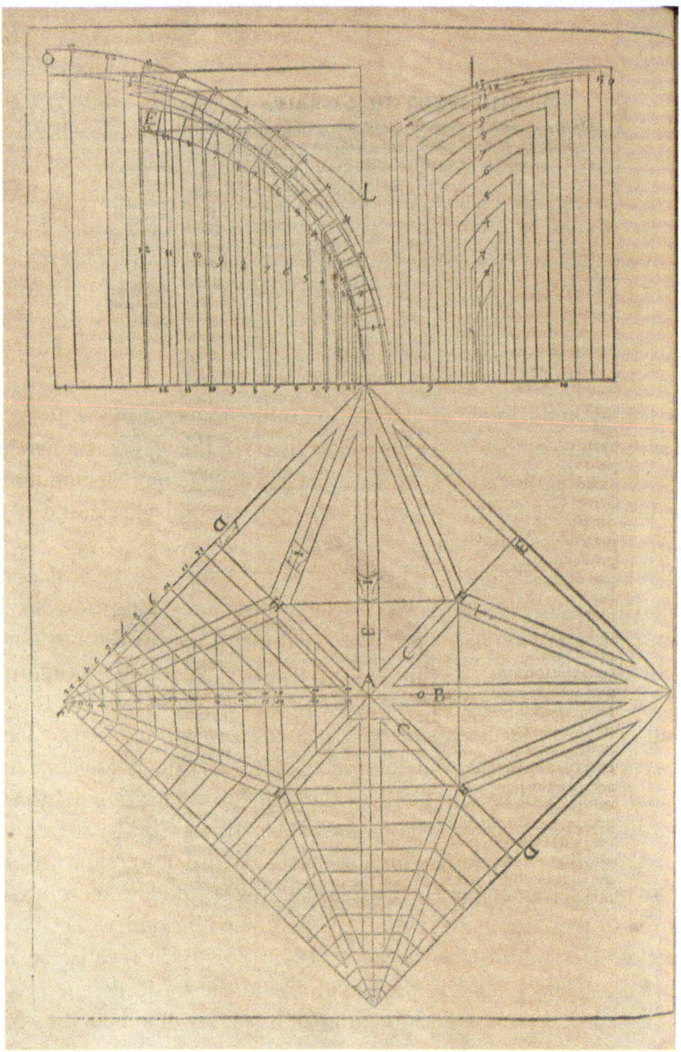

Fig. 10.15 Star vault (de l'Orme 1567: 108v)

l'Orme represents the width of the ribs and their division into voussoirs, to control the layout of the severies by means of templates. He draws the horizontal projections of the joints in each severy as parallels. However, since diagonal ribs, tiercerons and liernes do not feature equal lengths, the intrados surface of the severy is warped, and thus cannot be represented exactly by a template. Quite probably, he expected the mason to adjust this surface in the final phase of the carving process, as explained by Gelabert, but he does not say a word about it.

Spherical rib vaults. The manuscript of Vandelvira (c. 1585: 94v-97r) includes an example of a rib vault (Fig. 10.16), under the puzzling heading *De las jarjas* (About springers); it features diagonal ribs, tiercerons, liernes and two full circular ribs around the main keystone. The larger of these circles meets the tiercerons and liernes at their intersection; the smaller one passes through the midpoint of the liernes. This design follows closely a vault built by the author's father in the crossing of the Dominican convent in La Guardia de Jaén (Lázaro 1988; Gila 1992: 97–99; Galera 2000: 88–93), leaving aside some decorative mouldings. Both in La Guardia and in the manuscript, there are no bosses; rather, ribs directly intersect each other, as in many German vaults. In contrast to Rodrigo Gil, Hernán Ruiz and de l'Orme, the tiercerons in the manuscript do not follow the bisector of the diagonal rib and the wall arch; rather, their prolongations pass through the intersection of the vault axes with the opposite side of the enclosing square. As mentioned by Palacios ([1990] 2003: 290–301; 2009: 93–96) and Rabasa (1996a: 429–431), Vandelvira states that the vault is "modern" (that is, Gothic) since it includes ribs; however, he makes it clear that it does not follow mainstream Gothic design in the shape of the wall arches, which are round rather than pointed, so the general shape of the vault is hemispherical. All this is reflected in the elevations, which include true-size representations of all ribs, except the horizontal round ones; since the vault is hemispheric, the radius of the lierne equals that of the diagonal arch.

The lack of bosses demands a tight control over the shape of the bossless keystones that materialise the junction between ribs, called *crucetas* (little crosses). To address this problem, Vandelvira constructs intrados templates, utterly foreign to mainstream Gothic methods, using cone developments. Thus, a geometrical procedure originally conceived for Renaissance vaults is reused for ribbed vaults; such is the quick pace of technical evolution in the Spanish Renaissance. When applied to the *crucetas*, the problem is more complex. Vandelvira develops a vertical-axis cone to construct the horizontal section of the template for the *cruceta*, controlling the shape of other sections through triangulation. The slightly later manuscript by Guardia (c. 1600: 85r bis[9]) explains the procedure in greater detail; the use of a cone with a horizontal, oblique axis in order to construct the template for a whole tierceron attests to the sophistication of these methods in the early seventeenth century.

Tangencies. Derand's tierceron vault (1643: 392–395) recalls the manuscript of Hernán Ruiz in its graphical presentation and the basic geometrical concepts (Fig. 10.17). The wall arch is slightly pointed; it is clear that the radius of the lierne

[9]The drawing and the accompanying text are placed on an unnumbered sheet inserted between fol. 85 and fol. 86.

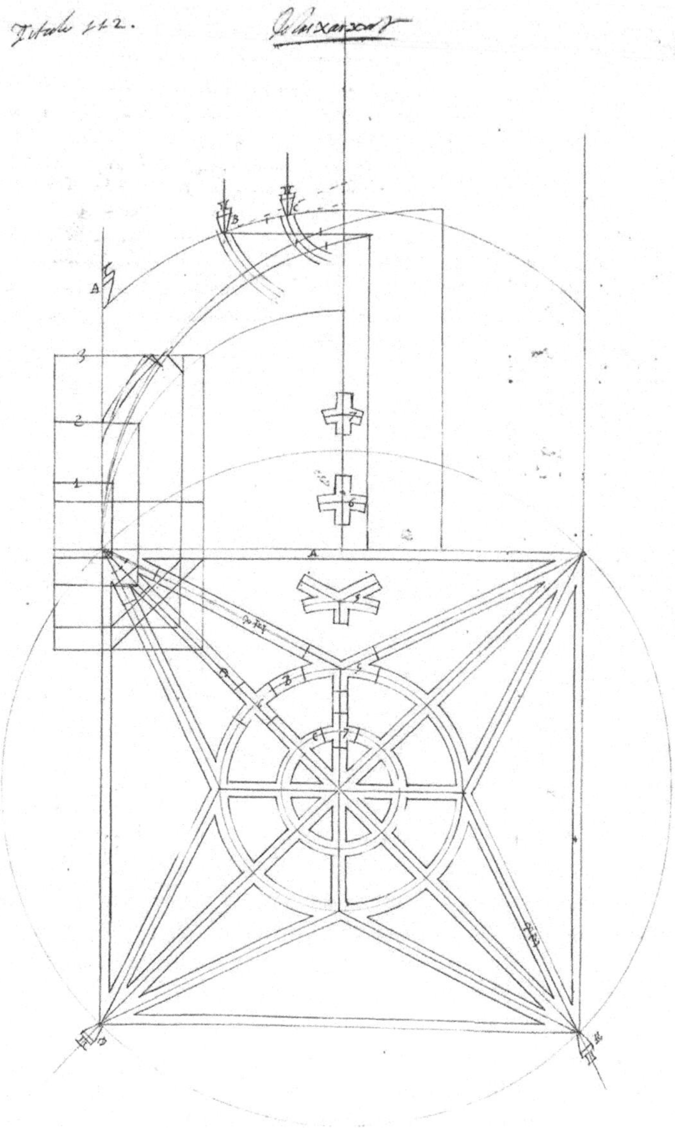

Fig. 10.16 Star vault (Vandelvira c. 1585: 96v)

is larger than that of the diagonal rib. Once again, the tierceron follows the bisector of the perimetral arch and the diagonal rib. Although some ribs are drawn in plan using two lines, the elevation depicts only the directrices of the ribs, showing their curvatures in true shape, as in Hernán Ruiz. Of course, this operation demands the disarticulation of the network of ribs; thus, an auxiliary horizontal line is used to

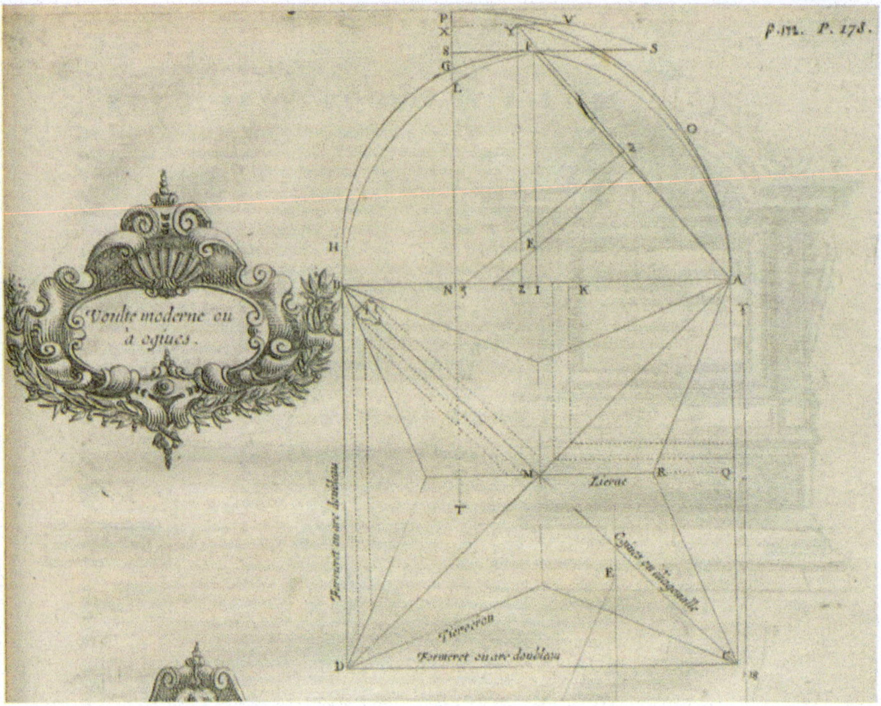

Fig. 10.17 Star vault (Derand [1643] 1743: pl. 172)

assure that the upper end of the tierceron and the starting point of the lierne are placed at the same level so both ribs meet neatly at the secondary keystone.

Up to here, the method closely follows the procedure used by Hernán Ruiz; the distinctive traits of de L'Orme's solution, such as the representation of the width of all ribs in the elevation or the diagonal ribs drawn parallel to the edges of the sheet, are nowhere to be seen. However, Derand uses several learned-geometry procedures to increase the precision of the result. In particular, to assure that the liernes "meet in a more pleasing way"[10] he advises the reader to place their centres on the vertical line passing through the main keystone. In other words, he draws the lierne with a horizontal tangent at its end, matching the tangent of the lierne coming from the other half of the vault and avoiding a salient point at the main keystone. Also, to guarantee that the tangent of the tierceron at the springing is vertical, he determines the ends of the tierceron, draws its chord, constructs its bisector, and finds the intersection of the bisector with the horizontal plane passing through the springings of the vault. All this furnishes the exact position of the centre of the tierceron, in contrast to Hernán Ruiz, who solved this problem by trial and error.

The general traits of Derand's procedure were taken up by Milliet ([1674] 1690: II, 680). However, the vault is spherical; this allows him to simplify the computation

[10]Derand (1643: 394): *se trouuent d'vne plus agreable rencontre.*

of the curvature of the tierceron, which is now a small circle of the intrados sphere. Also, the tierceron is drawn in plan by dividing into five parts the half-axis of the enclosing square and taking three parts, while the elevations are placed under the plan.

Structural functionalism. The encyclopaedic repertoire of Freziér's treatise includes a single rib vault (1737–39: III, 24–31; see also Rousteau 1996). The author explains that although such vaults are not usually built, this design should be known in order to carry out restorations in cloisters, churches and other buildings. His method is basically that of Derand; however, he critisises minor aspects, such as the position of the centres of diagonal ribs; at the same time, he adds ribs between secondary keystones, for didactic reasons (Fig. 10.18).

His comments about the advantages and disadvantages of Gothic constructive elements are more interesting. When explaining the rib vault, he points out that it offers remarkable advantages, although he finds the discontinuity of the severies quite unappealing. According to Frézier, the high slope of the severies allows their construction with light webs, only 5 or 6 inches thick, with a remarkable economy in labour and material. Further, these elements can be built using simple squared blocks, in contrast to the sterotomical techniques explained over and over in Frézier's treatise. At the same time, these vaults exert less thrust upon walls; this fact helps to avoid the need for massive buttresses, bringing about further economies in material. In a separate passage, Frézier (1737–39: II, 97–98) deals with the catenary, the ideal form

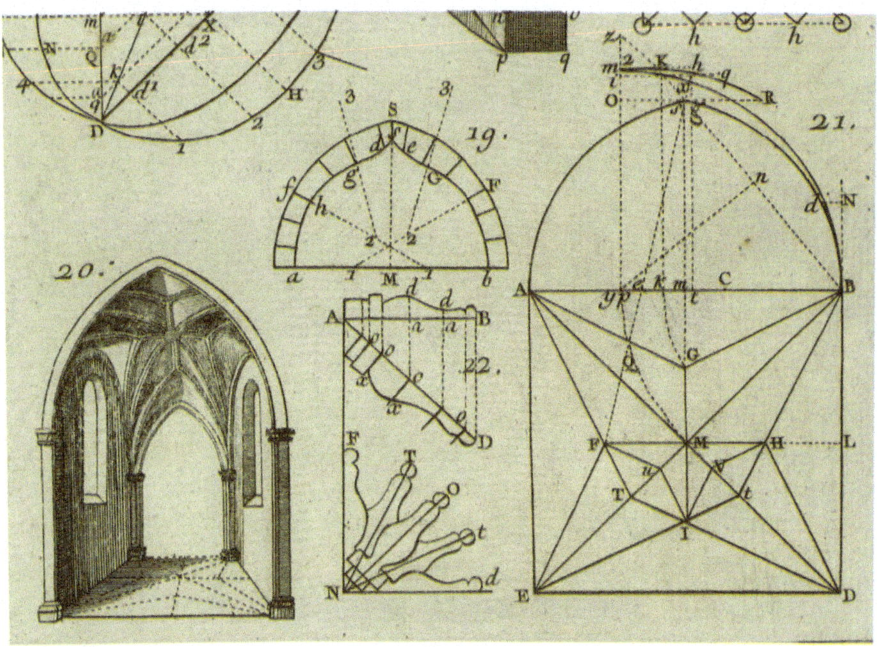

Fig. 10.18 Ribbed vault (Frézier [1737–1739] 1754–1769: pl. 71)

of arches which support only their own weight, studied some years before by Hooke and Couplet. Frézier explains that the arches that use the catenary as a directrix show an angle at their base, since the tangent to the catenary at this point is not vertical. He also explains that, among the arches usually employed in construction, the one that best approximates the catenary, while keeping vertical the tangent at the springings, is the Gothic or pointed arch. This marks the start of a new appreciation of Gothic geometry, based not only on its utility in restorations, but also as an example of structural rationalism; of course, a consideration of the implications of Frézier's passages lie out of are beyond the scope of this book.

10.1.5 Net Vaults and Other Complex Types

Net vaults and German methods. While Late Gothic masons in England, France and Spain frequently used a succession of tierceron vaults separated by transverse ribs, masons in the German Empire often eschewed transverse arches and tiercerons. The result of these design choices is a continuous network of ribs where primary divisions are hard to identify; the result is known in German as *Netzgewölbe* (net vault) (Fig. 10.19). In recent decades, following the pioneering work of Shelby and Mark (1979) and Müller (1989, 1990), a number of authors (Nussbaum and Lepsky 1999; Nussbaum 2000; Tomlow 2009; Martín Talaverano et al. 2012; Pliego 2017)

Fig. 10.19 Net vault over the nave and tribunes. Kutná Hora, Saint Barbara (Photograph by the author)

have pointed out that such elaborate vaults can be controlled using the elevation of a single arch, called *Prinzipalbogen* (main arch). In Jos Tomlow's graphical metaphor (2009: 197, note 15), the main arch can be folded like a Japanese screen as many times as necessary in order to generate a number of circular arcs placed on different vertical planes. Thus, all ribs have the same curvature; of course, such standardization brings about remarkable economies in arch squares, formwork and worksite organisation.

Many drawings of complex-plan vaults, either single exemplars (Böker 2005: 238) or belonging to manuscripts or compilations (Codex Miniatus c. 1565: 1r, 2v, 3v, 5r, 6r, 6v, 7r, 8r, 8v, 9r, 10v, 11r, 15v; Facht 1593: 5v, 17v) include the elevation of a single rib in the shape of a quadrant, usually showing the intrados and extrados. Vertical lines drawn accross the elevation, connected by letters with points in the plan, suggest that the elevation was used to determine the height of secondary keystones. This suggests that the height of the keystones was computed by measuring in the plan the horizontal distance from each secondary boss to the main keystone and transferring this distance to the elevation.

However, a detailed analysis of these drawings, in particular the neat ones in the manuscript by Facht von Andernach (1593; see also Shelby and Mark 1979) shows that the question is not so simple. In a few particular layouts, such as the uniform triangular network drawn by Facht von Andernach (1593: 11v-12r) all reasonable paths leading to a keystone have the same total length; thus, the problem can be solved consistently with a single *Prinzipalbogen*. However, in most cases, each keystone can be reached through several paths with different lengths (Fig. 2.8). If the height of the keystone were measured from a single "principal arch", these different paths would lead to different heights; in other words, one of the ribs meeting at the keystone would pass over other ribs. Masons may accept this fact and use keystones with a high cylindrical kernel, so that ribs can reach it at different heigths (Rabasa et al. 2015b: 1404–1405). In other cases, however, they used some adjustments to avoid this situation. The simplest one is a slight rotation of one of the ribs so that a circular arc with the standard curvature reaches the appropriate keystone; in some particular cases, this means that the tangent at the springing is not vertical (see for example Palacios 2009: 172–173; Tomlow 2009: 198) Another alternative is to use a different radius for one or several secondary paths, so all ribs will reach a secondary keystone at the same level (Facht 1593: 3v, 8v, 9v).

Bartel Ranisch's hypotheses. A different method is presented in Bartel Ranisch's book (1695) about the churches of Danzig, now Gdansk. As we have seen in Sect. 2.1.7, the book is not presented as an instruction manual or personal compilation, but rather as a description of all *extant* churches in Gdansk; however, they are explained in such detail that a separate diagram for rib curvature computation of each vault of every church is included. Ranisch's idiosyncratic method has been analysed by Pliego (2017) for three vaults in the book with different levels of complexity. Here, I will limit myself to the first and simplest one, a vault in the church of Saint Mary (Fig. 10.20) including tiercerons and double liernes for the short sides of the area and tierceron-like ribs starting from an intermediate point in the wall arches; this will allow us to grasp the essence of this method. It seems that Ranisch starts from the basis that all ribs feature the same curvature. He starts by unfolding the "Japanese

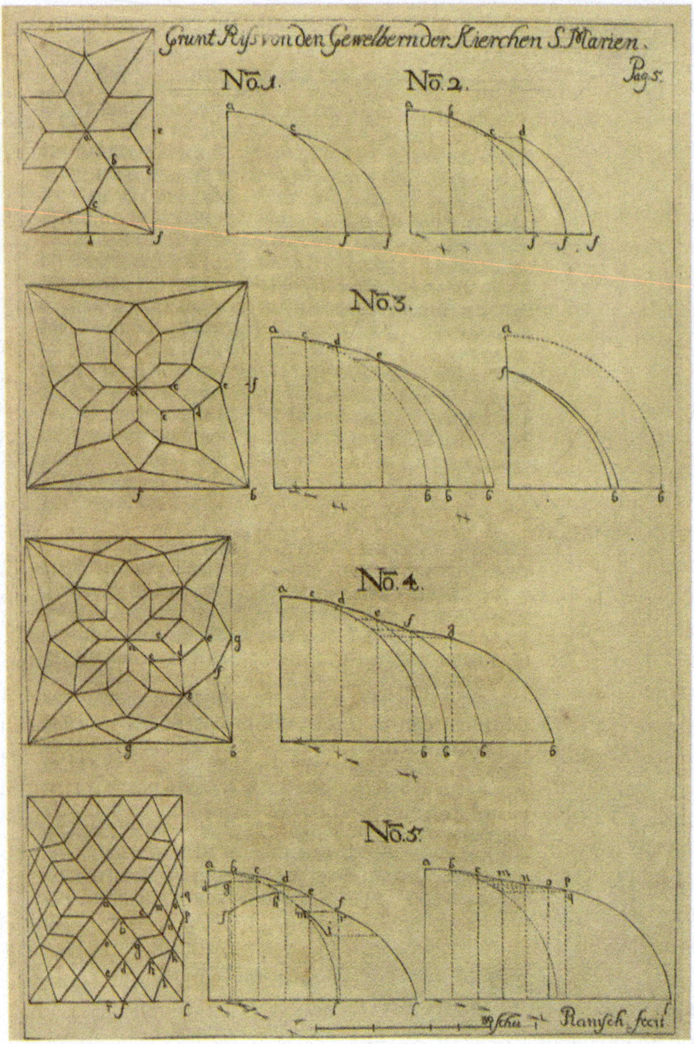

Fig. 10.20 Net vaults of the church of Saint Mary, Dantzig, now Gdansk (Ranisch 1695: 5)

screen", drawing a single quadrant representing the diagonal rib and reusing it to compute the height of the keystone of the perimetral arch over the larger side of the plan, that is, the end of the cross-section.

Then, in a separate drawing, he applies his singular procedure. To compute the height of the meeting point of the tiercerons placed near to the short side of the area, he measures its distance to the main keystone in the plan. Next, he applies this distance *twice* in the elevation. First, he draws an arc starting from the central keystone whose radius equals this distance, scoring its intersection with the quadrant that stands for the diagonal arch. Next, he marks the same distance at the springing line, drawing a vertical line from the intersection point, as done by Facht von Andernach and other writers. He then draws a horizontal line from the point marked on the quadrant until it reaches the vertical line, placing at this intersection the secondary keystone at the junction of both tiercerons. Since he takes it for granted that all ribs feature the curvature of the diagonal, he determines the centre of the arc connecting this point to the main keystone by drawing two arcs with this fixed radius from both ends of the arc; quite visibly, the intersection of both arcs and thus the centre of the rib, is placed below the springing line. Generally speaking, Ranisch uses the same method for other vaults, as explained by Pliego (2017).

Derivatives from the star vault. Vaults without a single vertical-plane rib, as in Kutná Hora, were frequently used in the Empire, but not outside it. In other countries, and particularly in the Iberian Peninsula, the typical Late-Gothic vault is a tierceron vault enriched with other elements, such as curved ribs or a rhombus joining the secondary keystones. Drawings in manuscripts and treatises (Alviz c. 1544; Gil de Hontañón and García [c. 1560] 1681: 25r; Vandelvira c.1585: 96v; Tornés and Juan [c. 1700] 2013: 56r-56v; Frézier 1737–1739: III, pl. 71) include these elements but do not furnish much information about methods for controlling their shape. Thus, we may surmise that these ribs were dressed on-demand once the main ribs were put in place, except for flat ribs such as the circles used by Gil de Hontañón and Vandelvira, and the rhombuses joining secondary keystones, all of which may be taken directly from the plan.

Other derivatives of the star vault pose interesting problems. Starting from the vault at Lincoln Cathedral crossing and a lost vault in Westminster Abbey, both built around 1300, tierceron vaults were sometimes grouped in sets of four or six vaults (Fig. 10.21), spanning large areas *without central piers.* Thus, two perimetral arches of each basic vault are transformed into axial ribs, while a diagonal rib from each unit is relegated to the role of lierne, the other diagonal rib is integrated in a large diagonal arch spanning the whole vault, and so on (Fig. 10.22). In order to guarantee a correct mechanical behaviour of the whole vault, the keystone of the main diagonal rib is lifted from the level of the springers (where it would belong if the central pier was kept) to the top of the vault. Of course, this brings about a general vertical displacement of the rest of the keystones and ribs in the vault, in order to guarantee geometrical consistency. In other words, while the plan of these vaults can be described as an assembly of four star vaults, the elevation is thoroughly transformed in order to assure efficient mechanical behaviour and consistent geometrical structure (Rabasa et al. 2017; Calvo et al. 2018b).

This design spread from England to the Continent, starting in Ornieta and Braniewo, then in Prussia and now in Poland, and appearing in the fifteenth and sixteenth centuries in France and the Iberian peninsula; in fact, the vaults in Hieronymite monastery in Belém, near Lisbon, which apply this system, are characterised by Frézier as "the most beautiful and well executed I have seen in (Gothic

Fig. 10.21 Crossing vault. Lincoln Cathedral (Photograph by the author)

vaulting)".[11] However, the Lincoln model was not adopted literally in the Continental examples. A team led by Enrique Rabasa and Ana López-Mozo has prepared photogrammetric surveys of fifteen combinations of star vaults, from Poland to Portugal, arriving to the conclusion that in each example, a particular rib controls the whole spatial layout of the vault, with other elements following suit in order to assure geometrical coherence. In Lincoln, the lead role is played by both axial ribs; in Belém, which includes six units, the controlling element is not the axial rib, which does not correspond to the perimetral ribs of a basic unit, but rather the transverse rib placed at one third of the longest side of the ensemble, that is, the perimeter of two of the basic units. In other cases, such as the crossing of the Royal Hospital in Santiago de Compostela, this function is entrusted to the main diagonal rib; in a small

[11] Frézier (1737-1739: III, 28): … *ce que j'ai vu de plus beau & de mieux exécuté dans ce genre, est au Monastere de Bethlehem, auprès de Lisbonne en Portugal, tant à l'Eglise qu'au Cloitre.*

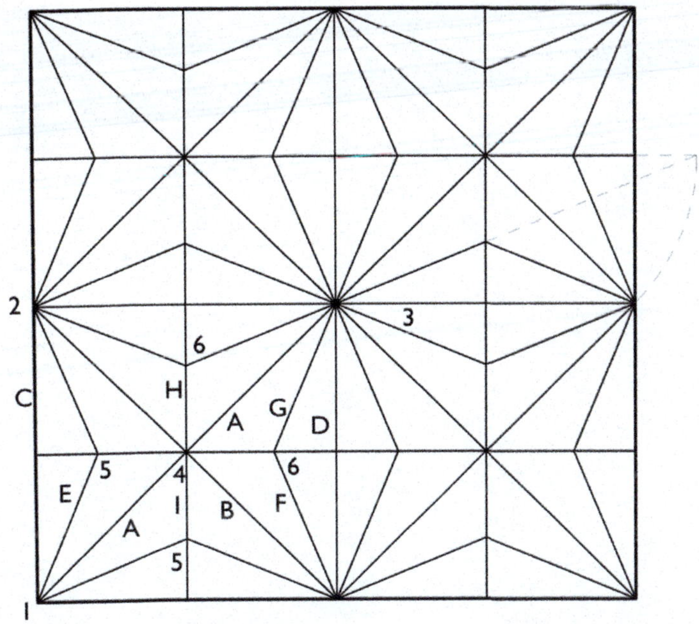

Fig. 10.22 Classification of ribs and keystones in the crossing vault of Lincoln Cathedral **A** Diagonal ribs **B** Diagonals of the individual units **C** Intersections of the severies with the perimetral walls **D** Axial ribs **E** Tiercerons starting from the corners **F** Tiercerons starting from the midpoints of perimetral walls **G** Tiercerons reaching the main keystone **H** Liernes starting at the perimeter **I** Liernes meeting the axial ribs **1** Corner springer **2** Middle springer **3** Main keystone **4** Secondary keystones **5** and **6** Tertiary keystones (Drawing by the author)

vault in the church of Santa María del Puerto in Santoña, encompassing six units, the directing element is not the diagonal of the whole plan, but rather an oblique rib formed by the diagonals of two units placed in staggered fashion. All these examples are self-contained forms; that is, they may not be enlarged adding successive units. In contrast, those in the chapel of the Bishops' Castle in Lidzwark Warminski (Poland) and Saint-Eustache in Paris are designed so that they could be indefinitely extended; in some way, they are ribbed barrel vaults (Rabasa et al. 2017; see also Calvo et al. 2018b).

Other derivatives of star vaults are the so-called "asymmetrical" vaults. When laid over a square plan, they are not actually asymmetrical, but they are symmetrical about a diagonal of the plan. However, this variant is not the usual one; as shown by López-Mozo et al. (2015; see also López-Mozo and Senent 2017), these vaults are used solve rectangular or slightly irregular plans. The simplest and earliest version is embodied in the "crazy" vaults of this powerful innovation centre, Lincoln Cathedral (Fig. 10.23; see Frankl 1953). They include two pair of ribs stemming from springings in the same side of the plan, meeting at a keystone on the longitudinal axis *but not at the centre of the plan.* Thus, there are two different keystones at the ridge rib; from each of them, an additional rib goes down to the nearest springers. In later examples,

Fig. 10.23 "Crazy" vaults in the nave. Lincoln Cathedral (Photograph by the author)

the solution is even more complex; for example, in Basel Cathedral, Bebenhausen Abbey or the parish church of Weißkirchen an der Traun (Austria), (Fig. 10.24), ribs starting from the springers meet at four different keystones, but not at the centre of the plan. Additional ribs connect these four keystones with a single diagonal rib;

Fig. 10.24 Asymmetrical vault. Weißkirchen an der Traun, Austria, parish church (Photogrammetric survey by Ana López-Mozo and Miguel Ángel Alonso)

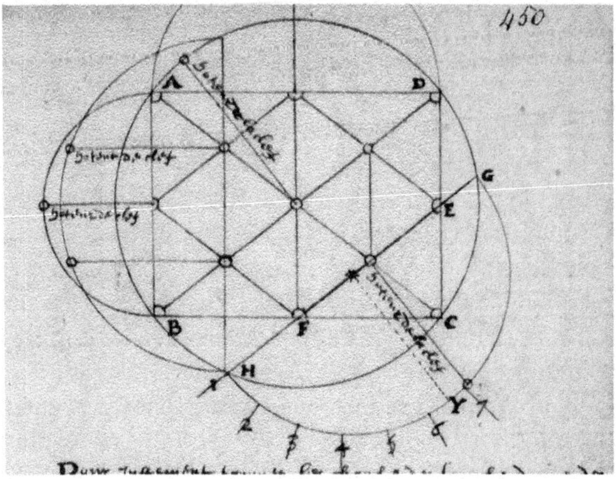

Fig. 10.25 Triangulated spherical vault (Gentillâtre c. 1620: 450r)

before reaching it, pairs of additional ribs meet at the other diagonal, which does not feature a rib. All in all, there are three kinds of keystones: the ones at the end of ribs starting from the springers; the ones where the additional liernes intersect at the ribless diagonal; and the ones where these liernes meet the ribbed diagonal; what is also lacking is the main keystone at the centre of the plan.

Jacques Gentillâtre's triangulated spherical rib vault. The stereotomic section of MS Fr. 14.727 in the National Library of France, attributed to Jacques Gentillâtre (c. 1620: 406r-451v, in particular 450r-451v) ends with a remarkable drawing of a rib vault, with an explanatory text, and twelve plans of Gothic vaults, without elevations or text. The first drawing, the one with the accompanying text, includes a rectangular-plan rib vault, divided into triangles by lines that are parallel to the short sides of the plan and the diagonals (Fig. 10.25). At first sight, it resembles some ceilings in Serlio (1537: 72v). However, several circles and semicircles suggest that the grid is projected onto a spherical surface, while the text includes the word *ogive*, the French term for diagonal ribs. Seen from this perspective, the vault recalls some drawings by Facht von Andernach (1593: 6r). In particular, a full circle seems to represent the elevation of a diagonal rib, rotated around its horizontal projection, as in Gil de Hontañón ([1560] 1681: 25r). Four semicircles stand for the elevations of the long and short perimetral arches, a rib laid out parallel to the short side of the enclosing rectangle, and a rib parallel to the main diagonal rib. There are no ribs parallel to the long sides of the plan, since the triangular grid makes them unnecessary. The result recalls some vaults in the upper storey of the Merchants' Exchange in Seville (see Palacios 1987; Minenna 2012: I, 156–159; Minenna 2014) and suggests connections with cartography, as Pinto (2002: 105–112, 127–142) pointed out for other Sevillian works. However, no hard evidence either supports or contradicts this hypothesis for the moment, since Gentillâtre draws only rib axes, as did Hernán Ruiz, eschewing the templates of Vandelvira and Guardia.

10.1.6 Fan Vaults

Architectural history manuals stress the contrast between ribbed Gothic construction and the single-tier Romanesque and Renaissance approach. However, it is widely accepted that ribbed vaults appeared in Europe in the Romanesque period, for example in the church of Rivolta d'Adda, Durham Cathedral or the Abbey of Jumièges (Bony 1984: 7–12). In contrast, it is usually forgotten that one-level construction reappeared inside the Gothic system in two remarkable, although geographically limited, schools: fan vaults in England and arrised vaults, built in brick in a number of locations in the German Empire and Poland and in ashlar in the city of Valencia.

The geometry of fan vaults (Fig. 10.26) is apparently quite different from mainstream ribbed vaults; they involve surfaces of revolution generated by a circular arc rotating around a vertical axis passing through the springings, which are usually placed regularly on two walls (Leedy 1978; Leedy 1980). However, vaults with duplicated or triplicated tiercerons are frequent in England, while cross-sections by planes passing through main keystones, both longitudinal and transversal, are

Fig. 10.26 Fan vault. Cambridge, King's College (Photograph by the author)

Fig. 10.27 Vault with multiple tiercerons. Winchester Cathedral, nave (Photograph by the author)

usually flat. Such vaults project the powerful image of several inverted pyramids,[12] with their apexes at the springings and the edges of their bases at the longitudinal and transversal ridges, where they meet the next pyramid. In some occasions, for example in the nave of Winchester Cathedral, tiercerons cross transversal ridges and progress until the longitudinal ridge; as a result, the pyramids are octagonal (Fig. 10.27). The next step, taken in the late fourteenth century in the cloister of Gloucester Cathedral, involves the transformation of the octagonal base into a half-circle, while the ribs starting from the springer are unified both in radius, width, and profile; thus, the distinction between diagonal and transverse ribs and tiercerons becomes secondary. The area around each springer therefore becomes a vertical-axis surface of revolution, called a "conoid" by Leedy.[13] Ribs starting from the springer act as meridians, while the half-circle at the top plays the role of parallel or directrix of the surface.

---

[12]From the church floor, the observer gets the impression of an inverted pyramid, since diagonal ribs and tiercerons, being laid on vertical planes, appear to the eye as straight segments. However, diagonals and tiercerons are actually curved, so the "pyramids" are actually an ensemble of cylindrical portions, or even more complex shapes.

[13]Leedy's terminology (1978: 207, 210, 211, 213; 1980, 1, 3, 5, 7) is at odds with the standard vocabulary of descriptive geometry and may be a bit confusing for readers coming from this field. His "conoids" are surfaces of revolution resembling cones, but they use circular arcs as generatrices, instead of straight lines. They are different from the "conoids" of descriptive geometry, which are usually ruled surfaces, applied for example by de la Rue and Frézier in annular groin vaults (see Sect. 9.5.3). Moreover, the term "ruled surface" used by Leedy (1980: 3) in connection with this surface is also misleading, since both the generatrices and the directrices of the surface are circular arcs; in fact, Leedy himself (1980: 3) remarks that horizontal sections of a rib vault are circular and all vertical "ribs" have the same curvature.

Between these half-circles, star-shaped areas known as *spandrels* remain; usually, they are covered by a flat section of the vault. In the retrochoir of Peterborough Cathedral or the chapel of King's College in Cambridge, in addition to the semi-circular rib at the top of each "conoid", other ribs in the shape of the half-circle materialise the directrices of the surface. Thus, each "conoid" is covered by a purely decorative grid pattern. The multiplication of ribs leads to thin profiles; rather than building an independent network of ribs, as in coffered French and Spanish vaults of the sixteenth century, masons integrated ribs and severies in the same voussoirs.[14] In the Gloucester cloister, meridional joints are placed at one side of the thin ribs, while horizontal ones are located at will. The independence of joints and ribs is more evident in Peterborough: meridional joints are placed half-way between ribs. This approach is also clear in the treatment of spandrels, which are usually carved from a single block of stone. Another significant detail, pointed out by Leedy (1978: 212–213; 1980: 6) concerns the orientation of the profile of horizontal ribs: rather than being generated by vertical lines, they are perpendicular to the surface of the vault.

As for the geometrical procedures used to control the dressing of the voussoirs and the layout of the vaults, no medieval or Early Modern book or manuscript dealing with this issue has been preserved. However, the detailed drawings of vault extradoses included in Robert Willis's ([1842] 1910: pl. 2, 3) pioneering book offer some clues. In Peterborough, some courses present horizontal extradoses, in the fashion of the surfaces of operation in keystones, mentioned by Willis in the same study ([1842] 1910: 24); this suggests they were dressed by squaring. Other courses in Peterborough, as well as most courses in the Henry VII Lady Chapel in Westminster Abbey feature conical extradoses. The voussoirs for these courses may have been carved as the voussoirs of a simple trumpet squinch, inverting the usual procedure and starting from the extrados. However, it is not easy to say if simple templates based on triangulation were used or if masons resorted to the squaring method.

10.1.7 Arrised Vaults

Just as fan vaults are a strictly English product, arrised[15] vaults are focalised in the city of Valencia, although they fostered a final, very late output in Assier, in Southwestern France. In fan vaults, ribs are transformed into decorative mouldings, but they are still quite visible. In arrised vaults, only a shadow of the ribs remains. The intrados surface

[14]There are some exceptions to this general rule. Leedy (1980: 2) mentions fan vaults with ribs and panels, as well as those with a combination of these elements and jointed masonry; later on (1980: 5, 7), he includes the use of jointed masonry as one of the preconditions for the emergence of the fan vault.

[15]This kind of vaults were first called *anervadas* (ribless) (Garín 1962); this denomination is not completely satisfactory, since most vaults used in the Romanesque and Renaissance periods do not use ribs. Later on, Zaragozá (2010) coined the term *aristadas*, which may be translated loosely as "arrised". This term may be misleading in Spanish, since *bóveda de arista* means "groin vault"; however, this confusion does not arise in English.

Fig. 10.28 Funerary chapel of Alphonse V of Aragon. Valencia, Dominican convent, now a military headquarters (Photograph by the author)

is folded at groins and creases following the layout of well-known Late Gothic types: skew quadripartite vaults in the Gates of Quart, star vaults in the present-day entrance to Valencia Cathedral bell tower, tierceron vaults ending in a semipolygonal vault at the funerary chapel of Alphonse V of Aragon in the Dominican convent (Fig. 10.28) and finally an octagonal-plan tierceron vault in the chapel of Galliot de Genouillac in Assier (Fig. 1.19, 10.29). At the same time, the groin vault reappears at the church of Saint Nicholas, in the sacristy of the Dominican convent, and in the Monastery of the Holy Trinity. Thus, the Valencian school represents a clear transition point between Gothic and Renaissance stonecutting (Garín 1962; Pérouse [1982a] 2001: 151–152, 212; Zaragozá 2008; Navarro and Rabasa 2018b).

The precision of the execution is generally very high, as shown by the intersection of the Holy Trinity groin vault with a sloping barrel vault, or the neat cut of the groin vaults in the sacristy of the Dominican convent by an oblique plane. However, the school lasted only for a short period, starting in 1440 with the Gates of Quart (Natividad and Calvo 2012c), and lasting no longer than the later decades of that century, except for the Assier vault, dated in 1540–1545 (Tollon 1989: 125, 133; Navarro et al. 2018a). These vaults show some points of contact with the diamond vaults of Central Europe (Acland 1972: 220–228; Zaragozá 2008: 20). However, Polish, Czech and German arrised vaults are built in brick, their layouts are generally different to those in Valencia and there is no evidence of direct contact between the two schools.

No book, manuscript or tracing gives information about the methods used to control the dressing and placement of such vaults. However, basing themselves on

Fig. 10.29 Chapel of Galliot de Genouillac. Assier, parish church (Photograph by the author)

precise surveys, Natividad (2010; 2012c) and Navarro (2018b: 105–166) have posited some plausible hypotheses. First, it is easy to surmise that some preliminary tracing was made at the start to determine the curvature of the groins, probably following the methods we have seen when dealing with tierceron vaults. Natividad (2010: 112–138) has analysed the geometry of the skew tierceron vault of the Gates of Quart, determining that groins and creases are circular arcs; in contrast, it seems that the webs are not formed by simple surfaces. This highlights the Gothic traits of the vault: simple curves are used in groins and creases, while complex surfaces resulting from geometrical operations starting from these lines materialise the web. The layout of the bed joints furnishes vital information about the geometry of the severies. Natividad has found that a fair number of courses—running up to more than half the total height of the vault—are laid out with horizontal bed joints, in *tas-de-charge* fashion. In the upper courses, the bed joints in each quarter of the vault (that is, the area between two intersecting liernes, at both sides of the diagonal groin) belong to the same sheaf of planes, with their common line placed at springing level. Taking all this into account, the first phase of the dressing process may have involved cutting the bed joints, either parallel for the springings, or concurrent for the upper courses; for the latter, the angle was probably measured from a cross-section through the vertical plane passing through the diagonal groin and transferred to the stone with a bevel. Once the bed joints were dressed, templets with the curvature of their intrados sides may be applied to both bed joints of the voussoir, upper and lower, carving marginal drafts. Starting from these drafts, other templets with the curvature of the diagonal ribs and the liernes may have been used to dress the intrados surface of each voussoir. All this leaves the issue of the computation of the curvature of the intrados side of bed joints open. Natividad mentions a study by Sánchez Simón (2009) stating that in the vault of the chapel in the Dominican convent, the angular measure of this curve for each severy is constant, namely 60º; this places a constraint in the position of the centres of these arcs and its radius; however, Natividad could not analyse this issue for the Gates of Quart, since the relatively short length of these joints in this construction does not make it possible to reach firm conclusions.

Navarro (2018b), Navarro and Rabasa (2018c), Navarro et al. (2018a) has carried out a comprehensive survey of the most significant vaults in Valencia as well as a separate study on the Assier vaults. First, he has confirmed the assertion by Sánchez Simón about the angular measure in the Dominican convent; however, he has also found that the vaults in Assier use different angles for each severy. Further, he has put forward an interesting hypothesis about the setting out and dressing operations. He has confirmed Natividad's findings about the division of the vaults in sections, for example between two symmetry axes, as well as the bed joints forming a sheaf of planes. In particular, the analysis of the precisely executed vaults at the Dominican convent (Navarro 2018c: 444–448) and Assier (Navarro et al. 2018a) shows that the axes of the sheaves of planes are orthogonal to the vertical plane passing through the diagonal arises, and that each portion or quarter of the vault is treated like an variable-width arch with its directing line placed at the diagonal groin. Thus, templates for the apparent bed joints may be constructed by either rabatting the different arrises for each quarter of the vault or drawing them in an independent scheme. Next, the diagonal groin should be divided into several portions; these divisions are then transferred to other arrises. This operation cannot be carried out with the bevel, since the slopes of the intersections of the planes in the sheaf with the vertical planes for each arris are different. Thus, Navarro (2018c: 129–164, 448–483) puts forward a likely hypothesis: the bed joint planes, shown in edge view, are represented in a number of auxiliary views, one for each arris (Fig. 10.30); the intersections of the planes with a vertical line passing through the springing are transferred from the diagonal groin to the other arrises; then the sheaf of planes is drawn in each auxiliary view, locating the intersection of each plane with the arris; then the distances of each of these intersections to the axis of the sheaf of planes are computed and brought to a true-shape construction representing the intersections of the sheaf of planes with the arrises. This allows the mason to construct a simplified bed-joint template; these

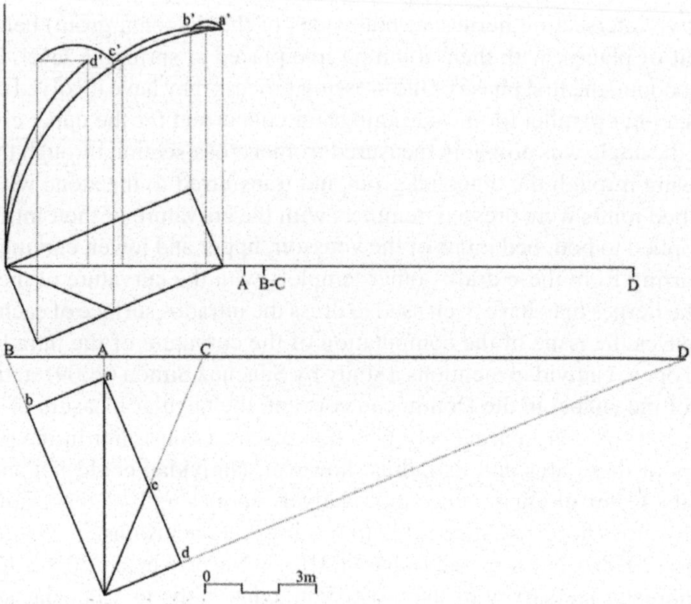

Fig. 10.30 Layout of the funerary chapel of Alphonse V of Aragon (Drawing by Pablo Navarro Camallonga)

points are connected either by using the standard 60° angle, as in the Dominican convent, or by drawing the curve between the arrises at will, as in Assier.

10.2 Coffered Vaults

While fan vaults and arrised vaults materialise Gothic formal schemes anticipating Renaissance one-tier construction, a fair number of vaults built in France, Spain and Portugal in the sixteenth century embody the opposite concept. Roman-looking coffered vaults were constructed using a network of ribs filled with severies.[16] The idea seems to have originated in France, during the second decade of the century, with such well-known constructions as the vault over the staircase in the courtyard of Blois castle and the halls around the central double staircase in Chambord. However,

[16]Of course, coffered vaults can be built using a single-level stonecutting scheme, where the protrusions and the coffers are materialised in the same voussoir, exactly as in fan vaults. In fact, such a system was widely used when coffered vaults using ribs and severies fell out of fashion: de l'Orme's vault at the chapel of Anet is an outstanding example of this approach (see Potié 1996: 114–120). However, from the stereotomic point of view, such vaults present no specific problems; they can be dressed using general methods, such as the standard procedure for a hemispherical vault, in the case of Anet. Thus, I will deal in this subsection exclusively with coffered vaults built using the rib-and-severies approach.

these vaults are quite scarce in France; neither de l'Orme (1567) nor Chéreau (c. 1567–74) mention them. In contrast, they were used quite frequently in Spain during the sixteenth century, although they vanished suddenly at the end of this period. Vandelvira (c. 1585) includes in his manuscript no fewer than sixteen vaults of this kind. In fact, the *Libro de trazas...* is the only written source for this kind of construction, with the exception of an isolated and problematical example in the manuscript of Guardia (c. 1600). Significantly, these sections were excluded in the mid-seventeenth century copy of Vandelvira's manuscript by Goiti (Vandelvira/Goiti 1646); perhaps he considered them outdated.

10.2.1 Coffered Cylindrical Vaults

Coffered barrel vaults. The cross-section of a coffered barrel vault can be a semi-circle, although surbased profiles are frequent, as in the four halls starting from the double staircase in Chambord in each storey. In the following decades, they are quite frequent in Spain and Portugal. Ribs follow the horizontal generatrices of the vault, as well as several evenly spaced directrices. As far as I know, no treatise or manuscript explains the setting out of these simple vaults. However, a few general principles can be deduced from existing examples and other vaults in the manuscript of Vandelvira. Ribs are divided into *crucetas*, that is, cross-shaped pieces including a part of a horizontal rib and a vertical one; junctions between *crucetas* are placed at the midpoint of each segment of the network of ribs. This choice may seem fanciful, but in fact, it is quite sensible; putting the unions at the nodes of the network would have involved four ribs meeting at each node, a difficult dressing problem. When dealing with the coffered octagonal vault, Vandelvira stresses that the horizontal ribs are *revirados*, meaning that their directing planes should not be vertical, as in traditional ribbed construction, but instead they should point to the centre of the vault. Extrapolating this advice to the barrel vault, the directing planes of the longitudinal ribs should pass through the axis of the vault; inspection of existing examples confirms that this idea was followed in practice.

Coffered octagonal vaults. Coffered octagonal vaults are formed by portions of cylinders with horizontal generatrices; their intersections are materialised by ribs in the shape of elliptical arcs, at least in theory. Thus, ribs are placed at the creases dividing each cylindrical portion from the next one; within each of these portions, horizontal ribs embody the generatrices, while another set of ribs are placed in vertical planes. However, in Vandelvira's (c. 1587: 104v-105r) solution, these vertical planes are not orthogonal to the generatrices of the cylinders, in order to avoid awkward intersections with the ribs placed between cylindrical portions. Rather, both the ribs dividing the vault into cylindrical portions and the internal ribs in each portion are placed in a sheaf of planes converging in a vertical line passing through the centre of the area (Fig. 10.31). Moreover, if the horizontal ribs were evenly spaced, the coffers near the top of the vault would be too long and narrow. Thus, Vandelvira employs a complex procedure to avoid this effect, controlling the spacing of the horizontal ribs.

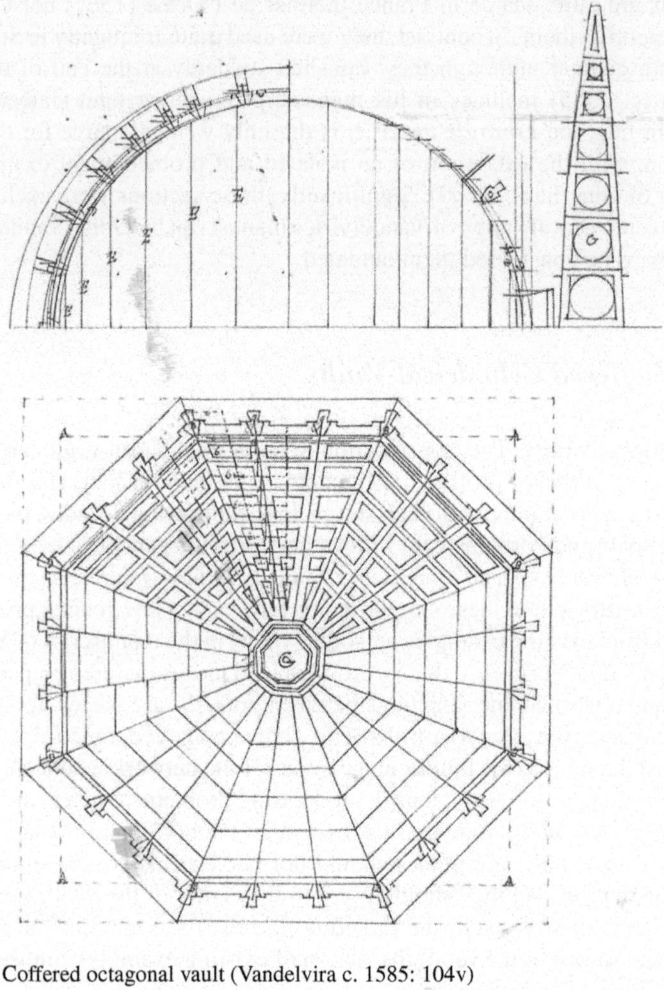

Fig. 10.31 Coffered octagonal vault (Vandelvira c. 1585: 104v)

First, he draws the octagon in plan using an efficient and exact method: after drawing a square, he measures the distance between the corner and the centre and brings it to the side of the square; this furnishes two corners of the octagon, placed at adjacent sides (see Serlio 1545: 19r and Sect. 3.1.2. Repeating this procedure, he can place all the vertexes of the octagon. Next, he draws the radii of the octagon, splits each side of the octagon into three parts and draws lines from these division points to the centre of the plan, so the vault is divided into 24 wedges. He then starts drawing the elevation; he does not give detailed instructions, but rather directs the reader to *la cabecera primera en ochavo*; this phrase alludes to the *cabecera ochavada*, a semi-octagonal vault for the chancel of a church on fol. 102v-103r. As we have seen in Sect. 8.3, Vandelvira makes a striking choice in this piece.

Rather than using circular arcs for the orthogonal section of the central portion, he uses a construction method leading to a raised section for the central portions. We should surmise that the same method is used in the coffered octagonal vault; in fact, Vandelvira traces two slightly different rib profiles in the elevation, standing for the intersections between cylindrical portions and the intermediate divisions. In the next step, he prepares a provisional development of the central third of a cylindrical portion. In this development, he draws the ribs, both meridional and horizontal, manipulating the spacing of the horizontal ribs so that a circle is inscribed inside each coffer. Of course, since the width of the coffers diminishes as they approach the top of the vault, the spacing of the horizontal ribs decreases accordingly.

The plan shows clearly that the ribs are divided into cross-shaped pieces or *crucetas*. In order to control their dressing, Vandelvira does not construct intrados templates; however, he gives a number of hints. He stresses that these pieces are skewed since vertical ribs cross the horizontal ones at an oblique angle. In the case of the crosses placed at the crease between two cylindrical portions, the horizontal ribs form equal angles at both sides of the vertical one, which should be controlled with the bevel. As for the horizontal ribs, he stresses that they are *revirados*, that is, their directing plane passes through the centre of the vault, rather than being vertical as usual in ribbed construction.

10.2.2 Coffered Spherical Vaults

The first coffered vault included in the Vandelvira manuscript (c. 1585; 62v-63r) is a hemispherical dome, with ribs laid out as parallels and meridians. The dressing procedure is explained in detail with a plan, a cross-section and two additional developments of a *lune* or *gore* (a spindle-shaped portion of a sphere between two meridians) (Fig. 10.32). As on other occasions, these stonecutting diagrams focus on the construction of the intrados templates, rather than providing an intuitive representation of the vault; thus, Vandelvira (c. 1585: 63v) adds two architectural drawings, a plan and a cross-section. The plan shows that ribs are rather thin, and the sides of the meridional ribs are parallel to their axes. As in coffered octagonal vaults, course heights are not equal, but rather diminish as they go up the surface of the vault, in order to keep the proportion of the coffers approximately constant. Thus, after drawing a plan and a cross-section of the vault, Vandelvira prepares a preliminary development of a lune, and then draws ribs while inscribing circles in them to keep the proportion of the coffers approximately square, as in the octagonal vault.[17]

Once the ribs have been drawn in the development and transferred to the cross-section and the plan, he constructs the intrados templates through cone developments, for both the ribs and the coffer panels. As we have seen, Vandelvira (c. 1585, 94v-97r; see also Sect. 10.1.4) uses cone developments for ribbed vaults. However, in this case, the problem is more complex since he must deal with cross-shaped pieces, rather

[17]More precisely, coffers are trapeziums where half the sum of the bases equals the height.

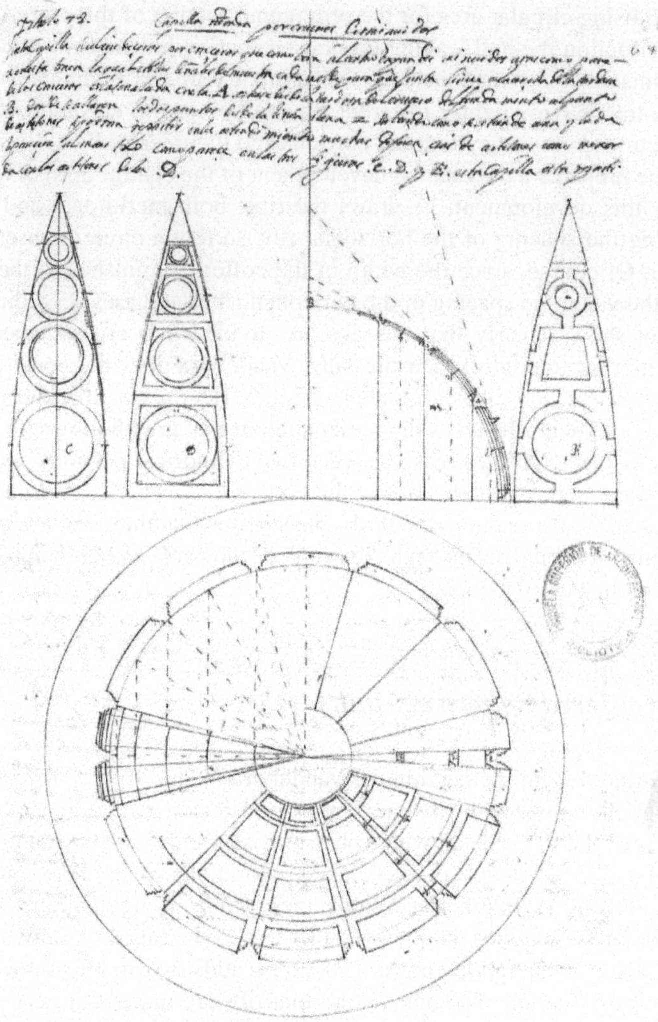

Fig. 10.32 Coffered spherical vault (Vandelvira c. 1585: 64r)

than simple spherical quadrilaterals. He explains it in few words, referring the reader to the preceding sections, that is, hemispherical and surbased non-coffered vaults. This means that, although the ribs are thin, Vandelvira draws generatrices of a cone passing through both sides of the horizontal rib, rather than trying to use a tangent, as he will do in the spiral vault (Vandelvira c. 1585: 65v-66r). Next, he develops a cone section as usual, drawing two arcs from the intersection of the generatrix with the axis of the vault; this furnishes the template for the horizontal arms of the *cruceta.* He then draws a meridional rib; although he does not explain the procedure, it seems clear that he performs this operation by tracing two parallels to the axis of the rib, starting from the intersection of the generatrix with the axis of the vault. Of course, the drawing makes it clear that both the horizontal and the meridional rib are part of the same piece. In the next phase, he develops intrados templates for the coffer panels, using the same method applied for the voussoirs of the hemispherical dome; it is worth remarking that he seems to take it for granted that the development does not alter the circles in the decoration.

Vandelvira also includes two variants of this scheme. In the first one (c. 1585: 64r), meridional ribs are tapered; their sides are traced so that their plan is given by two radii of the circle in the plan. He does not explain the details of the construction and, in fact, he does not draw the templates; however, it is easy to surmise that the sides of the meridional ribs should be drawn passing through the intersection of the axis of the rib and that of the vault.

The other variant (Vandelvira c. 1585: 64v-65r) involves deeper changes, since ribs are set following two families of vertical parallels. As in other pieces by Jousse (1642: 130–131) and Derand (1643: 364), this seems to be an adaptation to a hemispherical vault of a scheme originally designed for sail vaults. However, it is worth mentioning that the ribs do not meet at the intersections of the diagonals with the enclosing circle. Thus, rather than a square sail vault with the addition of identical caps, it may be understood as a rectangular sail vault complemented by four different caps, as in Derand (1643: 366–368).

The *ochavo de la Guardia*: toroidal or spherical? A puzzling problem is raised by the piece included by Vandelvira (c. 1585: 103v) that he calls *ochavo de La Guardia*. At least from the times of Fernando Chueca (1954: 28) it has been accepted that the archetype of this form is the vault over the chancel of the Dominican convent in La Guardia, near Jaén, covered by a vault of the same shape (Figs. 10.33, 10.34). This is a high honour; only four pieces in the Vandelvira manuscript take their names from Spanish towns. The name *ochavo* is hard to explain, since it does not involve an octagon or part of it, except for the impost, which is part of an irregular octagon. However, portions of octagons were placed frequently over chancels, as in the *cabecera ochavada* (Vandelvira c. 1585: 102v-103r); thus, the term *ochavo* reflects function rather than form.

Another remark from Vandelvira raises a difficult problem: "the *Ochavo de la Guardia* is the same design than half *Bóveda de Murcia*".[18] Now, as we have seen in Sect. 9.5.2, the "Murcia vault" stands for horizontal-axis torus vaults, but the horizontal section in Vandelvira's drawing of the *ochavo*, at springing level, includes two circular arcs at both ends, and both are part of the same circle, in contrast to the "Murcia vault", which includes a long rectangular section between two semicircles at the end of the plan; when rotated about a central axis, the semicircles generate

[18] Vandelvira (c. 1585: 103v):... *ochavo de La Guardia, la cual es la misma traza que media bóveda de Murcia*. Also, in the section about the *Bóveda de Murcia*, Vandelvira (c. 1585: 69v) states that *El ochavo de La Guardia y esta traza son todas una, excepto que el ochavo es la mitad de esta capilla con sus cruceros como la de adelante de ésta parece partida por el diámetro más largo*, that is, "the *ochavo de La Guardia* and this design are the same, except that the *ochavo* is half this vault, with coffers as the next one [which is a coffered version of the *Bóveda de Murcia*] divided by the longest diameter". Transcriptions are taken from Vandelvira and Barbé 1977.

Fig. 10.33 Vault over the chancel. La Guardia de Jaén, Dominican convent (Photograph by the author)

a torus surface. In contrast, rotating the arcs in the drawing of the *ochavo* would generate a portion of a sphere, rather than a torus, since both arcs are parts of a single circle.

This is in apparent contradiction of Vandelvira's assertion about the similarity between the *ochavo* and the "Murcia vault". In order to solve the problem, Salcedo et al. have surveyed the vault at the Dominican convent in La Guardia with a 3D laser scanner, reaching the conclusion that the intrados surface of the ribs of the vault is spherical, and thus consistent with Vandelvira's drawing. Was the author wrong, then? A careful reading of the text solves the issue: Vandelvira states that the *traza*–the setting out method–is identical in both the "Murcia vault" and the *ochavo de La Guardia*, but he does not mention the final shape of the two vaults; in other words, it is the process, not the product, which is similar in both archetypes.

Once this issue is settled, the piece may be understood easily. Although Vandelvira's explanations are quite brief, it may be appreciated at first glance that the drawing (Fig. 10.34) is similar to Vandelvira's drawing for the coffered dome, exchanging the roles of the plan and the elevation and, of course, cutting the sphere by two vertical parallel planes in addition to the usual section through the horizontal springing plane. Reversing the operations for the coffered dome, a provisional development of each gore between two meridional ribs is drawn; then, circles are inscribed in the gore, separated by transversal ribs; next, the distances between these transversal ribs are transferred to the horizontal section of the vault in order to place an unevenly spaced set of ribs playing the role of (vertical-plane) parallels. Once

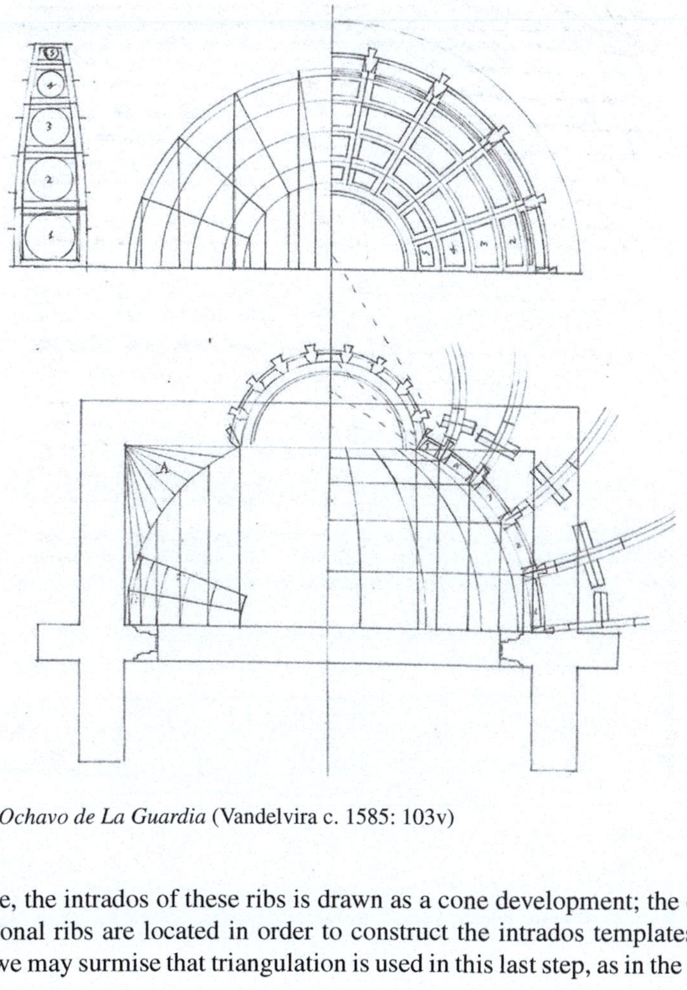

Fig. 10.34 *Ochavo de La Guardia* (Vandelvira c. 1585: 103v)

this is done, the intrados of these ribs is drawn as a cone development; the edges of the meridional ribs are located in order to construct the intrados templates for the *crucetas*; we may surmise that triangulation is used in this last step, as in the coffered dome.

10.2.3 Coffered Sail Vaults

Vandelvira eschews the coffered version of the sail vault with round courses. This is a sensible choice: the intersections of the ribs with the vertical planes enclosing the vault would have brought about a massacre of ribs. Thus, he presents five kinds of coffered sail vaults, with ribs parallel to the sides or the diagonals of the area, including both square and rectangular versions, and even two solutions for the square vault with diagonal ribs.

Square vaults with frontal ribs. Starting with the simplest one, the square vault with ribs parallel to the sides of the area (Fig. 10.35), Vandelvira (c. 1585: 97v-

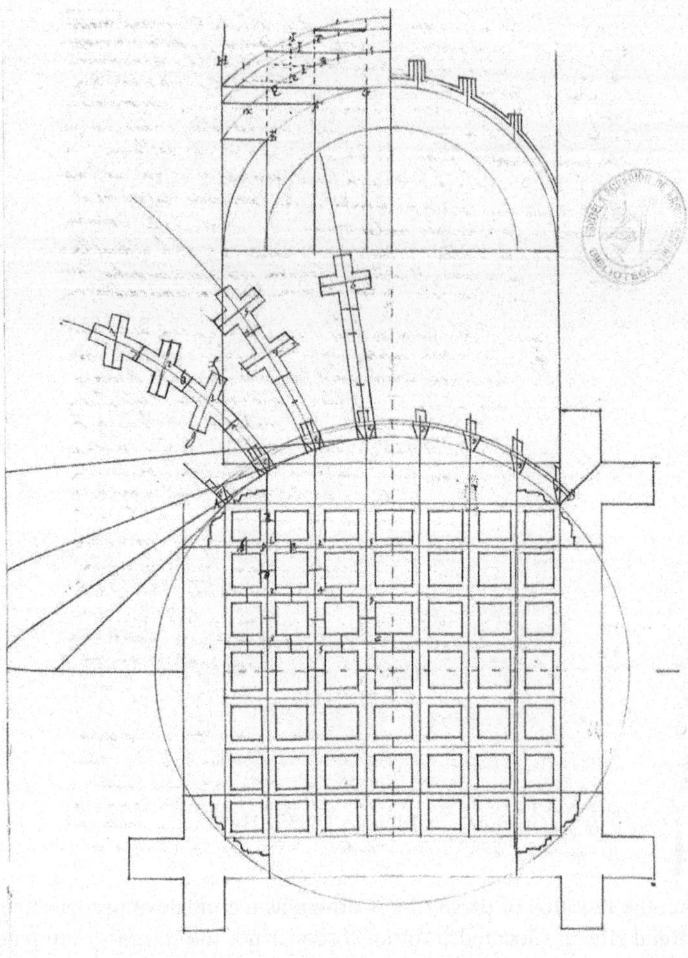

Fig. 10.35 Coffered square sail vault with frontal ribs (Vandelvira c. 1585: 98r)

98r; see also Palacios [1990] 2003: 302–315) introduces a sophisticated procedure for the construction of the intrados templates, combining cone developments and triangulation. To start with, he draws the square plan of the vault and an enclosing circle, used as a cross-section. Then, he divides the cross-section, from the keystone of one of the perimetral arches to the opposite one, into seven equal parts, placing a rib in each of these divisions; of course, he gets three ribs at the left side of the vault and three symmetrical ones at the right. He stresses that the section of each rib can be drawn using two alternative methods. In the first, and apparently simpler way,[19]

[19]This seemingly simple solution poses a complex problem. Although each individual section is symmetrical about the axis of the rib, axes at different points in the same rib are not parallel, since the curve that joins the starting points of the axes is a small circle. The axes, being the normals to

the axis of each rib is a radius of the intrados sphere of the vault, and the profile of the rib is symmetrical about the axis; this solution is called *moldes cuadrados*, (orthogonal templates). In the alternative solution, the axes of the ribs are vertical; in order to adapt to the spherical surface, the profile of the template is subject to an affine transformation (Rabasa 2000: 106) resulting in an asymmetrical shape; Vandelvira calls this solution *moldes revirados* (warped templates).

Using either square or warped templates, Vandelvira advises the reader to construct intrados templates using the cone development method he had applied in the hemispherical vault. However, the cross-shaped pieces used in coffered vaults raise the problem of the intersection of the front, side, and back ribs at awkward angles. To address this issue, Vandelvira constructs an elevation, used mainly to compute the spatial positions of rib intersections (Palacios [1990] 2003: 302–315; Natividad 2017: I, 108–112). First, he transfers the position of the three left-side ribs to the elevation. However, these ribs are also symmetrical about the diagonal to three back ribs. Thus, the intersections of the left-side ribs with the back perimetral arch stand at the same level as the intersections of the back ribs with the left perimetral arch. Taking this into account, Vandelvira draws schematic elevations of the three back ribs; they are depicted as arcs whose centres lie on the axis of the area, and thus are projected at the midpoint of the base of the cross-section; of course, these arcs pass through the intersections of each rib with the perimetral arch. These elevations, together with projecting lines drawn from the plan, allow Vandelvira to compute the height of all relevant points in the network of ribs, including the nodes at the centre of each cross and the junction points placed at the midpoints of the segments connecting the nodes. Using this information, Vandelvira sets the nodes and junction points along cone developments; to locate the ends of the crossing branches of each piece, he computes their distances to the points in the cone development, either nodes or ends. Given the symmetry of the piece, in some cases two points are placed at the same level, and their distance may be taken directly from the plan. In other cases, however, Vandelvira must compute the distance between two points by forming a right triangle with the horizontal distance and the difference in heights, taken from the cross-section. Once he knows the distance of a cross end to the centre of the cross and the lower point of the cross, he can place the end of the cross at the intersection of two arcs drawn from the centre and the lower points with the respective distances as radii.

However, not all built examples follow Vandelvira's tracing procedure. In particular, the parish church in Cazalla de la Sierra (Fig. 10.36), combines the spherical layout explained by Vandelvira with another solution, where ribs show a surbased design, in order to reduce the rise of the vault. Similar solutions may be found in the cloister of the monastery of Saint Jerome of Buenavista in Seville and the town hall

---

a spherical surface along a small circle, are the generatrixes of a cone; thus, the surface that passes through the axes is slightly bent. A third alternative is materialised by the ribs in the vault over the second story in the bell tower of Murcia cathedral: rib axes point to the centre of the sphere, but are laid along a great circle: as a result, the horizontal projections of the rib are elliptical, although they can be drawn as circular arcs with quite small errors (Calvo 2005a: 200–204).

Fig. 10.36 Coffered square sail vault with diagonal ribs. Cazalla de la Sierra, parish church (Photograph by Pau Natividad)

in the same city (Palacios [1990] 2003: 304, 309, 314; Natividad 2017: I, 238; 2017: II, 38, 186).

Square vaults with diagonal ribs. Vandelvira (c. 1585: 99v-100r, 101v-102r; see also Palacios [1990] 2003: 316–321; Natividad 2017; I, 117–121, 125–128) applies basically the same procedure in vaults with ribs laid in parallel to the diagonal of the area. However, there are some significant differences. The first cross-section, used to determine the position of the ribs, is drawn around the plan, as before, but now stands for a cross-section by the vertical plane passing through the diagonal of the plan. In contrast, the elevation is a frontal one. This is a striking choice; an engineer or architect trained in descriptive geometry would surely use an auxiliary view to show the diagonal ribs in true shape. Quite probably, Vandelvira uses a frontal view in order to draw the perimetral arches in true shape, as a first step. He then constructs a schematic elevation of each oblique rib taking into account its radius, taken from the plan, and its height, taken from the cross-section. This allows him to know the heights of all nodes and midpoints of the network and construct intrados templates as in the preceding case.

Rectangular vaults. Similar methods are used in rectangular vaults (Vandelvira c. 1585: 98v-99r; 100v-101v). However, in this case the vault is not symmetrical about the diagonal. This detail makes the problem even more complex. First, the radii of the frontal ribs are not equal to those of the left and right-side ribs; thus, Vandelvira needs elevations for both kind of ribs; in typical stonecutting fashion, he overlaps a frontal elevation and a side view (Natividad 2017: I, 112–116). Second, the number of different *crucetas* is bigger; while in the square-plan vault the problem can be solved with six different templates, the rectangular one requires nine. Third, there are no points at exactly the same level; thus, all distances must be computed by forming right triangles. In the case of the rectangular vault with diagonal ribs, all these difficulties are added to the complexities of the projections of diagonal ribs.

10.2.4 Oval, Annular and Conical Coffered Vaults

Oblate oval vault. Out of six oval vaults, Vandelvira includes two coffered pieces (c. 1585: 72v-73r, 74v-75r). The *capilla oval segunda* (second oval vault) is a coffered version of his first oval vault: an oblate ellipsoid whose height equals half the larger axis of the plan (Fig. 10.37). A set of ribs follows meridians rotating around the short

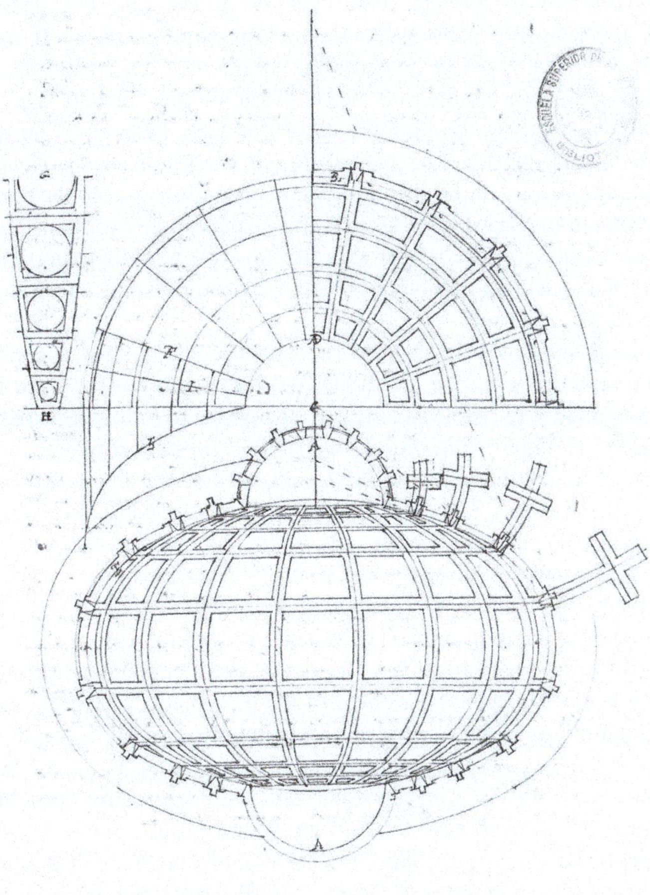

Fig. 10.37 Coffered oblate oval vault (Vandelvira c. 1585: 73r)

axis of the plan, while other ribs are set on vertical planes parallel to the longer axis. As we saw for Vandelvira's first oval vault in Sect. 9.4.2, there are no built examples, as far as I know, probably as a result of its excessive height. Vandelvira (c. 1585: 72v-73r) starts by tracing the oval or elliptical plan of the vault and a semicircle in the elevation, standing for a longitudinal section, as well as two oculi surrounding the ends of the smaller axis. Next, he divides the longitudinal section into an odd number of equal parts; starting from the division points, he will draw meridional ribs pointing to the centre of the vault. However, these ribs do not reach the centre of the section; in order to avoid a difficult junction and tiny coffers, meridional ribs end at the oculi at the ends of the short axis of the vault.

Vandelvira then marks a number of provisional divisions in the oval plan in order to draw the outline of a lune, as he did in the coffered dome (Vandelvira c. 1585: 62v-63r). He draws parallel ribs inside the lune, controlling their spacing with a circle inside each coffer. As a result, the layout of octagonal and hemispherical coffered vaults is reversed: the size of the coffers diminishes as they approach the circles placed at the poles of the network, that is, the ends of the short axis of the plan. Quite graphically, the lune shows a tiny base and a large top, which is left open since the central row of coffers spans the longitudinal plane of symmetry of the vault. Next, he transfers the division points to the base of the longitudinal section by means of projection lines; stressing the importance of this step, he goes as far as drawing a separate scheme of a quarter of the plan to explain this point clearly. The intersections of the projection lines with the springing plane provide a cue for the construction of the parallel ribs in the section. Of course, these ribs can be traced easily in the plan, since they are parallel to the longer axis of the oval. As a final step, he constructs intrados templates using a combination of cone developments and triangulation, as he did in the coffered dome.

Prolate oval vault. While Vandelvira's second oval vault is a coffered version of the first one, the fourth one (c. 1585: 74v-76r) translates the third into the rib-and-panel language (Fig. 10.38). Generally speaking, he follows the method he used in the prolate oval vault, using meridional ribs set in vertical planes and horizontal parallels, using an auxiliary semicircle to ensure that the angles between each rib and the next one are equal. He dutifully explains that the midpoints of panels, rather than the ribs, should be placed on the axes of the plan. Next, he draws a longitudinal section with the *cercha del fundamento*, that is, the springer outline, since he means to construct a prolate vault. In fact, Vandelvira could have generalised the procedure, raising or lowering the longitudinal section to construct a scalene oval vault, but he does not mention this possibility. Next, he divides the section into an arbitrary number of parts, constructs a provisional outline of the gore placed along the longer axis of the vault, and draws inside it a set of ribs and circles to control the proportions of the coffers, as on other occasions (Vandelvira c. 1585: 62v-63r). However, he cannot use the cone development technique, since parallels are not circular. Thus, he constructs a development of each different lune in the vault by triangulation, as he did in the third oval vault (Vandelvira c. 1585: 73v-74r), that is, the non-coffered prolate one.

Annular vaults. Vandelvira (c. 1585: 70v-71r) also includes a horizontal-axis coffered annular vault (Fig. 10.39), called the *bóveda de Murcia por cruceros*, since it

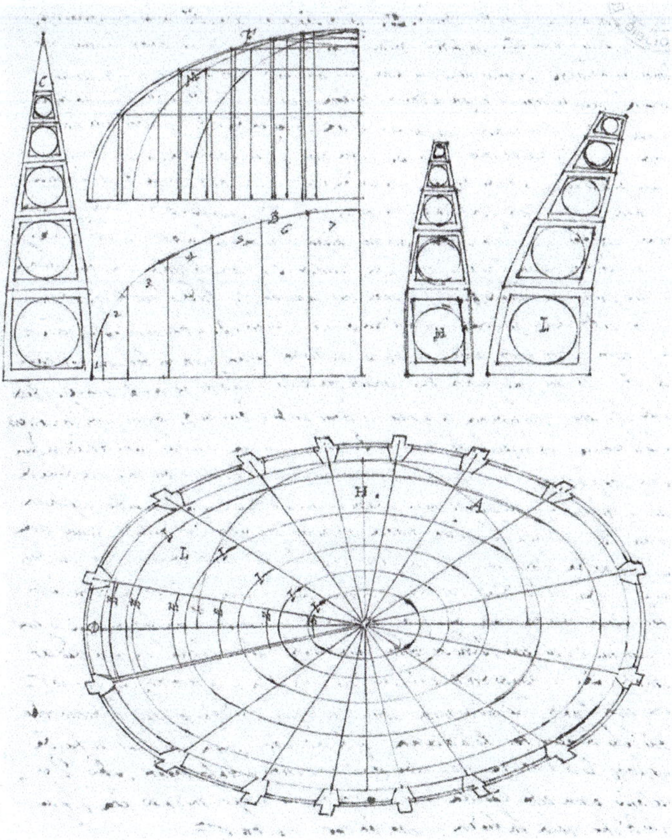

Fig. 10.38 Coffered prolate oval vault (Vandelvira c. 1585: 75r)

is a translation of the single-tier annular vault, exemplified by the chapel of Junterón in Murcia Cathedral. Generally speaking, the solution is similar to the one in the oblate oval vault: Vandelvira uses meridians set out in a sheaf of planes with a common intersection at the short axis of the vault and parallels in vertical planes following the direction of the longer axis; he develops the gore over the longer axis of the vault, while using cones in order to construct the intrados templates of the ribs and even the panels.

Conical vaults. Stonecutting manuscripts dealing with coffered vaults (Vandelvira c. 1585: 62r-65r, 70v-71r, 72v-73r, 74v-76r, 97v-102r, 103v, 104v-105r, 106v-107r, 125r; Guardia c. 1600: 89v) do not include coffered splayed arches or trumpet squinches. There are, however, a series of coffered conical vaults covering the passages between chancel and ambulatory in Granada Cathedral (Fig. 10.40), built under the supervision of Diego de Siloé around 1540. For liturgical and theological reasons, the chancel was originally wide open to the ambulatory; thus, a number of vaulted passages were opened in the thick wall supporting the spherical

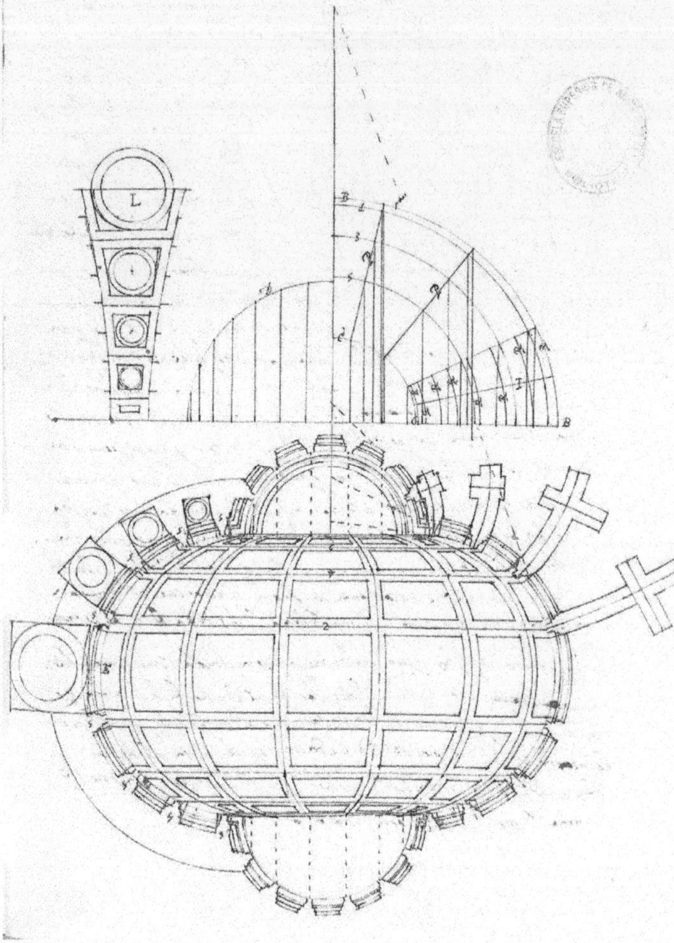

Fig. 10.39 Coffered annular vault (Vandelvira c. 1585: 71r)

vault over the chancel. It has been shown (Salcedo and Calvo 2018) that the central passages are conical coffered vaults with circular ribs parallel to the wall, and radial ribs acting as cone generatrices. The junction with the round chancel is solved by an arch with a concave face on the sanctuary side and a flat one furnishing the transition to the coffered vault. However, the problem in the last passages on each side is quite different, since at these points, the vaults must adapt at the same time to the radial geometry of the ambulatory and the orthogonal layout of the nave vaults. Therefore, the plans of the last vaults are asymmetrical, their spans at the ambulatory side are wider than in other passages, and the planes of the face arches are not parallel. All these problems are solved neatly. Diego de Siloé wished to keep the rise of all arches in the ambulatory side equal; thus, the face arch on the last passage is surbased, since its rise equals the rest of the arches, while the span is larger. In contrast, the arch

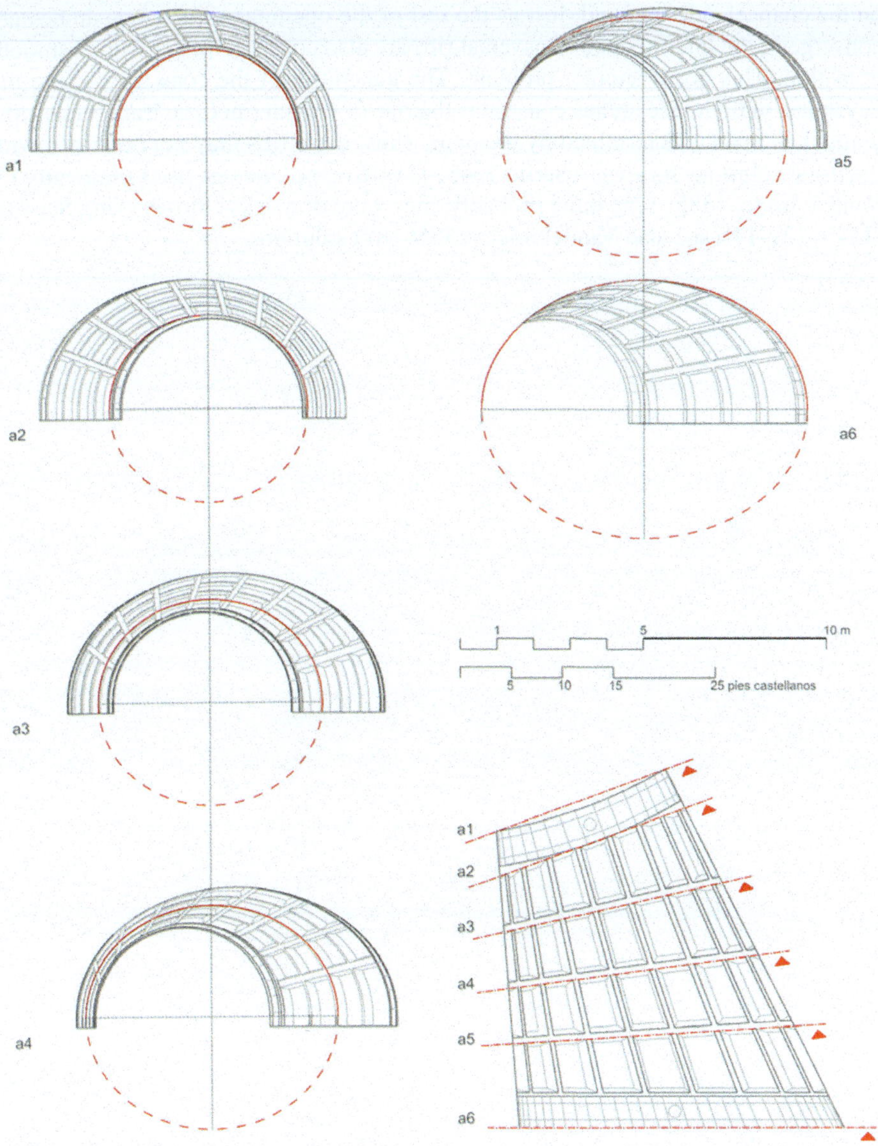

Fig. 10.40 Conical vaults in the passages between chancel and ambulatory. Granada Cathedral (Drawing by Macarena Salcedo)

on the chancel side is round, just as the rest of the openings. Since face arches are convergent, the intermediate transversal ribs are also convergent, to assure a smooth transition. This poses another problem. The generatrix of the cone connecting the keystones of both face arches constrains the rise of the intermediate transversal ribs, while their span is determined by the plan. Thus, these ribs may not be drawn with the usual solutions for ovals (Serlio 1545: 17v-18v[20]) or the one used frequently by Vandelvira (c. 1585: 18r); quite probably, they were drawn as ellipses using Serlio's (1545: 13v-14r; see also Vandelvira c. 1585: 18v) solution.

[20]Fol. 18 is numbered erroneously as fol. 20.

Chapter 11
Staircases

11.1 Spiral Staircases

11.1.1 Straight-Newel Spiral Staircases

The simplest type of spiral staircase in stone is known in Spanish as *caracol de husillo* (spindle stairway). A single member—the individual step—includes the tread, the riser, the cylindrical support known as a *newel* and the junction with the wall (Fig. 11.1); it even serves as scaffolding, since the mason stands on it while setting the pieces on their final position. Viollet-le-Duc (1854–1868: V, 297) stressed that this technique avoids the use of formwork, in contrast to twelfth-century vaulted staircases. Geometrically, this member is generally made up of two sections, both carved in the same stone: a wedge-shaped portion that includes the riser and the tread, extending inside the wall, and a circular end acting as a drum of the newel. Circular sections are placed one on top of the other, so the newel adopts a cylindrical shape. In many early examples, the wedge is treated as a simple extrusion, so the intrados of the member is stepped, just like the upper surface. This solution looks coarse; tall users may bump against the edges above the stair. Thus, usually the lower surface is dressed using a ruler sliding along two helixes, one at the newel and the other in the wall edge; of course, this operation is not applied in the part of the step that goes into the wall.

As usual, de l'Orme does not bother to explain such a simple example and includes little more than two outstanding staircases, the vaulted one known as *vis de Saint Gilles* and the cantilevered *quartier de vis suspendue* (de l'Orme 1567: 120r–126v). Thus we must start, as on other occasions, with Vandelvira (c. 1585: 49v–50r). He draws the plan of the staircase and a detail of the step, as well as two cylindrical developments showing the junctions of the steps with the newel and the wall (Fig. 11.2); he does not feel the need to draw an ordinary elevation. The upper edges of the step (that is, the sides of the tread) are traced radially, passing through the axis of the newel, while the lower edges (that is, the generatrices of the intrados surface) are tangent to the newel, in order to achieve a smooth union of the intrados with the newel (see

J. Calvo-López, *Stereotomy*, Mathematics and the Built Environment 4,
https://doi.org/10.1007/978-3-030-43218-8_11

Fig. 11.1 Straight-newel staircase. London, Saint Paul's Cathedral (Photograph by the author)

Benítez and Valiente 2015). Vandelvira's solution was repeated by other Spanish writers, with some variants. Martínez de Aranda (c. 1600: 227–229) sets the edges of the tread radially and the generatrices of the intrados tangentially, like Vandelvira, but he manages to shape the section where each step overlaps with the preceding one as a strip with parallel edges. Sanjurjo (2015: 71) has remarked that this leads to a narrow overlapping section; this small detail is important, since steps may be worn out through use, allowing light to pass between their edges. Another interesting detail in Martínez de Aranda is the simplification of the helix in the handrail as a series of circular arcs, constructed starting from two endpoints lying on the helix and an intermediate point. However, this is not an original invention by Martínez de Aranda; as we will see in the next paragraph, Vandelvira (c. 1585: 50v-51r) explains this method in his open-well staircase.

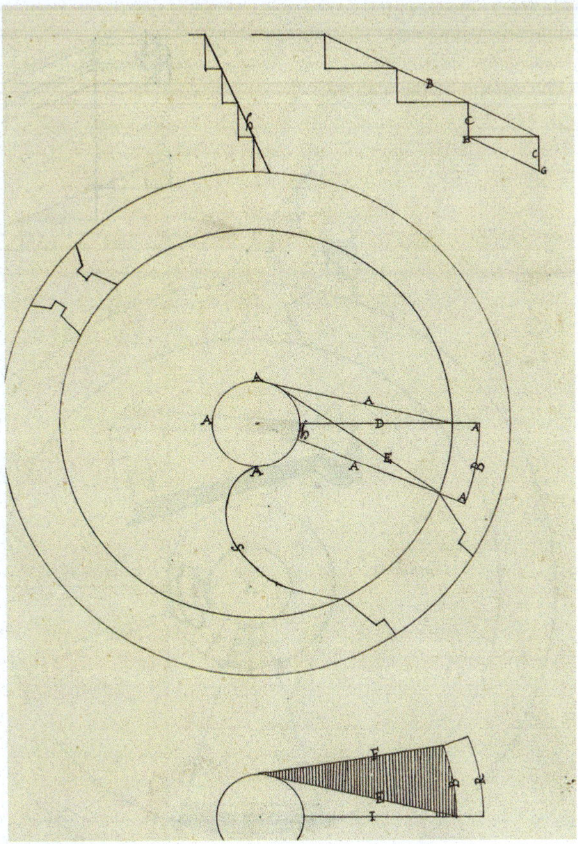

Fig. 11.2 Straight-newel staircase ([Vandelvira c. 1585: 50r] Vandelvira/Goiti 1646: 94)

11.1.2 Open-Well and Cantilevered Spiral Staircases

Open-well staircases with helical newel. Straight-newel staircases, particularly small ones, are usually dark and claustrophobic; when used as service stairs, they may be too narrow to carry large loads. An ingenious way of letting in more light and gaining space in a spiral staircase, without increasing the radius of the enclosing wall, is the design known as *caracol de Mallorca* in Spanish texts; according to Zaragozá (1992: 103-104) the name derives from an early example of this type in the corner turrets in Mallorca's Merchant Exchange (Fig. 11.3). The key to the solution is to place the end section not at the centre of the ensemble, but closer to the perimetral wall, usually about one foot off the axis of the staircase. In this case, the newel does not take the form of a cylinder, since its horizontal sections rotate as they go up; it assumes the shape of a vine shoot, creating a small open well.

De l'Orme (1567: 120r) makes a passing remark about this type of staircase, known as *vis a jour* in French, but furnishes no drawing. In contrast, Vandelvira (c. 1585: 50v–51r; see also Palacios [1990] 2003: 156–161; Sanjurjo 2007; Sanjurjo

Fig. 11.3 Open-well staircase. Palma de Mallorca, Merchants' Exchange (Photograph by the author)

2017: 99) explains the problem using, once again, a plan and a host of details, but no ordinary elevation (Fig. 11.4). In this case, he traces both the tread edges and the intrados joints radially; this gives as a result a crease between intrados and newel, but it is hidden by the elaborate moulding carved in the newel. As I have said, this line is controlled by substituting a series of circular arcs for the helixes in the mouldings; the same operation is performed for the handrail; in any case, other authors such as Martínez de Aranda (c. 1600: 246–247) and Jousse (1642: 180–181; see also Derand 1643: 401–406) eschew the mouldings and use a simple helical newel.

Spiral staircases cantilevered from the wall. The builder of Mallorca's Merchant Exchange, Guillem Sagrera, built a larger open-well staircase in the Castel Nuovo in Naples, apparently cantilevered from the wall; presently, it suffers from structural problems (Calvo and De Nichilo 2005b: 519–521; Tutton 2014: 37). The same idea reappears in the well-known oval staircase in the Convento della Caritá by Palladio (Fig. 11.5). Along the seventeenth century, some open-well stairs, known as "geometrical staircases", were built in England, such as the Tulip Staircase in the Queen's House, Greenwich, by Inigo Jones, and the Dean's Staircase in Saint Paul's

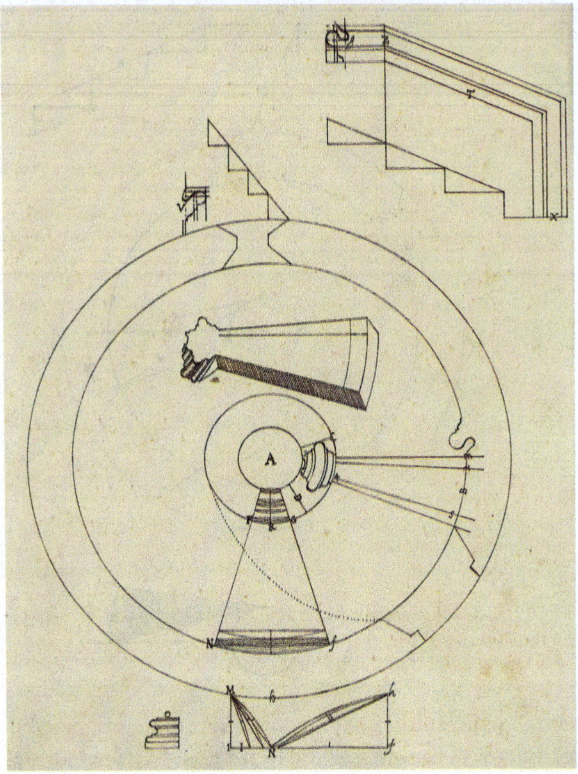

Fig. 11.4 Open-well helical-newel staircase ([Vandelvira c. 1585: 50v] Vandelvira and Goiti 1646: 95)

in London (Fig. 11.6; see also Campbell 2014a: 99-101 and Bergamo 2019). Some designs by Juan de Portor y Castro (1708: 20v, 21bis v, 22r) address this problem. The steps feature a complex cross-section and an arched longitudinal section used before by Martínez de Aranda in a staircase cantilevered from the newel, as we will see in the next paragraph. Portor's first solution (1708: 20r–20v) includes a round staircase and a circular well; it is half-way between a *caracol de Mallorca* and the wide-well English staircases. The second one features a square wall and a square well; however, Portor gives it the typical name associated in Spain with spiral stairs, *caracol* (snail), since the edges of the steps are traced radially. Finally, the third one includes a straight enclosure and a round well (Fig. 11.7).

Spiral staircases cantilevered from the newel. The opposite type, cantilevered from an isolated newel, without a perimetral wall, is used on some occasions to access pulpits. Such constructive stunt is not frequent, although some outstanding examples can be found in Saint Stephen cathedral in Vienna, Santa Maria Novella in Florence or Saint Étienne du Mont in Paris (See Battisti 1976: 292–293; de Rosa 2019, and Fig. 11.8). Philibert de l'Orme (1567; 120v–121v), presents a variant where only a quarter of the staircase is cantilevered from two supports at right angles, with a helical string, quite similar to a famous example in Autun Cathedral. Although his

Fig. 11.5 Oval staircase. Venice, Convento della Caritá, now Galleria della Accademia (Photograph by the author)

actual stonecutting explanations are quite concise, it may be guessed that the main geometrical problem addressed by de l'Orme is the warped nature of the upper and lower sides of the string voussoirs; also, in order to improve the stability of the piece, he uses a dog-tooth joint between voussoirs, as in the Mausoleum of Theodoric in Ravenna.

In contrast to this, Vandelvira (c. 1585: 51v–52r) offers a solution for a staircase fully cantilevered from an isolated newel (Fig. 11.9). The design of the step includes a broad wedge in the inner section, to be placed in a mortise cut in the newel; all edges are traced radially. The cross-section of the step is L-shaped so that each member overlaps the preceding one. Martínez de Aranda (c. 1600: 243–245) generally follows the same design, although he does not broaden the wedge at its junction with the newel. However, he uses a more complex cross-section, probably seeking to strengthen the union between steps, while the longitudinal section is arched. Although none of these authors engages in any computation of the structural feasibility of the piece, Martínez de Aranda's design shows a sound intuitive understanding of the mechanical problems that it raises.

Fig. 11.6 Cantilevered staircase. London, Saint Paul's Cathedral (Photograph by the author)

Fig. 11.7 Staircase cantilevered from the wall, detail (Portor 1708: 22r)

Fig. 11.8 Cantilevered staircase. Paris, Saint Etienne du Mont (Photograph by the author)

11.1.3 Vaulted Spiral Staircases

When Charles V of France tried to build a wide staircase in the Louvre, he could not find stones long enough, so he had to employ discarded tombstones (Viollet-le-Duc 1854–1868: V, 297–298). On other occasions, the problem is solved using a vault spanning the area between the outer wall and a central newel. The most usual solution is known as *vis de Saint Gilles*, from the priory in Saint Gilles, in Languedoc, where a remarkable example of this type stands in isolation following the destruction of the church in the Wars of Religion and the French Revolution (Pérouse 1985; Hartmann-Virnich 1999). However, there are earlier examples, in particular in Bab-el-Nasr, in the Fatimid walls of Cairo (Fig. 11.10), built between 1087 and 1092 by three Armenian monks coming from Edessa (Tamboréro 2006); another one at the Castello Maniace in Syracuse, although later, suggests a possible point of entry of this type into Europe (Bares 2007).

The intrados surface of such staircases is generated by a round arch that rotates around the axis of the ensemble while ascending, keeping as directrices two helixes

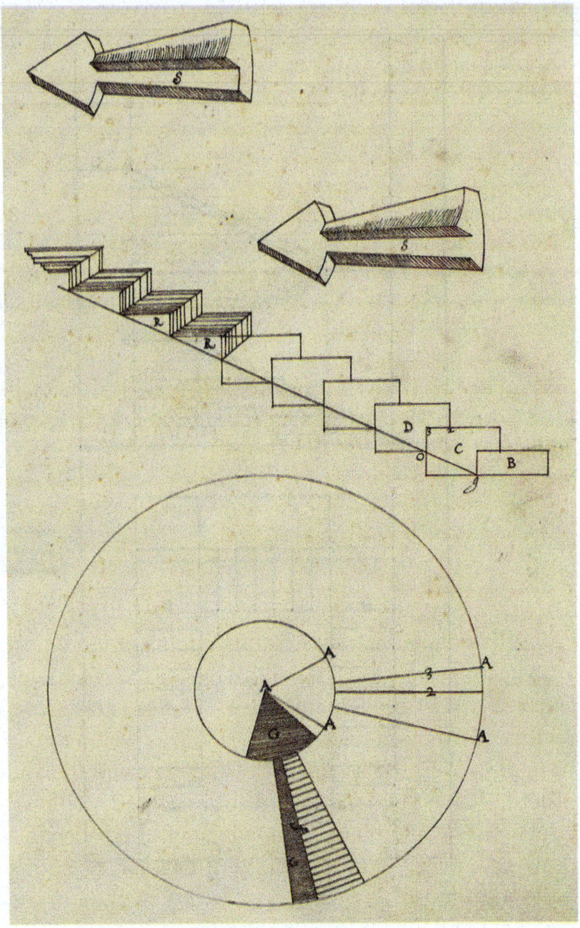

Fig. 11.9 Cantilevered staircase ([Vandelvira c. 1585: 52r] Vandelvira/Goiti 1646: 98)

which furnish the intersections with a usually massive newel and the exterior wall. The result is a singular double-curvature surface which is neither a revolution nor a ruled surface. The first explanation of the construction procedure is given by de l'Orme (1567: 123v–125v). As we have seen in Sect. 2.2.1, he boasts of his skill in dressing the voussoirs of the vault using merely the arch square and the bevel. However, he does not clearly explain some critical steps of the actual execution method; I will try to reconstruct them by comparison with later texts.

First, de l'Orme draws the plan of the wall and the newel (Fig. 11.11); next, he constructs a cross-section of the vault in the shape of a round arch, dividing it into seven voussoirs. Once this is done, he draws projection lines to bring the intrados and extrados joints of the round arch to the plan. These points provide cues for tracing the intrados and extrados joints of the entire vault in plan; next, radial joints are added in order to get voussoirs of reasonable size.

Fig. 11.10 Vaulted staircase. Cairo, Fatimid walls, Bab-el-Nasr (Photograph by the author)

De l'Orme points out that this tracing could be used to solve a ring-shaped vault with level springings. The vault at the palace of Charles V in Granada (see Salcedo and Calvo 2016: 54-55; Senent 2016: 537, 543) may serve as an example, although it was under construction when de l'Orme was writing his treatise and it is not mentioned in it. However, the ascension of the stair and the vault involves other problems. In order to solve them, de l'Orme draws three auxiliary schemes, marked P, Q and R, but he does not explain them, advising the reader to "learn the tracings reproducing drawings, rather than by long writings and discourses".[1] Later on, he embarks on a description of his own works in the open staircases or *perrons* in Fontainebleau and Anet and a well-known invective against Bramante's staircase in the Belvedere (Potié 1996: 133–134; Camerota 2006: 56–57; Fallacara 2009a: 140–144), but he gives no clue about diagrams P, Q and R.

Following de l'Orme's advice, in order to reproduce drawing P, we should draw projectors from the intersections of the steps and the newel in plan, until they reach horizontal lines placed at the level of each step, connecting the resulting points with a curved line; this will furnish a vertical projection of the helix at the intersection of the vault and the newel. Diagram R, which represents some lines converging in a point, is more intriguing; however, it bears a series of numbers from 0 to 7, which are also present in the plan, close to the intrados joints. These numbers make it clear that the scheme represents the length and slope of the intrados joints in each voussoir. These lines are computed by simply forming triangles with the horizontal projection of the

[1]De l'Orme (1567: 123v): *apprendre les traits plus en les contrefaisant, imitant et représentant, que par longues écritures et discours de paroles*. Transcription is taken from http://architectura.cesr.univ-tours.fr. Translation by the author.

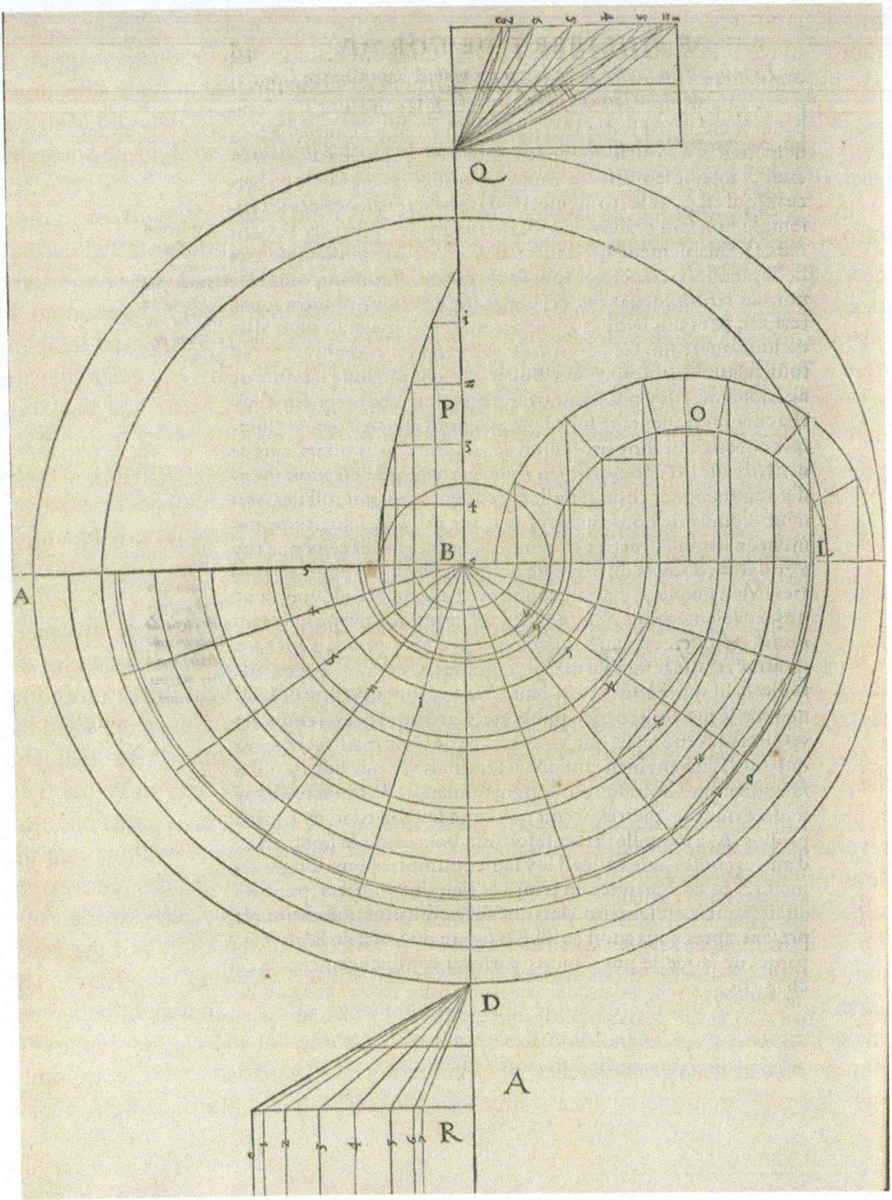

Fig. 11.11 Vaulted staircase (de l'Orme 1567: 125v)

chord, numbered from 0 to 7, and the height of the step, which is the same for all chords. Once this is clear, it is easy to interpret diagram Q: it reproduces diagram R, adding a templet for each intrados joint. This implies that circular arcs are substituted for portions of the helical intrados joints, as Vandelvira (c. 1585: 50v) and Martínez de Aranda (c. 1600: 213–215, 222–223) do for open-well and straight-newel stairs, respectively.

In any case, this leaves us with few hints as to the actual dressing procedure. In fact, the lines in diagram R seem at first sight to be of little use in the carving process; if materialised as a draft, they would go inside the finished voussoir in some cases, causing the member to be unusable.[2] Thus, lines in R seem to be used not in the chord of the actual voussoir, which appears to be controlled using the templets drawn in Q, but rather in an enclosing preliminary block. This suggests that the voussoirs were carved, at least in part, by squaring; the lines below the voussoirs in the cross-section, representing the lower part of the enclosing block, point in the same direction. However, Tamboréro (2006) identifies this solution with de L'Orme's reference to templates; this apparent contradiction may be solved considering that the main dressing strategy relies on orthogonal projections, which are usually identified with squaring, although some auxiliary templates, particularly those for the plan of the voussoir, are used.

De L'Orme (1567: 126r–126v) also offers an alternative solution to this piece (Fig. 11.12), presenting it as "the tracing of another sort of sloping staircase in the shape of the Vis de Saint Gilles".[3] He is even more concise than in the preceding variant; he only explains that the drawing shows how to measure the slopes and the *cerces ralongées*, that is, the templets, and that the execution can be controlled with the arch square. Moreover, he states that the issue is a complex one, and other lines could be included in the drawing, although this would cause great confusion; also, several long chapters could be devoted to its explanation, but he will shorten the description. In spite of the obscurity of the text, Tamboréro (2006) has put forward a likely interpretation of this accompanying drawing, which is clearly divided into three sections. The lower right repeats the construction of the chords of the voussoir, exactly as in the other solution presented by de l'Orme (1567: 125v). The upper right section presents some amazing shapes recalling at first sight helicopter blades; Tamboréro (2006) identifies them as the slopes of the chords, remarking that the engraver has mismatched the numbering. A comparison with de l'Orme's own "Cantilevered quarter of a staircase" (1567: 121v) and Martínez de Aranda's

[2]A modern reader may wonder if diagram R is merely a preliminary step for diagram Q, which involves exactly the same lines with the addition of the cerces. However, this sequential approach is typical of the assembly or process drawing of the Industrial Revolution and the technical training of the Enlightenment and is clearly at odds with the economy of stonecutting tracings of the Renaissance and Baroque periods, with the exception of Bosse (Bosse and Desargues 1643a: plates 12, 23, 53, 84–85), clearly an outsider.

[3]De l'Orme (1567: 126r): *Le traict d'une autre sorte de vis & montée rempante en façon de la vis Sainct Gilles*. Transcription is taken from http://architectura.cesr.univ-tours.fr. Translation by the author.

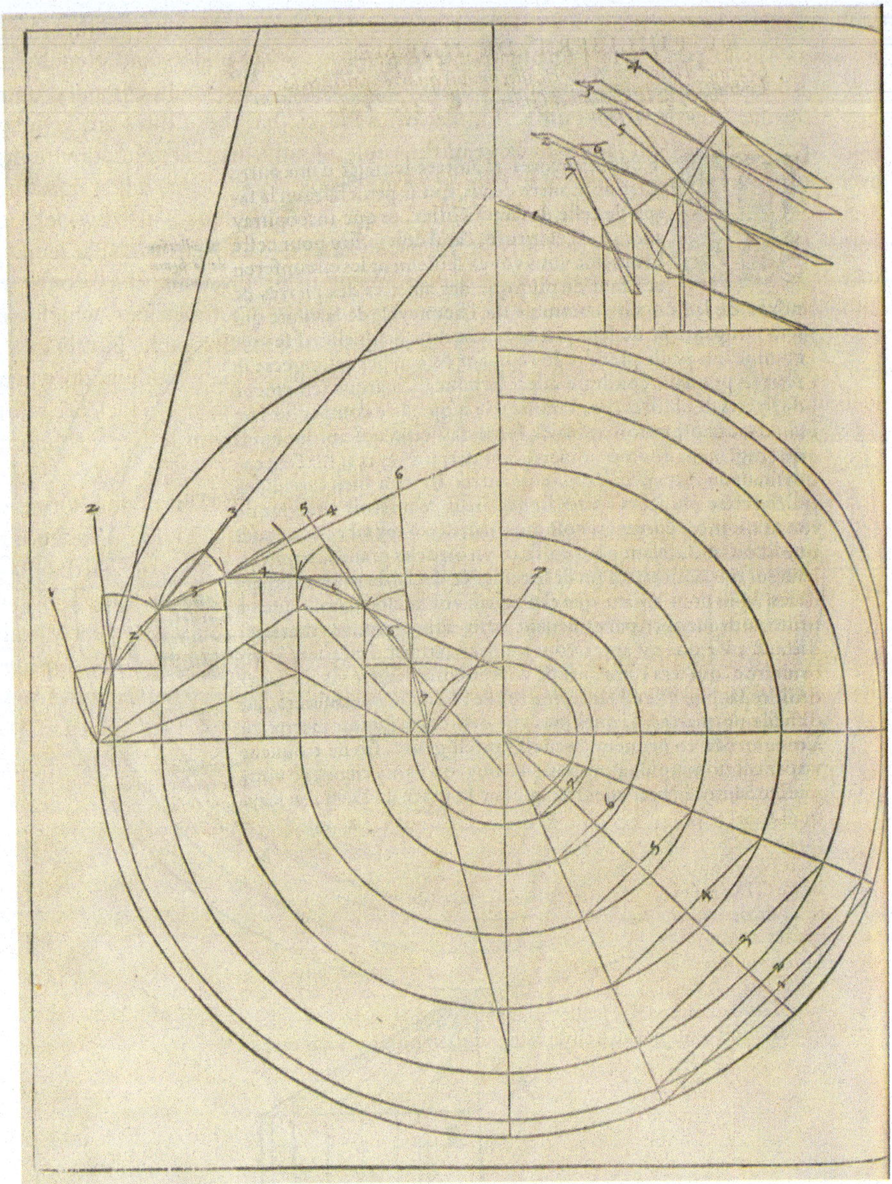

Fig. 11.12 Vaulted staircase, additional constructions (de l'Orme 1567: 126v)

explanation on how to dress a warped surface (c. 1600: 222–223) supports this interpretation. The left half of de l'Orme's scheme is not so easy to understand. It includes several lines starting from the semicircular cross-section of the vault, extended until they meet the axis of the drawing. Tamboréro remarks that these lines are generatrices of cones inscribed in the intrados of the vault, identifying this method with the arch squares and bevels mentioned by de l'Orme (1567: 123v–124r) when boasting about his skill. This is a plausible hypothesis since this technique is used widely in Renaissance stereotomy, particularly in spherical-intrados vaults, as we have seen in Sect. 9.1. However, there are some problems. When using these auxiliary cones in domes and other vaults, the conical surface is developed using directrices, which are represented by circular arcs with their centres in the axis of the vault. In this case, de l'Orme draws generatrices intersecting the vault axis, as many authors do when dealing with spherical vaults, but he does not use these intersections; in fact, arc radii are much shorter. I will come back to this issue when dealing with Jousse's solution below.

Sixteenth-century Spanish stonecutting manuscripts use a variant of de L'Orme's (1567: 123v–125v) first method, but they are much clearer (Fig. 11.13). Vandelvira (c. 1585: 52v–53r; see also Palacios [1990] 2003: 164–169 and Sanjurjo 2010: 640–641) draws clearly several schemes of the enclosing blocks, taking into account the rise of the vault from one radial joint to the next. In the text, he states clearly that the voussoirs should be dressed by squaring; he adds that the height of the starting block should equal the height of the enclosing rectangle in the cross-section, from

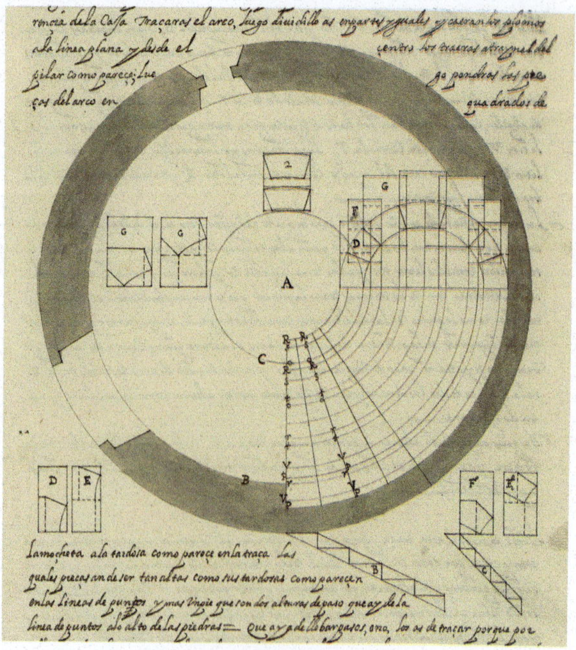

Fig. 11.13 Vaulted staircase ([Vandelvira c. 1585: 52v] Vandelvira/Goiti [c. 1585] 1646: 108)

the intrados to the extrados, plus one foot, that is, twice the rise of the vault between two consecutive radial joints; and he mentions that the mason can place the voussoirs in staggered fashion, in order to break the continuity of the radial joints. This detail explains why the height of the enclosing block should be increased twice the rise of the stair between radial joints.

These issues are even clearer in Martínez de Aranda (c. 1600: 230–233). Along with the usual plan, cross-section, and radial and longitudinal joints, he includes neatly drawn enclosing rectangles, both for longitudinal and cross-sections and the cerces we have seen in de l'Orme (Fig. 11.14). Further, he explains clearly the dressing method in these terms:

> to execute these voussoirs you will dress them first by squaring with the shape of the intrados templates and the width and height of the enclosing rectangle of the bed joint template between corners *5, 6, 7, 8, 9, 10, 11, 12*; after dressing the member by squaring, you should take off the wedges defined by the rise of the stair so that the surfaces are warped… so that it will have in the higher and lower joint the width and height of the enclosure of the voussoir

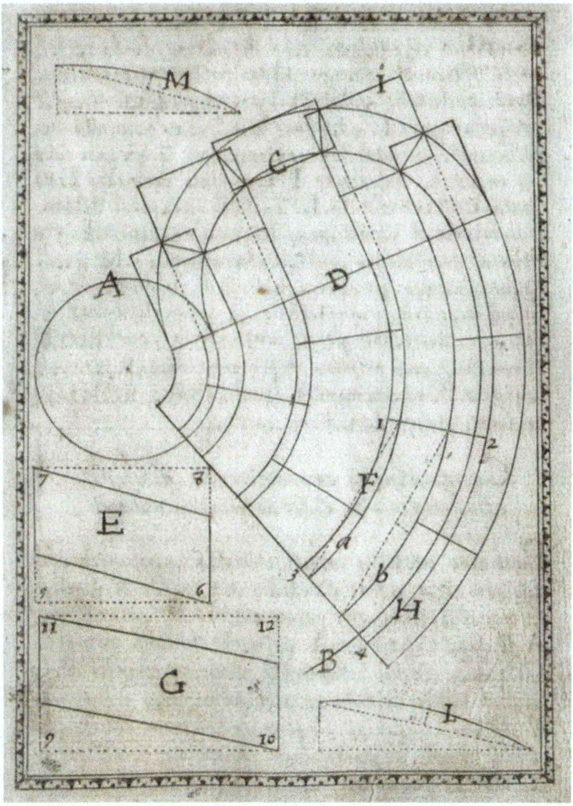

Fig. 11.14 Vaulted staircase (Martínez de Aranda c. 1600: 232)

as in figure *I*; next you will take off wedges in the intrados, bed joints and extrados taking as a cue figure *I* and the extended cerce *L*...[4]

In order to interpret this passage we must take into account that the intrados and bed joint templates mentioned at the start are not true-shape templates; rather, the intrados templates are horizontal projections, while the bed joint templates *E* and *G* are taken from projections on two vertical planes parallel to intrados joint chords. It is important to stress that the faces of the initial enclosing block should not coincide with the chords, in order to avoid portions of the final voussoir being left outside of the enclosing block. Thus, the dressing process involves three phases. In the first one, the mason should dress a block enclosing vertexes *1* and *3*, plus a face slightly displaced from chord *2–4* in order to encompass the whole voussoir. Its height should equal that of the rectangle *I* plus the rise of the stair between both radial joints. In the second phase, the mason should take off two wedges, one at the bottom and one at the top of the initial block, using templates *E* and *G* as guides, applying them to vertical planes passing through points *1* and *3* and slightly off points *2* and *4*. Finally, the mason should carve four wedges on the intrados, extrados, and bed joints, using the cross-section as a reference, to arrive at the final shape of the voussoir. All this coincides substantially with de l'Orme's first solution (1567: 123v–125v), which is based on orthogonal projections, including an auxiliary projection onto a plane parallel to the chord of the voussoir.

Jousse (1642: 187–192) devotes three sections of his book to the *vis de Saint-Gilles,* but his presentation is not easy to understand. The first section (1642: 186–187) basically follows De L'Orme's first solution (1567: 123v–125v; see also Vandelvira c. 1585: 52v–53r, and Martínez de Aranda c. 1600: 230–233); the title makes it clear that the mason is to dress the voussoirs by squaring with the aid of extended templets. He includes the plan, the elevation and the computation of the slopes of the helicoidal joints,[5] but instead of constructing the templets separately, starting from the slopes of the helicoidal joints, he groups them in a diagram resembling an intrados template of each voussoir (Fig. 11.15), although Jousse always refers to these figures as *serches*, that is, templets, and not as *panneaux* or templates.[6]

[4]Martínez de Aranda (c. 1600, 231–233): ... *y para labrar esta dicha pieza la labraras primero de cuadrado con la forma que tuviere la dicha planta por cara y con el ancho y alto que tuviere la dicha planta por cara y con el ancho y alto que tuvieren los cuadrados de las dichas plantas por lechos que están entre los ángulos 5 6 7 8 9 10 11 12 y después de labrada la dicha pieza de cuadrado con la forma arriba dicha la robarás primera vez por lo alto y por lo bajo con los robos que causare la subida del dicho caracol que quede la dicha pieza engauchida... que venga a quedar por las juntas alta y baja con el ancho y alto que tuviere el cuadrado de su bolsor que es la figura I y despues la robarás segunda vez por la cara lechos y tardós con los robos que causare su bolsor en el dicho cuadrado I robándole por la cara baja con la cercha extendida L ...*

[5]It is important to take into account that somebody drew some additional bright red lines in the ETH-Zurich library copy used as a source of the e-rara facsimile, including templets in the equivalent of de l'Orme's diagram R.

[6]The interpretation of these diagrams as actual templates is quite problematic. It involves the substitution of circular arcs for two portions of different helixes, connecting them with line segments standing for the circular arcs of the cross-sections. The whole ensemble would stand for a part of a

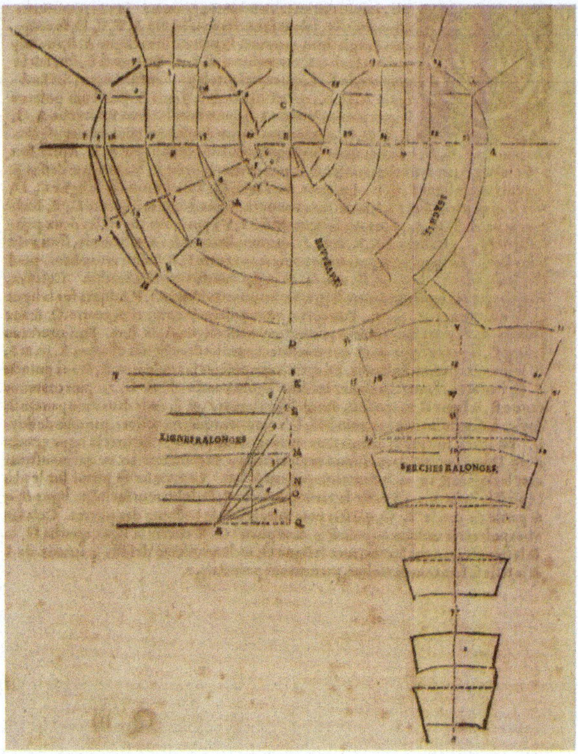

Fig. 11.15 Vaulted staircase (Jousse 1642: 186)

The second and third sections (Jousse 1642: 188–191) develop de l'Orme's second method in greater detail; he presents the problem as a *premier plan* furnishing the *cherches ralongées* or templets of the helicoidal joints and a *second plan* providing templets for the meridional joints. In both cases, the scheme is based on the template-constructing method for spherical vaults, based on cone developments. In the *premier plan* the generatrices of the cones pass through two consecutive helicoidal joints. As in the vertical-axis annular vault, those ones for the outer half intersect the axis over the springing plane, while some of the inner ones intersect it under the impost. However, in domes and other spherical pieces, intrados joints are horizontal and circular, and thus can be used as directrices of a cone. In contrast, the joints of the *vis de Saint Gilles* are cylindrical helixes, so they cannot be taken directly as directrices of a cone. Jousse tries to solve this problem with a coarse correction: he draws an auxiliary diagram with the slope of the chord of each joint, adding an orthogonal of the end with the length of the thickness of the vault and constructing a small triangle to measure the deviation of this line from the vertical passing through the end of the joint. Next, he transfers these deviations to the intrados templates in order to slant

double-curvature surface which is, of course, non-developable. Thus, it seems that the grouping of *serches* is a graphical device that tries to make it clear that two templets belong to the same voussoir.

them. Of course, this takes it for granted that meridional or side joints are placed on vertical planes.

The third section, dealing with bed joints, follows similar lines. Jousse constructs cones passing through the intrados and extrados joints. As in the vertical-axis annular vault, the centres of the cones in the inner half of the vault lie over the springing plane, while those of the outer section are placed below the impost. He draws single templets and full templates using circular arcs with their centres at these points. To slant these templates, he again constructs triangles at the end of the chords of the joints and transfers them to the templates, although in this case, the rationale of the operation is even more obscure.

Derand (1643: 406–413) puts forward similar methods. He expands the developments of helicoidal joints, including both intrados and extrados in several diagrams called *dispositifs*, which can be loosely translated as "preliminary constructions"; in this case, it is clear that the *dispositifs* are simplified oblique projections of the voussoirs on planes passing through the line joining two consecutive corners of the voussoir in the same helix; this recalls Martínez de Aranda's preliminary templates. Next, Derand uses the *dispositifs* to correct the intrados and bed joint templates constructed through cone developments; as in Jousse, the rationale of the operation remains unclear. In any case, neither the intrados surface nor the bed joints are developable, so the use of templates laid directly on the intrados or in the bed joints is problematical. De la Rue (1728: 134–139; see also Frézier 1737–1739: II, 419–427) takes strong exception against these methods, remarking that

> Father Derand's … method for dressing the voussoir is somewhat different from the one I am including. I do not know why this Father has proposed the use of intrados and bed joint templates to dress the voussoirs; he uses them in a way that is both mechanical and inexact … Is not there a method more precise than dressing the stone by trial and error, both in the intrados and the bed joints?[7]

As a consequence, he eschews the templates of Jousse and Derand, using only a set of diagrams in the fashion of Derand's *dispositifs* to construct one-sided templets (*not templates*) for the helicoidal joints, on both the intrados and extrados. In essence, he returns to de l'Orme's first method and even quotes the passage where he boasts of being able to dress the piece using only arch squares and templets. In contrast to this economy, he explains in great detail an issue other writers had not addressed. Rather than using a mortise to join the vault with the newel, as Vandelvira and Martínez de Aranda had done for cantilevered stairs, he includes a projecting fin in the drums of the newel, providing a support for the vault, drawing for it a *panneau de tambour* (newel template).

Up to this moment, all solutions in the literature use vertical joints between voussoirs in the same course. Although this solution is not ideal, it does not raise serious

[7]De la Rue (1728: 139): *Le P. Derand… sa méthode pour tailler les voussoirs, differe un peu de celle que nous donnons ici. Je ne sais comment ce Pere a pû proposer l'usage des panneaux de douelle & de joint pour couper les dits voussoirs, la façon dont il veut qu'on s'en serve étant aussi méchanique que peu exacte… Ne voilà-t-il pas un méthode bien certaine, que d'ôter de la pierre en tâtonnant, tantôt a la douelle & tantôt au lits.*

problems as long as these joints are laid out in staggered fashion. In contrast, most built examples, starting with the Saint-Gilles archetype, do not use vertical joints (Sanjurjo 2010: 637–639). Moreover, nineteenth-century treatises such as the one by Rondelet ([1802–1817] 1834: II, 208–211, see also pl. 54), or historical studies such as Choisy's *Art de Batir chez les Byzantins* (1883: 46–47) stress the use of sloping joints and end faces.

11.1.4 Double, Triple and Fourfold Spiral Staircases

On some rare occasions, two or more spiral staircases are built around the same newel or axis. An early example is a double spindle stairway in the Chapel of Alphonse V in Valencia. However, another double staircase, built around a hollow newel in the castle of Chambord, is better known; it was depicted, as a fourfold piece, in Palladio's (1570: I, 64–65) *Quattro libri*. Antonio da Sangallo the Younger built a double ramp in the Well of San Patrizio in Orvieto. It was used by mules to raise water while Pope Clement VII took refuge from the Sack of Rome; the unusual design prevented descending beasts from clashing with returning ones (Vasari 1568: III, 318). As far as I know, the only built example with more than two stairways is the triple one at the convent of Saint Dominic of Bonaval, in Santiago de Compostela; as we have seen, a full-scale tracing has been preserved right under the stair, the only known witness of this practice (Taín 2006: 3017–3019; Fernández Cabo et al. 2017). Although not frequent, the type did not die with the Industrial Revolution, as the examples in the Vatican Museums by Giuseppe Momo[8] and the dome in the Reichstag by Norman Foster attest.

Since stairways sharing the same axis do not intersect, their construction involves repeating the basic types and does not raise special problems, so most stonecutting treatises and manuscripts do not deal with this issue in any depth. De l'Orme (1567: 120r, 122r–123r) mentions a double vaulted stairway; he includes a drawing but no real explanation about stonecutting methods. Martínez de Aranda is more specific (c. 1600: 229–230, 241–243, 245–246, 248–250), particularly for straight-newel (Fig. 11.16), cantilevered, and open-well staircases. Two details are worth remarking. In the straight-newel example, each drum is divided into two parts, each one carved in the same piece as the step for each of the stairways; of course, this problem does not arise in the cantilevered or open-well variants. In contrast, Martínez de Aranda explains in detail the number of stairs in each revolution and the height of the risers of all these types. Surely, such fastidious explanations are justified by the difficult problems posed by double stairways: the combined risers of the steps in one

[8]The staircase by Momo is sometimes called the "Bramante stair". This is confusing, since in the Vatican Museums area there is another staircase actually designed by Bramante (see Sects. 2.3.5 and 11.1.3). While the real Bramante stair features a single stairway and is supported by columns placed on the interior edge of the stairway, Momo's one features two stairways around the same axis and is cantilevered from the wall. The main similarity between the two pieces, other than being spiral, is that neither one is strictly a stair; both are ramps.

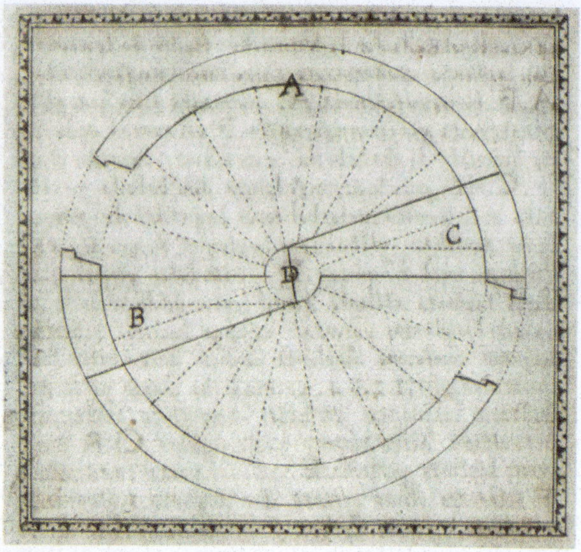

Fig. 11.16 Double straight-newel staircase (Martínez de Aranda c. 1600: 230)

revolution must allow for twice the height of the tallest user, plus twice the height of the riser and the thickness of the step (see Campbell 2014b: 254–256).

11.2 Straight-Flight Staircases

11.2.1 Straight-Flight Staircases with Curved Strings

Curved strings and joints orthogonal to the wall. Some of the earliest preserved stairs in hewn stone in the Iberian Peninsula are single-flight stairs, such as those in the Gates of Serranos and the Gates of Quart in Valencia, dating respectively from 1398 and 1451 (Fernández Correas 2007: 280). In these examples, there is no space to place multiple-flight stairways, so stairs are set along the interior face of the city wall. A similar solution is found in the main courtyard of the Palace of the Kings of Majorca in Perpignan, with two symmetrical stairs placed alongside the walls of the main court (Fig. 11.17). These stairs include two flights placed along the same wall and resting on buttresses; rampant vaults are built between each buttress and the next one. Intrados joints are horizontal and orthogonal to the wall; thus, the resulting intrados surface is a portion of a cylinder.

These stairs are placed in a huge courtyard, but the single-flight solution cannot be applied in the small courts of town palaces: there is not enough length to reach the first storey with a single flight and a reasonable tread-to-riser ratio. This problem is usually solved with two flights placed at a corner of the courtyard. In the earliest examples, such as the Queens' stair in the Palace of the Kings of Majorca in Perpignan

Fig. 11.17 Straight-flight staircase in the courtyard. Perpignan, Palace of the Kings of Majorca (Photograph by the author)

(Fig. 11.18), built before 1347, both flights rest on independent vaults with the second flight leaning on the string of the first one (Pérouse [1982a] 2001: 167–168). However, late-fifteenth century Valencian masons must have found such solution messy: it is hidden behind piers in the Scala and Generalitat palaces.

In the Scala stairway, the two flights are built as independent vaults (Gómez-Ferrer 2005: 118–127). The next step in the evolution of Valencian stairs was the junction of the vaults of two consecutive flights. An early example of this solution is the one in the Sancho palace, now Town Hall of Ontinyent (Fig. 11.19), dating from the early sixteenth century and attributed to Benoît (or Benet) Augier. As usual in the stairs in Eastern Spain, from Perpignan to Lorca, the horizontal projections of bed joints are perpendicular to the wall and thus parallel between them; however, the joints themselves are not parallel, and the intrados surface is warped. The space under the first flight is closed, along with a portion of the second flight, to provide a space for barley storage.

The existence of a long landing at the end of the second flight in Ontinyent raises the problem of the junction between flights. It solved in a practical and neat way: the masons manipulated the strings of the second flight to avoid a strange result. If the strings were parallel, the outer string of the final landing, set against the wall, would lie higher than the free edge; of course, this is not a good design for a cantilevered element. To avoid this, the builders of the Ontinyent stair traced the wall string of the second flight so that it actually goes down in the final portion, behind the landing. Also, the free string rises more steeply than that of the wall from its start at the barley storage. In this way, the intrados surfaces of the flight and the landing are tangent.

Fig. 11.18 Queen's staircase. Perpignan, Palace of the Kings of Majorca (Photograph by the author)

Fig. 11.19 Staircase. Ontinyent, Sancho Palace, now Town Hall (Photograph by author and Pau Natividad)

Fig. 11.20 Staircase known as *vis des archives*. Toulouse, Town Hall, known as Capitole. Photograph taken before demolition in 1885. Mediathéque de l'Architecture et du Patrimoine. Base de données Memorie. Número du negatif MH0005776

The junction of the voussoirs of both sections is solved in a practical way: rather than using V-shaped voussouirs, each course in the landing meets at its end a course in the flight, and this course in its turn meets at its end the next course in the landing.

Later on, Augier furnished drawings for the *vis des archives* at the Capitole or Town Hall at Toulouse, built between 1531 and 1541–42 by Sébastien Bougereau (Figs. 11.20, 11.21), which shows a neater, more advanced solution. Two centuries later, it had reached the status of an archetype: the eighteenth–century manuscript at Auxerre alludes to a "Square staircase in its plan in the shape of the Toulouse staircase".[9] (Pérouse [1982a] 2001: 84, 100, 168). The staircase was demolished in 1885 when a rear façade for the Capitole was built (Tollon 1992: 98, 99–101); however, several documents, such as a survey by Anatole de Baudot, a restoration project by Viollet-le-Duc, another set of drawings by an unknown draughtsman, and several exterior and interior photographs, have survived (Pérouse [1982a] 2001: 84; Tollon 1992: 101–102), allowing a reconstruction of the stair (Zaragozá et al. 2012). It was built within a square enclosure and included ten flights, although the last two were left unfinished. Bed joints are perpendicular to the walls; strings are traced as in the Sancho palace, so that the free edge of each flight is clearly steeper than the wall one; however, Bougier avoids the final downward section in the outer string.

[9] Auxerre, Bibliothéque Municipale, MS 388, without page numbers, *Escalier quarrés sur son plant en forme de l'escallier de Toulouse.*

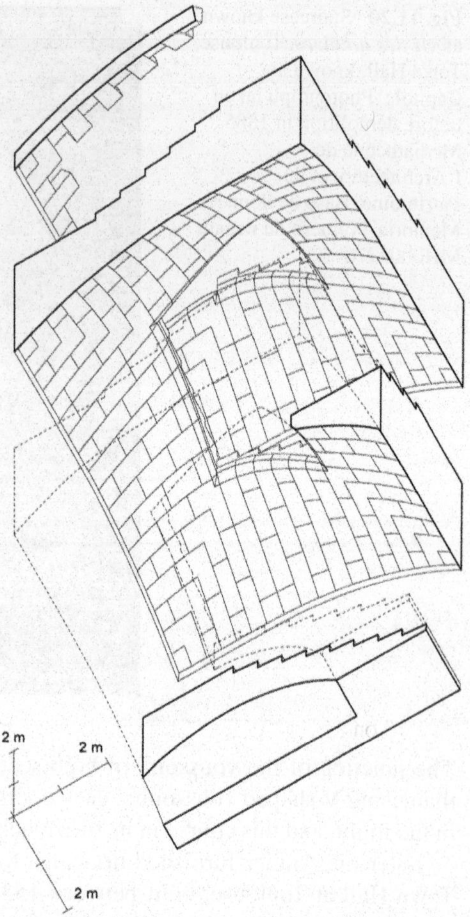

Fig. 11.21 *Vis des archives* (Drawing by Pau Natividad)

Both de Baudot's drawings and the interior photograph make it clear that the surface below the landings is tangent to both connecting flights, as in the Sancho palace.

Joints parallel to the wall. All the examples we have seen up to this moment, from Toulouse to Lorca, use bed joints orthogonal to the strings; of course, the voussoirs in each course are divided by joints parallel to the wall, set in a staggered fashion to avoid cracks. However, Vandelvira (c. 1585: 56v–59r; see also Palacios [1990] 2003: 176–181) reverses this layout, using bed joints parallel to the strings. He mentions two staircases in Granada, one at the convent of Santa María de la Victoria and the other one at the Real Chancillería (Courts of Justice) (Fig. 11.22). Both were built by Master Pedro Marín, who according to Vandelvira, "showed in both pieces his command of this art".[10] The staircase at Santa María de la Victoria is lost, but that of the Chancillería attests to Marín's skill. It includes three flights, with the first one

[10]Vandelvira (c. 1585: 59r):... *en las cuales bien enseñó su suficiencia en este arte*... Transcription is taken from Vandelvira and Barbé 1977.

Fig. 11.22 Straight-flight staircase. Granada, Real Chancillería, now Superior Court of Andalusia (Photograph by the author)

placed on a solid block. Thus, there is only a junction between flights, but it is solved neatly, without a trace of a groin and using V-shaped voussoirs; since the joints are parallel to the walls, the angle in the V points to the wall rather than the free edge, as usual in Valencian solutions. As if wishing to stress this solution, both bed joints and joints between voussoirs in the same course are clearly marked by indentations.

Vandelvira's explanation is far from clear, for many reasons. First, a present-day reader may mistake his drawings for double orthographic projections, that is, plans and elevations neatly connected by projection lines. On close inspection, it is clear that the plan is a standard one; however, in a typical modern multiview drawing, the middle flight would be shown in a lateral view, presenting the sides of the string frontally, while the starting and finishing flights would be depicted with the risers shown frontally. Instead, Vandelvira depicts all three flights in a lateral view, so that all strings are shown frontally. In other words, he is deconstructing the stair by rotating the starting and finishing flights, to control their shape. This procedure has some traits in common with the disarticulation of tierceron vaults in Late Gothic and Early Modern manuscripts and treatises (Rabasa 2000: 337–338; Calvo and Rabasa 2017: 81–82; see also Sect. 10.1.4).

Vandelvira uses three drawings to explain the problem. The first one (c. 1585: 56v), is pretty straightforward. He divides the side of the staircase into four parts; the stairwell will take two parts, that is, ten feet, while the flights will be five feet wide. Next, he divides the rise of the stair in four parts, giving a part and a half to the first and third flight, and a part to the second, traverse flight. The total height, twenty-two feet, is distributed in thirty steps so that each riser will be one span high.[11]

The following steps of the procedure are shown in the drawing on fol. 58r (Fig. 11.23), which is less clear: the problems raised by the disarticulation of the flights make themselves felt, and the lack of some letters in the notation does not help. Vandelvira instructs the reader to draw straight lines placed just beneath the steps. Under these lines, he draws both strings of the first flight. As in Valencian stairs, they are not parallel: the free edge ascends quickly, to compensate for its shorter overall length, while the wall string recovers the difference in heights in the first section of the second flight, that is, the portion of the vault placed below the corner landing. Thus, both strings start in the middle section of the second flight, placed between both corner landings, at the same height; they are traced as compound curves, including two or three circular arcs. Vandelvira does not explain a method to assure the tangency of the two arcs, so we may assume that approximate tangency was achieved by trial and error.

From this point on, the same solutions are reproduced in the second flight: the open string ascends quickly in the middle section, and the wall string recovers the difference in the corner. In the third flight, the free string ascends a bit higher than the wall one, although the difference is smaller than the one in the preceding flights. Anyway, Vandelvira states that this difference in heights is not really necessary: "the uppermost flight may be a segmental arch, without a rise in the free edge, since it does not match another element",[12] and in fact the drawing on fol. 59r (Fig. 11.24) depicts parallel strings placed at the same level.

Such a solution makes much practical sense. For example, in the first flight, strings rise sharply in the first portion, while the section of the wall string along the first wall of the corner portion is almost horizontal; in fact, both ends of the corner part are placed at the same height, although the profile shows a slight bulge. Thus, the corner rise of the wall string is materialised in the second wall of the corner area, belonging to the second flight. As a result, in each flight both the free edge and the wall string ascend quickly in the initial portion, while their final portions are almost horizontal; this guarantees that the surface of the flight, seen from below, is concave, as befits a vault. Were this layout reversed, with a flat initial section and a steep final stretch, it would be convex, an extremely unusual design in preindustrial construction.

After this, Vandelvira divides the flights into several courses and each course in voussoirs. These operations are carried out in the plan and then brought to the elevations; of course, in the first and third flight, the mason should transfer measurements individually, since he is not using conventional double orthogonal projection and their elevations are not correlated with the plan by projection lines, which are used only to separate the landings from the second flight.

[11] Vandelvira himself admits that these calculations are not exact. Since a Castilian handspan equals three-quarters of a foot, 22 feet equal 29 1/3 handspans, and thus each riser measures 0.977 handspans.

[12] Vandelvira (c. 1585: 57v): *la subida más alta puede también ser arco escarzano sin que haga capialzo a la parte de afuera, por no tener con quién cumplir…* Transcription is taken from Vandelvira and Barbé 1977.

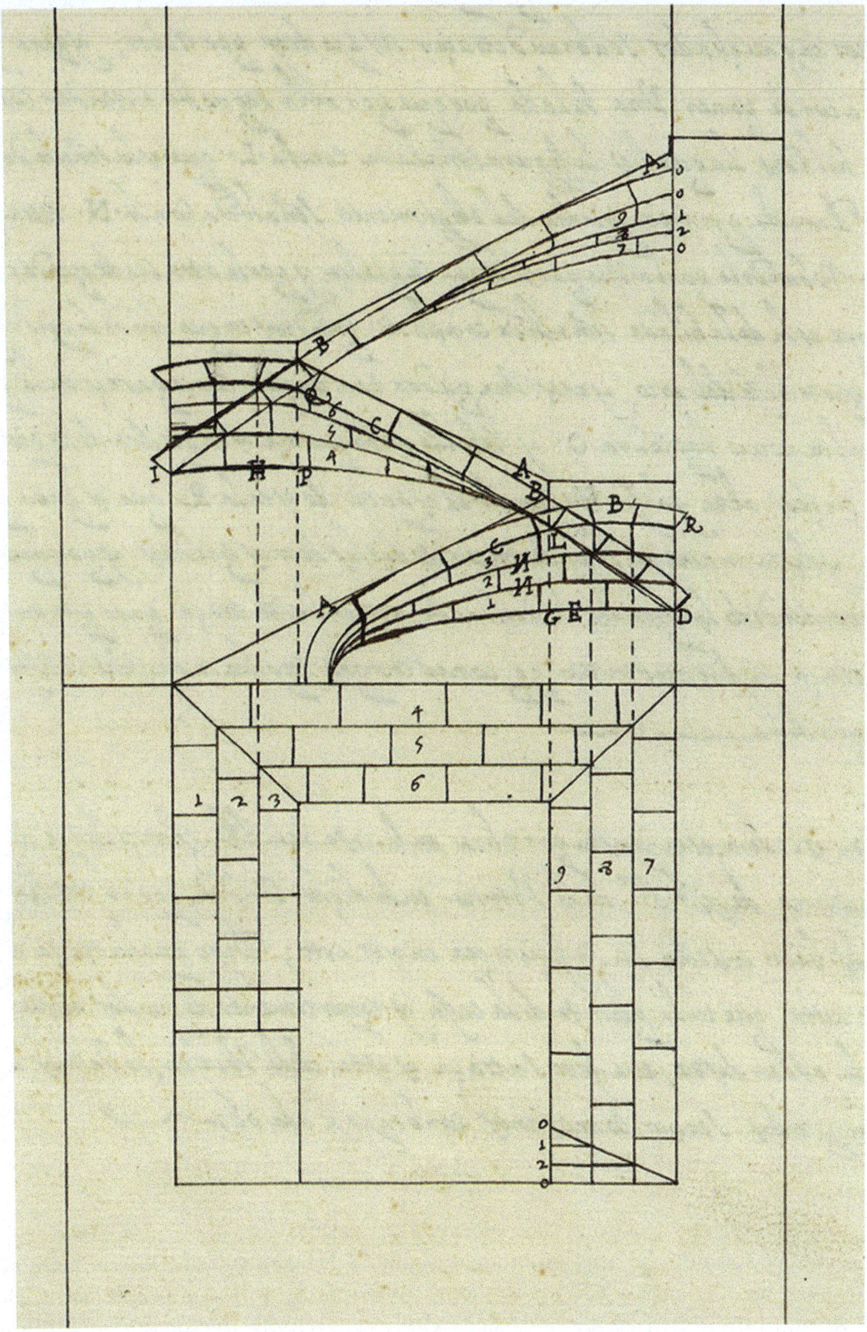

Fig. 11.23 Straight-flight staircase, curved strings ([Vandelvira c. 1585: 58r] Vandelvira/Goiti 1646: 102)

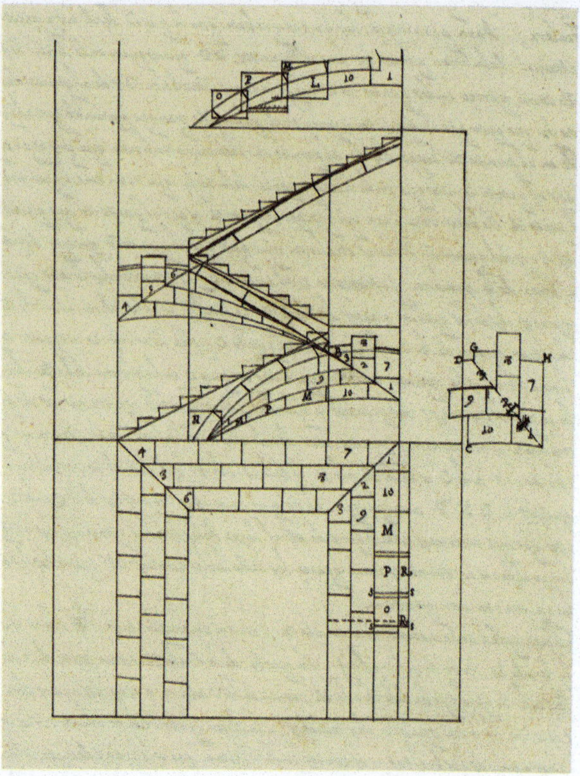

Fig. 11.24 Staircase with curved strings, landings ([Vandelvira c. 1585: 59r] Vandelvira/Goiti 1646: 104)

Next, Vandelvira gives some instructions for the dressing process, although they are not complete. He differentiates between the voussoirs in the middle section of each flight, which are *engauchidas* (warped), and those in the corners, which are *capialzadas* (raised at the head); in this context, the term *raised* is opposed to *warped*. This means that the outline of the middle section is a raised rectangle, with two corners at one level and the opposite ones at a higher level, but each side is parallel to the opposite edge. As a result, the voussoirs in the middle portion are to be dressed by squaring, and Vandelvira includes a diagram showing enclosing rectangles. In contrast, he states that the pieces in the corner sections are to be dressed using templates; however, it is not clear whether these templates are based on orthographic projection or if they take into account the difference in levels between voussoir corners.

Juan de Portor y Castro (1708: 19; see also Carvajal 2011b: 632–633, 643–644) includes a solution for this problem, essentially following Vandelvira's method; however, he adds cross-sections for some voussoirs as well as a profile along the diagonal of the corner area. As in Vandelvira's diagrams his tracing does not follow a standard double orthogonal projection; the elevations of some flights are rotated in order to show all strings in true shape. However, in another section of his manuscript, he includes an architectural drawing of a staircase (Portor 1708: 76r), following Fray Laurencio de San Nicolás (1639: 119r), drawn in standard orthographic projection. That is, still at the beginning of the eighteenth century, disarticulated diagrams stemming from the Gothic tradition were used in stonecutting diagrams since they offered clear advantages; solving this problem in a conventional orthographic drawing, with some strings seen in true shape and others shown frontally, would be most cumbersome.

11.2.2 Straight Strings

Straight strings at the wall and the free edge. Vandelvira includes an alternative to the preceding solution with straight strings. In this case, he does not mention any example of this type; however, San Nicolás (1639: 118v) brings up a stair in the convent of Saint Catherine, later known as Saint Prudentius, in Talavera de la Reina, probably built in the mid-sixteenth century, as well as another one in the convent of the military order of Saint James in Uclés. Juan de Portor y Castro mentions Talavera and the one in the Merchant's Exchange in Seville (Fig. 11.25), erected between 1609 and 1611 (Portor 1708: 15r; Carvajal 2011b: 642–644; Carvajal 2015; Pleguezuelo 1990: 32–34). Although Vandelvira had been in charge of the latter building, the straight-string staircase was built after he had left this post, and also many years after the death of Juan de Valencia in 1591, which furnishes a *terminus ante quem* for the

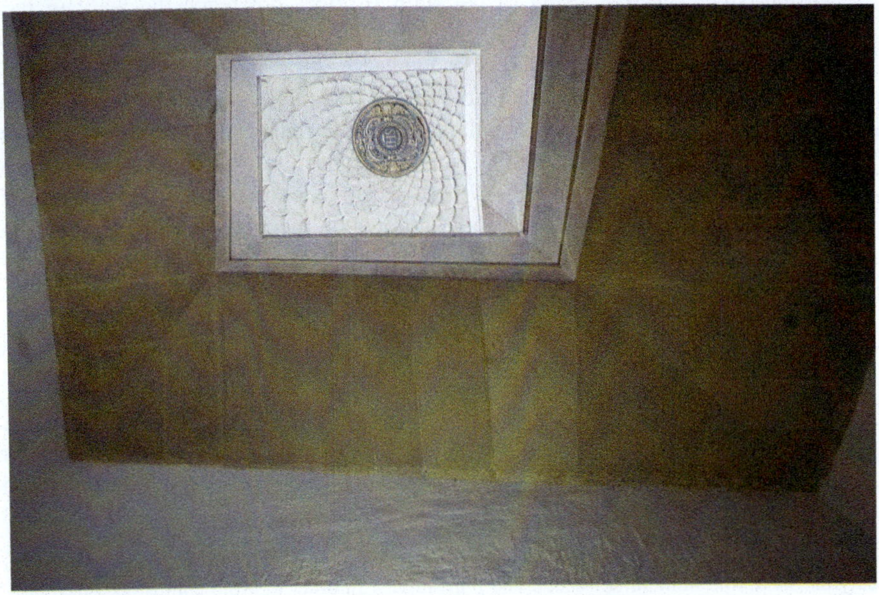

Fig. 11.25 Staircase with straight strings. Seville, Merchants' Exchange (Photograph by the author)

manuscript. Thus, Vandelvira could not have taken the Seville stair as a model; he may or may not have known the Talavera and Uclés examples. However, he may have discussed the design of a straight-string staircase with other masters in the Exchange workshop, and these masters may have executed the idea. In any case, there is no evidence to either prove or disprove these hypotheses, as far as I know.

At first sight, the only drawing included by Vandelvira in connection with this problem (Fig. 11.26) seems to be an adaptation of the preceding example to the case of straight strings; in particular, he shows all flights laterally in the elevation. There are several significant differences, however. First, the bed joints are set out orthogonally to the wall, as in the curved staircases of the Perpignan-Valencia-Toulouse genre. Second, there is another subtle but important detail. In the curved-string staircase, Vandelvira laid out the final section of each wall string so that the corners were at the same level; as a result, the wall string caught up with the free string in the initial section of the next string. Here, since the wall strings are straight, they feature the same slope in all portions of the same flight, although there are differences in slope between flights.[13] In any case, this means that the wall strings regain height on both sides of the corner section.

This fact must be taken into account to prevent the wall string from rising higher than the free edge. The procedure used by Vandelvira to guarantee this is striking at first sight. As in the curved staircase, he determines the number of steps of each flight; this allows him to construct his idiosyncratic elevations, drawing a line that shows the slope of each flight and the interior string of each flight. Next, he draws parallels to these lines, showing the thickness of the free edge. Then, he extends the lower edge of these free strings until they reach the wall. Of course, such extensions are virtual, since the free string of the staircase does not reach the wall; in fact, they are used only to construct the wall string. In particular, the extension of the second free string reaches the wall at *c*; Vandelvira draws a horizontal line from *c* until it arrives at point *G*, on the vertical line *E* passing through the intersection of the first and second free edges. Next, he draws the first wall string from the start of the stairway, *h*, passing through *G*, until it reaches the wall at *x*. For the second wall string, Vandelvira uses a similar method. He extends the third free edge until it reaches the wall at *c'*,[14] draws a horizontal which meets the free edge at *G'* and draws the second wall string passing through *x* and *G'*, until it reaches the wall at *M*.

---

[13]Vandelvira states that the stair shares its case with the preceding one. This suggests that the dimensions of the flights and the steps are identical; the drawing does not contradict this assumption. The first and fourth flights include ten steps, each rising one handspan; since a handspan equals a quarter of a yard or three quarters of a foot, the rise of the first flight equals 10 handspans, that is, 7.5 feet. The length of the flight equals the width of the well plus the width of a flight, 15 feet in total, so the slope of these flights is ½. For the second and third flights, the drawing shows six steps, so the rise equals six handspans or 4.5 feet. Since the length of these flights equals the width of the well, 10 feet, the slope equals 9/20.

[14]Vandelvira uses "c", and "G" for three different points each, perhaps to stress that these points play the same role in the first, second and third flights. In order to avoid confusion, I have used c' and G' for the second instances of these points, but the reader should be aware that the prime symbols are not used in Vandelvira's drawing.

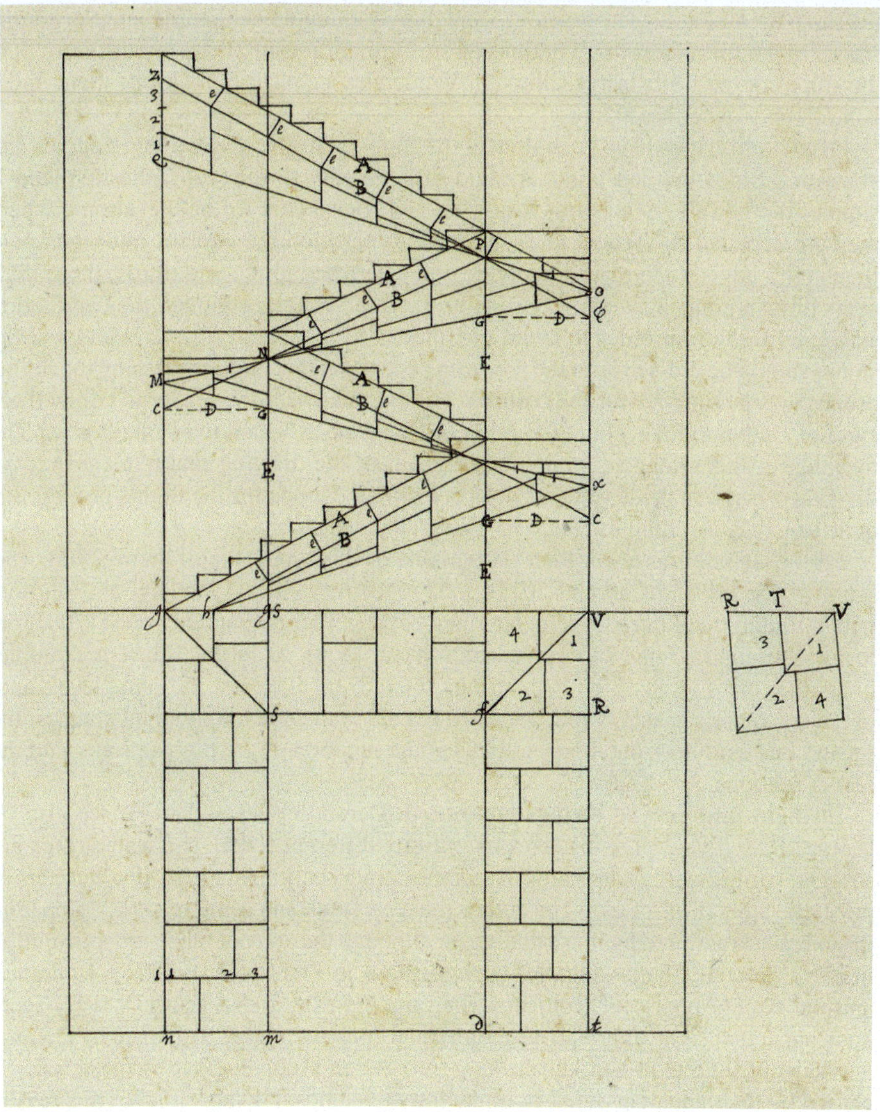

Fig. 11.26 Staircase with straight strings ([Vandelvira c. 1585: 60r] Vandelvira/Goiti 1646: 106)

At first sight, such operations resemble a magical incantation. However, there is some convoluted geometrical rationale behind them. Were all strings to feature the same length and slope, each wall string would ascend along the length of the flight plus two sides of the corner area. Since the rise of the wall string in each flight must equal the rise of the free string, and in fact, both strings reach the same level as I, N, etc. (that is, the final points of each flight) the slope of the wall string would be the same for all flights. However, in Vandelvira's example neither the lengths nor the slopes of the free edges of all flights are uniform; as a result, there are differences in the slopes of the wall strings, although Vandelvira's construction keeps them under control, up to a certain extent.

Once both strings have been drawn, the flights must be divided into courses and voussoirs. Since the bed joints are laid orthogonally to the walls, the first step is simple: the bed joints are drawn in plan and transferred to the elevations, where they are depicted as vertical lines. In order to partition the courses into voussoirs, Vandelvira advises the reader to divide one of the bed joints into thirds; these divisions furnish joints between voussoirs in the same course, although they are laid in staggered fashion in order to avoid continuous joints. Finally, he explains that the stones should be dressed as in the preceding example; this implies dressing the flights proper by squaring. As for the corner sections, he advises the reader to "draw them carefully, taking into account the point of each line at which the stone lies".[15] The meaning of this sentence is imprecise; however, the drawing makes clear that the division scheme in those sections does not follow the one for the flights proper; it is quite sensible, avoiding too large or too small stones.

Portor includes no fewer than six variants of this type. One of them (1708: 15r; see also 18r bis and Carvajal 2011b: 633–646) reproduces Vandelvira's solution, with the addition of intrados templates and bevel guidelines; another one (1708: 16r) features straight strings and curved transversal joints, as in the Talavera example; in another one (1708: 18 r), both strings seem to be placed at the same level; and, of course, other variants (1708: 19r–19v, 21r–21v) include longitudinal courses; the second one uses horizontal generatrices at the junction of the flights proper and the corner sections.

Straight and curved strings combined. Derand (1643: 429–433; see also de la Rue 1728: 151–152, 155) takes an entirely different route. The wall strings are straight, while the free edges are curved; the corner sections are treated as quarters of pavilion vaults (Fig. 11.27). The author makes a frank and acute remark: "This kind of staircase is admired by those that see it, although there is not much art or invention in it".[16] Certainly, built examples such as those in Premontré, Balleroy, Bourgueil and the Archbishopric of Bordeaux (Pérouse [1982a] 2001: 174–178, 197) seem amazing at first sight, although its geometrical construction, as explained by Derand, is relatively simple. In fact, the striking impression brought about by these stairs is caused by their apparent defiance of the laws of statics. Perhaps, it is for this reason that Derand added that "there are several [staircases of this kind] in Paris and other places, happily executed in the last years".[17]

[15] Vandelvira (c. 1585: 59v): ... *se trazarán con algún estudio teniendo cuenta en qué punto de cada línea se halla la piedra.* Transcription is taken from Vandelvira and Barbé 1977.

[16] Derand (1643: 429): *Ce sorte d'escalier donne plus d'admiration à ceux qui le voient mis en oeuure, qu'il n'ya d'art & d'inuention à le faire.*

[17] Derand (1643: 429): *Il s'en voit plusieurs à Paris, & ailleurs, executez heureusement depuis quelques années en çà.*

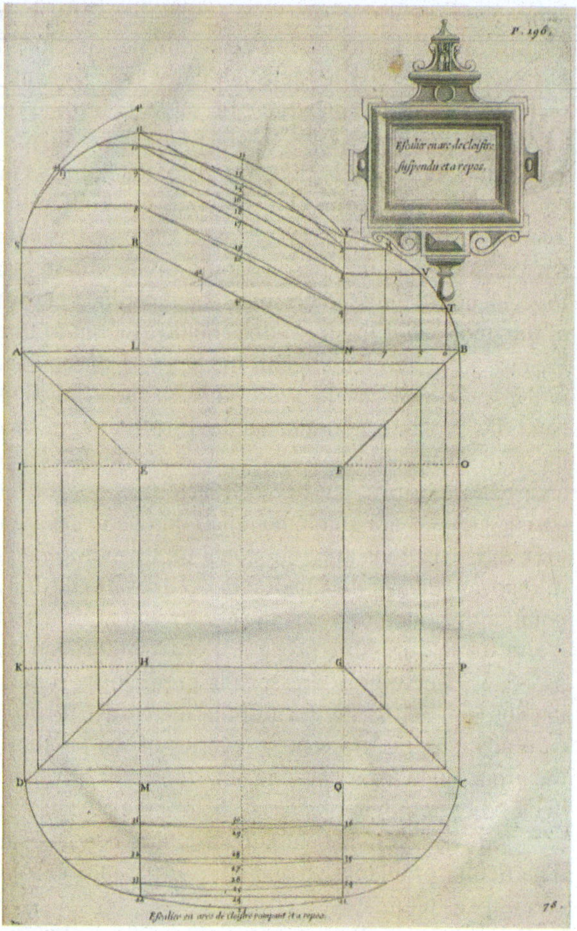

Fig. 11.27 Staircase with straight and curved strings (Derand [1643] 1743: pl. 196)

The mason should start by drawing the plan, which is quite similar to Vandelvira's schemes; in particular, Derand uses a square layout, with three equal–length flights, while the fourth side is used as the main landing, although the design may be modified easily. Next, he draws the profile of a flight, taking into account the number and rise of the steps, as well as the cross-section of the pavilion vault, in the shape of a quadrant. This also provides a section of the flights proper, although it is not strictly a cross-section since the cutting plane is not orthogonal to the axis; in any case, the rise of the flight, from the wall string to the free edge, equals its width. Derand then divides the quadrant into equal portions and transfers the divisions to the plane separating the pavilion vault section from the flights proper; from the resulting points, he draws several parallels to the profile of the vault, furnishing the first approximation of the longitudinal joints. However, Derand remarks that "it is beneficial for the solidity and firmness of this structure that the flights proper should be curved in the front, although in the back, where it meets the walls, it should be straight"[18] Thus, he advises the mason to choose an additional rise for the free edge so that it will not disturb the layout or the steps nor cause an ugly kink at the junction with the corner sections. However, since the end tangents in the sections of the pavilion vault are horizontal, and the baseline of the joints in the flights is sloping, the groin cannot be avoided simultaneously in both ends of the flight; in fact, Derand drawing shows that the bending of the joint almost avoids the kink in the upper end, while it amplifies it at the lower junction. In any case, once he has chosen an appropriate rise for the curved free edge, the mason should divide it into even parts and subtract a portion for each consecutive joint; in his example, he uses four joints, and thus the bending of the joint under the free edge should amount to three quarters of that of the edge, the next joint will feature half the bending of the edge, and so on, until the bending in the wall string reaches zero, that is, the string is straight, as planned by Derand. Next, he explains a similar method for the main landing, where the free edge and the wall string show no slope. He remarks his wish to construct a basket handle arch with no kinks, but his approach does not lead to this result. The edges of the corner sections are quadrants; their end tangents are horizontal; the end tangents of the central sections are orthogonal to the radii of a circle, and cannot coincide with the end tangents of the quadrants. As a result, the drawing shows a very slight salient point in the union between the corner and middle sections.

As for the dressing process, Derand refers the reader to pavilion vaults (1643: 344-348), where he addresses the problem by squaring. The main difficulty lies in the voussoirs crossing the groin in the corner sections, which must be dressed very carefully, starting from both severies of the pavilion vault. If the mason goes past the groin plane from one side, the stone will be ruined, since stonecutting technique is based on removing material; before the advent of modern filling techniques, any addition was excluded. In any case, this bending of the joints in the flights poses a particular problem; Derand advises the mason to use the arcs he has constructed carefully as templates.

Derand also includes two variants of this solution. One of them (1643: 434–436; see also de la Rue 1728: 152–155) replaces the quarter-pavilion vaults with corner trumpet squinches; in order to match the flights, the intrados joints are curved, so the author explains in detail how to draw the templets for them. The other variation (Derand 1643: 401, 420–423) is fit for a staircase without a well, or with a very narrow one, so the flights can rest on the wall on one side and two piers in the other (Fig. 11.28). Thus, flights are built as sloping barrel vaults, while the corner sections are full groin, rather than pavilion, vaults; Derand (1643: 25–29, 329–335) has addressed both problems before. Since there is no need to attain increased stiffness in the longitudinal direction, the bed joints are straight. The only new problem is brought about by the opening between two piers. Derand places a sloping arch between them, in order to provide support for the vault. However, rather than leaning

[18]Derand (1643: 430): *Et dautant qu'il est expedient pour plus de solidité & fermeté de ces ouurages, que cét arc de cloistre, en ce qu'il contient entre les parties de sa voûte, qui couurent les palliers, soit bombé sur le deuant, bien que sur le derriere, où il pose contre les murs, il soit reglé …*

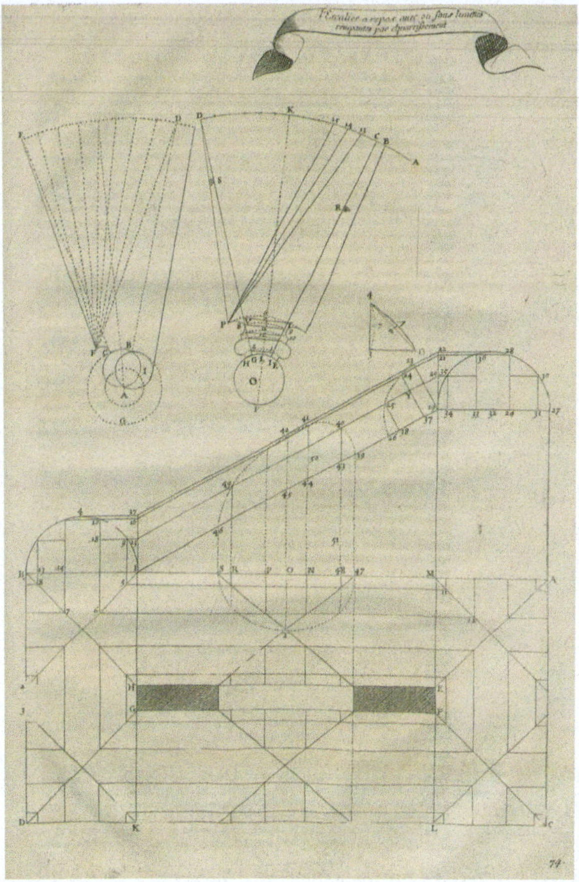

Fig. 11.28 Staircase with straight strings (Derand [1643] 1743: pl. 191)

the sloping vault directly on the arch, he uses a pair of pointed lunettes starting from the sloping arch. Derand explains the construction of the sloping arch using an affine transformation starting from a semicircle, bringing the heights of the intersections with the joints to the transformed arch; the same diagram serves as a cross-section of the lunettes.

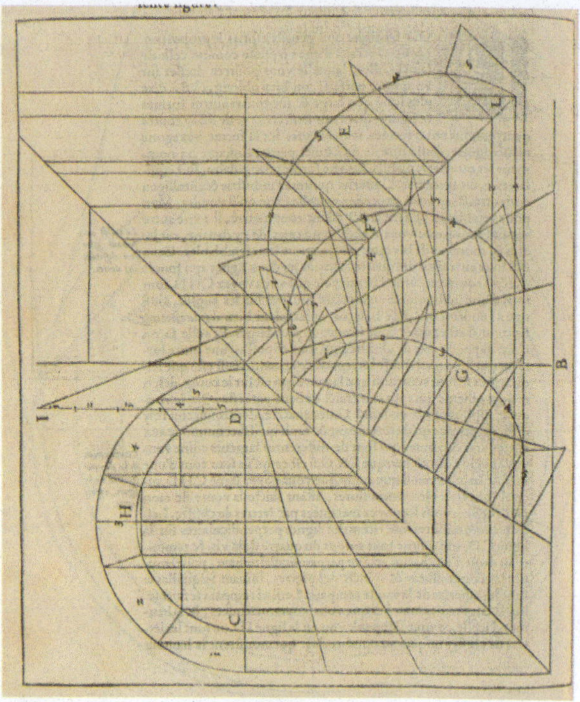

Fig. 11.29 Staircase with straight flights and radial strings (de l'Orme 1567: 127v)

11.2.3 Straight-Flight Staircases with Radial Joint Projections

Most French treatises, starting with de l'Orme, include a *vis de Saint Gilles quarrée*, that is, a staircase with three or more straight flights and two sets of joints, one parallel to the wall and the other one pointing to the centre of the staircase (Fig. 11.29). Spanish manuscripts explain similar solutions, although they do not associate them with the *vis de Saint Gilles* (Sanjurjo 2006: 2794).

After drawing the plan of the vault with the radial and parallel joints, de l'Orme (1567: 127–129) endeavours to construct cross-sections by radial joints, setting a constraint: the height of the intersections of the radial and parallel joints over the start of the radial joint (that is, the springings of the vault) should be equal for each radial joint. As a result, he uses a construction furnishing points of an ellipse to define the profiles of the joints. This procedure is similar to the one he had explained in his carpentry treatise for the intersection of two barrel vaults, alongside with the *trois points perdus* technique; thus we may surmise that Philibert means to connect the points with circular arcs (de l'Orme 1561: 13r–13v; de l'Orme 1567: 55v, 56v; see also Sect. 3.1.2). However, the problem here is different. Since the radial joints feature different spans but the same rise, the intrados joints are parallel in plan but not in elevation; thus, the resulting surface is a warped one known in the eighteenth century as *cylindroid*[19] where the intrados joints are parallel to the wall but not to one another (Sanjurjo 2006: 2803–2804). Wisely, de l'Orme does not attempt to build templates and limits himself to diagrams representing the slope of the intrados joints, as in the *vis de Saint Gilles*. Vandelvira (c. 1585: 55v, 56r) follows the same line, adding an elevation that shows clearly that interior joints need to ascend quickly and are therefore not parallel to exterior ones, as well as a host of diagrams illustrating the enclosing solids of each course. Jousse's (1642: 192–193) explanation is interesting for the extreme economy of the diagram: while most authors depict three or four flights, or at least, a quarter of the full staircase, he limits himself to half a flight. Of course, this is acceptable because all flights are equal, and the lower half of each flight may be obtained from the upper one as a product of symmetries. Derand (1643: 414–420) tries to construct intrados panels, in spite of the warped nature of the surface. De la Rue (1728: 140–145) limits himself to bed joint panels, using a more extensive version of de l'Orme's profiles, including a most detailed explanation of the centre pier and the vault support; he also focuses on the voussoirs crossing the groin between two consecutive flights.

Frézier (1737–39: III, 218) explains clearly, for the first time, the difference between a barrel vault and a cylindroid, as well as the main problem in straight staircases, remarking that

> Describing a *vis de Saint-Gilles quarrée* as M. de la Rue does, as a composite of *sloping barrel vaults, skew at both ends*, does not give a clear idea of its nature; it is necessary to add that these barrels are intrinsically irregular and completely different from ordinary barrel vaults ... the lines of both springings are straight and placed on parallel vertical planes, but they are not both in the same slanting plane ... the edges *a b* & *e f* cross at their midpoints at *m*, and therefore are not in the same plane. The reason for this is that the sides *a* & *b* of the long edge of the barrel, and those of the short one, *e* & *f*, must be placed at the same level, down at *a e* & up at *f b*, thus, the springings are unevenly inclined, in order for the shorter one to reach the same height as the longest one ...[20]

[19]Nowadays, the word "cylindroid" is used in English for different geometrical shapes, such as the elliptical cylinder or Plucker's conoid.

[20]Frézier (1737–39: III, 218): *Ce n'est pas donner une juste idée de la* vis Saint-Giles quarrée, *que de la représenter, iansi que M. de la Rue, comme un composé de* berceaux en descente, biais par les deux bouts; *il faut ajouter que ces berceaux sont d'une irrégularité intrinseque, & d'une espece toute différente des berceaux ordinaires ... les lignes des deux impostes sont bien droites, & dans des plans verticaux entre eux, mais elles ne sont pas toutes les deux dans un même plan incliné... les extrémités a b & e f se croisent a leur milieu en m, par consequént elles ne sont pas dans le même plan/La raison qui fait qu'elles se croisent, est que les extremités* a & b *du grand côté du berceau, & celles du petit* e & f *doivent être de niveau entre elles, en bas comme* a e, *& en haut comme* f b; *ainsi les impostes sont inégalment inclinées, afin que la plus courte parvienne a la même hauteur que la plus longue...*

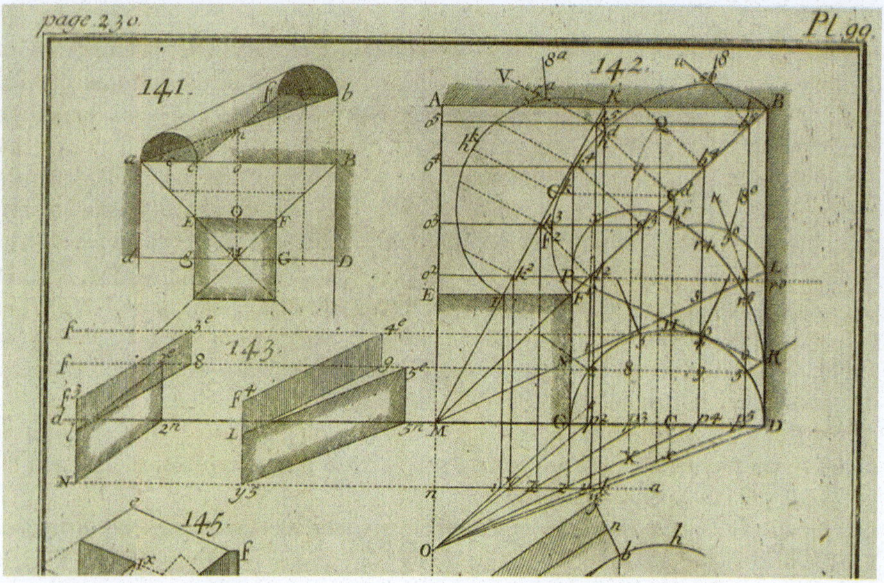

Fig. 11.30 Staircase with straight flights and radial strings (Frézier [1737-1739] 1754-1769: pl. 99)

In spite of departing from de la Rue's theory, he follows his solutions (Fig. 11.30): he uses sloping lines rather than full templates and includes a diagram to show the crossing of the springing and devotes attention to the support of the vault in the central pier.

Part III
Discussion

Chapter 12
Problems

Abstract The preceding chapters present stereotomy as an isolated science, but this vision does not match the real nature of the discipline. To avoid this, a few short essays about the connections of stonecutting with other branches of knowledge are included in this chapter. First, it addresses the influence of external factors in stereotomy, including functional constraints, mechanical behaviour, aesthetical ideals, patronal whim and artisanal pride. Next, it deals with the social standing of stonecutters or stereotomy theorists and the status of stereotomy as a branch of knowledge. The following section addresses the geographical distribution and the historical evolution of stereotomy, putting forward a polycentric vision of the origins of stereotomy and stressing its dependence on the availability of materials and architectural constraints. Then, it addresses the connections of stereotomy with learned science, in particular Euclidean and practical geometry, cosmography, perspective and gnomonics, as well as its links with the oral traditions in the artisanal trades. Finally, the chapter deals will the role of stonecutting in the formation of descriptive geometry, showing that this science, despite its purpose as a general tool, inherited a wide range of problems from stonecutting, as a result of the solid, spatial nature of this technology.

12.1 Reason and Caprice in Stonecutting

The section of Philibert de l'Orme's treatise dealing with stereotomy (1567: 65r-67r) starts with a parable. A gentleman has inherited from his great-grandfather a house with a typical irregular medieval plan (Fig. 12.1). He wishes to enlarge the mansion and adapt it to Renaissance taste. Rather than tearing down the existing wings, a legacy of the patron's ancestors, a good architect should add parts here and there to regularise the layout and solve practical problems. As a result, he needs to enter rooms through their corners or place studios outside the house walls or over a river. These constraints demand the use of skew arches, trumpet squinches, arches opened in curved walls, and of course staircases; this shows clearly the practical function of stonecutting.

Thus, in the next two books of his treatise, Philibert explains one by one all the constructive elements needed to carry out such programme, starting with sloping

J. Calvo-López, *Stereotomy*, Mathematics and the Built Environment 4,
https://doi.org/10.1007/978-3-030-43218-8_12

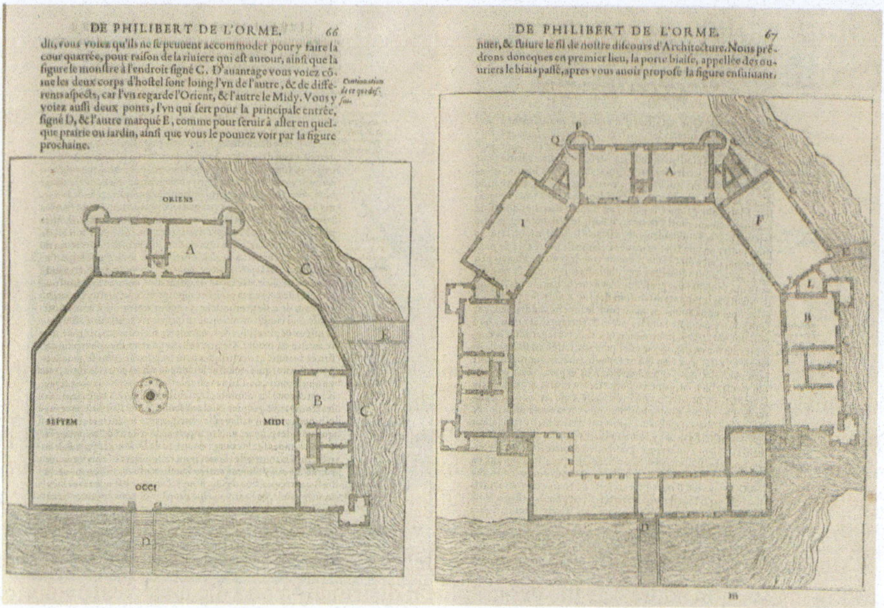

Fig. 12.1 House before and after refurbishing (de l'Orme 1567: 66r, 67r)

barrel vaults, rere-arches, skew arches, corner arches, arches in curved walls, trumpet squinches, vaults and staircases (de l'Orme 1567: 67v-128v). However, his repertoire also includes some pieces that cannot be justified by practical reasons, such as spiral vaults and sail vaults quartered in the shape of a fan vault (de l'Orme 1567: 112v, 119r-119v). Quite the contrary, these elements require tiresome tracing and dressing processes, just to solve problems that could be addressed with ordinary domes or sail vaults.

This shows that masonry construction is subject to the opposing forces of functionalism and caprice, practicality and whim. For example, the scriptorium of the monastery of Sanahin, in Armenia, built in 1063 (Strzygowski 1918: I, 67–68; II: 822; Cuneo 1988: 290–294; Maranci 2001: 149–151; Calvo et al. 2015c) features a complex and apparently unnecessary combination of different elements (Fig. 12.2). It is laid out on a square plan; however, a second, rotated square is inscribed in the first one, with its corners placed at the midpoints of the outer one. For no apparent reason, the triangles between the two squares are covered by different elements; two opposite triangles are spanned by trumpet squinches, while the other two are overlaid with halves of pavilion vaults, cut through their diagonals. Four arches materialise the inner square; typical Armenian pendentives, with planar bed joints, are laid over them. Using flat bed joints instead of the conical joints used in the West may seem a simplification. However, the intersections of these planar joints with the spherical surface of the pendentive lead to falling and rising apparent joints, rather than the horizontal ones used in the West; as a result, the thickness of the upper course in

Fig. 12.2 Scriptorium vaults. Sanahin, monastery (Photograph by the author)

the pendentive is variable, since it must provide a horizontal support for the vault over the pendentives. The vault itself is quite remarkable. At first sight, it looks like a typical Western octagonal vault made up from cylinder portions; however, close inspection and photogrammetric analysis have shown that its profile is raised, the section approximates a pointed arch, and the intrados joints go down and up in each severy. This last detail can be justified by the lack of trees in Armenia, which leads builders to use almost exclusively ashlar masonry and reduce centring and formwork to a minimum. These constraints may justify the curvature in the intrados joints of the pendentives and the vault severies; in this way, voussoirs tend to push towards the slightly lower midpoint of the course, rather than fall outside the wall, reducing the need for formwork, centring and shoring. In any case, there is no practical justification for the simultaneous use of trumpet squinches and half-pavilion vaults in equivalent positions.

A typical study case on functionalism versus caprice is provided by Gothic vaulting, which evolves from utmost rationalism in the twelfth and thirteenth centuries to the whimsical designs of Late Gothic (see Sects. 10.1.1–10.1.5). Any analysis of this issue must take into account that the word "functionalism" has different meanings in this context. Viollet-le-Duc advocated structural functionalism for High Gothic vaulting; in fact, he paid almost no attention to later, "degenerate" periods. However, Choisy's stance is slightly different; he refers to strictly constructive, rather than structural, rationalism:

> Blois, Saint-Germain do not present vaults other than those built by the Gothic system, "in the French fashion", as Philibert de L'Orme puts it; the only innovation is the use of brick in the severies. Builders use, as in the Middle Ages, forms that are easily built. It is not until the times of Philibert de l'Orme that the problem is inverted: builders arbitrarily impose upon themselves a more or less complex kind of vault and adapt to it a more or less expensive bond.[1]

Moreover, the Northern Gothic quest for light was based in part on functional reasons such as the dim light of the North; it also fostered a rational system of piers, responds, ribbed vaults, flying buttresses and vertical buttresses, while Mediterranean Gothic, either in Southern France, Spain, Italy or Cyprus, eschews large high windows. However, the search for light was also driven by theological reasons (Von Simson [1956] 1962: 55–58); when carried to the limit, it gave birth to glazed triforia. In this striking solution, builders invert the slope of aisle roofs in order to allow light to reach the triforium. Such roofs direct water against the wall, exactly at the base of the triforium, which is treated with stained-glass windows, built with a three-level structure of iron, lead and glass; of course, Gothic builders had no access to twentieth-century sealants. Thus, a non-optimal (to say the least) constructive solution is implemented in order to materialise the (Northern) Gothic ideals of light and linear consistency.

Further, the effect of World War I shellfire in Reims and Soissons Cathedrals showed that in some cases, severies could stand without supporting ribs, casting some doubts on Viollet's theories (Gilman 1920; Abraham 1934; Coste 2003); this opened the way for a mainly aesthetical understanding of Gothic architecture (Frankl 1962), which was in any case coldly received by other historians (Barnes 1965; Branner 1968). At the end, a century of debates on the structural behaviour and the general function of Gothic vaulting was summarised by Jacques Heyman (1995: 54), as we have seen at the start of Sect. 10.1 Paradoxically, what was highlighted by these debates is that Gothic vaulting, and in fact the entire Gothic construction system, is functional at many levels,—structural, geometrical, constructive or aesthetic,—and each school has stressed one of them. The same idea can be extrapolated to other architectural elements: their form is subject to structural, constructive, functional and aesthetical constraints; none of these, taken individually, determines its final shape. A designer should take all these constraints into account at the same time, although the wide range of reasonable solutions for most architectural elements shows that the designer retains a certain degree of creative freedom.

Using the rib vault again as an example, the relative weight of these constraints shifted over time. In the twelfth century, before the generalisation of the *tas de charge*, the sexpartite vault allowed the construction of square vaults, avoiding the difficulties of acute angles between ribs, while the central transversal rib in each vault

[1]Choisy (1899: II, 704): *Blois, Saint-Germain ne présentent que des voûtes bâties à la manière gothique, 'à la mode française', comme disait Ph. Delorme: la seule nouveauté consiste dans l'emploi de la brique pour les panneaux de remplissages. On s'attache, comme au moyen âge, aux formes qui se construisent simplement/Ce n'est guère qu'à l'époque de Ph. Delorme qu'on commence à renverser le problème: s'imposer arbitrairement une forme de voûte plus ou moins complexe, et lui adapter un appareil plus ou moins coûteux.*

was supported by intermediate piers connected with lesser, square vaults in the aisles. Later on, the use of the *tas de charge* allowed acute junctions between transverse and diagonal arches, fostering rectangular vaults in the nave. From the mid-thirteenth century on, the tierceron allowed the reduction of the size of the severies, simplifying formwork at the expense of dressing effort in ribs and additional centring. However, in this period constructive rationalism started its decline with glazed triforia, requiring gutters and increasing the need for maintenance. A further step in this path was the introduction of curved ribs in Late Gothic, in particular in the German Empire. Since this kind of member cannot withstand compressive efforts, it must be supported by the severies, reversing the constructive logic of Gothic architecture; in some occasions, curved ribs hang from a higher structure, such as the trusses in roofs.

This tension between practical, sensible constructive solutions and the subjection of design to aesthetical will or outright whim persisted in the Renaissance and Baroque periods. As we have seen, de l'Orme advocated the use of skew arches or trumpet squinches to address practical problems, but there is no conceivable practical justification for the use of spiral vaults; as remarked by Rabasa (2003: 1689), these elements, relatively frequent in sixteenth-century Spain, are "an early and atypical example of this unnecessary complication … perhaps … displays directed to the stone-cutting guild itself". Both Fray Laurencio de San Nicolás and Juan Caramuel remark the whimsical uses of corner and skew arches. While San Nicolás shows some hesitation, stating that "difficulties arise from the location where arches are to be built, sometimes as a result of construction constraints; in other cases, windows are placed by caprice, for example in a corner. I do not approve that, but neither do I condemn it",[2] Caramuel declares his admiration of whimsical stereotomic solutions in bombastic terms: "the architect did not design it because it was necessary for this spot, since this chapel is erected at a place without constraints, but rather to leave a testimony of his ingeniousness with an extraordinary piece".[3]

Rabasa (2009a) has charted other unnecessary complications in stonecutting. As we have seen in Sect. 9.4.2, ellipsoidal and oval vaults pose different problems. In elliptical-plan vaults all voussoirs in the same quarter are different; edges and corners, which are not orthogonal, may be easily damaged when dressing, moving or placing voussoirs. In contrast, using an oval plan and sections for the vault, there are only two different types of voussoir, while angles between faces are orthogonal. However, new difficulties appear; the junction of two sections over the longer axis of the plan poses tangency problems, as we have seen in Sect. 9.4.2. Notwithstanding all that, both ellipsoidal and oval ashlar vaults were used frequently during the Early Modern

---

[2]San Nicolás (1639: 69v): *De los sitios donde se han de hazer los arcos resultan dificultades, vnas vezes por pedirlo assi la obra, otras por elegir vna ventana por gala, como lo es elegirla en vna esquina. No la aprueuo, mas tampoco la reprueuo.*

[3]Caramuel (1678: II–VI, 21) … *que hizo assi el Architecto, no por necessidad del sitio (que cae esta Capilla en lugar libre) sino solo por su gusto, para con una obra extraordinaria dexar testimonio de su ingenio.* See also Rabasa (1994: 152).

Fig. 12.3 Flat vaults over the Grand Colonnade. Paris, Louvre (Photograph by the author)

period,[4] since they furnish a compromise between central and longitudinal vaults; an apparently whimsical choice hides much geometrical skill (Potié 2005: 75–76).

Other whimsical pieces mentioned by Rabasa (2009a: 66–68) are the flat vaults proposed by Abeille and Truchet in 1699 (see Gallon 1735: 159–164). Flat vaults had been built before. As we have seen in Sect. 4.4.2, a simple solution is to revolve the cross-section of a lintel divided into voussoirs. This operation generates a circular-plan vault, which may be cut to adapt to a square plan, as in, the medallions in the ceiling of the gallery behind the columnar screen in the east façade of the Louvre (Fig. 12.3). If necessary, pendentives are added to increase its strength, as done in the space below the high choir in the main church of the Escorial complex. However, Abeille and Truchet both took a different road, using wedges laid in two orthogonal directions. Their advantages over traditional solutions are not clear. The lack of practical sense in Abeille is evident since the upper surface of the ensemble shows holes which make walking on it difficult, while the lower face is flat; Abeille meant to cover the holes with ceramic tiles. Truchet tried to solve the problem with his fanciful curved surfaces; however, the implementation of Abeille's scheme in the cathedral of Lugo or the Casa de Mina de Limpia in the Canal de Isabel II near Madrid solved the problem simply by reversing the vault and placing the flat face above (Rabasa 1998; de Nichilo 2003).

Up to this moment I have been talking about these capricious pieces in general terms, but the forces fostering these solutions are quite different. On the one hand, patrons of all ages have shown a penchant for rare, strange or even exotic pieces,

[4]Such vaults were used quite frequently in the Baroque period, of course, but they were usually built in brick.

such as *Astwerk* in Late Gothic, nymphaea in the Renaissance, or follies in the Enlightenment. This peculiar psychology can explain such pieces as *Netzgewölbe* with curved ribs, spiral vaults, or the interwoven ribs in the entrance pavilion at Anet. We should remember that most of these works were only accessible to the patron, his family or his friends. The first and largest sixteenth-century Spanish spiral vault is placed at the entrance to the sacristy of Murcia cathedral, where only chapter members and their aides could see it (Calvo et al. 2005a).

Along with patronal whimsy, another force driving towards these pieces was artisanal pride. Other spiral vaults, such as those at Plasencia cathedral or the church of Santa María de la Coronada in Medina Sidonia are built in even more secluded parts of the structures, where only masons and the maintenance staff could see them (Rabasa 2003; Sanjurjo 2015: 95, 161–163, 181); perhaps they were conceived as a proof of skill for other masons. Another factor was the selection process of master masons of cathedrals, monasteries or King's Works in Spain from the sixteenth century on, which was based usually in an *oposición*, a contest between masters, where they were expected to present their qualifications. In many cases, these *oposiciones* led to bitter disputes between figurative artists and construction specialists, who boasted about their building skills. As we have seen in Sect. 2.2.5, an infrequent stonecutting problem helped Juan de Aranda Salazar to win an *oposición* in Granada cathedral, although he soon left for Jaén. In another *oposición*, also in Granada, Francisco del Castillo argued that he knew "stonecutting tracings that other contenders do not know since they have not put them into practice" (Moreno 1984: 358–359), as we will see in Sect. 12.2. It comes as no surprise that builders manoeuvred to introduce stonecutting exercises in other contests, while figurative artists tried to reduce their importance or submitted artistic works, such as monstrances, as legitimate stonecutting pieces (Blasco 1991: 166). That is, stereotomy was used as a showcase of the geometrical knowledge of the masons; of course, this leads us to another open problem, the social standing of stonemasons and their lore; I will address it in Sect. 12.2.

Although this may seem surprising, the presence of such elaborate problems as the scalene ellipsoid or the *arrière-voussure de Marseille* in nineteenth-century stereotomy treatises (see Sects. 2.4.3, 2.4.4, 7.3, and 9.4.2) derives from similar motivations: to exhibit the sophisticated geometrical knowledge of the author and, by extension, his school or the entire corps of military engineers, showcasing the power of science. Again, this takes us to other problems, in particular, the connections of stonecutting with learned science and its role in the formation of descriptive and projective geometry, which we will see in Sects. 12.4.1 and 12.5.

However, we should never forget that stonecutting practice is always subject to the tension between the opposing forces of practicality, economy, standardisation and rationality on the one hand, and patronal whim, artisanal pride, corporative scheming and excessive scientism on the other. The vault spanning the entire first storey of the Town Hall in Arles, designed by Jules Hardouin-Mansart and built by local masons in 1673–1674, offers a remarkable case study on these issues (Pérouse [1982a] 2001: 116–117; Tamboréro and Sakarovitch 2003; Etlin 2009; Fallacara et al. 2011). While the general layout shows much practical sense and structural skill, the use of a hidden central arch to support an exceptional low-rise, wide-span vault required by the use and placement of the entrance hall, the use of different shapes and combinations of lunettes, and the U-turns in the layout of joints reveal a desire to showcase the skill of the designer and builders.

12.2 The Social and Epistemological Standing of Stonecutting

In the introduction to *Cerramientos y trazas de montea*, Ginés Martínez de Aranda states that "if craftsmen are not rich, in a good position and well spoken, they are not granted authority in accordance to the effort of their studies, even when they assure the reader they are well-judged, and the public does not believe they know their lore".[5] Thus, he is demanding at the same time a fair social consideration for masons and a respectable intellectual standing for stonecutting knowledge. This is not mere rhetoric. We have seen in Sect. 2.2.1 that Catherine Wilkinson presented de l'Orme as the archetype of the "new professionals", an emergent group in sixteenth-century building trades, opposed at the same time to the pure artisans of medieval origin and the figurative artists that dominated the Central Italian scene. In contrast to traditional masters, the new professionals knew the emerging architectural literature fostered by the printing press and the smallest details of the classical language of architecture; in opposition with figurative artists, they mastered the technicalities of construction and knew how to handle workers. In the words of de l'Orme himself (1567: 81r).

> … as with some architects and masters, who do not understand the practice of full-scale tracings and geometry, and say when they find some constrained place, and they see some strange structure that is used to address the constraint, that it is not necessary to be so fanciful, that this is a masons' work. It is necessary to say, then, that by their own confession, masons know more than these architects, which is against reason, since the architect should be learned in order to direct and explain all works to master masons; however, in several countries the cart (as they say) drives the oxen, that is, in many places masons command and teach the masters. This should be said without offence to learned architects … and not those who abuse the patrons to rise to a status which does not fit their knowledge, and who know nothing, except what they have learned from master masons….[6]

---

[5]Martínez de Aranda (c. 1600: n. p. [ii]): … *los mismos artífices aunque prometan su prudencia si no son ricos y muy favorecidos y bien hablados no pueden alcanzar autoridad conforme a la industria de sus estudios para que se crea que saben aquello que profesan* …. The passage is included the second page of a prologue with the title "To the reader", written in three unnumbered pages at the beginning of the manuscript.

[6]De l'Orme (1567: 81r) … *voire de certains architectes et maîtres, qui par faute de n'entendre la pratique des traits, et la géométrie, disent quand ainsi ils rencontrent aucuns lieux de contrainte, et voient quelque étrange structure y être accommodée, qu'il n'était besoin de s'y amuser, et que c'est ouvrage de maçon. Il faut donc dire par leur confession, que les maçons savent plus que tels architectes, qui est contre raison, car l'architecte doit être docte pour bien commander et ordonner toutes oeuvres aux maîtres maçons; mais aujourd'hui en plusieurs pays, la charrette (comme l'on dit) conduit les boeufs, c'est-à-dire, les maçons en plusieurs lieux gouvernent et enseignent les maîtres, qui sera dit sans offenser les doctes, … et non ceux qui abusent les seigneurs pour se vouloir mêler d'un état q'ils n'entendent, et n'en savent autre chose, sinon ce qu'ils en ont ouï et*

Fig. 12.4 Capital in the cloister. Girona Cathedral (Photograph by the author)

Laboratores. However, the road leading to the formation of this group had been a long and difficult one. In medieval society, the borders between the categories of *bellatores* or noblemen, *oratores* or clerics, and *laboratores,* which included the rest of the population, were almost impermeable. A capital in the cloister of Girona cathedral shows two stonecutters dressing a stone under the vigilant gaze of a bishop (Fig. 12.4); after all, this word derives from the Greek *episkopos*, that is, "supervisor". This reflects, of course, the point of view of the clerics; it is easy to surmise that civil patrons held the same opinion. Such a state of events may explain the scarcity of drawings and tracings, as well as information about designers and technical supervisors in the Romanesque period. However, simple works such as small rural churches and dwellings may be built without specific means of formal control, but it is not easy to envisage the construction of such churches as the cathedral of Santiago de Compostela, Saint Sernin in Toulouse, Cluny III or the great Rhineland cathedrals without some kind of prior conception and the subsequent supervision.

Thus, such a situation was already outdated in the twelfth century; from this moment on, drawings on parchment, tracings in floors or walls and names of "architects" are increasingly frequent, regardless of the formal style of constructions. An early example of these named supervisors is Lanfranco, who is shown in a miniature inspecting work at the Romanesque cathedral of Modena; another one was Maestro Mateo, the artist to whom was entrusted the completion of the Cathedral in Santiago de Compostela. However, this evolution was not easily accepted; Nicolas de Biard, a cleric, showed his surprise and contempt in two well-known passages:

appris des maîtres maçons. Probably, De L'Orme is thinking about Primaticcio, who took control of Fontainebleau's works when Philibert was expelled from the direction of the Royal Works, or Serlio, who had been unable to control Gilles le Breton in the Salle de Bal in this palace. See Potié (1996: 42–43).

> In these great buildings it is customary to have a main master who gives orders only by word, and works with his hands seldom or never; however, he receives a higher salary than other workers…[7]

and

> The masters of the masons, with a bar and gloves in hand, say to the other workers "Cut this stone here", they do not work and receive a larger compensation; this is what many modern prelates do.[8]

This rich text tells us that the figure of the professional supervisor, as opposed to bishops and lords, first emerged in larger buildings and that these master builders were extracted from the same social medium as "the other workers"; that they instructed the others on how to cut stone and still worked themselves on some rare occasions; and that these changes were seen with surprise and with some disdain by contemporary society.

At the same time, a crucial innovation resurfaced, the architectural drawing on parchment. Slowly, the use of drawings allowed the differentiation of the functions of the designer and the supervisor. On many occasions, designers were required to spend a few months of the year at the working site or visiting the works a number of times. This is, of course, a first step in the separation of the designer and the supervisor; between site visits a lower-level supervisor, often chosen in agreement with the designer, ensured the correct execution of his plans. In the fifteenth and sixteenth centuries, contracts or payments for drawings not mentioning supervising duties were recorded with increasing frequence; in many of these cases, work was supervised by another master, and thus the differentiation between supervision and design was complete.

As a result of these developments, masons proper were slowly relegated to a secondary position in the evolution of stonecutting techniques. Significantly, the earliest documentary evidence for a large artisanal association, the *compagnons du devoir*, arises in the seventeenth century (Truant 1995) when the literature in the field was written mainly by clerics. At this moment, their historical task was accomplished: masons and Renaissance architects with artisanal instruction had created the full repertoire of classical stereotomy and most of its methods.

In any case, the organisation itself and some colourful traditions may date from earlier periods. One of these was the original *Tour de France*: apprentices were expected to travel around France, working in their crafts both to support themselves and progress on their training. After this, they prepared a *chef d'oeuvre* (see Sect. 3.1.4), usually in the form of an exquisitely crafted model, in order to be accepted as *compagnon fini*. In spite of many attacks against it, such as the *Loi Le Chapelier* in

---

[7] De Biard, attr. (c.1250: 30r): *In istis magnis edificiis solet esse unus magister principalis qui solum ordinat ipsa verbo, raro aut nunquam apponit manum, et tamen accipit majora stipendia aliis.* Transcription taken from Mortet (1906: 268).

[8] Sermon by Nicolas de Biard: *Magistri cementatiorum, virgam et cyrothecas in manibus habentes, aliis dicunt: "Par ci me la taille"et nihil laborant: et tamen majorem mercedem accipiunt, quod faciunt multi moderni prelati.* Transcription taken from Mortet (1906: 267–268).

the French Revolution, outlawing both guilds and *compagnons*, the organisation has survived to our times. It promises a thriving future, offering vocational instruction and even university credits in collaboration with another historical institution, the Conservatoire national des arts et métiers; it also manages an excellent specialised bookshop and publishing house, which has launched an ambitious *Éncyclopedie des métiers* with a comprehensive treatise on stonecutting (Compagnons du devoir: 2001–2007).

It has been stressed in recent decades that documents such as those of Villard and the geometrical, topographical and constructive schemata in fols. 18v-23r, drawn by a different master, known has Magister II or Hand IV, do not stem from the learned tradition, but rather from an oral one.[9] Such assertion sheds light on the purpose of these sheets, although it brings about the paradox of a written and drawn source standing for the oral tradition. Leaving aside Villard's drawings, in Hand IV's schemata we may surmise that actual instruction was, in fact, some kind of "learning by doing". In other words, the texts in the diagrams, such as *ar chu tail om vosure besloge*, play the role of headings in a textbook, while the master gave actual explanations through real practice with stonecutting or surveying tools, complemented by some lost verbal explanations. What remains in the sheets drawn by Hand IV is the shadow of this practical instruction, in the form of captions and mnemonic diagrams (Beffeyte 2004: 105; Potié 2008: 150).

The other key source on the instruction of medieval masons, Roriczer's booklets, reflects oral traditions in a completely different way in passages such as this one:

> Then make the square equal in size to the preceding; divide [the distance] from *a* to *b* into two equal parts and mark an *e* [at the midpoint]. Do the same from *b* to *d* and mark an *h*; from *d* to *c* and mark an *f*; from *c* to *a* and mark *a g*. Then draw lines from *e* to *h*, *h* to *f*, *f* to *g*, and *g* to *e*, as in the example of the figure drawn hereafter.[10]

This seems to reproduce the master's verbal explanations lacking in Hand IV's schemata. The accompanying diagrams are, as in Villard's portfolio, a reflection of actual tracings prepared during instruction, although in some cases, they seem to be divided sequentially, showing a number of phases in the execution of the tracing. This oral transmission was not haphazard or informal; the episode of the Ratisbone statutes (Frankl 1945: 46–47, 50; Roriczer, Schmuttermayer and Shelby 1977: 46–61; see also Sect. 2.1.2) banning the instruction of those who were not formally admitted into the apprenticeship system regulated by the lodges, should be proof enough. In the Renaissance, these barriers were lowered, mainly as a result of the emergence of the printing press, but the oral tradition still played an important role in the shaping of didactic methods, as we will see in the next paragraphs.

---

[9]See Beffeyte (2004) for a refreshing attempt to reconstruct such instruction method as an alternative to a standard scholarly study.

[10]Roriczer (1486: 3r): *Darnach mach dy vyervng gleich jn der voringen grosz vnd tail vom a pisz avf das b jn czway gleiche tail da seez ain e. Desgleichen vom b pisz czvm d da mach ain h vnd vom d pisz czvm c da mach ain f deszgleichen vom c pis czum a da mach ain g. Darnach czvich ain liny vom e jn das h und vom h jn das f vom f jn das g vom g jn das c des ain exempel jn der nach gemachten figvr.* Transcription and translation taken from Roriczer/Schmuttermayer/Shelby (1977: 84–85).

New professionals. However, at this moment another crucial change arose. The Renaissance was born in Central Italy; its architecture was built mainly in brick. While palace façades were overlaid with a not-so-thick veneer of rusticated stone, vaults were seldom executed with ashlar masonry. In other words, construction techniques were relatively simple. An artist with limited experience in actual construction, Michelangelo Buonarroti, called the model of Saint Peter by Antonio de Sangallo the Younger, an able builder, trained in the craft since his youth, as "meadow … for sheep and oxen".[11] Michelangelo did not actually displace Sangallo, who stayed in command of the works until his death, but his formal solution was finally adopted, although it was corrected by Giacomo della Porta, who introduced a pointed dome to reduce the thrust of the vault.

In France or Spain, figurative artists met with stronger resistance (see Sect. 2.2.1 about Gilles le Breton, Serlio and de l'Orme). Jacopo Torni and Pedro Machuca, painters, or Diego de Siloé, Jerónimo Quijano and Alonso de Covarrubias, sculptors, held key posts in Granada, Murcia and Toledo, but this is about all. The rest of the significant figures in the sixteenth-century Spanish architectural scene, such as Diego de Riaño, Martín de Gaínza, Hernán Ruiz II, Hernán González, Andrés de Vandelvira, Rodrigo Gil de Hontañón, Juan de Álava and Francisco del Castillo, were trained as masons.

Castillo's personality is fascinating; in an *oposición* for the post of master mason of Granada cathedral, held in 1577, he argued that:

> It is well known that I have worked in the most honourable place in the world … that is, in Rome, in a most rich building made by Pope Julius III, whom I served for three years … About theory and speculation, I stayed for nine years in Italy where this art flourishes, researching and measuring old buildings, practising with the most skilled masters in this nation and ours, studying old and modern books about this field, where everything is written in Italian …[12]

but also that

> The practice and skill I have in commanding officials and workmen of this works, since they respect me as an experienced man, because I know how to extract stone from the quarry, how to put them in carts and bring them to the worksite, even when they seem so large that no human force may transport them, how to dress them sparingly and economically in order not to waste the funds of the cathedral fabric, preparing machines in order to elevate materials to the top of the buildings, how to make templates and inverse templates, templets, arch squares, and other bevels and stonecutting techniques that my contenders do not know, since they had not practised nor exercised them …[13]

---

[11] Vasari (1568: III, 750): *… un prato … per le pecore & buoi …*

[12] *… como es público y notorio que yo las he hecho en el más honroso lugar del mundo … que fue en Roma en un edificio suntuosísimo que hizo el papa Julio Tercio a quien yo serví tres años … En lo que toca a la teoría y especulación tuve nueve años de ejercicio en Italia donde florece este arte viendo, escudriñando y midiendo las cosas antiguas, practicando con maestros habilísimos de aquella nación y de la nuestra, estudiando los libros modernos y antiguos que tratan de esta facultad que todo está escrito en lengua italiana …* (Archive of Granada Cathedral (AGC), file 667, 314, 6 and 7, fol. 118–119). Transcription is taken from Moreno (1984: 359–360).

[13] *Lo otro por el manejo y práctica que tengo yo en el gobierno de los oficiales y ministros para las dichas obras que por hombre ejercitado y experimentado me obedecen y respetan, porque yo sé*

This comprehensive curriculum was similar to the training of Philibert de L'Orme, the paradigm of the "new professionals" for Wilkinson ([1977b] 1986). In his youth, he had been in command of a substantial number of men in the fortifications of Lyons; later on, he stayed in Rome measuring ruins, where he attracted the attention of Cardinal De Bellay and landed his first important commissions. Thus, the "new professionals" were in the middle of a battlefield. They contended with ordinary masons on one side and with Italian or Italianate figurative artists on the other. Humanists such as Ronsard mocked their lack of instruction in classical matters, while Cristóbal de Rojas had to withstand the scorn of generals.

This situation led de l'Orme to an attempt to exalt the origin of stonemasonry by trying to connect it with Vitruvius and Euclid, although he remarked that "I have never heard that any architect, either ancient or modern, has written anything about stone-cutting tracings".[14] Taking this into account, Ginés Martínez de Aranda put forward a startling explanation in order to justify the classical pedigree of stonecutting, and indirectly its social acceptance: "Since a perpetual silence on these stonecutting traits has been kept, I reckon that the great ancient artisans have left their explanation aside so that future ones may practise this lore preparing tracings and models".[15]

In spite of these proclamations, stonecutting techniques were taught mainly at the worksite, adding practical examples to oral or written explanations. De L'Orme remarks that

> To be frank, if I wanted to explain everything in detail, I would need to undertake great work and excessive writing. And even if I wrote everything that I can think, there are many things in stonecutting practice that are not easy to explain, without showing with a finger how they must be done, either for tracing the stones, either for putting them in their proper place; thus I beg you to be satisfied with my drawings and tracings.[16]

---

y entiendo el sacar de las piedras en las canteras, el orden de cargarlas y traerlas aunque fueran de grandeza que pareciere imposible a fuerzas humanas el orden de labrarlas y trazarlas parca y escasamente para que no se gasten los bienes de la fábrica torpe ni demasiadamente, el hacer de las máquinas para subir los materiales a los edificios, cómo se hará los moldes y contramoldes, cerchas, barbeles y otras falsarreglas y cortes de piedras que de ninguno de mis opositores tienen noticias porque no lo han usado ni ejercitado ... (ACG, file 667, 314, 6 and 7, fol. 117–118). Transcription is taken from Moreno (1984: 358–359. Probably *barbeles* and *falsarreglas* are transcription errors for *baiveles* and *saltarreglas*, made by the ecclesiastical secretary recording the *oposición* proceedings.

[14]De l'Orme (1567: 87v) ... *traits de géométrie, lesquels je vous propose, sans jamais avoir entendu qu'il en ait été écrit aucune chose, soit par les architectes anciens ou modernes.* Transcription is taken from http://architectura.cesr.univ-tours.fr. The expression *traits de géométrie* must refer to stonecutting traits, which are the main subject of this chapter in de l'Orme's treatise, since he mentions architects at the end of the sentence, and also he knows, of course, Euclid, quoted in many passages of his treatise.

[15]Martínez de Aranda (c:1600: preface, [i]): *Siendo esto así como se ha tenido perpetuo silencio en estas dichas trazas de montea si no me engaño los grandes artífices antiguos las han dejado para que los venideros tengan continuo ejercicio en el trazar y contra hacerlas.* The passage is included in the introduction of the manuscript, written on unnumbered pages at the beginning.

[16]De l'Orme (1567: 112v): *Car, à dire vérité, qui voudrait par le menu expliquer le tout, il entreprendrait œuvre de grand labeur et excessive écriture. Et encore que j'eusse écrit tout ce que j'en pourrais penser, si est ce qu'il y a beaucoup de choses à la pratique des traits que l'on ne saurait faire entendre, sans montrer au doigt comme elles se doivent mettre en œuvre, soit pour*

Echoes of this position can be found in Martínez de Aranda, who stresses that perhaps Ancient (or modern) masters "have not found the definitions of their right terms to put them into writing",[17] or Rojas (1598: 88v), who states that "I will not explain in writing how to set out the arches, since I would need a ream of paper to address its difficulty, for this issue depends on experience".[18]

This means that the oral tradition was still in full force at this period. In fact, Rojas's strategy, furnishing only headings and drawings, is exactly the same one put forward by Hand IV, while Martínez de Aranda resorts to the didactic methods of Roriczer, trying to register every single word uttered by a master instructing an apprentice. De l'Orme and Vandelvira stand in an intermediate position: they include detailed explanations here and there, but they summarise and abbreviate other pieces; in particular, de l'Orme has a penchant for leaving aside tiresome stonecutting details all of a sudden, to embark on brilliant, apparently erudite discourse.

As in medieval times, this oral tradition was set inside a formal education system. In Spain, a fair number of apprenticeship contracts have been recorded in specialised literature. As we have seen in Sect. 3.1.4, Francisco de Goycoa agreed to instruct an apprentice on stonecutting, making models of different models in plaster and teaching him the construction of arch squares for vaults, trumpet squinches and rere arches. Similar provisions are included in a contract signed by Ginés Martínez de Aranda. In 1597 he undertook the instruction of Pedro Pablo de Ordóñez in stonecutting; however, if the apprentice wished to progress to the study of design and architecture, Martínez de Aranda was to instruct him in these advanced fields (Gila 1991: 276–277). Even more interesting is the contract between Juan de Aranda Salazar and Pedro del Portillo (Galera 1977: 146); the former was to instruct his pupil in this way:

> fourteen stonecutting problems, seven of them according to my will and the other seven according to the will and selection of the aforesaid Pedro del Portillo and also I should teach him arithmetic up to the cubic root and also as much drawing and geometry as the pupil can learn during four years.[19]

All this means that the medieval system of workshop education had evolved in the seventeenth century to encompass mathematical instruction. However, this looks

---

tracer les pierres, ou pour les appliquer en ladite œuvre. Pour ce est il que je vous prie de vous vouloir contenter, de ce que je vous en montrerai par figures et traits. Transcription is taken from http://architectura.cesr.univ-tours.fr. See also de l'Orme (1567: 19r, 124r).

[17] Martínez de Aranda (c. 1600: [i]) ... *o por no haber dado en las difinitiones que los términos de ellas tienen para ponerlas por escritura.*

[18] Rojas (1598: 88v): *no pondré por escrito la declaración de los cortes de los arcos, porque sería menester una resma de papel para poder declarar algo de su mucha dificultad, por ser cosa que consiste todo en experiencia.*

[19] Archivo Histórico Provincial de Jaén, file 1523, 43r-44v: ...*catorze trazas de montea las siete dellas a mi voluntad y las otras siete a boluntad y escoxenzia del dicho pedro Portillo. Y anssi messmo he de enseñar a el susso dicho contar hassta sacar la rayz cúbica y demás de ello el disseño y geometría lo que el sussodicho pudiere deprender de lass dichas trazas y antes conforme sube al saver y deprender durante el dicho tiempo de quattro años.*" My transcription. See also Galera (1977; 146).

like a desperate last stand; at this moment, other institutions were better fitted to carry out this task, as we will see soon. In fact, the empirical nature of masons' lore was underscored by the use of models as a didactic and research tool. As we have seen in Sect. 3.1.4, both Martínez de Aranda and Goiti stressed that they had tested the setting out procedures in their texts using models; the latter claims to have contrasted his manuscript with experience, the mother of all knowledge. All this recalls some passages in the first part of *Arte y uso de Arquitectura.* In one of these, San Nicolás instructs the reader to make plaster cuts "to master the difficulties" of stonecutting; but at the same time adds that "I have experimented these cuts before writing about them".[20] Thus, for San Nicolás, those models are used *both in instruction and research*; in one passage, he proposes to make plaster cuts, adding that *fuera del enterarte, conocerás ser así* (San Nicolás 1639: 65v). Although the difference between both terms is very subtle, the sentence can be translated as "in addition to *understanding* this issue, you will *ascertain* it". In another passage, this subtle difference between the comprehension of the concept and the test of its feasibility reappears: "From the designs that I bring to models I know the justification [of stereotomic solutions]; and ... I build safely, making good use of time and wasting less material". [21] In another sheet the contraposition between rational geometry and empirical test is explained clearly:

> If you want to put my lessons to good use and experiment my writings, you should make plaster cuts; and by means of these cuts you will know that practical knowledge is in concordance with speculative knowledge, as I have experimented with my hands before writing it.[22]

Thus, the same idea appears in Vandelvira, Martínez de Aranda, Goiti and San Nicolás. Vandelvira and San Nicolás remark that the reader should prepare models as a didactic resource, to gain a better understanding of their lessons. On the other hand, both Martínez de Aranda and Goiti and, again, San Nicolás, assert that they have made the models themselves to verify the exactitude of the tracing procedures explained in their manuscripts and treatises. In other words, the speculative constructions used to determine the true shapes of voussoir faces or the exact magnitude of the angles between the edges of these faces may be verified by the empirical, practical, use of models.

Jesuit science. New contenders appeared at the end of the sixteenth century. Francisco del Castillo was humiliated in the *oposición* for the post of master mason of Granada cathedral; Juan de Orea, a mason trained in the palace of Carlos V with

---

[20] San Nicolás (1639: 70v): *Importaria, que antes que hiziesses el arco, que le cortasses de yeso en pequeño, para que de su conocimiento resultasse el hazerte mas señor en las dificultades: mas los cortes dichos, antes los he experimentado, que llegasse a tratar dellos.*

[21] San Nicolás (1639: 69r): *Los cortes dichos ... por los deseños que obro en piecas de yeso, conozco su justificacion: y es obrar con seguridad, cuando lo que se obra es costoso, pues se aprouecha el tiempo, y se gasta menos.*

[22] San Nicolás (1639: 91r): *Si deseas aprouechar, y experimentar este mi escrito, haz cortes de yeso, y por ellos conoceràs ser cierto, y concordar lo practico con lo especulativo. Todo lo cual experimentè por mis manos antes de escriuirlo, siendo este mi exercicio, como en otras ocasiones he dicho.*

Machuca, and Lázaro de Velasco, the son of Jacopo Torni, who had followed an ecclesiastical career and was looking for a substantial stipend, received seven votes each (Rosenthal 1961, 205–206). Lázaro de Velasco was not the only cleric in the Spanish architectural scene of the sixteenth century. Bartolomé de Bustamante, a Jesuit, supervised the houses of the order and put forward alternative solutions to those of Alonso de Covarrubias for the Hospital of Cardinal Tavera, while Juan Bautista Villalpando, another Jesuit, worked on the Convent of Saint Hermenegild in Seville, the house of the Jesuits in Baeza, and the north crossing of the cathedral of the city, before embarking on *In Ezechielem explanationes*, an architectural project disguised as a theology treatise on the Temple of Salomon and the one described in the book of Ezekiel (Taylor 1991: 160–168; Morrison 2010).

As we have seen in Sect. 2.3, this led to a shift in the extraction of stonecutting authors from the field of masons and architects to that of clerics and learned gentlemen. Several Jesuit architects, such as Derand, François Aguillon or Étienne Martellange, the predecessor of Derand at the worksite of Saint-Paul-Saint-Louis, taught mathematics in the colleges of the Society of Jesus. This social change brought about an epistemic shift in stereotomy from artisanal lore to a specific part of mathematics. However, the transformation was not immediate. On the one hand, Derand performed a capillary revision of much material in traditional literature, polishing accepted procedures here and there and introducing powerful mathematical methods, such as the implicit use of tangencies for the control of unions in rib vaults, but most of these refinements were not obvious. On the other hand, the learned professor remarked that demonstrations were unnecessary in his treatise:

> Some people will surely ask for the geometrical proofs of my propositions. This would be hard work, and rather useless for our present subject. If those who take pains to read this book are versed in geometry … it will be annoying for them to spend much time reading my propositions, and just seeing the figures, they will understand that they are geometrically true, or at least exact enough for good and sound practice. Those that are ignorant of the principles, maxims and truths of geometry will see anything I can place on this book as a jargon they will not understand and will cause them more annoyance than pleasure in reading.[23]

Thus, Derand makes every effort to be understood by masons, recording artisanal practices that were probably in use for many centuries, such as drawing on stone surfaces, the most radical variant of the squaring method, and constructing templates for arches by measuring distances to a reference plane, as opposed to the perpendiculars used by Vandelvira. Also, he divided arc segments into two parts and substituted

[23]Derand (1643: 13): *Quelques-uns sans doute exigeront de moy les preuues geometriques de mes propositions. Mais outre que ce trauail seroit de longue haleine, ie le iuge d'ailleurs assez inutile quant au present suiet. Car [si] ceux qui se donneront la peine de lire cét Ouurage seront versez dans la Geometrie … ce leur sera chose ennuyeuse de consommer beucoup de temps à lire, ce qu'à l'ouuerture simple de mes propositions, & a la premiere veuë des figures ils connoissent geometriquement veritable, ou au moins autant exact qu'il en est de besoin pour en venir à vne bonne & solide pratique. Que s'ils sont ignorans des principes, & des maximes, & veritez de la Geometrie, ce que i'en pourrois icy coucher leur seroit vn iargon qu'ils n'entendroient point; & partant qui leur causeroit plutost du dégoust que du plaisir à le lire.*

the combined length of both chords for the rectification of the circumference. Of course, he knew this operation is an approximation, but he argued that the error is too small to be noticed in practice. A few years later, another cleric, the Theatine father Guarino Guarini, pointed out implicitly that this operation is inexact, but still recommended that "the parts of the circle or quadrant *F. 1. 10. D.* should be extended (on this line) measuring them with such small parts as possible".[24] The reader may be startled to find no mention to the number π in this debate, but we should take into account that stonecutters were not used to arithmetical operations, in particular with irrationals or decimal notation, which had been introduced in Europe a few decades before with Simon Stevin, without much success.

Another step for the incorporation of stonecutting into the mathematical realm was the publication of the fifteenth treatise of the *Cursus seu mundus mathematicus* by Milliet-Deschalles (1674), with the title *De Lapidum Sectione*, that is, "The section of stones". However, the transformation was far from complete. For consistency with the rest of his encyclopedic work, Milliet structured the subject in theorems, problems and corollaries. However, the treatise on stonecutting includes five theorems at the start, and this is about all; the rest are problems, and there is not much material that may be accepted as a proof.

In any case, the efforts of clergymen in the mathematisation of the discipline would be taken up by military engineers. This connection may seem strange, but treatises by Jesuits and other clerics were used in military academies in the eighteenth century. For example, the *Compendio matemático* by the Oratorian Father Tomás Vicente Tosca (1707–1715), which took much material from Milliet, was used as a textbook in the Academy of Military Engineering in Barcelona, while there were copies of Milliet and Tosca at the libraries of the Mathematical Military Society in Madrid, the Academy of Guardias de Corps and, probably, also in the Artillery Academy in Segovia (see Sect. 2.3.4).

Architectural and engineering instruction. The reasons for the involvement of military engineers with stonecutting are not as obvious as they may seem. Cristóbal de Rojas mentions "skew arches for the passages leading to the casemates of the fort, and also splayed arches, which are very appropriate for the gunports of the artillery ... "[25] (marked as *G* in Fig. 12.5) or asserts that "the engineer ... should know how to build the gunports of fortification ... that are called splayed in artillery ... and especially he should know the cuts and shapes of the vaults, for any architecture".[26] However, such explanations are not convincing. In Early Modern military constructions, vaults

[24] Guarini ([c. 1680] 1737: 197): *si estenderanno sopra la medesima le parti del circolo, o quadrante F. 1. 10. D misurandolo con parti piccole al possibile.*

[25] Rojas (1598: 97 v): *los arcos en viaje, para las entradas, o callejones de las casasmatas de la fortificación, y así mesmo arcos embocinados, que son muy a propósito para las troneras, o cañoneras del artillería.*

[26] Rojas (1613: 40v-41v): *... conviene mucho que el tal Ingeniero sea arquitecto práctico ... conviene saber hacer las troneras de la fortificación en los traveses en viaje, o en forma, que llaman embocinadas para el artillería, que son muy anchas, por la parte de adentro, y por de fuera angostas, y principalmente saber hacer los cortes, y cerramiento de las bóvedas, para toda arquitectura ...* Page numbers in this book are chaotic; fol. 41 is numbered as fol. 14.

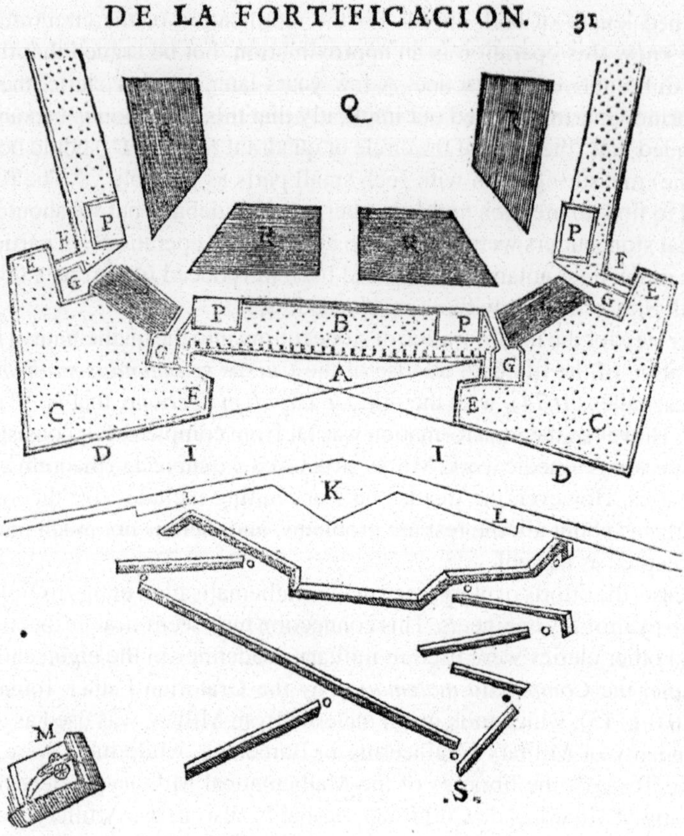

Fig. 12.5 Fortification plan (Rojas 1598: 31r)

were executed on many occasions in brick or concrete, while the stairs were usually small. Treatises also deal with arches in battered walls; however, these elements are not frequent in practice since they are at odds with the military function of these constructions. That is, knowledge of stonecutting is useful, certainly, for the military engineer, but does not seem to justify the importance granted by Rojas or the three volumes of Frézier's treatise.

Moreover, Belhoste's study about the exercises in engineering schools in the seventeenth and eighteenth centuries suggests that the teaching of stonecutting in military engineering schools was the result of an educational fad. As we have seen, stonecutting was taught in the "second hour" at the Royal Academy of Architecture in the late seventeenth century; later on, it was introduced in the École des Ponts et Chausées, a civil engineering school, and the military engineering school in Mézières followed suit (Belhoste et al. 1990b; see also de la Gournerie 1874: 114–115). Thus, a puzzling question arises: was the interest of engineers in the geometrical problems of stonecutting brought about by practical needs, didactic intentions, or esprit de corps? Military schools needed to instruct the aspiring engineer on the interpretation of drawings in double projection; the intricate problems of stonecutting provided a most suitable and almost endless source of exercises in this field. In fact, stonecutting instruction at the École Polytechnique was more abstract than its counterpart at the École des Ponts et Chauseés, although it still included practical exercises, in contrast to Monge's lessons in the École Normale (de la Gournerie 1874: 114–120, 152–156; Belhoste et al. 1990b: 54–72). All in all, the teaching of stonecutting theory in military schools focused not so much on the training of engineers for stone construction, but on the education of their spatial vision and the concept—almost the ideology—of the application of mathematics to practical problems (Sakarovitch 1998: 221–223).

In other words, the underlying mathematical concepts in stonecutting, pointed out by Desargues, Derand and Milliet, rather than its practical application, lured military engineers, like Frézier, or architects under the influence of engineers, like de la Rue, into taking further steps in the mathematisation of stonecutting. The nature of this operation is paradoxical: in order to fulfil it, de la Rue and Frézier needed to detach the subject from the particular configurations of constructive elements. As a result, their treatises included abstract sections on the intersections of cones, cylinders and spheres; this implies leaving all physical qualities aside, starting with gravity and therefore any mechanical analysis. For de la Rue, the strictly mathematical part was a small and dispensable appendix, the *Petit Traité de Stéréotomie*, placed at the end of his book. In contrast, Frézier turned the little appendix of de la Rue into the whole foundation of his monumental treatise, spreading it over the entire first volume and adding the theory of projections, developments, and angular measures to de la Rue's intersections. In the second and third volumes of his book, Frézier included many sections under the title *Explication demonstrative* (proving explanation) but these were not true proofs, but rather a mixed bag including fragments of an actual mathematical demonstration interwoven with a practical explanation (Sakarovitch 2002: 590).

This approach was followed in military engineering schools in the last decades of the Ancien Régime. The teaching of stonecutting theory by Gaspard Monge in the Mézières school addressed abstract topics such as the determination of the position of a point in horizontal and vertical projection starting from the angles formed by the lines connecting them to three given points. That is, the artisanal practice of stonecutting had evolved at this stage into an abstract science, which could be applied to any technical field. As Sakarovitch (1998: 243–244; see also de la Gournerie 1855: 15) stressed, the stonemason works on mass, starting from a three-dimensional object, in contrast to the carpenter and the coppersmith, who work on lines or developable surfaces. This suggests a parallel between the physical activity of the stonemason and the ideal position of the geometrician, shown by the rich repertoire of developable and warped surfaces that can be materialised by the stonemason, as well as their complex intersections.[27] This geometrical wealth is reflected in the wide variety of

---

[27] In descriptive geometry, a developable surface can be materialised through deformation (excluding extension, folding or cutting) of a flat sheet of paper, cardboard or metal. Examples are cones or cylinders with any directrix. In three-dimensional space, all developable surfaces are also ruled;

graphic techniques that can be used in stonecutting tracings: projections, developments, rabatments, rotations, changes of projection plane, in present-day parlance. Even the notion of generatrices and directrices introduced by Monge in his post-revolutionary lectures at the École Normale recalls the methods of stonemasons for defining a surface through drafts opened by the chisel; the ruler rests on these channels to materialise a ruled surface. And finally, the economy and elegance of stonecutting tracings had to be enticing to the military by force (Sakarovitch 1998: 244). All this explains, beyond any practical application, the interest of engineers in stonecutting.

The École Normale and the École Polytechnique. In any case, Monge did not draw on the full potential of Frezier's virtualisation of stonecutting procedures in Mézières, where military secrecy, or esprit de corps, prevented him from presenting all details of his research (Dupin 1819: 11), but rather in the École Normale and the École Polytechnique, two revolutionary institutions (see Sect. 2.4.3). If the foundation of stonecutting was an abstract science of surfaces, sections, developments and projections, why not reuse this science to the benefit of other building trades or even military issues such as the *défilement*? (see Sect. 2.4.3). After all, Frézier had extended the field of application of stereotomy to joinery. Following the same line of thought, it could be used in carpentry, metalwork or even in the resolution of abstract geometrical problems, such as the distance between two skew lines.[28]

Two factors crucially influenced the reception of Monge's theory. The first one was rather accidental: while the teachings of Monge at the École Normale were transcribed in shorthand and published in many languages, those at the École Polytechnique have survived only as a few drawings by pupils preserved in the school archives until they were unearthed in the late twentieth century. The other factor lies at a deeper level: while Blondel, de la Rue or Frézier were practising engineers or architects, Monge and his followers, such as Hachette and Leroy, built nothing; the latter stayed at their chairs at the École Polytechnique, with little or no connections with construction practice. All this led to a dismal conclusion: the teaching of descriptive geometry (the new name for the theoretical base of stonecutting) as well as stereotomy itself, stagnated in the endless repetition of Monge's exercises, in particular during the long tenure of Leroy in the descriptive geometry chair at the École Polytechnique. Even worse, such activities focused on issues selected in order to demonstrate abstract geometry concepts, such as the scalene ellipsoid or the *arrière-voussoure de Marseille*, along with their mechanical-engineering counterparts, such as the shadow of the screw with triangle-shaped thread.

The alarms signalling the weakness of such an approach were set off by the "Waterloo of skew bridges", in Sakarovitch's words (1998: 341; see also Sect. 8.1.2).

---

that is, through every point of the surface there is a straight line that lies entirely on the surface. Other surfaces, known as warped surfaces, cannot be materialised through deformation of a sheet, although they can be dressed in stone or modelled in clay. Warped surfaces may be ruled surfaces, such as some kinds of skew arches, or double curvature surfaces, such as the sphere, where no straight line lies in the surface (see Lawrence 2011).

[28]"Skew lines" is used here in its spatial geometry meaning, that is, two lines that are not parallel and do not intersect.

While the French had devised a solution that was mathematically unimpeachable but practically impossible to implement, the Englishmen suddenly turned up in the scene with a practical method. Two opposing voices were raised, calling for a comeback to the practical origins of descriptive geometry. Although Olivier and de la Gournerie (see Sect. 2.4.6) fought a bitter debate, seen from a distance it seems like a clash between different engineering specialities rather than different underlying conceptions. After all, de la Gournerie was a skilled civil engineer, and Olivier was at the heart of a movement to revitalise French industry. The former stated that changes of horizontal projection plane were useless in construction since they did not take into account the direction of the force of gravity; however, he forgot that this factor is not so relevant in industrial engineering, where rotating parts are subject to forces much stronger than gravity. In any case, this effort to revitalise stonecutting through a return to its centuries-old traditions was useless; the industrial revolution killed not only these traditions, but also stone construction itself, through the advent of cast iron, steel and concrete.

12.3 Stereotomy in History and Geography

Many present-day conceptions about stereotomy have been shaped by Jean-Marie Pérouse de Montclos's classic *L'architecture a la Française* (1982a), in particular by a section under the heading "History and Geography of [Early] Modern vaulting".[29] However, we must take into account the purpose and the historiographical framing of his book. Its main aim was to show that Early Modern French architecture, although using the classical language imported from Italy, added several specifically French traits or *gallicismes*, such as steeply sloping roofs, some particular idioms in decoration and mural composition, the emphasis on distribution and, above all, stereotomy. In fact, he devotes to stonecutting more than half the book.

The section on the history and geography of modern vaulting (Pérouse [1982a] 2001: 181–185), although intellectually honest, seems to be designed in order to support the central thesis of the book. First, Pérouse reviewed the classical antecedents of Renaissance stereotomy briefly, mentioning Rome, Syria and South-Eastern France, but dismissing them as isolated examples. This is true but not really meaningful (see, for instance, Etlin 2012: 8). Ashlar construction flourishes where geological conditions offer suitable stone, whether in Antiquity, the Middle Ages or the Early Modern period; I will come back to this issue in Sect. 12.3.3.

[29] The literal title of the section is *Histoire et géographie de la voûte moderne*. However, we should take into account that in French historiography, *moderne* means "Early Modern", a period ending with the French Revolution. Further, the full title of Pérouse book is *L'architecture a la Française. Du milieu du XVe a la fin du XVIIIe siècle*, that is, "Architecture in the French Manner. From the Mid-Fifteenth Century to the End of the Eighteenth Century". These details explain why Pérouse does not deal with the expansion of stereotomic theory to England, Italy and Germany in the nineteenth century: it just falls out of the scope of his book.

After this, Pérouse posited that the typology of Early Modern vaults derives from Romanesque precedents, such as the rere-arch, the lunette, the trumpet squinch, the pendentive and the sail vault, the ribbed dome, the annular vault, and, above all, the *Vis de Saint Gilles*, adding that Romanesque techniques were alive and well at least up to the seventeenth century, when the cathedral of Valence was literally reconstructed in imitation of the old Romanesque building.[30] Next, he dealt with Gothic architecture, pointing out the difference between ribs placed on a continuous surface and those marking the intersection of different surfaces, and the contrast between planar curves and warped surfaces in mainstream Gothic severies and developable[31] surfaces in Renaissance vaulting. Although he admitted that the complex geometry of (Late) Gothic vaulting might have been inherited by Renaissance stonecutting, he implicitly downplayed the influence of Gothic construction in Early Modern stereotomy,[32] moving forward to a page on "The Renaissance and the scientific description of space", where he mentions some connections between perspective and stereotomy, leading up to Desargues, to conclude that

> we can say that modern stereotomy comes out from the combination of Romanesque practice with an idea, an ambition, that of the rational conquest of the third dimension of space ... De l'Orme was the first one to take profit [from this combination]: initiated into the secrets of southern masons, he had the fortune to discover the Renaissance on the route to Italy[33]

Next, Pérouse included a lengthy section on the geographical distribution of vaults in France, both in ashlar masonry and other materials, and he added a chapter under the title "About other states adjoining France" dealing almost exclusively with Spain; his conclusion about the mutual standing of France and Spain can be summed up in a single phrase: "In all evidence, reciprocity" (Pérouse [1982] 2001: 212). In contrast, his attention to stereotomy in other European countries is almost nil. At a lesser scale, he noticed that most *appellations d'origine*, that is, stonecutting archetypes connected with a particular city, belonged to southern France, which boasts the *vis de Saint-Giles*, the rere-arches of Marseille and Montpellier, the squinch named after that city and the *Pendentif de Valence*, while Northern France can only present the

[30]For an extreme example of this position, see Potié (2005 :74–75); for a recent and nuanced one, see Galletti (2017a).

[31]Pérouse does not actually use the term "developable". In fact he states that *La révolution ogivale, c'est l'adoption d'intersection segmentaire, coplanaire, nervurée et de surfaces gauches; la contre-révolution, c'est le retour à des surfaces réglées et à des pénétrations elliptiques et gauches*, that is, "The Gothic revolution is based in the adoption of segmentary, coplanar and ribbed intersections, as well as warped surfaces; the counter-revolution is the comeback to ruled surfaces and elliptical and non-planar intersections". However, "gauche" is by no means the opposite of "réglée", since "réglée" or ruled surfaces can be either warped or developable. Thus, Pérouse seems to be thinking of Renaissance developable surfaces, such as in the lunette vault, as opposed to Gothic warped severies.

[32]For a later, balanced opinion, included in a not-so-well-know study, see Pérouse (1987: 15–16).

[33]Pérouse (1982:185): *Nous croyons pouvoir affirmer que la stéréotomie moderne est sortie de la rencontre de la pratique romane avec une idée, une ambition, celle de la conquête rationelle de la troisième dimension de l'espace ... De L'Orme fut le premier à en tirer publiquement parti: initié aux secrets des maçons méridionaux, il eut la chance de découvrir la Renaissance sur la route de l'Italie.*

lost rere-arch of Saint-Antoine in Paris. He also surveyed for classical precedents for Renaissance stereotomy and found only a few isolated examples. All this led him to place the origin of modern stereotomy in Languedocian Romanesque, remarking the quality of stonecutting in this school.

12.3.1 *The Historical Antecedents of Renaissance Stonecutting*

All in all, Pérouse described stereotomy as an essentially Franco-Spanish issue, with very slight precedents in Classical Antiquity, implying that the inspiration for the stereotomy of the Early Modern period was to be found in Romanesque stonecutting, leapfrogging over the Gothic phase. There are several problems with this position. First, it is not easy to understand why, for Pérouse, the restriction of classical stereotomy to three or four regions invalidates Antiquity as a precedent of Renaissance stonecutting, while Languedocian Romanesque is enough proof to place its roots in the High Middle Ages.

Admittedly, barrel vaults, trumpet squinches, annular vaults, lunettes, splayed arches or arches opened in curved walls, quarter-sphere and octagonal vaults, and vaulted spiral staircases are frequent both in Romanesque and Renaissance architecture. However, spiral and straight stairs pass from the Romanesque period to Gothic architecture and then to the Renaissance. Paradoxically, Pérouse himself (1985) casts doubts about the dating of the archetype of vaulted spiral staircases, the one at the priory of Saint Gilles, suggesting it may date from the fifteenth century. However, Hartmann-Virnich (1996; see also 1999) confirmed the Romanesque dating of the Saint Gilles stairway; moreover, the dating of another remarkable example, the one at the cathedral of Avignon, has not been put into discussion, as far as I know.

With other pieces of the repertoire, issues are more complex. Hemispherical domes can be found, although they are not frequent, both in Romanesque and Gothic architecture. In the Romanesque period they are associated with two particular epicentres, and neither of them has strong connections with the Languedoc: south-western France, in particular the cathedral of Angoulême and Saint-Front de Périgueux, and western Spain, with the group of domes over the crossings of the cathedral of Zamora, the old cathedral of Salamanca, and the collegiate church of Toro. Both epicentres may be traced to Byzantine roots, in particular, Saint-Front de Périgueux, which resembles Saint Mark of Venice and, indirectly, the Holy Apostles church in Constantinople, while the ribs and the placement of windows in Zamora, Salamanca and Toro recall other Byzantine examples (Erlande-Brandenburg and Mérel-Brandenburg 1995: 212–216; Torres Balbás 1922). However, there are also many small domes in Gothic architecture, in particular over spiral stairs. Corner arches flourish at the transition of Gothic to Renaissance, in particular in Trujillo and Ciudad Rodrigo, in western Spain (Hoyo 1976). Sail vaults in real ashlar are almost inexistent before the sixteenth century, and they seem to result from the confluence of Late

Medieval vaults built in brick or rubble and the influence of the Florentine Renaissance, in particular, the side aisles of San Lorenzo and Santo Spirito (Natividad 2017: I, 257–258; see also Molina and Arévalo 2014). True oval vaults are almost non-existent before the mid-sixteenth century; vaults over rectangular plans capped by two quarter-sphere vaults are a different problem. And of course, the coffered vaults of the sixteenth century are a derivative of Gothic technology adapted to Renaissance taste.

Moreover, Perouse's association of stereotomy with Romanesque architecture, and particularly with the Languedoc and Provence, focuses on the product, not on the process. Were Renaissance vaults built using the same technology and control methods as their Romanesque antecedents? The available evidence is clear: not at all. First, specific archaeological or documentary sources about control systems in Romanesque architecture are very scant, while mentions to professional architects, that is, designers and supervisors trained in the building crafts are all but non-existent. Quite to the contrary, such details as a bishop overseeing a stonemason carving a stone (Fig. 12.4), shown in a capital in the cloister of the cathedral of Girona dating before 1117, or the direction of the abbey church of Saint Denis directly by Abbot Suger, suggest that, as a rule, in the eleventh and early twelfth centuries supervision was performed directly by clients.

However, a quick change in these issues seems to have taken place in the late twelfth and early thirteenth century, as we have seen: named architects, orthographic projection, full-scale tracings, drawings in parchment and templates appear, at least in the historical and archaeological record, all at the same time. These innovations may have been brought about by the general complexity of Gothic architecture, although they are also adopted in late Romanesque buildings. Viollet le-Duc and Choisy made much about the coordination between the vertical structures such as piers and responds and the nervatures in vaults (see also Sakarovitch 1997: 16–17).[34] Thus, the need to envisage a full structure before building, in contrast to High Medieval *imago in mente conceptum* (the practice of building without drawings) brought about a real "inversion of constructive thinking", in Sakarovitch's words (2005b: 56–59), requiring professional builders to design and supervise these structures (see Figs. 10.3, 10.6 and 10.7 as examples, Fig. 12.6 as an exception).

There are also crucial differences in the dressing and placement stages. In particular, hoisting equipment in the Romanesque period was basic, to say the least, while Renaissance builders could use the huge cranes of the Escorial or Brueghel's paintings. As a result, blocks in unrestored Romanesque architecture are generally small, while their Renaissance counterparts are much larger. Further, as we have seen in

[34] There are exceptions to this principle, for example, the nave in Noyon cathedral, where the vertical structure seems to be prepared to receive sexpartite vaults, but the present-day building features quadripartite ones. This led Choisy (1899: II, 427) or Dehio (1894: IV, Atlas, pl. 361) to draw the vaults actually as sexpartite, perhaps assuming that the original ones had disappeared after a fire. Recently, Geraldine Victoir (2005) has shown that although the fire actually took place, the quadripartite vaults are original. However, Noyon is the exception that proves the rule: in ordinary cases, there is clear coordination between the vertical structure and the vaults.

Fig. 12.6 Quadripartite vaults over the nave with *a-b-a-b* rhythm. Noyon Cathedral (Photograph by the author)

Sect. 1.3, the specific surface of masonry increases as the size of the blocks diminishes; that is, for a given element, the total joint surface is much larger when built with small stones. Also, it is quite difficult to shape a small stone as a regular prism; percussion tools may topple the block, and the set square is simply too large. Thus, when hoisting equipment cannot lift great weights, builders generally resort to roughly dressed pieces, in order to avoid a huge amount of dressing work. Of course, the price they need to pay for the economy in dressing effort is a large quantity of mortar, arising both from the irregular shape of the stones and their small size. In contrast, when powerful lifting equipment is available, it is more efficient to dress large blocks in the form of precisely shaped prisms or voussoirs, in order to save mortar, improving at the same time the aspect of the masonry.

In short, Romanesque rubble masonry and Renaissance ashlar construction are the result of quite different, even opposing, building technologies. This is particularly evident in a key element, the groin vault. In the Romanesque period, it was used frequently in aisles, rather than naves. However, the shortcomings of Romanesque rubble groin vaults were one of the factors leading to the use of the rib vault from the twelfth century on, as Palacios and Martín (2011: 536–537) have pointed out: in groin vaults executed in rubble, it is almost impossible to give a precise elliptical shape to the groins. However, this had not always been the case; the groin vault at the lower chamber of the Mausoleum of Theodoric in Ravenna features a remarkably well-controlled execution, with neat groins and large, well-dressed voussoirs. There is no doubt that it was executed using the powerful lifting technology of Late Antiquity since the upper chamber is covered by a single stone slab measuring almost 11 m in diameter.

The Mausoleum of Theodoric brings us to another problem in Pérouse's hypothesis: the scarcity of complex ashlar vaults in Roman constructions. Pérouse himself ([1982a] 2001: 144, 181) mentions the annular, rising vault in Hadrian's Mausoleum in Rome, but dismisses this and other Roman pieces as isolated examples, while he makes passing remarks about Syria and southern France. The problem with this interpretation is that he focuses mainly on Rome, where the local stone, travertine, does not favour precise dressing. Quite significantly, the vault in Hadrian's Mausoleum was begun in ashlar masonry, but only the first portion, measuring about 20 m, was executed in hewn stone; after this, it was continued in concrete. Moreover, Pérouse set out to look specifically at vault intersections, but not so much at other complex stonecutting pieces, such as the dome in the tomb of Ummidia Quadratilla in Casino, the groin vault in Frasso Sabino, the skew arches in Hadrian's baths in Leptis Magna, the octagonal vault inside the arch of Marcus Aurelius in Oea (now Tripoli), the groin vault in the arch of Cáparra in Western Spain, or the sail vaults in Gerasa.

In spite of all these antecedents, Sakarovitch (1998: 105–107; 2005b: 53–54; 2009b: 7) located the cradle of stonecutting, or even learned architecture, in Christian Syria. Of course, in this period the use of complex constructive elements, in particular, quarter-sphere vaults, in such locations as Qalat' Simaan (the monastery commemorating Saint Simeon the Stylite) or the church of Qualb Lozeh, is frequent. However, he scarcely mentions the fact that in this particular historical period, Syria belonged to the Byzantine empire. This connects Syrian examples with other remarkable stonecutting pieces in countries under the sphere of influence of Byzantium, such as Armenia, where quarter-sphere vaults are also frequent in the fifth through the seventeenth centuries, for example in Saint Hripsime in Ejmiadzin, the church at Bayburt, Saint Kiriaki at Arzni, Saints Polos-Petros in Zovuni, the Ciranavor basilica at Astarak, Saint Grigor at Aruc, Saint Christopher at K'ristap'orivank, to name just a few examples (Cuneo 1988: 98–101, 118, 143, 166–167, 194, 212, 218). Although vaults in Byzantium itself were built in brick,[35] quarter-sphere vaults also played an essential role. In a broad generalisation, we could assume that the use of a set of geometrical shapes is a widely travelling cultural choice, while the selection of materials depends on local availability and cost.

The connection points between these Oriental sources and western European stereotomy have been located in a few significant areas. One of them is Fatimid Cairo: the gates in the city walls, in particular, Bab-el-Nasr and Bab-el-Futuh, hold a number of remarkable stonecutting pieces, in particular, different kind of lunettes, a groin vault, a sail vault, and a remarkable vaulted spiral staircase of the *vis de Saint Gilles* type (Tamboréro 2006). All of them are quite precisely executed, in contrast to contemporary Western examples. According to tradition, both gates were built under the direction of three Armenian monks coming from Edessa; this provides the connection with Armenia, and indirectly, Byzantium. The introduction in western

[35] Or rather in mortar filled with bricks, since in many occasions the width of joints exceeds the thickness of bricks. See Mainstone (1988: 70).

Europe of this archetype has been connected with another example in the Castello Maniace in Syracuse, dated to the thirteenth century (Bares 2007). However, taking into account the setting back of the date of the Saint-Gilles staircase to the twelfth century, this is not as relevant a connection step, independent of the high quality of the Sicilian staircase.

These connections can explain the appearance of many types of the stereotomic repertoire, such as trumpet squinches, vaulted staircases, sail vaults, pointed arches or rib vaults, but they cannot explain the full ensemble of elements and techniques used in European stonecutting. There is also the issue of the connections of Renaissance stonecutting with Gothic technology. Gothic and Renaissance architecture have nothing in common, at least in theory; in any case, Gothic vaulting typically involves a two-tier constructive system, while Renaissance uses a single-level method. Thus, Pérouse (1982a) and Sakarovitch (1998) barely mention this connection. However, since we have seen that Romanesque and Renaissance construction methods have nothing in common, did the Renaissance stonemasons start from scratch or did they profit from Late Medieval building technology? The answers to this question should be nuanced. What the Renaissance inherited from the Late Medieval period was basically the idea of a complex, specialised work organisation, which had been lost at the end of Antiquity and resurfaced gradually starting in the twelfth century. This involves professional designers and supervisors, the use of full-scale tracings as well as drawings in parchment, orthographic projection and, last but not least, powerful lifting equipment, allowing the use of medium-sized stones in the Late Middle Ages and large ones in the Renaissance.

Moreover, in the strictly architectural field, the idea of an impermeable frontier between Gothic and Renaissance does not hold outside Central Italy; in technological issues, it is entirely untenable. In particular, as we have seen in Sects. 10.1.6 and 10.1.7, there are at least three Late-Gothic schools that use ribless vaults: fan vaults in England, arrised vaults in Valencia, with an offspring in Assier, in southern France, and "diamond" vaults in Bohemia, Germany and Poland, usually built in brick (Leedy 1978; Leedy 1980; Zaragozá 2008; Navarro 2018c; Navarro and Rabasa 2018b; Navarro et al. 2018a; Acland 1972: 220–228). Although there is not much hard evidence about the formal control methods used in these schools, we may surmise that some of the knowledge brought about by these practices filtered down to Renaissance stonemasons. Moreover, the Gothic practice of drawing in the faces of stones in the dressing stage, in particular in keystones and springers, filtered down to Early Modern practice, as we have seen in Sect. 3.2.3 (Willis [1842] 1910: 24; Rabasa 2000: 96–121; Derand 1643: 124, 138, 164–166).

None of this means that Renaissance stonecutting derives entirely or mainly from Gothic sources. Quite to the contrary, there are many pieces of evidence of a rapid, almost frantic evolution of stonecutting techniques during the sixteenth century; so fast that typical Renaissance methods, such as those based on cone developments, are applied to Late Gothic vaults (see Sect. 10.1.4). At the initial stages of the introduction of the Renaissance in Spain and France, Italian or Italianate masters, most of them figurative artists, needed the help of local masons to materialise architectural works in the new idiom. Most probably, both groups were puzzled by the task they had to carry out. Italianate artists, although trained on sculpture in some occasions, were not familiar with advanced ashlar dressing techniques, since in Central Italy vaults were almost exclusively built in brick. Local masons, on the other hand, mastered two-tier Gothic stonecutting techniques, but were not acquainted with the new concepts of single-level Renaissance construction, except perhaps in England and Valencia. Under pressure from clients wishing to adopt the fashionable Renaissance language without renouncing the aristocratic connotations of ashlar masonry, the basic techniques of classical stonecutting were developed in a few decades. We may surmise that both groups, Italianate artists and local masons, collaborated in these advances; in any case, when De L'Orme's treatise (1567) left the presses, all these essential procedures were up and running.

From this moment on, sources are not lacking, and the evolution of stonecutting concepts and methods can be surely charted, as we have seen in the preceding chapters of this book. In the following centuries, open questions are not raised by the historical evolution of stonecutting, but rather by their geographical distribution; that is, if and why stereotomic advances were concentrated in France and Spain.

12.3.2 The Geographical Distribution of Early Modern Stonecutting

A quick glance at Chaps. 2–11 of this book may be sufficient to confirm Perouse's claim of the central role played by France and Spain in Early Modern stonecutting and the "reciprocity" of interchanges between both countries. However, such reciprocity does not extend to the whole period. During the sixteenth century, Spanish influence in France is undeniable; de l'Orme (1567: 72v-74r, 111v-115r; 119r-120v; see also Calvo 2015b; Galletti 2017b: 155) includes in his treatise several examples that may be found easily in Spain, such as corner windows or spiral and sail vaults, while they are infrequent or non-existent in France. At the same time, the rich and wide range of Spanish manuscripts of the period contrasts with the French production, essentially limited to de l'Orme's treatise and his reflection on Chéreau, both with a limited repertoire in comparison with Vandelvira (Pérouse 1982c). No French building site of the period can match the Escorial in sheer size or organisational complexity.

The picture changes suddenly in the middle years of the seventeenth century. The richest Spanish texts of the period, such as Portor, are derivatives of sixteenth- or early seventeenth-century manuscripts, or more or less disguised copies of French sources, such as the manuscript by Fray Francisco de Santa Bárbara or the treatise of Tosca. Most construction of the period in Spain is carried out in brick, and no building program, including the lost Alcázar and Buen Retiro palaces, both in Madrid, can match Versailles, the Louvre or the Invalides. And of course, such building activity, carried out in the fine stone of the Paris basin, is accompanied by the written works of Desargues, Jousse, Derand, de la Rue and Frézier.

All this seems to reflect political and economic conditions. While the sixteenth century is the period of Pavia and Saint-Quentin, the French reversed the situation the next century in Gravelines and Rocroi. It may be a coincidence, but a significant one, that this period of emergence of the French economic and political power overlaps the extraordinary outcome of stonecutting treatises in the 1640s. And of course, the overwhelming French influence in Spain after the War of the Spanish Succession is reflected in derivatives of French treatises such as those of Tosca and Bails.

After dealing with Spain, Pérouse ([1982a] 2001: 213–218) analysed stonecutting in Italy, Germany and England, stressing from the start that "in these countries the art of tracing, as I have defined it, is almost unknown".[36] To be fair, Pérouse mentions Guarini, Nicholson, the spiral staircases at the Ducal Palace in Urbino, the Convento della Caritá and Saint Paul's as well as the influence of de la Rue in George Dance the Younger and John Soane and other bits of information. In any case he ends this section with a quotation from Sauval, stating that (outstanding carpentry and) the art of tracing had not yet crossed the frontiers of France.[37]

Today, thirty years after Pérouse, thanks to the publication of a large number of case studies in this field, the time is ripe for a review of these positions. Beginning with Italy, the birthplace of classicism, the issue seems to be connected with the opposition between figurative artists and professional builders. The best representative of the latter, Antonio da Sangallo the Younger, built two remarkable asymmetrical splayed arches in the vestibule of Palazzo Farnese; however, such pieces are sometimes classified as "false perspective" rather than stereotomy, in order to frame them within the accepted narratives of Italian architecture. Even a figurative-artist-turned-builder, Brunelleschi, designed a cantilevered stairway for the pulpit at Santa Maria Novella. In the words of a writer not particularly interested in stereotomy, Eugenio Battisti (1976: 292), the stair "looks like a fallen sheet from De L'Orme's treatise or a Late-Gothic stonecutting wizard".[38] Palladio trained in his youth as a stonemason and used cylindrical lunettes, built in brick, in Villa Pisani at Montagnana and Il Redentore. More interesting for our purposes is Palladio's oval, cantilevered stone staircase at the Convento della Carità in Venice, which provided an important inspiration for English stereotomy, as we have seen in Sect. 11.1.2.

---

[36] Pérouse ([1982a] 2001: 213): ... *trois autres Etats voisins de la France, dans lesquels l'art du trait, tel que nous l'avons défini, est pratiquement inconnu: l'Italie, l'Allemagne et la Grande-Bretagne.*

[37] Sauval (1724: III, 44): *la necessité de bâtir à Paris en des lieux fort serrés ... ce qui a éveillé l'esprit des Charpentiers pour la coupe des bois & tout de même celui des Architectes pour la coupe des pierres. Deux découvertes inconnues aux Anciens, aussi bien qu'aux Etrangers, & qui cependant n'ont point encore passé nos portes, quelque merveilleuses qu'elles soient, & fassent tant d'honneur à l'Architecture* (the necessity of building in Paris in constrained places ... has awakened the ingenuity of carpenters on woodcutting, and that of the architects on stonecutting. Two discoveries unknown to the ancients, as well as foreigners, that have not passed our gates, although they are wonderful and bestow so much honour on architecture). Although this passage is inserted in a section on carpentry, Pérouse does not mention in his quotation the carpenters and the ancients.

[38] Battisti (1976: 292): *Sembra infatti di sfogliare il trattato d'architettura di Delorme, o di avere a che fare con un mago del taglio in pietra del tardo gotico ...*

However, these are isolated pieces. It is easier to find relevant examples in northwestern and southern Italy, two areas exposed to strong Spanish influence. Marco Rosario Nobile (2013) has been able to put together a collective volume on "Stereotomy in Sicily and the Mediterranean", a catalogue of stonecutting pieces such as trumpet squinches, domes and quarter-sphere vaults, octagonal vaults, spiral staircases, arches opened in curved walls, corner arches, skew arches and groin vaults, spread over the whole island. Perhaps in the coming years, comparable surveys will be carried out about stereotomy in Campania and even Lombardy, drawing a more complete picture of stonecutting in the peninsula. In the introduction to his study of Sicilian stonecutting, Nobile (2013: 37–42) mentions Malta, a first order stereotomic focus. The practical absence of construction materials other than stone, together with French and Spanish influence, both direct through the Knights of the Order of Malta and indirect through Sicily, fostered the construction of quite remarkable stereotomic pieces, such as the series of sail vaults in the Grand Master's Palace or the sloping coffered staircase at the Holy Infirmary.

This state of events explains the scarcity of stonecutting treatises in Italy. The exceptions are, of course, Caramuel and Guarini, but none of these books is strictly Italian. That of Caramuel, although published in Vigevano, is written in Spanish; the very short section dealing with stonecutting (Caramuel 1678: II, 20–22) is elementary and outdated, as we have seen in Sect. 2.3.5. It seems placed here just to show an application of oblique architecture and not to furnish a guide to stonecutters. Guarini, of course, is another issue. The stonecutting sections in *Euclides adauctus* (1671: 572–596) and *Architettura civile* ([c. 1680] 1737: 191–265) are remarkable for their geometrical correctness[39] and abstract approach, especially taking into account that the first one was written in the late seventeenth century, and the latter, published only in 1737, takes much material from the former. Pérouse stressed the passage stating that *ortografia getatta*, that is, developments, "is not well known in Italian Architecture, and only put in practice wonderfully in many occasions by the French".[40] However, we should not forget that Guarini trained in Paris, that the treatise, although written in Italian, was published in Turin, that the city was in this period the capital of Savoy, a buffer state between centralised France and Balkanized Italy, and that it was edited after Guarini's death by Bernardo Vittone and the Theatines of Turin. These factors may explain a slight influence of Guarini's treatise in France (cf. Guarini 1737: treatise 4, pl. 3 and Rondelet [1802–1817] 1834: pl. 34).

Pérouse ([1982a] 2001: 214) also mentioned writings by Bernardo Vittone and Francesco Milizia. However, Vittone included only a section of Saint Peter's in Rome and several methods for tracing arch and vault directrices and thicknesses, with no reference to the division of these pieces into voussoirs (Vittone 1740: 500–514; see also plates 88, 89). All this suggests that he was interested in mechanics, rather than

---

[39]There are isolated geometrical errors in *Architettura Civile* (1737) which was published by Bernardo Vittone and the theatines of Turin many years after the death of Guarini. See Borin and Calvo (2020: 51).

[40]Guarini ([c. 1680] 1737:191): … *abbenché poco conosciuta dalla Italiana Architettura, solamente dalla Francese in molte occasioni egregiamente adoperata.*

stereotomy; his praise of Frézier is brought about by a remark on the thickness of non-loadbearing vaults (Vittone 1740: 503). This is even clearer in Milizia (1781: III, 236): he mentions Derand, Dechalles, Blondel and de la Rue, stating that their rules had been corrected by de la Hire, Couplet, Bélidor and Frézier. Now, Couplet and Bélidor did not write about stonecutting, but rather on mechanical issues, while Derand (1643: 16–19), Blondel (1675–1683: IV, 419) and Frézier (1737–39: III, 342–427) included structural rules here and there in their writings, and de la Hire (1692; see also Lemmonier 1911–1929: II, 345–349) wrote a memoir on the thickness of buttresses. Later on, Milizia (1781: III, 267) mentions Derand, Dechalles, Blondel and De la Rue again, but only in connection with a well-known structural rule about arch buttressing. Thus, it seems that statical problems, rather than stereotomic ones, triggered Milizia's appreciation (or disregard) for these writers. Even his striking praise for Torija may be brought about by an interest in quantity surveying. Thus, we may surmise that Milizia's remark—"Neither the Ancients nor the Italians, who have built so many domes and vaults, have dealt with this subject in theory. In contrast, the French have addressed it with success, applying geometry"[41]—does not refer to actual stonecutting, but rather to structural issues. That is, Italian theorists, except the cosmopolitan Guarini, do not admit French superiority in this field, as posited by Pérouse; instead, they ignore the subject entirely.

Interest in actual stereotomy did not arise again until the nineteenth century, with Tramontini (1811), who included some elemental schemata in a general treatise on descriptive geometry, and Bordoni (1826) and Tatti (1839), the Italian representation in the controversy on oblique arches. As a summary, Italian writers, with the exception of Guarini, eschewed stereotomic theory up to the Industrial Revolution, while builders in the peninsula, Sicily and Malta constructed here and there some remarkable pieces, although admittedly not with the same frequency as in France or Spain.

A similar process took place in German-speaking countries. According to Werner Müller, German stereotomy derives from three sources: first, Late German Gothic, in particular Lechler and the Dresden sketchbook, known also as Codex Miniatus; second, the influence of Guarini and Vittone, and third, direct knowledge of French treatises (Müller 1969; Müller 1972; see also Pérouse [1982a] 2001: 214–215). However, these sources were alternative rather than additive. As we have seen, in the Late Gothic period, the virtuosity of the masters in the lodges of the whole Empire led to the preparation and circulation of a sizeable number of stonecutting manuals and even case studies, most of them in manuscript form (WG c. 1560; Codex Miniatus c. 1570; Facht 1593) although a few reached the presses (Ranisch 1695; see also Pliego 2017). This tradition was alive and kicking until the seventeenth century; from that moment on, the influx of Baroque architecture, imported directly from Italy, merged with German traditions, leading to a taste for painted surfaces. 'Stereotomic' forms, such as three-dimensional arches, are present in such well-known masterpieces of

[41] Milizia (1785: III, 236): *Nè gli Antichi, nè gl'Italiani, che eressero tante cupole e tante specie di volte, hanno trattato nai questo soggeto teoricamente. I Francesi all'incontro vi si sonon impegati con profitto, applicandovi la Geometria.*

German Baroque as the Vierzehnheiligen church, entirely painted; as Pevsner (1958) pointed out, the inclusion of curved arches may derive from Guarini. Later on, in the nineteenth century, the French influence fostered a large number of stonecutting manuals. Thus, external influence led German-speaking countries to leave aside their stonecutting traditions twice. First, they eschewed in part, although not totally, their own rich Late-Gothic tradition for Baroque architecture, which involved complex shapes, usually painted and materialised in brick. Later on, they again set aside this tradition for nineteenth-century French-inspired stonecutting manuals, which were put into practice in a limited number of locations, such as the additions to the Zwinger in Dresden by none other than the great theorist of the nineteenth century, Gottfried Semper.[42]

The picture in England is quite different; in fact, the introduction of stereotomy in England seems to have been driven mainly by Palladio's influence, as strange as it may seem. Inigo Jones built the Tulip Staircase in the Queen's House, Greenwich, following the model of the stair at the Convento della Carità, substituting a circular plan for Palladio's oval one (Campbell 2014b). De la Gournerie (1855: 24–25) suggests that Wren was able to apply geometrical methods to astronomy due to his long practice on stereotomy. Perhaps he knew about the striking stonecutting in Paris Observatory (Swanson 2002; Swanson 2003), or about Blondel's (1673) problems in architecture, which involved issues in conics, or its application to actual stonecutting in de la Hire's manuscripts (c. 1690).

In any case, Wren (or his collaborators) built several remarkable stonecutting pieces, such as the ox horns at both sides of the altar in Saint Lawrence Jewry, London. The Tulip staircase model was enlarged in the Dean's staircase at Saint Paul's (Campbell 2014); further, the entire nave of the cathedral is covered by sail vaults with spherical lunettes, a stereotomic tour de force that has passed mainly unnoticed. Nicholas Hawksmoor was probably aware of such solutions, if not personally involved; he used a carefully dressed sail vault in the entrance porch of Saint George in the East. After the parenthesis of Neo-Palladianism, sail vaults resurfaced, albeit rendered, in the work of George Dance the Younger and John Soane, who may have used de la Rue's treatise (Pérouse [1982a] 2001: 218)[43] either directly or through General Vallancey's interpretation. Later on, as we have seen, the skill of British engineers in stereotomy allowed them to find a practical solution for skew bridges, while

[42]According to Pérouse (1982: 215) Derand and Desargues were translated to German since Sturm had read them; in note 15 to Chap. 20, he admits that he has found no mention to these translations other than the one in Sturm's book. However, Goldmann and Sturm (1699; 12) quote Derand (1643) and Bosse and Desargues (1643a) by the French names of their books and do not allude to a translation into German. Page numbers in Goldmann and Sturm are taken from the electronic facsímile at https://digi.ub.uni-heidelberg.de.

[43]Pérouse mentions a brief remark by Summerson ([1953] 1969: 274–275): the use of a sail vault by Dance the Younger in the Council Chamber of the Guildhall, with no more than a thin moulding separating the pendentives from the cap, may have been fostered by his knowledge of De la Rue's solution to this problem. However, it is worthwhile to remark that in later editions (Summerson [1953] 1983: 452) he mentions Marie-Joseph Peyre as a source instead of De la Rue.

their French counterparts, hypnotised by the abstract power of descriptive geometry, had failed to do so.

Last but not least, Pérouse and other researchers in the field do not say a word about Portugal. This is rather unfair; Frézier (1737–39: III, 28) singled out the vaults in the crossing of the Hyeronimite Monastery at Belém as the foremost example of Gothic vaulting. Both in this period, at the beginning of the sixteenth century, and in the later classical epoch, for example in open-well staircases in the Claustro dos Felipes in Tomar, Portugal managed to absorb the Spanish influence to arrive to its own distinctive and brilliant solutions (Genin 2009; 2014; Delgado 2017).

12.3.3 A Polycentric Narrative

Thus, rather than looking for the origins of stonecutting in the Languedoc, Syria, or Armenia, the appearance of complex stonecutting techniques should be explained by a complex narrative, starting from three obvious facts. First, in pre-industrial architecture, transportation resources were scarce and expensive; as a rule, builders working on a particular region used locally available materials. This explains the almost exclusive use of hewn stone in Armenia and Malta, in contrast to the preference for brick in Toulouse, Zaragoza or Constantinople. Of course, there are exceptions to this principle, in particular when less expensive transportation by boat was used, as in the cathedrals of Canterbury and Seville and the Castel Nuovo in Naples, sourced respectively from Caen in Normandy, El Puerto de Santa María in the Atlantic, and Santañy in Majorca.

Second, as we have seen in Sect. 12.1, architecture is a strange craft, its logic lying half-way between the strictly practical rationale of engineering and different external influences, which can range from theological beliefs or cultural intention to pure and simple caprice. This set of complex and contradictory constraints fosters, of course, a wide range of architectural vocabulary and constructive solutions.

Now, twentieth-century architectural historiography has been written mainly by professional art historians. While their contributions have enlarged architectural history with essential concepts and useful methods, they are prone to apply several conceptions that fit the figurative arts reasonably well but may be misleading in the architectural field. One of them is their insistence on *divergent evolution* at the expense of *convergent evolution*. These concepts originated in biology; the latter has seldom been applied to art or architectural history. In *divergent evolution*, genes are transmitted from one living being to another, causing a similarity between a creature and its offspring. Gradually, as descendants adapt to different environments, the resemblance wanes, although some common traits are shared, as for example between hominids. All this is well known, both in biology and art history. Whole schools have been identified on the basis of the similarities between artists or architects apprenticed in the same workshop or under the same master, resembling one another like artistic siblings. However, two species or genera may show similar traits without common ancestors, simply because they have adapted to comparable environments.

This phenomenon, called *convergent evolution*, although less frequent than divergent evolution, is well known in biology. In contrast, it has been neglected in architectural history, although this can explain the development of similar architectural forms in different milieu. Maybe this is the result of a translation of mental habits stemming from the historiography of figurative arts to architecture. It is not frequent for painters or sculptors coming from different backgrounds to arrive at similar ways of drawing a hand. This is what justifies the method of attribution in the figurative arts. By comparison, attributions in architectural history are quite unreliable, since it is relatively easy to copy a set of mouldings from another building or from the sheets of Vignola. In fact, while there are potentially infinite ways to draw a hand, the available geometrical vocabulary in Western architecture before the twentieth century is somewhat limited, partly as a result of the constraints placed by pre-industrial setting out methods. Ropes, squares and compasses foster the use of straight lines, circles, ellipses and their derivatives—arcs, squares, rectangles, ovals, triangles—and that is about all. This explains the development of similar architectural forms in different milieu; a well-known example is the resemblance between Mycenean, Maya and Far Eastern corbelled arches. Builders from entirely different backgrounds in time and space arrive at comparable solutions through adaptation to a similar context or by pure chance.

Additionally, we should bear in mind that the visual image of a constructive element does not tell everything; apparently similar forms do not imply identical constructive technologies or underlying geometries. About the first factor, we have seen in Sect. 12.3.1 the radical differences between Romanesque and Renaissance technology in this field. For example, although Armenian half-domes may recall Western ones, the inspection of some ruins reveals that they are built with a rather thin veneer of stone acting as formwork for concrete, in contrast to the massive Western construction. Still another example stresses the contraposition of the apparent surface image and internal geometry: at first sight Armenian pendentives recall Western ones. On closer inspection, subtle details, such as slightly sloping pendentive joints, signal the use of planes for bed joints. Although this may seem counterintuitive, all this contrasts sharply with the horizontal joints generally used in the West, standing at the edge of conical surfaces (López-Mozo et al. 2013) (Fig. 12.7).

Taking these ideas into account, to explain the historical evolution and geographical distribution of complex stonecutting in Western Europe we should leave aside simple explanations and opt for a nuanced narrative. First, ashlar construction requires a specific, complex, skilled and expensive technology. It arises in locations with accessible natural formations of easy-to-dress stone, in particular limestone, although marble, sandstone and even granite have been used to materialise complex pieces. In areas where stone is the sole available construction material, such as Malta, Armenia or the Indian town of Fatehpur Sikri, ashlar masonry enjoys a monopoly. We should never forget that both rubble and brick masonry require much mortar, which in turn demands sand, water and wood to operate lime kilns; in contrast, ashlar masonry can be executed with little or no mortar. Thus, rather than looking for a single source of stereotomy in Syria, Armenia or the Languedoc, we should think about a polycentric origin, driven by adaptation to available materials and

Fig. 12.7 Pendentive with planar—as opposed to conical—bed joints in the monastery of Haghpat. Notice the decrease in width in the centre of the top course (Photograph by the author)

resources, social and economic conditions in each location and period, and the architectural needs, intentions and caprices; all three mechanisms may operate by way of divergent or convergent evolution.

About the first term of the equation, local material and resources, it is easy to mention Pharaonic Egypt, Classical and Hellenistic Greece, Early Christian and Byzantine Syria, Early and Medieval Armenia, Medieval Islamic architecture in different locations, Late Medieval Western Europe, Mughal India, Early Modern France, Spain, Malta and southern Italy, as well as Early Industrial Revolution England as places and periods making a specially intensive use of ashlar masonry. In contrast, Mesopotamia, Imperial Rome, the city of Byzantium, Northern Early Middle Ages or Central Italian Renaissance, used mainly adobe, pozzuolana concrete, thin bricks embedded in a mass of mortar, wood, and brick, respectively.

Before the Industrial Revolution and the relatively inexpensive transportation facilities of the twenty-first century, the choice of these materials has been influenced, although not determined, by a complex economic structure. It is easy to explain that the narrowness of the Nile valley made stone available for Pharaonic constructions, while the broader valley of the Tigris and the Euphrates hindered transportation of stone to Mesopotamia, leading the successive civilisations in the valley to use adobe. In other cases, however, the reasons are more complex. For example, Roland Bechmann ([1981] 1996: throughout, but especially 132–142) has connected the emergence of Gothic architecture to the depletion of forests and the lack of wood; this fact made economical formwork and centring schemes indispensable (see also Fitchen [1961] 1981: 175–196 and Sect. 3.3.2). This situation brought about a completely different approach to stone construction. The linear ribs and thin webs of Gothic construction require smaller volumes of stone; however, they demand much skilled labour in dressing the voussoirs of the ribs, in contrast to Romanesque rubble masonry. At the onset of the Renaissance, even greater skill was needed for dressing the complex pieces of Early Modern vaulting; this fact alone should be sufficient to make it clear that Romanesque and Renaissance stonecutting are based on entirely different technologies.

However, these factors do not explain all material and typological choices; thus, we must turn to the third member of the equation, architectural needs, constraints, intentions and caprice. For example, in Pharaonic Egypt and Classical Greece ashlar masonry was used as a core material, although both civilisations eschewed complex geometries; even arches and vaults, although present in a few isolated examples in both periods, were relegated to small, secondary elements. Complex shapes in ashlar masonry emerged in some particular locations in the Hellenistic and Roman world, such as the Baths in Gerasa or Leptis Magna, the theater of Sabratha, the tombs of Ummidia Quadratilla in Casino, the amphitheatres of Nîmes and Arles, the arch of Cáparra or the first section of the spiral ramp in the Mausoleum of Hadrian. Sakarovitch (1998:105–107) has described Christian Syria as no less than the cradle of learned architecture, but the typological repertoire focuses on the quarter-sphere vault and the trumpet squinch, exactly as in Early Christian Armenia: crossing arches and other ribbed vaults would appear much later, in the thirteenth century, in all evidence through the Islamic influence.

This brings us to pointed arches and rib vaults, developed in the Islamic world, although generally without diagonal ribs. The import of such solutions in the West has been explained as a result of the return of the Crusaders (Viollet-le-Duc 1854–1868: VI, 424–425; Choisy 1899: II, 24, 249, 512, 515; for a different opinion, see Sakarovitch 1997: 14–16; Sakarovitch 2009b: 13–14). This may be true, but we should also take into account that contacts with the Muslims of the southern shore of the Mediterranean, Spain and, up to the twelfth century, Sicily, were not infrequent.[44] Of course, the knowledge of these Islamic solutions, combined with the lack of wood and social changes, fostered the use of rib vaults in Gothic architecture, contributing new items to the stereotomic repertoire.

The school of Pérouse has made much, as we have seen, of another rich section of the catalogue: splayed arches, barrel, groin and pavilion vaults, and the like. However, these pieces are an interesting example of how architectural forms can travel from

[44]In fact, one of the first medieval translations of Euclid into Latin was prepared by Gerardo de Cremona from Arab sources in Toledo.

one material to another. Some of them—in particular vaults—were materialised frequently in lime and pozzolana concrete in the days of the Roman Empire. When the Mare Nostrum was no longer safe for maritime commerce, and pozzuolana was scarce or unavailable, vaults were built on some occasions in ashlar masonry, as in the Mausoleum of Theodoric. However, this solution demanded skilled labour, which was increasingly scarce; groin vaults were frequently built in rubble during the Romanesque period and revived again in ashlar masonry in the Renaissance.

Generally speaking, such knowledge transfers are driven by two sets of factors: while geological, economic and social circumstances constrain the choice of material, preference for circular, cylindrical or spherical shapes is the result of geometrical constraints—after all, these shapes were easily controlled with a rope in preindustrial times—but also, the offspring of a *meme* or cultural construct travelling from Late Roman to Early Christian, Byzantine, Romanesque and lately Renaissance architecture. In this context, Gothic architecture provides an excellent aporia against technological determinism: while the use of the rope as a means of formal control fostered round arches, barrel vaults and circular apses in Classical and Romanesque periods, Gothic architecture used the same control tool to fulfil its own interests, which brought about pointed arches, rib vaults, polygonal ones and radial chapels.

In any case, the classical repertoire was combined in the Renaissance with a rich set of techniques and processes created or rediscovered in the Gothic period, fostering the emergence of the ensemble of architectural types and geometrical methods that form the core of this book. A quick glance to Chap. 2 and Part II in this book will make it clear that this process was especially intense in France and Spain, although the latter was only active during the sixteenth century and entered a slow but unavoidable decline in the next century, while France maintained a high level of theoretical and practical production until the full onset of the Industrial Revolution pushed ashlar masonry out of the market, in the late nineteenth century. However, this French predominance does not mean that other countries were left out of the scene. The role of Britain, although focusing on a few types such as spiral staircases and skew arches, is altogether quite brilliant, as we have seen.

To summarise, trying to locate the specific origins or the central epicentres of stereotomy is akin to chasing a ghost. Comparable forms may be explained by divergent evolution, that is, knowledge travelling along long dynasties of masons or through structured training systems (see Sect. 12.2). However, as mentioned, convergent evolution, involving the use of similar forms to address similar constraints, should never be discarded; as mentioned, we should not forget that the geometrical vocabulary of architecture is infinitely more restricted than the endless repertoire of the figurative arts. Moreover, superficial (in both senses of the word) coincidences in the visual image of constructive elements in different locations or periods do not guarantee similar internal geometry or equivalent construction technology.

Additionally, the examination of the history of complex forms built in ashlar masonry should not be limited to Western Europe. First, we have seen how the Islamic tradition of rib vaulting was imported into Europe to develop Gothic architecture. In contrast, the image of Asian architecture in the West has focused, either positively or negatively, on the wooden constructions of China and Japan, while the rich tradition of ashlar masonry in India is frequently overlooked. Moreover, two traditions intermingle in Mughal India: an Oriental one, present for example in Fatehpur Sikri, and an Islamic one, derived from Iran and Turkestan, involving, for example, bulbous domes, which shines at the Tah Majal. Obviously, no single book or author can cover such a rich history; this is why this book endeavours modestly to furnish just a piece of this jigsaw puzzle and deals with Western Europe in a relatively limited time frame.

12.4 The Sources of Gothic and Early Modern Stonecutting Methods

12.4.1 Euclidean and Practical Geometry

The Middle Ages. A large number of popular publications take it for granted that medieval master masons had an exceptional command of geometry. In other cases, respectable scholars (Panofsky 1951) have suggested that masons shared the mental habits of the theologians of the period; in other words, they were in contact with the learned elite of the period. However, the available evidence points in the opposite direction. Generally speaking, the illiteracy of most medieval masons and the lack of translations of classical texts into the vernacular acted as strong barriers to the use of this knowledge by masons up to the Renaissance. Lon Shelby (1972) described the situation in a pivotal study that remains essentially valid today, in spite of many recent developments in the field. Medieval masons seem to have considered geometry as the basis of their craft. For example, Villard de Honnecourt mentions geometry twice as the foundation of the "art of portraiture"[45] (that is, drawing), while Hand IV inserts a caption in the sheet that includes his stereotomic diagrams, stating that "All these figures are taken from geometry".[46] However, in this context, the word "geometry" stands for something far different from classical geometry. An important English text on medieval masons, the Cooke manuscript (c. 1400: 19r-21r; see also Knoop et al. 1938: 30–38) includes a fantastic narrative in a section about "Articles and Points of Masonry": Abraham travelled to Egypt with his aide Euclid, a "worthy clerk", in order to teach geometry to the Egyptians, while Euclid instructed them on how to build walls to stop the floods of the Nile and divide the land between proprietors; as a result, the name "geometry" stands for the craft of masonry. Of course, such vision is incompatible with a clear knowledge of Euclidean geometry. Medieval masons had

---

[45] Villard (c. 1225: 1v): *Wilars de Honecort vos salve ... Et si troveres le force de le por ... traiture. les trais ensi come li ars/de iometrie le commande*, (Villard de Honnecourt salutes you ... [in this book] you will find the technique of representation as the discipline of geometry requires and instructs it [to be done]). Transcription and translation taken from Villard/Barnes 2009.

[46] Villard (c. 1225: 19v) *en ces. i .i .i .i fuelles a des figures de/Lart de iometrie* ...(On these four leaves are figures from the discipline of geometry ...) Transcription and translation taken from Villard/Barnes 2009.

no access to Euclid, since it was not translated into Latin until the twelfth century and there is no evidence about masons reading Latin or attending grammar schools or universities, the only effective road to classical literacy in this period. In fact, most masons could not read the vernacular languages until the sixteenth century.

Several scholars have tried to connect Villard with the medieval tradition of practical geometry, such as the treatises of Hugh of Saint Victor, Dominicus Gundissalinus and Leonardo di Pisa, known as Fibonacci, or the *Geometria Gerberti*. The first problem is that all these authors wrote in Latin, so their works were not accessible to the vast majority of masons. According to recent research, the main author of the Villard manuscript was not a mason, but rather some kind of agent for the Cambrai cathedral chapter; it is not clear whether he knew Latin (Villard/Barnes [c. 1225] 2009: 229–230). Anyway, both Villard and Hand IV may have used some practical geometry manual in the vernacular, such as *Pratike de geometrie*, written in Picard, the same language as most of the portfolio. However, both the Latin treatises and the *Pratike* deal mostly with planimetry, altimetry and cosmimetry, that is, the measure of planar areas, heights and volumes; they are more useful for surveyors than for masons. That is, the practical geometry treatises could have furnished material for the surveying diagrams in fol. 20r and 20v, but not for the stereotomic schemata in the same sheet. Only Dominicus Gundissalinus ([c. 1150] 1903: 109) makes a passing remark about masons in connection with practical geometry:

> ... The artificer of practice is he who uses [geometry] in working. There are two kinds of these, namely, surveyors and craftsmen. ... Artisans are those who exert themselves by working in the constructive or mechanical arts – such as the carpenter in wood, the smith in iron, the mason in clay and stones, and likewise every artificer of the mechanical arts – according to practical geometry. Each indeed forms lines, surfaces, squares, circles, etc., in material bodies in the manner appropriate for their art. These many kinds of craftsmen are distinguished according to the different materials in which and out of which they work. Any one of these thus has his proper materials and instruments. The instruments of the ... masons are the string, trowel, plumb, bob, and many others...[47]

However, he seems to put bricklayers and stonemasons into the same bag and, in any case, he does not include a single bit of geometrical information that could be useful to masons. All this led Shelby (1972: 409) to posit that

> ... stereotomical problems were solved by mediaeval masons primarily through the physical manipulation of geometrical forms by means of the instruments and tools available to the masons. These were rule-of-thumb procedures, to be followed step by step, and there

[47]Gundissalinus ([c. 1150] 1903: 109): *Artifex uero practice est, qui eam operando exercet. Duo autem sunt, qui eam operando exercent, scilicet mensores et fabri ... fabri uero sunt, qui in fabricando siue in mechanicis artibus operando desudant, ut carpentarius in ligno, ferrarius in ferro, cementarius in luto et lapidibus et similiter omnis artifex mechanicarum arcium secundum geometriam practicam. Ipse enim per semetipsum format lineas, superficies, quadraturas, rotunditates et cetera in corpore materiae, que subiecta est arti sue. horum autem fabrorum multe species esse dicuntur secundum diuersitatem materiarum in quibus et ex quibus operantur. Quorum unusquisque sicut habet materiam propriam sic et instrumenta propria. ... cementariorum (enim instrumenta sunt) uero linea, trulla, perpendiculum et multa alia ...*". Transcription and translation are taken from Shelby (1972: 403).

were virtually no mathematical calculations involved. We may thus characterize the practical geometry of Villard's *Sketchbook* more precisely as constructive geometry, by means of which technical problems of design and building were solved through the construction and physical manipulation of simple geometrical forms – triangles, squares, polygons, and circles.

Of course, this is consistent with the short extension and the evident errors of Roriczer's *Geometria Deutsch*, such as an incorrect construction for the regular pentagon, a problem which is solved flawlessly in Euclid (Roriczer c. 1490a: 2r; Euclid c. -300: 4.11; see also Meckspecker 1983 and Sect. 3.1.2). Shelby (1972: 412–416) pointed out the connections of this leaflet with a Latin manuscript, *De inquisitione capacitatis figurarum*, which uses the same techniques for the construction of the heptagon and the octagon; in neither case is a proof furnished. However, Shelby also remarked that *De inquisitione* explains the Archimedean method for the division of the circle as an arithmetical computation, while Roriczer transforms it into a graphic procedure; in other words, he takes material from the realm of learned geometry and carries it into the field of masons' constructive geometry. Much is lost along the journey: in fact, an Archimedean theorem states that π is smaller than 3+1/7 but larger than 3+10/71. Both *De Consolatione* and Roriczer convert this theorem in a problem with a single solution, arriving at inexact results. It is easy to mock Rorizcer's clumsiness, but we should remember that most forays into a new field of human activity are rather ungainly, to say the least; it is no wonder that, in his pioneering effort to apply learned geometry to practical construction, Roriczer choose as a guide a garbled account of classical science, either *De Consolatione* or a similar text. In any case, not all was lost: Roriczer's booklets mark the introduction in stonecutting literature of geometrical notation by letters, absent in Villard and ubiquitous in stonecutting literature from the late fifteenth century on.

At this point, the reader may be startled and ask: did medieval masons not assert that *ars sine scientia nihil est*? Well, this phrase is usually quoted without reference to context, and the context shows it is an exception that proves the rule. To summarise a complicated story, the cathedral of Milan was begun in 1386; in 1391 the architect, Nicolas Bonaventure, fled or was dismissed, the building committee decided to change the projected height of the nave, and the mathematician Gabriele Stornalocco was invited to give his opinion. After careful measurements, he proposed new heights for the nave and the aisles, determined using a procedure based on equilateral triangles, which he called *ad triangulum*, although he simplified the overall height of the cathedral to 84 *braccia*, in order to avoid irrationals. Shortly after, in 1392, Heinrich Parler was also asked for his opinion; he criticised Stornalocco and asked for the cathedral to be erected *ad quadratum*, that is, to equal the width and the height, rather than *ad triangulum*, as in Stornalocco's solution. A meeting of master masons was called; they rejected Parler's proposal and accepted Stornalocco's scheme only for the outer aisles, manipulating the height of the inner ones and fixing the height of the nave at 74 *braccia*, even lower than Stornalocco. Such decision involves both mathematical and architectural issues; it may be based on the Pythagorean triangle, but at the same time it seems to follow a penchant by Lombard, and generally Italian, architects for lower profiles. Later on, the French engineer or mason Jean Mignot was consulted; he held a number of bitter debates with the local masters. Most interesting for our purposes is his invective against the local masons, stating that *ars sine scientia nihil est*, that is, technical competence is useless without the support of scientific theory. His proposal, similar to that of Stornalocco in its scientific approach and the resulting height of the cathedral, won the support of Duke Gian Galeazzo Visconti initially, but it was finally discarded, and the cathedral was built according to the dimensions established in the 1392 meeting with some adjustments (Frankl 1945: 51–57; Ackerman 1949).

To summarise, medieval masons longed for material from Euclidean or practical geometry but were unable to use it for two reasons. On the one hand, the primary sources were just not available in the West until the twelfth century; even then, they were not accessible to masons, who generally could not read vernacular languages, not to speak of Latin. On the other hand, the learned geometry of the period, both Euclidean and practical, did not address the central problems in stonecutting, those related to projection and true shape. Thus, on the rare occasions when building committees sought the advice of mathematicians on these matters, masons strongly objected and ultimately, patrons followed the opinion of artisans.

The Renaissance. Only in the sixteenth century did masons start to use some concepts and methods coming from classical or practical geometry. Most Renaissance stonecutting texts followed this road, but the process was quite slow. De l'Orme states that "there is nothing so difficult or so strange that cannot be solved with the aid of tracings together with geometry, which is so rich that those who know it can make admirable things".[48] For Cristóbal de Rojas (1598: 1v-2v), geometry is the first and foremost branch of knowledge that the military engineer must command. Martínez de Aranda points out that "… the most excellent (stonecutting problem) is the one that hides many remarkable points of geometry in few words …".[49] However, we must ask ourselves whether such accolades are as devoid of actual meaning as the story of Euclid in the Cooke manuscript. The simplest way to answer this question is to analyse the geometrical propositions in these texts and their actual use in stonecutting. A few classical geometry constructions appear in books and manuscripts dealing with stonecutting, such as the determination of the centre of a circle when three points are known (de l'Orme 1561: 13r-13v; de l'Orme 1567: 55r-55v, 56v; Vandelvira c. 1585: 18v), the construction of the bisector of a segment, used as a perpendicular to the line (Vandelvira c. 1585: 3v) or the division of a segment in a given number of parts (de l'Orme 1567: 38v-39r; San Nicolás 1639: 24v).

However, we should not make too much of these bits of classical geometry in sixteenth-century stonecutting treatises. First, these propositions are placed at the introductory section of Vandelvira's manuscript or the beginning of the chapters devoted to stereotomy in de l'Orme's treatise. In contrast, the great majority of

[48]De l'Orme (1567: 71r) *De sorte qu'il ne se présentera chose tant étrange, ne tant difficile, qu'ils ne trouvent incontinent le moyen d'en venir à bout par l'aide de ces traits étant accompagnés de géométrie, qui est si riche que celui qui la connaît peut faire choses admirables.*

[49]Martínez de Aranda (c. 1600: preface [iii]): … *es más excelente traza la que en pocas palabras tiene encerrados muchos y notables puntos de geometría…*

the geometrical methods of Renaissance stonecutters, explained in the body of these texts, do not derive from classical science, nor are they a direct application of medieval stonecutting methods, since Renaissance architecture and single-tier construction brought about entirely new challenges, as we have seen at the end of Sect. 12.3.1. Second, the whole system of Renaissance stonecutting cannot be ascribed to rational geometry, since it seems to derive in large part from empirical methods, in particular, constructive geometry. As we have seen in Sect. 12.2, San Nicolás stressed he had "experimented these cuts before writing about them"; the phrase suggests links with the emerging science of the seventeenth century. At the same time, San Nicolás mentions the pair of terms *práctico* and *especulativo*, which derive from the tradition of medieval practical geometry of Hugh of Saint-Victor and his Spanish follower, Dominicus Gundissalinus. These concepts are relatively common in Early Modern Spanish architectural and engineering literature. In his introduction to the *Libro de Trazas de Cortes de Piedras*, Alonso de Valdelvira (c. 1585: 4r; see Barbé 1977: 26; Barbé 1993: 131–132) quotes *Fragmentos Matemáticos*, by Juan Pérez de Moya (1568: 5) who had also published *Geometría Práctica y Especulativa*. As the title suggests, in this book Pérez de Moya combines learned and practical geometry (see, for example, 1573: 20–28).

The Baroque period. The slow penetration of learned geometry into stonecutting texts continued during the seventeenth century. The controversy between Desargues and Curabelle (Desargues 1648; Curabelle 1644a, 1644b; see also Sakarovitch 1994b) signalled the beginning of a conceptual shift from practical craftmanship to abstract geometry. However, it did not bring about a significant transfer of learned geometry concepts to the field of stonecutting. Of course, Desargues tried to subject practical masonry to the rule of geometrical science, but, first, he was not waving the flag of true classical geometry, but rather a quite different science that, almost two centuries later, was called projective geometry; second, stonemasons strongly rejected his revolution. Another detail, practically unnoticed before now, is more significant for our purposes: Derand (1643: 394) applied the notion of tangency, without mentioning it, to the directrices of diagonal and tierceron ribs in order to smooth their junction. Also, the extension of flexible templates to most cylindrical vaults (Derand 1643: 153–161) seems to be fostered by the notion of cylindrical developments, although he deliberately simplifies the construction.

Remarkably, Milliet (1674), brought about almost no new transfers from learned science in his effort to include stonecutting into mathematics. In contrast, Guarini presented a consistent theory of projections and developments in his *Euclides adauctus* (1671); this was brought to architects and masons in *Architettura Civile* ([c. 1680] 1737). Blondel (1673: 56–59; see also Gerbino 2005), the first director of the Académie Royale d'Architecture, applied the theory of conic sections to sloping arches. De la Hire (c. 1668a: I, 71r-74r), a professor in the Académie, applied the notion of dihedral angle to groin vaults, perhaps following Florimond de Beaune (c. 1640; see also Tamboréro 2008: 73–78), although de Beaune deals mainly with trihedral angles. De la Rue (1728: 44–46) trod on the same path; he also addressed conic and cylindrical sections in the final part of his treatise, the *Petit traité de stéréotomie* (1728: 163–183).

However, as we have seen, Frézier (1737–39: I) turned de la Rue's strategy upside down: rather than being a final complement, and advanced course so to speak, abstract geometry was placed by Frézier at the base of stonecutting; his introductory volume includes in particular the theory of projections, surface intersections, developments and angular measures. None of these fields stems directly from classical, Euclidean geometry, but they comply, up to a certain extent, with the requisites of learned science. In particular, Frézier included in the second and third volume of his monumental treatise the proofs that Derand had omitted, although, in the form of the *Explanation Démonstrative*, that is, mixing scientific proofs with didactic explanations (see for example Frézier 1737–39: II, 81–82, 121). This supplied much material to nineteenth-century descriptive geometry; however, before addressing this issue, we must turn our heads to other, more plausible sources of stonecutting methods.

12.4.2 The Oral Tradition: Artisanal Instruction and Empirical Methods

As we have seen in the preceding sections, we may surmise that the medieval stonecutting tradition, in particular, Shelby's "constructive geometry", furnished at least a platform for Renaissance innovations. In fact, most basic methods of Renaissance stonecutting, such as the determination of the lengths of segments from their horizontal projection and the difference in heights between edges, triangulation or transfer of distances from a reference plane to intrados templates fit into Shelby's definition of constructive geometry as "the construction and physical manipulation of simple geometrical forms".

Since masons were generally illiterate up to the sixteenth century, we may assume that these practices were passed on from one generation of masons to the next one by oral transmission and, as a result, were not recorded until the time of de l'Orme. Thus, to discuss the formation of most Early Modern stonecutting methods is a risky business. In this section, I will try to prove that masons in the first half of sixteenth century *could* have put together most Renaissance stonecutting methods using constructive geometry, without substantial use of Euclidean and practical geometry; of course, most details of this process are unclear, due to the lack of recorded sources.

Triangulation. As we have seen in Sects. 2.1.1 and 6.2.2, Hand IV (Villard c. 1225: 20r, dr. 9) seems to control the shape of a voussoir of a skew arch through a true size and shape construction involving a triangle formed by a line in point view, a radial segment in the face plane and an intrados joint. While the length of the radial segment may be taken easily from an elevation, it is quite difficult to compute it through any other method. Thus, he used probably a double orthographic projection. This is a constant in most Renaissance stonecutting methods, which endeavour to construct a figure in true shape starting from orthogonal projections.

Triangulation, an important branch of Renaissance stereotomic methods, seems to stem from these roots, although surveying methods and practical geometry texts may have contributed to its development. In contrast to *planimetria*, the triangulation method used by Renaissance masons generally involves two phases.[50] First, the lengths of sides and diagonals of templates are measured or computed. Some of these segments may be horizontal and thus may be taken directly from the plan. On other occasions, masons construct right triangles with the horizontal projection of the segment and the difference of heights between its ends; the hypotenuse equals the real length of the segment. This operation recalls a surveying diagram, drawn by Villard's hand ([c. 1225] 2009: 20v, dr. 10) although the approach is different. In Villard's scheme the difference in heights is an unknown, so he must use the angle between the hypotenuse and the horizontal cathetus (Bechmann [1991] 1993: 156–157).[51] In contrast, for Renaissance masons the unknown is the hypotenuse; they do not use angles in this phase. This technique may be inspired by practical geometry or surveying treatises, although the details of such transfer are unclear. In the second phase, masons construct triangular templates using the real lengths of the three sides (de l'Orme 1567: 95r-96r; Vandelvira c. 1585: 7v) or quadrilateral ones starting from sides and diagonals (Alviz c. 1544: 7r, 8v; Vandelvira c. 1585: 26v-27r). This operation may derive from surveying practices, with much adaptation, or may have been developed independently in stonecutting workshops.

Another construction involving right triangles is applied to the ox horn (the splayed arch with a springing orthogonal to the face planes) and the *biais passé* (skew arch with bed joints orthogonal to the faces) (see Sects. 6.1.2 and 6.2.2). In both cases, masons construct a triangle using as catheti a line in point view whose length equals the thickness of the wall and the apparent distance between the vertical projections of the ends of the intrados joints. Although the latter sounds like a fairly advanced concept, for a mason it is an intuitive one since it represents a side of the wedge that must be taken off to shape the intrados of each voussoir. In any case, what masons are trying to compute with this operation is the angle between the intrados and face joints. The origin of the concept must have been purely empirical, since some form of it seems to be depicted in the skew arch drawn by Hand IV in Villard's manuscript (Villard [c. 1225] 2009: 20r, drawing 9; see Sect. 6.2.2); at this stage, no Euclidean or practical geometry text was in the hands of masons (see Sect. 12.4.1).

It is quite interesting to compare this procedure with Martínez de Aranda's interpretation (c. 1600: 11–12, 15–16). All other writers place these triangles near the springers, probably to avoid going back and forth on all fours. In contrast, Martínez de Aranda uses as a cathetus the horizontal projection of the line in point view; as a result, triangles are scattered throughout the length of the arch. This solution seems natural for an architect or engineer trained in nineteenth-century descriptive geometry, but nothing suggests that Martínez de Aranda knew the notion of revolution;

[50] Alviz (c. 1544: 3r, 7r, 8v); de l'Orme (1567: 92v-94r, 100v-101v); Vandelvira (c. 1585: 7v, 26v-27r). See also Palacios ([1990] 2003: 34–37, 96–99); García-Baño (2017: 104–113, 157–162); and Sect. 3.1.3.

[51] There is another surveying scheme in Villard's portfolio ([c. 1225] 2009, drawing 12) by Hand IV, which involves triangulation. However, it seems to be connected either with a right angle, as suggested by Bechmann ([1991] 1993, 154–155) or with similar triangles.

instead, it appears that he was trying to explain the procedure by assimilation with the method he used for quadrilateral templates, for greater clarity.

Auxiliary views and cylinder intersections. Renaissance texts, such as Alviz (c. 1544: 21r, 22r, 23r, 24r, 25r, 26r), de l'Orme (1567: 120v), Vandelvira (c. 1585: 13r, 23v, 80r, 81r) and Martínez de Aranda (c. 1600: 3, 7, 9, 10, 86) construct true-shape representations of vertical planar figures using a folding line parallel to the plane of the figure. Employing auxiliary lines, usually furnished by intrados joints, they raise perpendiculars from the folding line for each relevant point of the figure; next, they transfer the height of relevant points, taken from the original elevation, to the perpendiculars. It may be argued that most of these early examples are virtually indistinguishable from cylinder intersections, but there are a few cases of true-shape representations obtained by this method with no material cylinder present (Vandelvira c. 1585: 10v; Martínez de Aranda c. 1600: 76, 78, 88, 206). In any case, this observation suggests a likely hypothesis for the origin of this technique. Masons probably first conceived the idea of an intersection, using intrados joints as auxiliary lines, since the notion of a line or joint meeting a wall or plane is a strongly intuitive one; later on, they extended the technique to true-size representations where no cylinder was present.

A similar technique was used to tackle an apparently complex problem: the intersection of two cylinders. Provided that one of the cylinders is shown in edge view (that is, each generatrix is depicted as a line in point view), the projection of the cylinder appears as a single curve; thus, it is easy to determine the intersection of several generatrices of the other cylinder with the vault shown in edge view. As we have seen in Sect. 8.4, several variations of this method were applied to orthogonal and oblique intersections of horizontal-axis vaults, and even to the junctions of a sloping vault with a horizontal one, both orthogonal and oblique.

Rotational symmetry. Vandelvira (c. 1585: 7r), Rojas (1598: 99) Martínez de Aranda (c. 1600: 33–34, 66–67) and Jousse (1642: 78–79, 82–83) understood clearly that all voussoirs in symmetrical trumpet squinches or splayed arches are equal, provided that their directrices are circular and the apex is placed on the symmetry axis of the ensemble; as a result, the template for the springers, which can be drawn directly, may be used in all voussoirs. Vandelvira uses a more advanced notion when he states that in rhomboidal-plan skew arches, "since both springers are parallel, the same templates, turned around, can be used at the other end".[52] Thus, the templates for the left side are skew-symmetrical with the templates for the right one, so they may be reused placing the opposite face on the stone surface.

All this seems rather advanced. However, learned geometry of the period does not deal with such issues as symmetries and rotations in space. It seems most probable that masons discovered these notions empirically, observing the rotations of cartwheels, templates, or stones turned around with levers, as we have seen in Sect. 4.1.1, or the host of rotating devices used in the Middle Ages and the Early Modern period.

---

[52]Vandelvira (c. 1585: 28r) *estas propias plantas vueltas sirven para este otro lado por ser el viaje paralelo*. Transcription is taken from Vandelvira and Barbé 1977. See also Derand (1643: 155).

Distances to vertical reference planes. To construct four-sided templates by triangulation is tiresome, recursive and prone to the accumulation of errors. Martínez de Aranda and many other writers draw them using a procedure that has some traits in common with, and may be explained by means of, nineteenth-century revolutions, as we have seen in Sect. 6.2.1. However, it is important to remark that most writers do not explain this procedure using orthogonals to the axis of rotation and they rarely draw these orthogonals. Quite to the contrary, they measure distances to a vertical plane orthogonal to the so-called axis of rotation and transfer these distances to a parallel to the axis. This method features many variants, from the simplest to the most complex; this suggests that it appeared gradually along a slow historical process.

A remote antecedent on this procedure can be found in Hand IV's scheme for an arch opened in a round wall (Villard c. 1225: 20r, dr. 8-h). As we have seen in Sect. 6.4.2, a templet set tangentially to the wall and some marks between the templet and the wall suggest the instrument was used to measure horizontal distances to the masonry, probably to voussoir corners. However, given the schematical nature of Hand IV's diagrams, it is not safe to hypotesize beyond this point.

Jumping to the Renaissance, the simplest variant of the method is present in the trapecial-plan skew arch (Alviz c. 1544: 9; Martínez de Aranda c. 1600: 6–8; see also Sect. 6.2.1). When trying to construct true-shape templates for this piece, it is easy to conceive the idea of taking the width of bed joint templates from the length of the face joint, which may be measured on the elevation of the arch, since one of the sides of the template is orthogonal to the springings. In the same way, the width of the intrados joint may be taken from the distance between two successive intrados joints, which are shown in the elevation as lines in point view. All this furnishes three corners of the bed joint or intrados template. Masons took it for granted that the fourth corner maintains its distance to a vertical plane orthogonal to the intrados joints. This idea may be fostered by the observation of templates in symmetrical or skew-symmetrical pieces, stones in dressing, cartwheels or other rotating devices; most likely, it was tested using models (see Sects. 3.1.4 and 5.2). In order to reduce tracing effort, the springing line of the elevation is frequently reused as the footprint of an auxiliary vertical plane, thus acting as a remote antecedent of the *ligne de terre* of descriptive geometry.

The second step in this evolution involves a bold virtualisation. In circular-section, elliptical-face skew arches, there is no material orthogonal plane to use as a reference. However, masons used a virtual round arch as a cross section of the piece. Again, the width of the bed joint template may be taken from the length of face joint and that of the intrados template from the distance between two consecutive intrados joints, measuring both in the cross section. Also, the springing line of the cross section may be reused as a reference line in order to locate *both* ends of the extrados joint and the upper intrados joint. As an exception, Vandelvira (c. 1585: 19v, 20v-23v) actually draws orthogonals to intrados joints, although he explains them in the text as parallels to the reference line, rather than orthogonals to the intrados joints.[53]

[53] Vandelvira (c. 1585: 19v): *comienzan de sus plomos y acaban conforme el plomo de adelante toca en el arco llevado a trainel de la línea plana con aquella línea de puntos* (their plumb lines

The persistence of the mental habit of measuring distances to the springing line is shown in Martínez de Aranda's solution to the elliptical-section, circular-face skew arch (Martínez de Aranda c. 1600: 16–17; see also Rojas 1598: 99v, bottom, who may have inspired it). First, explains that he will draw an orthogonal to the springings as *juzgo* or auxiliary line. The use of this construction is remarkable: it furnishes a base for the cross section, an aid for the construction of intrados joints as parallels to the springings, and a vertical reference plane. Even more surprising is the fact that although Martínez de Aranda needs orthogonals to intrados joints, he does not explain this in so many words. Instead, he uses the word *galgar*, which applies to the construction of parallels, as we have seen in Sect. 6.2.1. Since the elevation represents in true size distances between intrados joint ends, he uses them to locate the upper corners of the template, drawing an arc with a radius equal to the distance between the ends of the upper and lower joints, from the end of the lower joint. Where the arc meets the orthogonal, he may place the end of the upper joint.[54] This is an interesting generalisation of the procedures we have seen in the preceding paragraphs. However, the fact that Martínez de Aranda does not mention the tracing of an orthogonal to the lower intrados joint, but rather a parallel to the *juzgo*, in apparent contradiction with the principle of tracing economy, hints that he had not fully grasped the potential of this new procedure. Consequently, he explained it by reference to usual practices, applying an empiric procedure rather than geometrical reasoning.

Another advanced example of these empirical methods is the one used by Jean Chéreau (c. 1567–74: 113r) and other writers to solve a trapecial or rhomboidal skew arch with elliptical faces. He noticed that the bed joints could be extended until they all meet at a single point, namely the intersection of the axis of the arch with the face plane. Thus, they can be drawn at a springer, transferring the distance from the lower end of the face joint to the springing; next, the mason should lean a straightedge on the common intersection and the transferred end, drawing the face joint starting from this end and dragging the scriber to the opposite side of the common intersection. Of course, this procedure is tantamount to a rotation of the face joint around the axis, but such theory was not explained in sixteenth-century learned geometry treatises. We may surmise, then, that it was inspired by the observation of wheels or other revolving artefacts, or by the rotation of templates and bevels, and quite probably tested and demonstrated using models.

Distances to horizontal reference planes. Interpretations of Hand IV's diagrams are still debatable; even the identification of the problems is not completely clear. It may be debated whether the triangle in the skew arch in was drawn in a full-scale tracing or inscribed directly on the stone while dressing it; moreover, the vertical

start and end where the front plumb meets the arch parallel to the level line with this dotted line). Transcription is taken from Vandelvira and Barbé 1977.

[54] If we read Martínez de Aranda's words literally, he should draw a parallel to the *juzgo* using two orthogonals to this line; one of these parallels is furnished by the lower intrados joint, but at this stage he has not yet drawn the upper intrados joint, since he needs to construct the arc and intersect it with the orthogonal to the intrados joint, that is, the parallel to the *juzgo* he is trying to draw. Thus, he needs an additional, auxiliary orthogonal to the *juzgo* to perform the *galgar* operation.

projection is not actually present in the manuscript. No details about the stonecutting technique of the arch opened in a round wall can be ascertained. In contrast, methods for the control of star vaults in Early Modern texts are clearer, as we saw in Sect. 10.1.4. Although details differ, masons did not try to construct an orthographic elevation and instead drew directly true shape representations. This brings about the issue of disarticulation: a set of independent true-shape depictions of wall arches, diagonal ribs, tiercerons and liernes does not help to verify that tiercerons and liernes meet at the secondary keystones. Thus, in order to guarantee the consistency of the system, Hernán Ruiz II (c. 1560: 46v; see also Vandelvira c. 1585: 96v or Derand 1643: 393) draws all ribs starting from a springing line and he uses horizontal lines joining the upper end of the tierceron and the lower one of the lierne. These horizontal lines are absent in Gelabert (1653: 147r; see also Ribas 1708: 109)[55]; quite probably, he checked that the tierceron and the lierne met at the secondary keystone by measuring the heights of their endpoints over the springing line; this method is akin to the concepts held by the German masters or Rodrigo Gil.

All this suggests a parallel with the distances to vertical reference planes used by Renaissance masters. Distances, which stand implicitly for orthogonals, are used by Facht von Andernach, Rodrigo Gil and de l'Orme for star or net vaults and most writers, up to Jousse and Derand, for arches; explicit orthogonals are used by Hernán Ruiz II, Vandelvira and Derand for the star vault and by Vandelvira again and de la Rue for arches. Thus, it may be that the practice of measuring distances to a vertical reference plane in arches may have derived from the distances to the springing plane used by the German masters and Rodrigo Gil for tierceron and net vaults, or the other way around.

However, there are several arguments against this hypothesis. First, this abstract transposition is at odds with the comment by Martínez de Aranda that masons are "men tied to matter". As remarked by Rabasa, the vertical segments used to measure the distances to a horizontal plane are known as *plomos* while there is no equivalent word in the masons' jargon for lines in point view. Second, the uses of the two methods are different. While horizontals or distances in star and net vaults are used to locate points in space and assure that different ribs will meet at a keystone, orthogonals and distances in arches are used to construct the planar shapes of templates, or at least the angles of bevel guidelines. Third, horizontals and distances in star and net vaults connect different true-size depictions of ribs, while orthogonals and measurements in arches always join a projection of the member with a true-shape template. At most, the two ideas may have been mutually reinforced, although the idea of independent origins seems more likely.

Thus, this section shows that the core geometrical methods used for squinches, arches and rere-arches in Renaissance stonecutting *may* have been conceived and tested empirically, without resort to the concepts and problems of learned and practical geometry. In contrast, they fit well into Shelby's definition of "constructive

[55]The pages of Ribas's manuscript are unnumbered, and there is, up to this moment, no published facsimile, either physical or electronic; thus, I am using the page numbers in a PDF copy furnished by the Biblioteca Nacional de Catalunya.

geometry", that is, the physical manipulation of geometrical forms by means of the instruments and tools available to the masons. However, to explain the formal control procedures used in spherical, annular and most oval vaults, we must turn to other scientific fields also dealing with other forms of projection, namely perspective, cartography and gnomonics.

12.4.3 Perspective, Cartography and Gnomonics

As we have seen, Pérouse included some remarks about perspective and stereotomy as "distinct, although constantly associated techniques",[56] pointing out that both are included in the manuscript of Jean Chéreau (c. 1567–74). However, the connection between perspective and stereotomy is not evident, since the former is based on central or conical projection and the latter on cylindrical, or specifically orthogonal, projection. An interesting study case of this issue is offered by *mazocchii*, a special kind of head garment used by Renaissance painters, in particular, Paolo Ucello and Piero della Francesca, as an exercise in perspective, and Spanish Renaissance annular vaults, which feature exactly the same geometrical shape. As remarked elsewhere (Calvo and Alonso 2010), when drawing *mazocchii* in perspective, Italian painters resorted to the use of double orthogonal projection in order to compute the position of points in the perspective. Further, Spanish stonemasons of the early sixteenth century solved the dressing of these pieces by squaring, without the use of true-shape templates, in contrast to Vandelvira (c. 1585: 69v-70r; see also Jousse 1642: 184–185; Derand 1643: 395–398; and Sect. 9.5.2), which used templates to solve the problem. Thus, in both perspective and stereotomy, orthogonal projection is used as a general-purpose method, since both have enough power to address the most difficult problems. Methods such as vanishing points and true-shape templates, although more economical, are used only where allowed by the nature of the problem.

However, there is another connection between perspective and stonecutting at a deeper level, although its real effects were only felt from the seventeenth century on. Some monastic scribes and a host of architects and masons had used orthogonal projection for centuries without knowing it; neither projectors nor projecting planes are mentioned at all in connection with this issue before the seventeenth century. In contrast, in the first written treatise about Renaissance perspective, Alberti's *De Pictura* (1435: book I, paragraphs 5–8, 12–13) the projection plane, mentioned as "the picture", is ubiquitous; the projectors are mentioned on several occasions as "rays" and even, one of them is described as "this ray, among all the strongest and most and lively, … may be called the prince of rays";[57] to cast aside any doubt,

[56]Pérouse ([1982a] 2001: 184–185): *La perspective et la stéréotomie sont deux techniques, certes bien distinctes, mais constamment associeés.*

[57]Alberti (1435: book 1, paragraphs 5–8, 12–13): *Questo uno razzo, fra tutti gli altri gagliardissimo e vivacissimo … si può dire prencipe de' razzi.*

Alberti specifies that the image of any point should be placed at the intersection of the corresponding ray with the picture, that is, the projection plane.

Thus, schematically but clearly, the theory of projection filtered down to cartographic literature in the mid-sixteenth and seventeenth centuries. Juan de Rojas Sarmiento ([1550] 1551; 25–26, 29), a cosmographer, explained orthographic projection applied to cartography as a particular case of perspective where the station point is placed at infinity. His system was still rather crude: he projected meridians as arcs of a circle rather than half-ellipses. Guidobaldo del Monte (1579: 91–92; see also Camerota 2005a: 82–84), pointed out this error mercilessly, while remarking that such use of perspective was repugnant. A clear explanation of orthographic projection can be found in Aguillon's treatise on optics (1613: 503–520). A few years later, Jean-Charles de la Faille wrote a short manuscript treatise, or rather a summary of the architectural teachings at the Imperial College in Madrid. It mentions stonecutting only briefly, but it bothers to explain that

> Orthography is a kind of perspective used frequently by architects in their drawings which can be classified into three genres. The 1st one is that of the plan. The 2nd one is that of the sides. The 3rd one is that of the profiles ... The difference between orthography and ordinary perspective is that perspective puts the view at an infinite distance, and that orthography puts it at a moderate distance.[58]

From this point on, de la Faille presents a short, but remarkably complete, theory of orthogonal projection. He explains that the projection of a line which is perpendicular to the projection plane[59] is a point, the projection of a line that is oblique to this plane is shortened, and the projection of a line that is parallel to the projection plane maintains its length. Next, he asserts that any figure that is parallel to the projection plane is represented by another figure that is equal to the original figure; that any surface or figure that is orthogonal to the projection plane is transformed in a straight line; that a circle that is oblique to the projection plane gives "an oval or ellipse", whose major axis is equal to the diameter of the original circle, while the minor axis can adopt different lengths depending on orientation. The oblique "oval" sometimes remains an oblique oval, while on other occasions it adopts the shape of a circle, while the oblique square is transformed in different kinds of trapeziums.[60] (de la Faille c. 1640: part 2, Chap. 1–2).

It is interesting to remark that both Aguillon and de la Faille were Belgian Jesuits, that Aguillon was the son of a civil servant in the entourage of Philip II of Spain, and although he was born in Brussels, he had studied theology at Salamanca, and he worked as an architect for the Society of Jesus, just as Derand had. Further,

---

[58]De la Faille (c. 1640: 5v-6r): *La orthographia es un genero de Perspectiva de que usan de ordinario los architectos en los dibuxos que se reduccen a tres generos. El 1º es de la planta. El 2º de los lados. El 3º de los perphiles ... La differençia de la orthographia y de la Perspectiva ordinaria que se llama Sçenographia, consiste en que la primera pone la vista en una distançia infinita, y la otra en una distançia moderada.* Transcription and translation by the author.

[59]He uses the phrase *plano de la proieçcion* (projection plane) explicitly; in contrast, the projection itself is called *el lugar* (the locus).

[60]It seems that De la Faille's definition of a trapezium encompasses the rhomboid.

the Imperial College, was, up to a certain extent, a successor of the Academy of Mathematics founded by Juan de Herrera (Esteban Piñeiro 2002–2003: 14).

There are other associations between stonecutting and cartography at a different level. Francisco Pinto (2002: 105–142) has suggested the existence of connections between stonecutters and cartographers, starting from the fact that both disciplines must address the problem of the development of the sphere. This issue leapt suddenly to the centre of the stage at the beginning of the sixteenth century. Classical cartography and, in particular, Ptolemy's conical projection, was known in the late Middle Ages. However, mainstream cartographic production in this period was based on portolan charts, designed for navigation and including a small portion of the terrestrial surface, most frequently the Mediterranean Sea or part of it (Campbell 1987). The main concern of portolan mapmakers, based mainly in Majorca and several Italian harbours, was the precise representation of rhumb lines, reported by pilots; generally speaking, this task was addressed empirically. At the same time, hemispherical domes were built, in particular over spiral staircases. They were small, about 2 m in diameter, and their layout was probably solved by dressing the voussoirs by the squaring method; the lack of true-shape templates in the tracing for a sail vault in the sacristy of Murcia cathedral points this way (Calvo et al. 2013b).

However, this vault, together with an earlier one at the Merchants' Exchange in Valencia, signals a remarkable change in the use of spherical vaults. While the vaults over staircases—for example, the one in Alphonse V's chapel in Valencia—measured about 2 m in diameter, the Merchants' Exchange and the Murcia sacristy vaults cover areas about 6 m square, and their diameters equal the diagonal of the area, roughly 10 m, that is, five times the ordinary vault over a staircase (Natividad 2012a; Calvo et al. 2013b). Moreover, in contrast to the rough finish of the vault over the staircase in Alphonse V's chapel, both the Merchants' Exchange and the Murcia sacristy vaults feature a most precise finish. Surely, this approach demanded new methods of formal control, probably based in squaring. However, two decades later, even small vaults over staircases were dressed using true-shape templates, as attested by the tracing for a vault over a stairway leading to the rooftops of Seville cathedral (Ruiz de la Rosa and Rodríguez 2002), which implements in diagrammatic form a construction resembling cone developments.

In the cartographic field, an even greater change was brought about by the (re)discovery of America. The empirical approach of portolan charts was useless in oceanic routes; a new cartographic system had to be devised. Meanwhile, the cartographic centre had shifted from Mediterranean ports to the Casa de Contratación or Contract House in Seville, where all American trade was centralised by Charles V. The institution was also charged with the control of all maps of America. When pilots were to sail for the New World, the House furnished them with copies of a *Padrón Mayor* or Main Map, prepared by the Higher Pilot; in return, when coming back, pilots had to report discoveries in order to update the Main Map (Sandman 2007). Pinto pointed out a revival of interest in Ptolemy's projections, in particular "conic" projection, also mentioning the work of the cosmographer Fernán Pérez de la Oliva, as well as spindle-shaped developments, used in gores and illustrated by

Dürer (1525: Miiii ter r), and their similarities with some sections of the manuscript by Alonso de Vandelvira (c. 1585: 63r).[61]

Such ideas are quite enticing; Ptolemy's projection shares its general traits with the general method used by the masons who prepared the tracing in Seville's cathedral rooftops, de l'Orme (1567: 113v-115r) and Vandelvira (c. 1585: 60v-61r), while the spindle-shaped developments respond literally to a particular technique used by Vandelvira (c. 1585: 63r, 64r, 78r, 102v, 104r, 104v; see also Sects. 10.2.2 and 10.2.4) for hemispherical and oval domes. Moreover, other pieces of factual evidence point in this direction. Hernán Ruiz II himself was involved in some repairs in the Casa de Contratación (Morales 1996: 120–121; Pinto 2002: 110). He was probably in contact with the Main Pilot or the Cosmographer, and he may have exchanged information about conical and cylindrical developments, a common interest, although there is no factual evidence to either prove or refute this. As we have seen in Sect. 2.1.4, he prepared a personal notebook including a mixed bag of architectural drawings, stereotomic schemes, geometrical methods and idiosyncratic perspective drawings; this recalls Pérouse comments about the presence of perspective and stereotomy in the manuscript of Jean Chéreau. He did not include spherical vaults, but instead, he used cylindrical developments to control the decoration of barrel vaults; also, his perspectival schemata seem to be connected to cylindrical developments (Ruiz c. 1560: 52r; Calvo 2019). All this is quite relevant since after its Ptolemaic phase, Sevillian cartography evolved to cylindrical projections, perhaps an antecedent of Mercator's projection.

Another interesting study case is provided by a coffered sail vault in the second stage of the bell tower of Murcia cathedral, used traditionally to store liturgical ornaments and recently as an archive. The piece was built under the supervision of Jerónimo Quijano, an innovative master mason; however, the layout of the ribs in this vault departs from other solutions in his extensive work and usual practice in sixteenth-century Spanish stonecutting, since the ribs do not lie on vertical planes. Rather, they follow diametral planes, so their centres are set at the midpoint of a square placed at springing level; their horizontal projections follow ellipses, rather

[61] The phrase "conic projection" is used both in cartography and descriptive or projective geometry, although with different meanings. In these latter disciplines, "conic" or "central" projections involve projectors passing through a single point, known as centre of projection; the projection of a point is given by the intersection of its projector with a projecting surface, usually a plane. Particular cases of this concept are linear perspective and stereographic projection, used in cartography. However, in this science, the phrase "conic projection" is applied to a procedure used to represent a limited portion of the earth, usually contained between two parallels and two meridians. Elementary conic projections are furnished by a specific procedure where a cone is placed with its apex at a particular point on the earth's axis, so that it is tangent to the earth's surface at a particular parallel or, more frequently, it cuts the surface at two parallels of the earth surface. First, the geographical accidents in the surface of the earth are projected onto the conical surface by different methods, depending if the projection is intended to preserve areas, angles or distances; none of them is strictly a "conical projection", in the sense of the term used in descriptive geometry. Next, the cone is developed to represent parallels, meridians and geometrical accidents on a planar surface. Of the different methods used in order to project geographical accidents onto the conical surface, the only one relevant for us is the one described by Ptolemy, where the distance between two parallels is preserved in the final developed surface (For a detailed explanation of this subject, see Snyder 1987: 97, 111).

than straight lines, as usual in Medieval and Renaissance ribbed vaults (Calvo et al. 2005a: 197–210). Although there are some debates about the dating of the vault, it was most probably built during the tenure of Juan Martínez Guijarro, known as Siliceus, at the bishopric of Cartagena, whose cathedral is located at Murcia (Gutiérrez-Cortines 1987: 130, 137; Vera 1993: 109–111). Siliceus was a remarkable mathematician; in fact, he was entrusted with the diocese as a reward for his tutoring of the future Philip II of Spain in mathematics and cosmography. His *Ars Arithmetica* (Martínez Guijarro: 1519), published in Paris, was corrected by Oronce Finé, author of a well-known cosmography treatise (1542). Moreover, Siliceus had supervised construction in the University of Salamanca during the incumbency as Rector (President) of no less than Fernán Pérez de la Oliva (Pereda 2000: 69; see also Pinto 2002) mentioning also his connections with Hernán Ruiz II's father). Thus, perhaps Silíceus asked Quijano for a new layout of sail vaults showing the advances in cosmography of the period, with Quijano following his patron's wishes, leaving aside his usual methods.

Another connection between stereotomy and cartography is given by the remarkable vaulting of the second storey of the Merchants' Exchange in Seville. This institution is different from, but closely connected to, the Contract House. The traders who assembled in the Exchange had to go to the Contract House in order to pay duties; in modern terms, the Exchange was the equivalent to the Chamber of Commerce while the Contract House performed the functions of a Customs House. The second storey was to be erected under the supervision of Alonso de Vandelvira; however, funds were scarce, and he accepted a more rewarding job at Sanlúcar de Barrameda, so the vaults were finally built by Miguel de Zumárraga (Pleguezuelo 1990; Cruz Isidoro 2001: 96–100; see also Sect. 2.2.3). In contrast to the severe first storey, designed by Juan de Herrera and directed by Juan de Minjares, Zumárraga used a startling combination of square and rectangular sail vaults with clearly marked parallels and meridians, as well as other great and small circles, set at oblique planes. On many occasions, the decoration does not follow the joints between courses and voussoirs, adding a new set of circles on the spherical surface of the vault. Like a three-dimensional Portuguese emblem, it seems that this stereotomic tour de force alludes to the sea-faring traders that had erected the building.

However, if we try to delve deeper in order to identify more precisely the mechanisms of knowledge transfer between architects and cartographers, it is almost impossible to reach a firm conclusion. Two interconnected facts obscure our knowledge of the actual methods used by the Main Pilots. Spain and Portugal had agreed in the Treaty of Tordesillas to divide their areas of exploration in America along a meridian placed 370 leagues west of the Cape Verde islands. However, at that moment nobody knew how to measure longitudes exactly and further, Spanish and Portuguese relations were not altogether friendly until 1580, when Philip II inherited the Kingdom of Portugal. Thus, general maps of America were kept secret or falsified (Goodman 1988: 72–81). No instance of the Main Map held at the Contract House has survived; the nearest proxy is a copy preserved in the Vatican archives, sent to the Pope to substantiate the dispute with Portugal. The extent to which the Vatican copy responds to actual practice in Seville is open to discussion. In any case, this map seems to use an approximate cylindrical projection, an antecedent of

Mercator's projection, but it is not entirely accurate; this may derive from discussions between theoretical cosmographers and practical pilots, who wanted to use empirical charts, the modern versions of the portolans they were used to (Turnbull 2000: 105–110; Sandman 2007). Thus, to summarize this point, the interest of cartographers in Ptolemy's conical projection could have fostered the use of cone developments in stereotomy; the gores used in the construction of globes, illustrated by Dürer, might be reflected in Vandelvira's special method for coffered hemispherical domes, also used in oval vaults; and cartographical cylindrical projections, used as empirical antecedents of Mercator's projections in the Contract House, could have been adapted by Hernán Ruiz II in order to control the decoration of barrel vaults; in any case, the hard evidence for these connections is not conclusive.

Things are a bit clearer with gnomonics, the science of sundials. Architects such as Jean Bullant were keenly interested in the issue: he wrote a treatise on gnomonics (1561) and a shorter summary on the subject, as an appendix to a geometry treatise (1562); by his own confession, he had taken much material from no other scholar than Oronce Finé (see Tamboréro 2008: 25–40 and Manceau 2009). Hernán Ruiz II (c. 1550: 29v-33r), Chéreau (c. 1567–74: 107v-108r) or Gentillâtre (c. 1620: 509r-511v) include sundials or armillary spheres in their miscellaneous manuscripts, while Desargues (1640: 4) added a paragraph on the subject at the end of his stonecutting leaflet; some years later this short passage was enlarged into a full-fledged book by Bosse (1643b). Desargues uses the same word, *essieu*, to indicate both the shadow-casting *gnomon* in sundials and the axis in barrel vaults; this gives an important clue about his stonecutting method. First, the *plan sous-essieu* is the equivalent of the projecting plane of the gnomon; this hints that sundials may have given Desargues the idea of an oblique projection with slanting projectors, although he did not use the term; all this is essential for the right understanding of Desargues's system.

As Luc Tamboréro pointed out (2008: 38–40), the construction of sundials involves the determination of the angle between the projecting or shadow plane and the plane of the sundial itself; it is thus connected with dihedral angles. Two years before Desargues, Florimond de Beaune wrote his pioneering *Doctrine de l'angle solide* (c. 1640), which remained unpublished until the twentieth century. However, this treatise inspired Philippe de la Hire to make use of this notion as a new, powerful tool for stereotomy; this idea was taken up by de la Rue (1728: 45–46, pl. 24, 24bis) to streamline Jousse's method for the groin vault, based on folding templates; Frézier would bring about further applications of this concept. Another issue raised by Tamboréro (2012) is that of the connections of optical corrections and stonecutting. According to Frézier (1737–39: II, 400–404), de la Hire delivered a lecture at the Académie Royale d'Architecture dealing with the use of raised elliptical vaults to correct the optical image of hemispherical domes. Frézier mentions Saint Peter's in Rome and the churches of the Sorbonne, Val-de-Grace abbey and the Invalides in Paris as examples; Tamboréro (2012), adds another interesting case: some singular arches in the side chapels of the Invalides.

12.5 The Role of Stonecutting in the Formation of Descriptive and Projective Geometry

Pérouse de Montclos took it for granted that "It is certainly not French geometry that has made up descriptive geometry; instead, thanks to Desargues, de la Hire, Frézier, and Monge, French stereotomy has produced descriptive geometry".[62] This idea may stem from a short work by René Taton (1954). Before Taton, de la Gournerie had underscored the role of stonecutting, along with other trades, in the formation of descriptive geometry; however, he criticised the shortcomings of Monge's method, based exclusively on geometry, when applied to stonecutting (de la Gournerie 1855: 6–15, 25–31; de la Gournerie 1874: 114; see also Sect. 2.4.6). Gino Loria (1921: 84–87) may have taken data from him for his classic *Storia della Geometria Descrittiva*, but he paid no attention to his comments on Monge. Loria mentioned de l'Orme, stressing the empirical nature of his treatment of double orthogonal projection and putting in the same bag Jousse, Derand and Milliet. For Loria, Desargues was the first to "endeavour to treat this essential part of engineering with scientific criteria".[63] He admitted Bosse and Guarini in this scientific realm; however, he stated that while Desargues and Bosse had not succeeded in putting stereotomy on a genuinely scientific path, Frézier had achieved this goal. Thus, he included a relatively long section on Frézier, summarising his first, abstract volume, while ignoring completely the second and third ones, those dealing with practical stonecutting.

Later on, Taton (1954) published a short paper on the history of descriptive geometry, stemming from a lecture in the Palais de la Découverte, the Parisian Museum of the History of Science. He chose a reversed historical narrative, starting with Monge's teachings, both in the École Normale and the École Polytechnique, and thus differentiating between the theoretical orientation of the former and the emphasis on applications of the latter. From this point, Taton went backwards to Monge's stay at the school in Mézières, looking for the genesis of the idea of descriptive geometry, stressing his involvement with the problem of the *défilement*. In a further step back, he analysed stonecutting literature, starting with Villard; about de l'Orme, he remarked that proofs are lacking and graphical constructions are too elaborate and in many occasions, inexact. He admitted that "A number of stonecutting and architectural treatises published during the seventeenth century show clear progress from a theoretical point of view; however, graphical procedures remain too elaborate, and each problem requires the use of particular methods";[64] he was probably thinking about Jousse, Derand, or perhaps Milliet. In the next lines, he turned to Desargues,

[62]Pérouse ([1982a] 2001: 185): *Ce n'est certainement pas la géométrie française qui a fait la stéréotomie; mais au contraire, grâce a Desargues, a La Hire, a Frézier, a Monge, la stéréotomie française qui a produit la géométrie descriptive.*

[63]Loria (1921: 85): *…chi per primo si propose di trattare anche quest'importantissima parte dell'arte dell'ingegnere con criteri veramente scientifici è ancora Desargues…*

[64]Taton (1954: 14): *Plusieurs traités de coupe des pierres et d'architecture publiés au XVIIe siècle sont en net progrès du point de vue théorique, mais les procedés graphiques y demeurent très compliqués, chaque problème nécessitant l'emploi de méthodes particulières.*

remarking his general approach and asserting that some of his constructions show a clear understanding of several descriptive geometry methods. Then, he followed on to the eighteenth century, stressing further progress in the work of de la Rue and, particularly, Frézier.

Joël Sakarovitch (1998: 157–166, 218–246) followed Taton's path, presenting much information about the role of Monge as Professor of the Theory of Stonecutting in the school at Mézières and his predecessors in stonecutting treatises. However, a fair understanding of Sakarovitch's position requires a careful evaluation. At first sight, his analysis of the different solutions to the sloping, skew barrel vault in his longest work (Sakarovitch 1998: 157–170; 362–384) may suggest that he downplays the evolution that Taton had underscored, giving the impression that it was stonecutting literature as a whole which furnished the primary basis for descriptive geometry. However, one year later, in a very short paper, Sakarovitch (1999) made it clear that a deliberate desire for scientific neutrality had led him to apply in these particular sections the "Rashomon method", alluding to a film by Akira Kurosawa whose plot revolves around the different reports about a rape and a murder given by different characters. In other words, he allowed de l'Orme, Jousse, Desargues, Derand and Frézier to express themselves. In an early work, Sakarovitch stressed the contraposition between practitioners and theorists (Sakarovitch 1992a: 531; see also Potié 2008: 155–158, adding a third line revolving around Desargues and Monge). But perhaps Sakarovitch himself felt that this dichotomy is too coarse; in an appendix to his main work, he writes down the cast of his version of *Rashomon*: De l'Orme the precursor, Jousse the stonecutter, Desargues the geometer, Derand the educator, Frézier the engineer (Sakarovitch 1988: 362–384; see also 137–140, 145–147).

All in all, the plot is even more tangled than that of Kurosawa. Desargues the geometer takes more material from traditional sources that we may suspect at first sight (see the end of Sect. 8.5). Of course, Derand does this systematically in order to reach his audience, but the treatment of tangencies in rib vaults and cylindrical developments, as well as his remarks about the lack of proofs, betray the mathematics teacher. Milliet and Guarini are also mathematicians, but they generally follow Derand, and Milliet's list of theorems is less than elementary.[65] De la Rue has been classified as another traditionalist, but this is clearly unfair in the light of his treatment of groin vaults, both the basic version and the annular one, his reticence about the classical cone-development method and his renovation of "drawing in stone" practices, extending them to space. Apparently Frézier explains traditional

[65] In addition ot his scant list of theorems in the treatise on stonecutting Milliet (1690: II, 622–623) includes some propositions in *ichnographia projecta*, *ichnographia* and *orthographia* in the treatise in perspective included in the third volume of *Cursus seu mundus mathematicus* (Milliet [1674] 1690: III, 502–515, 542–545). However, this is not relevant to our interests for several reasons. First *ichnographia projecta* is not really orthographic projection, but rather a *central projection* of the footprint of an object, used as a preparatory stage in the construction of a linear perspective. Second, the propositions on orthographic projection in pp. 542–545 are not abstract theorems, but rather problems such as how to draw plans and elevations of slanted prisms and pyramids. And third and most important, since these propositions are presented in the third volume in the *Cursus,* it seems that Milliet himself takes it for granted that they not necessary in stonecutting, which is explained in the second volume.

methods to attack them; however, one may suspect that he presents standard procedures as an operating alternative to his own impractical solutions, such as the skew arch with elliptical joints or the deceptively simple flat templates for spherical vaults. That is, the branches of practitioners and theoreticians interweave so tightly that the attempt to ascribe the source of descriptive geometry to the latter, while excluding the former, is not consistent with the factual evidence.

In other words, the analysis of particular stonecutting methods in Part II in this book shows clearly that de l'Orme, Jousse, Derand and de la Rue cannot be treated as a compact, homogeneous whole, and that Spanish treatises add further, essential information. In arches, sixteenth-century triangulations and true-shape constructions were replaced by approximate cylindrical developments in the times of Jousse (1642: 46–47) and Derand (1643: 155–157). In vaults, sixteenth- and seventeenth-century cone developments applied to spherical surfaces were attacked by de la Rue (1728: 50–52) as inexact; this led him to put forth his idiosyncratic method, drawing in space, while Frézier (1737–1739: II, 312–331) went back to traditional methods, including de la Rue's invention, as options in a comprehensive portfolio of solutions to spherical vaults. Desargues's approach to skew arches and sloping vaults (1640) was rejected by practitioners; however, it fostered a systematic approach to oblique arches, which led to the treatment of the problem through conics theory by Frézier (1737–39: II, pl. 49). The list could go on, but the conclusion is clear: the mass of knowledge leveraged by Monge in his teaching about the theory of stonecutting in Mézières was the result of a slow historical evolution involving many debates and controversies, both implicit and explicit. In order to get a clearer picture from this quandary, we must ask ourselves: in which sense was stonecutting an antecedent of descriptive geometry? Did it furnish the problems descriptive geometry attempts to solve, the methods of the discipline, or its fundamental concepts?

12.5.1 Problems

The answer to these questions is complex. First, as stated by Monge in his lessons at the École Normale,

> [Descriptive geometry] has two main purposes.
>
> The first one is to represent exactly three-dimensional objects in drawings with just two dimensions, provided that these objects are susceptible of rigorous definition. From this point of view, it is a necessary language for the engineer who must conceive a project, for those entrusted the supervision of its execution, and for the artisans who must execute its different parts.
>
> The second aim of descriptive geometry is to deduce from the exact depiction of bodies everything that follows necessarily from their shapes and their respective positions. In this sense, it is a way to seek truth; it offers continual examples of the passage from the known to the unknown … It is not only useful for the exercise of the intellectual faculties of a great

people, contributing to the improvement of the human species; it is also indispensable for all workers whose aim is to give to physical bodies certain determined shapes ...[66]

In other words, the discipline has two faces. The first one is extroverted: the engineer uses it to convey orders to supervisors and executors. The second one is introverted: workers and, implicitly, engineers, may use it to determine magnitudes or shapes that are not shown explicitly by orthographic projection, such as lengths of segments, angular values and planar shapes, when they are not parallel to projection planes. Even more, in Monge's opinion, even the general public can use this science as an intellectual exercise.

Both uses of orthogonal projection had been employed by stonecutters for centuries; however, its extension to other crafts and abstract problems was a relative novelty. Admittedly, Frézier (1737–39: II, 290–309) had included wood in his stereotomy treatise, thus furnishing a cue for Monge's surprising enlargement of the field of application of the discipline; however, Frézier dealt with joinery, not carpentry, and only in a few isolated examples. This "special relationship" of stonecutting with descriptive geometry has been explained persuasively by Sakarovitch (1998: 243–244): the stonemason works on volume, starting from a three-dimensional object, while the carpenter deals with lines and the coppersmith with developable surfaces. A parallel can be established between the physical activity of the stonemason and the abstract reasoning of the geometrician, shown by the rich repertoire of developable and warped surfaces and complex intersections that can be materialised by the stonemason.

As a result, from stonecutting have emerged many of the practical problems of descriptive geometry. First, of course, is that of a systematic representation of volumes in space. A single projection—whether orthographic, axonometric or linear perspective—cannot provide a precise representation of the position of a point in space since all points in the same projector share the same projection. Double orthogonal projection offers a simple, elegant and efficient solution to this problem, as explained by Monge in his lessons at the École Normale (Laplace, Lagrange and Monge 1989: 308–312). In theory, axonometrics or even linear perspective may supply such precise information about the position of a point through the addition of a horizontal projection, but this is an elaborate nineteenth-century construct; even with this support, problems such as the determination of distances, true shapes and

[66]Laplace, Lagrange and Monge ([1795] 1992: 305–306): *(La géométrie descriptive) a deux objets principaux/Le premier est de représenter avec exactitude, sur des dessins qui n'ont que deux dimensions, les objets qui en ont trois, et qui sont susceptibles de définition rigoureuse/Sous ce point de vue, c'est une langue nécessaire a l'homme de génie qui conçoit un projet, à ceux qui doivent en diriger l'exécution, et enfin aux artistes qui doivent eux-mêmes en éxecuter les différentes parties/Le second objet de la géometrie descriptive es de déduire de la description exacte des corps tout ce qui se suit nécessairement de leurs formes et de leurs positions respectives. Dans ce sens, c'est un moyen de rechercher la vérité; elle offre des exemples perpétuels du passage du connu à l'inconnu ... Elle est non-seulement propre à exercer les facultés intellectuelles d'un grand peuple, et à contribuer par-là au perfectionnement de l'espece humaine, mais encore elle est indispensable à tous les ouvriers dont le but es de donner aux corps certaines formes déterminées ...*

angles are carried out much more easily and efficiently through double orthogonal projection.

These distances, shapes and angles are the "truths" mentioned by Monge: information implicit in orthographic projections but not shown explicitly in them. In fact, plans represent exactly the lengths of horizontal line segments and the shape and area of horizontal figures. However, they distort the lengths and shapes of sloping lines and figures, even to the extent of showing vertical lines as points and vertical figures as line segments, as we have seen in Sect. 1.5. As for elevations, they are even more selective; they reproduce the lengths of segments and the shape of figures laid on vertical planes parallel to the *projection plane of the elevation*, not all vertical planes. Such problems can arise in any technical field; for example, it is easy to understand that the determination of the length of sloping, oblique segments is important in carpentry. The treatise on carpentry by Diego López de Arenas (1633: 15r-16r; see also Nuere 1985: 131–138) also poses the problem of the determination of angles between segments, between planes and between lines and planes. However, given the three-dimensional nature of Early Modern stonecutting, it raises all kinds of problems, as stressed by Sakarovitch (1998: 243–244): lengths of segments are essential in the triangulation techniques used by Alviz, de l'Orme, Vandelvira and many other authors; shape determination is the central problem raised by rigid templates; angles between segments are used as bevel guidelines by most writers in different forms; angles between planes are used from the period of de la Hire, de la Rue and Frézier. Taking into account Monge's teaching experience at Mézières, it seems clear that these problems were mainly introduced to descriptive geometry through stonecutting.

The issue with developments is more complex. A rough approximation to cylindrical developments seems to be present, as we have seen in Sects. 3.2.6 and 6.4.1, in de L'Orme (1567: 74v-77r) and Vandelvira (c. 1585: 21v-22v), although the technique does not appear explicitly until Jousse (1642: 46–47) and Derand (1643: 155–157). However, such developments were solved approximately, using two chords of the directrix, in order to avoid the intractable problem of the rectification of the circumference. Conical developments also appear in sixteenth-century stonecutting, in squinches and splayed arches (see Chap. 5 and Sect. 6.1 and Tamboréro 2009: 81–82) although they were present much earlier in Ptolemy's conical projection. In any case, neither the text of Ptolemy's geography nor sixteenth-century stonecutting treatises and manuscripts mention cones. Thus, we may ask ourselves whether cartographers and stonecutters had a clear understanding of these methods. Fortunately, Vandelvira (c. 1585, 61v) included in his manuscript an additional explanation to the section dealing with hemispherical vaults solved by means of cone developments. In this section, probably the only text resembling a proof in Renaissance stonecutting literature, the author does not mention cones, but rather trumpet squinches; this means that at least he understood that he was substituting the shape of a squinch for that of a section of a dome, and in fact he uses the word *esférica* (spherical).

At the same time, stonecutting writers, in particular Martínez de Aranda (c. 1600: 12–14, 15–16, 34–38, 41–16, etc.), tried here and there to develop warped surfaces, although they were perfectly aware of the difference between these surfaces and developable ones (Martínez de Aranda c. 1600: 222–223). The modern notion of developable surface, tied to analytical geometry, derives from these problems, together with spherical developments in cartography. In fact, it stems from two seminal papers, one by Euler (1772), fostered by cartographic needs, and another one, closer to stereotomy, by Monge (1796). In any case, the concept underwent an important transformation: for stonecutters, a voussoir face was *gauche* if and when two intrados joints were neither parallel nor convergent; the modern notion states that a ruled surface is non-developable when two generatrices placed at an infinitely small distance are skew lines.

Tangency problems arise here and there in stonecutting; a good example is the direction of the end tangents in tierceron and diagonal ribs, placed by Hernán Ruiz II (c. 1560: 46v; see also Rabasa 1996) by trial and error and determined exactly by Derand (1643: 393–394). However, the issue of tangent planes was brought to descriptive geometry from other fields: on the one hand, topography and military engineering, with the problem of the *défilement*; on the other hand, shadow theory, where Monge used them to determine the dividing line between light and shadow on a cone. Later on, the problem of the continuity of tangent planes was posed by the classical solution to the *arrière-voussure de Marseille*, but rather than a new problem stemming from practical needs, this constructive type was used as an illustration of a theoretical concept.

12.5.2 Methods

Orthogonal projection. It seems clear that stonecutting, along with cartography and topography, furnished most of the practical problems that descriptive geometry deals with. The issue with methods is not as clear. The first examples of the emergence of orthographic projection appear in clerical sources, such as the illustrations to Richard of Saint Victor's commentary on Ezekiel or such miniatures as those representing the tower of Tábara (see Cahn 1994 and Galtier 2001: 374–389. However, Ackerman (1997: 47) has put forward, although with some reservations, the hypothesis that Villard may have taken information for his elevations of Reims cathedral, drawn before the construction of the clerestory, from drawings for the cathedral shop. Thus, these drawn orthographic representations in clerical sources may be contemporary with lost mason's drawings or large-scale tracings; we should recall that parchment, an expensive material (Erlande-Brandenburg 1993: 74), was not usually in the hands of masons. In any case, if orthographic projection arose almost at the same time in clerical and building sources, masons used it extensively; in fact, Dürer (1525: Oiii ter r) attributed it to stonecutters when using it to determine the shadow of a cube (Fig. 12.8).

It has been argued that most of these representations include evident geometrical errors, but we should take into account three factors. First, as the English say, in order to learn to ride, you must fall seven times from the horse; of course, the first steps of any human activity are rather clumsy. Second, some of the blatant errors, such as the incorrectly drawn projection of an oblique circumference in an elevation

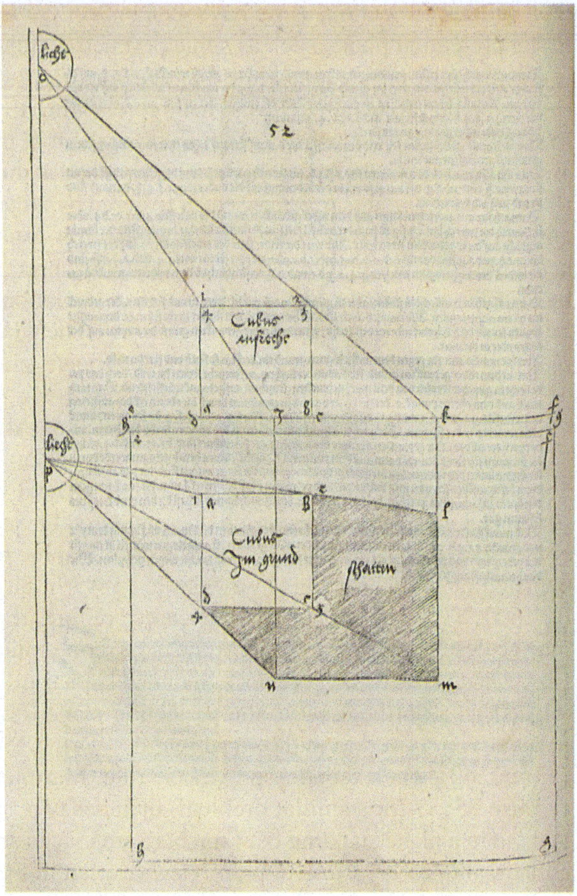

Fig. 12.8 Projections of a cube as a preparation for a shadowed perspective drawing (Dürer 1525: Oiii ter v)

of the spires in Strasbourg cathedral)[67] arose when stonecutters addressed problems that nobody was able to solve in this period; even learned geometricians knew very little about conic sections before the publication of Commandino's translation of Apollonius of Perga (1566). Third, a correct orthographic projection of the circular mouldings in the oblique sides of the Strasbourg spire would have been useless for a medieval mason, while a true-shape representation furnished a valuable tool for formal control. That is, masons used orthographic projection when it suited their needs and turned to other methods when necessary.

In any case, while horizontal projection was used at all scales, from general plans to detail drawings, vertical orthographic projection appears mainly in façade elevations. In contrast, in stonecutting tracings, ribs were usually represented in true shape rather

[67] Sanabria (1992: 168); see also Recht (1995: 63), or also Böker (2005: 175) for a similar problem in the archivolts of a doorway. The Strasbourg drawing is lost, but a nineteenth-century copy is included both in Sanabria (1992) and Recht (1995).

than vertical projection, as we have seen in Sect. 10.1.4. Of course, only ribs parallel to the projection plane, such as some perimetral arches and liernes, are represented in true form by projection, while diagonal and tierceron ribs are distorted by vertical orthographic projection. Thus, when representing tierceron vaults or *Netzgewölbe*, masons disassembled ribs in order to show them in true size and shape. In treatises and manuscripts, ribs frequently appear grouped at the corners of the plan, or around a central axis, but this practice seems fostered by didactic reasons. In one of the rare, and very late, full-scale tracings for a tierceron vault which have survived, in Tui cathedral, ribs are placed where the available space allows (Taín et al. 2012). The most striking example of this medieval disarticulation of stonecutting elevations can be found in de l'Orme's treatise (1567: 108v). There is no coordination at all between plan and elevation; the keystone in the elevation is placed above the springer in plan and the other way around.

In any case, sixteenth-century writers such as Alviz (c. 1544: 7, 8, 13, 15), de l'Orme (1567: 69r, 70v; see also Sakarovitch 1992a: 532) and Vandelvira (c. 1585: 73r, 76r, for example) extended the use of vertical orthographic projections from general elevations to stonecutting diagrams. Again, an objection may be raised to the consideration of such schemata as true orthographic projections: oblique shapes are depicted in some drawings by Alviz (c. 1544: 7r, 8r, 13r) and Martínez de Aranda (c. 1600: 81, 90, 92, 95), but these cases are rare. In most occasions, the diagrams may be described as overlays of true-shape representations rather than projections.[68] However, these overlays are not assembled haphazardly. Quite to the contrary, points aligned on an orthogonal to the projection plane (or, in American parlance, a line in point view) overlap in the elevation (de l'Orme 1567: 70v; Martínez de Aranda: c. 1600: 40). The exception proves the rule: Martínez de Aranda (c. 1600: 46–48; see also 35, 37 for a similar problem) includes two variants of an arch with a groin laid out in a plane standing between both arch faces. In the straight version, voussoir corners aligned on a line in point view overlap in the elevation. In contrast, in the skew version, points overlapping in the elevation are placed at oblique parallel lines (Fig. 12.9). In other words, Martínez de Aranda uses orthogonal projection for the straight case, but he resorts to oblique projection, for the skew one.[69] The whole issue seems connected with the squaring method, which is used in the straight case, encouraging the use of orthogonal projection; in contrast, in the skew case, the ruler or square should follow an oblique direction, applying a variant of the standard squaring method. All this strongly suggests that the use of the squaring method fostered the transfer of vertical orthographic projections from architects' drawings to stonecutters' diagrams; after all, stonecutters had used horizontal projections for three centuries.

---

[68] Alviz (c. 1544: 6r, 18r); L'Orme (1567: 69r, 70v); Vandelvira (c. 1585: 70r, 71r, 72r); Martínez de Aranda (c. 1600: 11, 13, 35, 40), to quote just a few examples.

[69] Aranda's oblique projection may seem utterly strange to our eyes, but it had a long tradition in the Antiquity and the Middle Ages. It is based in horizontal, oblique projectors, so that the resulting projection preserves the shape of frontal figures while showing the lateral faces of the object.

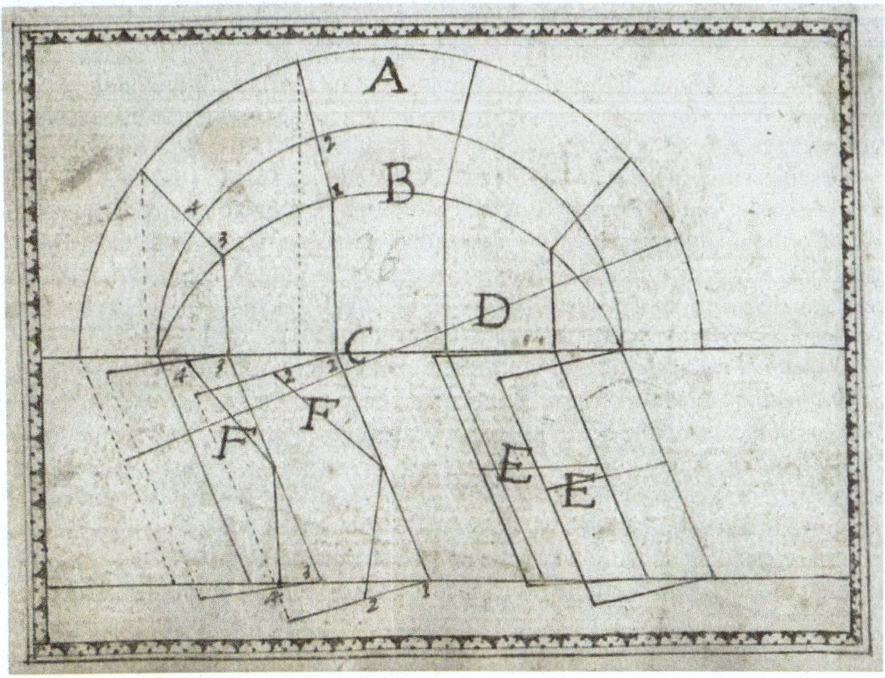

Fig. 12.9 Skew "groin arch" (Martínez de Aranda c. 1600: 47)

All this shows that for these writers, orthogonal projection is a method, rather than a rule; it may be used where adequate and left aside when other means are more suitable. This approach is also applied by Serlio (1544: ix, xx, xxvii) and Hernán Ruiz II (c. 1560: 90, 97, 101) to perspective: in section drawings, they mix perspective and orthogonal projection freely. Later on, the use of orthogonal projection is codified more strictly: for example, Palladio (1570: IV, 53; see also Sakarovitch 1990: 99–100) represents the temple of Vesta near the Tiber in a strictly orthographic elevation, without the perspectival contaminations of Serlio. Derand (1643: 35, 123, 165; Alonso et al. 2011b: 659–661) noticed that orthographic projections do not furnish a clear, intuitive rendering of the complex volumes used in stonecutting, so he resorted to oblique projections as a didactic supplement of his utterly economical stereotomic diagrams. Later on, de la Rue (1728: pl. 9, 18, 21, 26, 35, etc.) and Frézier (1737–39: II, pl. 29, 31, 34, 63, etc.) followed this path, using not only axonometrics but also linear perspective with instructional, rather than operative intentions.

However, Monge reacted against this practice and ripped axonometrics from the copy of de la Rue used in the École Polytechnique; in fact, he did not write a single line about axonometry. It seems clear, then, that he was trying to convert orthographic projection into a universal technical language, as stated in the introduction to his lectures in the École Normale (Laplace, Lagrange and Monge [1795] 1992, 305; see also de la Gournerie 1855: 37–38) and in fact he succeeded, at least in Continental

Europe; in Anglo-Saxon countries the use of orthographic drawing was brought about by other sources (see Lawrence 2003).

In any case, Monge had still a hurdle to leap: he tried to overcome the lack of intuitive volume representation of orthographic projection through the use of shading and shadows, following a long tradition that stretches back to a didactic manuscript on the subject used at the school at Mézières and, indirectly to the *Traité des pratiques perspectives et géométrales* by Bosse (1665: 80–89). Thus, it seems clear that the use of orthographic projection by descriptive geometry stems from stonecutters' traditions; however, it must be stressed, as with other issues, that this tradition was the result of quite complex developments and Monge's abstract concept of projection had scarcely anything in common with the medieval pragmatic approach.

Rotations, rabatments and changes of projection plane. The issue of the so-called "methods" of descriptive geometry is even more complex. First, classical manuals of the discipline differentiate between rotations and rabatments; the former applies to points and lines, while the latter involve figures and planes. However, both operations are in essence identical. For example, a point on the axis of rotation does not move in a revolution or rabatment; thus, a line or figure that intersects the axis before the rotation will still meet the axis after the revolution at the same point, and so on.

However, the key question is whether stonecutters knew the foundations and properties of these geometrical operations, or, if they were at least aware that they were performing a rotation. As we have seen in Sect. 12.4.2, most writers use distances to a reference plane, rather than orthogonals to a rotation axis, and none of them uses the concept of a "revolution". Quite probably, this operation was one of the many that masons could not explain by words; as stated by Cristóbal de Rojas "to know how to build vaults it is essential to have much experience, so I will not say anything (in writing) about this subject, which is hidden very deep".[70]

These workshop practices may have been encouraged by the intuition, rather than the concept, of radial symmetry in space. One of these techniques is used by Chéreau (c. 1567–74: 113r) without explanation, but Jousse (1642: 10–11; see also 18–23) is a bit more explicit. Noticing that the extensions of all face joints meet at the intersection of the axis of the arch with the face plane, they draw the joints passing through this point. Of course, they need another point to determine the direction of the joint; this second point is placed at the intersection of an orthogonal to the axis drawn[71] through the horizontal projection of the voussoir with the springing. It is difficult to explain this operation without some kind of rotation; in modern terms, it may be described as a revolution of the face joint around the axis of the arch. Again,

[70]Rojas (1598, 97v-101r): *consiste el saber hacer las bóvedas en el mucho uso y experiencia que se tendrá de ellas, y así no diré su declaración, por ser la materia que la tiene dentro en sí muy escondida.* See also de l'Orme (1567: 112v, 119r, 124r).

[71]In fact, in Chereau's manuscript this point must have been located with the square, leaving no material mark on the sheet, in contrast with other drawings in this manuscript, where drypoint lines are found easily.

masons probably conceived the notion[72] while moving templates from one voussoir face to another; the notion, rather than the concept, since the basis of these operations is never explained in words.

Other methods of descriptive geometry clearly seem to stem from stonecutters' practice: in particular, changes of projection plane (or, in American parlance, auxiliary views), constructed by transferring heights to several orthogonals to a folding line. As stated by de la Gournerie (1860: vi; see also Rabasa 2011 and Tamboréro 2009: 76–77) changes of vertical projection plane are frequent in stonecutting treatises and manuscripts. However, this sentence condenses a complex evolution in a single line. As we have seen in Sect. 12.4.2, Renaissance masons routinely constructed auxiliary views determining the intersection of several intrados joints with a vertical plane, raising perpendiculars from a folding line, and transferring the heights of the intersection points, taken from an elevation or cross section to the perpendiculars. Such a method has points in common with the changes of projection plane of descriptive geometry, but the central concept is lacking: there is no projection since there are no points in different planes involved.

Later on, actual changes of projection plane appear when other writers from Derand (1643: 63, 67, 89, 197, to quote just a few extreme examples) to Frézier (1737–39: II, pl. 40, 41) extended this method to three-dimensional forms. This method underwent a momentary eclipse in Monge and his immediate followers since they were more interested in showing that two, and only two, projections were sufficient to determine the position of a point in space unambiguously. However, as we have seen, they resurfaced with Théodore Olivier (1832: 344–348), together with changes of the horizontal projection plane.

This latter method was not an offspring of mainstream stonecutting practice. De la Gournerie remarked that it had points in common with Desargues's solution for skew sloping vaults (1640). This procedure was all but forgotten for two hundred years. In the nineteenth century, Olivier, a geometrician teaching at the École Centrale, a school of industrial or mechanical engineering, introduced changes of horizontal projection plane as a generalisation of the vertical plane changes used for centuries by builders. However, the similarities between Desargues's method and Olivier's horizontal plane changes are generic: both Desargues and Oliver manipulated the orientation of the horizontal plane, but the actual method used by Desargues has traits in common with triangulations, rather than the procedures used by Olivier. All in all, horizontal plane changes did not stem from building practice, in contrast to vertical ones, but rather from the abstract concepts of a mathematician and the needs of mechanical engineering.

[72]I am using "notion" here in two meanings of the word: "a general understanding; vague or imperfect conception or idea of something" and "an ingenious article, device, or contrivance", and not as "a fanciful or foolish idea; whim".

12.5.3 Concepts

It appears that most problems and some methods of descriptive geometry were anticipated by stonecutting literature. However, this idea cannot be extended to the core concepts of this discipline, such as point, line, plane and surface; generally speaking, they stem from classical learned geometry. In an apparently trivial passage, Vandelvira translates these Euclidean notions into stonecutters' language:

> Point is something so small that cannot be divided into different parts; it is the starting point of stonecutting tracings. Line, which is called in Spanish 'stroke', and in stonecutting parlance 'trait', is something that is imagined with its length and without breadth, the ends of which are two points. This line is divided into two types: straight and curved; we call 'straight' the line that goes by the shortest way, as illustrated above; curved line, also called templet, is the one that does not go by the shortest way. These kinds of lines are like genera, since each one is divided into many categories, such as the parallels or *a trainel*, as stonecutters say, the orthogonal or plumb line, and the flat or level line, (and the two create squaring) and the diagonal and the diametral. Curved lines include the spiral, which turns like a snail, and the helical, which surrounds a body, and the circular, which is called circumference in the circle, as shown in the illustrations. Parallels are straight lines that are equidistant from one another and never meet, even if extended to the infinite; among stonecutters, they are called lines *a trainel*, and we will use this word in this book.[73]

Such detailed explanations hint that the terms and the concepts of classical geometry—point, line, parallel, orthogonal, and so on—were not known by stonemasons up to the sixteenth century. It seems, in any case, that Vandelvira was trying to tie together two branches of the same tree, as de l'Orme had done a few decades before: in Latin *linea* means both "line" and "rope"; the word derives from *lineus* or *linum*, meaning "flax".[74] Also, the notion of projection in Vandelvira is purely empirical: he refers to horizontal projections as *plomos* or "plumb bobs"; of course, the bob materialises the projection and the plumb line the projector. In contrast, his understanding of vertical projection seems rather vague: generally, he makes no difference between a given point and its vertical projection.

All this contrasts with the distinction between the *raggi* or projecting rays, the *pittura* or picture plane and the *intersegazione* or intersection of ray and picture

[73]Vandelvira (c. 1585: 3r-3v): *Punto es una cosa tan pequeña que no puede ser partido en mas partes, el cual es principio de la traza. Línea que en español llamamos raya y entre canteros trazo es una cosa que se imagina según longura y sin anchura, los extremos y fines de la cual son dos puntos. Divídese este línea en línea recta y línea curva; la línea recta decimos a la que va por más breve camino, que es la de arriba; línea curva que llamamos cercha es la que no va por más breve camino. Estas dos líneas son como géneros porque de cada una de ellas se sacan muchas; de la recta salen las que dicen paralelas o a trainel [como] llaman los canteros, y la perpendicular o a plomo y la línea plana o a nivel, las cuales causan la escuadría y la diagonal y la diametral. De la curva sale la espiral que es la que va rodeando a manera de caracol y la eliaca que es la que va rodeando algún cuerpo y la circular que es la que dicen circunferencia en el círculo; las cuales se figuran de esta manera. Las paralelas rectas son las que son equidistantes una de otra, que nunca se junta aunque procedan en infinito; llámanse entre canteros líneas a trainel, del cual vocablo usaremos en este libro.* Transcription taken from Vandelvira and Barbé 1977. (See also Bosse and Desargues 1643a: 14–16).

[74]Isidore of Seville (c. 630: 19.18.3): *Linea genere suo appellata, quia ex lino fit.*

plane, that is, projection, in Leon Battista Alberti's *De Pictura* (1435: book I, paragraphs 5–8, 12–17), written almost one hundred and fifty years before. In other words, the modern notion of central projection, including projectors and projection plane, appears clearly in the first treatise about perspective. As we have seen, its transfer to orthographic projections was slow: it appears first in cartography with Rojas Sarmiento ([1550] 1551: 25–26, 29; see Camerota 2005: 82–84), although its application is quite insecure, as remarked by Del Monte (1579: 91–92). A few decades later, the notion is quite clear in the optics treatise by Aguillon (1613: 503–520), who had worked as an architect for the Society of Jesus, and in a short manuscript about the didactic program of the Colegio Imperial in Madrid, dealing with the orders, stereotomy and drawing, by Jean Charles de la Faille (c. 1640), another Jesuit, born in Antwerp.

Remarkably, in spite of the widespread use of projection in stereotomic literature, there are few mentions of the concept in stonecutting treatises before Frézier. Derand explains horizontal projections in this way:

> The plan of a vault … is nothing other than the space on earth, or the ground that it covers … The architects, in order to draw the plan of an arch or vault geometrically, use perpendicular lines, known by masons as plumb lines, as we have seen before; these lines start in fact, or ideally, from all or several parts of the arches or vaults they mean to build, and are extended until they meet one or several straight lines, which the masons know as lines of direction, placed under the arches and the curved lines, both interior and exterior, of the vaults; in this way they form or find the tracings. Once these plumb lines are created, they mark and determine on the line of direction the plan of the corresponding arch, so that only the part of this line of direction between the plumb lines is taken as the plan of the arch, the excess being excluded. It does not matter whether this line of direction is close to the arches, or distant from them, as long as it is parallel to their diameter, or orthogonal to the plumb lines … The same idea must be applied to the proportions of the plans of all the surface of vaults: if we conceive a vault suspended in the air, and a flat surface under it receiving the plumb lines coming from all its parts, or several different ones, in particular, those that surround and end the vault, these plumb lines form the real plan.[75]

This passage may strike the reader as inconsistent: it starts talking about direction lines and finishes with "a flat surface under (the vault)" which plays clearly the role

---

[75]Derand (1643, 10): *Le plan d'vne voûte … n'est autre chose que l'espace sur la terre, ou bien le sol qu'elle couure … Or les architectes pour designer geometriquement le plan de quelque arc ou voûte que ce soit, se seruent de lignes perpendiculaires, nommées com dit est cy-dessus par les ouurieurs, des aplombs, qu'uis font partir par effet, ou par idée, de toutes, ou de plusiers & differentes parties des arcs, ou des voûtes qu'ils desirent tracer, & les produisent iusques à la rencontre d'vne ou plusiers lignes droites, qu'ils appellent lignes de direction, lesquelles ils posent au dessous des arcs, & des cherches tant interieures, qu'exterieures des voûtes, desquelles ils forment ou recherchent les traits. Et ces aplombs ainsi produits, marquent & determinent sur la ligne de direction, le plan de l'arc d'où ils procedent; en sorte que la partie seule de cette ligne de direction qui est conprise entre ces aplombs, est prise pour le plan de l'arc, le surplus en estant exclus. Or il n'importe que cette ligne de direction soit ionte aux arcs, ou separée d'iceux, pourueu qu'elle soit parallelle à leur diametre, ou perpendiculaire aux aplombs … Le mesme se doit entendre par proportions des plans de toute la superficie des voûtes: Car si on conçoit vne voûte soustenuë en l'air, & au dessous d'icelle vne superficie plane receuant des aplombs procedans de toutes, ou de plusiers & differentes de ses parties, particulierement de celles que la bornent, & la terminent, ces aplombs en formeront ou determineront le vray plan.*

of projection plane. On closer inspection, it seems that Derand clearly understands the notion of projection of a point as the intersection of projectors with a projection plane; however, he is trying to introduce it gradually, starting from a method known by masons. In particular, he begins with a procedure for the construction of the plan of built elements, in particular arches, starting from their vertical projection. This technique, which inverts Roriczer's method of extracting the elevation from the plan, it is the simplest way of finding the plan of a round arch or a hemispherical dome. The voussoirs of the arch are equal, but their horizontal projections are different; thus, the shortest method to construct the plan starts by drawing the voussoirs in the elevation and projecting them onto the plan, using projecting lines intersecting a "direction line". This line is akin to the *ligne de terre* of nineteenth-century descriptive geometry treatises, introduced by Monge, or the "folding line" of American manuals. However, it is essential to note that Derand mentions one or several direction lines and stresses that they may be close to or far from the actual arch; that is, the position of the folding plane, and thus the horizontal projection plane is not fixed, as Derand makes clear. Only in the final part of the passage does Derand consider a flat surface as a projection plane; this hints he was aware of the abstract concept of projection; after all, he was a Jesuit and, like Aguillon, an architect or at least a building supervisor for the Society.

Milliet includes in the introduction to the section on stereotomy of his *Cursus seu mundus mathematicus* a variant of the theorem that was later known in French treatises as *théoreme des trois perpendiculaires* (theorem of the three orthogonal lines). The accompanying figure shows clearly projectors and projections, but the Latin vocabulary sounds odd for a modern reader: projections are *vestigia*, that is, footprints (Fig. 12.10). This is appropriate since Milliet only deals with the case where the *vestigia* are placed on a horizontal plane, while there is no specific word for projectors or vertical projections.

Not a single word about projection planes, projectors, or projections can be found in de la Rue (1728); in contrast, Frézier (1737–39: I, ix), states at the start of his treatise that ichnography and orthography are important sections of his "stereotomy"; in this context, the word designates the theoretical foundation of stonecutting, as we have seen. However, the subject appears as a section of a chapter about "The drawing of sections of bodies that should not, or cannot, be drawn other than in concave or convex surfaces" (Frézier 1737–39: I, 206–211). This strange placement hints that Frézier did not think about projection as a general method for the representation of all geometric entities, as Monge would do fifty years later; rather, he conceived it as a specialised tool for figures that cannot be depicted in any other way. This is why he placed at the beginning of the contents of his book (1737–39:1, viii-ix) the *Tomomorphie* (shape of intersections) and the *Tomographie* or (drawing of sections), before *Ichnographie* (horizontal projections) and *Ortographie* (vertical projections).

In any case, his approach to this issue is remarkably precise and acute. First, he explains projection in the following terms:

> The word "projection" has several meanings. It may be applied to the action of throwing; however, we shall limit it here to the depiction of a body formed on a plane by the perpendiculars to such plane, or, if you want to extend the notion, by several parallels passing through

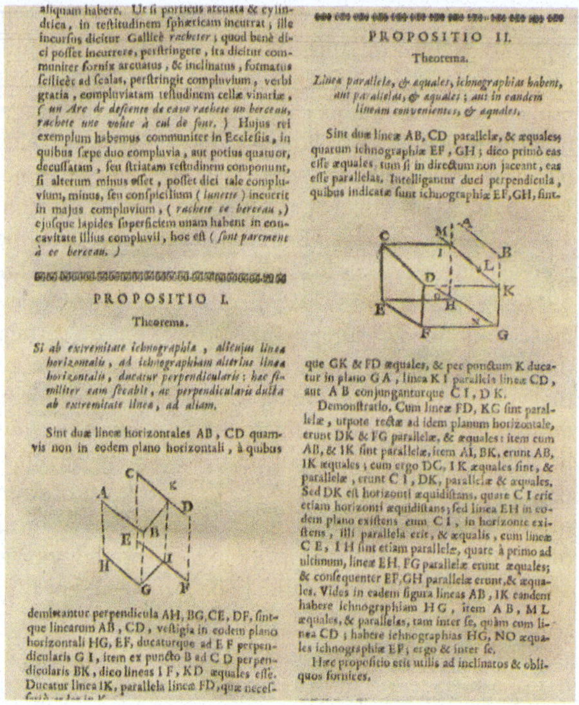

aliquam habere. Ut ſi porticus arcuata & cylindrica, in teſtitudinem ſphæricam incurrat; ille incurſus dicitur Gallicè *racheter*; quod benè dici poſſet incurrere, perſtringere, ita dicitur communiter fornix arcuatus, & inclinatus, formatus ſcilicèt ad ſcalas, perſtringit compluvium, verbi gratia, compluviatam teſtudinem cellæ vinariæ, (*un Arc de deſcente de cave rachete un berceau, rachete une voûte à cul de four.*) Hujus rei exemplum habemus communiter in Eccleſiis, in quibus ſæpe duo compluvia, aut potius quatuor, decuſſatam, ſeu ſtriatam teſtudinem componunt, ſi alterum minus eſſet, poſſet dici tale compluvium, minus, ſeu conſpicilium (*lunette*) incurrit in majus compluvium, (*rachete ce berceau*,) ejuſque lapides ſuperficiem unam habent in concavitate illius compluvii, hoc eſt (*font parement à ce berceau.*)

PROPOSITIO I.

Theorema.

Si ab extremitate ichnographiæ, alicujus lineæ horizontalis, ad ichnographiam alterius lineæ horizontalis, ducatur perpendicularis: hæc ſimiliter eam ſecabit, ac perpendicularis ducta ab extremitate lineæ, ad aliam.

Sint duæ lineæ horizontales AB, CD quamvis non in eodem plano horizontali, à quibus demittantur perpendicula AH, BG, CE, DF, ſintque linearum AB, CD, veſtigia in eodem plano horizontali HG, EF, ducaturque ad E F perpendicularis G I, item ex puncto B ad C D perpendicularis BK, dico lineas I F, KD æquales eſſe. Ducatur linea IK, parallela lineæ FD, quæ neceſ-

PROPOSITIO II.

Theorema.

Lineæ parallelæ, & æquales, ichnographias habent, aut parallelas, & æquales; aut in eandem lineam convenientes, & æquales.

Sint duæ lineæ AB, CD parallelæ, & æquales; quarum ichnographiæ EF, GH; dico primò eas eſſe æquales, tum ſi in directum non jaceant, eas eſſe parallelas. Intelligantur duci perpendicula, quibus indicatæ ſunt ichnographiæ EF, GH, ſintque GK & FD æquales, & per punctum K ducatur in plano G A, linea K I parallela lineæ CD, aut A B conjunganturque C I, D K.

Demonſtratio. Cum lineæ FD, KG ſint parallelæ, utpote rectæ ad idem planum horizontale, erunt DK & FG parallelæ, & æquales: item cum AB, & IK ſint parallelæ, item AI, BK, erunt AB, IK æquales; cum ergo DC, I K æquales ſint, & parallelæ, erunt C I, DK, parallelæ & æquales. Sed DK eſt horizonti æquidiſtans, quare C I erit etiam horizonti æquidiſtans; ſed linea EH in eodem plano exiſtens cum C I, in horizonte exiſtens, illi parallela erit, & æqualis, cum lineæ C E, I H ſint etiam parallelæ, quare à primo ad ultimum, lineæ EH, FG parallelæ erunt æquales; & conſequenter EF, GH parallelæ erunt, & æquales. Vides in eadem figura lineas AB, IK eandem habere ichnographiam HG, item A B, M L æquales, & parallelas, tam inter ſe, quàm cum linea CD; habere ichnographias HG, NO æquales ichnographiæ EF; ergo & inter ſe.

Hæc propoſitio erit utilis ad inclinatos & obliquos fornices.

Fig. 12.10 Projections of line segments (Milliet [1674] 1690: II, 622)

the corners of this body or through several points of its contour onto this plane, in whatever situation it is placed. It is sufficient for our purposes, however, to consider vertical and horizontal lines, since lines that are inclined in respect to the horizon must be referred to these constant positions, which may always be determined. Taking into account this restriction, we can say, in mason's terms, that the projection of a body is the mark of several *plumb lines*, going down from the corners or the contours of this body in order to construct the *plan* or Ichnography, or several same-level lines starting from its angles or its contour on a perpendicular surface, in order to make the *profiles* or the *elevations*.[76]

---

[76]Frézier (1737–39: I, 206–207): *Le mot de projection a plusieurs significations, il peut s'appliquer à l'action de jetter, mais nous la resserrons ici à la description d'un corps formée sur un plan par des perpendiculaires à ce plan, où si l'on veut l'étendre encore davantage, par des paralleles menées des angles de ce corps ou de plusieurs points de son countour sur ce plan en quelque situation qu'il soit à son égard / Il suffit cependant à l'usage que nous en devons faire, de considérer les lignes verticales & les horisontales, parce que c'est à ces deux genres de situations constantes, & que l'on peut toujours déterminer, qu'on doit rapporter les lignes inclinées a l'horison. Selon cette restriction nous pouvons dire, pour nous accomoder aux termes de l'Art, que la projection d'un corps est la trace de plusiers* à-plombs, *abaissées de leurs angles ou de leurs contours pour en faire* le plan *ou Ichnographie, ou de plusieurs lignes de niveau tirées de même de ses angles, ou de son contour sur une surface* à plomb, *pour en faire les* profils *ou les* élévations.

It is not easy to translate this passage preserving Frézier's subtle distinctions between learned terminology and masons' jargon. In particular, *a-plomb* means a vertical line, but it is associated with the plumb line or the plumb bob; *ligne de niveau* means horizontal line, but it seems to allude also to the ropes used to guarantee the horizontality of courses in walls or barrel vaults; *surface à plomb* means "a surface with vertical generatrices" or, more specifically, "vertical plane", but

Thus, Frézier presents at the start a general notion of projection, where projectors may follow any direction. Moreover, if the phrase "in whatever situation it is placed" is to be interpreted broadly, projectors are not necessarily orthogonal to the projection plane; in this case, Frézier is considering a generalised notion of projection enclosing both orthogonal and oblique projections. However, in the second paragraph, he consciously restricts the kind of projections used in stonecutting to horizontal and vertical orthogonal ones. Of course, in the first variant projectors are vertical lines, and it is implied that the projection plane is horizontal, while in the second one he stresses that projectors are horizontal and the projection plane vertical.

Next, Frézier (1737–39: I, 207–211) extracts several useful conclusions from these definitions: (orthographic) projections on vertical and horizontal planes reduce the length or breadth of all lines and surfaces that are not parallel to the projection plane; curved lines belonging to a plane which is orthogonal to the projection plane are projected as straight lines; the projection of a circle that is not parallel to the projecting plane is an ellipse and, conversely, an ellipse can be projected as a circle; projections of ellipses, parabolas or hyperbolas are curves of the same kind, although more or less extended or reduced. That is, Frézier furnishes some fundamental tenets of Monge's science of projection, although he does not try to construct a universal system where everything should be projected, nor limit the number of projections to two.

Similar arguments may be put forward about the concepts of generatrix and directrix of a surface. As shown by Sakarovitch (1993: 136), the notions of directrices and generatrices of a surface may stem from the standard technique for dressing ruled or double-curvature surfaces (see Sect. 3.2.1). The stonemason starts by dressing two drafts playing the role of directrices; next, he carves the actual surface controlling the operation by means of a ruler leaning on both marginal drafts; each position of straightedge furnishes a generatrix. In any case, it is important to bear in mind that, for the stonemason, this is an implicit notion, rather than an explicit concept; no manuscript or treatise explains the issue it in so many words. Another descriptive geometry concept anticipated, but not embodied, in stonecutters' literature is the distinction between developable and warped surfaces. Masons clearly differentiate between warped surfaces (known as *gauche* in French or *engauchida* in Spanish) and "regular" or developable ones. However, the selection criterion seems to rest in the parallelism or convergence of intrados joints, rather than two infinitely close generatrices, as in modern geometry; also, some writers as Martínez de Aranda and Portor, do not extract the relevant conclusion and try to dress warped surfaces using templates.

---

again is explained using terms that are familiar to masons. Moreover, *plan*, *profil* and *élévation* are italicised by Frézier to stress that they are masons' terms opposed to such learned words as "ichnography".

Chapter 13
A Provisional Summary

Pre-industrial ashlar construction raises complex geometrical problems. First, an ashlar element, either block or voussoir, must be made to conform to a previously conceived shape. This shape is, up to a certain extent, the result of structural concerns, since the mechanical behaviour of the element, in particular, the thrust exerted on its supporting elements, depends more on the shape and weight of the element than on its material (see Sect. 1.2). In any case, the wide repertoire of arches used during pre-industrial periods shows clearly that these constraints are rather loose: although the ideal form of an arch is the catenary or the parabola,[1] round, segmental, basket handle, pointed, and many other shapes have been used for arches in western architecture during the second millennium. Regarding vaults, a similar argument may be put forward; even flat vaults—that is, slabs divided into wedges—work as ordinary vaults, although they exert very strong thrusts and are usually placed between the bases of towers or in building foundations.

Thus, other constraints are usually taken into account when choosing the shape of a construction element. Strictly functional[2] considerations demand special pieces: for example, skew arches are used to enter a room through a corner; splayed arches or rere-arches to distribute evenly the light coming through an opening; lunettes to cast light from above; and corner arches to control two streets visually (see Chap. 7 and Sects. 6.1, 6.2, 6.3 and 8.4). Aesthetical ideals brought back round arches and spherical vaults in the Renaissance and fostered oval vaults in the late Renaissance and the Baroque era; architectural syntax brings about further difficulties, such as arches opened in curved walls or sloping vaults. Moreover, feasibility and worksite organisation are essential concerns in the choice of constructive solutions. For example, a round arch can be controlled easily with a rope, which gives the shape of

---

[1]In theory, the ideal form of a self-supporting arch is the catenary, while the most efficient shape for an arch that supports a uniform load per unit of horizontal projection is the parabola.

[2]"Functional" is used here in opposition not only to "aesthetical", but also to "structural" or "mechanical". That is that is, the term is used here in the sense of *utilitas* (utility) and connected with circulation, lighting and other practical issues, but not with *firmitas* (stability).

J. Calvo-López, *Stereotomy*, Mathematics and the Built Environment 4,
https://doi.org/10.1007/978-3-030-43218-8_13

the arch directrix and the direction of the bed joints (see Sect. 4.3.1). However, the pointed arch helps to standardise voussoirs and formwork; different openings can be spanned by arches with the same curvature, and thus with the same kind of voussoirs, using the same arch squares and formwork. In addition to all this, there is caprice (see Sect. 12.1); masons everywhere, from Armenia to Portugal, have used particularly complex solutions, just to display their skill and the power of their clients.

Most of this applies to all kinds of masonry: concrete, rubble, pisé, adobe, brick, rough-hewn stone, and ashlar. However, depending on the material, different strategies are used to ensure that the final element conforms to the desired shape. In some cases, such as concrete, rubble or pisé, full formwork is essential; in addition to support, casing provides formal control. In arches and vaults built in brick or half-squared stone, more or less complete formwork is usually necessary, but the final form is generated by the use of wedge-shaped mortar joints. As an extreme example, timbrel vaults show that it is possible to build in brick without formwork; in this case, the role usually played in formal control by falsework is entrusted to additional devices, such as a guiding bar (see Sect. 1.1).

Ashlar construction is based on a completely different strategy. In this case, special-purpose pieces, the voussoirs, take responsibility for formal control, in lieu of formwork or the small mortar wedges in brick or rough-hewn stone. However, such an approach raises an additional problem: voussoirs must fit one another precisely in order to fill the three-dimensional space occupied by the member. If this condition is not fulfilled, the mechanical behaviour of the piece may be compromised, unless substantial quantities of mortar are used. In this case, the construction process would revert to a particularly uneconomical variety of the brick scheme.[3] Thus, formal control of the shape of the voussoir is essential. It serves two purposes: it insures that voussoirs will fit snugly with the next ones and it furnishes the desired final shape of the entire piece. At the start of the process, this final shape is ideally divided into voussoirs. This scheme is usually complex, and therefore must be represented graphically; full-scale tracings on site are used frequently for this purpose, to prevent errors derived from scale changes and, in the preindustrial period, to avoid the use of expensive materials such as parchment and paper (see Sect. 3.1.1). Voussoir shapes—or more precisely their faces, the lengths of their edges and the angles between them—must be transferred from the tracing to the stones during the dressing process by means of a host of geometrical instruments such as the ruler, the gauge, the templet, the bevel, the arch square and, above all, templates (see Sect. 3.2.1). Formwork is still necessary, due to the comparatively large size of the voussoirs; however, its role in formal control its limited to the final assembly.

Except when representing simple, almost two-dimensional pieces such as simple arches, these full-scale tracings use orthogonal projection, either simple, double or multiple. Horizontal projections are used systematically, while vertical projection planes are employed if and when the need arises. If voussoir faces or edges are

[3] As seen in Sect. 1.3, ashlar construction requires the use of comparatively large voussoir sizes, in order to keep the dressing effort within reasonable limits. If these large pieces do not fit well, the amount of mortar needed to fill the holes may exceed that used in a comparable brick element.

parallel to the projection plane, their shapes, lengths and angles are preserved and may be transferred directly from the tracing to the block of stone while dressing it. However, in many cases, voussoir faces or edges are neither horizontal nor frontal. In these cases, some special operations must be performed in order to deduce the true sizes and shapes of faces, edges and angles from their orthographic projection (see Sect. 1.3). This applies to templates used for planar surfaces, such as the faces and bed joints of voussoirs, or flat surfaces used as an intermediate stage in the dressing process of a curved face. Such templates are usually made of wood, although other materials can be used; in any case, they are conceptually rigid, since they stand for planar figures. In other cases, masons apply flexible templates directly on curved surfaces, such as cones and cylinders; of course, such templates must be executed in flexible materials, such as cloth, leather, tin, cardboard or paper (see Sect. 3.2.1). They are constructed using conical and cylindrical developments, although in many cases drastic simplifications are used; on some occasions, two chords of an arch are substituted for its real length. Such a method cannot be applied directly to spherical vaults since the sphere is a non-developable surface. To circumvent this constraint, masons inscribed cones into the intrados of spherical vaults; this is, of course, another simplification, albeit an effective one, as practice has shown.

It has been said (Pérouse [1982a] 2001: 184) that geometry did not furnish the methods of stonecutting; rather, it was stonecutting that brought about the essentials of descriptive geometry. This is an attempt to condense in a single sentence a story encompassing at least eight centuries, several architectural periods, writers and practitioners from quite different social and professional backgrounds, different media of knowledge transmission ranging from masons' personal notebooks to scientific journals and, above all, a tangled succession of two-way exchanges between learned geometry and artisanal practices (see Chap. 2 and Sects. 12.2, 12.4 and 12.5).

Regrettably, our actual knowledge about construction procedures in the Romanesque period—in particular, formal control methods and instruments—is rather scarce. There are few details about architects or master masons or their names; almost no drawings, whether on parchment, floors or walls, have been preserved. Only a few manuscripts such as the one by Herrad von Landsberg (c. 1160) or the *Biblia Sancti Petri Rodensis* (c. 1020: III, 89v; see also Colombier [1953] 1973), or even sculpted capitals, like those in the cloister of Girona cathedral (Fig. 12.4) offer some information on dressing, transportation, elevation and placement. Construction in this period seems to have been directed by ecclesiastical, monastic or civil patrons, while masons appear as mere executors. However, this may be an image deliberately projected by clients; gradually, artists known by name, such as Lanfranco in Modena or Maestro Mateo in Santiago de Compostela emerge as construction directors. Dressing seems to be carried out mainly by means of the pick and the chisel. Transportation by hand prevails, although carts are used here and there, for example in the *Biblia Rodensis* (c. 1020: III, 89v); also, hoisting equipment seems to be quite scarce (see Sect. 3.3.3). This explains the use of rubble and rough-hewn stone in the first phases of Romanesque architecture. Later on, medium-sized ashlar is used in elevations of the great works of mature Romanesque, such as some pilgrimage churches, Cluny, or the great Rhineland cathedrals. The main architectural element in these constructions, the barrel vault, either circular or pointed, as in the Cluny school, can be controlled easily without sophisticated formal control methods; in contrast, other elements, such as the annular vault in Montmajour or the archetype of the *vis-de-Saint-Gilles*, require complex control processes; unfortunately, any hypothesis about these methods must remain just that, a hypothesis, due to the lack of precise information.

Gothic architecture brought about a complete rethinking of these construction procedures. The depletion of forests fostered the use of economical formwork and centring schemes (Bechmann [1981] 1996: 42–51, 141–142), leading to a two-tier vault construction system, based on a linear network of ribs supporting web surfaces. It has been noted on many occasions that both rib vaults and pointed arches have Islamic antecedents (see for example Bony 1984: 12–17). The school of Viollet-le-Duc ascribed the transfer of this knowledge to the Crusaders, taking into account the central place of Île-de-France and Picardy in the origin of Gothic architecture; modern scholarship favours a diffuse transmission scheme, taking into account such early Christian ribbed vaults as those in Durham or Rivolta d'Adda. In any case, rib directrices were controlled rather precisely, while severies were materialised as ruled or double curvature surfaces (see Sect. 10.1.1 and 10.2.2) leaning on ribs, which cannot be reduced to the usual surfaces in descriptive or analytical geometry.

Formal control of early Gothic vaulting—in particular, sexpartite and quadripartite vaults—can be carried out through relatively simple methods, since the symmetry of square and rectangular vaults leads diagonal ribs to a precise meeting point in space: the main keystone. As a result, such vaults can be controlled without a complete full-scale tracing, using templates for the springers and axes marked on operating surfaces on the top face of keystones (see Sect. 10.1.1).

From the thirteenth century on, the introduction of the tierceron raised other problems: nothing guarantees that tiercerons and liernes will meet at secondary keystones. Quite probably this problem was solved through full-scale tracings, frequently executed on planks laid on scaffoldings, as pointed out by Rodrigo Gil de Hontañón (c. 1560: 24v-25r; see also Sect. 3.3.4). This practice allowed both the determination of the curvature of the ribs and the location of secondary keystones prior to dressing, as well as the verification of the placement of dressed voussoirs in the final stages of construction. However, because these drawings were executed on scaffoldings, few or no medieval vault tracings have been preserved, with the exception of a huge one in Szydłowiec, Poland (Fig. 3.3), probably prepared for the patrons rather than for actual execution, and two very late examples in northwestern Spain. In contrast, many examples of rib vault tracings are included in stonecutting manuscripts and treatises (for a summary, see Rabasa 2007b or Calvo 2017). Comparing them with actual tracings and keeping in mind that texts may have been embellished by didactic intentions, it seems clear that the basis of this system was a horizontal orthographic projection of the vault. The elevations of the ribs were generally disassembled in order to represent them in true size, eschewing projection; an orthodox vertical projection of diagonal ribs and tiercerons would distort them and be useless for construction purposes (see Sects. 10.1.4 and 12.4.2).

In theory, German net vaults raise the same problems, but their sheer complexity led masons to standardise rib radii as much as possible. This led to the idea of the *Prinzipalbogen* or main rib, a master arch providing the curvature for all ribs in a vault. In some cases, such as vaults with a tightly knit triangular network (Facht 1593: 6, 12), this goal can be accomplished without problems, since all reasonable pathways from the springings to any keystone have the same length. In this case, the height of the keystone, measured in the *Prinzipalbogen* on account of the combined length of the ribs going from the springing to the keystone, leads to a single result. On other occasions, however, a keystone can be reached by several pathways starting in the springings, with different lengths. As a result, the *Prinzipalbogen* method cannot be applied literally, since different pathways will attain different heights when reaching the same keystone, and thus the last ribs of each pathway would not meet in space. This led sometimes to the use of different curvatures in a limited number of ribs, in order to guarantee the geometrical consistency of the scheme (see Sects. 2.1.7 and 10.1.5).

Two aspects of medieval construction should be briefly commented on here. Medieval society was segmented into airtight compartments such as *oratores* or clerics, *bellatores* or noblemen and *laboratores*, that is, the rest of the society. Masons, of course, belonged to *laboratores* and, generally speaking, had no access to grammar schools and universities, the only effective places where classical science, and in particular learned geometry, was transmitted in the period. In particular, Euclid was translated into Latin in the twelfth century, but vernacular translations were not available until the late fifteenth century. Further, masons did not read Latin; in fact, up to the sixteenth century, many of them could not read the vernacular. It comes as no surprise, then: that masons' conception of Euclid and Euclidean geometry was strictly mythical: as we have seen, they took Euclid for a direct disciple of Abraham and, as late as the fifteenth century, Roriczer did not know an exact construction for the regular pentagon (see Sects. 12.2 and 12.4.1).

Moreover, neither Euclidean nor medieval practical geometry did provide useful tools for the main problems faced by masons, spatial representation and formal control of masonry. The main instrument for these purposes, orthogonal projection, seems to have been devised mainly by masons, although the influence of clerical sources cannot be discarded. In fact, the older witnesses of this technique appear in clerical circles aware of practical geometry, or at least in the hands of figurative artists connected with cathedral chapters, particularly in the manuscripts of Richard of Saint Victor, a student of Hugh of Saint Victor and the sketchbook of Villard de Honnecourt's. However, we should take into account the scarcity and high price of parchment and that Villard may have gathered information from plans prepared by or for the architects of Reims (see Erlande-Brandenburg 1993: 74; Ackerman 1997: 42, 47, note 10; and Sects. 2.1.1 and 3.1.3); this suggests at least that these early instances of orthogonal projection may have appeared at the same time in clerical and artisanal media.

However, the concept of *double* orthogonal projection seems strongly connected to architectural and building activities. In contrast to the combination of horizontal projection and disarticulated elevations in stonecutting tracings, masons used vertical projections for other purposes. General elevations appear in the sketchbook of Villard de Honnecourt (c. 1225: 31v, 32v; see also Sects. 2.1.1 and 3.1.3), but recent studies (Villard/Barnes [c. 1225] 2009: 229–230) show that he was not an architect or mason. However, remarkable orthographic elevations from the next centuries are present in the collections in Vienna (Böker 2005), Strasbourg (Recht et al. 2014) and Segovia (Ruiz Hernando 2003). While these are single orthogonal projections, a drawing for a bell tower in Sienna, probably a copy of Giotto's project for the *campanile* in Florence, shows a remarkable octagonal upper stage, with correctly drawn oblique sides (Recht 1995: 57–63; Ascani 1999: 266–272). Quite probably, the projected widths of the oblique sides were taken from a plan drawn at the same scale. In any case, double orthogonal projection is clearly explained by Mathes Roriczer (1486: 5r-5v) when dealing with pinnacle elevations; his method shares a number of traits with the one explained in the well-known letter to Leo X (Sanzio and Castiglione [c. 1518] 2003: 79–80).

It is essential to take into account that in the Middle Ages, and even in the mid-sixteenth century, orthogonal projection was a procedure and not at all a rule. In some occasions, draughtspersons preparing an orthographic elevation found that representing in true shape elements such as elevation responds, circles in oblique sides of spires, or vault ribs, fit their purposes better than depicting them in orthogonal projection. In these cases, they set consistency aside and flattened this particular element, regardless of its integration in an orthographic drawing. The use of linear perspective in Renaissance orthographic drawings by Serlio and Hernán Ruiz II stems from the same notion, that is, the lack of concern for the modern notion of consistency in representation systems.

Another important factor is the slow but steady development of transportation and hoisting technology during the Gothic period. In representations of Romanesque construction, stones are usually carried by workmen on their shoulders; later on, in a Chartres stained-glass window, two workers carry a stone in a handbarrow. At the same time, windlasses operated by two workers appear, allowing the lifting of medium-sized stones. The size and complexity of such devices increased gradually; in the sixteenth century, they are powered by men treading in a cage, as in the well-known painting of the Tower of Babel by Pieter Breughel the Elder in Vienna. All this led to an increase in the size of the stones used in ribs and, particularly, in the web. In southern Spain, the carefully dressed stones in the severies are sometimes quartered in round courses, exactly as in spherical domes (Pinto 2002). This process blurs the lines between ribbed and non-ribbed construction; if both ribs and severies are carefully dressed, why not forget the distinction and advance to single-tier construction? This step was taken in two different and, as far as we know, independent epicentres: England, with fan vaults (Leedy 1980), and Valencia (Zaragozá 2008), with arrised vaults.

Ashlar vaults, in the modern sense, were born in this period. Spanish and French Renaissance masons made good use of them. Patrons in both countries demanded ashlar, deemed to be an aristocratic material, while in Italy vaults were usually built in brick. Thus, masons had to devise, in a short period, a number of graphical procedures to control the execution of the new Renaissance forms in ashlar. Except for coffered vaults, this required the use of single-tier construction, which differs radically from mainstream Gothic vault construction. Builders resorted to orthographic projection and devised several ancillary procedures, such as auxiliary views, triangulation and several methods based on distances to a reference plane, prefiguring modern rotations and developments. The use of vertical projections may be obscured by the fact that masons, driven by an economy principle, usually aligned vertical projection planes with the most complex figures of each problem, so they were represented in true shape; in contrast, they placed other, less complex, shapes in oblique position. However, when necessary, and in particular for architectural drawings as opposed to stonecutting diagrams, masons used orthogonal projection competently.

The repertoire of classical stonecutting (see Sect. 1.4) was all but fixed at this moment. While de l'Orme's (1567: 58v-64r, 67v-128v) catalogue is relatively short, as a consequence of its inclusion in a general architectural treatise, the specialized manuscript by Vandelvira (c. 1585) includes trumpet squinches, skew, corner and sloping arches, rere-arches, spiral and straight staircases and spherical, sail, groin, pavilion, octagonal, torus, oval, rib and coffered vaults. Later treatises (Jousse 1642; Derand 1643; de la Rue 1728; Frézier 1737–1739; Hachette 1822; Leroy 1844; see also de la Gournerie 1855: 7; Sakarovitch 2003a: 21; Sakarovitch 2009a: 301) brought about much refinement in geometrical methods, but little or no additions to this repertoire.

The Renaissance started a shift in the social standing of stonecutting writers from masons to architects, from manual workers to intellectual designers. The ruling classes did not immediately accept such change. Poets and generals derided de l'Orme and Rojas, while ordinary masons were wary of a change in their own standing that would place them under the command of these "new professionals". This explains why de l'Orme was attacked by journeymen few days after his dismissal as the architect of the King's Works (see Sect. 2.2.1).

In any case, the "new professionals" could not understand Latin, (as shown by Ronsard's jokes about de l'Orme) but they could read and write vernacular languages competently, and thus had access to the new editions of Euclid. This new situation led to a slow integration of learned geometry concepts and methods in stonecutting literature. In the sixteenth century, this presence is still scant: Thales' theorem, the procedure to find the centre of a circle given three points, and the vocabulary of classical geometry are dutifully explained, while accolades of the power of geometry are routinely included in introductions or the first sections of stonecutting literature. In contrast, the main body of these works systematically uses projections and empirical procedures playing the role of rotations and developments; such methods do not belong to classical geometry. In any case, connections with other learned sciences, such as cosmography and cartography seem to have furnished an essential tool for stonecutting methods: conical developments applied, by approximation, to portions of a spherical surface.

Another social shift took place in the seventeenth century (see Sect. 2.3). The masons-turned-into-architects of the previous period were replaced by clerics, most of them practising designers for their respective orders or other patrons, such as

San Nicolás, Derand and Guarini. Many of them had a solid knowledge of mathematics and even had taught the subject in the schools of their orders, like Derand. However, the need to be understood by practitioners coloured their books, in particular those of San Nicolás and Derand. Such a stance may have also been influenced by the controversies brought about by Desargues. A gentleman geometer, rather than a practising architect, he made a fundamental contribution to the theory of conic sections, including a theorem on homological triangles that is now seen as the foundation of projective geometry. He also wrote two short leaflets putting forward new methods in perspective and stonecutting, in particular for skew arches, sloping vaults and arches in battered walls. Masons were infuriated by this intrusion of an outsider in their craft; the best stonecutter in Paris, Jacques Curabelle, reacted with a series of fierce attacks on Desargues. A contest involving the execution of arches according to the methods of both contenders was scheduled. Ultimately, it did not take place, since the rivals did not agree on the standards that should be taken into account. Curabelle took it for granted that the quality of the final work should be the ruling criterion of the contest, while Desargues stated that it was the correctness of the geometrical reasoning that should be held paramount; in other words, he put geometry and geometricians on a superior level than that of masons and artisanal practice (Sakarovitch 1994b; see also Sect. 2.3.2).

Desargues's methods did not exert much influence in the evolution of stonecutting in the next centuries, at least up to the Industrial Revolution, when stonecutting was being supplanted by new technologies. However, Derand chose a different, indirect strategy. In his treatise (1643; see Sect. 2.3.3), he addressed stonecutters, leaving aside mathematical demonstrations and trying to explain everything using masons' terms and concepts. However, here and there he introduced small tweaks, as the systematic placement of arc centres in rib vaults; the use of bisectors and tangencies is evident, although not explicitly mentioned. The same approach is applied to cylindrical developments: Derand makes the length of an arc portion equal to the combined length of two chords, but he is aware that this is a simplification and leaves the door open to using more than two chords. These are specific issues, but Derand applies the same strategy when dealing with a central concept of stonecutting methods: orthogonal projection. He first explains the operating procedure in empirical terms, and only at the end of this passage does he present the concepts of projectors—as opposed to that of projection line—and projection plane. This is quite remarkable since the operating methods of orthogonal projection had been connected with masons and stonecutting from the beginning. However, the theoretical concept of projection, and the name itself, had taken shape in other fields (see Sect. 12.5.3), starting with perspective and Alberti's rays, and following with cosmography, cartography and optics, with Rojas Sarmiento and Aguillon. An isolated attempt by Jean-Charles de la Faille to introduce these concepts and a correct theory of orthogonal projection in architectural instruction at the Imperial College in Madrid, seems to have fostered no derivatives.

A further effort to bring stonecutting into the realm of learned geometry was attempted by Claude Milliet-Dechales, another Jesuit that worked as an engineer but had devoted his career to teaching mathematics. He included a treatise on *De Lapidum Sectione* in his multivolume *Cursus seu mundus mathematicus* (1674: II, 619–692);

obviously, it was not directed to masons. The nature of this work led him to classify his material in theorems and problems; however, in the stonecutting treatise, theorems are notable for their absence, except for a few introductory sections dealing with the theorem of the three perpendiculars and other issues (see Sect. 2.3.4). Guarino Guarini's contemporary treatise on mathematics (1671) has drawn less attention in stonecutting studies (see D'Amato and Fallacara 2005: 71–72 as an exception); however, it is quite interesting for our purposes. It includes one treatise (Guarini 1671: 444–452) stating clearly some fundamental tenets of orthographic projection, as well as another about developments that may be applied to stonecutting procedures, as explained by Guarini himself (1671: 572–596, in particular 573; see also Sect. 2.3.5); this mathematical work provided the basis for two substantial chapters on projections and developments in his posthumous *Architectura Civile* (1737).

Another engineer, François Blondel, was placed during this period at the direction of the Royal Academy of Architecture in Paris. Although he did not write a general treatise on stonecutting, he stressed the importance of this discipline in architectural instruction in his well-known *Cours d'Architecture* (1675–1683) and included the directrices of sloping arches and the shape of the faces of rampant, skew and other kinds of arches as two of the four most important problems of architecture making good use of the theory of conic sections (Blondel 1673). Two writers in his entourage followed the same path: Philippe de la Hire applied advanced theorems on conic sections and their tangents in his unpublished *Architecture Civile*, while Jean-Baptiste de la Rue, an architect connected with the Academy, who had carried out some engineering projects, published finally his *Traité de la Coupe des Pierres* (1728), including the essential repertoire of stonecutting (see Sect. 2.4.1). He tried to improve tracing and, in particular, dressing methods, using new procedures that leverage rotational symmetry and the notion of dihedral angle. In other passages, he took exception to the use of conical developments in spherical vaults and proposed a new method, using triangulations on the dressed spherical surface of the stone in order to construct the exact shape of the intrados face of each voussoir; it does not seem that such an idiosyncratic method had much success among practitioners. However, de la Rue's most significant contribution, together with the excellent quality of his engravings, was the inclusion of a *Petit Traité de Stéréotomie* at the end of his book, providing an abstract treatment of the intersections of cylinders and cones with planes. Such placement suggests that de la Rue thought that most problems in stonecutting could be solved by the methods explained in the body of his book, while abstract geometry should be used as an advanced method to tackle the most difficult problems.

Amedée-François Frézier (see Sect. 2.4.2) reversed de la Rue's didactic strategy and placed the theoretical section not at the end, but rather at the beginning of his three-volume set, providing a firm foundation for stonecutters' lore. This led, finally, to the implementation of a full theory of projection, including surface intersections, angular measures and a host of abstract geometry propositions, filling the first volume of his treatise; only at the beginning of the second volume did he deal with actual stonecutting. Frézier also noticed that stonecutting methods could be applied to other disciplines, in particular joinery, or *menuisserie* in French; this is why his book is entitled "The theory and practice of stone and wood cutting ...". When dealing with stonecutting problems, Frézier applied a somewhat encyclopaedic, even eclectic strategy. He explained all or most known methods for dealing with a specific stonecutting problem carefully, although, in many places, he introduced sections under headings such as *Remarque sur la falseté de l'ancien trait* (Remarks on the falsehood of ancient tracings), and the like. However, old methods are explained in such detail that the reader has the impression that Frézier is presenting them as an option. Another example of this stance is his treatment of multiple projections: at first, Frézier takes exception to the use of elevations placed sideways or head down, but in the end, he accepts that such practice shows more clearly the connections between plans, elevations and profiles.

It is important to stress that the evolution of stonecutting literature, from Jousse to Frézier, brought about no new constructive types and few original methods, and even the proposals of Desargues for the use of slanting reference planes, Blondel for the use of conic sections and de la Rue for constructing templates directly on spherical surfaces do not seem to have been accepted in practice. What the ever-increasing introduction of learned geometry in this field brought about was a slow, capillary revision of stonecutters' procedures, gradually excluding such practices as templates for warped surfaces, rectification of the circumference using chords, placement of rib centres by trial and error, and so on. However, this evolution was quite remarkable: in less than a century, Frézier's procedures have little in common with those of Jousse.

In any case, Gaspard Monge was to move forward into a different phase (see Sect. 2.4.3). Stonecutting had an important, although somewhat fluctuating, place in the didactic programs of French engineering schools in the eighteenth century, both in the École des Ponts et Chausées, concerned with the instruction of civil engineers, and the École de Genie de Mézierès, entrusted with the formation of military engineers. Exercises at the Ponts et Chausées, where stonecutting had some weight (although not so much as mechanics, algebra, geometry and architecture) dealt with sophisticated problems. One of them involved no less than

> a trumpet squinch in a curvilinear angle, formed by two towers with different curvatures, one of them elliptical and the other one circular. This squinch supports a third tower, forming at the same time the crown of a door which is supported by a cylinder[4]

The practical application of such a conundrum is hard to guess. In other words, the schools' interest in stonecutting does not seem to be fostered mainly by its practical application (for a different opinion, see Sakarovitch 1995: 206–207); rather, it seems that stereotomy was a vehicle for the formation of the spatial vision of the engineer. The same approach is shown in the *Traité des ombres dans le dessin géométral* by Nicolas-Françoise -Antoine de Chastillon (1763; see Belhoste 1990a; Sakarovitch 1995: 208; Sakarovitch 2006b: 2; Carlevaris 2014: 634–636), the first director of the School of Mézières, stating that:

[4]Belhoste et al. (1990b: 63) ... *une trompe dans un angle curviligne formé par deux tours de différentes courbures. L'une d'elle est elliptique el l'autre circulaire. Ce trompe soutient une troisème tour, et forme en même temps, para un cylindre qui la soutient, une porte à laquelle elle sert de couronnement.* Translation by the author.

> We have found nothing more proper for them [the engineers] than to procure that perfect knowledge of design through the study of stone and wood cutting. Independent of the advantages which result from this study, relative to constructions of which the offices of engineering have the direction, one conceives easily that when one knows how to develop all the faces and knows all the angles of any stone used in a vault, a squinch, etc ... one has easily the facility to develop a bastion, a demi-lune, a cavalier retrenchment, a battery, etc.[5]

Monge was entrusted with the task of developing these ideas; the approach of his Theory of Stonecutting courses in Mézières seems to have been even more abstract than those in Ponts et Chauseés (Belhoste et al. 1990b: 74–87; Sakarovitch 1998: 85–89, 218–229). According to Monge himself (Dupin 1819: 11), he was not allowed to explain his theories in full until the French Revolution. At that stage, they had taken the shape of a new science, descriptive geometry, which took much material from the problems and methods accumulated on stonecutting lore during the previous three centuries. However, he added a number of new ideas. First, rather than being restricted to stonemasonry and joinery, the field of application of this new science could be extended to any field, both technical, as shown by its use to solve artillery problems, and abstract, such as finding the distance between two skew (as opposed to parallel or convergent) straight lines. Second, since two projections are sufficient to determine the position of a point in space unambiguously, additional projections are unnecessary and thus are excluded from the system. Third, the surviving projection planes, horizontal and vertical, were fixed; this allowed Monge to represent any plane by means of its intersections with the projection ones, in order to solve abstract problems; also, the intersections of straight lines with these fixed planes were useful for many problems, starting with the determination of their intersections with planes.

The contributions of masons' lore to this new science are outstanding, but by no means exclusive (see Sect. 12.5). Stonecutting brought about the problem of the representation of complex three-dimensional bodies, including the projection of elements oblique to the projection planes and, inversely, the determination of the true size of these elements when only their projection is known; the computation of lengths of straight lines and angles in oblique planes; cylindrical and conical developments; and intersections of cylinders, cones, spheres and planes. In theory, such problems arise in any construction technology. However, as explained by Sakarovitch (1998: 243–244), although the linear nature of carpentry or the two-dimensional nature of metalwork raise several of these problems, only the solids of classical ashlar construction are involved with them all.

Regarding methods, stonecutting was the main factor in the development of orthographic projection and particularly, auxiliary views, which resurfaced later in

[5]Chastillon (1763: Avant-propos) *On n'a rien trouvé de plus propre pour leur [les ingénieurs] procurer cette connaissance parfaite du dessin que de leur faire suivre des cours de coupe des pierres et des bois ... Indépendamment des avantages qui résultent de cette étude, relativement aux constructions dont les officiers du génie ont la direction, on conçoit facilement que, quand on fait développer toutes les faces et connaître tous les angles plans ou solides d'une pierre quelconque employée dans une voûte, une trompe, etc, ou d'une pièce de charpente employée dans un comble, un dôme, un escalier, etc., qu'on a bien de la facilité à développer un bastion, une demi-lune, un cavalier de tranchée, une batterie, etc...* Transcription is taken from Belhoste (1990a: 111); translation is based on Sakarovitch (1995: 208).

the work of Théodore Olivier (1843–1844: I, 18–22). In particular, vertical plane changes are ubiquitous in stonecutting literature, as we have seen; in contrast, the use of sloping reference planes was proposed by Desargues (1640) but rejected by practitioners. Computation of the lengths of segments by forming a right triangle with their horizontal projection and the difference in heights between their edges was known by stonecutters at least as early as the sixteenth century. The issue of rotations and rabatments is less clear: in the same period, stonecutters were able to determine exactly the true size and shape of an oblique planar figure by measuring the distance of its corners to a reference plane or, less frequently, tracing an orthogonal to a horizontal line through the projection of a point. However, it is not clear whether they understood the foundation of this method, or if they were aware they were performing a rotation or, even less, an affine transform. Cylindrical developments were performed, from the sixteenth century on, through a grossly simplified method; only in the eighteenth century were refined methods, although by no means exact ones, proposed by Guarini ([c. 1680] 1737) and Frézier (1737–1739).

Other methods and problems of descriptive geometry, however, arose from different fields of knowledge. Conical developments applied to approximate representations of the sphere seem to have been transferred from cartography and cosmography to stonecutting, although some feedback in the opposite direction cannot be discarded for the moment. The problem of *défilement*, at the intersection of artillery and fortification, posed the issue of tangent planes, which is also present in the theory of shadows (Sakarovitch 2007: 49–51; see also de la Gournerie 1855: 25). This field of knowledge was quite relevant in the school of Mézierès, probably for two reasons. On the one hand, there was a tendency on the part of military engineers to eschew all kinds of perspective and focus on orthographic representation, as stated by Muñoz (2015: 73–75) for Spain. Now, orthogonal projection does not offer by itself an intuitive representation of volume, unless it is complemented by shadows or other devices. On the other hand, shadow theory provided powerful exercises in order to train the spatial vision of the engineer, just as stonecutting did, since it involves the intersection of straight lines, cylinders and cones with other surfaces, and so on; in other words, it furnished additional problems and some methods of descriptive geometry.

Moreover, shadows themselves may be seen as projections, cylindrical in the case of natural light and conical when artificial sources are involved. Thus, shadows cast by the sun embody the concept of parallel projection; remarkably, this concept is absent from stonecutting literature. While masons made intensive use of orthogonal projections from the Middle Ages on, the introduction of the concept in stonecutting treatises (Derand 1643: 10; Frézier 1737–1739: I, 206–207) was remarkably slow. In contrast, the notions of projector and projection plane appear explicitly in the seminal text of perspective theory, *De pictura* by Leon Battista Alberti (1435); later on, they are applied to orthographic projection by cosmography, cartography and optics (Rojas Sarmiento [1550] 1551: 25–26, 29; Aguillon 1613: 503–520).

All in all, the notions of stereotomy as "applied science" and descriptive geometry as "crystallised practice" are remarkably reductive. What we have in front of us is a mixed bag of exchanges in both directions between the realms of artisanal practice and learned science. If masons' practice brought to the fore the ideas (or rather the practices) of projection, change of projection plane, true size and shape, development, triangulation, and intersection of surfaces, a slow refining process, lasting for a century and a half and using the concepts and tools of learned geometry, led to the formation of descriptive geometry, a generalized science applicable to other crafts and even to abstract problems. Moreover, other sciences and practices, such as perspective, cartography, cosmography, artillery and shadow theory, made important contributions to the mix. Similar considerations may be applied to social groups or countries. While the empirical approach of masons and architects was essential to the development of this branch of knowledge, the intervention of clerics, engineers and, later, scientists was crucial for the generalisation and systematisation that turned an artisanal practice into a learned science. For several historical and geographical reasons, starting with its centrality in Western Europe, France played a leading role in this process; however, the contributions of Germany in regard to complex rib vaults, Italy with the creation of the Renaissance and Guarini's systematization, Spain with its rich sixteenth-century contributions, and England with staircases and skew bridges, should never be put aside.

Glossary

abacus The flat slab on top of a capital.
absolute orientation In photogrammetry, an operation that determines the scale of a model and its orientation in relation to verticals.
acanthus A plant in the family Acanthaceae with thick, fleshy, scalloped leaves used in the carved ornament of Corinthian and Composite capitals and also in some mouldings.
altimetria (Lat) A branch of medieval practical geometry dealing with the measurement of the height of inaccessible objects.
analogical photogrammetry See *photogrammetry, analogical.*
analytical photogrammetry See *photogrammetry, analytical.*
angle de talus (Fre) Angle between the face plane of an arch or vault and a horizontal plane. In a battered wall, it is not a right angle.
angle du biais (Fre) In Desargues' terminology, angle between the horizontal projection of the axis of an arch or vault and a horizontal line in the face plane of the arch or vault. In a skew arch, it is not a right angle.
angle du pente du chemin (Fre) In Desargues' terminology, angle between the axis of a vault and a horizontal plane. In a sloping vault, it is not null.
annular vault See *vault, annular.*
appelation d'origine (Fre) Designation of a constructive element by reference to a city where a famous example or archetype stands or stood; for example, *vis de Saint-Gilles, arrière-voussure de Marseille* or *caracol de Mallorca.*
arc de cloître (Fre) A reinforcement arch at the intersection of two barrel vaults forming an L-shaped vault.
arch Constructive element formed by a series of wedge-shaped pieces, called voussoirs, set along a curved line or directrix. The combination of the wedge shape of the voussoirs and the curvature of the directrix guarantees that the pieces are subject only to compressive stress, excluding tensile stress. This allows building arches in a wide variety of materials that can withstand compression, such as stone, brick, wood, concrete, iron or steel.

J. Calvo-López, *Stereotomy*, Mathematics and the Built Environment 4,
https://doi.org/10.1007/978-3-030-43218-8

arch in a concave wall Arch opened in wall with a concave and a planar face. Both the intrados and one of the faces are cylinders and thus, one of the intrados edges is a warped curve.

arch in a convex wall Arch opened in wall with a convex and a planar face. Both the intrados and one of the faces are cylinders and thus, one of the intrados edges is a warped curve.

arch in a round wall Arch opened in a wall that has at least a curved face. Variants are the arch in a convex wall, arch in a concave wall, arch in a wall with a concave and a convex face and, more rarely, arch in a wall with two convex or two concave faces. Both the intrados and, at least, one of the faces are cylinders. One or both of the intrados edges are warped curves.

basket handle arch Arch whose directrix is a three-centre oval, that is, a combination of three circular arcs. Usually, the centre of each lateral arc, that of the central arc and the junction of the lateral and the central arc are aligned in order to guarantee tangency at the junction point.

corner arch Arch opened in the union of two walls. The intrados is a single cylinder cut by four oblique planes and thus its horizontal projection is V-shaped. Usually, the intersections of this cylinder with the walls are ellipses.

groin arch Arch with a groin on a plane parallel to both faces; it is extremely rare and only to be found in Ginés Martínez de Aranda (c. 1600).

ox horn Splayed arch where one of the springings is orthogonal to the faces while the other one is horizontal but oblique to both faces. Usually its bed joints are orthogonal to the faces to facilitate dressing by squaring.

perimetral arch In a rib vault, the arches that strengthen the edges of the vault. They may be wall arches, when embedded in a wall or framing windows, or transverse arches, when dividing a rib vault from the adjacent one.

pointed arch Arch whose directrix is given by two circular arcs, their radius being larger than half the span of the arch. It is usually symmetrical about a vertical plane. Its rise is larger than half the span and both arcs are not tangent in the uppermost point or apex.

round arch Arch whose directrix is a semicircle. Its rise equals half its span.

segmental arch Arch whose directrix is a single circular arc smaller than a semicircle. Its rise is less than half its span.

skew arch Arch whose springers are parallel between themselves but oblique to one or both faces. Three different solutions may be used: (a) the intrados may be a circular cylinder, in which case, the intersection with the oblique face(s) gives a surbased ellipse as a result; (b) the intrados surface may be generated from semicircular directrices in the faces; since the generatrices are oblique to these directrices, the intrados surface is a raised elliptical cylinder and the cross-section is a raised ellipse; (c) two circular directrices are cut by planes orthogonal to the faces; intrados joints connect the intersections of these planes with both directrices; the resulting intrados is a warped surface and the element is called *biais passé*.

splayed arch Arch whose springers are not parallel. Thus, one or both springers are not orthogonal to the face planes. Three cases may arise: (a) both springers are oblique to the face planes but symmetrical around an axis; (b) both springers are oblique to the face planes and asymmetrical between themselves; (c) one of the springers is orthogonal to the face plane and the other is oblique to them; the last variant is called *ox horn*.

transverse arch In a building covered with rib vaults, an arch placed between two adjacent vaults.

Tudor arch Arch whose directrix is an ensemble of four circular arcs. Usually, the centre of each lateral arc, the centre of the adjacent internal arc and the junction of the both arcs are aligned in order to guarantee tangency at the junction point. By contrast, the internal arcs are not tangent between themselves, causing a usually slight apex.

wall arch In a rib vault, one of the arches embedded in a wall or framing a window that may strengthen one of the edges of the vault.

arch square Geometrical stonecutting instrument in the shape of a square with a straight and a curved branch or, more rarely, both curved branches.

archetype A particular construction that stands, through metonymy, for a constructive type, such as the *Vis de Saint Gilles*, which stands for vaulted staircases.

archipenzolo (Ita) A kind of level based on a three-sided square in the shape of an isosceles triangle, with a plumb line hanging from the corner of the square placed at its symmetry axis. If the opposite side is levelled, the plumb line should cross this side at its midpoint.

area Portion of ground covered by a building or a vault.

arrière-voussure de Marseille (Fre) Rere-arch whose front and back edges are arcs. Usually the lower edge is a semicircle and the higher one takes the shape of a segmental arch. The intrados surface is formed by three smoothly joined sections, the central one spanning both edges, while the lateral ones go from the lower edge to a curve in the intersection of the intrados surface and the jambs of the supporting walls.

arrière-voussure de Montpellier (Fre) Rere-arch whose upper edge is straight while the lower one is curved.

arrière-voussure de Sainte-Antoine (Fre) Rere-arch with a lower straight edge and an upper curved edge, whose intrados is a double-curvature surface and, thus, intrados joints are curved.

arrised vault *See vault, arrised.*

ashlar (1) A cuboid-shaped block in hewn stone, used for example in the construction of wall. Ashlars are usually grouped into courses, although they may stand alone at their level, for example in pillars. (2) See *ashlar masonry*.

ashlar masonry Construction executed in hewn stone, in particular with ashlars and voussoirs.

astwerk (ger) Late Gothic decorative motifs in the shape of tree branches and boughs, replacing ribs on some occasions.
automated photogrammetry See *photogrammetry, automated.*
auxiliary view In orthographic projections, a projection in addition to the usual horizontal and frontal projections. It is usually constructed by means of a change of projection plane.
axe Mechanical stonecutting instrument with a handle and a cutting piece with a linear edge parallel to the handle. It may be used orthogonally, tangentially or obliquely to the surface being dressed.
axonometric drawing Drawing representing a three-dimensional object constructed using three orthogonal axes in the three-dimensional space, which are represented as three mutually oblique axes in the paper plane. It may be based on orthogonal or oblique parallel projection. In the first case, variants include isometric, dimetric or trimetric drawing; in the second one, cavalier and military perspective.
axonometric projection See *projection, axonometric.*
axonometry See *axonometric drawing.*
barrel vault See *vault, barrel.*
base Member at the lower end of a column, connecting it with a stylobate or podium.
basket handle arch See *arch, basket handle.*
batiente (Spa) Strip in a rere-arch where a window frame is attached.
beat The working mass of a mallet or dummy, as opposed to the handle.
bed joint Horizontal or nearly horizontal joint in a masonry element dividing one course from the upper and lower ones. An ashlar or voussoir usually has a lower bed joint, that leans on the underlying block or voussoir, as well as an upper bed joint, where the next block or voussoir lies.
bed joint template Template representing a bed joint, usually in true size and shape.
bellatores (Lat) A group of medieval society formed by noblemen; the term literally means "warriors".
bevel Geometrical stonecutting instrument in the shape of a compass, used to transfer angles, for example from a tracing to a block being dressed.
bevel guideline A line in stonecutting tracings, representing either an intrados or a face joint, used to represent an angle together with another line.
biais passé A skew arch with bed joints orthogonal to the faces and forming a sheaf of planes whose common intersection is a line orthogonal to the faces.
bisector (1) A straight line orthogonal to a given segment, passing through its midpoint. (2) The line that passes through the vertex of an angle and divides it into two equal angles.
biveau de la nivelée en face (Fre) In Desargues' terminology, angle between the face of a barrel vault and the horizontal plane. In a battered wall, it is not a right angle.
block (1) An approximately cuboid piece of stone, resulting from extraction in the quarry; it should be dressed in order to get an ashlar or voussoir (2) Ashlar.

boaster Chisel with a wide edge.
bob A moderately heavy mass hanging from the end of a plumb line in order to materialise verticals. Usually made from lead, but examples in wood where used when metals were scarce.
bogenaustragung (Ger) Diagram used to represent or compute the curvature of ribs and the height of keystones in a rib vault, especially in complex Late-Gothic German vaults.
bond The specific layout of blocks or voussoirs in a wall or vault.
boning Control operation performed while dressing ashlars or voussoirs to verify that four points, for example the four corners in a face of an ashlar, are placed in a single plane. It is usually carried out checking that two straightedges overlap to the eyesight of a mason while closing an eye; the term derives from the French *borgne*, "one-eyed man".
boning block Small blocks used to perform boning when dressing hard stones such as granite; using them, the mason avoids carving a marginal draft before boning.
boss An ornamental element placed below a keystone and used to cover the intersection of ribs in a ribbed or coffered vault, often carved with foliage or religious or heraldic imagery. It may be carved from the same block as the keystone or be executed in a different material, for example wood, and attached to the keystone.
bóveda de Murcia (Spa) Horizontal-axis annular vault; takes its name from the one in the funerary chapel of Gil Rodríguez de Junterón in Murcia cathedral.
bricklayers' scaffold Scaffold resting in a single series of supports, separated from the wall. Compare with masons' scaffold, built on two series of supports.
bundled pier A pier with several shafts (2), or half shafts attached to a central nucleus, covering the full circumference of the pier.
bush hammer Mechanical stonecutting instrument in the shape of a hammer, with a cuboid beat including diamond-shaped points.
cantilevered staircase See *staircase, cantilevered.*
canting machine Stonecutting machine used to shape bevels.
cantonnée pillar A pier or pillar with several half shafts attached to a central nucleus, but not covering it completely, so the nucleus is partially visible.
cap The upper part of a sail vault, placed above a circle joining the keystones of the perimetral arches.
capilla de Cuenca (Spa) Square-plan coffered vault, with ribs parallel to the sides of the area; takes its name from the funerary chapel of the Muñoz family in Cuenca cathedral.
capital The upper part of a column, supporting the entablature.
caracol de Mallorca (Spa) Helical-newel open-well staircase; probably takes its name from a staircase placed at the corner turrets of Mallorca Merchants' Exchange.
cavalier perspective (1) Drawing based on parallel oblique projection; it preserves the shape of figures parallel to the projection plane. (2) Specifically, drawing based on a parallel projection where the projection plane is vertical and

the projectors are oblique to the projection plane. It preserves the shape of vertical figures that are parallel to the projection plane.

central projection See *projection, central.*

centre of projection In central projection, a point where all projectors converge. In linear perspective, it is called also “station point”. Compare with *direction of projection.*

centring Part of the falsework for a vault, arch, trumpet squinch or rere-arch. It takes the shape of a structure intended to hold an arch or vault while the voussoirs are being placed, before the arch or vault is completed and the arch behaviour comes into effect. Voussoirs are supported by formwork resting on the centring, supported in turn by underpinning.

change of the horizontal projection plane Descriptive geometry operation which involves replacing the horizontal projection plane with a sloping projection plane, in order to show the true size and shape of a slanting planar shape.

change of projection plane Descriptive geometry operation which involves replacing one of the projection planes with a different one, in order to show the true size and shape of a planar figure.

change of the vertical projection plane Descriptive geometry operation which involves replacing the vertical projection plane with a different vertical plane, in order to show the true size and shape of a vertical planar shape.

chef d'oeuvre (Fre) Object made by a compagnon in order to prove his or her competence and reach the status of *compagnon fini.* Aspiring masons usually present stonecutting models as *chef d'oeuvres.* The word probably derives from the use of models in order to convey instructions to masons, in lieu of drawings.

chisel Mechanical stonecutting instrument in the shape of a small bar with a cutting edge at one of the ends. In order to cut stone, it is hit with a mallet or a dummy. It is used to dress small voussoirs, to materialise complex shapes, for sculptoric work or to open marginal drafts. Variants of the chisel include the point, the punch, the driver, the boaster, the waster and the gouges.

cimbra (Spa) (1) A set of face templates (2) Spanish for “centring”.

cintre primitif (Fre) In Frezier's terminology, the cross-section of an arch or vault. It may adopt the shape of a semicircle, if the setting out of the arch or vault begins with the cross-section; however, if the setting out procedure starts with the face arch, it will be shaped as an ellipse.

ciseau boucharde (Fre) A chisel with teeth arranged as a square, in the fashion of a small bush hammer. In contrast with the bush hammer, it must be hit with a mallet or dummy.

cloister vault See *vault, pavilion.*

coffered vault See *vault, coffered.*

column Architectural supporting element in the approximate shape of a cylinder. Not to be confused with pillars, that are square or rectangular in section, with piers that involve different shafts, or with pilasters and semicolums, which are attached to walls.

compagnon (Fre) An artisan has completed the Tour de France and presented a *chef d'oeuvre* and as a result has reached the first stage of the *Compagnons du devoir.*
Compagnons du devoir (Fre) An artisanal organisation encompassing many trades, including masons, originated in the 17th century, if not earlier. It has its nucleus in France but has extended to other countries. The access to the status of *compagnon* involves such passage rites as the Tour de France and the preparation of a *chef d'oeuvre.*
compressive force Equal and aligned forces that point to the inside of a structural member. Compressive forces cause a (usually slight) shortening of the member. If they exceed an admissible stress, they may cause structural failure, either by deformation or crushing.
conical projection See *projection, conical.*
conoid See *right conoid.*
contramolde A negative template, that is, one that surrounds a face of a voussoir or block, instead of covering it.
contre-essieu (Fre) In Desargues' terminology, intersection of a plane orthogonal to the axis of a barrel vault and the plan sous-essier, that is, the plane orthogonal to the face that passes through the vault axis.
corner arch See *arch, corner.*
cosmimetria A branch of medieval practical geometry, dealing with the computation of volumes.
course A set of ashlars or voussoirs placed at the same level. Usually, it rests on an underlying course and supports another course. The contact surfaces between courses are called bed joints. A given course has both upper and lower bed joints. The lower bed joint of a course is the upper bed joint of the underlying course.
cross section The shape resulting from the intersection of a three-dimensional object with a vertical plane. Strictly speaking, in a cross-section the vertical plane should be orthogonal to the longest horizontal dimension of the three-dimensional object; if the vertical plane is parallel to that dimension, the intersection is called longitudinal section.
cross-cut saw A saw capable of cutting through the whole section of a stone element. It may be adapted for use by one man or two men; the latter is also known as a whipsaw. As all saws, it is only useful when dressing very soft stones.
crossing-image photogrammetry See *photogrammetry, crossing image.*
cruceta (Spa) A member for a coffered vault in the shape of a cross. It has four arms so that each arm connects a node in the network of ribs with a point half-way between two nodes.
cuboid A solid with six mutually orthogonal faces (1). Generally, its three dimensions (length, height and breadth) are not equal.
curved-string staircase See *staircase, curved-string.*
cylindrical lunette 1. A constructive element in the shape of a small barrel vault abutting on another, larger-diameter barrel vault. 2. The line resulting from the intersection of two cylinders with different radii, which appears at the intersection of a cylindrical lunette (1) with the larger vault.

cylindrical projection See *projection, parallel.*
cylindroid A ruled surface whose directrixes are two circles in non-parallel planes.
défilement (Fre) The problem of determining the required height of a wall that should protect soldiers or material from enemy gunfire coming from a particular point.
develop To transform a surface into a planar figure without stretching, shrinking or cutting it.
developable surface A surface that can be developed. In three-dimensional geometry, all developable surfaces are ruled ones, but not viceversa. In fact, there are only three kinds of developable surfaces: cones, cylinders and tangential surfaces, that is, surfaces generated by the tangents to a three-dimensional curve.
development The result of the transformation of a surface portion into a planar figure.
diagonal rib See *rib, diagonal.*
diamond vault A vault, usually built in brick, involving many groins and creases. Most examples are to be found in Germany, the Czech Republic and Poland.
digital photogrammetry Surveying technique based on the use of digital photographs. Variants are crossing-image photogrammetry and automated photogrammetry.
direction of projection In parallel projection, the common direction of all projectors. In conical projection, projectors have different directions and the concept does not apply. Compare with *centre of projection.*
directrix (1) A line that sets a constraint on the movement of the generatrix during the generation of a surface. For example, a cylindrical surface is generated by a line that rotates while keeping contact with a circular directrix (2) A line that controls the shape of an arch, for example the edge at the intersection of the front face and the intrados, or the intersection between the longitudinal symmetry plane with the intrados.
distance meter A surveying instrument, usually based on a laser beam, which furnishes precise distance measures. It can be a self-standing instrument or be integrated in another instrument such as a total station.
dome A curved structure with rotational symmetry designed to cover an area.

- **hemispherical dome** A dome with a hemispherical intrados surface.
- **oval dome** See *vault, oval.*
- **surbased dome** A dome with a surbased profile, either segmental, oval or elliptical.

double-curvature surface A surface that is not a ruled surface. No straight line entirely included in the surface can be drawn through all or some of the points of the surface. The name derives from the fact that two planar sections through all or some of the points of the surface will feature non-zero curvature.
draft Linear cut in a block (2) which is being dressed to obtain an ashlar or voussoir. It can be an end in itself or a preliminary stage in the dressing process of a face of the block or voussoir; in this latter case, it is called a *marginal draft.*
drafter Chisel with a medium sized edge, used for cutting marginal drafts.
drag Toothed steel plate used to materialise a smooth surface in soft limestones.

dress To cut a rough-shaped stone block (1) in order to adjust it to the predefined shape and dimensions of an ashlar or voussoir.

dressing by templates Dressing a stone block in order to materialise a voussoir making heavy use of true size and shape templates.

dressing by the direct method See *dressing by templates.*

dressing by squaring Dressing method based on orthographic projections of the faces of each voussoir. Starting from these projections, scored on the faces of a block, the mason materialises planes or cylinders that form the faces of the voussoirs.

driver See *boaster.*

drum (1) Cylindrical architectural element serving as the base for a dome (2) Cylindrical (or approximately cylindrical) constructive member used as a part of the shaft of a column.

drypoint See *stylus.*

dummy Mechanical stonecutting instrument with a metal beat, used to hit instruments in the chisel family.

elevation Architectural or engineering drawing based on the orthographic projection of a construction on a vertical projection plane.

ellipsoid A closed surface shaped so that all planar sections are ellipses or circles. The ellipsoid features bilateral symmetry about three orthogonal planes; the intersections of these planes are called the axes of the ellipsoid. Variants of the ellipsoid are the sphere, the oblate ellipsoid, the prolate ellipsoid and the scalene ellipsoid.

axes of an ellipsoid 1) The intersections between the symmetry planes of an ellipsoid 2) The segments of the axes of the ellipsoid between their intersections with the ellipsoid.

oblate ellipsoid Ellipsoid generated by rotation of an ellipse around its shorter axis. Two of the axes (2) of the ellipsoid are equal, while the third one is shorter. One the sections through the symmetry planes of the ellipsoid is a circle, while the other two sections are ellipses.

prolate ellipsoid Ellipsoid generated by rotation of an ellipse around its longer axis. Two of the axes of the ellipsoid (2) are equal, while the third one is longer. One the sections through the symmetry planes of the ellipsoid is a circle, while the other two sections are ellipses.

scalene ellipsoid Ellipsoid where the lengths of the three axes are different. Neither of the three sections through the symmetry planes of the ellipsoid is a circle.

elliptical vault See *vault, elliptical.*

entablature A horizontal structure in a classical temple or other buildings derived from it; it rests on columns and includes the architrave, the frieze and the cornice.

entasis The curve resulting from the longitudinal section of a column shaft. It can adopt slightly different shapes, such as a circular or elliptical arc, a conchoid of Nicomedes or a mechanical curve resulting from the deformation of a thin, flexible lamina.

equinus The middle part of a capital, placed between the necking and the abacus.

essieu (Fre) In Desargues' terminology, axis of a barrel vault. In a sloping vault, it is not horizontal; in a skew vault or a vault in a battered wall, it is not orthogonal to the face.

extrados The upper surface of an arch, vault, trumpet squinch or rere-arch.

extrados template Template representing the true size and shape or the development of the intrados of an arch or vault.

face 1) One of the enclosing surfaces of an ashlar or voussoir 2) One of the opposing vertical surfaces in an arch or rere-arch, intersecting the intrados and the extrados 3) A vertical surface in a trumpet squinch, intersecting the intrados and the extrados.

face joint An edge of a bed joint of an arch, rere-arch or trumpet squinch, showing in the face (2) of the arch. The concept does not generally apply to vaults.

face template Template representing the true size and shape or the development of the face of an arch, trumpet squinch or rere-arch.

falsework Provisional structure intended to hold an arch or vault while the voussoirs are being placed, before the arch or vault is completed and the arch behaviour comes into effect. Typically, it includes the formwork, a surface directly supporting the voussoirs; the centring, a loosely horizontal structure supporting the formwork; and the underpinning, a loosely vertical structure supporting the centring.

fan of planes See *sheaf of planes*.

fan vault See *vault, fan*.

file A stonecutting instrument with small teeth, used to finish the surface of soft stones, in particular limestones.

flight A section between two landings of a non-spiral staircase, usually rectangular in plan.

folding line The intersection of two projection planes. The concept may include the intersection of two slanting planes, which is not horizontal and does not fit into the definition of ground line.

formwork A part of the falsework that supports directly an arch or vault under construction. It takes the form of a convex surface corresponding exactly to the concave intrados surface of the arch or vault. It is supported by the centring, which is in turn supported by the underpinning.

french scraper A mechanical stonecutting instrument including several iron blades held together by a small wooden block, used to finish soft stones, in particular limestones.

frontal plane In descriptive geometry, plane parallel to the vertical projection plane.

frontal projection See *projection, frontal*.

frontal segment In descriptive geometry, segment parallel to the vertical projection plane.

gallicisme (Fre) According to Jean Marie Pérouse de Montclos, one of several specific traits of early modern French architecture, such as the use of complex ashlar vaulting, mansard roofs, the emphasis on distribution or the use of wide segmental pediments.

gauge (1) An instrument used to transfer a measure from one point to another, specifically a wooden strip where notches are made (2) To transfer a measure from one point to another, for example to construct a parallel to an existing straight line.
generatrix A line that moves or rotates following some constraints. The set of the points of the generatrix in all its successive positions forms a surface.
geometrical staircase See *staircase, geometrical.*
gin (1) A hoisting machine based on two struts joined at their tops and held in place by several ropes. A pulley hangs from the intersection of the struts, so weights can be lifted pulling a rope (2) A similar machine, built with three struts and no sustaining ropes; it is quite inefficient.
glyptography A branch of knowledge dealing mainly with stonecutters' marks; the conferences in the field occasionally deal with other kinds of marks and stonecutters' tracings.
gnomon An iron bar used to project a shadow on a sundial; the position of the shadow indicates the hours and occasionally the months or the phases of the Zodiac.
goniographie (Fre) A part of stereotomy dealing with the measure of angles.
gore A part of a sphere enclosed between two meridians; terrestrial and celestial globes where usually constructed using paper gores glued to a spherical base. See also *lune.*
gouge A curved chisel used to dress mouldings.
groin 1) A sharp edge formed at the intersection of vault surfaces 2) Specifically, one of the two edges formed at the intersection of two semicylinders in a groin vault 3) The set of L-shaped voussoirs placed along the groins.
groin arch See *arch, groin.*
groin vault See *vault, groin.*
ground line In descriptive geometry, the intersection of a vertical and a horizontal projection plane. Compare with *folding line.*
gunport An opening in a wall or parapet through which a gun can be fired. Usually, it features an intrados surface in the shape of a portion of a right or oblique cone. The axis of the cone is usually slanting in order to direct arrows or gunfire downwards, although stonecutting treatises include frequently gunports with a horizontal axis.
hand saw A saw that can be operated by a mason with just one hand; it is used to cut very soft stone.
helicoidal bonding A bond used in 19th-century skew vaults, with helical intrados joints.
helix A three-dimensional curve generated by the movement of a point that rotates around an axis while moving in the direction of the axis, so that the rotation angle between two positions of the point is proportional to the movement in the direction of the axis.
helical-newel staircase See *staircase, helical-newel.*
hemispherical dome See *dome, hemispherical.*
hemispherical vault See *dome, hemispherical.*

hoist (1) A placement instrument consisting of a rotating cylinder whose axis is attached to a support secured to the ground and furnished with handles. A rope winds around the cylinder; masons rotate the handles to lift medium-sized stones. (2) To lift medium or heavy weights, such as ashlars or voussoirs, either from the ground to a cart or from the ground or a cart to its position in a wall, arch, trumpet squinch or vault or more rarely, a staircase.

horizontal projection See *projection, horizontal.*

horizontal-axis annular vault See *vault, horizontal-axis annular.*

ichnographie (Fre) In 18th-century parlance, horizontal projection.

impost The surface below the starting voussoir or springer of an arch or vault.

intrados The lower surface of an arch, trumpet squinch, rere-arch, vault or staircase; it is usually concave.

intrados joint The edge of a bed joint showing in the intrados of the constructive element.

intrados template The template representing the intrados of an arch, trumpet squinch, rere-arch, vault or staircase, either in true size and shape or developed.

jamb The vertical surface of an archway, doorway or window; it usually intersects the impost, that is, the horizontal surface below an arch or rere-arch.

joint The surface dividing one ashlar or voussoir from adjoining ones. In walls, vaults and vaulted staircases, bed joints are horizontal or nearly horizontal and divide the masonry in courses, while side joints divide a course into blocks or voussoirs. The distinction between bed and side joints does not apply to arches and most trumpet squinches and rere-arches.

juzgo (Spa) Auxiliary geometrical construction, such as a straight line orthogonal to a given line, used to draw parallels to the given line.

keystone The central voussoir at the crown of an arch, trumpet squinch, rere-arch or vault. Usually it is the last one to be put in place, although there are exceptions. In rib vaults, it usually includes a boss with carved decoration; in some occasions, the boss is an independent piece carved in other material such as wood.

laboratores (Lat) A group of medieval society formed by anybody who was not a king, a nobleman or a cleric; the word means "worker".

lagging A series of short boards joining twin frames in the centring of an arch.

laser total station See *total station.*

lathe An instrument capable of cutting stone while rotating, used to dress revolution surfaces such as bases, shafts and balusters.

level ridge A horizontal rib at the top of a rib vault.

lewis An instrument used to hoist stones formed by three metallic pieces. The end pieces are dovetail shaped in both horizontal directions; the central one is flat; all them have a hole in the upper part. In order to use it, the mason should open previously a dovetail mortise in the upper face of the ashlar he intends to hoist. Then he introduces both end pieces of the lewis in the mortise; next, he places the central piece between both end pieces; finally, he passes a rope through the aligned holes of all three pieces. Then the rope can be attached to a hoist or crane in order to securely lift the ashlar. It is rarely used to lift voussoirs; pincers are more suitable to this end.

lierne A rib in a (typically Late-Gothic) rib vault that is neither a perimetral rib, a diagonal rib or a tierceron.
line in point view In descriptive geometry, a straight line orthogonal to the vertical projection plane; as an exception to the general rule, its projection is a point. Vertical lines are also projected as a point in horizontal projection; however, they are not usually called "lines in point view".
lintel A linear, horizontal constructive element bridging an opening. It can be made from a single block of stone or divided into several wedge-shaped pieces.
longitudinal section The shape resulting from the intersection of a three-dimensional object with a vertical plane parallel to the longest dimension of the object.
L-shaped vault See *vault, L-shaped.*
lune A part of a sphere included between two meridians. See also *gore.*
lunette 1. A small vault abutting on another, larger-diameter vault. Usually, the small vault is either a barrel vault, a splayed arch or a pointed lunette, while the larger vault may be a barrel vault or a dome. 2. The line resulting from the mutual intersection of two cylinders or a cylinder and a cone.
mallet A stonecutting instrument used to hit instruments in the chisel family, including a wooden beat and a handle. Compare with *dummy.*
mandorla A shape enclosed between two symmetrical circular arcs, resembling an almond.
marginal draft A linear incision opened, usually with the chisel, in the position of an edge of an ashlar or voussoir face, as a preliminary step to the actual carving of the face. The planarity of the face is then controlled leaning a straightedge on two marginal drafts.
masons' scaffold A provisional structure used to place—but usually not support—ashlars in a wall, or to dress or clean them after placement. In opposition to bricklayers' scaffolds, it includes two rows of supports, one placed along the wall and another one at a distance from it.
mastel (Spa) A rotating cylinder. It can be used as the nucleus of a hoist or as the axis of a crane.
meridian Each of the successive positions of the generatrix of a surface of revolution.
military perspective Drawing based on a parallel projection where the projection plane is horizontal and the projectors are oblique to the projection plane; it preserves the shape of horizontal shapes parallel to the projection plane.
mitre Special file with a curved tip used to clean difficult sections, in particular in mouldings.
model Three-dimensional object reproducing an artistic, architectural or engineering work, or a constructive element, at a reduced scale. It is usually prepared before actual construction in order to test feasibility, ask for the client's approval or serve as a guide during construction.
mold See *template.*
molde cuadrado (Spa) Applied to ribs whose cross-section is symmetrical and not subject to *revirado.*

molde revirado (Spa) See *revirado.*
moulding A continuous protrusion or groove, used with aesthetic purposes and sometimes to throw away rain from a wall.
multi-image photogrammetry See *photogrammetry, crossing image.*
net vault See *vault, net.*
newel A central strut in a spiral staircase. It is usually straight, although open-well staircases with helical newels are also frequent.
newel template A template representing the development of the surface of a newel, in order to carve a fin that supports a vault in staircases of the *Vis de Saint Gilles* type.
numerical control machine A machine able to cut stone in the shape of a digital model. Variants include three- and five-axes models.
oblate ellipsoid See *ellipsoid, oblate.*
oblique cone A cone with a circular base and the apex placed outside the orthogonal straight line passing through the centre of the directrix. It can be proved that this figure is congruent with a cone with an elliptical directrix and the apex placed at an orthogonal to the directrix passing through its centre. This cone is not a surface of revolution.
oblique projection A kind of parallel projection where the projectors are oblique to the projection plane. Variants include cavalier and military perspective.
ochavo (Spa) (1) An octagonal vault, solved by means of eight cylindrical portions (2) A semioctagonal vault used frequently in chancels of small churches, covered by three or five cylindrical portions (3) by extension, any other vault with the same function, such as the *Ochavo de La Guardia,* which is actually spherical.
ochavo de La Guardia (Spa) A spherical coffered vault, cut through two parallel vertical planes; takes its name from the presbitery of the church of the Dominican convent in La Guardia de Jaén.
octagonal vault See *vault, octagonal.*
oculus A round opening in a wall; usually it is splayed, so the intrados surface is conical. As an exception, the intrados cannot be described as the lower part of the member since it is placed in the interior of the oculus.
oposición (Spa) A procedure used to select master masons in Spanish cathedrals in the Early Modern period, among many other positions. Aspiring masons should present a portfolio or explain how they would continue the work in the cathedral; they were also allowed to criticise the presentations by other candidates, thus the name *oposición,* that is, opposition. After this, the cathedral chapter or a specialised panel was to vote to choose the most suitable candidate.
oratores (Lat) A group of medieval society formed by clerics, including archbishops, bishops, priests, monks and friars. It literally means "those who pray".
orientation The quality of all planes that are mutually parallel.
orthogonal bonding A bond used in 19th-century skew vaults, with intrados joints orthogonal to the faces.
orthogonal projection See *projection, orthogonal.*
orthographic projection See *projection, orthographic.*

orthography A part of stereotomy dealing with vertical projections.
orthophotograph A transformation of one or several photographs to furnish an orthogonal projection. Not to be confused with a rectified photograph, which is still based on conical projection.
oval dome See *vault, oval.*
oval vault See *vault, oval.*
ox horn A variant of the skew arch where the bed joints are set in planes orthogonal to the faces, forming one or two sheaves of planes with their common axes orthogonal to the faces.
parallel projection See *projection, parallel.*
patent axe A bush hammer with interchangeable heads, so the mason can use larger diamond points, for rough and efficient dressing, or small diamond points for a finer finish.
pavilion vault See *vault, pavilion.*
pencil of planes See *sheaf of planes.*
pendentif de Valence (Fre) A sail vault with vertical courses and intrados joints laid on planes parallel to the sides of the area. Takes its name from the funerary monument of Canon Nicolas Mistral in Valence.
pendentive 1) In a sail vault, one of the four spherical triangles below the circle that connects the keystones of the perimetral arches; compare with *cap* 2) One of four spherical triangles laid below a hemispherical dome or a drum (1) supporting a dome, allowing the transition from a square plan to the circular impost of the dome.
pendentive vault See *vault, pendentive.*
pénétration extradosée (Fre) A variant of the cylindrical lunette (1) where the full thickness of the lunette, including the extrados edge, is visible in the intrados of the main vault.
pénétration filée (Fre) A variant of the cylindrical lunette (1) where the intrados joints of the lunette and the main vault are coordinated so they meet at the groin between the lunette and the main vault; in this case the full thickness of the lunette does not show in the intrados of the main vault. Usually, this solution requires the manipulation of the joints of either the lunette or the main vault.
perimetral arch An arch placed at the perimeter of a vault, in particular in sail and rib vaults, in order to support the vault.
perron (Fre) An exterior staircase formed by two symmetrical flights; usually, each flight is semicircular, although more complex shapes are occasionally used.
photogrammetry A surveying technique that uses photographs to get parallel projections, in particular plans, elevations and sections. These can be presented as line drawings or as representations with color, texture, shades and shadows, based on the transformation of one or more photographs, known as orthophotographs. A variant of photogrammetry that uses two images from analogical cameras. Using these images, two virtual images are generated by means of optical and mechanical devices and projected in two eyepieces, so that an operator can position points or lines in space; the system furnishes two-dimensional projections of the three-dimensional object.

analytical photogrammetry A variant of photogrammetry that uses two images from analogic or digital cameras. Using these images, two virtual images are generated by means of computer software and projected in two eyepieces, so that an operator can position points or lines in space. The system furnishes a digital model that can be used to obtain two-dimensional projections of the three-dimensional object.

automated photogrammetry A variant of digital photogrammetry where the operator furnishes a program with several digital photographs. From this information the program computes directly the relative positions of the camera used in each photograph and the position of a large number of points in the object. Then, the program may construct a three-dimensional mesh representing the object and project a number of photographs on this mesh, obtaining a representation of the three-dimensional object with colour and texture. As a final step, ortophotographs in orthogonal projection (plans, elevations, sections or axonometrics) can be obtained from the three-dimensional model.

crossing-image photogrammetry A variant of digital photogrammetry where the operator furnishes a program with several digital photographs and marks on each one of them the images of a number of relevant points (for example, voussoir corners). From this information the program computes the relative positions of the camera used in each photograph and the position of the relevant points.

digital photogrammetry Surveying technique based on the use of digital photographs. Variants are crossing-image photogrammetry and automated photogrammetry.

multi-image photogrammetry See *photogrammetry, crossing image.*

pick A dressing instrument with two points in both ends, used for rough dressing.

pincers A hoisting instrument in the shape of two S-shaped arms tied by a hinge at their middle part. Holes must be opened in the sides of the ashlar of voussoir being lifted; the weight of the load causes the arms to close, increasing the pressure on the sides of the ashlar or voussoir and thus securing the load.

plan Drawing based on a horizontal orthographic projection. In architecture, it is usually a section cutting through a specific storey of a building, except in the case of roof plans.

plan de chemin (Fre) In Desargues' terminology, plan of the floor of a barrel vault. It is usually parallel to the plan of the springings of the vault. In a sloping vault, it is not horizontal.

plan de face (Fre) Plan of the face of a vault. In a vault opened in a battered wall, it is not vertical.

plan droit a l'essieu (Fre) In Desargues' terminology, plane orthogonal to the axis of a barrel vault that passes through its intersection with the face plane. In a sloping vault, it is not vertical.

plan droit au face et au niveau (Fre) In Desargues' terminology, a plane orthogonal to the horizontal lines of a barrel vault. In a skew wall, it is not orthogonal to the axis of the vault.

plan sous-essier (Fre) In Desargues' terminology, the plane orthogonal to the face of a barrel vault that passes through the vault axis. In a sloping vault, it is not horizontal.

plane in edge view A plane orthogonal to the vertical projection plane; as an exception to the general rule, its projection is a straight line. Horizontal planes are also projected as a straight line in horizontal projection; however, they are not usually called "planes in edge view".

planer A mechanical stonecutting element used to materialise planar faces.

planimetria (Lat) A branch of medieval practical geometry, dealing with the computation of areas.

plantar al justo (Spa) In Martínez de Aranda's terminology, to place a true-size template on the face of a block in order to dress a voussoir by the direct method.

plantar de cuadrado (Spa) In Martínez de Aranda's terminology, to place a template representing the orthogonal projection of an intermediate stage of the dressing process on a side of a block.

Plucker's conoid See *cylindroid*.

plumb line A placement instrument formed by a string and a mass, called bob or plummet, which materialises a vertical line.

plummet The mass at the end of a plumb line. It is usually made from metal, specifically lead, although wood was also used during the Middle Ages.

pointed and raised lunette The most usual solution to the pointed lunette. The apex of the round arch in one of the edges is placed at a lower level that the intersection of two elliptical arcs in the opposite edge. As a result, intrados joints are slanting lines.

pointed arch*See arch, pointed.*

pointed lunette A lunette (1) whose intrados surface spans the area between an arch, usually round, and the intersection of the intrados of a barrel vault with two oblique planes, which takes the shape of two elliptical arcs meeting at an acute angle, therefore the attribute "pointed". It usually takes the shape or a ruled surface, but not a cylinder, since the projection of the elliptical arcs in the plane of the arch does not adjust to a circle. Compare with *cylindrical lunette*.

polisher Mechanical stonecutting instrument used to finish or polish a stone surface.

polka (Fre) A dressing instrument with two cutting edges; one is parallel to the handle, as in the axe, while the other cutting edge is orthogonal to it.

polygonal vault See *vault, polygonal.*

portal d'apotecari (cat) A lintel with parallel joints in the front face and converging ones in the back face, described by Josep Gelabert (1653).

pressure polygon A virtual polygon used to test the equilibrium of an arch; if the pressure polygon falls entirely within the arch, the arch is stable.

prinzipalbogen (ger) An arch used in German tracings for rib or specifically net vaults, representing the curvature of all or most of the ribs in the vault.

projection Operation used to construct a bidimensional representation of a three-dimensional object. In order to do that, a line called projector is drawn through each point of the three-dimensional object following some rules and constraints; its intersection with a projection plane is the projection of the point.

axonometric projection Any kind of cylindrical projection used as the base of an axonometric drawing. This encompasses orthogonal projections, excluding orthographic ones, which furnish isometric, dimetric or trimetric drawings, as well as oblique projections used in cavalier and military perspectives.

central projection Projection in which all the projectors are convergent in a single point, known as centre of projection or station point. It is the basis of linear perspective and stereographic projection.

conical projection See *projection, central.*

frontal projection Orthogonal projection on a vertical plane used as a reference in a double-projection system, taking into account that the ensemble of a single vertical projection and a horizontal projection is sufficient to represent unambiguously a point in space. Other vertical projections are possible, but they are considered auxiliary views. Planes, shapes, lines and segments parallel to the frontal projection plane are called frontal; notice that other vertical planes, shapes, lines and segments are not frontal.

horizontal projection Orthogonal projection onto a horizontal plane; the resulting drawing is called a plan.

oblique projection Parallel projection where the projectors are oblique to the projection plane. It is the foundation of cavalier and military perspective.

orthogonal projection Parallel projection where the projectors are orthogonal to the projection plane. It encompasses orthographic projections and orthogonal axonometry, which is in turn divided into isometric, dimetric and trimetric axonometry.

orthographic projection Orthogonal projection where the projection plane is horizontal or vertical and usually is associated with some important surface of the object depicted in the projection. Such drawings as plans, elevations, cross-sections and longitudinal sections are based in orthographic projections.

parallel projection Projection in which all the projectors are parallel. Projectors may be orthogonal or oblique to the projection plane. The first case encompasses orthographic projections and orthogonal axonometry. The second one, cavalier, military and ordinary perspective; in the latter, the projection plane is slanted and the projectors are oblique to the projection plane.

projection line Straight line connecting the horizontal and vertical projections of a point.

projection plane Plane whose intersection with projectors determines the projection of a point or a line. It can be assimilated to the plane of the drawing paper or the supporting surface of a large-scale tracing.

vertical projection An orthogonal projection on a vertical plane; resulting drawings are called elevations, cross-sections or longitudinal sections.

projector Straight line joining a point with its projection.

prolate ellipsoid See *ellipsoid, prolate.*
punch A dressing instrument in the chisel family, heavier than the chisel and with a point at the end.
quadripartite vault See *vault, quadripartite.*
quarter-sphere vault See *vault, quarter-sphere.*
quartier-de-vis suspendu (Fre) A staircase or a section of it whose plan is a quarter of a circle; it is supported by walls or pillars at the ends of the quarter of circle, with no intermediate supports.
rabatment Descriptive geometry operation consisting in the rotation of a plane or a planar shape until it reaches a horizontal or frontal plane, so it is represented in true size and shape.
raised Arch or vault whose rise is larger than half its span. Compare with *surbased.*
raised ellipse An ellipse whose vertical axis is larger than its horizontal one.
raised oval An oval whose vertical axis is larger than its horizontal one.
rampante llano (Spa) The cross-section of a rib vault when it adopts a horizontal or nearly horizontal shape.
rampante redondo (Spa) The cross-section of a rib vault when it adopts a circular shape, in particular when its radius equals that of the diagonal ribs.
rear arch See *rere-arch.*
relative orientation In photogrammetry, an operation that determines the relative station points of the photographs. It does not determine, however, neither the absolute values of the distances between station points nor its inclination.
rere-arch A constructive element used to enlarge an opening, usually in the interior of a building. Its edges may be given by a horizontal line and an arc or by two arcs. Spanish manuscripts include a third variant with horizontal edges at both faces; this variant is called *capialzado* in Spanish but is not included in sources from other countries.
revirado (Spa) An affine transformation applied to the cross-sections of ribs in a rib or coffered vault so that it adapts to the spherical or cylindrical surface of the severies.
revolution (1) Rotation (2) The part of a helix included between two consecutive intersections of the helix with a generatrix of the supporting cylinder.
rib An arch or lintel protruding from a vault and supporting the severies, at least during construction. Ribs are usually materialised by independent pieces, although in some cases they may be included in the same voussoir than the severies, particularly in fan vaults.

> **diagonal rib** In a square or rectangular rib vault, a rib connecting the opposed corners of the area covered by the vault.
>
> **ridge rib** A rib at the top of a rib vault; it follows usually the longitudinal axis of a set of vaults, although it can be also transversal.

transverse rib See *arch, transverse.*
rib vault See *vault, rib.*
ridge rib See *rib, ridge.*
rifler A special file with a pointed tip used to dress or clean difficult features, such as scotias and other mouldings.

right cone A cone with a circular directrix and the apex placed at some point of a straight line passing through the centre of the directrix and orthogonal to the plane of the directrix. This cone is a surface of revolution.
right conoid A ruled surface generated by straight lines intersecting both a generic directrix and a fixed straight line so that the generatrices are always perpendicular to the fixed straight line.
rise The difference in heights between the imposts of an arch or vault and the highest point of its intrados.
rooster's feet A particular rib in rib vaults, usually placed near the keystones of perimetral arches; in contrast with most ribs, its horizontal projection is curved, with a relatively small radius.
rotation (1) A mechanical movement in which a piece turns around an axis (2) In descriptive geometry, an operation in which a point or line turns around an axis, in order to bring it to a horizontal or frontal position, so that distances between points, lines and planes or angles between lines and planes can be measured.
round arch See *arch, round.*
ruled surface A surface generated by the movement or rotation of a straight line. As a result, at least a straight line completely included in the surface may be drawn through any point of the surface. Compare with *double-curvature surface.*
ruler See *straightedge.*
ruling line In a ruled surface, one of the straight lines entirely included in the surface.
sail vault See *vault, sail.*
Saint Peters' keys See *lewis.*
saw One of a group of stonecutting instruments used to cut planar faces in a stone block, consisting in a metal blade with many small teeth. In contrast with other stonecutting tools, it is operated moving the saw back and forth, instead of hitting the stone surface.
scaffold Provisional structure erected along a wall or under a vault, allowing the access of masons or bricklayers to the wall to place bricks or small stones, control placement of large stones, retouch the stone surface or the mortar joints after placement, and store bricks or small stones, but usually not full ashlars, which are hoisted by specialised equipment. It differs from the falsework, since it is not used to support the structure during construction.
scalene ellipsoid See *ellipsoid, scalene.*
scanning A surveying operation performed using a 3D laser scanner; this instrument can measure horizontal and vertical angles and distances very quickly, furnishing the coordinates of a large number of points, usually in the range of millions, in a very short time span, usually in the range of minutes. However, in contrast to a total station, it does not allow the operator to choose the scanned points. Some instruments, known as multistations, combine a total station and a slow scanner.
scotia Concave moulding, used for example in Attic bases.

score To make an incision on a stone or wood surface, representing a dimension, an angle, or the shape of a face of an ashlar or voussoir, either directly or using a template.
scriber An instrument akin to a stylus used to mark or score incisions, either directly on stone faces or in a large- or full-scale tracing executed on walls and floors.
secondary keystone In star or net vaults, a keystone that is not placed at the highest point of the vault; it is usually placed at the intersection of tiercerons and liernes.
section 1) The intersection of a three-dimensional solid with a plane known as cutting plane 2) An orthographic projection representing a section (1) of a solid and, usually, other parts of the solid behind the cutting plane, but not those in front of the cutting plane.
segmental arch See *arch, segmental.*
semidome See *vault, quarter-sphere.*
semioctagonal vault See *vault, semioctagonal.*
semipolygonal vault See *vault, semipolygonal.*
set out To prepare preliminary tracings for a construction in ashlar.
severy A part of a rib vault in the shape of a portion of a surface, leaning on previously built ribs.
sexpartite vault See *vault, sexpartite.*
shaft (1) The central and larger part of a column, placed between the base and the capital, or, in the Greek Doric order, between the crepidoma and the capital 2) A broad semycilindrical moulding in a bundled or composite pier or attached to a wall in a respond.
sheaf of planes A set of planes intersecting at a single straight line known as the axis of the sheaf.
shoring See *underpinning.*
shuttering See *formwork.*
skew arch See *arch, skew.*
skew lines Two lines are called skew lines if they do not belong to a single plane. Skew lines are neither convergent nor parallel.
skew vault See *vault, skew.*
sloping barrel vault See *vault, sloping barrel.*
sloping vault A barrel vault with an inclined axis and imposts.
sous-essieu (Fre) In Desargues' terminology, intersection of the face plane and the *plan sous-essier*. It can be understood as an orthogonal projection of the axis onto the face plane.
span The free distance between the supports or imposts of an arch, rere-arch or vault.
spandrel (1) A portion of a wall between two arches, or between an arch and a pilaster (2) The flat portions of a fan vault between the upper edges of adjoining curved sections.
spiral staircase See*staircase, spiral.*

splayed An opening with convergent jambs, such as a splayed arch, a gunport, most rere-arches, or some oculi.
splayed arch See *arch, splayed.*
springer (1) The lower voussoir in an arch or vault, resting on the impost (2) The edge of the impost facing the opening of an arch or vault.
springing The group of the lower voussoirs in an arch or vault. The term is frequently applied to rib vaults, where these voussoirs include different ribs and arches, in particular wall arches, transverse ribs, diagonal ribs, and occasionally tiercerons, all carved in the same stone. In this case, the springing ends where these ribs depart from one another.
square Geometrical stonecutting instrument including two branches at right angles; it can be used both in the setting out and dressing phases. Another variant, with three branches, guarantees precision in tracing, but is useless in the dressing phase.
squaring See *dressing by squaring.*
squinch Constructive element that spans the space between two walls set as an angle. It may be built as an arch, a corbel, a lintel or a trumpet squinch. Except in the case of the trumpet squinch, it requires a platform spanning the space between the arch, lintel or corbels and the walls.
staircase A constructive element used to ascend between two levels in a building, by means of a series of small horizontal surfaces set at different levels, known as treads. Another element, known as riser, may close the gap between successive treads. When built in ashlar, each tread and the underlying riser are usually built in a single piece known as step. When the staircase connects more than two levels, or when the levels to be connected are placed at fairly different heights, intermediate landings or flat sections are included.

cantilevered staircase (1) Spiral staircase resting on a wall, with no newel (2) Spiral staircase resting on a newel, with no supporting wall.
curved-string staircase Straight-flight staircase whose strings are curved. Compare with *straight-string staircase.*
geometrical staircase Spiral staircase cantilevered from the enclosing wall, with a large open well. It is relatively frequent in early modern England.
helical-newel staircase Spiral staircase resting on a wall and a helical newel. Compare with *straight-newel staircase,* and *cantilevered staircase.*
spiral staircase Staircase whose outer edge takes the shape of a helix. It rests usually in a central straight newel and a wall; however, there are variants resting in a helical newel and the wall, a straight newel alone, or the wall alone. Compare with *straight-flight staircase.*
straight-flight staircase Staircase where the horizontal projections of the inner and outer edges are one or more straight lines. Usually, the outer edge rests on a wall and the outer one is free. Compare with *spiral staircase.*
straight-newel staircase Spiral staircase with a central cylindrical newel. Steps rest on the newel and an enclosing wall. When the wall is absent, the element is known as a *cantilevered staircase.*
straight-string staircase Straight-flight staircase whose strings are straight.

vaulted staircase A relatively wide spiral staircase that requires to be covered with a vault with several courses, divided by helical intrados joints; each of the courses is divided into voussoirs. Steps are independent pieces placed over the vault, in contrast with other spiral staircases. The most frequent type is the *Vis de Saint Gilles*, although other solutions have been put forward in treatises.

stairway staircase, especially when it is not enclosed by a wall.

straight-flight staircase See *staircase, straight-flight.*

straight-newel staircase See *staircase, straight-newel.*

straight-string staircase See *staircase, straight-string.*

star vault See *vault, star.*

stellar vault See *vault, star.*

stereotomy A science dealing with the division of three-dimensional solids, including the shape of their sections, the representation of solids and sections, the measure of angles and the practical execution of stonecutting in order to materialise predefined solids. From the late 18th century on, most of these subjects were grouped under the name Descriptive Geometry and the term "stereotomy" was restricted to the practical execution of stonecutting.

stonecutters' hammer A mechanical stonecutting instrument akin to a hammer, with a heavy head in the shape of a cuboid. It is used to dress stones roughly, either as a definitive finish or as a preliminary step before smoother dressing with the axe, bush hammer or other instruments.

straightedge A geometrical dressing instrument materialising a straight line.

straight-newel staircase See *staircase, straight-newel.*

stretched template (1) A curve constructed by locating several independent points and joining them by hand or by means of circular arc; such curve may represent an ellipse or a helix (2) A kind of bed joint template used in the springings of a rib vault. Usually, a single, standard template is used in the springings; however, in some case, a basic template is subject to partial modifications so each template in thee springings is slightly different.

string The lateral edge of a flight in a staircase. Strings can be open to a well or adjoining a wall. They can be also straight or curved, although the horizontal projections are usually straight.

stylus A metallic instrument used to mark blank lines on paper or parchment. Used in stonecutting manuscripts to construct auxiliary lines that are only visible under flush light.

surbased Arch or vault whose rise is less than half its span. Compare with *raised.*

surbased dome See *dome, surbased.*

surbased ellipse An ellipse whose vertical axis is shorter than its horizontal one.

surbased oval An oval whose vertical axis is shorter than its horizontal one.

surface of operation A vertical surface in top of a keystone, where marks such as rib axes are inscribed, usually taking their angles from a general plan or tracing of the vault. If these marks were inscribed on a sloping surface, important errors would occur.

surface of revolution Surface generated through the rotation of a line or curve called generatrix around an axis of rotation. The successive positions of the generatrix are called meridians, while the set of the positions of a single point of the generatrix is a circle known as parallel.
surmounted An arch placed over vertical stilts. The result is similar, but not exactly equal to a raised arch.
tabula gracilis (Lat) A thin, flexible wooden lamina used to draw large-radius curves.
tape measure A strip of metal, plastic, or cloth with numbers marked on it, used for measuring distances.
tas-de-charge A group of stones at the lower sections of a rib vault that include in a single stone sections of the perimetral arches, the diagonal rib and, sometimes, the tiercerons or the perimetral arches of adjacent vaults. In order to reduce complexity and the need for falsework, bed joints are usually horizontal.
template A geometrical stonecutting instrument consisting in a complex shape materialised in wood, tin, paper, cardboard, cloth, leather or other materials. It represents either the true size and shape of a face of a voussoir, as a rigid template, or its development, as a flexible template. Wood templates can only be used as rigid templates, unless they are very thin. The rest of materials are suitable for flexible templates; however, they can be used also as rigid templates, applying them to a previously dressed flat surface. Not to be confused with *templet*.
by templates Dressing method that relies heavily on true-shape templates.
templet A ruler with a curved side. Not to be confused with *template*.
tensile force Equal and aligned forces that point to the outside of a structural member. Tensile forces cause a (usually slight) lengthening of the member. If the result of tensile force exceeds admissible stress, the element may develop cracks and cause structural failure, either by deformation or splitting.
theorem of the three perpendiculars Theorem stating that if two straight lines are mutually orthogonal and at least one of them is orthogonal to the projection plane in an orthogonal projection, the projections of the lines are orthogonal. The theorem does not apply in oblique or conical projections.
three lost points A geometrical construction that allows determining the centre of a circle when three points are known. A draughtsperson should draw two lines joining pairs of points; then he or she should construct the bisectors of these lines; the centre of the circle will be placed at the intersection of the bisectors.
three-piece lewis See *lewis*.
thrust Lateral force exerted by an arch or vault over its supports. If this force is not adequately counteracted by the weight of the support, another adjoining arch or vault, or an additional element such as a flying buttress, the support may topple and the arch or vault, deprived of one of its supports, will probably collapse.
tierceron A rib that, together with a lierne, divides a quarter of a quadripartite vault in three sections. The result is a vault with twelve sections called star, stellar or tierceron vault. However, in some occasions, two or three pairs of tiercerons are used, resulting in a vault with a larger number of portions. Also, additional liernes or curved ribs may be added to the scheme.

tierceron vault See *vault, star*.
timbrel vault See *vault, timbrel*.
tomographie (Fre) A part of stereotomy dealing with the drawing of sections of solids, and thus with the construction of true size and shape templates.
tomomorphie (Fre) A part of stereotomy dealing with the shape of sections of solids.
tomotechnie (Fre) A part of stereotomy dealing with the construction of particular templates and dressing procedures.
torno encarcelado (Spa) A particular kind of hoist, set among three struts.
torus (1) A surface of revolution generated by a circle revolving around an axis placed on the same plan than the circle but not passing through its centre. Usually, the axis does not intersect the torus. If it does, the torus may be described as a self-intersecting torus 2) A convex moulding, used for example in Attic bases.
torus vault See *vault, annular.*
total station A surveying instrument including a telescope, angle sensors in a vertical and a horizontal plane and a laser distance meter, capable of furnishing the coordinates of any point visible from the station with high precision, typically in the range of millimetres.
tour de France (Fre) A rite and a formation practice of *compagnons du devoir*. An aspiring compagnon should travel around France for a given period, working in his craft to make a living and to master the techniques of the craft, usually living in lodging houses managed by *compagnons*; at the same time he should visit important examples of his craft, such as the *Vis de Saint Gilles* in the case of masons.
tracing Large- or full-scale drawing executed on a floor or wall as a preliminary operation in masonry construction. It depicts the element to be built in orthogonal projection. It may include its division into voussoirs or true-shape templates.
transverse arch*See arch, transverse.*
transverse rib See *arch, transverse.*
trapezial vault See *vault, trapezial.*
traversieu In Desargues' terminology, intersection of the face plane and the *plan droit a l'essieu.* It is orthogonal to the axis.
triangular vault See *vault, triangular.*
trompe de Montpellier (Fre) A trumpet squinch opened in a curved wall; voussoir corners are placed projecting centrally, rather than orthogonally, a round arch on the curved wall.
trompillon (Fre) A single stone placed in the portion of a trumpet squinch adjoining the apex of the conical intrados surface. This layout eschews the use of bed joints in this section, avoiding the appearance of very acute angles, which are prone to cause dents.
trumpet squinch A constructive element that spans the space between two walls set at an angle by means of a conical intrados surface.
Tudor arch See *arch, Tudor.*
twisted column A column whose shaft includes helical arrises.

underpinning The vertical portion of the provisional structure or falsework supporting an arch or vault under construction. It may include inclined struts.

vault A curved structure designed to cover an area. When featuring rotational symmetry, it is usually called a dome.

annular vault Vault whose intrados is part of a torus (1). Usually, if the axis of the torus is vertical, the upper half of the torus is used. If the axis is horizontal, the upper and outer quarter of the torus is used.

arrised vault Vault without ribs but with folds and creases in the position where ribs would be placed in an ordinary rib vault. It requires careful coordination of the voussoirs placed along the groins and creases. Examples are to be found mainly in the city of Valencia, with two isolated examples in Assier, Southern France.

barrel vault Vault whose intrados surface is the upper half of a cylinder.

cloister vault See *vault, pavilion*.

coffered vault Vault taking the shape of Roman vaults with coffers. However, it is built using the Gothic scheme of a combination of independent ribs and coffer caps.

elliptical vault Vault with its intrados in the shape of an ellipsoid, either prolate, oblate or scalene. It is difficult to tell apart built examples of this type from oval vaults, and thus elliptical vaults are frequently described as oval, and viceversa.

fan vault Typically English vault generated by the rotation of circular arcs around vertical axes. Individual members end in horizontal semicircles; the spaces between these semicircles are filled by planar spandrels. Mouldings resembling ribs are carved along the parallels and meridians of the surface. However, they are usually too thin to be executed as independent ribs, so they are included in the same voussoirs as the webbing of the vault.

groin vault Vault whose intrados is formed by two cylinders with the same radius and intersecting axes. Only the portions with generatrixes orthogonal to the sides of the covered area are used. Compare with *vault, pavilion*.

hemispherical vault See *dome, hemispherical*.

horizontal-axis annular vault Annular vault with a horizontal axis; only the outer upper half of the torus is materialised. Meridians are used as bed joints, while portions of parallels are used as side joints.

L-shaped vault Vault used to span two intersecting corridors or cloister galleries. Its intrados is formed by two intersecting semicylinders, cut at the vertical plane acting as bisector of the axes of the semicylinders; a crease appears at this intersection. Generatrices of the cylinders act as intrados joints.

net vault Rib vault with multiple ribs but without transverse arches, so the division in bays is not materialised. It was used mainly in German Late Gothic.

octagonal vault Vault with an octagonal plan. It is covered usually with eight cylindrical portions, belonging to four cylinders with the same radius and intersecting axes. As an alternative, it can be built also with a network of ribs covered by severies.

oval vault Vault with an oval impost. Its intrados surface may be made up from two spherical portions at the ends and two toroidal portions in the middle section, but this layout causes problems. See also *vault, elliptical.*

pavilion vault Vault whose intrados is formed by two semicylinders with the same radius and intersecting axes. Only the portions with generatrixes parallel to the sides of the covered area are used. Compare with *vault, groin.*

pendentive vault The ensemble of a dome and the pendentives (2) that support it; it may also include a drum (1) between the dome and the pendentives.

polygonal vault Rib vault covering a polygonal area, usually in the shape of a regular polygon. It is frequent in English chapter houses or in central plan churches.

quadripartite vault Rib vault with square or rectangular plan, four perimetral arches that may be wall arches or transverse ribs and two diagonal ribs.

quarter-sphere vault Vault whose intrados is a spherical surface cut by two planes, a horizontal one at the impost and a vertical one at the face; thus, the intrados is a quarter of a sphere.

rib vault Vault built using a combination of linear elements, the ribs, and surface elements, the severies, resting on the ribs.

sail vault Vault with a spherical or approximately spherical intrados, cut through three or more, and usually four, vertical planes. It is used to cover a square, rectangular or generally polygonal area with a spherical vault.

semioctagonal vault Vault whose impost is a portion of an octagon. The impost may include three or five sides of a regular octagon, or in some occasions three full sides and two half-sides at the ends; in this case it is exactly semioctagonal. It is usually covered by cylindrical portions, although it can be also built as a coffered vault.

semipolygonal vault Rib vault covering an area shaped as half a regular polygon, or a slightly larger one. It is used frequently in church chancels.

sexpartite vault Rib vault including four perimetral arches, either transverse or wall arches; two diagonal ribs; and an intermediate transverse arch. These ribs divide the area in six portions, covered with severies; therefore the term *sexpartite vault*.

skew vault Vault in which the springers are mutually parallel, but they are oblique to one or both faces.

sloping barrel vault Barrel vault with a slanting axis and imposts.

star vault Rib vault with tiercerons, in addition to perimetral arches, diagonal ribs and liernes connecting main and secondary keystones.

stellar vault See *vault, star*.

surbased vault Vault whose rise is less than half its span.

tierceron vault See *vault, star.*

timbrel vault Vault built in brick so that the larger face of each brick is a portion of the intrados surface. Thus, the thickness of each layer equals the smallest dimension of the brick. To strenghten the vault, it is usually built with two or three layers, with staggered joints in each successive layer. These vaults can be built without any kind of provisional falsework, using thin bricks and quick-setting gypsum mortar in the first layer; then, the second and succesive layers use the self-standing first layer as formwork.
trapezial vault Vault covering a trapezial area, used frequently in ambulatories.
triangular vault Vault covering a triangular area, used in triangular rooms, ambulatories, or the spaces between an octagonal vault and a square enclosing wall. It can be covered with a simple web; alternatively, it can be divided in three portions by ribs starting from the corners of the triangle.
vertical-axis annular vault Annular vault with a vertical axis; only the upper half of the torus is materialised. Parallels are used as bed joints, while portions of meridians are used as side joints.

vaulted staircase See *staircase, vaulted.*
vertical projection See *projection, vertical.*
vertical-axis annular vault See *vault, vertical-axis annular.*
vesica piscis See *mandorla.*
vis de Saint-Gilles A particular kind of vaulted staircase, with an enclosing wall and a thick newel. The intrados surface is generated by a semicircle rotating and ascending at the same time so that its ends are always placed at two helixes, a broader one in the wall and another one in the newel. Bed joints are helixes passing through the successive positions of a particular point in the semicircle, while side joints are portions of a particular position of the semicircle.
volute An ornamental element at the sides of the capital of the Ionic and Composite orders in the shape of a spiral.
voussoir A wedge-shaped member used to build arches, trumpet squinches, rere-arches and vaults.
wall A vertical two-dimensional constructive element used to enclose a building, providing shelter from cold, heat, rain or snow and/or support for a vault, roof or other horizontal construction. Its plan can be straight or curved. When built in half-squared stone, ashlar masonry or brick, it is divided into horizontal courses, and each course is divided into well-defined pieces such as ashlars, half-squared stones or bricks. The surface between courses is called bed joint; that between pieces in the same course is called side joint. Walls can be also built in rammed earth, rubble masonry or concrete, and in this case the concept of course does not apply.
wall arch See *arch, wall.*
warped surface A ruled surface where two consecutive positions are never in the same plane. As a result, the surface is non-developable.
warped template See *revirado.*
waster A chisel with a toothed edge.

web The ensemble of the severies in a rib vault. The web may be interrupted if the ribs have a fin on top, or either it may pass over the rib. In the second case, the web may stay in place even if the ribs have fallen as a result of war or earthquake damage.
webbing See *web*.
well (1) A hole dug in the ground in order to collect water or other purposes. In many occasions, a wall is built .in the perimeter of the hole in order to prevent the ground from slipping (2) A central portion in a staircase which is not covered by flights or steps in order to let light come through.
whipsaw A saw formed by a toothed blade with two handles at both ends, operated with both hands or, frequently, by two carpenters or masons.
windlass See *hoist*.

Image Sources

Agustín Castellanos. 10.4.
Alonso Rodríguez, Miguel Ángel, Pau Natividad Vivó, and José Calvo-López. 3.17, 1.20.
Alonso, Miguel Ángel, and José Calvo-López. 3.2.
Bayerische StaatsBibliothek. 4.4.
Biblioteca de Cultura Artesana. Mallorca. 2.6, 4.2, 4.9, 6.15.
Biblioteca de la Escuela de Arquitectura de la Universidad Politécnica de Madrid. 1.1, 2.11, 2.12, 2.17, 2.18, 2.24, 3.6, 4.6, 5.6, 6.2, 6.9, 6.23, 8.4, 8.24, 8.27, 9.24, 9.30, 9.32, 10.16, 10.31, 10.32, 10.34, 10.35, 10.37, 10.38, 10.39, 12.5.
Biblioteca della Universitá IUAV. Venezia. 1.3, 1.7, 1.9, 2.25, 3.41, 4.5, 5.5, 5.7, 6.22, 6.26, 6.51, 7.15, 7.20, 8.10, 8.11, 8.12, 9.36, 9.6.
Biblioteca Nacional de España. Madrid. 2.14, 2.15, 2.4, 2.5, 3.18, 3.24, 3.25, 3.37, 5.2, 5.10, 5.13, 6.13, 6.20, 6.34, 6.40, 6.48, 7.2, 7.5, 7.8, 8.2, 8.7, 8.13, 8.15, 8.19, 8.21, 8.25, 9.2, 9.3, 9.4, 9.10, 9.11, 9.14, 9.18, 9.19, 9.26, 9.28, 9.34, 11.2, 11.4, 11.7, 11.9, 11.13, 11.23, 11.24, 11.26.
Bibliothéque Nationale de France. 2.1, 2.10, 3.1, 3.15, 3.29, 6.27, 9.31,10.25.
Bibliothéque Universitaire de Lyon. 2.22.
Calvo-López, José. 1.2, 1.8, 1.11, 1.12, 1.13, 1.14, 1.15, 1.16, 1.17, 1.19, 1.21, 3.3, 3.27, 3.32, 3.33, 3.34, 3.39, 3.40, 3.43, 4.1, 4.11, 4.12, 4.7, 4.8, 5.1, 5.12, 5.14, 5.8, 6.1, 6.5, 6.14, 6.17, 6.19, 6.33, 6.37, 6.47, 7.1, 7.6, 7.13, 7.16, 8.6, 8.18, 8.20, 8.22, 8.26, 9.1, 9.12, 9.13, 9.16, 9.21, 9.27, 9.8, 9.9, 10.1, 10.3, 10.10, 10.11, 10.12, 10.13, 10.19, 10.2, 10.21, 10.22, 10.23, 10.26, 10.27, 10.28, 10.29, 10.33, 11.3, 10.5, 10.6, 10.7, 10.8, 11.1, 11.10, 11.17, 11.18, 11.22, 11.25, 11.5, 11.6, 11.8, 12.2, 12.3, 12.4, 12.6, 12.7.
Calvo-López, José, and Pau Natividad Vivó. 11.19.
Camiruaga Osés, Idoia, Miguel Taín Guzmán and José Calvo-López. 3.7.
Colegio Oficial de Arquitectos de Murcia. 9.35.
e-rara. ETH-bibliothek. Zurich. 1.4, 2.16, 2.19, 2.20, 2.21, 2.23, 2.26, 2.27, 2.28, 3.10, 3.11, 3.12, 3.13, 3.14, 3.22, 3.26, 3.31, 3.38, 3.42, 3.45, 3.49, 3.9, 4.13, 4.3, 5.11, 5.3, 5.4, 6.4, 6.10, 6.11, 6.25, 6.31, 6.32, 6.36, 6.41, 6.42, 6.43, 6.44, 6.46, 6.50, 7.10, 7.11, 7.12, 7.14, 7.19, 7.21, 8.1, 8.5, 8.8, 8.9, 8.16, 8.17,

J. Calvo-López, *Stereotomy*, Mathematics and the Built Environment 4,
https://doi.org/10.1007/978-3-030-43218-8

8.23, 8.29, 8.30, 8.31, 8.32, 9.7, 9.15, 9.29, 9.33, 10.9, 10.17, 10.18, 11.15, 11.27, 11.28, 11.30, 12.10, 12.8.

e-rara. Stiftung Bibliothek Werner Oechslin. 3.48.

e-rara. Universitätsbibliothek Bern. 2.9, 3.21, 3.23, 3.28, 3.30, 3.4, 3.46, 3.5, 3.8, 5.15, 5.16, 5.17, 5.18, 5.9, 6.6, 6.18, 6.28, 6.35, 6.39, 7.7, 7.17, 8.3, 9.17, 9.20, 9.23, 9.25, 10.15, 11.11, 11.12, 11.29, 12.1.

España. Ministerio de Defensa. Biblioteca Central Militar, MS-457. 2.13, 3.19, 3.20, 3.35, 3.36, 3.44, 6.12, 6.16, 6.21, 6.24, 6.29, 6.3, 6.30, 6.38, 6.45, 6.49, 6.7, 6.8, 7.3, 7.4, 7.9, 7.18, 8.14, 8.28, 11.14, 11.16, 12.9.

Herzog August Bibliothek. Wolfenbüttel. 2.7, 3.47.

Historisches Archiv Köln. 2.2, 2.3, 2.8, 3.16.

Internet Archive. 1.6.

López Mozo, Ana, and Miguel Ángel Alonso Rodríguez. 10.24.

Médiatheque de l'Architecture et du Patrimoine. Base de données Mémoire. 11.20.

Rabasa Díaz, Enrique. 1.5, 4.10, 9.5, 10.14.

Salcedo Galera, Macarena. 10.40.

Taín Guzmán, Miguel. 1.10.

Navarro Camallonga, Pablo. 10.30.

Natividad Vivó, Pau. 1.18, 10.36, 11.21, 9.22.

References

Sources and Relevant Editions

Adhémar, Joseph-Alphonse. 1861. *Traité des ponts biais en pierre et en bois.* Paris: E. Lacroix.

Alberti, Leon Battista. 1435. De Pictura.

Alberti, Leon Battista. 1485. *De Re Aedificatoria.* Firenze: Nicolai Laurentii Alamani.

Alberti, Leon Battista, and Cosimo Bartoli. 1550. *L' architettura di Leonbattista Alberti.* Firenze: Lorenzo Torrentino.

Alberti, Leon Battista, Joseph Rykwert, Neil Leach, and Robert Tavernor. [1485] 1991. *On the art of building in ten books.* Cambridge (MA): MIT Press.

Alviz, Pedro de, attr. c. 1544. Manuscrito de cantería. Madrid: Biblioteca Nacional de España. MS 12686.

Aguillon, François. 1613. *Opticorum libri sex.* Antwerpen.

Apollonius Pergaei, Federico Commandino, Pappus of Alexandria, Eutocii Ascalonitae and Sereni Antinsensis. 1566. *Apollonii Pergaei conicorum libri quattuor ...* Bononiae: Ex Officina Alexandri Benatii.

Bachot, Ambroise. 1598. *Le gouvernail d'Ambroise Bachot, ... lequel conduira le curieux de géométrie en perspective dedans l'architecture des fortifications ...* Melun-Paris: L'auteur—La Croix Blanche.

Bails, Benito. 1779-1790. *Elementos de Matemática ...* Madrid: Joaquín Ibarra.

Biblia Sancti Petri Rodensis. c. 1020. Bibliothèque Nationale de France. Ms. lat. 6.

Bails, Benito. 1779-1790. *Elementos de Matemática...* Madrid: Joaquín Ibarra.

Blanchard, Edme. 1729. *Traité de la coupe des bois pour le revêtement des Voutes ...* Paris: Josse et Jombert.

Blondel, François-Nicolas. 1673. *Résolution des quatre principaux problèmes d'architecture ...* Paris: Imprimerie Royale.

Blondel, François-Nicolas. 1675-1683. *Cours d'architecture.* Paris: L'auteur.

Bordoni, Antonio. 1826. *Nota di stereotomia sopra i cunei dei ponti in isbieco ...* Milano: Paolo Emilio Giusti.

Bosse, Abraham. 1643b. *La Manière universelle de M. Desargues, ... pour poser l'essieu et placer les heures et autres choses aux cadrans au soleil, par A. Bosse.* Paris: Pierre des Hayes.

Bosse, Abraham. 1648. *Manière universelle de M. Desargues, pour pratiquer la perspective par petit-pied, comme le géométral ...* Paris: Pierre des Hayes.

Bosse, Abraham. 1665. *Traité des pratiques géométrales et perspectives enseignées dans l'Académie royale de la peinture et sculpture ...* A Paris: chez l'auteur.

Bosse, Abraham, and Girard Desargues. 1643a. *La practique du traict a preuues de M. Desargues ... pour la coupe des pierres en l'Architecture ...* Paris: Pierre des Hayes.

J. Calvo-López, *Stereotomy*, Mathematics and the Built Environment 4,
https://doi.org/10.1007/978-3-030-43218-8

Bullant, Jean. 1561. *Recueil d'horlogiographie, contenant la description, fabrication et vsage des horloges solaires*. Paris.

Bullant, Jean. 1562. *Petit traité de géométrie et d'horlogiographie pratique*. Paris: Guillaume Cavellat.

Caramuel y Lobkowitz, Juan. 1678. *Arquitectura civil recta y oblicua* ... Vigevano: Imprenta obispal.

Cattaneo, Pietro. 1567. *L'Architettura di Pietro Cataneo senese. A le quale oltre ... primi quattro libri ... sono aggiunti di piu il Quinto, Sesto, Settimo o Ottavo libro ...*

Cheréau, Jean. c. 1567-1574. Livre d'Architecture. Gdansk: Polska Akademia Nauk Biblioteka Gdańska. MS 2280.

Codex Miniatus 3. c. 1570. Oesterreiche Nationalbibliothek, Wien.

Compagnons du Devoir et du Tour de France. 2001-2010. *Encyclopédie des métiers. La maçonnerie et la taille de pierre*. Paris: Librairie du compagnonnage.

"Cooke" Manuscript. c. 1400. London. British Library. Add MS 23198.

Curabelle, Jacques. 1644a. *Foiblesse pitoyable du sr G. Desargues employée contre l'examen fait de ses oeuures*. Paris.

Curabelle, Jacques. 1644b. *Examen des oeuvres du Sieur Desargues*. Paris: M. & I. Henault—F. L'Anglois dit Chartres.

De Biard, Nicholas, attr. 1250, c. Distinctiones. Paris: Bibliothéque Nationale de France.

De Beaune, Florimond. c. 1640. Doctrine de l'angle solide construit sur trois angles plans. Paris: Acadèmie des Sciences. Papiers de Gilles Personne de Roberval.

De Beaune, Florimond, and Pierre Costabel. [c. 1640] 1975. *Doctrine de l'angle solide. Inventaire de sa bibliothèque*. Paris: J. Vrin.

Del Monte, Guidoubaldo. 1579. *Planisphaeriorum Universalium Theorica*. Pisauri: Hyeronimum Concordiam.

Derand, François, S. I. 1643. *L'Architecture des voûtes ou l'art des traits et coupe des voûtes ...* Paris: Sébastien Cramoisy.

Derand, François, S. I. 1743. *L' Architecture des voûtes, ou l'art des traits et coupe des voûtes ...* Paris: A. Cailleau.

Desargues, Girard. 1636. *Exemple de l'une des manieres universelles du S.G.D.L. touchant la pratique de la perspective sans emploier aucun tiers point, de distance ny d'autre nature, qui soit hors du champ de l'ouvrage*. Paris: Jacques Dugast.

Desargues, Girard. 1639. *Brouillon project d'une atteinte aux evenemens des rencontres du cone avec un plan*. Paris.

Desargues, Girard. 1640. *Brouillon project d'exemple d'une manière universelle du S.G.D.L. touchant la practique du trait a preuues pour la coupe des pierres en l'architecture ...* Paris: Melchoir Tavernier.

Desargues, Girard. 1644. *Recit au vray de ce qui a este la cause de faire cet escrit*. s. l.: s. n.

Desargues, Girard. 1648. Reconnoissance. In *Maniere universelle de Mr. Desargues, pour pratiquer la perspective par petit-pied, comme le geometral*. Abraham Bosse. Paris: Pierre Des Hayes.

Dürer, Albrecht. 1525. *Underweysung der messung mit dem zirkel und richtscheyt ...* Nuremberg: s. n.

Dupuit, J. 1870. *Traite de l'equilibre des voûtes et de la construction des ponts en maçonnerie*. Paris: Dunod.

Euclid. c. -300. Elements.

Euclid. 1482. *Elementa Geometriae*. Venezia: Erhard Ratdolt.

Euclid and Rodrigo Zamorano. 1576. *Los seis libros primeros de la geometría de Euclides traducidos en lengua española ...* Sevilla: Alonso de la Barrera.

Euler, Leonhard. 1772. De solidis quorum superficiem in planum explicare licet. *Novi Commentarii academiae scientiarum Petropolitanae* 16:3–34.

Facht von Andernach, Jacob, Mathes Roriczer and Lorenz Lechler. 1593. Musterbuch. Köln. Historisches Archiv Köln. Ms. 7020 (W*) 276.

Faventino, Marcus Cetius. c. 250. De diversis fabricis.

Faventino, Marcus Cetius. [c. 250] 1540. *De diversis fabricis*. Paris: Vascosan.

Fernández Sarela, Francisco. 1740. Algunos cortes de arquitectura: Archivo Provincial Franciscano de Santiago de Compostela.

Fernández Sarela, Francisco Antonio, and Miriam Elena Cortés López. 2015. *O manuscrito 'Algunos cortes de arquitectura' de Francisco Antonio Fernández Sarela.* Santiago de Compostela: Xunta de Galicia.

Finé, Oronce. 1542. *Orontii Finaei Delphinatis... De mundi sphaera, sive cosmographia.* Paris: Simon de Colines.

Frézier, Amédée–François. 1737–1739. *La théorie et la pratique de la coupe des pierres et des bois pour la construction des voutes et autres parties des bâtiments ... ou traité de stéréotomie ...* Strasbourg–Paris: Jean Daniel Doulsseker-L. H. Guerin.

Frézier, Amédée François. 1754-1769. *La théorie et la pratique de la coupe des pierres et des bois, pour la construction des voutes ..., ou traité de stereotomie ...* Paris: Charles-Antoine Jombert.

Gallon, M. 1735. *Machines et Inventions Approuvees par l'Academie Royale des Sciences.* Paris: Gabriel Martin, Jean-Baptiste Coignard fils, Hippolyte-Louis Gérin.

García, Simón, and Rodrigo Gil de Hontañón. 1681. Compendio de arquitectura y simetría de los templos, conforme a la medida del cuerpo humano. Madrid: Biblioteca Nacional de España. MS 8884.

García, Simón, Rodrigo Gil de Hontañón, and Antonio Bonet Correa. [1681] 1991. *Compendio de arquitectura y simetría de los templos, conforme a la medida del cuerpo humano.* Valladolid: Colegio de Arquitectos.

García Berruguilla, Juan. 1747. *Verdadera práctica de las resoluciones de la geometría sobre las tres dimensiones para un perfecto arquitecto ...* Madrid: Francisco Mojados.

Gelabert, Joseph. 1653. Verdaderes traces de l'art de picapedrer. Palma de Mallorca: Biblioteca del Consell Insular de Mallorca.

Gelabert, Joseph, and Pep Alcover. [1653] 1977. *De l'art de picapedrer.* Palma de Mallorca: Instituto de Estudios Baleáricos.

Gelabert, Joseph, Enrique Rabasa Díaz, Ramon Garriga, and Yuka Irie. [1653] 2011. *El manuscrito de cantería de Joseph Gelabert.* Palma de Mallorca—Madrid: Col.legi Oficial d'Arquitectes de les Illes Balears—Fundación Juanelo Turriano.

Gentillâtre, Jacques, attr., c. 1620. Manuel d'un ingénieur-architecte de la première moitié du XVIIe siècle. Paris: Bibliothèque Nationale de France. MS fr. 14727.

Gil de Hontañón, Rodrigo. c. 1560. Manuscrito. Included in García 1681. Madrid: Biblioteca Nacional. MS 8884.

Guardia, Alonso de. c. 1600. Rasguños de arquitectura y cantería. Written and drawn in a factitious copy of Pittoni 1562 and Pittoni 1566. Madrid: Biblioteca Nacional. ER/4196.

Guarini, Guarino. 1671. *Euclides adauctus et methodicus mathematicaque universalis.* Augustae Taurinorum.

Guarini, Guarino. [c. 1680] 1737. *Architettura Civile del padre D. Guarino Guarini chierico regolare opera postuma ...* Torino: Gianfrancesco Mariesse.

Gundissalinus, Dominicus. c. 1150 Liber de divisione philosophiae in partes suas ...

Gundissalinus, Dominicus, and L. Baur. [c. 1150] 1903. Liber de divisione philosophiae in partes suas ... In *Beiträge zur Geschichte der Philosophie und Theologie des Mittelalters*, 4:2, 3–142.

Hachette, Jean–Nicolas–Pierre. 1822. *Traité de géométrie descriptive: comprenant les applications de cette géométrie aux ombres, à la perspective et à la stéréotomie.* Paris: Guillaume et Corby.

Hachette, Jean–Nicolas–Pierre. 1828. *Traité de géométrie descriptive... par M. Hachette,...* 2ème ed. Paris: Corby.

Hammer, Hans. c. 1500. Schizzenbuch. Herzog August Bibliothek Wolfenbüttel. Cod. Guelf. 114.1 Extrav.

Herrera, Juan de. c. 1577. Architectura y machinas. Archivo General de Simancas. Casa y Sitios Reales. Leg 258, fol. 488.

Hooke, Robert. 1676. *A description of helioscopes and some other instruments ...* London: John Martyn.

Isidore of Seville. c. 630. Etymologiarum.

Jousse, Mathurin. 1642. *Le secret d'architecture découvrant fidélement les traits géométriques, couppes et dérobements nécessaires dans les bastimens.* La Flèche: Georges Griveau.

La Faille, Jean–Charles de. 1640 c. De la arquitectura. Madrid: Biblioteca del Palacio Real. MS 3729.

La Gournerie, Jules–Antoine–René Maillard de. 1855. *Discours sur l'art du trait et la Géometrie Descriptive* ... Paris: Mallet–Bachelier.

La Gournerie, Jules–Antoine–René Maillard de. 1860. *Traité de géométrie descriptive* ... Paris: Mallet-Bachelier.

La Gournerie, Jules-Antoine-René Maillard de. 1874. Mémoire sur l'enseignement des Arts graphiques. *Journal de Mathématiques Pures et Apliquées* 2ème serie, 19:113–156.

La Hire, Philippe de. 1688a. Traité de la coupe des pierres. Paris: Bibliothèque de l'Ecole Nationale des Ponts et Chaussées.

La Hire, Philippe de. c. 1688b. Traité de la coupe des pierres. Paris: Bibliothèque de l'Institut de France.

La Hire, Philippe de. c. 1688c. Traité de la coupe des pierres. Rennes: Bibliothèque Municipale.

La Hire, Philippe de. c. 1688d. Traité de la coupe des pierres. Langres: Bibliothèque Municipale.

Landsberg, Herrad von. c. 1160. Hortus Deliciarum. Bibliothèque Nationale et Universitaire de Strasbourg, now lost.

Laplace, Pierre-Simon de, Joseph-Louis Lagrange and Gaspard Monge [1795] 1992. *Leçons de mathématiques*... ed. Bruno Belhoste, Amy Dahan-Dalmédico, Jean Dhombres, Joël Sakarovitch. Paris: Dunod.

La Rue, Jean-Baptiste de. 1728. *Traité de la coupe des pierres où par méthode facile et abrégée l'on peut aisément se perfectionner en cette science.* Paris: Imprimerie Royale.

Lastanosa, Pedro Juan de, or Juanelo Turriano, attr. c. 1570. Los veintiún libros de los ingenios y las máquinas. Madrid: Biblioteca Nacional de España. MS 3772–3776.

Lechler, Lorenz. 1516. Unterweisungen für seynen Son Moritzen. Included in Facht et al. 1593. Köln. Historisches Archiv Köln. W 276.

Leroy, Charles–François–Antoine. 1834. *Traité de géométrie descriptive, avec une collection d'épures composée de 60 planches.* Paris: Carilian–Goeury.

Leroy, Charles–François–Antoine. 1844. *Traité de stéréotomie comprenant les applications de la géométrie descriptive a la théorie des ombres, la perspective linéaire, la gnomonique, la coupe des pierres et la charpente.* Paris: Bachelier, Carilian–Goeurt et Dalmont.

Leroy, Charles-Félix-Auguste, and E. Martellet. 1877. *Traité de stéréotomie comprenant les applications de la géométrie descriptive la théorie des ombres, la perspectiva linéaire, la gnomique, la coupe des pierres et la charpente* ... 7ème ed. Paris: Gauthier-Villars.

Llaguno y Amírola, Eugenio. 1829. *Noticias de los arquitectos y arquitectura de España* ... Madrid: Imprenta Real.

López de Arenas, Diego. 1633. *Breve compendio de la carpintería de lo blanco, y tratado de alarifes.* Sevilla: Luis Estupiñán.

L'Orme, Philibert de. 1561. *Nouvelles inventions pour bien bastir a petits frais.* Paris: Federic Morel.

L'Orme, Philibert de. 1567. *Le premier tome de l'Architecture.* Paris: Federic Morel.

L'Orme, Philibert de, and Jean–Marie Pérouse de Montclos. 1988. *Traités d'architecture.* Paris: Léonce Laget.

Manuscrito de arquitectura y cantería. 1650 c. Madrid: Biblioteca Nacional de España. MS 12744.

Martínez de Aranda, Ginés. 1600 c. Cerramientos y trazas de montea. Madrid: Biblioteca Central Militar. MS 457.

Martínez de Aranda, Ginés, and Antonio Bonet Correa. 1986. *Cerramientos y trazas de montea.* Madrid: Servicio Histórico Militar—CEHOPU.

Martínez Guijarro, Juan, known as Silíceo. 1519. *Arithmetica Ioannis Martini Scilicei in theoricen et praxim scissa / nuper ab Orontio Fine, Delphinate, summa diligentia castigata*... Paris: Henri Stephane.

Mazarrasa, Andrés Julián. 1750 c. Tratado de arquitectura.

Milliet–Dechalles, Claude–François, S. I. 1674. *Cursus seu mundus mathematicus*. Lugduni: Officina Anissoniana.

Milliet–Deschalles, Claude–François, S. I., and Aimé Varcin. 1690. *Cursus seu mundus mathematicus... Editio altera ex manuscriptis authoris aucta et emendata* ... Lugduni: Annison, J. Posuel et C. Rigaud.

Milizia, Francesco. 1781. *Principj di Architettura Civile*. Bassano: A spese Remondini di Venezia.

Miscelanea secundum usum ordinis cluniacensis. c. 1200. Paris: Bibliothèque Nationale de France. MS lat. 17716.

Monge, Gaspard. 1796 [An IV]. Des lignes de courbures de la surface de l'Ellipsoïde. *Journal de l'Ecole Polytechnique* (2):145–165.

Monge, Gaspard. 1799 [An VII]. *Géométrie descriptive, leçons données aux Écoles normales, l'an 3 de la République* ... Paris: Baudouin.

Moseley, Henry. 1843. *The Mechanical Principles of Engineering and Architecture*. London: Longman, Brown, Green and Longmans.

Nicholson, Peter. 1827. *A popular and practical treatise on masonry and stone–cutting*. London: Thomas Hurst, Edward Chance and Co.

Nicholson, Peter. 1839. *The guide to railway masonry, containing a complete treatise on the oblique arch*. London–Newcastle: John Weale–Emerson Charnley, Currie and Bowman–Finlay and Charlton.

Olivier, Théodore. 1832. Mémoire de Géométrie Descriptive sur la construction des Tangentes en un point multiple d'une Courbe plane ou a double courbure, dont l'Equation n'est pas connue. *Journal de l'Ecole Polytechnique* 13 (21): 303–355.

Olivier, Théodore. 1843–1844. *Cours de géométrie descriptive*.

Olivier, Théodore. 1851. *Mémoires de géométrie descriptive, théorique et appliquée* ... Paris: Carillan–Goeury et Dalmont.

Palladio, Andrea. 1570. *I quatri Libri d' Architettura*. Venezia: Domenico de Francesci.

Paris, Matthew. 1250–1259. Liber additamentorum. British Library. Cotton MS Nero D I.

Pérez de Moya, Juan. 1568. *Obra intitulada fragmentos mathemáticos. En que se tratan cosas de Geometría y astronomía y geographia y Philosophia natural y sphera y astrolabio y navegación y reloxes*. Salamanca: Juan de Cánova.

Pérez de Moya, Juan. 1573. *Tratado de Geometria Practica, y Speculativa por el bachiller Iuan Perez de Moya natural de Santesteban del Puerto*. Alcalá de Henares: Juan Gracián.

Pittoni, Battista. 1562. *Di Battista Pittoni pittore vicentino Imprese di diversi prencipi, dvchi, signori, e d'altri personaggi ... con alcune stanze del Dolce che dichiarano i motti di esse impresse*. Venezia.

Pittoni, Battista. 1566. *Imprese di diversi Pricipi, duchi, signori, e d'altri personaggi et huomini illustri. Libro secondo*. Venezia.

Portor y Castro, Juan de. 1708. Cuaderno de arquitectura. Madrid: Biblioteca Nacional. MS 9114.

Rabasa Díaz, Enrique. 2007. *Guía práctica de la estereotomía de la piedra*. León: Editorial de los Oficios.

Rabelais, François. 1534. *Gargantua*. Paris.

Rabelais, François. ([1534] 1978) *Gargantua*. Paris: Gallimard.

Ranisch, Bartel. 1695. *Beschreibung Aller Kirchen–Gebäude der Stadt Dantzig* ... Dantzig.

"Regius" or "Hallywell" Manuscript. London. British Library. Royal MS 17 A 1.

Ribes i Ferrer, Joseph. 1708. Llibre de trasas de vias y muntea. Barcelona. Biblioteca Nacional de Catalunya. MS 096–Fol–398.

Rixner, Wolfgang. 1467–1500. Bauhuttenbuch. Vienna. Graphische Sammlung Albertina. Cim Kasten, Fach VI, Nr. 55.

Rojas, Cristóbal de. 1598. *Teórica y práctica de fortificación, conforme a las medidas de estos tiempos* ... Madrid: Luis Sánchez.

Rojas, Cristóbal de. 1613. *Compendio y breve resolución de fortificación*. Madrid: Juan de Herrera.

Rojas Sarmiento, Juan de. 1550. *Illustris uiri D. Ioannis de Roias Commentariorum in Astrolabium quod Planisphaerium uocant libri sex. Nunc primum in lucem editi. His additus est index capitum ac rerum, quae toto opere continentur, locupletissimus...* Paris: Vascosan.

Rojas Sarmiento, Juan de. [1550] 1551. *Illustris uiri D. Ioannis de Roias commentariorum in Astrolabium quod Planisphaerium uocant libri sex nunc primum in lucem editi...* Paris: Vascosan.

Rondelet, Jean–Baptiste. 1802–1817. *Traité theorique et pratique de l'art de batir*. Paris: L'auteur.

Rondelet, Jean-Baptiste. 1834. *Traité theorique et pratique de l'art de batir*. Paris: Firmin-Didot.

Roriczer, Mathes. 1486. *Büchlein von der fialen Gerechtigkeit*. Regensburg.

Roriczer, Mathes. c. 1490a. *Geometria Deutsch.*

Roriczer, Mathes. c. 1490b. *Wimpergbüchlein.*

Roriczer, Mathes, Hannes Schmuttermayer, and Lon R. Shelby. 1977. *Gothic Design Technics: The fifteenth-century design booklets of Mathes Roriczer and Hans Schmuttermayer*. Carbondale: Southern Illinois University Press.

Ruiz II, Hernán. c. 1560. Libro de Arquitectura. Madrid: Biblioteca de la Escuela de Arquitectura de la Universidad Politécnica de Madrid.

Ruiz II, Hernán, and Pedro Navascués Palacio. 1974. *Libro de Arquitectura de Hernán Ruiz El Joven*. Madrid: Escuela de Arquitectura de Madrid.

Ruiz II, Hernán, Alfonso Jiménez Martín, and Consuelo Álvarez Márquez. 1998. *Libro de Arquitectura*. Sevilla: Fundación Sevillana de Electricidad.

San Nicolás, Fray Laurencio de. 1639. *Arte y uso de Arquitectura*. Madrid?: Juan Sánchez? Imprenta Real?

San Nicolás, Fray Laurencio de. 1665. *Segunda parte del Arte y uso de Arquitectura*. Madrid?.

Santa Bárbara, Fray Francisco de. 1766. Secretos de arquitectura. Tratado geométrico que comprehende lo más usual y corriente de la montea y arte de la cantería. Xàtiva. Arxiu Municipal. Llibre 995.

Sanzio, Rafaello and Baldassare Castiglione, attr. c. 1518. Lettera a Leone X.

Sanzio, Rafaello and Baldassare Castiglione, attr. and Francesco Paolo di Teodoro. 2003. *Raffaello, Baldassar Castiglione e la Lettera a Leone X*. Bologna: Minerva.

Sauval, Henri, and Bernard Rousseau [c. 1654–1676] 1724. Histoire et recherches des Antiquités de la ville de Paris.

Savot, Louis, and François Blondel [c. 1654–1676] 1685. *L'Architecture française des bâtiments particuliers*. 2nd ed.

Schmuttermayer, Hans. c. 1500. *Fialenbüchlein.*

Serlio, Sebastiano. 1537. *Regole generali di architetura sopra le cinque maniere degli edifici ...* Venetia: F. Marcolini da Forli.

Serlio, Sebastiano. 1545. *Il primo [-secondo] libro d'architettura ... Le premier [-second] livre d'architecture*. Paris: Iehan Barbé.

Serlio, Sebastiano, and Giovanni Domenico Scamozzi. 1584. *Tutte l'opere d'architettura di Sebastiano Serlio ...* Venetia: F. de Franceschi.

Serlio, Sebastiano. 1600. *Tutte le opere di architettura et prospetiva ... dove si mettono in disegno, quelle cose che sono più necessarie a sapere gli Architetti ...* Venezia: Heredi di Francesco de' Franceschi.

Sigüenza, Fray José de. 1600–1605. *Historia de la Orden de San Jerónimo*. Madrid: Imprenta Real.

Sigüenza, Fray José de. [1600–1605] 1907. *Historia de la Orden de San Jerónimo*. Madrid: Bailly–Bailliére.

Simonin, and Delagardette. 1792. *Traité élémentaire de la coupe de pierres, ou art du trait ouvrage dans lequel on enseigne à tracer et à tailler ... mis au jour par M. Delagardette ...* . Paris: Jombert.

Tatti, Luigi. 1839. Sul taglio dei cunei in isbieco, osservazioni dell'ingegnere Francesco Colombani. *Il Politecnico* 1 (1):67–69.

Torija, Juan de. 1661. *Breve tratado de todo tipo de bóvedas así regulares como irregulares ...* Madrid: Pablo de Val.

Tosca, P. Thomas Vicente. 1707–1715. *Compendio mathemático, en que se contienen todas las materias más principales de las Ciencias, que tratan de la cantidad....* Valencia: Antonio Bordazar—Vicente Cabrera.

Tosca, P. Thomas Vicente. 1757. *Tratados de Arquitectura civil, montea, canteria, y reloxes.* Valencia: Joseph García.

Tornés family and Natalia Juan García. [c. 1700] 2013. *Trazas y diseños: el manuscrito de la familia Tornés, su aportación al arte de la Edad Moderna y su vinculación con la tratadística arquitectónica.* Sariñena, Huesca: Nalvay.

Tornés family and Natalia Juan García. [c. 1700] 2015. *Más trazas y nuevos diseños. Otro cuaderno de taller de la familia Tornés. La transmisión de ideas artísticas en el Barroco a través de libros y tratados.* Sariñena: Nalvay.

Tramontini, Giuseppe. 1811. *Delle projezione grafiche e delle loro principali applicazioni ...* Modena: Società tipografica.

Vallancey, Charles. 1766. *A Practical Treatise on Stone–Cutting. In five parts. Containing, Part I. Of Plan Gates, and of voussures, or Gates revealed, splayed and vaulted ... Part the first.* Dublin: Thomas Ewing.

Vallée, Louis-Léger. 1819. *Traité de la géométrie descriptive* ... Paris: Vve. Courcier.

Vallée, Louis-Léger. 1825. *Traité de la géométrie descriptive* ... Paris: Bachelier.

Vandelvira, Alonso de. c. 1585. Libro de trazas de cortes de piedras. Madrid: Biblioteca de la Escuela de Arquitectura de la Universidad Politécnica de Madrid. MS Raros 31.

Vandelvira, Alonso de, and Geneviève Barbé Coquelin de Lisle. 1977. *El tratado de arquitectura de Alonso de Vandelvira ...* Albacete: Caja de Ahorros Provincial.

Vandelvira, Alonso de, and Felipe Lázaro de Goiti. 1646. Libro de cortes de cantería de Alonso de Vandeelvira, arquitecto ... Madrid: Biblioteca Nacional de España. MS 12719.

Vasari, Giorgio. 1550. *Le vite de' piú eccelenti architetti, pittori et scultori italiani, da Cimbaue insino a'tempi nostri ...* Firenze: Lorenzo Torrentino.

Vasari, Giorgio. 1568. *Le vite de' piú eccelenti pittori, scultori et architettiori italiani.* Firenze: Giunti.

Vignola, Iacomo Barozzi de. 1562. *Regola delli Cinque ordini d'Architettura.* Roma.

Villanueva, Juan de. 1827. *Arte de albañilería.* Madrid: Francisco Martínez Dávila.

Villard de Honnecourt. c 1225 c. Sketchbook. Paris: Bibliothèque Nationale de France.

Villard de Honnecourt and H. R. Hahnloser. [1935] 1972. *Villard de Honnecourt. Kritische Gesamtausgabe des Bauhuttenbuches, ms. fr 19093 der Pariser Nationalbibliothek.* Wien: Schroll.

Villard de Honnecourt and Carl F. Barnes. [c. 1225] 2009. *The Portfolio of Villard de Honnecourt, A New Critical Edition and Color Facsimile.* Farnham: Ashgate.

Villard de Honnecourt, Alain Erlande-Brandenburg, Régine Pernoud, Jean Gimpel, and Roland Bechmann. 1986. *Carnet de Villard de Honnecourt.* Paris: Seuil.

Villard de Honnecourt, Jean-Baptiste-Antoine Lassus, and Alfred Darcel. 1858. *Album de Villard de Honnecourt, architecte du XIIIeme siecle [...].* Paris: Imprimerie Nationale.

Villard de Honnecourt, Meister WG, and François Bucher. 1979. *Architector. The Lodge Books and Sketchbooks of Medieval Architects.* New York: Abaris Books.

Viollet le Duc, Eugène. 1854–68. *Dictionnaire raisonné de l'architecture française du XIe au XVIe siècle.* Paris: B. Bauce.

Vitruvius Pollio, Marcus. c. –25. De Architectura libri decem.

Vitruvius Pollio, Marcus, and Fra Giocondo Veronese. 1511. *M. Vitruvius per Iocundum solito castigator factus...* Venezia: Tacuino.

Vitruvius Pollio, Marcus, and Lázaro de Velasco. 1564. Traducción de los diez libros de arquitectura de Vitrubio. Cáceres: Biblioteca Pública del Estado.

Vitrubio Polion, Marco, and Quirico Viviani. 1830. *L'architettura di Vitruvio tradotta in italiano da Quirico Viviani illustrata con note critiche ed ampliata di aggiunte* ... Udine: Fratelli Matiuzzi.

Vittone, Bernardo Antonio. 1740. *Instruzioni elementari per indirizzo de' giovanni allo studio dell'architettura civile* Lugano: Agnelli.

Warland, E.G. [1929] 2015. *Modern practical masonry*. Oxford: Routledge.
WG, Meister. 1560-1572. Steinmetzbuch. Frankfurt am Main. Städelsches Kunstinstitut und Städtische Gallerie. 8–494.
Zabaglia, Niccola. [1743] 1824. *Castelli e ponti ... con la descrizione del transporto dell'obelisco Vaticano*. Roma: Crispino Puccinelli.

Literature

Abraham, Pol. 1934. *Viollet-le-Duc et le rationalisme medieval*. Paris: Vincent, Fréal & Cie.
Ackerman, James S. 1949. Ars sine scientia nihil est. Gothic Theory of Architecture at the Cathedral of Milan. *The Art Bulletin* 31 (2): 84–111.
Ackerman, James S. 1997. Villard de Honnecourt's Drawings of Reims Cathedral: A Study in Architectural Representation. *Artibus et Historiae* 18 (35): 41–49.
Acland, James H. 1972. *Medieval Structure: The Gothic Vault*. Toronto: University of Toronto Press.
Adam, Jean Pierre. 1984. *La construction Romaine, materiaux et tecniques*. Paris: Picard.
Aita, Danila. 2003. Between geometry and mechanics: a re–examination of the principles of stereotomy from a statical point of view. In *First International Congress on Construction History*, ed. Santiago Huerta Fernández, 161–170. Madrid: Instituto Juan de Herrera.
Aldana Fernández, Salvador. 1988. *La Lonja de Valencia*. Valencia: Consorci d'Editors Valencians.
Alexander, Jennifer S. 2004. Villard de Honnecourt and Masons' Marks. In *Villard's Legacy: Studies in Medieval Technology, Science, and Art in Memory of Jean Gimpel*, ed. Marie-Thèrese Zenner, 53–69. Aldershot-Burlington (VT): Ashgate.
Alonso Rodríguez, Miguel Ángel, José Calvo López, and Pau Natividad Vivó. 2013. The Vault of the Chapel of Our Lady of the Alcázar in the Church of Saint Patrick in Lorca. Geometrical Experimentation in Spanish Renaissance Architecture. *Revista EGA Expresión Gráfica Arquitectónica* 18 (22): 122-131.
Alonso Rodríguez, Miguel Ángel, Elías Hernández Albadalejo, José Calvo López, and Pau Natividad Vivó. 2011a. La capilla del canónigo Jerónimo Grasso y el desplome de la fachada renacentista de la catedral de Murcia. In *XXII Jornadas de Patrimonio Cultural de la Región de Murcia*, ed. José Antonio Melgares Guerrero et al., 13-22. Murcia: Ediciones Tres Fronteras.
Alonso Rodríguez, Miguel Ángel, Ana López Mozo, José Carlos Palacios Gonzalo, Enrique Rabasa Díaz, José Calvo López, and Alberto Sanjurjo Álvarez. 2009. Functionalism and caprice in stonecutting. The case of the Nativity Chapel in Burgos Cathedral. In *Proceedings. Third International Congress on Construction History*, 31-38. Cottbus: Brandenburg Technical University.
Alonso Rodríguez, Miguel Ángel, Elena Pliego, and Alberto Sanjurjo Álvarez. 2011b. Graphical tools for an epistemological shift. Contribution of Protoaxonometrical Drawing to the Development of Stonecutting Treatises. *Nexus Network Journal* 13 (3): 649-670.
Álvarez Márquez, Consuelo. 1998. Transcripción de los textos. In Hernán Ruiz II, *Libro de arquitectura*, 43-80. Sevilla: Fundación Sevillana de Electricidad.
Antón Solé, Pablo. 1975. La catedral vieja de Santa Cruz de Cádiz. Estudio histórico y artístico de su arquitectura. *Archivo Español de Arte* 48 (189): 83-96.
Aranda Alonso, María. 2017. La trompa de Montpellier. Origen, uso, desarrollo y evolución a partir del tratado de cantería de Alonso de Vandelvira. In *Actas del Décimo Congreso Nacional y Segundo Iberoamericano de Historia de la Construcción*, ed. Santiago Huerta Fernández, 27-36. Madrid: Instituto Juan de Herrera.
Ascani, Valerio. 1989. Le dessin d'architecture medieval en Italie. In *Les bâttisseurs des cathédrales gothiques*, ed. Roland Recht, 255-277. Strasbourg: Editions Les Musées de la Ville de Strasbourg.
Azambuja Varela, Pedro de, and J. P. Sousa. 2018. Variable Casting of Voussoirs for a Stereotomic Shell. *Nexus Network Journal* 20 (3): 627-644.

Babelon, Jean. 2006. Jousse, Mathurin. Le secret d'architecture. In *Architectura. Books on Architecture*, eds. Fréderique Lemerle, and Yves Pauwels. Tours: Centre d'etudes superieures de la Renaissance. http://architectura.cesr.univ-tours.fr (Accessed December 12, 2019).

Banda y Vargas, Antonio de la. 1974. *El arquitecto andaluz Hernán Ruiz II*. Sevilla: Universidad de Sevilla.

Barbé-Coquelin de Lisle, Geneviève. 1977. Introducción. In *Tratado de arquitectura de Alonso de Vandelvira*, 1-36. Albacete: Caja de Ahorros Provincial.

Barbé-Coquelin de Lisle, Geneviève. 1981. Noticia del 'Breue tratado de todo tipo de bóvedas'. In *Breue tratado de todo tipo de bóvedas así regulares como irregulares* ..., 11-19. Madrid: Albatros.

Barbé-Coquelin de Lisle, Geneviève. 1993. Progresos de la cantería y nivel científico en España en la época de Juan de Herrera. In *Juan de Herrera y su influencia*, 129-136. Santander: Universidad de Cantabria.

Barberio, Maurizio. 2014. Building the stereotomic design. In *Stereotomic Design*, eds. Giuseppe Fallacara, and Vincenzo Minenna, 50-55. Bari: Dicar-Pimar.

Bares, María Mercedes. 2007. La Vis de Saint Gilles del Castello Maniace di Siracusa: un'audace sperimentazione di stereotomia. *Lexicon* no. 4: 15-23.

Barnes, Carl F., Jr. 1965. Gothic Architecture by Paul Frankl. *Journal of the Society of Architectural Historians* 24 (2): 174-176.

Barnes, Carl F., Jr. 1972. The gothic architectural engravings in the cathedral of Soissons. *Speculum* 47 (1): 60-64.

Battisti, Eugenio. 1976. *Filippo Brunelleschi*. Milano: Electa.

Becchi, Antonio. 2009. Pregnant Columns. From Word to Shape. In *Creating shapes in civil and naval architecture. A cross-disciplinary comparison*, eds. Horst Nowacki, and Wolfgang Lefevre, 279-296. Leiden-Boston: Brill.

Becchi, Antonio, and Federico Foce. 2002. *Degli archi e delle volte. Arte del costruire tra meccanica e stereotomia*. Venezia: Marsilio.

Bechmann, Roland. [1981] 1996. *Les racines des cathédrales. L'architecture gothique, expression des conditions du milieu*. Paris: Payot.

Bechmann, Roland. [1991] 1993. *Villard de Honnecourt. La penseé technique au XIIIe siécle et sa communication*. Paris: Picard.

Bermejo, Elisa. 1954. Bartolomé de Zumbigo, arquitecto del siglo XVII. *Archivo Español de Arte* 27 (108):291–302.

Beffeyte, Renaud. 2004. The Oral Tradition and Villard de Honnecourt. In *Villard's Legacy: Studies in Medieval Technology, Science, and Art in Memory of Jean Gimpel*, ed. Marie-Thèrese Zenner, 93-119. Aldershot: Ashgate.

Belda Navarro, Cristóbal. 1971. La obra de rejería de la catedral de Murcia. *Anales de la Universidad de Murcia*, 207-243.

Belhoste, Bruno. 1990a. Du dessin d'ingenieur a la geometrie descriptive. L'enseignement de Chastillon a l'Ecole royale du genie de Mézières. *In Extenso* no. 13: 103-135.

Belhoste, Bruno, and René Taton. 1992. L'invention d'une langue de figures. In *L'Ecole normale de l'an III. Leçons de Mathematiques. Laplace, Lagrange, Monge*, ed. J. Dhombres, 269-303. Paris: Dunod.

Belhoste, Bruno, Antoine Picon, and Joël Sakarovitch. 1990b. Les exercices dans les ecoles d'ingenieurs sous l'Ancien Regime et la Revolution. *Histoire de l'Education* no. 46: 53-109.

Benítez Hernández, Patricia, and Mercedes Valiente López. 2015. The Tangent Solution to the Late Gothic Helical Staircase: A Justified Oblivion. *Nexus Network Journal* 17 (2): 379-398.

Bergamo, Francesco. 2020. La scala elicoidale a sbalzo di Andrea Palladio al convento della Carità a Venezia e la sua eredità geometrico-costruttiva. In *Geometria e costruzione. Steretomia e configurazione in architettura*, eds. José Calvo-López, Alessio Bortot, and Giulia Piccinin, 71-83. Roma: Aracne.

Bérchez, Joaquín. 2005-2006. Visiones y mentalidad arquitectónica de un maestro del siglo XVIII. La 'Descripción breve de las medidas y magnificencia ...' del convento de Santa Clara de Játiva, por Fray José Alberto Pina. *Ars Longa. Cuadernos de Arte* no. 14-15: 195-216.

Bessac, Jean-Claude. 1984. Tracés et épures gravés dans l'ancienne cathédrale Saint-Just de Narbonne (Aude). In *Actes du Colloque international de glyptographie de Cambrai*, 33-55. Braine-le-Château.

Bessac, Jean-Claude. 1986. *L'outillage traditionnel du tailleur de pierre de l'Antiquité á nos jours*. Paris: Editions du CNRS.

Bischoff, Franz. 1989. Les maquettes d' architecture. In *Les bâttisseurs des cathédrales gothiques*, ed. Roland Recht, 287-295. Strasbourg: Editions Les Musées de la Ville de Strasbourg.

Blasco Esquivias, Beatriz. 1991. Sobre el debate entre arquitectos profesionales y arquitectos artistas en el Barroco Madrileño. Las posturas de Herrera, Olmo, Donoso y Ardemans. *Espacio, Tiempo y Forma. Serie VII: Historia del Arte* no. 4: 159-194.

Block, Philippe, Thierry Ciblac, and John Ochsendorf. 2006. Real-time limit analysis of vaulted masonry buildings. Computers and Structures 84 (29-30):1841–1852

Blunt, Anthony. 1958. *Philibert de l'Orme*. London: A. Zwemmer.

Boccadoro, Niccola, and Maurizio Barberio. 2014. Parametrization of form. In *Lithic Tree. A Search for Natural Stereotomy*, eds. Giuseppe Fallacara, and Claudia Calabria, 108-118. Paris: Presses des Ponts.

Böker, Johann Joseph. 2005. *Architektur der Gotik. Gothic Architecture. Catalogue of the World-Largest Collection of Gothic Architectural Drawings (bequest Franz Jäger) in the … Academy of Fine Arts, Vienna*. Salzburg: Verlag Anton Pustet.

Bonet Correa, Antonio. 1966. *La arquitectura en Galicia durante el siglo XVII*. Madrid: Instituto Diego Velázquez.

Bonet Correa, Antonio. [1979] 1991. Simón García, tratadista de arquitectura. In Simón García, *Compendio de arquitectura y simetria de los templos*, 13-18. Valladolid: Colegio de Arquitectos.

Bonet Correa, Antonio. 1986. Ginés Martínez de Aranda, arquitecto y tratadista de cerramientos y arte de montea. In Ginés Martínez de Aranda, *Cerramientos y trazas de Montea*. Madrid: Servicio Histórico Militar.

Bonnet Saint-Georges, Dominique. 1993. *Philibert de L'Orme, lyonnais*. Lyon: Archives municipales.

Bony, Jean 1984. *French Gothic Architecture of the Twelfth and Thirteenth Centuries*. Berkeley: University of Californa Press.

Borin, Paolo, and José Calvo López. 2020. Notes on Guarini's stereotomy: spherical vaults. In *Geometria e costruzione. Steretomia e configurazione in architettura*, eds. José Calvo-López, Alessio Bortot, and Giulia Piccinin, 45-57. Roma: Aracne.

Bork, Robert. 1997-1998. Holy Toledo: Art-Historical Taxonomy and the Morphology of Toledo Cathedral. *AVISTA Forum* 10 (2): 31-37.

Bortot, Alessio, and José Calvo-López. 2020. De la Rue's Traité de la coupe des pierres: a crucial step in the articulation of material stonecutting and abstract stereotomy. In *Geometria e costruzione. Steretomia e configurazione in architettura*, eds. José Calvo-López, Alessio Bortot, and Giulia Piccinin, 19-31. Roma: Aracne.

Boscaro, Cristian. 2016. La *Manière universelle* de Girard Desargues per la comprensione delle strutture stereotomiche. PhD diss., Università IUAV.

Branner, Robert. 1957. Three problems from the Villard de Honnecourt manuscript. *The Art Bulletin* 39 (1): 61-66.

Branner, Robert. 1958. Drawings from a Thirteenth-Century Architect's Shop: The Reims Palimpsest. *Journal of the Society of Architectural Historians* 17 (4): 9-21.

Branner, Robert. 1960. *Burgundian Gothic Architecture*. London: Zwemmer.

Branner, Robert. 1960. Villard de Honnecourt, Archimedes, and Chartres. *Journal of the Society of Architectural Historians* 19 (3): 91-96.

Branner, Robert. 1963. Villard de Honnecourt, Reims and the origin of gothic architectural drawing. *Gazette des Beaux-Arts* 61: 129-146.

Branner, Robert 1968. Gothic Architecture by Paul Frankl. *The Art Bulletin* 50 (2): 199.

Brunet, E. 1928. La restauration de la cathédrale de Soissons. *Bulletin Monumental* 87: 65-99.

Bruno, Mathias. 2016. Fori di olivela nel Colosseo. In *I Tracciati di cantiere: disegni esecutivi per la trasmissione e diffusione delle conoscenze tecniche*, eds. Carlo Inglese, and Antonio Pizzo, 159-165. Roma: Gangemi.

Brykowska, Maria. 1992. Quadratur des spätgotischen Gewölbes im Chorraum der Pfarrkirche zu Szydłowiec/Polen. *Architectura* 22: 101—108.

Bucher, François. 1972. The Dresden sketchbook of vault projection. In *22nd International Congress on Art History*, 527-537. Budapest.

Burns, Howard. 1991. Building and construction in Palladio's Vicenza. In *Les Chantiers de la Renaissance*, 191-226. Paris: Picard.

Bustamante García, Agustín. 1994. *La octava maravilla del mundo. Estudio histórico sobre el Escorial de Felipe II*. Madrid: Alpuerto.

Cabezas Gelabert, Lino. 1992. Los modelos tridimensionales de arquitectura en el contexto profesional y en las teorías gráficas del siglo XVI. In *Dibujo y Arquitectura. Investigación Aplicada. IV Congreso Internacional de Expresión Gáfica Arquitectónica*, 105-115. Valladolid: Grapheus.

Calvo López, José. 2007. Piezas singulares de cantería en la ingeniería y la arquitectura militar de Cartagena en el siglo XVIII. In *Actas del Quinto Congreso Nacional de Historia de la Construcción*, eds. Miguel Arenillas Parra et al., 167–176. Madrid: Instituto Juan de Herrera.

Cahn, Walter. 1994. Architecture and Exegesis: Richard of St.-Victor's Ezekiel Commentary and Its Illustrations. *The Art Bulletin* 76 (1): 53-68.

Cajigal Vera, Miguel Ángel, Miguel Ángel Alonso Rodríguez, José Calvo-López, and Miguel Taín-Guzmán. 2016. The Full-Scale Tracings of the Parish Church of Nogueira do Miño. In *I Tracciati di cantiere: disegni esecutivi per la trasmissione e diffusione delle conoscenze tecniche*, eds. Carlo Inglese, and Antonio Pizzo, 108-117. Roma: Gangemi.

Calabria, Claudia. 2014. The Lithic Tree as a 'New Generation of Branched Structures'. Historical Antecedents and Evolution. In *Lithic Tree. A Search for Natural Stereotomy*, eds. Giuseppe Fallacara, and Claudia Calabria, 48-59. Paris: Presses des Ponts.

Calandriello, Antonio. 2020. Dal piano allo spazio: il linguaggio grafico di de l'Orme nelle sue tre dimensioni. In *Geometria e costruzione. Steretomia e configurazione in architettura*, eds. José Calvo-López, Alessio Bortot, and Giulia Piccinin, 29-68. Roma: Aracne.

Calvo-López, José. 1998. Los trazados de cantería en la 'Teórica y práctica de fortificación' de Cristóbal de Rojas. In *Actas del Segundo Congreso Nacional de Historia de la Construcción*, eds. Fernando Bores Gamundi et al., 67-75. Madrid: Instituto Juan de Herrera.

Calvo-López, José. 2000a. 'Cerramientos y trazas de montea' de Ginés Martínez de Aranda. PhD diss., Universidad Politécnica de Madrid.

Calvo López, José. 2000b. Lunetas y arcos avanzados. El trazado de un elemento constructivo en los siglos XVI y XVII. In *Actas del Tercer Congreso Nacional de Historia de la Construcción*, eds. Amparo Graciani García et al., 165-175. Madrid: Instituto Juan de Herrera.

Calvo-López, José. 2002. Arquitectura oblicua y trazas de montea. *Revista de Expresión Gráfica en la Edificación (EGE)* (2): 38-51.

Calvo López, José. 2003. Orthographic projection and true size in Spanish stonecutting manuscripts. In *Proceedings of the First International Congress on Construction History*, ed. Santiago Huerta Fernández, 461-471. Madrid: Instituto Juan de Herrera.

Calvo-López, José. 2006. Sixteenth-Century Spanish Cranes and Lázaro de Velasco's Translation of Vitruvius. In *Second International Congress on Construction History*, eds. Malcolm Dunkeld et al, 493-507. Cambridge: Construction History Society.

Calvo-López, José. 2009. El manuscrito 'Cerramientos y trazas de montea', de Ginés Martínez de Aranda. *Archivo Español de Arte* 82 (325): 1-18.

Calvo-López, José. 2015a. Los rasguños de Alonso de Guardia y la práctica de la cantería española en la Edad Moderna. In *Teoría y Literatura Artística en España. Revisión historiográfica y estudios contemporáneos*, eds. Miguel Taín Guzmán, and Nuria Rodríguez Ortega, 411-457. Madrid: Real Academia de Bellas Artes de San Fernando.

Calvo-López, José. 2015b. Philibert de l'Orme and Spanish Stereotomy. In *Philibert de l'Orme. Un architecte dans l'histoire*, eds. Fréderique Lemerle, and Yves Pauwels, 199-213. Turnhout: Brepols.

Calvo-López, José. 2017. The Geometry of the Rib Vault in Early Modern Iberian and French Literature. In *Techniche construttive nel Mediterraneo: dalla stereotomia ai criteri antisismici*, ed. Marco Rosario Nobile, 27-45. Palermo: Edizioni Caracol.

Calvo-López, José. 2019. Teaching, Creating, and Using Perspective in Sixteenth-Century Spain: The Architectural Notebook of Hernán Ruiz II. In *Renaissance Cultures of Optics and Practices of Perspective*, ed. Sven Dupré, 301-332. Turnhout: Brepols.

Calvo López, José, and Miguel Ángel Alonso Rodríguez. 2010. Perspective versus stereotomy: from Quattrocento polyhedral rings to Sixteenth-Century Spanish torus vaults. *Nexus Network Journal* 12 (1): 75-111.

Calvo-López, José, and Eliana De Nichilo. 2005b. Stereotomia, modelli e declinazioni locali dell'arte del costruire in pietra da taglio tra Spagna e Regno di Napoli nel XV secolo. In *Teoria e Pratica del costruire: saperi, strumenti, modeli*, 517-526. Ravenna: Università di Bologna-Fondazione Flaminia.

Calvo-López, José, and Enrique Rabasa Díaz. 2017. Construcción, dibujo y geometría en la transición entre Gótico y Renacimiento. *Artigrama* (31): 67-86.

Calvo López, José, and Miguel Taín Guzmán. 2018a. *Las monteas del convento de Santa Clara de Santiago de Compostela. Un repertorio de trazados, tanteos y dibujos del Barroco español.* Santiago de Compostela: Consorcio de Santiago-Guiverny.

Calvo-López, José, and Macarena Salcedo Galera. 2017. Geometric Proportion in the Sixteenth Century. Methods and Constraints. *Nexus Network Journal* 19: 155-178.

Calvo-López, José, Miguel Ángel Alonso Rodríguez, Enrique Rabasa Díaz, and Ana López Mozo. 2005a. *Cantería renacentista en la catedral de Murcia.* Murcia: Colegio Oficial de Arquitectos de Murcia.

Calvo-López, José, Miguel Ángel Alonso Rodríguez, Enrique Rabasa Díaz, Ana López Mozo, Carmen Pérez de los Ríos, and Pau Natividad Vivó. 2015c. Geometry and Capriciousness in 11th-century Armenian Architecture. The Scriptorium of the Monastery of Sanahin. In *5th International Congress on Construction History.* Chicago: Construction History Society of America.

Calvo López, José, Enrique Rabasa Díaz, Ana López Mozo, and Miguel Ángel Alonso Rodríguez. 2018b. The Vault over the Crossing Tower in Lincoln Cathedral in the Context of European Gothic Architecture. In *Fourth Construction History Society Annual Conference*, eds. James W. P. Campbell, et al., 221-232. Cambridge: Construction History Society.

Calvo-López, José, Miguel Angel Alonso Rodríguez, Miguel Taín Guzmán, and Pau Natividad Vivó. 2013a. La construcción en cantería en la Galicia barroca. Las monteas del coro alto de la capilla de San Telmo de Tui. *Informes de la Construcción* 65 (Extra-2): 127-140.

Calvo-López, José, Juan Carlos Molina-Gaitán, Pau Natividad-Vivó, Miguel Ángel Alonso-Rodríguez, Enrique Rabasa-Díaz, Ana López-Mozo, Miguel Taín-Guzmán, and José Antonio Sánchez-Pravia. 2013b. The Tracing for the Sail Vault at the Murcia Cathedral Vestry: Surveying a 16th-Century Full-Scale Working Drawing. *International Journal of Architectural Heritage: Conservation, Analysis, and Restoration* 7 (3): 275-302.

Calvo-López, José, Miguel Taín-Guzmán, and Idoia Camiruaga-Osés. 2016. The eighteenth-century full-scale tracings in the church of Saint Clare in Santiago de Compostela: execution drawings or design sketches? *Construction History* 31: 81-106.

Calvo López, José, et al. 2020. *An International Bibliography on Stonecutting and Stereotomy*, forthcoming.

Cámara Muñoz, Alicia. 1981. La arquitectura militar y los ingenieros de la monarquía española: aspectos de una profesión. (1530-1650). *Revista de la Universidad Complutense* (3): 255-269.

Cámara Muñoz, Alicia. 1998. *Fortificación y ciudad en los reinos de Felipe II*. Madrid: Nerea.

Cámara Muñoz, Alicia. 2014. Cristóbal de Rojas. De la cantería a la ingeniería. In *Ingenieros del Renacimiento*, 135-161. Madrid: Fundación Juanelo Turriano.

Cámara Muñoz, Alicia. 2015. Cantería e ingeniería del Renacimiento en el puente de Zuazo en Cádiz. *Lexicon* (15): 7-20.

Camerota, Filippo. 2005a. 'L'oeil du Soleil': la projection orthographique selon Galilée et Pietro Accolti. In *Perspective, projections, projet: techniques de la représentation architecturale*, 81-90. Paris: Monum.

Camerota, Filippo. 2005b. Prospettiva e stereotomia: la visualizzazione dello spazio geometrico. In *Prospettiva e geometria dello spazio*, ed. Marco Franciosi, 83-100. Sarzana (La Spezia): Agorà Edizioni.

Camerota, Filippo 2006. Architecture as mathematical science: the case of 'Architectura Obliqua'. In *Practice and science in early modern Italian building. Towards an epistemic history of architecture*, ed. Hermann Schlimme, 51-60. Milano: Mondadori.

Camón Aznar, José María. 1941. La intervención de Rodrigo Gil en el manuscrito de Simón García. *Archivo Español de Arte* 14 (45): 300-305.

Campbell, James W. P. 2014a. The British Staircase. In *Staircases. History, Repair and Conservation*, eds. James W. P. Campbell, and Michael Tutton, 75-140. London-New York: Routledge.

Campbell, James W. P. 2014b. The Geometry of Staircases: Their Comfort, Safety and Setting Out. In *Staircases. History, Repair and Conservation*, eds. James W. P. Campbell, and Michael Tutton, 245-265. London-New York: Routledge.

Campbell, Tony 1987. Portolan Charts from the Late Thirteenth Century to 1500 In *The History of Cartography. Vol I. Cartography in Prehistoric, Ancient, and Medieval Europe and the Mediterranean*, eds. J. B. Harley, and David Woodward, 371-463. Chicago-London: University of Chicago Press.

Capel, Horacio, Joan Eugeni Sánchez, and Omar Moncada. 1988. *De Palas a Minerva. La formación científica y la estructura institucional de los ingenieros militares en el siglo XVIII*. Madrid: CSIC.

Carazo Lefort, Eduardo. 1994. El modelo de la Catedral de Valladolid. *Revista EGA Expresión Gráfica Arquitectónica* (2): 95-100.

Carlevaris, Laura. 2000. Le volte di de L'Orme. Problemi di ricostruzione di alcuni Traits. In *Il disegno e la pietra*, ed. Riccardo Migliari, 81-92. Roma: Gangemi.

Carlevaris, Laura. 2014. Nicolas-François-Antoine de Chastillon: The Défilement of Fortifications at the Roots of Descriptive Geometry. *Nexus Network Journal* 16 (3): 631-652.

Carriero, Alessandra, and Fabio Sabbadini. 2016. Note sulla mechanica del sollevamento mediante 'olivella'. In *I Tracciati di cantiere: disegni esecutivi per la trasmissione e diffusione delle conoscenze tecniche*, eds. Carlo Inglese and Antonio Pizzo, 170-175. Roma: Gangemi.

Carvajal Alcaide, Rocío. 2011a. Estructura y singularidad del Cuaderno de Arquitectura de Juan de Portor y Castro (1708-1719). In *Actas del Séptimo Congreso Nacional de Historia de la Construcción*, eds. Santiago Huerta Fernández et al., 211-220. Madrid: Instituto Juan de Herrera.

Carvajal Alcaide, Rocío. 2011b. Stairs in the Architecture Notebook of Juan de Portor y Castro: An Insight into Ruled Surfaces. *Nexus Network Journal* 13 (3): 631-648.

Carvajal Alcaide, Rocío. 2015. La iglesia del monasterio jerónimo de Santa Catalina en Talavera de la Reina. Piezas singulares de cantería documentadas en la tratadística española. In *Actas del Noveno Congreso Nacional y Primer Congreso Internacional Hispanoamericano de Historia de la Construcción*, eds. Santiago Huerta Fernández and Paula Fuentes, 357-365. Madrid: Instituto Juan de Herrera.

Carvajal Alcaide, Rocío, and Miriam Elena Cortés López. 2013. Aportaciones gallegas para la historia del corte de la piedra en España: Los cuadernos de Juan de Portor y Francisco Sarela. In *Actas del Octavo Congreso Nacional de Historia de la Construcción*, eds. Santiago Huerta Fernández and Fabián López Ulloa, 162-170. Madrid: Instituto Juan de Herrera.

Casaseca Casaseca, Antonio. 1988. *Rodrigo Gil de Hontañón (Rascafría, 1500—Segovia, 1577)*. Valladolid: Junta de Castilla y León.

Cascione, Vito. 2014. From Conception to Finished Product: Digital Fabrication and Stone Techniques of a Stone Prototype. In *Lithic Tree. A Search for Natural Stereotomy*, eds. Giuseppe Fallacara and Claudia Calabria, 90-107. Paris: Presses des Ponts.

Ceccarelli Pellegrino, Alba. 1996. *Le "Bon architecte" de Philibert de l'Orme. Hypotextes et anticipations.* Fasano-Paris: Schena-Nizet.

Celtibère, M. 1860. *Monographie de Notre-Dame de Paris et de la Nouvelle Sacristie de MM. Lassus et Viollet-le-Duc.* Paris: A. Morel.

Chaboud, Marcel. 1996. *Girard Desargues, bourgeois de Lyon, mathématicien, architecte.* Lyon: IREM de Lyon-Aléas.

Chanfón Olmos, Carlos. [1979] 1991. Simón García y la antropometría. In *Compendio de Arquitectura y Simetría de los Templos*, 31-42. Valladolid: Colegio de Arquitectos.

Chatêlet-Lange, Liliane. 2006. Jacques Gentillâtre. Traité d'architecture. In *Architectura. Books on Architecture*, eds. Fréderique Lemerle and Yves Pauwels. Tours: Centre d'etudes superieures de la Renaissance. http://architectura.cesr.univ-tours.fr (Accessed December 12, 2019)

Checa, Fernando. 1989. El estilo clásico. In *Arquitectura del Renacimiento en España, 1488-1599*, 251-380. Madrid: Cátedra.

Choisy, Auguste. 1883. *L'art de bâtir chez les bizantins.* Paris: Librairie de la Societé Anonyme de Publications Périodiques.

Choisy, Auguste. 1899. *Histoire de l'Architecture.* Paris: Gauthier-Villars.

Chueca Goitia, Fernando. 1954. *Andrés de Vandelvira.* Sevilla-Madrid: Laboratorio de Arte de la Universidad de Sevilla- Instituto Diego Velázquez.

Chueca Goitia, Fernando. 1971. *Andrés de Vandelvira, arquitecto.* Jaén: Diputación de Jaén.

Clark, William W. 2004. Reims Cathedral in the Portfolio of Villard de Honnecourt. In *Villard's Legacy: Studies in Medieval Technology, Science, and Art in Memory of Jean Gimpel*, ed. Marie-Thèrese Zenner, 23-51. Aldershot: Ashgate.

Claval, Florence. 1988. Les épures de la cathedrale de Clermont-Ferrand. *Bulletin Archéologique du Comité des travaux historiques et scientifiques*, new series, no. 20-21: 184-224.

Clifford, Brandon, Wes McGee, and Mackenzie Muhonen. 2018. Recovering Cannibalism in Architecture with a Return to Cyclopean Masonry. *Nexus Network Journal* 20:583-604.

Coffin, David R. 1956. Padre Guarino Guarini in Paris. *Journal of the Society of Architectural Historians* 15 (2): 3-11.

Colchester, L. S., and John Harvey. 1974. Wells Cathedral. *Archaeological Journal* 131: 200-214.

Colella, Michaela. 2014. Stereotomic design exhibition. Evolution: from the hand to the robot. In *Stereotomic Design*, eds. Giuseppe Fallacara, and Vincenzo Minenna, 38-49. Bari: Dicar-Pimar.

Colombier, Pierre du. [1953] 1973. *Les chantiers des cathédrales.* Paris: Picard.

Comberousse, Charles Jules Félix de. 1879. *Histoire de l'Ecole Centrale des Arts et Manufactures depuis sa fondation jusqu'a ce jour.* Paris: Gauthier-Villars.

Coope, Rosalys. 1972. *Catalogue of the Drawings Collection of the Royal Institute of British Architects: Jacques Gentilhâtre.* Farnborough: Gregg International Publishers.

Coste, Anne. 2003. Gothic vaults: A rationalist or a tectonic track for the role of the ribs? In *First International Congress on Construction History*, ed. Santiago Huerta Fernández, 657-664. Madrid: Instituto Juan de Herrera.

Cruz Isidoro, Fernando. 2001. *Alonso de Vandelvira (1544-ca. 1626/7) Tratadista y arquitecto andaluz.* Sevilla: Universidad de Sevilla.

Cuneo, Paolo. 1988. *Architettura armena.* Roma: De Luca.

D'Amato, Claudio, and Giuseppe Fallacara. 2005. Tradizione e innovazione nella progettazione / costruzione dell'architettura: ruolo del 'modello' e attualità della steretotomia. In *L'arte della stereotomia: i Compagnons du devoir e le meraviglie della costruzione in pietra*, eds. Claudio d'Amato, and Giuseppe Fallacara, 52-89. Paris: Compagnons du devoir.

De Nichilo, Eliana. 2002. Sisteme architettonici e cupolati nei secoli XVI-XVII in Spagna. Sequenze, nodi tettonici e technique stereotomiche a confronto. PhD diss., Politecnico de Bari.

De Rosa, Agostino. 2020. Principi costruttivi e rigore geometrico: architetture in equilibrio tra gravità e immaginazione. In *Geometria e costruzione. Steretomia e configurazione in architettura*, eds. José Calvo-López, Alessio Bortot, and Giulia Piccinin. Roma: Aracne, forthcoming.

Delgado, Ricardo Jorge Conduto Rodrigues. 2017. A geometria na estereotomia da pedra na arquitectura religiosa portuguesa entre 1530 e 1580. PhD diss., Universidade de Lisboa.

Díaz Moreno, Félix. 2008. Introducción. In Fray Laurencio de San Nicolás, *Arte y uso de arquitectura. Edición anotada.* Madrid: Instituto de Estudios Madrileños.

Diles, Justin. 2018. Lightweight Stereotomy with Glass-Fiber Reinforced Plastic. *Nexus Network Journal* 20:645-669.

Dupin, Charles. 1819. *Essai historique sur les services et les travaux scientifique de Gaspard Monge.* Paris: Bachelier.

Duvernoy, Sylvie. 2015. Baroque Oval Churches: Innovative Geometrical Patterns in Early Modern Sacred Architecture. *Nexus Network Journal* 17 (2): 425-456.

Errede, Francesco. 2014. SHN Energy: Energy analysis and environmental comfort. In *Stereotomic Design*, eds. Giuseppe Fallacara, and Vincenzo Minenna, 77-83. Bari: Dicar-Pimar.

Erlande-Brandenburg, Alain. 1993. *Quand les cathédrales étaient peintes.* Paris: Gallimard.

Erlande-Brandenburg, Alain, and Anne-Bénédicte Mérel-Brandenburg. 1995. *Du Moyen Age à la Renaissance.* Paris: Mengés-CNMHS.

Esteban Piñeiro, Mariano, and María Isabel Vicente. 1991. El nivel atribuido a Juan de Herrera y su fundamento geométrico. *Llull* 14 (26): 31-57.

Esteban Piñeiro, Mariano. 2002-2003. Las academias técnicas en la España del siglo XVI. *Quaderns d'Historia de l'Enginyeria* 5:10-19.

Etlin, Richard. 2009. Génesis y estructura de las bóvedas de Arles. In *Actas del Sexto Congreso Nacional de Historia de la Construcción*, eds. Santiago Huerta Fernández et al., 425-434. Madrid: Instituto Juan de Herrera.

Etlin, Richard A. 2012. Stereotomy: The Paradox of an Acrobatic Architecture. In *Stereotomy: Stone Architecture and New Research*, eds. Giuseppe Fallacara, and Marco Stigliano, 18-32. Paris: Presses des Ponts.

Etlin, Richard, Giuseppe Fallacara, and Luc Tamboréro. 2008. *Plaited Stereotomy: Stone Vaults for the Modern World.* Roma: Aracne.

Ettlinger, Leopold D. [1977] 1986. The emergence of the Italian architect in the fifteenth century. In *The Architect. Chapters in the history of the profession*, ed. Spiro Kostof, 96-123. New York: Oxford University Press.

Evans, Robin. 1988. La trompe d'Anet. *Eidos* no. 2: 50-57.

Evans, Robin. 1995. *The Projective Cast.* Cambridge, Mass.: MIT Press.

Falcón Márquez, Teodoro. 1994. El nombramiento de Ginés Martínez de Aranda como maestro mayor de las diócesis de Cádiz y Santiago de Compostela. In *Tiempo y espacio en el arte. Homenaje al profesor Antonio Bonet Correa*, 461-468. Madrid: Universidad Complutense.

Fallacara, Giuseppe. 2003a. The formal unity of aerial vaults texture: The 'trompes'…'. In *First International Congress on Construction History*, ed. Santiago Huerta Fernández, 839-850. Madrid: Instituto Juan de Herrera.

Fallacara, Giuseppe. 2003b. Il paradigma stereotomico nell'arte di construire. Dalla natura sincretica della modellazione digitale alla proggetazione / produzione di elementi in pietra di taglio. PhD diss., Politecnico de Bari.

Fallacara, Giuseppe. 2006. Digital Stereotomy and Topological Transformations. Reasoning about Shape Building. In *Second International Congress on Construction History*, eds. Malcolm Dunkeld, James W. P. Campbell, Hentie Louw, Michael Tutton, Bill Addis, and Robert Thorne, 1075-1092. Cambridge: Construction History Society.

Fallacara, Giuseppe. 2009a. Philibert de l'Orme e l'*invenzione.* In *Nouvelles inventions pour a bien bastir et à petit frais. Edizione critica e traduzione integrale* ed. Maria Rita Campa, 129-148. Bari: PolibaPress.

Fallacara, Giuseppe. 2009b. Toward a Stereotomic Design: Experimental Constructions and Didactic Experiences. In *Third International Congress on Construction History*: Brandenburg Technical University.

Fallacara, Giuseppe. 2014a. Lithic Tree. A Search for Natural Stereotomy. Fossilization vs. Litization. In *Lithic Tree. A Search for Natural Stereotomy*, eds. Giuseppe Fallacara and Claudia Calabria, 19-47. Paris: Presses des Ponts.

Fallacara, Giuseppe. 2014b. New stereotomic prototypes: Pi.Mar for research. In *Stereotomic Design*, eds. Giuseppe Fallacara, and Vincenzo Minenna, 56-69. Bari: Dicar-Pimar.

Fallacara, Giuseppe, and Maurizio Barberio. 2018a. Stereotomy 2.0. The Rebirth of a Discipline that Never Died. *Nexus Network Journal* 20:509-514.

Fallacara, Giuseppe, and Maurizio Barberio. 2018b. An Unfinished Manifesto for Stereotomy 2.0. *Nexus Network Journal* 20:519-543.

Fallacara, Giuseppe, and Marco Stigliano. 2014. Solar natural house. In *Stereotomic Design*, eds. Giuseppe Fallacara, and Vincenzo Minenna, 68-73. Bari: Dicar-Pimar.

Fallacara, Giuseppe, and Luc Tamboréro. 2007. *Verso una progettazione stereotomica. Towards an Sterotomic Design*. Roma: Aracne.

Fallacara, Giuseppe, Fiore Resta, Niccoleta Spalucci, and Luc Tamboréro. 2011. The vault of the Hôtel de Ville in Arles. *Nexus Network Journal* 13 (3): 599-629.

Fergusson, Peter J. 1979. Notes on two engraved cistercian drawings. *Speculum* 54:1-17.

Fernández Cabo, Miguel Carlos, Ana González Uriel, and Manuel de Miguel Sánchez. 2017. La triple escalera de caracol en el Convento de Santo Domingo de Bonaval (Santiago, España): hipótesis de diseño y construcción. *Informes de la Construcción* 69 (546): 1-10.

Fernández Correas, Lorena. 2007. La ingeniería al servicio de las puertas de la ciudad: el caso del Portal de Quart, Valencia. In *Actas del Quinto Congreso Nacional de Historia de la Construcción*, eds. Miguel Arenillas Parra et al., 227-284. Madrid: Instituto Juan de Herrera.

Filangieri, Riccardo. 1937. Rassegna critica delle fonti per la storia di Castel Nuovo. Parte seconda. Il castello aragonese. *Archivio storico per le province napoletane* 23: 267-333.

Fitchen, John. [1961] 1981. *The construction of Gothic Cathedrals. A study of medieval vault erection*. Chicago: Chicago University Press.

Fitchen, John. [1986] 1989. *Building Construction Before Mechanization*. Cambridge, Mass.: MIT Press.

Fleury, Philippe. 1993. *La mécanique de Vitruve*. Caen: Presses Universitaires de Caen.

Frankl, Paul. 1945. The Secret of the Mediaeval Masons. *The Art Bulletin* 27 (1): 46-60.

Frankl, Paul. 1953. The 'Crazy' Vaults of Lincoln Cathedral. *The Art Bulletin* 35 (2): 95-107.

Gadaleta, Roberta. 2018. New Stereotomic Bond for the Dome in Stone Architecture. *Nexus Network Journal* 20: 707-722.

Galera Andreu, Pedro. 1977. *Arquitectura de los siglos XVII y XVIII en Jaén*. Granada: Caja General de Ahorros de Granada.

Galera Andreu, Pedro. 1978. Una familia de arquitectos jiennenses: los Aranda. Estudio genealógico. *Boletín del Instituto de Estudios Giennenses* (95): 9-19.

Galera Andreu, Pedro. 2000. *Andrés de Vandelvira*. Madrid: Akal.

Galletti, Sara. 2017a. Stereotomy and the Mediterranean: Notes toward an Architectural History. *Mediterranea. International Journal on the Transfer of Knowledge* (2): 73-120.

Galletti, Sara. 2017b. From Stone to Paper. Philibert de L'Orme, the Premier tome de l'architecture (1567), and the Birth of Stereotomic Theory. *Aedificare. Revue internationale d'histoire de la construction* (2): 143-162.

Galindo Diaz, Jorge Alberto. 1996. El conocimiento constructivo de los ingenieros militares españoles del siglo XVIII. PhD diss., Universidad Politécnica de Cataluña.

Galtier Martí, Fernando. 2001. *La iconografía arquitectónica en el arte cristiano del primer milenio. Perspectiva y convención; sueño y realidad*. Zaragoza: Mira.

Gambús Sáiz, Mercedes. 1989. De l'art de picapedrer de Josep Gelabert: un testimonio literario de la arquitectura mallorquina del siglo XVII. *Mayurqa. Annals de Ciencies Historiques* 22: 777-785.

García Baño, Ricardo. 2017. El manuscrito de cantería Mss. 12686 de la Biblioteca Nacional de España. PhD diss., Universidad Politécnica de Cartagena.

Garcia Baño, Ricardo, 2019. *Arcos de esquina en el Renacimiento español. Forma, función y construcción*. Murcia: Editum.

García Baño, Ricardo, and José Calvo López. 2015. About an Early 16th-century Stonecutting Manuscript in the National Library of Spain and the Origins of Modern Stereotomy. In *5th*

International Congress on Construction History. Chicago: Construction History Society of America.

García Tapia, Nicolás. 1990. *Ingeniería y arquitectura en el Renacimiento español*. Valladolid: Universidad de Valladolid.

Garín Ortiz de Taranco, Felipe María. 1962. Una posible escuela hispanolevantina de crucerías anervadas. In *Homenaje al profesor Cayetano de Mergelina*, 431-439. Murcia: Universidad.

Gasperuzzo, Francesca. 2020. La facciata obliqua di Juan Caramuel de Lobkowitz: un caso di circulaciòn architettonica. In *Geometria e costruzione. Steretomia e configurazione in architettura*, eds. José Calvo-López, Alessio Bortot, and Giulia Piccinin, 85-95. Roma: Aracne.

Genin, Soraya. 2014. Voûtes à Nervures Manuélines. PhD diss., KU Leuven.

Genin, Soraya, Krista De Jonge, and José Carlos Palacios Gonzalo. 2009. Portuguese Vaulting Systems at the Dawn of the Early Modern Period. Between Tradition and Innovation. In *Third International Congress on Construction History*, 671-678. Cottbus: Brandenburg Technical University.

Gentil Baldrich, José María. 1996. La traza oval y la Sala Capitular de la catedral de Sevilla. Una aproximación geométrica. In *Quatro edificios sevillanos*, 73-147. Sevilla: Colegio de Arquitectos.

Gentil Baldrich, José María. 1998. El libro de perspectiva. In *Libro de arquitectura,* Hernán Ruiz II, 215-234. Sevilla: Fundación Sevillana de Electricidad.

Gerbino, Anthony. 2005. François Blondel and the Résolution des quatre principaux problèmes d'architecture (1673). *Journal of the Society of Architectural Historians* 64 (4): 498-521.

Gila Medina, Lázaro. 1988. Ginés Martínez de Aranda. Su vida, su obra y su amplio entorno familiar. *Cuadernos de Arte de la Universidad de Granada* no. 19: 65-81.

Gila Medina, Lázaro. 1991. *Arte y artistas del Renacimiento en torno a la Real Abadía de Alcalá la Real*. Granada-Alcalá la Real: Universidad-Ayuntamiento.

Gila Medina, Lázaro, and Vicente M. Ruiz Puentes. 1992. Andrés de Vandelvira: aproximación a su vida y obra. In *Arquitectura del Renacimiento en Andalucía. Andrés de Vandelvira y su época*, 79-118. Sevilla: Consejería de Cultura.

Gilman, Roger. 1920. The Theory of Gothic Architecture and the Effect of Shellfire at Reims and Soissons. *American Joumal of Archeology* 24 (1): 37-74.

Gimpel, Jean. [1958] 1980. *Les bâttiseurs des cathédrales*. Paris: Seuil.

Gómez Martínez, Javier. 1998. *El gótico español de la Edad Moderna. Bóvedas de Crucería*. Valladolid: Universidad de Valladolid.

Gómez-Ferrer Lozano, Mercedes. 2005. Patios y escaleras de los palacios valencianos del siglo XV. In *Historia de la Ciudad IV. Memoria urbana*, 113-142. Valencia: ICARO CTAV-COACV.

Gómez-Moreno, Manuel. 1925. Sobre el Renacimiento en Castilla. Notas para un discurso preliminar. *Archivo Español de Arte y Arqueología* 1 (1): 1-40.

Gómez-Moreno, Manuel. 1941. *Las águilas del Renacimiento español*. Madrid: Consejo Superior de Investigaciones Científicas.

Gómez-Moreno, Manuel. 1949. *El libro español de Arquitectura*. Madrid: Instituto de España.

Gómez-Moreno, Manuel. 1963. *Diego Siloé. Homenaje en el IV centenario de su muerte*. Cuadernos de Arte y Literatura. Granada: Universidad de Granada.

Goodman, David. 1988. *Power and Penury. Government, technology and science in Philip II's Spain*. Cambridge: Cambridge University Press.

Gutiérrez Hernández, Alexandra María. 2017. Monteas en Jaén. *Boletín del Instituto de Estudios Giennenses* (215): 75-106.

Gutiérrez-Cortines Corral, Cristina. 1987. *Renacimiento y arquitectura religiosa en la antigua diócesis de Cartagena* ... Murcia: Colegio de Aparejadores y Arquitectos Técnicos.

Hara, Takashi, Kenichiro Hidaka, and Reiko Seta. 2013. Structural characteristics of ancient dome in Jerash. In *The 2013 World Congress on Advances in Structural Engineering and Mechanichs (ASEM13)*. Jeju, Korea.

Hartmann-Virnich, Andreas. 1996. L'escalier en vis voûté et la construction romane: exemples rhodaniens. *Bulletin Monumental* 152 (2): 113-128.

Hartmann-Virnich, Andréas. 1999. La "vis" de Saint-Gilles. In *Congrès archéologique de France*, 293-299. Gard: Societé Française d'Archéologie.
Harvey, John. 1954. *English mediaeval architects*. London: B. T. Batsford.
Harvey, John. 1968. The Tracing Floor in York Minster. *Annual Report of the Friends of York Minster* no. 40: 1-8.
Haselberger, Lothar. 1983. Die Bauzeichnungen des Apollontempels von Didyma. *Architectura* 13 (1): 13-26.
Haselberger, Lothar 1994a. Ein Giebelriss der Vorhalle des Pantheon. Die Werkrisse vor dem Augustusmausoleum. *Mitteilungen des Deutschen Archäologischen Instituts. Römische Abteilung* (101): 279-308.
Haselberger, Lothar. 1994b. The Hadrianic Pantheon—a Working Drawing Discovered. *American Journal of Archaeology* 98 (2): 327.
Hastings, Maurice. 1955. *St Stephen's Chapel and its place in the development of perpendicular style in England*. Cambridge: Cambridge University Press.
Hawkings, John Sydney, and John Tomas Smith. 1807. *Antiquities of the City of Westminster; the Old Palace; St. Stephen's Chapel* ... London: J. T. Smith.
Heyman, Jacques. 1969. The safety of stone arches. *International Journal of Mechanical Sciences* no. 11: 363-385.
Heyman, Jacques. 1982. Calculation of abutment sizes for masonry bridges. In *Colloquium on history of structures*, 115-120. London: Institution of Structural Engineeers.
Heyman, Jacques. 1995. *The stone skeleton*. Cambridge: Cambridge University Press.
Hoag, John D. 1958. Rodrigo Gil de Hontanon: His work and writings. Late medieval and Renaissance architecture in Sixteenth century Spain. Ph D diss., Yale University.
Holton, Alexander. 2006. The Working Space of the Medieval Master Mason: the Tracing Houses of York Minster and Wells Cathedral. In *Second International Congress on Construction History*, eds. Malcolm Dunkeld et al., 1579-1597. Cambridge: Construction History Society.
Horn, Walter, and Ernerst Born. 1979. *The Plan of St. Gall: A Study of the Architecture and Economy of, and Life in a Paradigmatic Carolingian Monastery*. Berkeley-Los Angeles-London: University of California Press.
Hoyo Alonso-Martínez, Paloma del. 1976. Las ventanas de ángulo del Renacimiento español. *Goya: Revista de arte* (130): 228-233.
Huerta Fernández, Santiago. 2002. The medieval 'scientia' of structures: the rules of Rodrigo Gil de Hontañón. In *Towards a History of Construction*, eds. Antonio Becchi, Massimo Corradi, Federico Foce, and Orietta Pedemonte, 567-585. Basel-Boston-Berlin: Birkhäuser.
Huerta Fernández, Santiago. 2004. *Arcos, bóvedas y cúpulas. Geometría y equilibrio en el cálculo tradicional de estructuras de fábrica*. Madrid: Instituto Juan de Herrera.
Huerta Fernández, Santiago. 2007. Oval Domes: History, Geometry and Mechanics. *Nexus Network Journal* 9 (2):211-248.
Huerta Fernández, Santiago, and Enrique Rabasa Díaz. 2001. Estereotomía y cálculo de estructuras en la reconstrucción de una boveda del siglo XVIII: Pazo de Antequeira (A Coruña). *OP* (57): 60-69.
Inglese, Carlo. 2000. *Progetti sulla pietra*. Roma: Gangemi.
Inglese, Carlo. 2013. Il tracciato di cantiere dell'Augusteo in Roma: integrazione di metodologie di rilievo. *Disegnare Idee Immagini* 23 (46): 64-73.
Inglese, Carlo, and Antonio Pizzo. 2014. *I tracciati di cantiere di epoca romana*. Roma: Gangemi.
Íñiguez Almech, Francisco. 1963a. Los ingenios de Juan de Herrera. In *El Escorial. 1563-1963*, 181-214. vol. 2. Madrid: Patrimonio Nacional.
Íñiguez Almech, Francisco. 1963b. Los ingenios de Juan de Herrera. Notas marginales. *Revista de Archivos, Bibliotecas y Museos* 71 (1-2): 163-170.
Irles Mas, Ricardo, and Ramón Maestre López-Salazar. 2002. *Homograf. LSP. Aplicación infográfica para arquitectura sobre AutoCAD*. Alicante: Universidad de Alicante.
Jiménez Martín, Alfonso. 1998a. Contexto de la presente edición. In *Libro de arquitectura*, Hernán Ruiz II, 15-22. Sevilla: Fundación Sevillana de Electricidad.

Jiménez Martín, Alfonso. 1998b. Anatomía del manuscrito. In *Libro de arquitectura*, Hernán Ruiz II, 23-41. Sevilla: Fundación Sevillana de Electricidad.

Jiménez Martín, Alfonso. 1998c. El libro de las portadas. In *Libro de arquitectura*, Hernán Ruiz II, 237-256. Sevilla: Fundación Sevillana de Electricidad.

Juan García, Natalia. 2014. Traces of stereotomy: Descriptive geometry drawings in the workshop notebook of the Tornes family from Jaca in north-eastern Spain (c.1650-1750). *Construction History* 29 (2): 21-37.

Juan García, Natalia. 2015. La impronta de los tratados de estereotomía en los cuadernos de taller: los diseños localizados en el manuscrito de la familia de arquitectos Tornés de Jaca. In *Teoría y Literatura Artística en España. Revisión historiográfica y estudios contemporáneos*, eds. Miguel Taín Guzmán, and Nuria Rodríguez Ortega, 537-572. Madrid: Real Academia de Bellas Artes de San Fernando.

Kitao, Timothy. 1974. *Circle and oval in the square of Saint Peter. Bernini's Art of Planning*. New York: New York University Press.

Knoop, Douglas, and G. P. Jones. 1938. *The Two Earliest Masonic MSS*. Manchester: University Press.

Kostof, Spiro. [1977] 1986. The architect in the Middle Ages, East and West. In *The Architect. Chapters in the history of the profession*, ed. Spiro Kostof, 59-95. New York: Oxford University Press.

Kubler, George Edward. 1957. *Arquitectura de los siglos XVII y XVIII*. Madrid: Plus Ultra.

Kubler, George Edward. 1982. *Building the Escorial*. Princeton: Princeton University Press.

Lachauer, Lorenz, Mathias Rippmann, and Philippe Block. 2010. Form Finding to Fabrication: A digital design process for masonry vaults. In *IASS Symposium 2010. Spatial Structures—Permanent and Temporary*. Shanghai.

Lalbat, Claude, Gilbert Margueritte, and Jean Martin. 1987. De la stéréotomie médiévale: la coupe des pierres chez Villard de Honnecourt. *Bulletin Monumental* 145 (4): 387-406.

Lalbat, Claude, Gilbert Margueritte, and Jean Martin. 1989. De la stéréotomie médiévale: La coupe des pierres chez Villard de Honnecourt (II). *Bulletin Monumental* 147 (1): 11-34.

Lassaulx, Johann Claudius von, and William Whewell. 1830-1831. Description of a Mode of erecting light Vaults over Churches and similar Spaces. *Journal of the Royal Institution of Great Britain* 1 (2): 224-240.

Lawrence, Snezana. 2003. History of Descriptive Geometry in England. In *Proceedings of the First International Congress on Construction History*, ed. Santiago Huerta Fernández, 1269-1282. Madrid: Instituto Juan de Herrera.

Lawrence, Snezana. 2011. Developable Surfaces: Their History and Application. *Nexus Network Journal* 13 (3):701-714.

Lázaro Damas, María Soledad. 1988. El Convento de Santa María Magdalena de la Cruz, de la Guardia. Programa Constructivo. *Boletín del Instituto de Estudios Giennenses* 34 (136): 115-143.

Le Boeuf, François. 2001a. Mathurin Jousse, maître serrurier à La Flèche et théoricien d'architecture (vers 1575-1645). *In situ, revue de l'Inventaire général*. https://www.researchgate.net/ (Accessed December 12, 2019).

Le Boeuf, Patrick. 2001b. La Bibliothèque de Mathurin Jousse. Une tentative de reconstitution. *In situ, revue de l'Inventaire général*. https://www.researchgate.net/ (Accessed December 12, 2019).

Le Moël, Michel. 1994. Jacques Curabelle et le monde des architectes parisiens. In *Desargues en son temps*, 389-395. Paris: Blanchard.

Lemonnier, Henry. 1911-1929. *Proces-verbaux de l'Academie Royale d'Architecture (1671-1793)*. Paris: Jean Schemit.

Leedy, Walter C. 1978. The Origins of Fan Vaulting. *The Art Bulletin* 60 (2): 207-213.

Leedy, Walter C. 1980. *Fan vaulting: A study of Form, Technology and Meaning*. London: Scholar Press.

Lemerle, Fréderique. 2016. Jean Chéreau. Livre d'Architecture. In *Architectura. Books on Architecture*, eds. Fréderique Lemerle, and Yves Pauwels. Tours: Centre d'etudes superieures de la Renaissance. http://architectura.cesr.univ-tours.fr (Accessed December 12, 2019).

Lenz, Martina. 2009. The Squinch of Anet. In *Creating shapes in civil and naval architecture. A cross-disciplinary comparison*, eds. Horst Nowacki, and Wolfgang Lefevre, 321-341. Leiden-Boston: Brill.

López Martínez, Celestino. 1932. *Notas para la historia del Arte: desde Martínez Montañés hasta Pedro Roldán*. Sevilla.

López-Mozo, Ana. 2005. Bóvedas cilíndricas en el Monasterio de El Escorial: dos ejemplos de lunetos. In *Actas del Cuarto Congreso Nacional de Historia de la Construcción*, ed. Santiago Huerta Fernández, 667-677. Madrid: Instituto Juan de Herrera.

López-Mozo, Ana. 2008. Tres monteas escurialenses. *Revista EGA Expresión Gráfica Arquitectónica* no. 13: 190-197.

López-Mozo, Ana. 2009. Bóvedas de piedra del Monasterio de El Escorial. PhD diss., Universidad Politécnica de Madrid.

López-Mozo, Ana. 2011. Ovals for Any Given Proportion in Architecture: A Layout Possibly Known in the Sixteenth Century. *Nexus Network Journal* 13 (3): 569-597.

López-Mozo, Ana, Miguel Ángel Alonso Rodríguez, José Calvo López, and Enrique Rabasa Díaz. 2013. Sobre la construcción de pechinas de cantería. El caso de Armenia. In *Octavo Congreso Nacional de Historia de la Construcción*, eds. Santiago Huerta Fernández, and Fabián López Ulloa, 555-564. Madrid: Instituto Juan de Herrera.

López-Mozo, Ana, and Rosa Senent-Domínguez. 2017. Late Gothic Asymmetrical Diamond Vaults in Spain. *Nexus Network Journal* 19 (2): 323-343.

López-Mozo, Ana, Rosa Senent Domínguez, Miguel Ángel Alonso Rodríguez, José Calvo López, and Pau Natividad Vivó. 2015. Asymmetrical vaults in late European Gothic: Basel and Bebenhausen as case studies. In *5th International Congress on Construction History*, eds. Brian Bowen, Donald Friedman, Tom Leslie, and John Ochsendorf. Chicago: Construction History Society of America.

Lorda Iñarra, Joaquín. 1997. Herrera y las grúas de la basílica de El Escorial. *Revista de Obras Públicas* (3367): 81-104.

Lorda Iñarra, Joaquín, and Angélica Martínez Rodríguez. 2000. Las grúas de Juan de Herrera. In *Actas del Tercer Congreso Nacional de Historia de la Construcción*, eds. Amparo Graciani García et al. Madrid: Instituto Juan de Herrera.

Loria, Gino. 1921. *Storia della Geometria Descrittiva, dalle origini sino ai giorni nostri*. Milano: Ulrico Hoepli.

Lotz, Wolfgang. 1955. Die Ovalen Kirchenraume des Cinquecento. *Römische Jahrbuch für Kunstgeschichte* (6):7-99.

Lotz, Wolfgang. [1974] 1995. *Architecture in Italy 1500-1600*. New Haven: Yale University Press.

Maira Vidal, Rocío. 2015. Bóvedas sexpartitas. Los orígenes del gótico. PhD diss., Universidad Politécnica de Madrid.

Maltese, Corrado. 1994. Elissi ed ellissografi. 'Querelles' semi-scientifiche e applicazioni pratiche negli anni di Desargues. In *Desargues en son temps*, 101-110. Paris: Blanchard.

Manceau, Jean-Pierre. 2015. La culture mathématique de Philibert De l'Orme. In *Philibert de l'Orme. Un architecte dans l'histoire*, eds. Fréderique Lemerle, and Yves Pauwels, 191-198. Turnhout: Brepols.

Maranci, Christina. 2001. *Medieval Armenian architecture: constructions of race and nation*. Leuwen: Peeters.

Marías, Fernando. 1983-86. *La arquitectura del Renacimiento en Toledo (1541-1631)*. Toledo: Instituto Provincial de Investigaciones y Estudios Toledanos.

Marías, Fernando. 1991. Piedra y ladrillo en la arquitectura española del siglo XVI. In *Les Chantiers de la Renaissance*, ed. Jean Guillaume, 71-84. Paris: Picard.

Marías, Fernando. 1988. Juan de Torija. Breve tratado de todo género de bóvedas. In *Los tratados de arquitectura*, ed. Dora Wiebenson and Juan Antonio Ramírez, 110-113. Barcelona: Hermann Blume.

Marías, Fernando. 1989. *El largo siglo XVI*. Madrid: Taurus.

Marías, Fernando. 1992. Trazas, trazas, trazas. Tipos y funciones del diseño arquitectónico. In *Juan de Herrera y su influencia*, 351-360. Santander: Universidad de Cantabria.

Mariátegui, Eduardo. 1880. *El Capitán Cristóbal de Rojas, ingeniero militar del siglo XVI*. Madrid: Memorial de Ingenieros.

Martín, Fèlix. 1981. *Els picapedrers i la indùstria de la pedra a la Floresta*. Barcelona: Fundació Salvador Vives Casajuana.

Martín González, Juan José. 1989. Noticia del Arte y Uso de Architectvra. In Fray Laurencio de San Nicolás, *Arte y uso de Arquitectura*. Valencia: Albatros.

Martín Talaverano, Rafael, Carmen Pérez de los Ríos, and Rosa Senent Domínguez. 2012. Late German Gothic Methods of Vault Design and Their Relationships with Spanish Ribbed Vaults. In *Nuts and Bolts of Construction History. Culture, Technology and Society*, eds. Robert Carvais et al., 83-90. Paris: Picard.

Martines, Giangiacomo. 2016. Le olivele di Erone. In *I Tracciati di cantiere: disegni esecutivi per la trasmissione e diffusione delle conoscenze tecniche*, eds. Carlo Inglese and Antonio Pizzo, 154-158. Roma: Gangemi.

Marzal Martínez, Amparo. 1991. La ingeniería militar en la España del XVIII, nuevas aportaciones a la historia de su legado científico y monumental. PhD diss., Universidad Complutense, Madrid.

Mazarrasa Mowinckel, Olav, and Fernando Fernández Herrera. 1988. *Mazarrasa, maestros canteros y arquitectos de Trasmiera*. Santander: Colegio de Arquitectos.

Meckspecker, Cord. 1983. Über die Fünfeckkonstruktion bei Villard de Honnecourt und im späten Mittelalter. *Architectura* 13 (1): 31-40.

Meek, Harold Alan. 1988. *Guarino Guarini*. New Haven: Yale University Press.

Menéndez Pelayo, Marcelino. 1883-1889. *Historia de las ideas estéticas en España*. Madrid: Pérez Dubrull.

Merino de Cáceres, José Miguel. 2005. La torre de la iglesia de San Millán de Segovia y su construcción. In *Actas del Cuarto Congreso Nacional de Historia de la Construcción*, ed. Santiago Huerta Fernández, 771-779. Madrid: Instituto Juan de Herrera.

Millon, Henry A. 1997. Models in Renaissance Architecture. In *The Renaissance from Brunelleschi to Michelangelo. The Representation of Architecture*, ed. Henry A. Millon, 19-73. New York: Rizzoli.

Minenna, Vincenzo. 2012. Forma e struttura dei sistemi voltati complessi nell' opera *vandelviresca*, prolegomeni alla relazione tra la prattica della stereotomia e l'apparato decorativo del 'Libro de trazas de cortes de piedra' di Alonso de Vandelvira. PhD diss., Politecnico di Bari.

Minenna, Vincenzo. 2014. La mappa del nuovo mondo: dall'Archivo delle Indie agli spazi stereotomici generativi. In *Stereotomic Design*, eds. Giuseppe Fallacara, and Vincenzo Minenna, 30-37. Bari: Dicar-Pimar.

Moisy, Pierre. 1950. L'architecte François Derand, jésuite lorrain. *Revue d'Histoire de l' Église de France* 36 (148): 149-167.

Moisy, Pierre. 1952. Martellange, Derand et le conflit du baroque. *Bulletin Monumental* 110: 237-261.

Molina Rozalem, José Francisco, and Federico Arévalo Rodríguez. 2014. Analysis of the Towers built by the Council of Seville for the Defense of the Moorish Band. *Revista EGA Expresión Gráfica Arquitectónica* 19 (24): 164-175.

Moore, Norman. 1885-1900. Vallancey, Charles. In *Dictionary of National Biography*. London: Smith, Elder & Co.

Morales, Alfredo J. 1993. El proyecto arquitectónico en la Sevilla del Renacimiento. Elementos y condicionantes. In *Juan de Herrera y su influencia*, 341-350. Santander: Universidad de Cantabria.

Morales, Alfredo J. 1996. *Hernán Ruiz 'El Joven'*. Madrid: Akal.

Moreno, Arsenio. 1984. *Francisco del Castillo y la arquitectura manierista andaluza*. Jaén: Fundación Pablo de Olavide.

Morgan, B.G. 1961. *Canonic design in English medieval architecture. The origins and nature of systematic architectural design in England. 1215-1515*. Liverpool: Liverpool University Press.

Morrison, Tessa. 2010. Juan Bautista Villalpando and the Nature and Science of Architectural Drawing. *Nexus Network Journal* 12 (1): 63-74.

Mortet, Victor. 1906. La maîtrise d'oeuvre dans les grandes constructions du XIIIe siècle et la profession d'appareilleur. *Bulletin Monumental* 70:263-270.

Mula Cruz, Francisco José. 2016. Reconstrucción de modelos a partir de perspectivas cónicas. PhD diss., Universidad Politécnica de Cartagena.

Müller, Werner. 1969. Stereotomie und Barockbaukunst. *Deutsches Museum Abhandlungen und Berichte*:41-42.

Müller, Werner. 1972. Vittone ed il modo stereotomico. In *Bernardo Vittone e la disputa fra classicismo e barocco nel Settecento. Convegno internazionale ...*, 81-117. Torino: Accademia delle Scienze.

Müller, Werner. 1989. Le dessin technique a l'époque gothique. In *Les bâttisseurs des cathédrales gothiques*, 237-254. Strasbourg: Editions Les Musées de la Ville de Strasbourg.

Müller, Werner. 1990. *Grundlagen gotischer Bautechnik. Ars sine scientia nihil est.* München: Deutscher Kunstverlag.

Muñoz Cosme, Alfonso. 2015. El proyecto y su representación en la arquitectura y la ingeniería militar entre los siglos XVI y XVIII. In *Ingeniería de la Ilustración*, 71-92. Madrid: Fundación Juanelo Turriano-UNED.

Natividad Vivó, Pau. 2010. Levantamiento y análisis estereotómico de las Torres de Quart de Valencia. Master's thesis, Universidad Politécnica de Valencia.

Natividad Vivó, Pau. 2012a. La bóveda en la cárcel de comerciantes de la Lonja de Valencia. *Revista P+C Proyecto y ciudad* (3): 71-86.

Natividad Vivó, Pau. 2012b. Las pechinas de las bóvedas baídas en el manuscrito de Alonso de Vandelvira. In *XI Congreso Internacional de Expresión Gráfica aplicada a la Edificación*, 321-328. Valencia: APEGA.

Natividad Vivó, Pau. 2014. Estereotomía renacentista en el Levante: la Capilla Cruzada de Orihuela. In *Bóvedas valencianas: arquitecturas ideales, reales y virtuales en época medieval y moderna*, ed. Juan Carlos Navarro Fajardo, 108-133. Valencia: Universidad Politécnica de Valencia.

Natividad Vivó, Pau. 2017. Bóvedas baídas de cantería en el Renacimiento español. PhD diss., Universidad Politécnica de Cartagena.

Natividad Vivó, Pau, and José Calvo López. 2013. Precisión del trazado de plantillas para pechinas de baídas por hiladas redondas segun el manuscrito de Vandelvira. In *Jornadas de Introduccion a la Investigación de la UPCT*, 6: 16-18. Cartagena: AJICT.

Natividad Vivó, Pau, and José Calvo López. 2014. El despiece helicoidal de la baída del acceso de la iglesia de San Juan de Caballeros (Jerez de la Frontera). In *Anuario de Jóvenes Investigadores UPCT*, 7: 123-125. Cartagena: AJICT.

Natividad Vivó, Pau, José Calvo López, and Gaspar Muñoz Cosme. 2012c. The Ribless Lierne Vault at the Gates of Quart in Valencia. *Revista EGA Expresión Gráfica Arquitectónica* 17 (19): 190—199.

Navarro-Camallonga, Pablo. 2018c. Arcos, bóvedas de arista y bóvedas aristadas de cantería en el círculo de Francesc Baldomar y Pere Compte. PhD diss., Universidad Politécnica de Valencia.

Navarro-Camallonga, Pablo, and Enrique Rabasa Díaz. 2018b. Vaulted ceilings built like arches in the architectural work of Baldomar. *Revista EGA Expresión Gráfica Arquitectónica* 23 (32): 142-153.

Navarro-Camallonga, Pablo, José Calvo-López, and Miguel Ángel Alonso-Rodríguez. 2018a. Knowledge transfer in vaulting. The Assier church and Valencian stonecutting. *Building Knowledge, Constructing Histories*, eds. Ine Wouters et al., 991-998. Brussels: CRC Press-Balkema-Taylor and Francis.

Navascués Palacio, Pedro. 1974. Estudio. In *El libro de arquitectura de Hernán Ruiz el Joven*, ed. Hernán Ruiz el Joven. Madrid: Escuela de Arquitectura de Madrid.

Navascués Palacio, Pedro. 1986. La obra como espectáculo: el dibujo de Hatfield. In *Monasterio de El Escorial. Las casas reales (El palacio)*, 55-67. Madrid: Patrimonio Nacional.

Nobile, Marco Rosario. 2013. *La stereotomia a Sicilia e nel Mediterraneo*. Palermo: Caracol.

Nuere, Enrique. 1985. *La carpintería de lo blanco. Lectura dibujada del primer manuscrito de Diego López de Arenas*. Madrid: Ministerio de Cultura.

Ousterhout, Robert. [1999] 2008. *Master Builders of Byzantium*. Philadelphia: University of Pennsylvania.

Palacios Gonzalo, José Carlos. 1986. La estereotomía en el Renacimiento. El Escorial. In *Fábricas y orden constructivo (La Construcción) IV Centenario del Monasterio del Escorial*, 97-107. Madrid: Comunidad de Madrid.

Palacios Gonzalo, José Carlos. 1987. La estereotomía de la esfera. *Arquitectura* 68 (267): 54-65.

Palacios Gonzalo, José Carlos. [1990] 2003. *Trazas y cortes de cantería en el Renacimiento Español*. 2nd ed. Madrid: Munilla-Llería.

Palacios Gonzalo, José Carlos. 2006. The Gothic Ribbed Vault in Rodrigo Gil de Hontañón. In *Second International Congress on Construction History*, eds. Malcolm Dunkeld et al., 2415-2431. Cambridge: Construction History Society.

Palacios Gonzalo, José Carlos. 2009. *La cantería medieval. La construcción de la bóveda gótica española*. Madrid: Munilla-Llería.

Palacios, José Carlos, et al. 2015. *Taller de Construcción Gótica I*. Madrid: Munilla -Lería.

Palacios Gonzalo, José Carlos, and Rocío Maira Vidal. 2014. The Sexpartite Vault in the School of Architecture of Madrid. In *Lithic Tree. A Search for Natural Stereotomy*, eds. Giuseppe Fallacara, and Claudia Calabria, 136-152. Paris: Presses des Ponts.

Palacios Gonzalo, José Carlos, and Rafael Martín Talaverano. 2009. La construcción de una bóveda de crucería en la Escuela Técnica Superior de Arquitectura de Madrid. *Informes de la Construcción* 61 (515): 49-58.

Palacios Gonzalo, José Carlos, and Rafael Martín Talaverano. 2011. Geometric Tools in Juan De Álava's Stonecutting Workshop. *Nexus Network Journal* 13 (3): 535-568.

Panofsky, Erwin. 1951. *Gothic Architecture and Scholasticism*. Latrobe, Pa.: Archabbey Press.

Paré, Ronald, R. O. Loving, I. L. Hill, and R. C. Paré. 1996. *Descriptive Geometry*. Upper Saddle River (NJ): Prentice Hall.

Parisi, Nico, and Giuseppe Fallacara. 2009. Nuove prospettive della costruzione in pietra: innovando la tradizione tra Occidente e Oriente. In *El arte de la piedra. Teoría y práctica de la cantería*, 279-308. Madrid: CEU Ediciones.

Perdue, Martin C. 1982. François Nicolas Blondel. Cours d'architecture enseigné dans l'Academie Royale d'Architecture. In *Architectural Theory and Practice from Alberti to Ledoux*, ed. Dora Wiebenson. Chicago: Architectural Publications.

Pereda, Felipe. 2000. *La arquitectura elocuente. El edificio de la Universidad de Salamanca bajo el reinado de Carlos V*. Madrid: Sociedad para la conmemoración de los centenarios de Carlos V y Felipe II.

Pérez de los Ríos, Carmen, and Elena García Alías. 2009. Mathurin Jousse, preludio de la estereotomía moderna. In *Actas del Sexto Congreso Nacional de Historia de la Construcción*, eds. Santiago Huerta Fernández et al., 1041-1050. Madrid: Instituto Juan de Herrera.

Pérez de los Ríos, Carmen, and Enrique Rabasa Díaz. 2014. Stretched templates in Gothic tas-de-charge construction. In *First Conference on Construction History*, 333-342 Cambridge: Construction History Society.

Peroni, Marco. 2016. Valutazione sul peso sollevabile mediante 'olivella'. In *I Tracciati di cantiere: disegni esecutivi per la trasmissione e diffusione delle conoscenze tecniche*, eds. Carlo Inglese and Antonio Pizzo, 166-169. Roma: Gangemi.

Pérouse de Montclos, Jean-Marie. [1982a] 2001. *L'Architecture a la française*. Paris: Picard.

Pérouse de Montclos, Jean-Marie. 1982b. Juan de Torija. Breve tratado de todo género de bóvedas. In *Architectural Theory and Practice from Alberti to Ledoux*, ed. Dora Wiebenson. III-C-10. Chicago: Architectural Publications.

Pérouse de Montclos, Jean-Marie. 1982c. Alonso de Vandelvira. Libro de traças de cortes de piedras. In *Architectural Theory and Practice from Alberti to Ledoux*, ed. Dora Wiebenson. III-C-3. Chicago: Architectural Publications.

Pérouse de Montclos, Jean-Marie. 1982d. Mathurin Jousse. Le secret d'architecture ... In *Architectural Theory and Practice from Alberti to Ledoux*, ed. Dora Wiebenson. III-C-6. Chicago: Architectural Publications.

Pérouse de Montclos, Jean-Marie. 1982e. Father François Derand. L'architecture des voûtes. In *Architectural Theory and Practice from Alberti to Ledoux*, ed. Dora Wiebenson. III-C-9. Chicago: Architectural Publications.

Pérouse de Montclos, Jean-Marie. 1982f. Jean-Baptiste de la Rue. Traité de la coupe des pierres. In *Architectural Theory and Practice from Alberti to Ledoux*, ed. Dora Wiebenson. III-C-16. Chicago: Architectural Publications.

Pérouse de Montclos, Jean-Marie. 1985. La vis de Saint-Gilles et l'escalier suspendu dans l'architecture française du XVIe. siècle. In *L'escalier dans l'architecture de la Renaissance*, 83-92. Paris: Picard.

Pérouse de Montclos, Jean-Marie. 1987. Le théâtre de l'art de construire. In *L'idee constructive en architecture*, ed. Xavier Malverti. 15-20. Paris: Picard.

Pérouse de Montclos, Jean-Marie. 1988. Présentation des traités. In *Traités d'architecture*, ed. Philibert de L'Orme, 3-22. Paris: Leonce Laget.

Pérouse de Montclos, Jean-Marie. 2000. *Philibert de l'Orme, architecte du roi, 1514-1570*. Paris: Mengès.

Pérouse de Montclos, Jean-Marie. 2001. Serlio à Fontainebleau. *Annali di Architectura* no. 13: 71-79.

Pérouse de Montclos, Jean-Marie. 2009. Derand, François. L'architecture des voûtes. In *Architectura. Books on Architecture*, eds. Fréderique Lemerle, and Yves Pauwels. Tours: Centre d'etudes superieures de la Renaissance. http://architectura.cesr.univ-tours.fr (Accessed December 12, 2019)

Pevsner, Nikolaus. 1958. The Three-Dimensional Arch from the Sixteenth to the Eighteenth Century. *Journal of the Society of Architectural Historians* 17 (4): 22-24.

Piccinin, Giulia, and Pau Natividad Vivó. 2020. Stereotomy in ancient Rome. The Mausoleum of Ummidia Quadratilla. In *Geometria e costruzione. Steretomia e configurazione in architettura*, eds. José Calvo López, Alessio Bortot, and Giulia Piccinin, 33-43. Roma: Aracne.

Pinault-Sørensen, Madeleine. 2013. Biographie. In *Philippe de la Hire 1640-1718. Entre architecture et sciences*, eds. Antonio Becchi, Helène Rousteau-Chambon, and Joël Sakarovitch, 17-23. Paris: Picard.

Pinto Puerto, Francisco. 1998. El libro de cantería. In *Libro de arquitectura*, Hernán Ruiz II, 199-214. Sevilla: Fundación Sevillana de Electricidad.

Pinto Puerto, Francisco. 2002. *Las esferas de piedra. Sevilla como lugar de encuentro entre arte y ciencia en el Renacimiento*. Sevilla: Diputación.

Pinto Puerto, Francisco, and Alfonso Jiménez Martín. 1993. Monteas en la Catedral de Sevilla. *Revista EGA Expresión Gráfica Arquitectónica* (1): 79-84.

Pinto Puerto, Francisco, and Alfonso Jiménez Martín. 2016. Geometric Working Drawing of a Gothic Tierceron Vault in Seville Cathedral. *Nexus Network Journal* 18 (2): 439-466.

Pizzo, Antonio. 2016. Observaciones sobre los trazados de obra de época romana de Lusitania. In *I Tracciati di cantiere: disegni esecutivi per la trasmissione e diffusione delle conoscenze tecniche*, eds. Carlo Inglese, and Antonio Pizzo, 55-71. Roma: Gangemi.

Pleguezuelo Hernández, Alfonso. 1990. La Lonja de Mercaderes de Sevilla: de los proyectos a la ejecución. *Archivo Español de Arte* 63 (249): 15-42.

Pliego de Andrés, Elena. 2017. The Germ of the Prinzipalbogen Concept in Bartel Ranisch. *Nexus Network Journal* 19 (2): 405-425.

Potié, Philippe. 1984. Le projet constructif de Philibert de l'Orme. In *L'idee constructive en architecture*, ed. Xavier Malverti, 21-31. Paris: Picard.

Potié, Philippe. 1996. *Philibert de l'Orme. Figures de la pensée constructive*. Marseille: Parenthèses.

Potié, Philippe. 2005. Géométrie savante et écriture baroque. In *Perspective, projections, projet. Techniques de la représentation architecturale*, 71-79. Paris: Monum.

Potié, Philippe. 2008. Le tracé d'épure, des carnet médiévaux aux traités de stéréotomie. In *La construction savante: les avatars de la littérature technique*, eds. Jean-Philippe Garric, Valérie Nègre, and Alice Thomine-Berrada, 149-160. Paris: Picard.

Puppi, Lionello. 1973. *Andrea Palladio*. Milano: Electa.

Quicherat, Jules Etienne Joseph. 1849. Notice sur l'album de Villard de Honnecourt, architecte du XIIIe siècle. *Revue archéologique* 6 (1): 65-80, 164-188, 209-226; pl. 116-118.

Rabasa Díaz, Enrique. 1994. Los arcos oblicuos en la traza de cantería. *Revista EGA Expresión Gráfica Arquitectónica* (2): 145-153.

Rabasa Díaz, Enrique. 1996a. Técnicas góticas y renacentistas en el trazado y la talla de las bóvedas de crucería españolas del siglo XVI. In *Actas del Primer Congreso Nacional de Historia de la Construcción*, eds. Antonio de las Casas Gómez et al., 423-434. Madrid: Instituto Juan de Herrera.

Rabasa Díaz, Enrique. 1996b. Arcos esviados y puentes oblicuos. El pretexto de la estereotomía en el siglo XIX. *OP* (38): 30-41.

Rabasa Díaz, Enrique. 1998. La bóveda plana de Abeille en Lugo. In *Actas del Segundo Congreso Nacional de Historia de la Construcción*, eds. Fernando Bores Gamundi et al., 409-415. Madrid: Instituto Juan de Herrera.

Rabasa Díaz, Enrique. 2000. *Forma y construcción en piedra. De la cantería medieval a la estereotomía del siglo XIX*. Madrid: Akal.

Rabasa Díaz, Enrique. 2003. The single coursed ashlar vault. In *Proceedings of the First International Congress on Construction History*, ed. Santiago Huerta Fernández, 1679-1689. Madrid: Instituto Juan de Herrera.

Rabasa Díaz, Enrique. 2005. Construcción de una bóveda de crucería en el Centro de los Oficios de León. In *Actas del Cuarto Congreso Nacional de Historia de la Construcción*, ed. Santiago Huerta Fernández, 909-917. Madrid: Instituto Juan de Herrera.

Rabasa Díaz, Enrique. 2007. Principios y construcción de las bóvedas de crucería. *Loggia: Arquitectura y restauración* (20): 86-97.

Rabasa Díaz, Enrique. 2008. La construcción medieval de bóvedas. In *Ars Mechanicae: ingeniería medieval en España*, ed. Pedro Navascués Palacio, 119-129. Ministerio de Fomento-CEDEX-CEHOPU-Fundación Juanelo Turriano.

Rabasa Díaz, Enrique 2009a. Soluciones innecesariamente complicadas en la estereotomía clásica. In *El arte de la piedra. Teoría y práctica de la cantería*, 49-68. Madrid: CEU Ediciones.

Rabasa Díaz, Enrique. 2011. La Gournerie versus Monge. *Nexus Network Journal* 13 (3): 715-735.

Rabasa Díaz, Enrique. 2015a. Traza, descripción, razón. Lenguaje y grafismo en los tratados de corte de piedras. In *Teoría y Literatura Artística en España. Revisión historiográfica y estudios contemporáneos*, eds. Miguel Taín Guzmán, and Nuria Rodríguez Ortega, 459-494. Madrid: Real Academia de Bellas Artes de San Fernando.

Rabasa Díaz, Enrique, Miguel Ángel Alonso Rodríguez, Ana López Mozo, Tomás Gil López, and José Calvo López. 2012. The 100 Ft Vault: The Construction and Geometry of the Sala dei Baroni of the Castel Nuovo, Naples. In *Nuts and Bolts of Construction History. Culture, Technology and Society*, eds. Robert Carvais et al., 53-59. Paris: Picard.

Rabasa Díaz, Enrique, Miguel Ángel Alonso Rodríguez, and Elena Pliego de Andrés. 2015b. Trazado de bóvedas en las fuentes primarias del tardogótico: configuración tridimensional. In *Actas del Noveno Congreso Nacional y Primer Congreso Internacional Hispanoamericano de Historia de la Construcción*, eds. Santiago Huerta Fernández and Paula Fuentes, 1399-1408. Madrid: Instituto Juan de Herrera.

Rabasa Díaz, Enrique, and José Calvo López. 2009b. Gothic and Renaissance Design Strategies in Stonecutting. In *Creating shapes in civil and naval architecture. A cross-disciplinary comparison*, eds. Horst Nowacki and Wolfgang Lefevre, 167-191. Leiden-Boston: Brill.

Rabasa Díaz, Enrique, José Calvo López, and Rafael Martín Talaverano. 2017. Bóvedas de crucería que se proyectan en planta según una matriz de estrellas. Transmisión de conocimiento técnico en el tardogótico europeo. In *Décimo Congreso Nacional y Segundo Iberoamericano de Historia de la Construcción*, ed. Santiago Huerta Fernández, 1335-1344. Madrid: Instituto Juan de Herrera.

Recht, Roland. 1995. *Le dessin d' architecture. Origine et fonctions*. Paris: Adam Biro.

Recht, Roland, et al., 2014. *Dessins. Cathédrale de Strasbourg*. Strasbourg: Editions des Musées de Strasbourg.

Rippmann, Mathias, and Philippe Block. 2010a. New Design and Fabrication Methods for Freeform Stone Vaults Based on Ruled Surfaces. In *Design Modeling Symposium*, eds. C. Gengnagel, A. Kilian, N Palz, and F. Scheurer, 181-189. Berlin: Springer.

Rippmann, Mathias, and Philippe Block. 2010b. Rethinking Structural Masonry: Unreinforced, stone-cut shells. *Proceedings of the Institute of Civil Engineers. Construction Materials* https://www.block.arch.ethz.ch (Accessed December 12, 2019).

Rippmann, Mathias, and Philippe Block. 2011. Digital Stereotomy: Voussoir geometry for freeform masonry-like vaults informed by structural and fabrication constraints. In *IABSE-IASS Symposium 2011*. London. https://www.block.arch.ethz.ch/brg/publication (Accessed December 12, 2019).

Rippmann, Mathias, and Philippe Block. 2018. Computational Tessellation of Freeform, Cut-Stone Vaults. *Nexus Network Journal* 20: 545-556.

Rockwell, Peter. 1993. *The art of stoneworking: a reference guide*. Cambridge: Cambridge University Press.

Rodríguez Ruiz, Delfín. 2001. Las Trazas del Palacio de Carlos V en la Alhambra de Granada. In *Las trazas de Juan de Herrera y sus seguidores*, 417-448. Madrid: Patrimonio Nacional-Fundación Marcelino Botín.

Rodríguez Estévez, Juan Clemente 2010. La construcción de la catedral de Sevilla (1433-1537). In *Arquitectura en construcción en Europa en época medieval y moderna*, ed. Amadeo Serra Desfils, 103-146. Valencia: Universitat de València.

Rosenthal, Earl E. 1961. *The Cathedral of Granada. A Study in the Spanish Renaissance*. Princeton: Princeton University Press.

Rosenthal, Earl E. 1985. *The Palace of Charles V in Granada*. Princeton: Princeton University Press.

Rousteau, Hélène 1996. A. F. Frézier, ou le regard d'un ingénieur du XVIIIe siècle sur le gothique. *Cahiers de recherches médiévales et humanistes* (2): 119-125.

Rousteau-Chambon, Hélène. 2003. *Le gothique des Temps modernes. Architecture religieuse en milieu urbain*. Paris: Picard.

Rousteau-Chambon, Hélène. 2013. Un savant chez les hommes de l'art. In *Philippe de la Hire 1640-1718. Entre architecture et sciences*, eds. Antonio Becchi, Hélène Rousteau-Chambon, and Joël Sakarovitch, 107-117. Paris: Picard.

Ruiz de la Rosa, José Antonio. 1987. *Traza y simetría de la arquitectura en la Antigüedad y el Medievo*. Sevilla: Universidad de Sevilla.

Ruiz de la Rosa, José Antonio. 1996. Giralda-Catedral gótica. In *Quatro edificios sevillanos*, 17-71. Sevilla: Colegio de Arquitectos.

Ruiz de la Rosa, José Antonio. 1998. El libro de geometría. In *Libro de arquitectura*, ed. Hernán Ruiz II, 97-141. Sevilla: Fundación Sevillana de Electricidad.

Ruiz de la Rosa, José Antonio, and Juan Clemente Rodríguez Estévez. 2002. 'Capilla redonda en vuelta redonda' (sic): Aplicación de una propuesta teórica renacentista para la catedral de Sevilla. In *IX Congreso Internacional Expresión Gráfica Arquitectónica. Re-visión: Enfoques en docencia e investigación*, 509-516. A Coruña: Universidad de A Coruña.

Ruiz Hernando, José Antonio. 2003. *Las trazas de la catedral de Segovia*. Segovia: Diputación Provincial de Segovia-Caja Segovia.

Saint Aubin, Jean-Paul. 1994. Les enjeux architecturaux de la didactique stéréotomique de Desargues. In *Desargues en son temps*, 363-370. Paris: Blanchard.

Sakarovitch, Joël 1990. Architecture et représentation, de la projection à la double projection. *In Extenso* 13:63-101.

Sakarovitch, Joël. 1992a. La coupe des pierres et la géometrie descriptive. In *L'Ecole Normale de l'an III Leçons de Mathématiques, Laplace-Lagrange-Monge*, ed. Jean Dhombres, 530-540. Paris: Dunod.

Sakarovitch, Joël. 1992b. La geometrie descriptive apres Monge. In *L'École normale de l'an III. Leçons de mathematiques. Laplace-Lagrange-Monge*, ed. Jean Dhombres, 583-590. Paris: Dunod.

Sakarovitch, Joël. 1993. La taille des pierres et la geometrie descriptive. In *La figure et l'espace*, 117-138. Lyon: IREM.

Sakarovitch, Joël. 1994a. La géométrie descriptive, une reine déchue. In *La formation polytechnicienne, deux siècles d'histoire*, eds. Bruno Belhoste, Amy Dahan Dalmedico, and Antoine Picon, 77-93. Paris: Dunod.

Sakarovitch, Joël. 1994b. Le fascicule de stéréotomie; entre savoir et metiers, la fonction de l'architecte. In *Desargues en son temps*, eds. Jean Dhombres, and Joël Sakarovitch, 347-362. Paris: Blanchard.

Sakarovitch, Joël. 1994c. Theodore Olivier, Professeur de Geometrie descriptive. In *Les professeurs du Conservatoire national des arts et metiers, dictionnaire bibliographique, 1794-1955*, eds. A. Grelon, and C. Fontanon, 326-335. vol. 2. Paris: INRP-CNAM.

Sakarovitch, Joël. 1995. The Teaching of Stereotomy in Engineering schools in France in the XVIIIth and XIX centuries: an Application of Geometry, an 'Applied Geometry', or a Construction Technique? In *Entre mécanique et architecture / Between Mechanics and Architecture*, eds. Patricia Radelet-de Grave, and Edoardo Benvenuto, 204-218. Basel-Boston-Berlin: Birkhäuser.

Sakarovitch, Joël. 1997. La construction du géométral. *Cahiers de la recherche architecturale* (40): 9-18.

Sakarovitch, Joël. 1998. *Épures d'architecture*. Basel-Boston-Berlin: Birkhäuser.

Sakarovitch, Joël. 1999. Aux origines de la géométrie descriptive: entre histoire des sciences et histoire de l'architecture. *Bulletin d'information de l'EAPV* no. 30: 31-33.

Sakarovitch, Joël. 2002. Entre mécanique et géométrie: penser l'architecture clavée, l'exemple de Frézier. In *Towards a History of Construction*, eds. Antonio Becchi, Massimo Corradi, Federico Foce, and Orietta Pedemonte, 587-600. Basel-Boston-Berlin: Birkhäuser.

Sakarovitch, Joël. 2003a. Mesure, précision et stéréotomie. In *Mesure et grands chantiers, 4000 ans d'histoire*, ed. Joël Sakarovitch, 16-21. L'Ordre des Géomètres-experts.

Sakarovitch, Joël. 2003b. Stereotomy, a multifaceted technique. In *Proceedings of the First International Congress on Construction History*, ed. Santiago Huerta Fernández, 67-79. Madrid: Instituto Juan de Herrera.

Sakarovitch, Joël. 2005a. Gaspard Monge: Géométrie descriptive, first edition (1795). In *Landmark Writings in Western Mathematics 1640-1940*, ed. I. Grattan-Guinness, 225-241. Amsterdam: Elsevier Science.

Sakarovitch, Joël. 2005b. Géométrie pratique, géométrie savante. In *Géométrie, mesure du monde*, eds. T. Paquot, and C. Younès, 45-60. Paris: Editions La Découverte.

Sakarovitch, Joël. 2006a. Construction History and Experimentation. In *Proceedings of the Second International Congress on Construction History*, eds. Malcolm Dunkeld et al., 2777-2792. Cambridge: Construction History Society.

Sakarovitch, Joël. 2006b. La géométrie descriptive dans les projets pédagogiques de Monge. In *Histoires de géométries*, ed. D. Flament, 1-12. Paris: Fondation Maison des Sciences de l'Homme.

Sakarovitch, Joël. 2007. Gaspard Monge, un géomètre de la transformation douce. *Bulletin de la Sabix, Ecole polytechnique* no. 41: 47-59.

Sakarovitch, Joël. 2008. La voûte plate. In *Les Grands Ateliers*, 78—81. Paris: Jean-Michel Place.

Sakarovitch, Joël. 2009a. From One Curve to Another or the Problem of Changing Coordinates in Stereotomic Layouts. In *Creating shapes in civil and naval architecture. A cross-disciplinary comparison*, eds. Horst Nowacki, and Wolfgang Lefevre, 297-319. Leiden-Boston: Brill.

Sakarovitch, Joël 2009b. From the Stone Carver's Techniques to Descriptive Geometry. In *El arte de la piedra*, eds. José Carlos Palacios Gonzalo, and Alberto Sanjurjo Álvarez, 7-26. Madrid: CEU Ediciones.

Sakarovitch, Joël. 2009c. Gaspard Monge Founder of 'Constructive Geometry'. In *Proceedings. Third International Congress on Construction History*, 1293-1300. Cottbus: Brandenburg Technical University.

Sakarovitch, Joël. 2010. Le fascicule de coupe des pierres de Girard Desargues. In *Encyclopédie des métiers: la maçonnerie et la taille de pierre* 2: 121-147. Paris: Presses du Compagnonnage.

Sakarovitch, Joël. 2013. Philippe de La Hire (1640-1718) et les sciences de la construction. In *Philippe de la Hire 1640-1718. Entre architecture et sciences*, eds. Antonio Becchi, Helène Rousteau-Chambon, and Joël Sakarovitch, 9-15. Paris: Picard.

Salcedo Galera, Macarena, and José Calvo López. 2015. 'Los primeros lunetos en cantería de los tiempos modernos': sobre la bóveda de la cripta del Palacio de Carlos V en Granada. In *Actas del IX Congreso Nacional y I Congreso Internacional Iberoamericano de Historia de la Construcción*, ed. Santiago Huerta Fernández, 1551-1560. Madrid. Instituto Juan de Herrera.

Salcedo Galera, Macarena. 2017. Construcción pétrea en la Granada del Renacimiento. PhD diss., Universidad Politécnica de Cartagena.

Salcedo Galera, Macarena, and José Calvo López. 2016. The Ring Vault in the Palace of Charles V in Granada. Survey and Geometric Analysis. *Revista EGA Expresión Gráfica Arquitectónica* 21 (28): 52-59.

Salcedo Galera, Macarena, and José Calvo López. 2018. From Constructive Practice to Masonry Literature: The Vaulted Passages in the Presbytery of the Granada Cathedral. *Revista EGA Expresión Gráfica Arquitectónica* 23 (32): 154-163.

Salmerón Avellaneda, Carmen. 2015. Desarrollos teóricos en el manuscrito de cantería atribuido a Juan de Aguirre (Biblioteca Nacional de España, Mss/12744). In *Teoría y Literatura Artística en España. Revisión historiográfica y estudios contemporáneos*, eds. Miguel Taín Guzmán, and Nuria Rodríguez Ortega, 597-619. Madrid: Real Academia de Bellas Artes de San Fernando.

Salvatore, Marta. 2011a. Prodromes of descriptive geometry in the *Traité de stéréotomie* by Amédée François Frézier. *Nexus Network Journal* 13 (3): 671-699.

Salvatore, Marta. 2011b. *La stereotomia scientifica in Amédée François Frézier. Prodromi della geometria descrittiva nella scienza del taglio delle pietre.* Firenze: Firenze University Press.

Sanabria, Sergio Luis. 1982. The mechanization of design in the XVIth century: The structural formulae of Rodrigo Gil de Hontañón. *Journal of the Society of Architectural Historians* 41 (4): 281-293.

Sanabria, Sergio Luis. 1984. The evolution and late transformations of the Gothic mensuration system. PhD diss., University of Princeton.

Sanabria, Sergio Luis. 1989. From Gothic to Renaissance Stereotomy. *Technology and Culture* 30 (2): 266-299.

Sanabria, Sergio Luis. 1992. A Late Gothic Drawing of San Juan de los Reyes in Toledo at the Prado Museum in Madrid. *Journal of the Society of Architectural Historians* 51 (2): 161-174.

Sánchez Simón, Ignacio. 2011. Traza y montea de la bóveda de la Capilla Real del convento de Santo Domingo de Valencia. La arista del Triángulo de Reuleaux entre las aristas de la bóveda. In *Actas del Séptimo Congreso Nacional de Historia de la Construcción*, eds. Santiago Huerta Fernández et al., 1301-1309. Madrid: Instituto Juan de Herrera.

Sandman, Alison D. 2007. Spanish Nautical Cartography in the Renaissance. In *The History of Cartography. Vol III. Cartography in the European Renaissance*, ed. D. Woodward, 1: 1095-1142. Chicago-London: University of Chicago Press.

Sanjurjo Álvarez, Alberto. 2006. The Vis Saint Gilles Quarrée or the Caracol de Emperadores Cuadrado. A Model Frequently Encountered in Spanish-French Architectural Treatises from the Modern Period. In *Second International Congress on Construction History*, eds. Malcolm Dunkeld, James W. P. Campbell, Hentie Louw, Michael Tutton, Bill Addis, and Robert Thorne, 2793-2814. Cambridge: Construction History Society.

Sanjurjo Álvarez, Alberto. 2007. El caracol de Mallorca en los tratados de cantería españoles de la edad moderna. In *Actas del Quinto Congreso Nacional de Historia de la Construcción*, eds. Miguel Arenillas Parra et al., 835-845. Burgos: Instituto Juan de Herrera.

Sanjurjo Alvarez, Alberto 2010. La Vis-de-Saint-Gilles: analyse du modèle dans les traités de coupe des pierres et de son influence sur les traités espagnols de l'âge moderne. In *Edifice & Artifice. Histoires constructives*, ed. J. Sakarovitch, 633-643. Paris: Picard.

Sanjurjo Álvarez, Alberto. 2015. La escalera de caracol en los tratados de cantería españoles de la Edad Moderna y su presencia en el Patrimonio Construido hispánico: estudio geométrico y constructivo. PhD diss., Universidad Politécnica de Madrid.

Schneegans, F. E. 1901. Ueber die Sprache des Skizzenbuches von Villard de Honnecourt. *Zeitschrift für Romanische Philologie* 25:45-70.

Schneider, Mark Edward. 1983. Girard Desargues, the architectural and perspective geometry. PhD diss., Virginia Polytechnic Institute and State University.

Sené, Alain. 1970. Un instrument de précision au service des artistes du Moyen Age: l'equerre. *Cahiers de civilisation médiévale* no. 4: 349-358.

Senent Domínguez, Rosa. 2016. La deformación del tipo. Construcción de bóvedas no-canónicas en España (siglos XVI-XVIII). Tesis doctoral, Universidad Politécnica de Madrid.

Shelby, Lon R. 1961. Medieval mason's tools: the level and the plumb rule. *Technology and Culture* 2 (2): 127-130.

Shelby, Lon R. 1964. The Role of the master mason in Medieval English Buildings. *Speculum* 39 (3): 387-403.

Shelby, Lon R. 1965. Medieval mason's tools: compass and square. *Technology and Culture* 6 (2): 236-248.

Shelby, Lon R. 1969. Setting out the keystones of Pointed Arches: A note on medieval 'Baugeometrie'. *Technology and Culture* 10 (4): 537-548.

Shelby, Lon R. 1971. Medieval masons' templates. *Journal of the Society of Architectural Historians* 30 (2): 140-154.

Shelby, Lon R. 1972. The Geometrical Knowledge of Medieval Master Masons. *Speculum* 47 (3): 395-421.

Shelby, Lon R., and Robert Mark. 1979. Late Gothic Structural Design in the 'Instructions' of Lorenz Lechler. *Architectura* no. 12: 113-131.

Sobrino González, Miguel, and Carlota Bustos Juez. 2007. Cimbras para bóvedas: Noticia de algunos casos. In *Actas del Quinto Congreso Nacional de Historia de la Construcción*, eds. Miguel Arenillas Parra et al., 907-914. Madrid: Instituto Juan de Herrera.

Strzygowski, Josef. 1918. *Die Baukunst der Armenier und Europa.* Wien: Anton Schroll.

Summerson, John. [1953] 1969. *Architecture in Britain 1530-1830.* Harmondsworth: Pelican.

Summerson, John. [1953] 1983. *Architecture in Britain 1530-1830.* Harmondsworth: Pelican.

Swanson, Randy S. 2002. Practical and Theoretical Applications of Geometry at Claude Perrault's l'Observatoire de Paris (1667-1672). In *Nexus 2002 International Conference*, eds. Jose Rodrigues, and Kim Williams, 237-251. Torino: Kim Williams Books.

Swanson, Randy S. 2003. Late XVIIth century practice of stereotomy prior to the establisment of Engineering Schools in France. In *Proceedings of the First International Congress on Construction History*, ed. Santiago Huerta Fernández, 1875-1885. Madrid: Instituto Juan de Herrera.

Taín Guzmán, Miguel. 2003. The drawings on stone in Galicia: Types, uses and meanings. In *First International Congress on Construction History*, ed. Santiago Huerta Fernández, 1887-1898. Madrid: Instituto Juan de Herrera.

Taín Guzmán, Miguel. 2006. Fifteen Unedited Engraved Architectural Drawings Uncovered in Northwest Spain. In *Second International Congress on Construction History*, eds. Malcolm Dunkeld et al., 3011-3023. Cambridge: Construction History Society.

Taín Guzmán, Miguel, and Pau Natividad Vivó. 2011. La montea para las bóvedas de horno de Santa Columba de Carnota. In *Actas del Séptimo Congreso Nacional de Historia de la Construcción*, eds. Santiago Huerta Fernández et al., 1387-1399. Santiago de Compostela: Instituto Juan de Herrera.

Taín-Guzmán, Miguel, Miguel Ángel Alonso-Rodríguez, José Calvo-López, and Pau Natividad-Vivó. 2012. Stonecutters' literature and construction practice in Early Modern Gothic: the tracings for a rib vault at the Cathedral of Tui in Galicia. *Construction History* no. 27:1-21.

Tamboréro, Luc. 2006. The Vis Saint-Gilles, Symbol of Compromise between Practice and Science. In *Second International Congress on Construction History*, eds. Malcolm Dunkeld et al., 3025-3039. Cambridge: Construction History Society.

Tamboréro, Luc. 2008. De Delorme à de La Hire, la recherche d'une méthode universelle dans les traités de stéréotomie. Opérations géométriques et emprunts. Diploma thesis, École des Hautes Études en Sciences Sociales, Paris.

Tamboréro, Luc. 2009. Les traites du trait, Les lires avec des outils. In *El arte de la piedra. Teoría y práctica de la cantería*, 69-98. Madrid: CEU Ediciones.

Tamboréro, Luc. 2012. Curves Mastery at the Royal Academy of Architecture: The Case of Jules Hardouin-Mansart. In *Nuts and Bolts of Construction History. Culture, Technology and Society*, eds. Robert Carvais et al., 1: 243-248. Paris: Picard.

Tamboréro, Luc. 2013. Le *Traité de la coupe des pierres*. In *Philippe de la Hire 1640-1718. Entre architecture et sciences*, eds. Antonio Becchi et al., 191-202. Paris: Picard.

Tamboréro, Luc, and Joël Sakarovitch. 2003. The vault at Arles City Hall: A carpentry outline for a stone vault? In *Proceedings of the First International Congress on Construction History*, ed. Santiago Huerta Fernández, 1899-1907. Madrid: Instituto Juan de Herrera.

Taton, René. 1950. *Gaspard Monge*. Basel: Birkhäuser.

Taton, René. 1954. *L' Histoire de la géométrie descriptive*. Paris: Université de Paris.

Taylor, René. 1991. Juan Bautista Villalpando y Jerónimo Prado: de la arquitectura práctica a la reconstrucción mística. In *Dios Arquitecto*, 153-211. Madrid: Siruela.

Tellia, Fabio. 2018. Stereotomy and Architectural Design at Foster + Partners. *Nexus Network Journal* 20:605-626.

Tellia, Fabio, and José Carlos Palacios Gonzalo. 2012. The Squinch Vaults in Joseph Ribes' *Llibre De Trasas De Viax y Muntea*. In *Nuts and Bolts of Construction History. Culture, Technology and Society*, eds. Robert Carvais et al., 647-654. Paris: Picard.

Tellia, Fabio, and José Carlos Palacios Gonzalo. 2015. Las bóvedas de crucería del manuscrito Llibre de Trasas de Viax y Muntea de Joseph Ribes. *Locus Amoenus* (13): 29-41.

Tibuzzi, Edoardo. 2018. Revisiting Stereotomic Principles in Contemporary AEC Practice. *Nexus Network Journal* 20:693-705.

Todisco, Leonardo, Elizabeth Stocks, Javier Leon, and Hugo Corres. 2018. Enhancing the Structural Performance of Masonry Structures by Post-Tensioning. *Nexus Network Journal* 20:671-691.

Tollon, Bruno. 1989. L'église d'Assier. In *Congrès archéologique de France*, 125-136. Quercy: Société Française d'Archéologie.

Tollon, Bruno. 1992. *L'escallier de Toulouze* ou la vis des archives revisité. *Mémoires de la Société archéologique du Midi de la France* (52): 97-106.

Tomlow, Jos. 2009. On Late-Gothic Vault Geometry. In *Creating shapes in civil and naval architecture. A cross-disciplinary comparison*, eds. Horst Nowacki, and Wolfgang Lefevre, 193-219. Leiden-Boston: Brill.

Torres Balbás, Leopoldo. 1922. Los cimborios de Zamora, Salamanca y Toro. *Arquitectura* (36): 137-156.

Trevisan, Camillo. 1998. Le proporzioni nascoste nel trait della trompe di Anet. *Disegnare Idee Immagini* (16): 59-66.

Trevisan, Camillo. 2000. Sulla stereotomia, il CAD e le varie trompe d'Anet. In *Il disegno e la pietra*, ed. Riccardo Migliari, 27-53. Roma: Gangemi.

Trevisan, Camillo. 2011. *Per la storia della stereotomia. Geometrie, metodi e costruzioni. Apparato critico.* Roma: Aracne.

Truant, Cynthia Maria. 1995. *The Rites of Labour: Brotherhoods of Compagnonage in Old and New Regime France*. Ithaca: Cornell University Press.

Turnbull, David. 2000. *Masons, Tricksters and Cartographers: Makers of Knowledge and Space.* Taylor and Francis.

Tutton, Michael. 2014. History and Development of the Staircase. In *Staircases. History, Repair and Conservation*, eds. James W. P. Campbell, and Michael Tutton, 13-73. London-New York: Routledge.

Vera Botí, Alfredo. 1993. *La Torre de la Catedral de Murcia. De la teoría a los resultados*. Murcia: Academia Alfonso X el Sabio.

Victoir, Géraldine. 2005. La polychromie de la cathédrale de Noyon et la datation des voûtes quadripartites de la nef. *Bulletin Monumental* 163 (3): 251-254.

Vigo Trasancos, Alfredo. 1996. El arquitecto jiennense Ginés Martínez de Aranda y la iglesia de San Martín Pinario en Santiago de Compostela. *Norba-Arte* (16): 103-129.

Viollet le Duc, Eugène. 1854-1868. *Dictionnaire raisonné de l'architecture française du XIe au XVIe siècle*. Paris: B. Bauce.

Von Simson, Otto George. [1956] 1962. *The Gothic Cathedral: Origins of Gothic Architecture and the Medieval concept of Order*. New York: Pantheon Books.

Webb, Alfred. 1878. *A Compendium of Irish Biography*. Dublin: M. H. Gill and Son.

Wendland, David, María Aranda Alonso, and María José Ventas Sierra. 2015. El corte de la piedra en bóvedas tardogóticas complejas a la luz de los primeros tratados modernos de estereotomía. In *Actas del Noveno Congreso Nacional y Primer Congreso Internacional Hispanoamericano de Historia de la Construcción*, eds. Santiago Huerta and Paula Fuentes, 1781-1791. Madrid: Instituto Juan de Herrera.

Wilkinson [Zerner], Catherine. 1977a. The Hospital of Cardinal Tavera in Toledo. A documentary and stylistic study on Spanish architecture in the mid-sixteenth century. PhD diss., Yale University.

Wilkinson [Zerner], Catherine. [1977b] 1986. The new professionalism in the Renaissance. In *The Architect. Chapters in the history of the profession*, ed. Spiro Kostof, 124-153. New York: Oxford University Press.

Wilkinson [Zerner], Catherine. 1985. Proportion in Practice: Juan de Herrera's Design for the Façade of the Basilica of the Escorial. *The Art Bulletin* 67 (2): 229-242.

Wilkinson [Zerner], Catherine. 1991. Building from drawings at the Escorial. In *Les Chantiers de la Renaissance*, ed. Jean Guillaume, 263-278. Paris: Picard.

Wilkinson Zerner, Catherine. 1993. *Juan de Herrera. Architect to Philip II of Spain*. New Haven: Yale University Press.

Willis, Robert. [1842] 1910. *On the construction of the vaults of the Middle Ages*. London: Royal Institute of British Architects.

Wilson Jones, Mark. 2003. *Principles of Roman Architecture*. New Haven—London: Yale University Press.

Wilson Jones, Mark. 2009. Ancient Greek and Roman Architects' Approach to Curvature. The Corinthian Capital, Entasis and Amphitheaters. In *Creating shapes in civil and naval architecture. A cross-disciplinary comparison*, eds. Horst Nowacki, and Wolfgang Lefevre, 93-116. Leiden-Boston: Brill.

Wölfflin, Heinrich. 1888. *Renaissance und Barock*. München: Theodor Ackermann.

Zaragozá Catalán, Arturo. 1992. El arte del corte de piedras en la arquitectura valenciana del cuatrocientos. Francesch Baldomar y el inicio de la esterotomía moderna. In *Primer Congreso de Historia del Arte Valenciano*, 97-105. Valencia: Generalitat.

Zaragozá Catalán, Arturo. 1997. La Capilla Real del antiguo Monasterio de Predicadores de Valencia. In *La Capella Reial d'Alfons el Magnànim de l'antic monestir de predicadors de València*, 14-59. Valencia: Conselleria de Cultura.

Zaragozá Catalán, Arturo. 2008. *El arte del corte de piedras en la arquitectura valenciana del cuatrocientos. Un estado de la cuestión* ... Valencia: Real Academia de Bellas Artes de San Carlos de Valencia.

Zaragozá Catalán, Arturo, José Calvo López, and Pau Natividad Vivó. 2012. Stereotomic Exchanges between Iberia and France in the 16th Century: Benoît Augier, Valencian Stairways and the 'Escalier de Toulouse'. In *Nuts and Bolts of Construction History. Culture, Technology and Society*, eds. Robert Carvais et al., 385-392. Paris: Picard.

The manufacturer's authorised representative in the EU is Springer Nature Customer Service Centre GmbH, Europaplatz 3, 69115 Heidelberg, Germany. If you have any concerns regarding our products, please contact ProductSafety@springernature.com

Printed and bound by CPI Group (UK) Ltd, Croydon, CR0 4YY
30/09/2026
02225454-0016